SIGNALS AND SYSTEMS

RISHABH ANAND

B.E (Hons), MBA (IT), M.Tech (ECE), IEEE,CSI, IETE

Global Service Delivery Manager (IT Operations)

ITIL V4, Prince(2) and PMP Certified Professional

Visiting Faculty (Assistant Professor)

Global Institute of Technology and Management

Farrukh Nagar, Gurgaon

KHANNA BOOK PUBLISHING CO. (P) LTD.

Publisher of Engineering and Computer Books

4C/4344, Ansari Road, Darya Ganj, New Delhi-110002

Phone : 011-23244447-48 **Mobile:** +91-9910909320

E-mail : contact@khannabooks.com

Website : www.khannabooks.com

Dear Readers,

To prevent the piracy, this book is secured with HIGH SECURITY HOLOGRAM on the front title cover. In case you don't find the hologram on the front cover title, please write us to at contact@khannabooks.com or whatsapp us at +91-99109 09320 and avail special gift voucher for yourself.

Specimen of Hologram on front Cover title:

Moreover, there is a SPECIAL DISCOUNT COUPON for you with EVERY HOLOGRAM.

How to avail this SPECIAL DISCOUNT:

Step 1: Scratch the hologram
Step 2: Under the scratch area, your "coupon code" is available
Step 3: Logon to www.khannabooks.com
Step 4: Use your "coupon code" in the shopping cart and get your copy at a special discount
Step 5: Enjoy your reading!

Thanks,
Khanna Publishing House

Price: ₹725.00

SIGNALS AND SYSTEMS

by **Rishabh Anand**

ISBN: 978-93-86173-53-9

Edition: 2025

To view complete list of books, please scan the QR Code:

Published by:
Khanna Book Publishing Co. (P) Ltd.
CIN: U22110DL1998PTC095547

Visit us at: www.khannabooks.com
Write us at: contact@khannabooks.com

Printed in India.

Dedicated to

my **Mother** *whose love and affection made me capable of writing this book*

and

Shri Guru Maharaj Ji *of* **Shri Anandpur**

PREFACE

When something can be read without effort, great effort has gone into its writing.

—Enrique Jardiel Poncela

Signals and Systems is a textbook designed for undergraduate students of electronics and communication engineering, telecommunication engineering, computer engineering, electronics and instrumentation engineering, and electrical and electronics engineering. The book will also be useful to AMIE and IETE students. Written with student-centered, pedagogically driven approach, the text provides a self-contained introduction to the theory of signals and systems. This book looks at the concepts of systems, and also examines signals and the way that signals interact with physical systems. It covers topics ranging from basic signals and systems to signal analysis, properties of continuous-time and discrete-time Fourier transforms including Fourier transforms of standard signals, signal transmission through linear systems, relation between convolution and correlation of signals, sampling theorems and techniques, transform analysis of LTI systems, z-transform, Random Signals and Systems. All the solved and unsolved problems in this book are designed to illustrate the topics in a clear way. It includes numerous solved examples with step-by-step solutions for easier understanding of the theoretical concepts.

One approach to learning a new language is to become conversant in the core vocabulary and to start using it right away. At first, you may only know how to ask for directions, order coffee, and so on, but once you become confident with the core, you will start engaging others in "conversations" and rapidly learn how to get things done. This step-by-step approach works equally well in many areas of science and engineering.

With this introduction, welcome to The Essential Guide to Signals and Systems! The reader will find that this Guide covers introductory, intermediate and advanced topics of Signals and Systems, and is intended to be highly accessible for those entering the field or wishing to learn about the topic for the first time. As such, the Guide can be effectively used as a classroom textbook.

The authors hope that the book will fulfill the pressing need of interested readers and welcome any suggestions towards the improvement of the book.

New Delhi **Rishabh Anand**

ACKNOWLEDGEMENT

No one walks alone and when one is walking on the journey of life just where I start to thank those who joined me, walked besides me, and helped me along the way.

Over the years, those that I have met and worked with have continuously motivated me to write this book. So at last, here it is. So, perhaps this book and its pages will be seen as "thanks" to the tens of thousands of you who have helped bringing out this book in the form what is today.

There was, in and always will be the love of my students whom I taught, for what is written in this book has been the fruit of experience that I taught them. Finally, thanks to my mother and to rest of my family for their patience and support during the long hours of writing this book and above all there is the one almighty whose humble children we are. It is his blessings we cherish and pray for. It is the blessing I wish for you.

My special acknowledgment to all the authors whose books, journals and paper are referred while writing this book.

I deeply express my heartful thanks to the Management Team, Mukul Seth and Editorial Staff of **Khanna Book Publishing Co. (P) LTD.** for publishing the book in such a beautiful get-up and well in time.

When you finish the journey through this book, having variety of experiences, ideas and thoughts. You are welcome to express your opinion at *rishabhanand009@gmail.com*.

Rishabh Anand

Contents

1

Introduction to Signals and Systems

1.1 INTRODUCTION

The concepts of signals and systems arise in diverse fields such as science, engineering, communications, aeronautics, circuit design, acoustics, seismology, biomedical engineering, chemical process control, speech processing, image processing, economics and politics etc. Physical nature of the signals and systems varies with disciplines but they all have some very basic features in common. The signals are functions of one or more independent variables which convey information about the behavior or nature of some phenomenon. Systems respond to particular signals by producing other signals. Electrical circuit is an example of system where voltage and current are signals (which are function of time). A computer program for the automated diagnosis of electrocardiograms is an example of system in which a digitized electrocardiogram is an input signal and produces an output signal in the form of estimates of parameters such as heart rate. Another example of a system is camera which receives light from different sources and reflected from objects and it produces a photograph (as output). In the next few articles, first we discuss about the signals then systems separately.

1.2 SIGNALS

A signal can be defined as a function of one or more independent variable(s) which conveys information about the behavior or nature of some phenomenon. Independent variables may be time, space etc. depend on the type of signals.

Examples of signals are speech signals, picture signals, electrocardiogram (ECG) signals etc. A speech signal is represented mathematically as a function of time where time is an independent variable. A picture signal is represented as a brightness function of two spatial variables (x and y) where x and y are independent variables. Fig. 1.1 shows a waveform of speech signal.

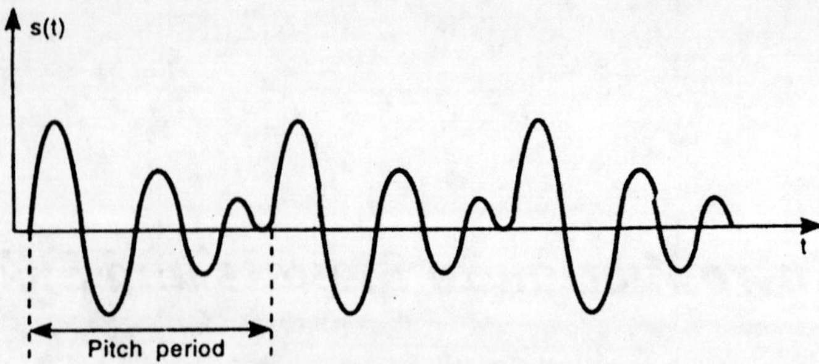

Fig. 1.1 *Speech signal.*

Signals are represented by its mathematical model

Mathematical models are nothing but mathematical equations representing signals. For example, the mathematical signal model for a household voltage is given by equation

$$v(t) = 115\sqrt{2}\cos[2\pi(50)t] \quad -\infty < t < \infty \qquad ...(1.1)$$

1.3 CLASSIFICATION OF SIGNALS

Any investigation in signal and system analysis is started with a classification of signals involved in the specific applications.

Signals can be classified in the manner given below.

1.3.1 Classification of Signals Based on Number of Dimensions of a Signal

Signals are categorized in two groups on the basis of dimensions of a signal. The dimension of signal is defined as number of independent variable(s) on which value of signal, *i.e.*, dependent variable depends.

1. One dimensional signals
2. Multidimensional signals.

One Dimensional Signal. A signal is called one dimensional signal if it is a function of single independent variable.

For example, speech signal is a one dimensional signal because amplitude of signal [dependent variable] depends upon signal independent variable, *i.e.*, time.

Multidimensional Signal. A signal is called multidimensional signal if it is a function of more than one independent variable. For example, black and white (B/W) TV picture signal is an example of two-dimensional signal because brightness [dependent variable] of the signal at each point is function of two spatial independent variables, namely x and y. Variables x and y are width and height of the picture element [pixel].

Similarly, colour TV picture signal is an example of three-dimensional signal because brightness of the signal at each point is a function of three independent variables, namely x, y and time (t).

1.3.2 Classification of Signals Based on Number of Channels of a Signal

Signals are categorized in two groups based on the number of channels of a signal. The channels of a signal are defined as the number of sources from which it is generated. They can be

1. Single channel signals.
2. Multichannel signals.

Single Channel Signals. Signals which are generated by a single source or a single sensor are called *single channel signals*.

Example of a single channel signal is the speech signal as it is generated by a single source or by using a single sensor.

Multichannel Signals. Signals which are generated by more than one source or sensor are called multichannel signals. These signals are represented by vector

$$s(t) = \begin{bmatrix} s_1(t) \\ s_2(t) \\ s_3(t) \end{bmatrix} \qquad ...(1.2)$$

Eq. 1.2 represents a three-channel signal. In electrocardiography, 3-lead and 12-lead electrocardiograph is often used which results in, 3-channel and 12-channel signals respectively.

1.3.3 Classification of Signals based on Condition of Independent Variable of a One-Dimensional Signal

In this classification of signals, time is considered as an independent variable. On the basis of condition of time, signal is classified into two groups

1. Continuous-time signals
2. Discrete-time signals.

Continuous-time Signal. A signal that varies continuously with time is called continuous-time signal. A continuous-time signal has a value defined as each point in time. A continuous-time signal is denoted by a function of time such as $s(t)$ where t is continuous-time independent variable. For example, speech signal and temperature of the room are continuous-time signals. Continuous-time signal is shown in Fig. 1.2.

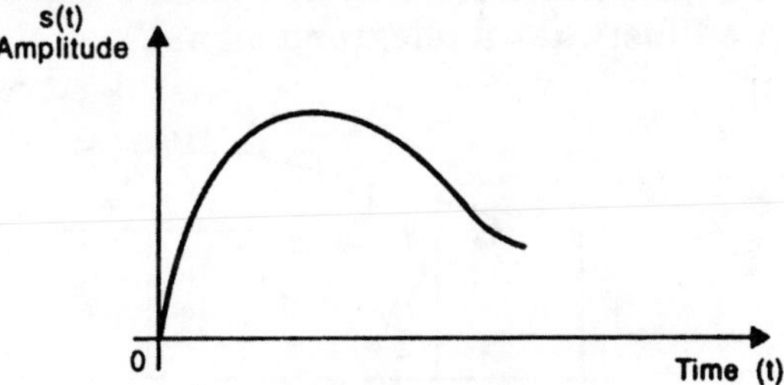

Fig. 1.2 *Continuous-time signal.*

Discrete-time Signal. Discrete-time signals are signals which are defined at discrete times. It is shown in Fig. 1.3. These are represented by sequences of numbers. A discrete-time signal has a value defined only at discrete points in time. Discrete-time signals can be recovered by periodic sampling of continuous-time signals. A discrete-time signal is denoted by $s(n)$ where n is discrete-time independent variable. For example, demographic data in which various attributes, such as average budget, crime rate, are tabulated against such discrete variables as family size, total population respectively.

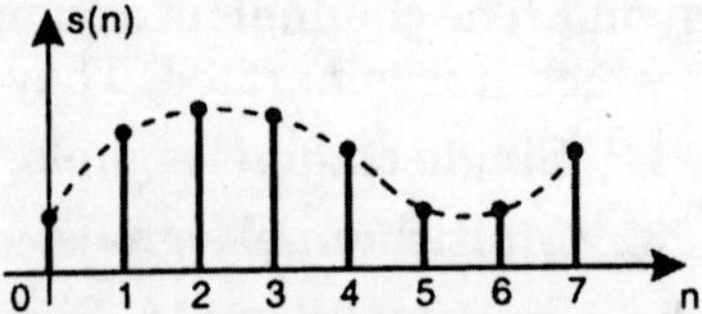

Fig. 1.3 *Discrete-time signal.*

1.3.4 Classification of Signals based on Condition of Dependent Variable and Independent Variable(s) of a Signal

On this basis, signals are classified in two groups

1. Analog signals
2. Digital signals.

Analog Signals. Analog signals are signals whose both dependent variable and independent variable(s) are continuous in nature. Analog signals arise when a physical waveform is converted into an electrical signal. This conversion is performed by means of a transducer. For example, telephone speech signal, TV signals etc., are very common types of analog signals.

Telephone Speech Signals. A telephone message comprises speech sounds having vowels and consonants. These sounds produce an audio signal. These sound waves are converted into analog electrical signals by means of a transducer (microphone).

Transducer is a device which converts non-electrical quantity into electrical signals.

Continuous-amplitude, continuous-time signals are called *analog signals*. Analog signal is shown in Fig. 1.2.

Digital Signals. Digital signals are signals whose both dependent variable and independent variable(s) are discrete in nature. Digital signals comprise pulses occurring at discrete intervals of time. Telegraph and teleprinter signals are examples of digital signals. Fig. 1.4 illustrates a telegraph signal.

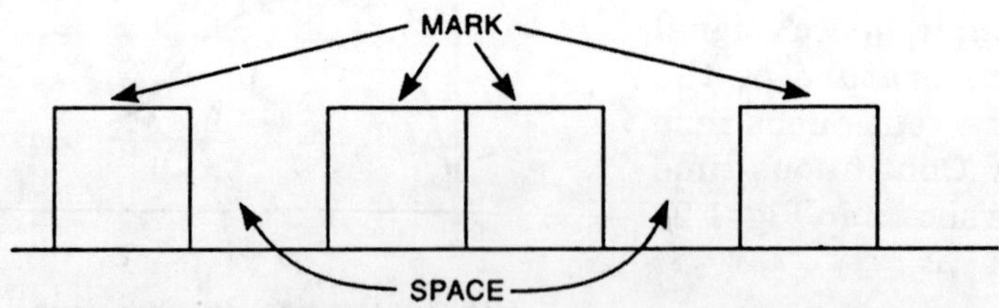

Fig. 1.4 *Telegraph signal.*

1.4 GENERAL SIGNAL CHARACTERISTICS

Continuous-time and discrete-time signals can be further categorised in terms of various characteristics they possess.

1. Deterministic and random signals
2. Piecewise-defined and simply-defined signals
3. Periodic and non-periodic signals
4. Power and energy signals
5. Even and odd signals.

1.4.1 Deterministic and Random Signals

A deterministic signal is one which has no uncertainty with respect to its value at any value of independent variable, namely, time. A deterministic signal behaves in a fixed known way with respect to time. A deterministic signal can be modeled by a known function of time, t, (or a sample number, n) for continuous-time signals (or discrete-time signals). For example, continuous-time signal given by

$$s(t) = \frac{2}{(1+t^2)} \qquad -\infty \le t \le \infty$$

is a deterministic signal and shown in Fig. 1.5. Similarly, discrete-time signal given by

$$s(nT_s) = s(n) = \frac{3n}{2(4+n^2)}$$

is a deterministic signal and shown in Fig. 1.6.

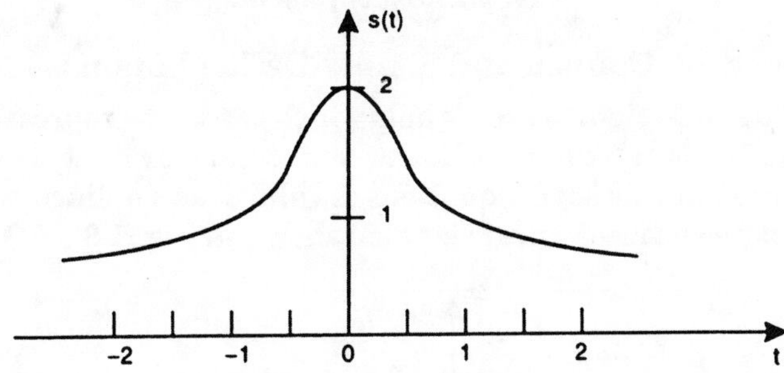

Fig. 1.5 *Continuous-time deterministic signal* $s(t) = \left(\frac{2}{1+t^2}\right) -\infty \le t \le \infty.$

If no range of values for t or n is shown, then the function is valid for all values of t or n.

A random signal is a signal which has some degree of uncertainty with respect to its value at any value of independent variable, namely, time. A random signal takes on one of several possible values at each time for which a signal value is defined. Random signals are represented

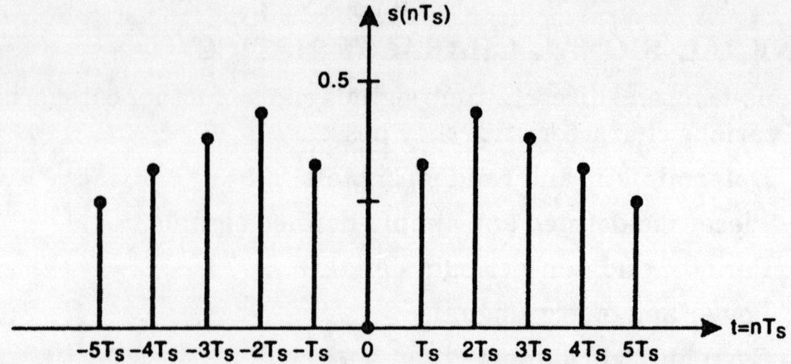

Fig. 1.6 *Discrete-time deterministic signal* $s(n) = s(nT_s) = 3n/2(4 + n^2)$

by their probabilistic models because there is a probability of occurrence associated with each of the resulting time functions or sequence of signal values that can occur. For example, thermal agitation noise in conductors is a random signal. A random signal is shown in Fig. 1.7.

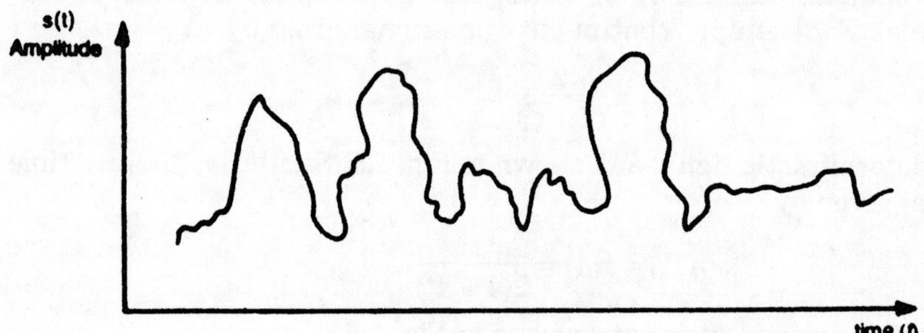

Fig. 1.7 *Random signal.*

1.4.2 Piecewise-Defined and Simply-Defined Signals

Piecewise-defined signals are signals which cannot be represented by a single equation for all time but can be represented by a set of equations. Each equation in this set of equations is valid over a defined time span. Example of piecewise-defined signal is shown in Fig. 1.8.

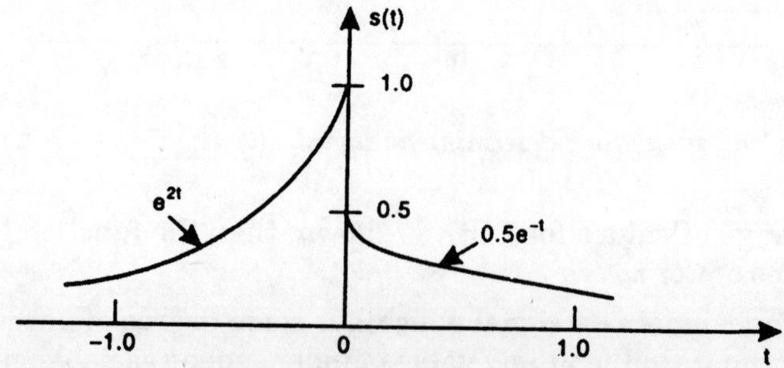

Fig. 1.8 *Example of piecewise-defined continuous-time signal*

$$s(t) = \begin{cases} 1.0\, e^{2t}, & t < 0 \\ 0.5\, e^{-t}, & t \geq 0 \end{cases}$$

Simply defined signals are signals which are represented by a single equation for all times. A few signals that can be defined by a single equation are more conveniently analyzed if a piecewise definition is used.

A continuous-time signal is given by a single equation,

$$s(t) = e^{-|t|} \quad -\infty \le t \le \infty \qquad \text{...(1.3)}$$

Its piecewise-defined model is given as

$$s(t) = \begin{cases} e^{t}, & t \le 0 \\ e^{-t}, & t > 0 \end{cases} \qquad \text{...(1.4)}$$

Simply defined signals are shown in Fig. 1.9.

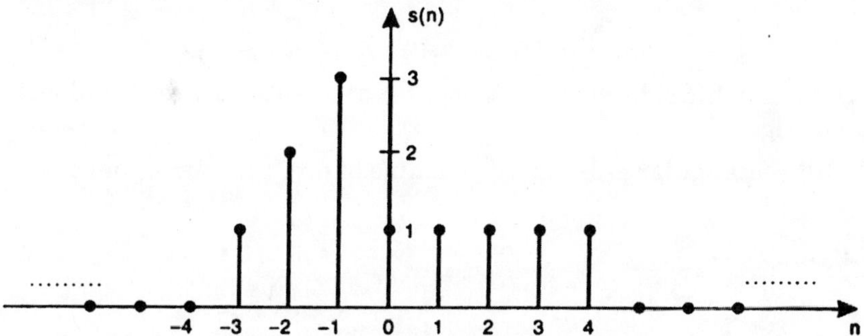

Fig. 1.9 *Example of piecewise-defined discrete-time signal*

$$s(n) = \begin{cases} 1, & 0 \le n \le 4 \\ n + 4, & -4 \le n < 0 \\ 0, & elsewhere \end{cases}$$

1.4.3 Periodic and Non-Periodic Signals

A signal which repeats its waveform after a fixed period of time is called a periodic signal. This fixed time is called *time period* (T_0).

A periodic signal has a waveform that repeats over and over, with the time between repeats defined as the period of the signal.

The continuous-time signal $s(t)$ is periodic if and only if

$$s(t + T_0) = s(t) \quad \text{for all } t \qquad \text{...(1.5)}$$

where T_0 is the period of the signal measured in units of time.

The discrete-time signal $s(n)$ is periodic if and only if

$$s(n + N_0) = s(n) \quad \text{for all } n \qquad \text{...(1.6)}$$

where N_0 is the period of the signal measured in units of number of sample spacings.

The smallest value of T_0 or N_0 is called the fundamental period of the periodic signal. For example, sinusoidal signals are example of periodic signals shown in Fig. 1.10.

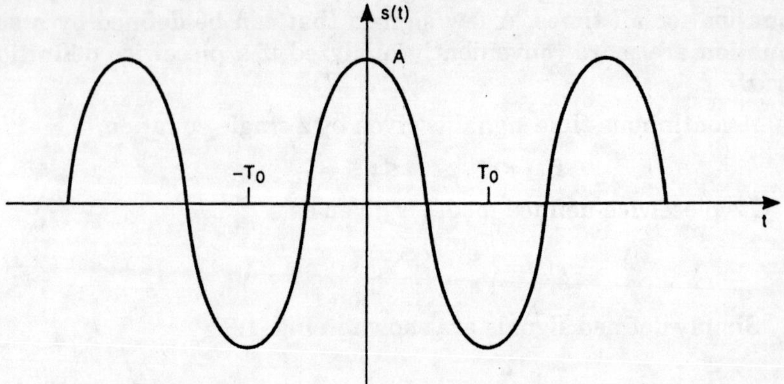

Fig. 1.10 *Sinusoidal signal* $s(t) = A \cos \omega t$, $-\infty \le t \le \infty$.

Signals which does not satisfy above conditions are called non-periodic signals.

Unit rectangular pulse is the example of non-periodic signal shown in Fig. 1.11.

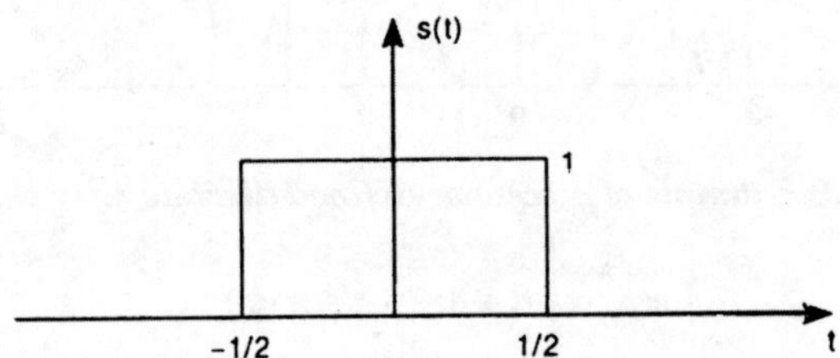

Fig. 1.11 *Unit rectangular pulse* $s(t) = \begin{cases} 1, & |t| < \frac{1}{2} \\ 0, & \textit{otherwise} \end{cases}$

In general, a periodic continuous-time signal can be expressed as

$$s(t) = \sum_{k=-\infty}^{\infty} s_p(t - kT_0) \qquad \text{...(1.7)}$$

where $$s_p(t) = \begin{cases} s(t), & t_1 \le t < t_1 + T_0 \\ 0, & \text{otherwise} \end{cases} \qquad \text{...(1.8)}$$

t_1 is an arbitrary time

and a periodic discrete-time signal can be expressed as

$$s(n) = \sum_{k=-\infty}^{\infty} s_p(n - kN_0) \qquad \text{...(1.9)}$$

where $$s_p(n) = \begin{cases} s(n); & n_1 \le n \le n_1 + N_0 \\ 0, & \text{otherwise} \end{cases}$$

n_1 is arbitrary sample index.

The sum of N periodic continuous-time signals is not necessarily periodic. The condition of periodicity for the sum of N periodic continuous-time signals is given by

$$\frac{T_0}{T_{oi}} = n_i\,, \quad 1 \le i \le N \qquad \text{...(1.10)}$$

where T_0 is the period of sum of N periodic continuous-time signals

T_{0i} is the period of ith signal in the sum

and n_i is an integer.

$$\frac{T_{01}}{T_{0i}} = \frac{n_i}{n_1}\,, \quad 2 \le i \le n \qquad \text{...(1.11)}$$

Eq. 1.11 is an equivalent necessary and sufficient condition for a sum of periodic continuous-time signals to be periodic. This condition states that the sum of N periodic continuous-time signals is periodic. If all the ratios of the period of first signal in the sum to the period of another signal in the sum be rational then it means it is the ratio of integers.

Determination of Period of Sum Signal if it is Periodic First convert each ratio from Eq. 1.11 to a ratio of integers. If this ratio is not rational then the sum signal is not periodic. If it is rational, then factor the highest common divisor from numerator and denominator of each individual ratio. If the least common multiple of the denominators of the resulting ratios in n_1,

Then we can determine the period of sum signal by

$$T_0 = T_{01}\,.\,n_1 \qquad \text{...(1.12)}$$

Eqs. 1.11 and 1.12 can also be applicable to the sum of discrete- time periodic signals if T_0 is replaced by N_0 and each T_{0i} is replaced by N_{0i}.

$$\frac{N_{01}}{N_{0i}} = \frac{n_i}{n_1}\,, \quad 2 \le i \le n \qquad \text{...(1.13)}$$

$$N_0 = N_{01}\,.\,n_1 \qquad \text{...(1.14)}$$

Sum of discrete-time periodic signals is always periodic because the period ratios N_{01}/N_{0i} are always rational.

Example 1.1 Determine periodicity for the following signals and compute the period of sum signals.

(*i*) $s(t) = s_1(t) + s_2(t) + s_3(t)$, where $s_1(t)$, $s_2(t)$, $s_3(t)$ are continuous-time periodic signals with 8/3, 1.26 and $\sqrt{2}$ seconds respectively.

(*ii*) $s(n) = s_1(n) + s_2(n)$, where $s_1(n)$, $s_2(n)$ are discrete-time periodic signals with periods 90, 54, respectively.

Solution. (*i*) A necessary and sufficient condition for sum of periodic continuous-time signals to be periodic is given by

$$\frac{T_{01}}{T_{0i}} = \frac{n_i}{n_1}, \quad 2 \le i \le n$$

$i = 2,$ $$\frac{T_{01}}{T_{02}} = \frac{n_2}{n_1} \text{ and } i = 3, \ \frac{T_{01}}{T_{03}} = \frac{n_3}{n_1}$$

Given $T_{01} = \frac{8}{3}$ seconds

$$T_{02} = 1.26 \text{ second and } T_{03} = \sqrt{2}$$

$$\frac{T_{01}}{T_{02}} = \frac{(8/3)}{1.26} = \frac{800}{378} \text{ rational}$$

and $$\frac{T_{01}}{T_{03}} = \frac{(8/3)}{\sqrt{2}} = \frac{8}{3\sqrt{2}} \text{ not rational}$$

The signal $s(t)$ is not periodic since T_{01}/T_{03} is not rational.

(*ii*) A necessary and sufficient condition for sum of periodic discrete-time signals to be periodic is given by

$$\frac{N_{01}}{N_{0i}} = \frac{n_i}{n_1}, \quad 2 \le i \le n$$

$i = 2,$ $$\frac{N_{01}}{N_{02}} = \frac{n_2}{n_1}$$

Given $N_{01} = 90$ and $N_{02} = 54$

$$\frac{N_{01}}{N_{02}} = \frac{90}{54} = \frac{5}{3} \text{ rational}$$

The signal $s(n)$ is a periodic signal since $\frac{N_{01}}{N_{02}} = \frac{5}{3}$ is a rational.

n_1 = least common multiple of denominators = 3

Period of sum signal $N_0 = N_{01} \,.\, n_1 = 90 \times 3 = 270$.

1.4.4 Power and Energy Signals

In many, but not all, applications, the signals we considered are directly related to physical quantities capturing power and energy in a physical system. A signal is called *energy signal* only if its total energy is finite. For example, rectangular pulse is an energy signal.

A signal is called power signal only if its average power is finite. For example, sinusoidal waves are power signals.

The energy signals have zero average power and power signals have infinite energy. It means energy and power signals are mutually exclusive.

Instantaneous power across a resistor is

$$p(t) = v(t)\, i(t) \qquad ...(1.15)$$

But $$i(t) = \frac{v(t)}{R} \qquad ...(1.16)$$

From Eqs. 1.15 and 1.16, we get

$$p(t) = v(t)\, i(t) = v(t)\, \frac{v(t)}{R}$$

or
$$p(t) = \frac{v^2(t)}{R} \qquad ...(1.17)$$

where $v(t)$ is the voltage across the resistor

$i(t)$ is the current across the resistor.

The total energy expanded over the time interval $t_1 \le t \le t_2$ is

$$E = \int_{t_1}^{t_2} p(t)\, dt = \int_{t_1}^{t_2} \frac{v^2(t)}{R}\, dt \qquad ...(1.18)$$

The average power in the interval, $t_1 \le t \le t_2$ is

$$P = \frac{1}{t_2 - t_1} \int_{t_1}^{t_2} p(t)\, dt$$

$$= \frac{1}{t_2 - t_1} \int_{t_1}^{t_2} \frac{v^2(t)}{R}\, dt \qquad ...(1.19)$$

The total energy over the time interval $t_1 \le t \le t_2$ in a continuous-time signal $s(t)$ is given by

$$E = \int_{t_1}^{t_2} |s(t)|^2\, dt \qquad ...(1.20)$$

where $|s(t)|$ denotes the magnitude of the signal $s(t)$ and $s(t)$ is possibly a complex signal.

The average power in the time interval, $t_1 \le t \le t_2$ in a continuous-time signal $s(t)$ is given by

$$P = \frac{1}{t_2 - t_1} \int_{t_1}^{t_2} |s(t)|^2\, dt \qquad ...(1.21)$$

where $t_2 - t_1$ is the time interval.

Similarly, the total energy in a discrete-time signal $s(n)$ over the time interval $n_1 \le n \le n_2$ is given by

$$E = \sum_{n=n_1}^{n_2} |s(n)|^2 \qquad ...(1.22)$$

The average power in the time-interval, $n_1 \le n \le n_2$ in a discrete-time signal $s(n)$ is given by

$$P = \frac{1}{n_2 - n_1 + 1} \sum_{n=n_1}^{n_2} |s(n)|^2 \qquad ...(1.23)$$

where $n_2 - n_1 + 1$ are the number of points in the interval, $n_1 \le n \le n_2$.

The total energy in a continuous-time signal over an infinite time interval, $-\infty < t < \infty$.

$$E_\infty \triangleq \lim_{T \to \infty} \int_{-T}^{T} |s(t)|^2 \, dt$$

$$= \int_{-\infty}^{\infty} |s(t)|^2 \, dt \qquad \text{...(1.24)}$$

The total energy in a discrete-time signal over an infinite time interval, $-\infty < n < \infty$.

$$E_\infty \triangleq \lim_{N \to \infty} \sum_{n=-N}^{N} |s(n)|^2$$

$$= \sum_{n=-\infty}^{\infty} |s(n)|^2 \qquad \text{...(1.25)}$$

For some signals, integral in Eq. 1.24 or sum in Eq. 1.25 does not converge. Such signals have infinite energy. If integral in Eq. 1.24 or sum in Eq. 1.25 converges then these signals have finite energy, $E_\infty < \infty$.

The time-averaged power in a continuous-time signal over an infinite interval, $-\infty < t < \infty$ is given by

$$P_\infty \triangleq \lim_{T \to \infty} \frac{1}{2T} \int_{-T}^{T} |s(t)|^2 \, dt \qquad \text{...(1.26)}$$

The time-average power in a discrete-time signal over an infinite interval, $-\infty < n < \infty$ is given by

$$P_\infty \triangleq \lim_{N \to \infty} \frac{1}{2N+1} \sum_{n=-N}^{N} |s(n)|^2 \qquad \text{...(1.27)}$$

On the basis of total energy and average power over an infinite time-interval, signals are categorized in three classes given as follows.

First of these is the class of signals which have finite total energy and zero average power. These are called *energy signals*.

Second of these is the class of signals which have finite average power and infinite total energy. These are called *power signals*.

Third of these is the class of signals whose neither average power nor total energy is finite. These are neither power signals nor energy signals.

1.4.5 Even and Odd Signals

Another set of useful properties of signals relates to their symmetry under time reversal. A signal $s(t)$ or $s(n)$ is called even signal if it is identical to its folded (time- reversed) counter part. Time-reversal of a signal means its reflection about the origin ($t = 0$ or $n = 0$).

For continuous-time signal $s(t)$,

If $s(-t) = s(t)$, then $s(t)$ is even signal

If $s(-t) = -s(t)$, then $s(t)$ is odd signal.

For discrete-time signal $s(n)$,

If $s(-n) = s(n)$, then $s(n)$ is even signal

If $s(-n) = -s(n)$, then $s(n)$ is odd signal.

An odd signal must be zero at $t = 0$ or $n = 0$. Even and odd signals are also called *symmetric and asymmetric signals* respectively. Examples of even and odd continuous-time signals are shown in Fig. 1.12.

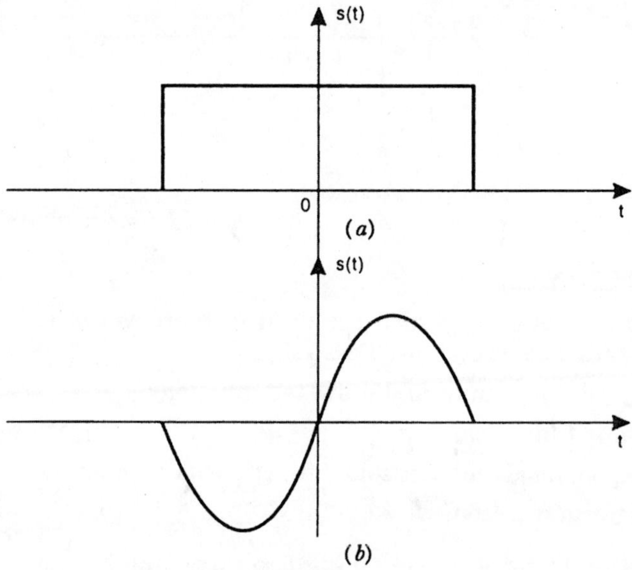

Fig. 1.12 *(a) An even continuous-time signal. (b) An odd continuous-time signal.*

Any signal can be decomposed into a sum of two signals, one of which is even (symmetric) and other is odd (asymmetric).

$$\text{Ev}\ [s(t)] = \frac{1}{2}\ [s(t) + s(-t)] \qquad \text{...(1.28)}$$

$$\text{Od}\ [s(t)] = \frac{1}{2}\ [s(t) - s(-t)] \qquad \text{...(1.29)}$$

where Ev $[s(t)]$ and Od $[s(t)]$ are called even part and odd part of $s(t)$ respectively.

Examples of even and odd discrete-time signals are shown in Fig. 1.13.

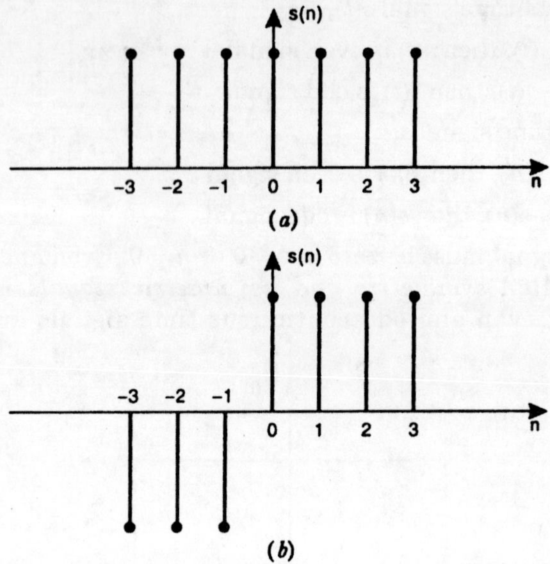

Fig. 1.13 (*a*) *An even discrete-time signal.* (*b*) *An odd discrete-time signal.*

1.5 BASIC SIGNALS

The signals or sequences which are used to represent other complex signals are called basic signals. These signals are

1. Impulse functions and unit-step functions.
2. Ramp functions.
3. Exponential and sinusoidal functions or signals.
4. Signum function.

1.5.1 Impulse Function and Unit-Step Functions

These two functions are studied under a common heading because both are related to each other. Discrete-time functions are also called *discrete-time sequences*. In this article, we will study the following functions

1. Continuous-time impulse function.
2. Continuous-time unit-step function.
3. Discrete-time impulse sequence.
4. Discrete-time unit-step sequence.

We will discuss each and every function one by one in the next sections.

Continous-time Impulse Function. It is a simplest basic continuous-time signal. It is also called *Dirac-Delta function* and it is denoted by $\delta(t)$. This function is defined as

$$\delta(t) = \begin{cases} 1, & t = 0 \\ 0, & t \neq 0 \end{cases} \qquad ...(1.30)$$

Dirac-Delta function $\delta(t)$ is graphically shown in Fig. 1.14.

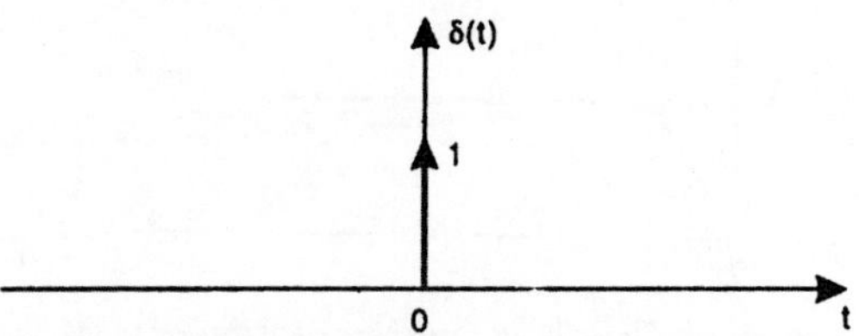

Fig. 1.14 *Graphical representation of* $\delta(t)$.

Continuous-time Unit-step Function. It is also a basic continuous-time signal and is denoted by $u(t)$. This function is defined as

$$u(t) = \begin{cases} 1, & t > 0 \\ 0, & t < 0 \end{cases} \qquad ...(1.31)$$

It is graphically shown in Fig. 1.15.

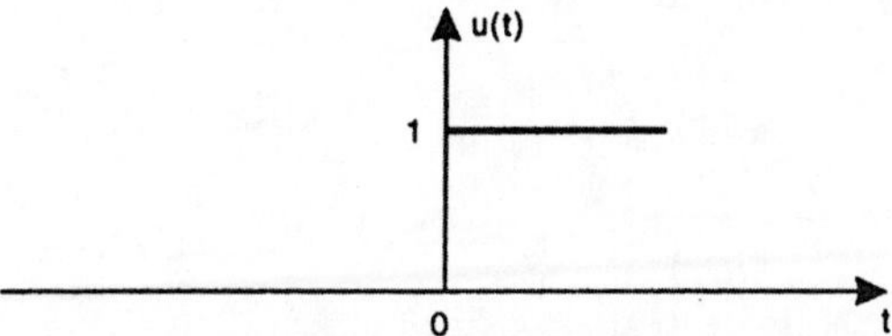

Fig. 1.15 *Graphical representation of* $u(t)$.

Unit-step function $u(t)$ is discontinuous at time, $t = 0$.

Relationship between unit impulse function $\delta(t)$ and unit step function $u(t)$.

In particular, the unit-step function $u(t)$ is the running integral of the unit-impulse function $\delta(t)$.

$$u(t) = \int_{\tau=-\infty}^{t} \delta(\tau)\, d\tau \qquad ...(1.32)$$

In other words, we can say that unit impulse function $\delta(t)$ is the first derivative of unit-step function $u(t)$.

$$\delta(t) = \frac{du(t)}{dt} \qquad ...(1.33)$$

Formally, $u(t)$ is not differentiable because $u(t)$ is discontinuous at $t = 0$ and therefore, there is some formal difficulty with this equation $\delta(t) = \frac{du(t)}{dt}$ as a representation of $\delta(t)$.

Now, we are considering an approximation to the unit-step function and it is denoted by $u_\Delta(t)$. Approximation of $u(t)$ is shown in Fig. 1.16. $u_\Delta(t)$ rises from the value 0 to the value 1 in short time

Fig. 1.16 *Graphical representation of* $u_\Delta(t)$.

interval to length Δ.

Formally, unit step function $u(t)$ is the limit of $u_\Delta(t)$ as Δ tends to zero.

$$u(t) = \lim_{\Delta \to 0} u_\Delta(t) \qquad \text{...(1.34)}$$

Therefore $\delta_\Delta(t)$ is given by

$$\delta_\Delta(t) = \frac{d\, u_\Delta(t)}{dt} \qquad \text{...(1.35)}$$

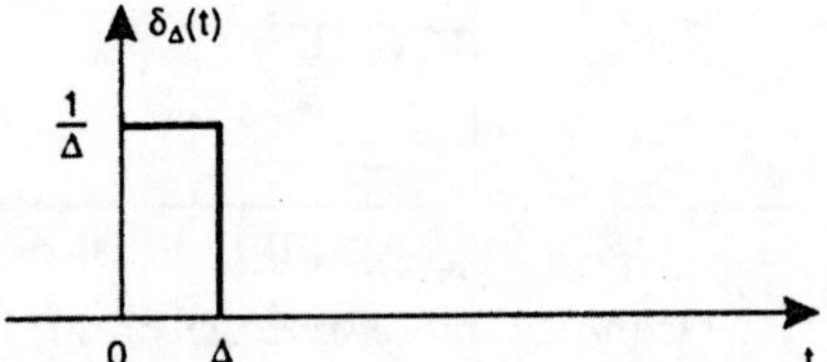

Fig. 1.17 *Graphical representaiton of derivative of* $u_\Delta(t)$.

$\delta_\Delta(t)$ is the derivative of $u_\Delta(t)$ and it is shown in Fig. 1.17.

Limiting form of $\delta(t)$ is given by

$$\delta(t) = \lim_{\Delta \to 0} \delta_\Delta(t) \qquad \text{...(1.36)}$$

$\delta(t)$ has no duration but unit area.

Properties of Unit-impulse Function $\delta(t)$. $\delta(t)$ has following properties

1. Time scaling. $\delta(a.t) = \frac{1}{|a|}\,\delta(t)$
2. Time-reversal. $\delta(-t) = \delta(t)$
3. $\int_{t_1}^{t_2} s(t)\,\delta(t - t_0) = s(t_0)\,, \quad t_1 < t < t_2$
   $= 0, \quad$ otherwise

4. $s(t)\,\delta(t-t_0) = s(t_0)\,\delta(t-t_0)$

Discrete-time Impulse Function. It is a simplest basic discrete- time signal. It is also called *unit-impulse* or *unit-sample sequence.* This sequence is defined as

$$\delta(n) = \begin{cases} 1\,, & n = 0 \\ 0\,, & n \neq 0 \end{cases} \qquad ...(1.37)$$

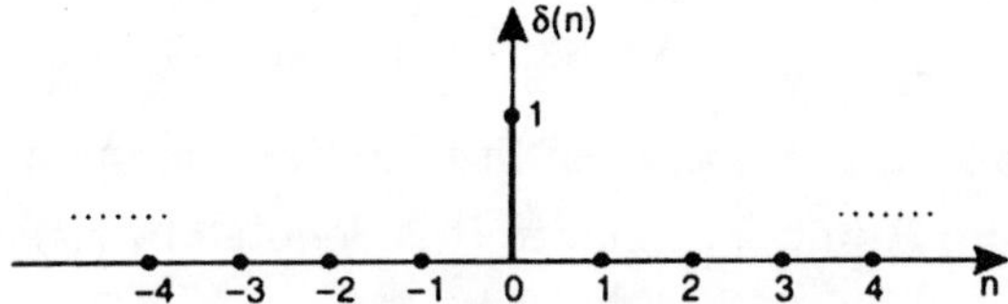

Fig. 1.18 *Graphical representation of* $\delta(n)$.

and it is graphically shown in Fig. 1.18.

Discrete-time Unit-step Function. It is also a basic discrete- time signal. It is also called *unit-step sequence.* This sequence is defined as

$$u(n) = \begin{cases} 1\,, & n \geq 0 \\ 0\,, & n < 0 \end{cases} \qquad ...(1.38)$$

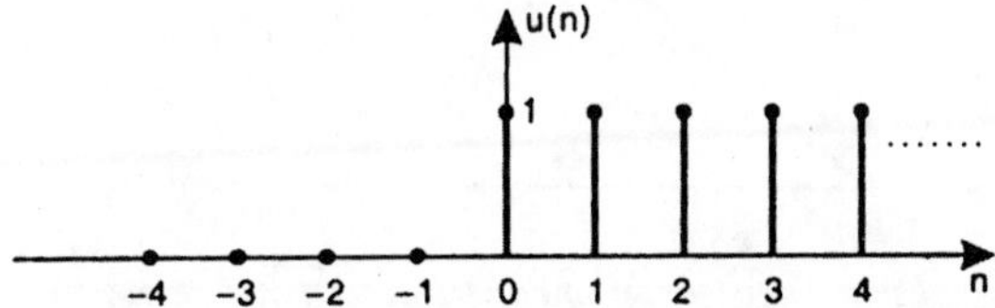

Fig. 1.19 *Graphical representation of* $u(n)$.

and it is shown in Fig. 1.19.

Relationship between Unit-Sample Sequence, $\delta(n)$ and Unit-Step Sequence, $u(n)$

In particular, unit-sample sequence, $\delta(n)$ is the first difference of the unit-step sequence, $u(n)$

$$\delta(n) = u(n) - u(n-1) \qquad ...(1.39)$$

In other words, the unit-step sequence, $u(n)$ is the running sum of the unit-sample sequence, $\delta(n)$.

$$u(n) = \sum_{k=-\infty}^{n} \delta(k) \qquad ...(1.40)$$

The unit-step sequence, $u(n)$ can also be written in terms of $\delta(n)$.

$$u(n) = \sum_{m=0}^{\infty} \delta(n-m) \qquad \text{...(1.41)}$$

The unit-sample sequence $\delta(n)$ can be used to sample the value of signal at $n = 0$. Any sequence $s(n)$ can be represented in terms of $\delta(n)$.

$$s(n) = \sum_{k=-\infty}^{\infty} s(k)\, \delta(n-k) \qquad \text{...(1.42)}$$

1.5.2 Ramp Function

These are either continuous time or discrete-time ramp functions.

Continuous-time Ramp Function. It is denoted by $r(t)$ and defined as

$$r(t) = \begin{cases} t\,, & t > 0 \\ 0\,, & t < 0 \end{cases} \qquad \text{...(1.43)}$$

The relationship between continuous-time ramp function $r(t)$ and unit-step function $u(t)$ is given as

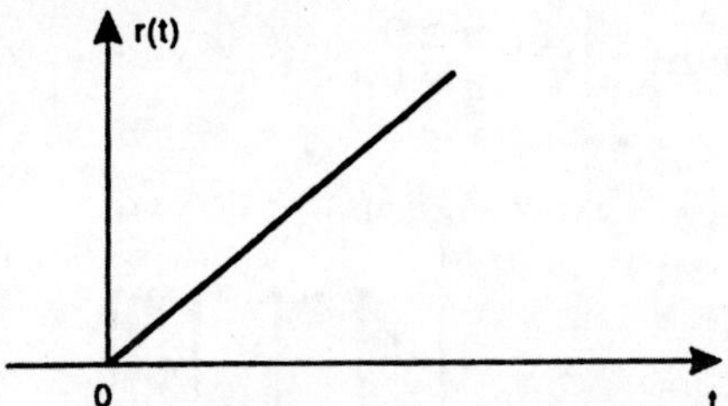

Fig. 1.20 *Graphical representation of continuous-time ramp function r(t).*

$$r(t) = t\, u(t) \qquad \text{...(1.44)}$$

Continuous-time ramp function is shown in Fig. 1.20.

Discrete-Time Ramp Sequence. It is denoted by $r(n)$ and defined as

$$r(n) = \begin{cases} n\,, & n \geq 0 \\ 0\,, & n < 0 \end{cases} \qquad \text{...(1.45)}$$

The relationship between $r(n)$ and $u(n)$ is given as

$$r(n) = n\, u(n) \qquad \text{...(1.46)}$$

Fig. 1.21 *Graphical representation of r(n).*

Discrete-time ramp sequence is shown in Fig. 1.21.

1.5.3 Exponential and Sinusoidal Signals

These signals serve as basic building blocks. These are used to construct many other signals. These are two types

1. Continuous-time complex exponential and sinusoidal signals.
2. Discrete-time complex exponential and sinusoidal signals.

Continuous-time Complex Exponential and Sinusoidal Signals. The continuous-time complex exponential signal is given by

$$s(t) = be^{at} \qquad \text{...(1.47)}$$

where b and a are, in general, complex numbers. These complex exponentials can have several different characteristics that depend upon the values of parameters b and a.

Real Exponential Signals. If both the parameters b and a are real then it is called *real exponential signal*. In this case exponential signal shows two types of behavior. If parameter a is positive, then independent variable (t) increases signal $s(t)$ is a growing exponential. This growing exponential is used to describe many physical processes such as chain reactions in atomic explosions etc.

If parameter a is negative, then signal $s(t)$ is a decaying exponential. This signal is also used to describe many physical processes such as the responses of *RC* circuits etc.

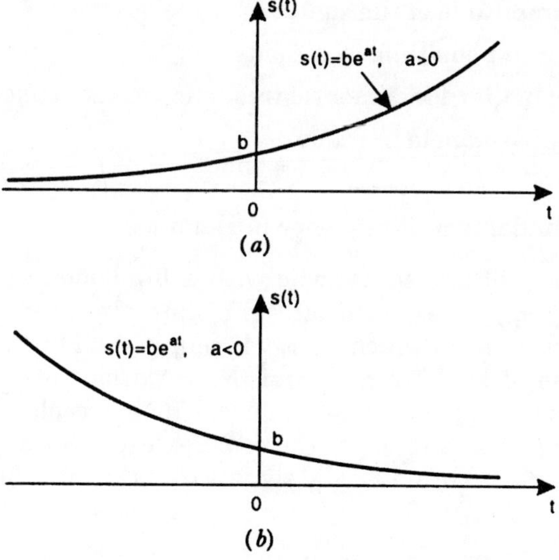

Fig. 1.22 *Continuous-time real exponential (a) growing exponential signal (b) decaying exponential signal.*

If parameter a is zero, then signal $s(t)$ is a constant.

Periodic Complex Exponential and Sinusoidal Signals. Purely imaginary exponential signal is given by

$$s(t) = e^{j\omega_0 t} \qquad ...(1.48)$$

where $\sqrt{-1} = j$

This is a periodic signal.

Verification. Above signal will be periodic with period T

if $\qquad e^{j\omega_0 t} = e^{j\omega_0(t+T)}$

or $\qquad e^{j\omega_0 t} = e^{j\omega_0 t} \cdot e^{j\omega_0 T}$

or $\qquad e^{j\omega_0 T} = 1$

If $\omega_0 = 0$ then signal $s(t) = 1$, which is periodic for any value of T.

If $\omega_0 \neq 0$, then the fundamental period T_0 of signal $s(t)$, which is the smallest positive value of T is given by

$$T_0 = \frac{2\pi}{|\omega_0|} \qquad ...(1.49)$$

Both the exponential signals $e^{j\omega_0 t}$ and $e^{-j\omega_0 t}$ have the same fundamental period.

Sinusoidal signal is closely related to periodic complex exponential signal and it is given as

$$s(t) = A \cos(\omega_0 t + \phi) \qquad ...(1.50)$$

where A is the amplitude of the signal

t is the continuous-time in seconds

ω_0 is the angular fundamental frequency in radian/sec

ϕ_0 is the phase angle in radian

$$\omega_0 = 2\pi f_0$$

f_0 is the fundamental frequency in Hertz.

Sinusoidal signal is also periodic with a fundamental period T_0. Sinusoidal and complex exponential signals are also used to describe many physical systems in which energy is conserved. For example, the natural response of an LC circuit and the variations of the acoustic pressure corresponding to a single musical tone are also sinusoidal signals. By using Euler's relation, the complex exponential signal can be represented in terms of sinusoidal signals with the same fundamenal period (T_C).

$$e^{j\omega_0 t} = \cos \omega_0 t + j \sin \omega_0 t \qquad ...(1.51)$$

Similarly, sinusoidal signal can also be represented in terms of periodic complex exponential signals with same fundamental period.

$$s(t) = A \cos(\omega_0 t + \phi) = A\left[\frac{e^{j(\omega_0 t + \phi)} + e^{-j(\omega_0 t + \phi)}}{2}\right]$$

$$= \frac{A}{2} e^{j\phi} e^{j\omega_0 t} + \frac{A}{2} e^{-j\phi} e^{-j\omega_0 t} \quad ...(1.52)$$

Note that the two exponentials given in above Eq. 1.52 have complex amplitude.

Sinusoidal signal can also be expressed in terms of complex exponential signal as

$$A \cos[\omega_0 t + \phi] = A \text{ Re } [e^{j(\omega_0 t + \phi)}] \quad ...(1.53)$$

and $$A \sin[\omega_0 t + \phi] = A \text{ Im } [e^{j(\omega_0 t + \phi)}] \quad ...(1.54)$$

We know that the fundamental period T_0 of a continuous- time sinusoidal signal or a periodic complex exponential is inversely proportional to the fundamental frequency $|\omega_0|$. If the magnitude of fundamental frequency ω_0 decreases then rate of oscillation decreases and therefore, period increases. Exactly opposite effects occur if magnitude of fundamental frequency increases. A constant signal ($\omega_0 = 0$) has a zero rate of oscillation.

Periodic complex exponential signals will play a central role in much of our treatment of signals and systems. These signals will serve as extremely useful constitutents or building blocks for many other signals. A necessary and sufficient condition for a complex exponential $e^{j\omega t}$ to be periodic with period T_0 is given by

$$e^{j\omega T_0} = 1$$

where $\omega T_0 = 2\pi k$, $k = 0, \pm 1, \pm 2, ...$

General Complex Exponential Signals. Now, consider a complex exponential given by

$$s(t) = be^{at}$$

where b is expressed in polar form and
a in rectangular form.

$$b = |b|\, e^{j\theta} \text{ and } a = r + j\omega_0$$

Substituting the value of b and a in signal

$$s(t) = be^{at} = |b|\, e^{j\theta} e^{(r + j\omega_0)t}$$

$$= |b|\, e^{rt} e^{j(\omega_0 t + \theta)} \quad ...(1.55)$$

Using Euler's relation, we can expand Eq. 1.55 as

$$s(t) = be^{at} = |b|\, e^{rt} [\cos(\omega_0 t + \theta) + j \sin(\omega_0 t + \theta)]$$

$$= |b|\, e^{rt} \cos(\omega_0 t + \theta) + j\, |b|\, e^{rt} \sin(\omega_0 t + \theta) \quad ...(1.56)$$

For $r = 0$, Eq. 1.56 becomes

$$s(t) = |b| \cos(\omega_0 t + \theta) + j\, |b| \sin(\omega_0 t + \theta) \quad ...(1.57)$$

and the real and imaginary parts of complex exponential are sinusoidal.

For $r > 0$, the signal $s(t)$ corresponds to sinusoidal signals multiplied by growing exponential, and

For $r < 0$, the signal $s(t)$ corresponds to sinusoidal signals multiplied by decaying exponential.

These two cases are shown in Fig. 1.23.

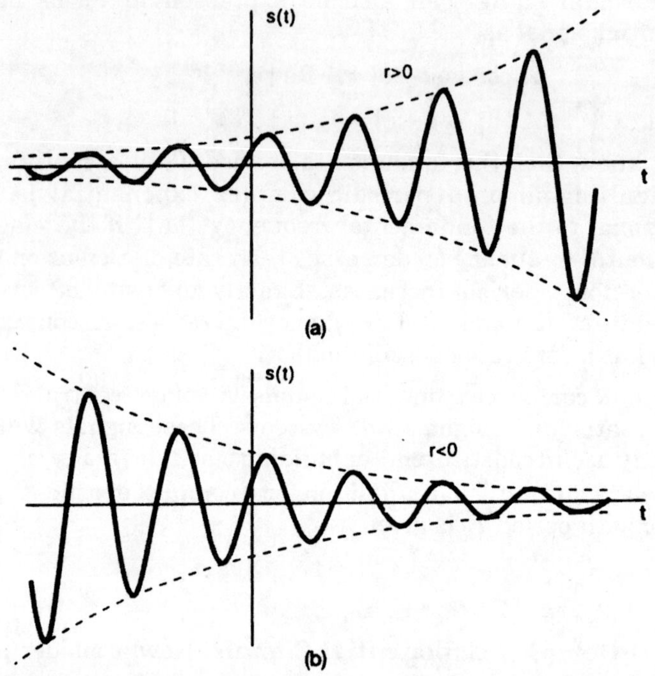

Fig. 1.23 (*a*) *Growing sinusoidal signal* (*b*) *decaying sinusoidal signal.*

If sinusoidal signals are multiplied by decaying exponential then it is called *Damped sinusoidal signal*. For example, such type of signals arise in the response of *RLC* circuit etc. These damped systems dissipate energy with oscillations that decay with time.

Discrete-Time Complex Exponential and Sinusoidal Signals. An important signal in discrete-time is the complex exponential sequence. It is given by

$$s(n) = b(a)^n \quad \text{...(1.58)}$$

where a and b are, in general, complex numbers in Eq. 1.58.

Above complex exponential sequence can also be expressed in the form

$$s(n) = be^{\alpha n} \quad \text{...(.159)}$$

where $a = e^{\alpha}$

We note that Eq. 1.59 is more analogous to the form of the continuous-time exponential.

Real Exponential Signals. If both a and b are real then Eq. 1.59 represents a real exponential signal. If $|a| > 1$ the magnitude of the signal grows exponentially with n.

While if $|a| < 1$ the magnitude of the signal decreases exponentially with n.

Real-value discrete-time exponentials are often used to describe population growth as a function of generation.

Sinusoidal Signals. Discrete-time sinusoidal signal or sequence is defined as

$$s(n) = A \cos (\omega_0 n + \phi) \qquad \text{...(1.60)}$$

If n is dimensionless, then both ω_0 and ϕ have units of radians.

Complex exponential signals are related with sinusoidal signals by using Euler's relation.

$$e^{j\omega_0 n} = \cos \omega_0 n + j \sin \omega_0 n \qquad \text{...(1.61)}$$

and
$$s(n) = A \cos (\omega_0 n + \phi) = A\left[\frac{e^{j(\omega_0 n + \phi)} + e^{-j(\omega_0 n + \phi)}}{2}\right]$$

$$= \frac{A}{2} e^{j\omega_0 n} e^{j\phi} + \frac{A}{2} e^{-j\omega_0 n} e^{-j\phi} \qquad \text{...(1.62)}$$

The signals $s(n) = e^{j\omega_0 n}$ and $s(n) = A \cos (\omega_0 n + \phi)$ are examples of discrete-time signals with finite average power but infinite total energy.

The real exponential signal is shown in Fig. 1.24.

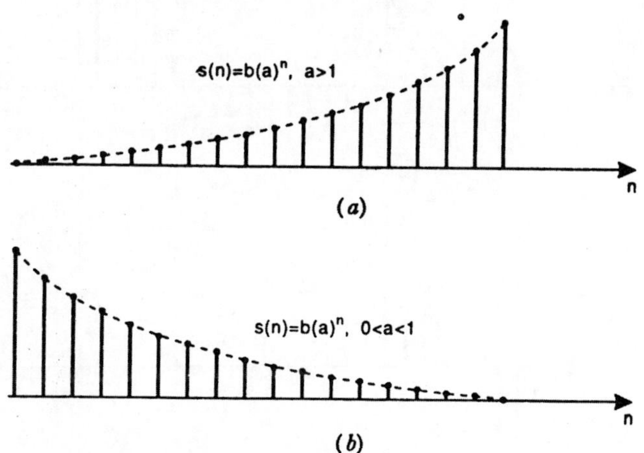

Fig. 1.24 *The real exponential signal* $s(n) = b(a)^n$ (a) $a > 1$ (b) $0 < a < 1$.

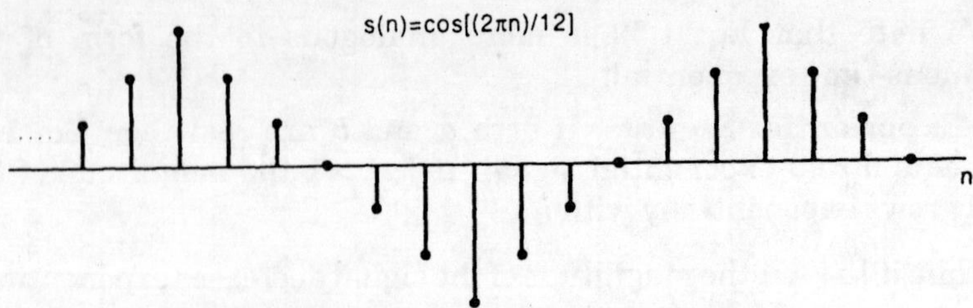

Fig. 1.25 *Discrete-time sinusoidal signal.*

General Complex Exponential Signals. The general discrete-time complex exponential signals can be written in terms of real exponentials and sinusoidal signals.

Substituting $b = |b|e^{j\theta}$ and $a = |a|e^{j\omega_0}$ in Eq. $s(n) = b(a)^n$, we get

$$s(n) = b(a)^n = |b|\ e^{j\theta}\ [|a|\ e^{j\omega_0}]^n = |b|\ |a|^n\ e^{j(\omega_0 n + \theta)}$$

$$= |b|\ |a|^n \{\cos(\omega_0 n + \theta) + j \sin(\omega_0 n + \theta)\}$$

$$= |b|\ |a|^n \cos(\omega_0 n + \theta) + j\ |b|\ |a|^n \sin(\omega_0 n + \theta) \quad ...(1.63)$$

$$s(n) = b(a)^n = |b|\ |a|^n \cos(\omega_0 n + \theta) + j\ |b|\ |a|^n \sin(\omega_0 n + \theta)$$

There will be the following three cases :

Case I : for $|a| = 1$, the real and imaginary parts of a complex exponential sequence are sinusoidal.

Case II : for $|a| < 1$, they correspond to sinusoidal sequences multiplied by decaying exponential.

Case III : for $|a| > 1$, they correspond to sinusoidal sequences multiplied by growing exponential.

Examples of these signals is shown in Fig. 1.26.

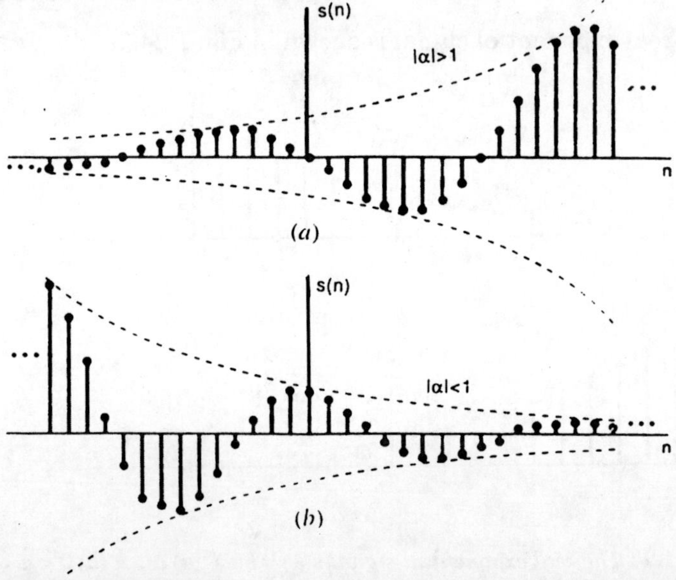

Fig. 1.26 (*a*) *Growing discrete-time sinusoidal signals* (*b*) *Decaying discrete-*

1.5.4 Signum Function

Signum function is denoted by $sgn(t)$ and it is defined as

$$sgn(t) = \begin{cases} 1\,, & t > 0 \\ 0\,, & t = 0 \\ -1\,, & t < 0 \end{cases} \qquad \text{...(1.64)}$$

Relationship between signum function, $sgn(t)$ and unit-step function $u(t)$ is given by

$$sgn(t) = 2u(t) - 1 \qquad \text{...(1.65)}$$

Signum function $sgn(t)$ is graphically shown in Fig. 1.27.

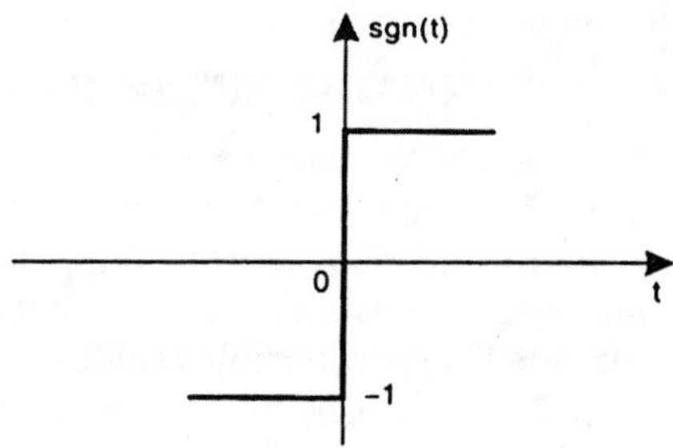

Fig. 1.27 *Graphical representation of signum function, sgn(t).*

1.6 SYSTEMS

Physical systems in general sense are an interconnection of components, devices or subsystems. A system can be viewed as a process in which input signals are transformed by the sytsem and resulting in other signals as outputs. In other words, a system responds to particular signals by producing other signals having some desired behavior. Examples of systems are audio amplifier system, electric circuit system, wheel suspension system. An audio amplifier system contains microphone, amplifier and speakers. These are subsystems of audio amplifier system. Microphone subsystem converts acoustic signals to electrical signals, amplifier subsystems amplifies the electrical signals and speaker subsystem converts the electrical signals to acoustic signals.

The electrical circuit is a system which comprises various system components such as resistors, capacitors, inductors, transistors, and so on. Voltages and currents in the circuit are signals.

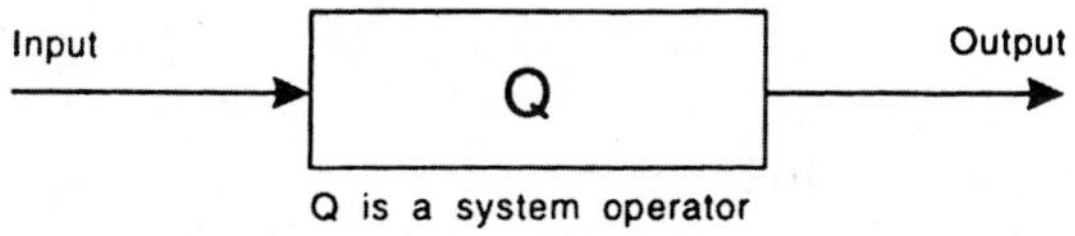

Fig. 1.28 *Block diagram representation of a system.*

The wheel suspension is the system. It comprises various system components such as wheel, tyre, spring, shock absorber, and so forth. The position and velocity of the various system components are the signals.

Signals that enter a system from some external source are called inputs signals. Signals produced by the system in its response to the input signals are called the *output signals*. Signals that occur within a system are called *internal signals*.

Signals are classified in two classes on the basis of type of signals (continuous-time or discrete-time) to be applied an input and recovered from output of a system.

1.7 CONTINUOUS-TIME AND DISCRETE-TIME SYSTEMS

Continuous-time systems are the systems for which both input and output are continuous-time signals. In other words, in continuous-time systems we apply continuous-time input signals and result in continuous-time output signals. Systems may be represented by block diagrams that feature symbols representing various components in the system. This convention is often used for continuous-time electrical network systems, and referred to as a circuit diagram . The input and output relation of a continuous-time system is given by

$$y(t) = Q_c\,[s(t)] \qquad ...(1.66)$$

where $s(t)$ and $y(t)$ are inputs and outputs of the system, respectively. and Q_c is a continuous-time operation, such as a derivative, an integral, or a multiplication by a constant output.

Block diagram of continuous-time system is shown in Fig. 1.29.

Fig. 1.29 *Continuoust-time system.*

Now we will discuss some of the simple examples of systems given as follows.

1. Example of an Electrical Network System. A simple *RC* circuit with source voltage $v_s(t)$ and capacitor voltage $v_c(t)$ is shown in Fig. 1.30.

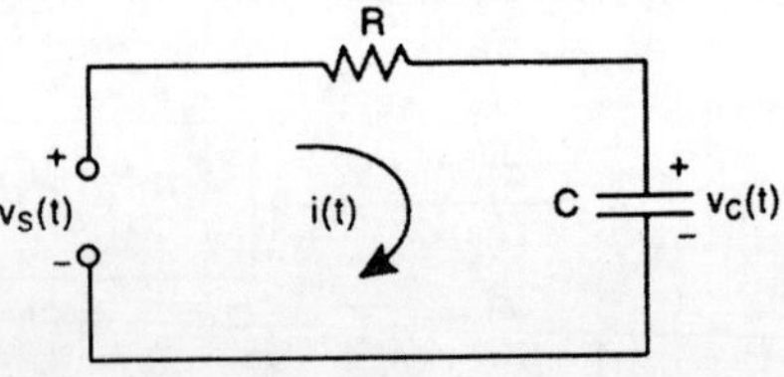

Fig. 1.30 *Electrical network system.*

In this system, $v_s(t)$ and $v_c(t)$ are input and output signals respectively. Now, we can use simple circuit analysis to derive an equation describing the relationship between the input and output. According to Ohm's law, the current $i(t)$ through the resistor is directly proportional to the voltage drop across the resistor.

Current through resistor $\propto$ voltage drop across the resistor.

or $$i(t) \propto v_s(t) - v_c(t)$$

or $$i(t) = \frac{1}{R}\,[v_s(t) - v_c(t)] \qquad ...(1.67)$$

Similarly, current $i(t)$ will be equal to the rate of change of voltage with time across the capacitor

$$i(t) = C\,\frac{dv_c(t)}{dt} \qquad ...(1.68)$$

By equating Eqs. 1.67 and 1.68, we obtain a differential equation which describes the relationship between the input $v_s(t)$ and output $v_c(t)$.

$$C\,\frac{dv_c(t)}{dt} = \frac{1}{R}\,[v_s(t) - v_c(t)]$$

or $$\frac{dv_c(t)}{dt} + \frac{1}{RC}\,v_c(t) = \frac{1}{RC}\,v_s(t) \qquad ...(1.69)$$

This is the characteristic equation of a continuous-time system.

Discrete-time systems are the systems for which both input and output are discrete-time signals. In other words, in discrete- time systems, we apply discrete-time input signals and result in discrete-time output signals. The input and output relation of a discrete-time system is given by

$$y(n) = Q_d\,[s(n)] \qquad ...(1.70)$$

where $s(n)$ and $y(n)$ are inputs and outputs of the system, respectively and Q_d is a discrete-time operation, such as multiplication, addition or delaying operations.

The block diagram of a discrete-time system is shown in Fig. 1.31.

Fig. 1.31 *Discrete-time system*.

2. Examples of a Discrete-time System. A simple example of a discrete-time system is a simple model for the balance in a bank account from month to month.

The balance at the end of nth month is given by the difference equation. This difference equation relates the balance $y(n)$ with the net deposit during nth month.

$$y(n) = 1.02\, y(n-1) + s(n) \qquad ...(1.71)$$

or $$y(n) - 1.02\, y(n-1) = s(n) \qquad ...(1.72)$$

where $s(n)$ represents net deposit during nth month

net deposit = deposits – withdrawls

and the term $1.02\, y(n-1)$ models the fact that we accrue 2% interest each month.

1.8 INTERCONNECTIONS OF SYSTEMS

Many real systems are built as interconnection of several subsystems. One example of such a system is an audio system. It is an interconnection of several subsystems such as a radio receiver, compact disc (CD) player or tape deck with an amplifier and one or more speakers. These subsystems of a system are also called its *component systems*. By viewing such a system as an interconnection of its components, we can use our understanding of the subsystems (component systems) and of how they are interconnected to analyse the behavior of the overall system.

Now we can construct a variety of system interconnections that are frequently encountered.

- Series or Cascade Interconnection of subsystems.
- Parallel Interconnection of subsystems.
- Series-Parallel Interconnection of subsystems.
- Feedback Interconnection of subsystems.

Series or Cascade Interconnection of Subsystems. In this system interconnection of two or more subsystems (component systems) are cascaded in each other. It means in this interconnection of subsystems, output of one component system is used as input of next component system. A series interconnection of two subsystems is shown in Fig. 1.32.

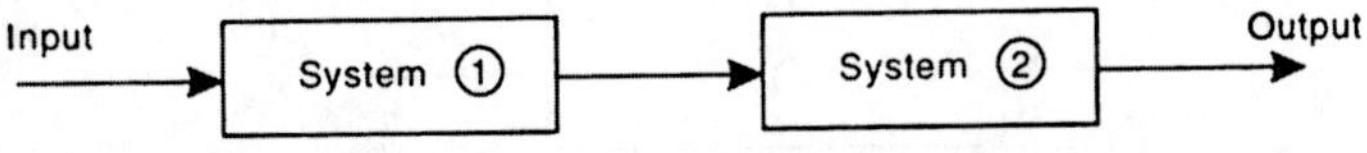

Fig. 1.32 *Series interconnection of two systems.*

Interconnection of radio receiver and an amplifier is an example of series interconnection of two subsystems and this interconnection is shown in Fig. 1.33.

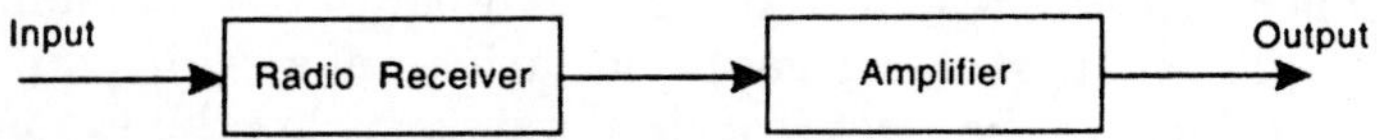

Fig. 1.33 *Example of series interconnection of two subsystems. [Radio receiver and amplifier].*

Parallel Interconnection of Subsystems. In parallel interconnection of subsystems, a common input is used in all subsystems and output is the sum of outputs of all subsystems. A parallel interconnection of two subsystems is shown in Fig. 1.34.

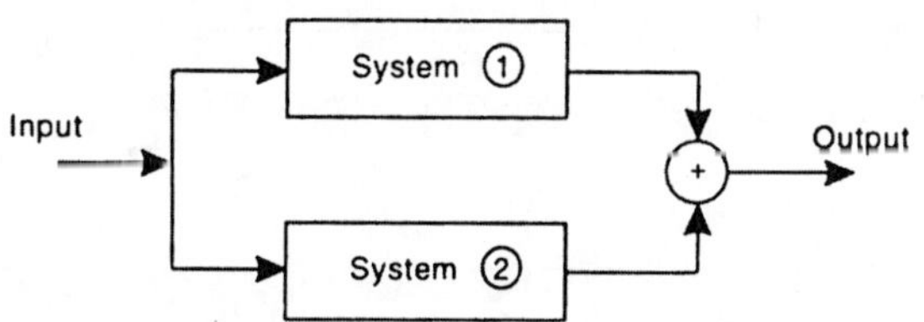

Fig. 1.34 *Parallel interconnection of two systems.*

Interconnection of simple audio system with several microphones feeding into a single amplifier and speaker system is an example of parallel interconnection of subsystems.

Series-Parallel Interconnections of Subsystems. It is the combination of both series and parallel interconnections of subsystems. A series-parallel interconnection of five subsystems is shown in Fig. 1.35.

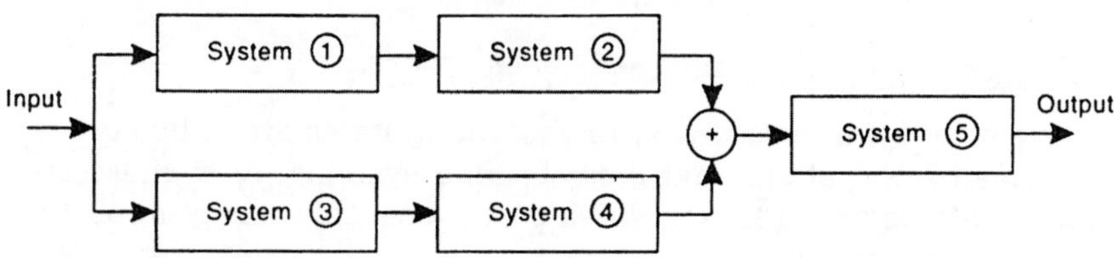

Fig. 1.35 *Series-parallel interconnection of five subsystems.*

Feedback Interconnection of Subsystems. It is a very important interconnection of subsystems. It is shown in Fig. 1.36.

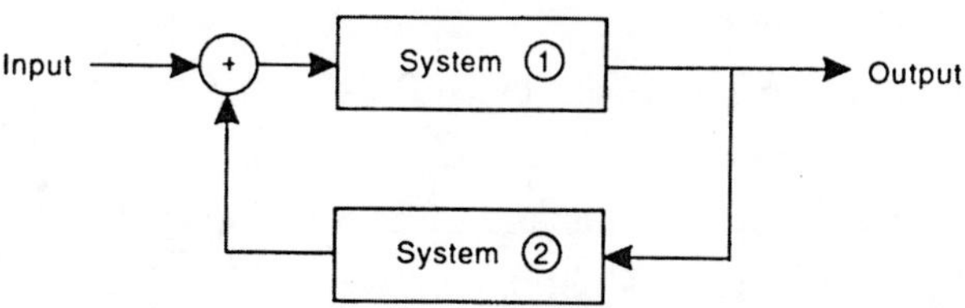

Fig. 1.36 *Feedback interconnection of two systems.*

In this interconnection of subsystems, the output of system 1 is the input to the system 2 and output of system 2 is fed back. The output of system 2 is added to the external input to produce the actual input to system 1. These systems (feedback) arise in a wide variety of applications. Most of the electrical circuit systems are the examples of feedback interconnection of subsystems. One electrical circuit system is shown in Fig. 1.37. It is a feedback interconnection of two subsystems.

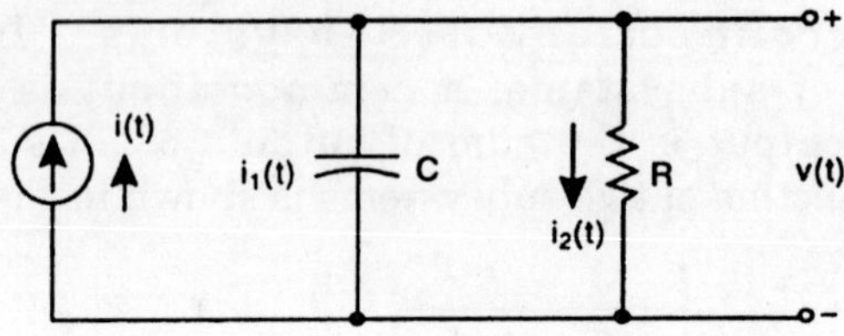

Fig. 1.37 *Simple electrical circuit system.*

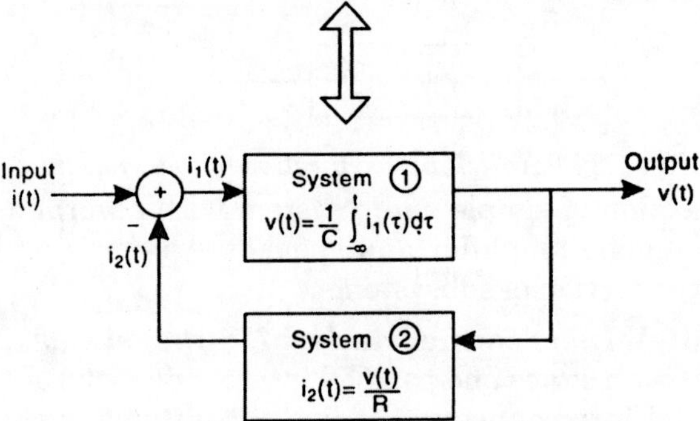

Fig. 1.38 *Block diagram showing the feedback interconnection of two circuit subsystem (elements).*

1.9 CLASSIFICATION OF SYSTEMS

We have already discussed that in general, systems are of two types : continuous-time systems and discrete-time systems. Now these two types of systems are again classified on the basis of system properties in the following groups.

1. Distributed-Parameter systems and Lumped-Parameter systems.
2. Static systems and Dynamic systems.
3. Inverse systems.
4. Causal systems and Non-causal systems.
5. Linear systems and Non-linear systems.
6. Time-invariant systems and Time-varying systems.
7. Stable systems and Unstable systems.

1.9.1 Distributed-Parameter Systems and Lumped-Parameter Systems

Distributed-Parameter Systems. In the most general sense, all physical systems contain distributed parameters because of the physical size of the components. For example, resistor is a distributed-parameter system because its resistance is distributed throughout its volume. The distributed-parameter systems are modeled with partial differential equations for continuous-time systmes or with partial difference equations for discrete-time systems.

If the size of the component is large with respect to the wavelength of the highest frequency component present in the signals associated with it then it is called *distributed- parameter system.*

Lumped-Parameter Systems. If the action of a system component occuring at a point and the size of the component is small with respect to the wavelength of the higest-frequency component present in the signals associated with it and it is true for all components in a system, then the system is called *Lumped-parameter system.* The lumped-parameter systems are modeled with ordinary differential equations for continuous-time systems or with ordinary difference equations for discrete-time systems.

Example of Distributed-Parameter Systems and Lumped-Parameter Systems. 50 Hz electric power system is an example of a distributed-parameter system as well as a lumped-parameter systems depends upon the wavelength of the signal and size of the component. The wavelength of the signal in this case is 6000 kms. The electrical system inside a building can be treated as a lumped-parameter system and same system is treated as distributed-parameter system for long-distance transmission lines. In this book, we will consider only lumped-parameter systems.

1.9.2 Static Systems and Dynamic Systems

On the basis of memory requirement, systems are categorized as static or dynamic systems.

Static Systems. These systems are also called memoryless sytems. Physically, these systems contain no energy-storage elements. The equation relating its output signal to its input signal contains no derivative, integrals, or signal delays.

A static or memoryless system is a system with an output signal that at any specific time depends on the value of the input signal only at that time.

Examples of static systems are resistor, identity system etc. System $y(n) = 3\,s(n) - s^2(n)$ is memoryless (static) system because the value of output $y(n)$ at any particular time n_0 depends only on the value of input $s(n)$ at that time.

Resistor is a memoryless system because its output $y(t)$ at any particular time t_0 depends only on the value of input $s(t)$ at that time.

$$y(t) = Rs(t) \quad ...(1.73)$$

where $y(t) = v(t)$, is the voltage across the resistor

$s(t) = i(t)$, is the current that flows through resistor

R is the resistance.

Another simple memoryless (static) system is the identity system whose both output and input are equal.

Continuous-time identity system is described by the following input-output relation

$$y(t) = s(t) \quad ...(1.74)$$

and discrete-time identity system is described by

$$y(n) = s(n) \quad ...(1.75)$$

Dynamic Systems. A dynamic system or system with memory is a system with an output signal that at any specified time depends on the value of the input signal at both the specified time and at other times. These systems have one or more energy storage elements. Input-output relationship of a dynamic continuous-time system is described by its differential equation. Similarly, input-output relationship of a dynamic discrete-time system is described by its difference equation.

In dynamic systems, stored energies define the initial conditions. These initial conditions are used to solve the differential or difference equations.

Examples of dynamic systems or systems with memory are given below.

A capacitor is an example of a continuous-time system with memory and its input-output relationship is given by

$$y(t) = \frac{1}{C}\int_{-\infty}^{t} s(\tau)\, d\tau \quad ...(1.76)$$

where $y(t)$ is the output voltage

$s(t)$ is the input current

and C is the capacitance of the capacitor.

An accumulator is an example of a discrete-time system with memory and its input-output relationship is given by

$$y(n) = \sum_{k=-\infty}^{\infty} s(k) \quad ...(1.77)$$

A delay element is also an example of dynamic system and its input-output relationship is given by

$$y(n) = s(n-1) \quad ...(1.78)$$

1.9.3 Invertible System and Inverse System

A system is said to be invertible if there is a one-to-one correspondence between its input and output signals.

If a system is invertible, then an inverse system exists. The cascading of an invertible system and its inverse system is equivalent to the identity system.

The frequency response of an inverse system is basically reciprocal of the frequency response of the original system [or invertible system].

An example of an invertible continuous-time system is given by

$$y_1(t) = 3\,s(t) \qquad ...(1.79)$$

and its inverse system will be given by

$$y_2(t) = \frac{1}{3}\,y_1(t) \qquad ...(1.80)$$

These invertible system and inverse systems are cascaded and shown in Fig. 1.39.

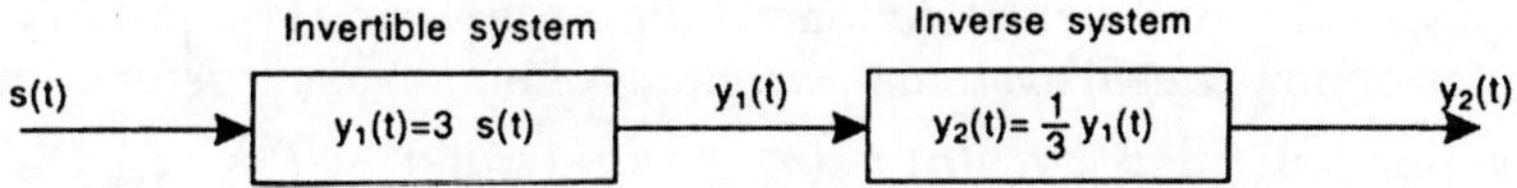

Fig. 1.39 *Inverse system for the invertible system described by* $y_1(t) = 3\,s(t)$.

An example of an invertible discrete-time system is the accumulator and it is described by

$$y_1(n) = \sum_{k=-\infty}^{\infty} s(k) \qquad ...(1.81)$$

and its inverse system is given by

$$y_2(n) = y_1(n) - y_1(n-1) \qquad ...(1.82)$$

These invertible and its reverse systems are cascaded. These are shown in Fig. 1.40.

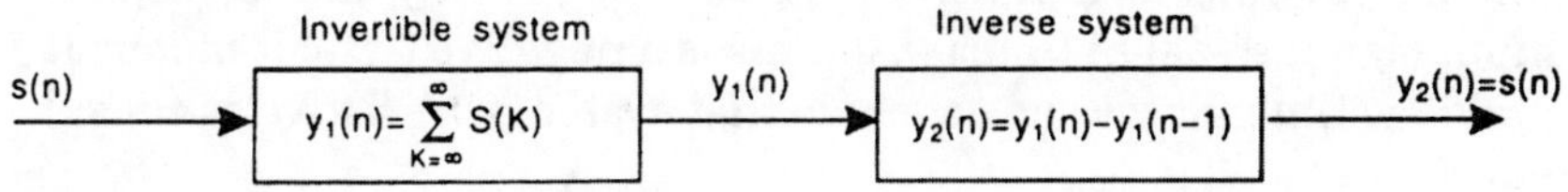

Fig. 1.40 *Inverse system for the invertible system described by* $y_1(n) = \sum_{k=-\infty}^{\infty} s(k)$.

For lossless coding, the input to the encoder must be exactly recoverable from the output. It means that the encoder must be invertible.

1.9.4 Causal Systems and Non-Causal Systems

Causal Systems. A system is called causal system if the output at any time depends only on values of the input at the present and previous time. These systems are often referred to as being non-anticipative because system output does not anticipate future values of the input. All the static systems are always causal systems. All the real-time systems are also causal systems because these systems cannot know future values of the input signal when it constructs output signal. Examples of causal systems are :

Resistor is described by $y(t) = R\, s(t)$. It is causal continuous-time system because output $y(t)$, *i.e.*, voltage depends only on the input $s(t)$, *i.e.*, current at the present time.

Delay element is described by $y(n) = s(n-1)$. It is a causal discrete-time system because output $y(n)$ depends only on the past input $s(n-1)$, not on future input.

Non-Causal System. A system is called non-causal system if its output at a given time depends on the input at future time. Some of the non-real time systems are non-causal systems. Image processing systems are the examples of non-causal systems. Examples of non-causal systems are :

Systems defined by $y(n) = s(n) - s(n+1)$ and $y(t) = s(t+1)$ are non-causal systems because output at time t or n depends on the input at future values.

Example 1.2 Check for causality of the following systems :

(*a*) $y(n) = s(-n)$

(*b*) $y(t) = s(t) \cos(t+1)$

Solution. (*a*) Discrete-time system is defined by

$$y(n) = s(-n) \qquad ...(i)$$

Put $n = n_0$, where n_0 is a positive.

$$y(n_0) = s(-n_0) \qquad ...(ii)$$

The output $y(n_0)$ at a positive time n_0 depends only on the value of the input signal $s(-n_0)$ at time $(-n_0)$. This is a negative time. It means it is a previous time. So we can conclude that system $y(n) = s(-n)$ is causal system.

Again put $n = -n_0$ in above Eq. (*i*), we get

$$y(-n_0) = s(n_0) \qquad ...(iii)$$

so that the output at this time depends on future values of input. Hence the above system is a non-causal system.

(*b*) Given a system defined by

$$y(t) = s(t) \cos(t+1) \qquad ...(iv)$$

The output of above system at any time t is equal to the input at the same time multiplied time-varying function cos $(t + 1)$.

In this system, only the present value of the input $s(t)$ influences the present value of the output $y(t)$. Now we can conclude that the above system is a causal system.

Order of a System. The order of a continuous-time system corresponds to the highest derivative of the output signal that appears in input-output differential equation.

Similarly, the order of a discrete-time system corresponds to largest number of units of delay of the output discrete-time signal appearing in the input-output difference equations.

The order of a system is equal to the number of energy storage elements in the system when the system is reduced to configuration with the fewest-energy storage elements that produces the same input-output equation. For example, Parallel capacitors reduced to one capacitor.

Example 1.3 Find the order of the system which are described by following input-output relations :

(*a*) Input-output equation of a continuous-time system is given by

$$\frac{d^3y(t)}{dt^3} + 3\frac{dy(t)}{dt} + 2y(t) = \frac{ds(t)}{dt} + s(t)$$

where $y(t)$ is the output of the system and $s(t)$ is the input of the system.

(*b*) Input-output equation of a discrete-time system is given by

$$y(n) = y(n-2) - 4y(n-1) + s(n) - s(n-1)$$

where $y(n)$ and $s(n)$ are output and input of the system.

Solution. (*a*) The continuous-time system is a third order system since the highest order derivative of the output $y(t)$ in differential equation is three.

(*b*) The discrete-time system is a second order system since the largest number of units of delay of the output $y(n)$ in difference equation is two.

1.9.5 Linear Systems and Non-Linear Systems

On the basis of superposition principle, system are two types : linear systems or non-linear systems.

Superposition Principle. It is stated as response of the system to weighted sum of input signals be equal to the corresponding weighted sum of responses of the system to each of the individual input signals.

Linear Systems. A linear system (continuous-time/discrete-time) is a system that possesses the important property of superposition.

Superposition property of a system is given by two properties. These are additivity property and scaling or homogeneity property.

For a continuous-time system, let $y_1(t)$ be the response (output) of the system to an excitation (input) $s_1(t)$ and $y_2(t)$ be the response corresponding to the input $s_2(t)$. Then the system is linear if it satisfies two properties separately.

1. Additivity property and it is stated as the response to $s_1(t) + s_2(t)$ will be equal to $y_1(t) + y_2(t)$.
2. Homogeneity property and it is stated as the response of $A\, s_1(t)$ will be equal to $A\, y_1(t)$, where A is any complex constant.

Above two properties defining a linear system can be combined into a single statement.

$$T\,[A\, s_1(t) + B\, s_2(t)] = A\, y_1(t) + B\, y_2(t) \qquad ...(1.83)$$

where A and B are arbitrary constants.

$s_1(t)$ and $s_2(t)$ are the arbitrary inputs and $y_1(t)$ and $y_2(t)$ are the responses of the continuous-time system to $s_1(t)$ and $s_2(t)$ respectively.

Superposition property for discrete-time system is given by

$$T\,[A\, s_1(n) + B\, s_2(n)] = A\, y_1(n) + B\, y_2(n) \qquad ...(1.84)$$

where T is system operator

A and B are arbitrary constants

$s_1(n)$ and $s_2(n)$ are arbitrary inputs

$y_1(n)$ and $y_2(n)$ are the responses of the discrete-time system to $s_1(n)$ and $s_2(n)$ respectively.

Non-Linear System. A non-linear system (continuous-time/discrete-time) is a system which does not satisfy the superposition property.

Note that a system can be linear without being time-invariant and it can be time-invariant without being linear.

We can understand linear and non-linear systems with the following examples.

Example 1.4 Check the continuous-time systems for linearity

(*i*) $y(t) = t\, s(t)$

(*ii*) $y(t) = s^2(t)$

Solution. (*i*) Given input-output relationship of a continuous-time system by $y(t) = t\, s(t)$

The corresponding outputs (responses) for two continuous-time inputs $s_1(t)$ and $s_2(t)$ are

$$y_1(t) = t\, s_1(t)$$
$$y_2(t) = t\, s_2(t)$$

A linear combination of two inputs results in the response (output)

$$y_3(t) = T\,[s_3(t)] = T\,[A\, s_1(t) + B\, s_2(t)]$$
$$= t\,[A\, s_1(t) + B\, s_2(t)]$$
$$= A\, t\, s_1(t) + B\, t\, s_2(t) \qquad ...(1)$$

A linear combination of the two outputs results in the output

$$A\, y_1(t) + B\, y_2(t) = A\, t\, s_1(t) + B\, t\, s_2(t) \qquad ...(2)$$

Since both responses (outputs) in Eqs. 1 and 2 are equal, the system is linear.

(*ii*) Given input-output relationship of a continuous-time system be $y(t) = s^2(t)$

The corresponding outputs for two continuous-time inputs $s_1(t)$ and $s_2(t)$ are

$$y_1(t) = s_1^2(t)$$
$$y_2(t) = s_2^2(t)$$

A linear combination of two inputs results in output (response)

$$y_3(t) = T\,[s_3(t)]^2 = T\,[A\, s_1(t) + B\, s_2(t)]^2$$
$$= 1\,[A\, s_1(t) + B\, s_2(t)]^2$$
$$= A^2\, s_1^2(t) + B^2\, s_2^2(t) + 2AB\, s_1(t)\, s_2(t) \qquad ...(3)$$

A linear combination of the two outputs results in the output

$$A\, y_1(t) + B\, y_2(t) = A\, s_1^2(t) + B\, s_2^2(t) \qquad ...(4)$$

Since both outputs (responses) in Eqs. 3 and 4 are not equal, so the system is non-linear.

Example 1.5 Check the discrete-time system for linearity

$$y(n) = 3\, s(n) + 5$$

where $y(n)$ and $s(n)$ are output and input of a discrete-time system respectively.

Solution. Given input-output relationship of a discrete- time system is

$$y(n) = 3\, s(n) + 5$$

The corresponding outputs for two discrete-time inputs $s_1(n)$ and $s_2(n)$ are

$$y_1(n) = 3\, s_1(n) + 5$$
$$y_2(n) = 3\, s_2(n) + 5$$

A linear combination of two inputs results in output

$$y_3(n) = T\,[s_3(n)] = T\,[A\, s_1(n) + B\, s_2(n)]$$

$$= 3\,[A\, s_1(n) + B\, s_2(n)] + 5$$

$$= 3A\, s_1(n) + 3B\, s_2(n) + 5 \qquad \text{...(1)}$$

Linear combination of two outputs results in the output

$$A\, y_1(n) + B\, y_2(n) = A\,[3\, s_1(n) + 5] + B\,[3\, s_2(n) + 5]$$

$$= 3A\, s_1(n) + 5A + 3B\, s_2(n) + 5B$$

$$= 3A\, s_1(n) + 3B\, s_2(n) + 5A + 5B \qquad \text{...(2)}$$

Since both outputs (responses) given in Eqs. 1 and 2 are not equal, so the system is non-linear.

Alternate Method. We have given discrete-time system

$$y(n) = 3\, s(n) + 5$$

To determine the zero-output response of the above discrete-time systems.

Put $s(n) = 0$

Zero-input response, $y_0(n) = [3\, s(n) + 5]_{s(n)=0}$

$$= 3 \times 0 + 5 = 5$$

This system violates the "zero-in/zero-out" property of the linear systems, so it is a non-linear system. System $y(n) = 3\, s(n) + 5$ is a non-linear system even if it is characterized by linear difference equation. The output of a non-linear system is represented as the sum of the output of a linear system and zero-input response of the system.

Above non-linear system $y(n) = 3\, s(n) + 5$ is represented by one linear system

$$y(n)_{\text{linear}} = 3\, s(n)$$

and zero-input response, $y_0(n) = [y(n)]_{s(n)=0} = [3\, s(n) + 5]_{s(n)=0} = 5$

Some of the definitions regarding linear and non-linear systems. These terms are :

1. Zero-initial-energy response.
2. Zero-input response, and
3. Decomposable system response.

Zero-Initial-Energy Response. The zero-initial-energy response of a system is the system output to an input signal when all initial stored energies are zero. It is also called zero-state response in state space techniques.

Zero-Input Response. The zero-input response of a system is the system output (response) to initial stored energies when the input signal is zero.

Decomposable System Response. A decomposable system response can be written as the sum of the system's zero-initial energy response and zero-input response.

Decomposable system response = zero-initial energy response + zero-input response

---

1.9.6 Time-Invariant Systems and Time-Varying Systems

On the basis of time-invariance property, systems are two types :

Time-Invariant Systems. Conceptually, a system is called time-invariant system if the behavior and input-output characteristics of the system does not change with time.

If the response to a delayed input and the delayed response are equal then the system is called *time-invariant system.*

The response to a delayed input and delayed response of a system are denoted by $y(n, n_0)$ [or $y(t, t_0)$] and $y(n - n_0)$ [or $y(t - t_0)$] respectively.

If $y(n, n_0) = y(n - n_0)$ then the system is called *time-invariant discrete-time system*.

If $y(t, t_0) = y(t - t_0)$ then the system is called *time-invariant continuous-time system*.

Time-Varying Systems. If the input-output characteristics of a system changes with time then the system is called *time- varying*. In other words, if the response to a delayed input and delayed response are not equal then the system is referred as time-varying system.

For example, the *RC* circuit shown in Fig. 1.37 is a time- invariant continuous-time system if the values of resistance *R* and capacitance *C* are constant over time.

If the values of *R* and *C* change with time then this system will be time-varying.

Differentiator $y(n) = s(n) - s(n - 1)$ is a time- invariant discrete-time system and time multiplier $y(n) = n\, s(n)$ is a time-varying discrete-time system.

Algebraic, differential and difference equations which are the mathematical models for time-invariant systems all have constant coefficients for the signal terms in them.

These constant coefficients are produced by physical systems either continuous-time or discrete-time type with components that do not vary with time.

Example 1.6 Check the following continuous-time system for time-invariance

(*i*) $y(t) = \sin [s(t)]$

(*ii*) $y(t) = t\, s(t) + 4$

(*iii*) $y(t) = s(3t)$

where $y(t)$ and $s(t)$ are input and output of the continuous-time system respectively.

Solution. (*i*) Given input-output relationship for a continuous-time system is

$$y(t) = \sin [s(t)]$$

The response to a delayed input is

$$y(t\ , t_0) = \sin [s(t - t_0)] \quad ...(1)$$

The delayed response is

$$y(t\ , t_0) = \sin [s(t - t_0)] \quad ...(2)$$

Both the responses given in Eqs. 1 and 2 are equal and therefore, this system is time-invariant.

(*ii*) Given input-output relationship for a continuous-time system is

$$y(t) = t\ s(t) + 4$$

The response to a delayed input is

$$y(t\ , t_0) = t\ s(t - t_0) + 4 \quad ...(3)$$

The delayed response is given by

$$y(t - t_0) = (t - t_0)\ s(t - t_0) + 4 \quad ...(4)$$

Both the responses given in Eqs. 3 and 4 are not equal in each other and therefore, this system is time-varying.

(*iii*) Given input-output relationship for a continuous-time system is

$$y(t) = s(3t)$$

This system represents a time scaling.

That is, $y(t)$ is a time compressed version of input $s(t)$ by a factor 3. Intuitively, then, any time shift in the input will also be compressed by the factor of 3. That is why this system is not time-invariant.

Example 1.7 Check the following discrete-time systems for time-invariance

(*i*) $y(n) = n\ s(n)$

(*ii*) $y(n) = s(n) - s(n - 1)$

Solution. (*i*) Given input-output relationship for a discrete-time system is

$$y(n) = n\ s(n)$$

It is a time-multiplier.

The response to a delayed input is given by

$$y(n\ , n_0) = n\ s(n - n_0) \quad ...(1)$$

The delayed response is given by

$$y(n - n_0) = (n - n_0)\ s(n - n_0) \quad ...(2)$$

Both the responses given in Eqs. 1 and 2 are not equal in each other and therefore, this system is time-varying.

(*ii*) Given input-output relationship for a discrete-time system is

$$y(n) = s(n) - s(n-1)$$

It is a discrete-time differentiator.

The response to a delayed input is given by

$$y(n\ , n_0) = s(n - n_0) - s(n - n_0 - 1) \quad ...(3)$$

The delayed response is given by

$$y(n - n_0) = s(n - n_0) - s(n - n_0 - 1) \quad ...(4)$$

Both the responses given in Eqs. 3 and 4 are equal in each other and therefore, this system is a time-invariant.

1.9.7 Stable Systems and Unstable Systems

On the basis of system stability, systems are categorized as stable or unstable systems. Here we will study only Bounded-Input Bounded-Output (BIBO) stability. A bounded signal has an amplitude that remains finite. We want a physical system to have a bounded output for any bounded input so that its output does not grow unreasonably large. In other words, we can say that such a system has bounded-input, bounded-output (BIBO) stability.

BIBO Stable System. An initially relaxed system (continuous-time/discrete-time) is said to be bounded-input, bounded-output (BIBO) stable only if every bounded-input produces a bounded-output. Informally, a stable system is one in which small inputs lead to responses (outputs) that do not diverge.

BIBO Unstable System. A system which does not produce a bounded output from a bounded input is termed as BIBO unstable system.

Stability of a physical system generally results from the presence of mechanism that dissipates energy.

Example of stable system is a pendulum shown in Fig. 1.41. In the pendulum, the input is the applied force $s(t)$ and the output is the angular deviation $y(t)$ from the vertical position. In this case, gravity applies a restoring force. This restoring force tends to return the

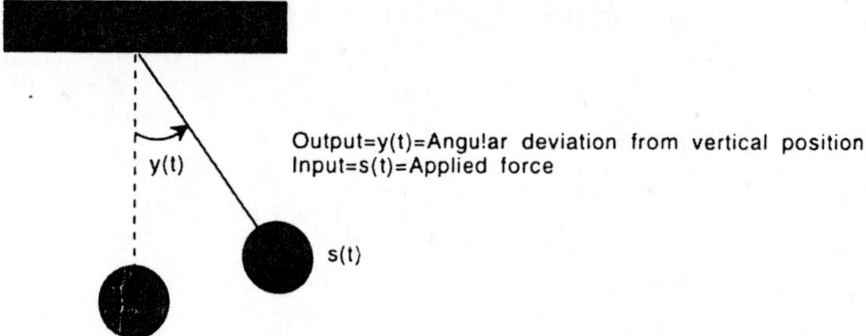

Fig. 1.41 *Stable normal pendulum.*

pendulum to the vertical position. Frictional losses due to drag tend to slow it down.

Example of an unstable system is an inverted pendulum shown in Fig. 1.42. In this case effect of gravity is to apply a force that tends to increase the deviation from vertical position. So in this system, a small applied force tends to a large vertical deflection causing the pendulum to topple over but there is no retarding force due to friction.

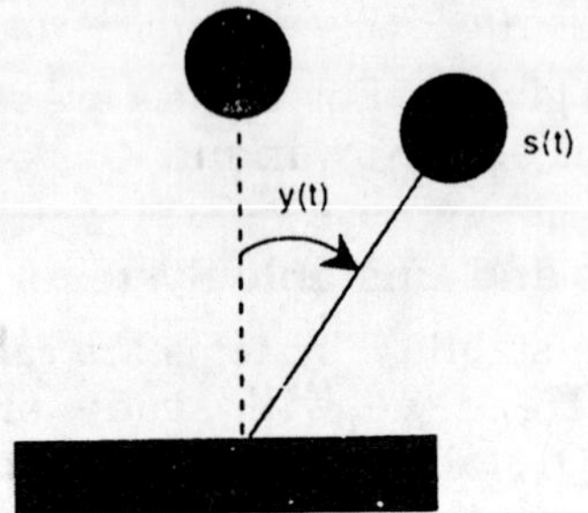

Fig. 1.42 *An unstable inverted pendulum.*

Other examples of stable systems are : Resistor in *RC* circuit is a stable system because it dissipates energy. Automobile system is also a stable system because it also dissipates energy through friction.

Other examles of unstable systems are : Model of chain reactions is an example of unstable system.

The model for a bank account balance is an example of an unstable system.

An accumulator is an unstable system because its output $y(n)$ grows continually even if input $s(n)$ is bounded.

For example, if the input $s(n)$ to the accumulator is a unit step sequence $u(n)$, the output will be given by

$$y(n) = \sum_{k=-\infty}^{\infty} s(n) \qquad ...(1.85)$$

After putting input $s(n) = u(n) = \begin{cases} 1, & n \geq 0 \\ 0, & n < 0 \end{cases}$, we get

$$y(n) = \sum_{k=-\infty}^{\infty} u(n) = (n+1)\, u(n)$$

or

$$y(n) = (n+1)\, u(n) \qquad ...(1.86)$$

$n = 0 \qquad y(0) = (0 + 1)\, u(0) = 1 \times 1 = 1$

$n = 1 \qquad y(1) = (1 + 1)\, u(1) = 2 \times 1 = 2$

$n = 2 \qquad y(2) = (2 + 1)\, u(2) = 3 \times 1 = 3$

and so on.

and $y(n)$ grows without bound. So accumulator is an unstable system.

Example 1.8 Check the following systems for stability :

(*i*) $y(t) = t\, s(t)$

(*ii*) $y(t) = e^{s(t)}$

Solution. (*i*) We have a system which is characterised by

$$y(t) = t\, s(t) \qquad ...(1)$$

Now we are going to check the stability of the above system by using simple bounded inputs such as a unit step function or constant.

Substituting input $s(t) = 1$ = constant in Eq. 1 yields

$$y(t) = t\, s(t) = t \qquad ...(2)$$

Here $y(t)$ is unbounded.

So we can say that this system is BIBO unstable because Bounded input, $s(t) = 1$, produces unbounded output $y(t) = t$.

(*ii*) We have a system which is characterised by

$$y(t) = e^{s(t)} \qquad ...(3)$$

Now we are assuming that input $s(t)$ is a bounded input *i.e.*,

$$|s(t)| < B < \infty \quad \text{for all time } (t) \qquad ...(4)$$

Then output of above system will be given by

$$|y(t)| < e^{B} < \infty \quad \text{for all time } (t) \qquad ...(5)$$

where B is arbitrary positive number.

We conclude that above system is a BIBO stable because it produces bounded output from bounded input.

Example 1.9 Find out which of the following signals are periodic

(*a*) $s(t) = \sin 15\pi t$

(*b*) $s(t) = \sin 20\pi t$

(*c*) $s(t) = \sin \sqrt{2}\pi t$

(*d*) $s(t) = \sin 5\pi t$

(*e*) $s(t) = \sin 15\pi t + \sin 20\pi t$

(*f*) $s(t) = \sin 20\pi t + \sin 5\pi t$

Solution. (*a*) Signal $s(t) = \sin 15\pi t$ is periodic. The fundamental period is

$$T_0 = \frac{2\pi}{\omega} = \frac{2\pi}{15\pi} = 0.1333 \text{ seconds}$$

(*b*) Signal $s(t) = \sin 20\pi t$ is periodic. The fundamental period is

$$T_0 = \frac{2\pi}{\omega} = \frac{2\pi}{20\pi} = 0.1 \text{ seconds}$$

(*c*) Signal $s(t) = \sin \sqrt{2}\pi t$ is periodic. The fundamental period is

$$T_0 = \frac{2\pi}{\omega} = \frac{2\pi}{\sqrt{2}\pi} = 1.414 \text{ seconds}$$

(*d*) Signal $s(t) = \sin 5\pi t$ is periodic. The fundamental period is

$$T_0 = \frac{2\pi}{\omega} = \frac{2\pi}{5\pi} = 0.4 \text{ seconds}$$

(*e*) Signal $s(t) = \sin 15\pi t + \sin 20\pi t$ suppose that

$$s(t) = s_1(t) + s_2(t)$$

with $s_1(t) = \sin 15\pi t$

and $s_2(t) = \sin 20\pi t$

Fundamental period of

$$s_1(t) = T_{01} = \frac{2\pi}{15\pi} = 0.1333 \text{ seconds}$$

Fundamental period of

$$s_2(t) = T_{02} = \frac{2\pi}{20\pi} = 0.1 \text{ seconds}$$

$\frac{T_{01}}{T_{02}} = \frac{0.1333}{0.1}$, this ratio cannot be expressed as a ratio of integers.

Hence, signal $s(t) = \sin 15\pi t + \sin 20\pi t$ is not periodic.

(*f*) Signal $s(t) = \sin 20\pi t + \sin 5\pi t$ support that

$$s(t) = s_1(t) + s_2(t)$$

with $s_1(t) = \sin 20\pi t$ and $s_2(t) = \sin 5\pi t$

Fundamental period of $s_1(t)$

$$T_{01} = \frac{2\pi}{20\pi} = 0.1 \text{ seconds}$$

and fundamental period of $s_2(t)$

$$T_{01} = \frac{2\pi}{5\pi} = 0.4 \text{ seconds}$$

$\frac{T_{01}}{T_{02}} = \frac{0.1}{0.4} = \frac{1}{4}$, this ratio is the ratio of integers, hence signal $s(t) = \sin 20\pi t + \sin 5\pi t$ is periodic.

Example 1.10 Check the following systems described by differential equations for linearity

(*a*) $\frac{d\,y(t)}{dt} + 2\,y(t) = s(t)$ (*b*) $\frac{d\,y(t)}{dt} + y(t) + 4 = s(t)$

Solution. (*a*) Let the response of the continuous-time system to input $s_1(t)$ be $y_1(t)$ and the response of the system to input $s_2(t)$ be $y_2(t)$.

Thus, for the input $s_1(t)$, the describing equation is

$$\frac{d\,y_1(t)}{dt} + 2\,y_1(t) = s_1(t) \qquad ...(1)$$

and for the input $s_2(t)$,

$$\frac{d\,y_2(t)}{dt} + 2\,y_2(t) = s_2(t) \qquad ...(2)$$

Multiplying Eqs. 1 and 2 by A_1 and A_2, respectively and adding yields

$$A_1 \frac{d\,y_1(t)}{dt} + 2A_1y_1(t) + A_2 \frac{d\,y_2(t)}{dt} + 2A_2y_2(t) = A_1s_1(t) + A_2s_2(t)$$

i.e., $$\frac{d}{dt}\,[A_1y_1(t) + A_2y_2(t)] + 2\,[A_1y_1(t) + A_2y_2(t)] = A_1s_1(t) + A_2s_2(t) \qquad ...(3)$$

The response of the system to the input $A_1s_1(t) + A_2s_2(t)$ is $A_1y_1(t) + A_2y_2(t)$.

Thus, the superposition condition is satisfied and hence the system is linear.

(*b*) Thus, for input $s_1(t)$, the describing equation is

$$\frac{d\,y_1(t)}{dt} + y_1(t) + 4 = s_1(t) \qquad ...(1)$$

and for input $s_2(t)$,

$$\frac{d\,y_2(t)}{dt} + y_2(t) + 4 = s_2(t) \qquad ...(2)$$

Multiplying Eqs. 1 and 2 by A_1 and A_2 respectively, and adding yields,

$$\frac{d}{dt}\,[A_1y_1(t) + A_2y_2(t)] + [A_1y_1(t) + A_2y_2(t)] + y\,[A_1 + A_2]$$
$$= A_1s_1(t) + A_2s_2(t) \qquad ...(3)$$

This equation can not be put into the same form as the original differential equation describing the system. Hence the system is non-linear.

Example 1.11 Write down the corresponding equation for the given signal.

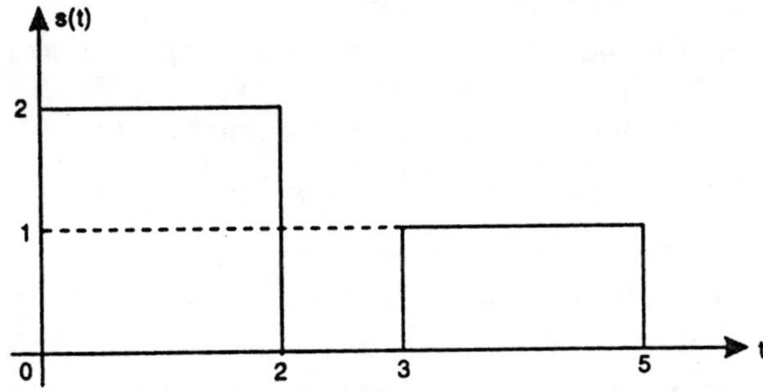

Fig. 1.43

Solution. Representation through addition of two unit step functions. Thus signal $s(t)$ can be obtained by adding both the pulses, *i.e.*,

$$s(t) = 2\,[u(t) - u(t-2)] + [u(t-3) - u(t-5)]$$

Representation through multiplication of two unit step functions

$$s(t) = 2\,[u(t)\,u(-t+2)] + [u(t-3)\,u(-t+5)]$$
$$= 2[u(t)\,u(2-t)] + [u(t-3)\,u(5-t)]$$

Example 1.12 Draw the single-sided and double-sided amplitude and phase spectra of continuous-time signal

$$s(t) = 8\sin\left(20\pi t - \frac{\pi}{6}\right), \quad -\infty, t < \infty$$

Solution. The single-sided spectra is plotted by expressing signal $s(t)$ as the real part of the rotating phasor.

Now using the trignometric identity $\cos\left(u - \frac{\pi}{2}\right) = \sin u$

$$s(t) = 8\sin\left(20\pi t - \frac{\pi}{6}\right)$$
$$= 8\cos\left(20\pi t - \frac{\pi}{6} - \frac{\pi}{2}\right)$$
$$= 8\cos\left(20\pi t - \frac{2}{3}\pi\right) \quad \text{...(1)}$$

The single-sided amplitude and phase spectra is shown in Fig. 1.44 (*a*).

The signal has an amplitude of 8 units at $f = 10$ Hz.

To plot the double-sided spectrum, the signal is converted into exponential form using Euler's relation.

$$s(t) = 8\cos\left(20\pi t - \frac{2}{3}\pi\right)$$
$$= 8\left[\frac{e^{j[20\pi t - (2/3)\pi]} + e^{-j[20\pi t - (2/3)\pi]}}{2}\right]$$
$$= 4\,e^{j[20\pi t - (2/3)\pi]} + 4\,e^{-j[20\pi t - (2/3)\pi]} \quad \text{...(2)}$$

The double-sided amplitude and phase spectra are shown in Fig. 1.44 (*b*). The signal has two components at $f = 10$ Hz and $f = -10$ Hz. The amplitude of these components are 4 units each and the phase of these components are $-\frac{2\pi}{3}$ and $\frac{2\pi}{3}$ radians, respectively.

Example 1.13 Verify whether signal $s(t) = A\,e^{-\alpha t}\,u(t)$, $\alpha > 0$ is an energy signal or not.

Solution. Given : $s(t) = Ae^{-\alpha t}\,u(t)$, $\alpha > 0$, where A and α are constants.

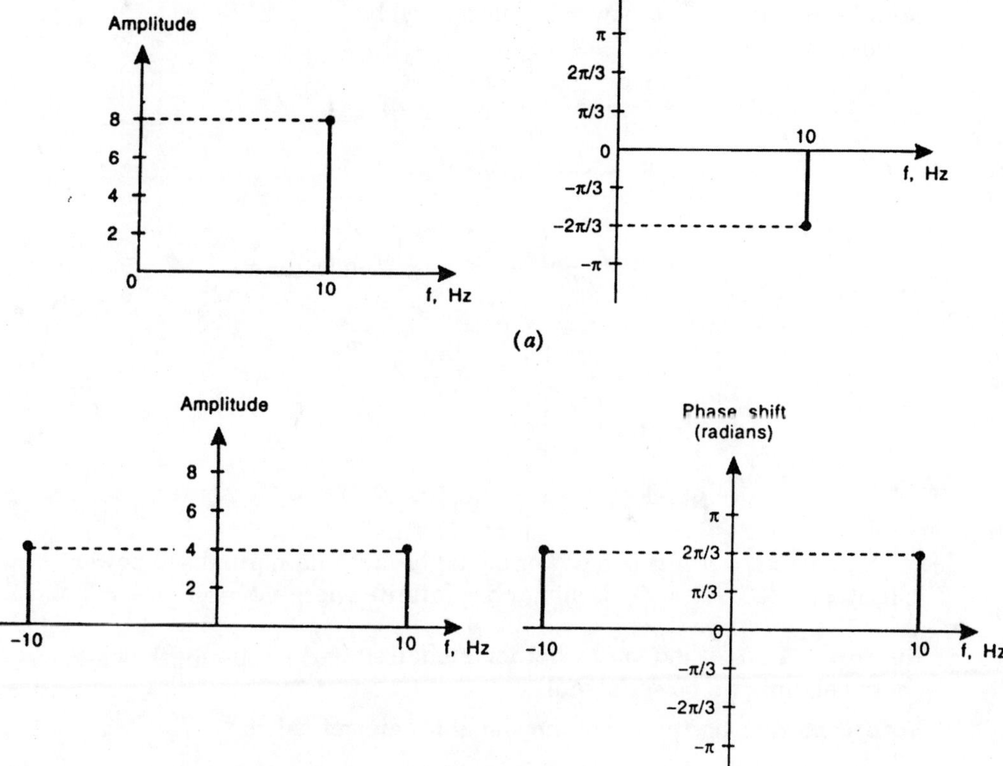

Fig. 1.44 *Amplitude and phase spectra (a) single-sided, and (b) double-sided.*

Total normalized energy is given by

$$E = \lim_{T \to \infty} \int_{-T}^{T} |s(t)|^2 \, dt = \int_{-\infty}^{\infty} |s(t)|^2 \, dt$$

$$= \int_{-\infty}^{\infty} [Ae^{-\alpha t} u(t)]^2 \, dt \qquad \left[\text{Since } u(t) = \begin{cases} 1, & t > 0 \\ 0, & t < 0 \end{cases} \right]$$

$$= \int_{0}^{\infty} A^2 e^{-2\alpha t} \, dt = A^2 \left[\frac{e^{-2\alpha t}}{-2\alpha} \right]_0^{\infty} = \frac{A^2}{2\alpha}$$

or $$E = \frac{A^2}{2\alpha}$$

$s(t)$ is an energy signal, since $0 < E < \infty$, from the above result obtained. Here average power is zero.

Example 1.14 Verify whether signal $s(t) = A\,u(t)$, is a power signal or not.

Solution. Normalized power is determined by

$$P = \lim_{T \to \infty} \frac{1}{2T} \int_{-T}^{T} |s(t)|^2 \, dt$$

$$= \lim_{T \to \infty} \frac{1}{2T} \int_{-T}^{T} [Au(t)]^2 \, dt$$

$$= \lim_{T \to \infty} \frac{1}{2T} \int_{0}^{T} A^2 \, dt \qquad \left[\text{Since } u(t) = \begin{cases} 1, & t > 0 \\ 0, & t < 0 \end{cases} \right]$$

$$= \lim_{T \to \infty} \frac{A}{2T} [\, T \,]_0^T = \lim_{T \to \infty} \frac{A}{2T} T$$

$$= \lim_{T \to \infty} \left[\frac{A}{2} \right]$$

or $$P = \frac{A}{2}$$

Signal $s(t) = A\,u(t)$ is a power signal because its normalised power is finite, *i.e.*, $0 < P < \infty$. This signal has infinite energy, $E = \infty$.

Example 1.15 Find out whether signal $s(t) = A\cos(\omega_0 t + \theta)$ is an energy signal or a power signal.

Solution. Average power of the signal is determined as

$$P = \frac{1}{T_0} \int_{t_0}^{t_0 + T_0} |s(t)|^2 \, dt$$

$$= \frac{1}{T_0} \int_{t_0}^{t_0 + T_0} [A \cos(\omega_0 t + \theta)^2] \, dt$$

$$= \frac{1}{T_0} \int_{t_0}^{t_0 + \frac{2\pi}{\omega_0}} A^2 \cos^2(\omega_0 t + \theta) \, dt$$

$$= \frac{1}{T_0} \int_{t_0}^{t_0 + \frac{2\pi}{\omega_0}} A^2 \left[\frac{1 + \cos 2(\omega_0 t + \theta)}{2} \right] dt$$

$$= \frac{\omega_0}{2\pi} \int_{t_0}^{t_0 + \frac{2\pi}{\omega_0}} \frac{A^2}{2} \, dt + \frac{\omega_0}{2\pi} \int_{t_0}^{t_0 + \frac{2\pi}{\omega_0}} \frac{A^2}{2} \cos 2(\omega_0 t + \theta) \, dt$$

or $$P = \frac{A^2}{2}$$

Signal $s(t) = A\cos(\omega_0 t + \theta)$ is a power signal because its average is finite and total energy is infinite.

Example 1.16 Determine the normalized energy and power of the signal

$$s(t) = 8e^{j4\pi t}\, u(t)$$

Solution. Normalized energy E is given by

$$E = \int_{-\infty}^{\infty} |s(t)|^2\, dt$$

$$= \int_{-\infty}^{\infty} |8e^{j4\pi t}\, u(t)|^2\, dt \qquad \left[\text{Since } u(t) = \begin{cases} 1, & t > 0 \\ 0, & t < 0 \end{cases}\right]$$

$$= \int_{0}^{\infty} |8e^{j4\pi t}|^2\, dt = 64 \int_{0}^{\infty} e^{j4\pi t}\, e^{-j4\pi t}\, dt$$

$$= 64 \int_{0}^{\infty} 1\, dt = 64\,[\,t\,]_{0}^{\infty} = \infty$$

or $$E = \infty$$

Average power P of the signal $s(t)$ is given by

$$P = \lim_{T\to\infty} \frac{1}{2T} \int_{-T}^{T} |s(t)|^2\, dt$$

$$= \lim_{T\to\infty} \frac{1}{2T} \int_{-T}^{T} |s(t)|^2\, dt$$

$$= \lim_{T\to\infty} \frac{1}{2T} \int_{-T}^{T} |8e^{j4\pi t}\, u(t)|^2\, dt$$

$$= \lim_{T\to\infty} \frac{1}{2T} \int_{0}^{T} 64\, |e^{j4\pi t}|^2\, dt$$

$$\left[\text{Since } u(t) = \begin{cases} 1, & t > 0 \\ 0, & t < 0 \end{cases}\right]$$

$$= \lim_{T\to\infty} \frac{1}{2T} \int_{0}^{T} 64\, e^{j4\pi t}\, e^{-j4\pi t}\, dt$$

$$= \lim_{T\to\infty} \frac{64}{2T} \int_{0}^{T} dt$$

$$= \lim_{T\to\infty} \frac{64}{2T}\,[\,t\,]_{0}^{T} = \lim_{T\to\infty} \frac{64}{2T}\, T = 32$$

or Average power $P = 32$

Example 1.17 Show that the signal $s(t) = t^{-1/4}\, u(t-1)$ is neither an energy nor a power signal.

Solution. Normalized energy is determined by

$$E = \int_{-\infty}^{\infty} |s(t)|^2\, dt = \int_{-\infty}^{\infty} |t^{-1/4}\, u(t-1)|^2\, dt$$

$$= \int_{1}^{\infty} |t^{-1/4}|^2\, dt \qquad \left[\text{Since } u(t-1) = \begin{cases} 1, & t > 1 \\ 0, & t < 1 \end{cases}\right]$$

$$= \int_{0}^{\infty} [t]^{-1/2}\, dt = \left[\, 2[t]^{1/2}\right]_{0}^{\infty} = \infty$$

or $\qquad E = \infty$

Average power is determined by

$$P = \lim_{T \to \infty} \frac{1}{2T} \int_{-T}^{T} |s(t)|^2\, dt$$

$$= \lim_{T \to \infty} \frac{1}{2T} \int_{-T}^{T} |t^{-1/4}\, u(t-1)|^2\, dt$$

$$= \lim_{T \to \infty} \frac{1}{2T} \int_{1}^{T} t^{-1/2}\, dt$$

$$= \lim_{T \to \infty} \frac{1}{2T} [2t^{1/2}]_{1}^{T} = \lim_{T \to \infty} \left[\frac{T^{1/2} - 1}{T}\right]$$

$$= \lim_{T \to \infty} \frac{\left[1 - \frac{1}{\sqrt{2}}\right]}{\sqrt{T}} = 0$$

or Average power $P = 0$

Here average power $P \to 0$, when total energy $E \to \infty$, which means that the condition $0 < E < \infty$ is not satisfied. Hence signal $s(t) = t^{-1/4}\, u(t-1)$ is not an energy signal.

When $E \to \infty$, the value of $P \to 0$, which means that the condition $0 < P < \infty$ is not satisfied. Therefore, signal $s(t) = t^{-1/4}\, u(t-1)$ is not a power signal. Therefore, the given signal $s(t) = t^{-1/4}\, u(t-1)$ is neither an energy signal nor a power signal.

Example 1.18 Determine the values of P_∞ and E_∞ for the following signals

(*a*) $s(t) = e^{-2t}\, u(t)$ $\qquad$ (*b*) $s(t) = \cos(t)$

(c) $s(n) = \left(\frac{1}{2}\right)^n u(n)$ $\qquad$ (*d*) $s(n) = \cos\left(\frac{\pi}{4} n\right)$

(e) $s(n) = e^{j[(\pi/2n) + (\pi/8)]}$

Solution. (a) $E_\infty = \int_{-\infty}^{\infty} |s(t)|^2 \, dt$

$$= \int_{-\infty}^{\infty} |e^{-2t} u(t)|^2 \, dt \qquad \left[\text{Since } u(t) = \begin{cases} 1, & t > 0 \\ 0, & t < 0 \end{cases} \right]$$

$$= \int_0^{\infty} e^{-4t} \, dt = \left[\frac{e^{-4t}}{-4} \right]_0^{\infty}$$

or $$E_\infty = \frac{1}{4}$$

$$P_\infty = 0 \text{ because } E_\infty < \infty$$

(b) $$E_\infty = \int_{-\infty}^{\infty} |s(t)|^2 \, dt = \int_{-\infty}^{\infty} |\cos(t)|^2 \, dt$$

$$= \int_{-\infty}^{\infty} |\cos^2(t)| \, dt$$

or $$E_\infty = \infty$$

$$P_\infty = \lim_{T \to \infty} \frac{1}{2T} \int_{-T}^{T} |s(t)|^2 \, dt$$

$$= \lim_{T \to \infty} \frac{1}{2T} \int_{-T}^{T} |\cos(t)|^2 \, dt$$

$$= \lim_{T \to \infty} \frac{1}{2T} \int_{-T}^{T} \left(\frac{1 + \cos(2t)}{2} \right) dt = \frac{1}{2}$$

(c) $$E_\infty = \sum_{n=-\infty}^{\infty} |s(n)|^2$$

$$= \sum_{n=-\infty}^{\infty} \left| \left(\frac{1}{2} \right)^n u(n) \right|^2 \qquad \left[\text{Since } u(n) = \begin{cases} 1, & n \geq 0 \\ 0, & n < 0 \end{cases} \right]$$

$$= \sum_{n=0}^{\infty} \left(\frac{1}{2} \right)^{2n} = \frac{1}{1 - \frac{1}{4}} = \frac{4}{3}$$

or $$E_\infty = \frac{4}{3}$$

$$P_\infty = 0 \text{ because } E_\infty < \infty$$

(*d*)
$$E_\infty = \sum_{n=-\infty}^{\infty} |s(n)|^2 = \sum_{n=-\infty}^{\infty} \left| \cos\left(\frac{\pi}{4} n\right) \right|^2$$
$$= \sum_{n=-\infty}^{\infty} \cos^2\left(\frac{\pi}{4} n\right) = \infty$$
$$P_\infty = \lim_{N\to\infty} \frac{1}{2N+1} \sum_{n=-N}^{N} |s(n)|^2$$
$$= \lim_{N\to\infty} \frac{1}{2N+1} \sum_{n=-N}^{N} \left| \cos\left(\frac{\pi}{4} n\right) \right|^2$$
$$= \lim_{N\to\infty} \frac{1}{2N+1} \sum_{n=-N}^{N} \left[\frac{1+\cos\left(\frac{\pi}{2} n\right)}{2} \right] = \frac{1}{2}$$

(*e*) Given $s(n) = e^{j[(\pi n/2) + (\pi/8)]}$
$$E_\infty = \sum_{n=-\infty}^{\infty} |s(n)|^2 = \sum_{n=-\infty}^{\infty} \left| e^{j[(\pi n/2) + (\pi/8)]} \right|^2 = \infty$$
$$\left[\text{Since } \left| e^{j[(\pi n/2) + (\pi/8)]} \right|^2 = 1 \right]$$
$$P_\infty = \lim_{N\to\infty} \frac{1}{2N+1} \sum_{n=-N}^{N} |s(n)|^2$$
$$= \lim_{N\to\infty} \frac{1}{2N+1} \sum_{n=-N}^{N} \left| e^{j[(\pi n/2) + (\pi/8)]} \right|^2$$
$$= \lim_{N\to\infty} \frac{1}{2N+1} \sum_{n=-N}^{N} 1 = 1$$

Example 1.19 Check the periodicity of the following signals

(*a*) $s(t) = 2e^{j[t + (\pi/4)]}\, u(t)$

(*b*) $s(n) = u(n) + u(-n)$

Solution. (*a*) Signal $s(t) = 2e^{j[t + (\pi/4)]}\, u(t)$
$$= \begin{cases} 2e^{j[t + (\pi/4)]}, & t > 0 \\ 0, & t < 0 \end{cases}$$

The above signal is not periodic because it is zero for negative time, *i.e.*, $t < 0$.

(*b*) Signal $s(n) = u(n) + u(-n)$
$$= \begin{cases} 1, & n \geq 0 \\ 1, & n < 1 \end{cases}$$
$$= \{1, \text{ for all values of } n$$

Therefore, it is periodic with a fundamental period of 1.

Example 1.20 Find out the fundamental period of the signal

$$s(t) = 2\cos(10t + 1) - \sin(4t - 1)$$

Solution. Given a continuous-time signal

$$s(t) = 2\cos(10t + 1) - \sin(4t - 1)$$

It can be written as a combination of two signals as

$$s(t) = s_1(t) + s_2(t)$$

Fundamental period of first signal $s_1(t) = 2\cos(10t + 1)$,

$$T_{01} = \frac{2\pi}{\omega} = \frac{2\pi}{10} = \frac{\pi}{5}$$

Fundamental period of second signal $s_2(t) = \sin(4t + 1)$,

$$T_{02} = \frac{2\pi}{\omega} = \frac{2\pi}{4} = 0.5\,\pi = \frac{\pi}{2}$$

$$\frac{T_{01}}{T_{02}} = \frac{\frac{\pi}{5}}{\frac{\pi}{2}} = \frac{2}{5}$$

Therefore, the overall signal is periodic because the ratio of fundamental periods of first signal and second signal is the ratio of integers.

The fundamental period of the overall signal is the least common multiple (LCM) of the periods of the first and second signals. This is equal to π.

Example 1.21 Find out the fundamental period of the discrete- time signal

$$s(n) = 1 + e^{(j4\pi n)/7} - e^{(j2\pi n)/5}$$

Solution. Given a discrete-time signal

$$s(n) = 1 + e^{(j4\pi n)/7} - e^{(j2\pi n)/5}$$

It can be written as a combination of three signals (or parts) as

$$s(n) = s_1(n) + s_2(n) + s_3(n)$$

Fundamental period of first part, $s_1(n) = 1$,

$$T_{01} = 1$$

Fundamental period of second part, $s_2(n) = e^{(j4\pi n)/7}$,

$$T_{02} = m\left(\frac{2\pi}{\omega}\right) = 2\left(\frac{2\pi}{4\pi/7}\right) = 7, \text{ when } m = 2$$

Fundamental period of third part $s_3(n) = e^{(j2\pi n)/5}$,

$$T_{03} = m\left(\frac{2\pi}{\omega}\right) = 1\left(\frac{2\pi}{2\pi/5}\right) = 5, \text{ when } m = 1$$

$$\frac{T_{01}}{T_{01}} = \frac{1}{7}, \quad \frac{T_{01}}{T_{03}} = \frac{1}{5}$$

The overall signal $s(n)$ is periodic because the above ratios are the ratio of integers. The period of the overall signal will be the least common multile (LCM) of the periods of the three terms in $s(n)$. The fundamental period of the overall signal will be 35.

Example 1.22 Calculate the value of E_∞ for the signal

$$y(t) = \int_{-\infty}^{t} s(\tau)\, d\tau$$

where $s(t)$ is a continuous-time signal given by $s(t) = \delta(t + 2) - \delta(t - 2)$

Solution. Given : $y(t) = \int_{-\infty}^{t} s(\tau)\, d\tau$ ...(1)

Putting $s(t) = \delta(t + 2) - \delta(t - 2)$ in Eq. 1, we obtain

$$y(t) = \int_{-\infty}^{t} [\delta(\tau + 2) - \delta(\tau - 2)]\, d\tau$$

$$= \begin{cases} 0, & t < -2 \\ 1, & -2 \le t \le 2 \\ 0, & t > 2 \end{cases}$$

Therefore, $E_\infty = \int_{-2}^{2} (1)\, dt = [\, t\,]_{-2}^{2}$

$$= 2 - (-2) = 4$$

Example 1.23 The input-output relationship for a discrete-time LTI system is given by

$$y(n) = s(n)\, s(n - 2)$$

(*a*) Whether above system is memoryless or not

(*b*) Find out the output of the system when the input is $A\,\delta(n)$, where A is any real or complex number

(*c*) Whether above system is invertible or not.

Solution. (*a*) System described by $y(n) = s(n)\, s(n - 2)$ is not memoryless because its output $y(n)$ depends on past values of input $s(n)$.

(*b*) Given input-output relationship for a discrete-time system

$$y(n) = s(n)\, s(n - 2)$$

Output for input $s(n) = A\,\delta(n)$ is determined as

$$y(n) = A\,\delta(n)\, A\,\delta(n - 2) \qquad ...(1)$$

$$\left[\text{Since } \delta(n) = \begin{cases} 1, & n = 0 \\ 0, & n \neq 0 \end{cases} \text{ and } \delta(n - 2) = \begin{cases} 1, & n - 2 = 0 \rightarrow n = 2 \\ 0, & n - 2 = 0 \rightarrow n \neq 2 \end{cases} \right]$$

$$y(n) = A^2\, \delta(n)\, \delta(n - 2) = 0$$

(*c*) From the result of part (*b*), we have seen that the output of the system is always zero for inputs of the form $\delta(n-k)$, k is any integer number. Therefore, the system is not invertible.

Example 1.24 Check the system described by $y(t) = s(\sin t)$ for (*a*) causality, (*b*) linearity.

Solution. (*a*) The system is not-causal because its output $y(t)$ at some time may depend on future values of input $s(t)$.

(*b*) The corresponding outputs for two continuous-time input signal $s_1(t)$ and $s_2(t)$ are given as

$$\left.\begin{aligned} s_1(t) &\to y_1(t) = T\,[s_1(t)] = s_1(\sin t) \\ s_2(t) &\to y_2(t) = T\,[s_2(t)] = s_2(\sin t) \end{aligned}\right] \quad ...(1)$$

A linear combination of two input signals results in the output

$$y_3(t) - T\,[s_3(t)] \quad ...(2)$$

where

$$\begin{aligned} s_3(t) &= As_1(t) + Bs_2(t) \\ y_3(t) &= T\,[As_1(t) + Bs_2(t)] \\ &= T\,[As_1(t)] + T\,[Bs_2(t)] \\ &= AT\,[s_1(t)] + BT\,[s_2(t)] \\ &= As_1(\sin t) + Bs_2(\sin t) \quad ...(3) \end{aligned}$$

A linear combination of two outputs results in the output

$$Ay_1(n) + By_2(n) = As_1(\sin t) + Bs_2(\sin t) \quad ...(4)$$

Since both outputs are equal, therefore, the system is linear.

Example 1.25 Input-output relationship for some continuous-time systems is given as follows :

(*a*) $y(t) = s(t-2) + s(2-t)$ (*b*) $y(t) = \cos(3t)\, s(t)$

Check for the following properties :

1. Memoryless or with memory

2. Causality

Solution. (*a*) Given a continuous-time system

$$y(t) = s(t-2) + s(2-t)$$

1. It is not a memoryless system because its output $y(t)$ at some time depends on both future and past values of inputs. So it is a system with memory.

2. It is a non-causal system because its output $y(t)$ at some time depends on future values of input $s(n)$.

(*b*) Given a continuous-time system

$$y(t) = [\cos(3t)]\, s(t)$$

1. It is a memoryless system because its output $y(t)$ at some time depends only on that time. Output of this system does not depend on past and future values of input.
2. It is also a causal system because its output $y(t)$ at some time does not depend on future values of input $s(t)$.

Example 1.26 Input-output relationship for some discrete-time systems is given as

$$(a) \quad y(n) = s(n-2) - 2s(n-8) \qquad (b) \quad y(n) = \begin{cases} s(n)\,, & n \geq 1 \\ 0\,, & n = 0 \\ s(n+1)\,, & n \leq -1 \end{cases}$$

Check above systems for following properties :

1. Memoryless or with memory system
2. Causality.

Solution. (*a*) Given a discrete-time system described by following input-output relationship

$$y(n) = s(n-2) - 2s(n-8)$$

1. It is not a memoryless system because its output $y(n)$ at some discrete-time index depends on past values of time-index. So there is a need of memory to store past values of input signal.
2. It is a causal system because its output $y(n)$ at some time index does not depend on future values of inputs.

(*b*) Given a discrete-time system which is described by

$$y(n) = \begin{cases} s(n)\,, & n \geq 1 \\ 0\,, & n = 0 \\ s(n+1)\,, & n \leq -1 \end{cases}$$

1. This sytem is memoryless for $n \geq 1$ becasue its output $y(n) = s(n)$ and its output at some value of time index depends on the input at that time index. Same system for $n \leq -1$ is described by $y(n) = s(n+1)$ is not memoryless system its output $y(n)$ depends on future values of input. For example :

$$y(n) = s(n+1)$$

$$n = -1, \quad y(-1) = s(-1+1) = s(0)$$

$$n = -2, \quad y(-2) = s(-2+1) = s(-1)$$

$n = -3, \quad y(-3) = s(-3+1) = s(-2)$ and so on.

2. Above system $y(n) = s(n)$ for $n \geq 1$ is causal because its output $y(n)$ does not depend on future values of input $s(n)$. But same system is described by $y(n) = s(n+1)$ for $n \leq -1$ is non-causal because its output at some time index depends on future values of input.

Example 1.27 A continuous-time signal is shown in Fig. 1.45 (*a*). Sketch the following signals :

(*a*) $s(t-1)$ (*b*) $s(2-t)$

(*c*) $s(2t+1)$ (*d*) $s[\,4-(t/2)\,]$

(*e*) $[s(t)+s(-t]\,u(t)$ (*f*) $s(t)\,[\delta(t+1.5)-\delta(t-1.5)]$

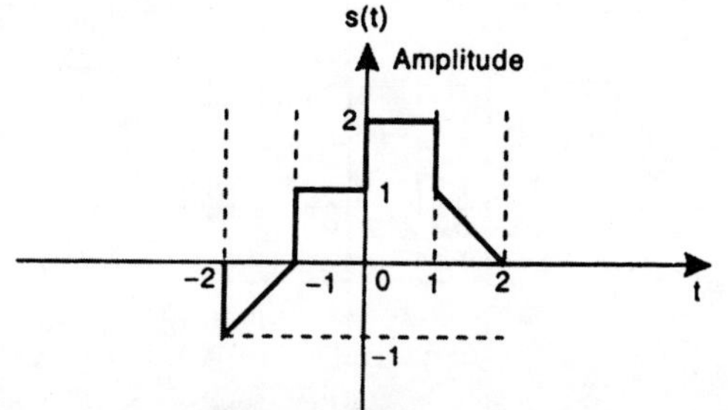

Fig. 1.45 (*a*) *Waveform of a continuous-time signal.*

Solution. (*a*) Signal $s(t-1)$ is a delayed version of $s(t)$ by one unit of time. It is shown in Fig. 1.45 (*b*).

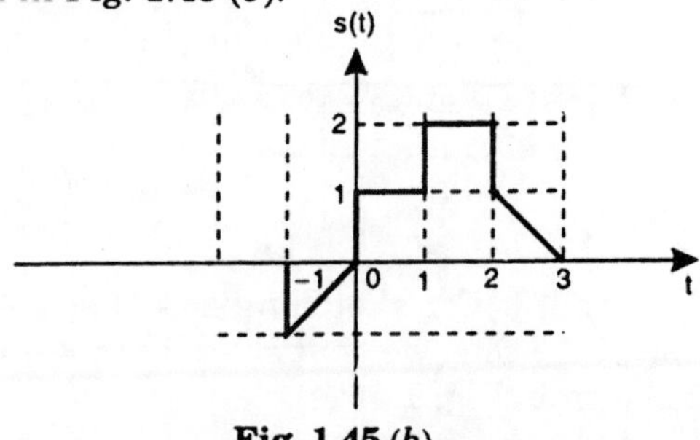

Fig. 1.45 (*b*)

(*b*) Continuous-time signal $s(2-t)$ is sketched as follows and shown in Fig. 1.45 (*c*). First of all we take mirror image of signal $s(t)$ shown in Fig. 1.45 (*a*). Taking of mirror image of a signal is also called folding of a signal. Therefore, we shift folded version of signal $s(t)$ by two units of time from left to right.

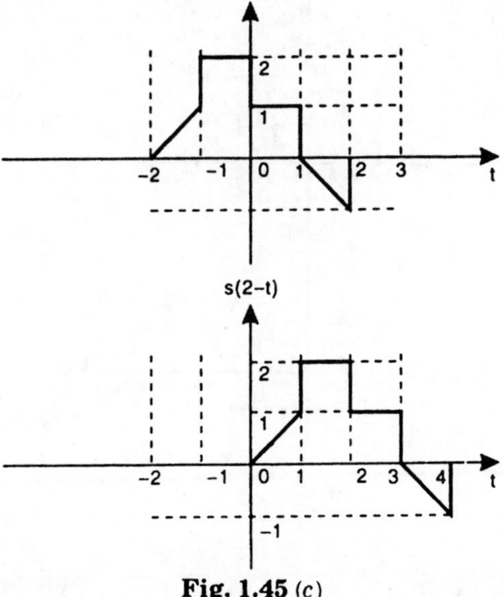

Fig. 1.45 (*c*)

(*c*) Signal $s(2t + 1)$. For sketching signal $s(2t + 1)$, we first sketch $s(2t)$ and them we shift $s(2t)$ by one unit of time from right to left. It is shown in Fig. 1.45 (*d*).

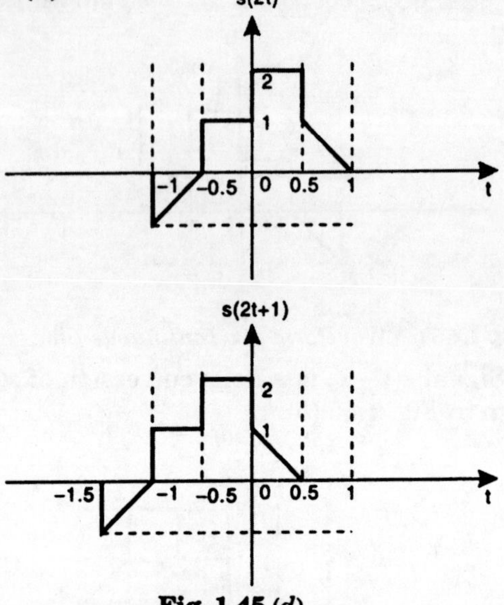

Fig. 1.45 (*d*)

(*d*) Signal $s[\,4 - (t/2)\,]$. For sketching signal $s[\,4 - (t/2)\,]$, we first sketch $s(-t)$ then $s(-t/2)$ then we shift $s(-t/2)$ by 4 units of time from left to right. It is shown in Fig. 1.45 (*e*).

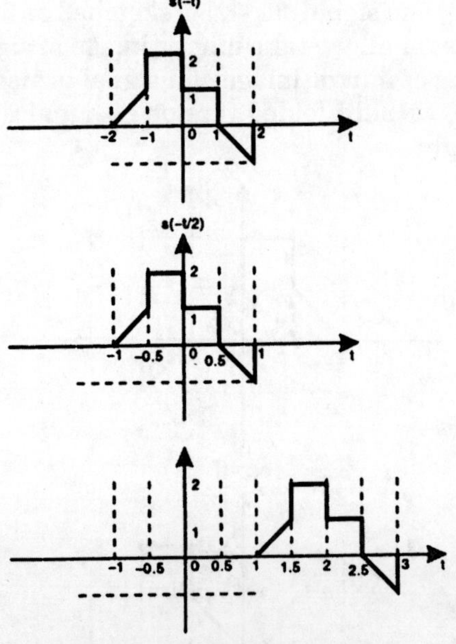

Fig. 1.45 (*e*)

(*e*) Signal $[s(t) + s(-t)]\ u(t)$ can be sketched by adding signals $s(t)$ and $s(-t)$ then multiplying $u(t)$. It is shown in Fig. 1.45 (*f*).

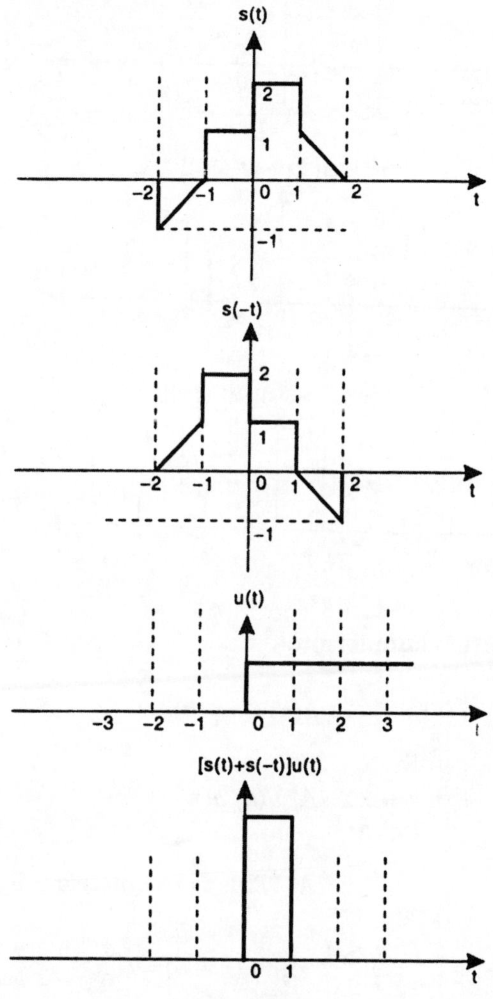

Fig. 1.45 (*f*)

(*f*) Signal $s(t)\left[\delta\left(t+\frac{3}{2}\right)-\delta\left(t-\frac{3}{2}\right)\right]$ can be drawn by multiplying $s(t)$ by $\left[\delta\left(t+\frac{3}{2}\right)-\delta\left(t-\frac{3}{2}\right)\right]$. It is shown in Fig. 1.45 (*g*).

Example 1.28 Determine the unit step response of a discrete-time LTI system when unit sample response is $h(n) = A^n\ u(n)$, $|A| < 1$.

Solution. For computing the unit-step response, we put input sequence $s(n) = u(n)$, where $u(n)$ is a causal sequence and the system is also causal.

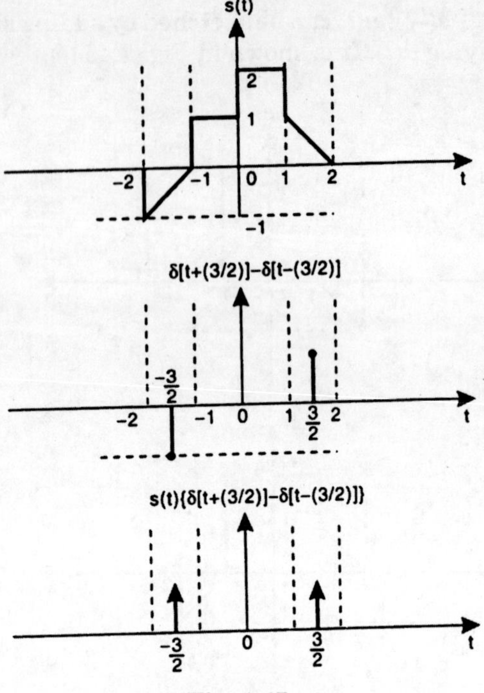

Fig. 1.45 (*g*)

From convolution sum formula

$$y(n) = \sum_{k=-\infty}^{\infty} h(k)\, s(n-k)$$

$$= \sum_{k=-\infty}^{\infty} A^k\, u(k)\, u(n-k)$$

$$= \sum_{k=0}^{n} A^k \text{ (This is Geometric progression)}$$

$$\left[\text{Since } (u(k) = \begin{cases} 1, & k \geq 0 \\ 0, & k \leq 0 \end{cases} \text{ and } u(n-k) = \begin{cases} 1, & n-k \geq 0 \rightarrow k \leq n \\ 0, & n-k < 0 \rightarrow k > n \end{cases} \right]$$

or $$y(n) = \frac{1 - A^{n+1}}{1 - A}, \quad n \geq 0 \qquad \text{...(1)}$$

and $$y(n) = 0, \quad n < 0 \qquad \text{...(2)}$$

Now combining Eqs. 1 and 2, we obtain

$$y(n) = g(n) = \left(\frac{1 - A^{n+1}}{1 - A} \right) u(n)$$

It is the unit step response of the discrete-time LTI system.

REVIEW QUESTIONS

1. Define signal. Give some examples of signals.
2. Give classification of signals.
3. What do you mean by dimensions of signal ?
4. What do you mean by number of channels of signal ?
5. Discuss dependent and independent variables of a signal.
6. Discuss the following :
   (*a*) One-dimensional and multi-dimensional signals
   (*b*) Single-channel and multi-channel signals
   (*c*) Continuous-time and discrete-time signals
   (*d*) Analog and digital signals.
7. Classify signals in accordance with signal characteristics.
8. Discuss the following :
   (*a*) Deterministic and random signals
   (*b*) Piece wise-defined and simply-defined signals
   (*c*) Periodic and non-periodic signals
   (*d*) Power and energy signals
   (*e*) Even and odd signals.
9. Describe various basic signals.
10. Discuss following basic signals :
    (*a*) Continuous-time impulse function $\delta(t)$
    (*b*) Continuous-time unit step function $u(t)$
    (*c*) Discrete-time impulse function $\delta(n)$
    (*d*) Discrete-time unit step function $u(n)$
    (*e*) Ramp function (continuous-time/discrete-time).
11. Write short notes on the following :
    (*a*) Exponential functions or signals
    (*b*) Sinusoidal signals.
12. What is Signum function ?
13. Define fundamental period in periodic signals.
14. With suitable examples distinguish a deterministic signal from a random signal.
15. Describe the procedure used to determine whether the sum of two periodic signals is periodic or not.
16. What are singularity functions ?
17. What is unit step function ? How can it be obtained from a unit-step function ?
18. How are systems classfieid ?
19. What is system ? Give examples.

20. Distinguish the following systems :
(*a*) Continuous-time and discrete-time systems
(*b*) Distributive-parameter and lumped-parameter systems
(*c*) Static and dynamic systems
(*d*) System and inverse system
(*e*) Linear and non-linear system
(*f*) Time-invariant and time-varying systems
(*g*) Causal and non-causal systems
(*h*) Stable and unstable systems
(*i*) System with memory and memoryless system.

21. What do you mean by interconnection of systems ?

22. Describe the following types of interconnection of systems.
(*a*) Series or cascade interconnection
(*b*) Parallel interconnection
(*c*) Feedback interconnection.

23. What is a causal system ? Why are non-causal systems unrealizable ?

24. What is BIBO stability ?

25. Define superposition condition for a linear system.

NUMERICAL PROBLEMS

1. Find the normalized power and the energy for the following signals

(*a*) $s(t) = 4\sin\left(6\pi t + \frac{\pi}{6}\right)$ (*b*) $s(t) = e^{-\alpha t}\,u(t)$

(*c*) $s(t) = e^{\alpha t}\,u(-t)$ (*d*) $s(t) = [\alpha^2 + t^2]^{-1/4}$

(*e*) $s(t) = e^{-\alpha|t|}$ (*f*) $s(t) = \cos(5\pi t)\,u(t)$

Ans. (*a*) $P = 8$ and $E = \infty$, Power signal; (*b*) $P = 0$, $\alpha > 0$ and $E = \frac{1}{2\alpha}$, Energy signal; (*c*) $P = 0$ and $E = \frac{1}{2\alpha}$, Energy signal; (*d*) $P = \infty$, $E = \infty$, it is neither energy nor power signal; (*e*) $P = 0$, $E = \frac{1}{\alpha}$, Energy signal; (*f*) $P = \frac{1}{2}$, $E = \infty$, Power signal

2. Find whether the signal $\Pi(t/\tau) + \Pi(t/2\tau)$ is an energy signal or a power signal. **Ans.** Energy signal

3. A continuous-time signal is shown in Fig. 1.43. Sketch following signals
(*a*) $s(t + 1)$ (*b*) $s(-t + 1)$
(*c*) $s\left(\frac{3}{2}t\right)$ (*d*) $s\left(\frac{3}{2}t + 1\right)$

4. Whether following continuous-time signal is periodic or not

$$s(t) = \begin{cases} \cos(t), & \text{if } t < 0 \\ \sin(t), & \text{if } t \geq 0 \end{cases}$$

Ans. Signal $s(t)$ is not periodic

5. Express the sum of two complex exponentials as the product of a single complex exponential and a magnitude of $s(t)$. Given signal $s(t) = e^{j2t} + e^{j3t}$.

Ans. $s(t) = 2e^{j2.5t} \cos(0.5t)$; $|s(t)| = 2|\cos(0.5t)|$

6. Determine the fundamental period of the discrete-time signal

$s(n) = e^{j(2\pi/3)n} + e^{j(3\pi/4)n}$ **Ans.** Fundamental period $N_0 = 24$

7. Check the causality for the following systems

(*a*) $y(t) = s(-t)$ (*b*) $y(n) = s(n) \cos(n+1)$

8. Find the order of the systems which are described by following input-output relations :

(*a*) A continuous-time system is described by

$$\frac{d^4 y(t)}{dt^4} + 4\frac{d^3 y(t)}{dt^3} + 3\frac{d^2 y(t)}{dt^2} + 2\frac{d\,y(t)}{dt} + y(t) = \frac{d\,s(t)}{dt} + s(t)$$

(*b*) A discrete time system is described by

$$y(n) = y(n-3) + 2y(n-2) - 4y(n-1) + s(n) - s(n-1)$$

Ans. (*a*) 4 ; (*b*) 3

9. Check the following discrete-time system for linearity

$y(n) = 2s(n) + 3$ **Ans.** Linear

10. Determine whether or not following signal is periodic

$$s(n) = \sum_{k=-\infty}^{\infty} \{\delta(n-4k) - \delta(n-4k-1)\}$$

Ans. Periodic with period of 4

11. Determine whether or not each of the following signals is periodic. If a signal is periodic, specify its fundamental period

(*a*) $s(t) = je^{j10t}$ (*b*) $s(t) = e^{(-1+j)t}$

(*c*) $s(n) = e^{j7\pi n}$ (*d*) $s(n) = 3e^{j3\pi/5[n+(1/2)]}$

(*e*) $s(n) = 3e^{j3/5[n+(1/2)]}$

Ans. (*a*) Periodic with period of $\pi/5$; (*b*) It is not periodic; (*c*) Periodic with period of 2; (*d*) Periodic with period of 10; (*e*) It is not periodic

12. Express the real part of the following continuous-time signals in the form $Ae^{-\alpha t} \cos(\omega t + \phi)$, where A, α, ω and ϕ are real numbers with $A \geq 0$ and $-\pi < \phi < \pi$.

(*a*) $s(t) = -2$ (*b*) $s(t) = e^{-t} \sin(3t + \pi)$

(*c*) $s(t) = je^{(-2+j100)t}$

Ans. (*a*) $\text{Re}\{s(t)\} = 2e^{0t} \cos(0t + \pi)$; (*b*) $\text{Re}\{s(t)\} = e^{-t} \cos\left(3t + \frac{\pi}{2}\right)$;

(*c*) $\text{Re}\{s(t)\} = e^{-2t} \cos\left(100t + \frac{\pi}{2}\right)$

13. A discrete-time system is described by following difference equation

$$y(n) = \sum_{k=n-n_0}^{n+n_0} s(k)$$

where n_0 = finite positive number

(*a*) Is this system linear ?

(*b*) Is this system time-invariant ?

Ans. (*a*) It is a linear system; (*b*) It is a time-invariant system

14. Find whether following systems are linear, time-invariant or both.

(*a*) $y(t) = t^2\, s(t-1)$ (*b*) $y(n) = s(n+1) - s(n-1)$

Ans. (*a*) Linear system, not time-invariant system; (*b*) Linear, time-invariant system

15. A discrete-time signal is shown in Fig. 1.46. Sketch the following signals

(*a*) $s(n-4)$ (*b*) $s(3-n)$

(*c*) $s(3n)$ (*d*) $s(3n+1)$

(*e*) $s(n)\,u(3-n)$ (*f*) $s(n-2)\,\delta(n-2)$

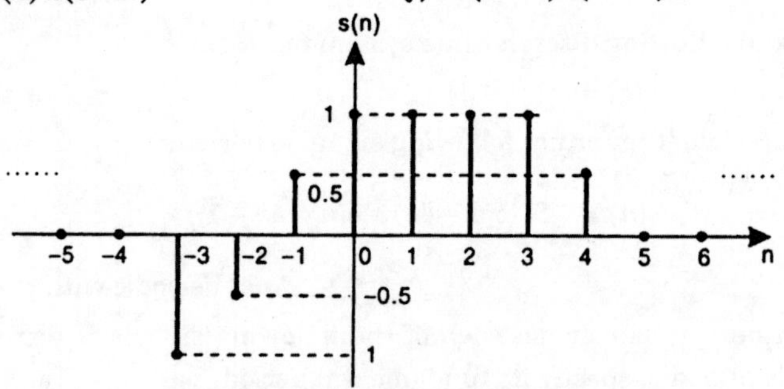

Fig. 1.46 *Discrete-time signal s*(*n*).

16. Find whether or not each of the following continuous-time signals is periodic. If the signal is periodic, find its fundamental period.

(*a*) $s(t) = 3\cos\left(4t + \frac{\pi}{3}\right)$ (*b*) $s(t) = e^{j(\pi t - 1)}$

(*c*) $s(t) = \left[\cos\left(2t - \frac{\pi}{3}\right)\right]^2$

Ans. (*a*) Periodic, period = $\pi/2$; (*b*) Periodic, period = 2;
(*c*) Periodic, period = $\pi/2$

17. Find whether or not each of the following discrete-time signals is periodic. If the signal is periodic, find the fundamental period.

(*a*) $s(n) = \sin\left(\frac{6\pi}{7}n + 1\right)$ (*b*) $s(n) = \cos\left(\frac{n}{8} - \pi\right)$

(*c*) $s(n) = \cos\left(\frac{\pi}{8}n^2\right)$ (*d*) $s(n) = \cos\left(\frac{\pi}{2}n\right)\cos\left(\frac{\pi}{4}n\right)$

(*e*) $s(n) = 2\cos\left(\frac{\pi}{4}n\right) + \sin\left(\frac{\pi}{8}n\right) - 2\cos\left(\frac{\pi}{2}n + \frac{\pi}{6}\right)$

Ans. (*a*) Periodic with period of 7; (*b*) It is not periodic; (*c*) Periodic with period of 8; (*d*) Periodic with period of 8; (*e*) Periodic with period of 16

18. Find whether following systems are invertible or not. If system is invertible, construct the inverse system. If it is not invertible, find two input signals to the system that have the same output.

(*a*) $y(t) = s(t - 4)$ (*b*) $y(t) \cos [s(t)]$

(*c*) $y(n) = n\, s(n)$ (*d*) $y(t) = \int_{-\infty}^{t} s(\tau)\, d\tau$

(*e*) $y(n) = \begin{cases} s(n-1), & n \geq 1 \\ 0, & n = 0 \\ s(n), & n \leq -1 \end{cases}$ (*f*) $y(n) = s(n)\, s(n-1)$

(*g*) $y(t) = s(2t)$ (*h*) $y(n) = s(2n)$

Ans. (*a*) Invertible, inverse system $\rightarrow y(t) = s(t + 4)$; (*b*) Non-invertible, the signals $s(t)$ and $s(t) + 2\pi$ give the same result; (*c*) Non-invertible, the input $\delta(n)$ and $2\delta(n)$ give the same result; (*d*) Invertible, inverse system $\rightarrow$ $y(t) = \frac{d\, s(t)}{dt}$; (*e*) Invertible, inverse system $y(n) = s(n + 1)$ for $n \geq 0$ and $y(n) = s(n)$ for $n < 0$; (*f*) Non-invertible, input $s(n)$ and $-s(n)$ give the same result; (*g*) Invertible, inverse system $\rightarrow y(t) = s\left(\frac{t}{2}\right)$; (*h*) Non-invertible, input $s(n) = \delta(n) + \delta(n - 1)$ and $s(n) = \delta(n)$ give $y(n) = \delta(n)$

19. Check the linearity for the following systems :

(*a*) $y(n) = 2s(n - n_o)$, where $n_o > 0$

(*b*) $y(n) = (n + 3)\, s(n - 3)$

(*c*) $y(n) = 5n\, s^2(n)$ **Ans.** (*a*) linear (*b*) linear (*c*) non-linear

20. Check the following systems for causality

(*a*) $y(n) = s(n) \sin \omega n$ (*b*) $y(n) = k[s(n + 1) - s(n)]$

(*c*) $y(n) = k[s(n) - s(n - 1)]$ (*d*) $y(n) = s(n + 1)\, e^{-n}$

Ans. (*a*) causal (*b*) non-causal (*c*) causal (*d*) non-causal

21. Check the following systems for time-invariance

(*a*) $y(n) = s(n) \sin \omega n$ (*b*) $y(n) = 2\, [s(n + 1) - s(n)]$

Ans. (*a*) non time-invariant (*b*) time-invariant

2

Linear Time-Invariant (LTI) Continuous-Time and Discrete-Time Systems

2.1 INTRODUCTION

A system which possesses two properties namely—linearity and time-invariance is called *linear time-invariant* (LTI) system. These two properties, linearity and time-invariance, play a fundamental role in signal and system analysis. Many physical processes have these properties and thus can be modeled as linear time-invariant (LTI) system. LTI systems can be analyzed in adequate detail with the help of a set of powerful tools which are already available. These tools are the basis of signal and system analysis. LTI systems are of two types viz., continuous- time and discrete-time. One of the primary reasons why linear time-invariant (LTI) systems are amenable to analysis is that any such system has the *superposition property.* LTI systems are characterized by their impulse response in both continuous-time and discrete-time and these are represented as linear combinations of delayed impulses. The complete characterization of any LTI system in terms of its impulse response is performed by convolution integral in the continuous-time case and by the convolution sum in discrete-time case. A continuous-time LTI system is described by linear constant-coefficient *differential equations*. Similarly, a discrete-time (LTI) system is described by linear constant-coefficient *difference equations.*

2.2 CONTINUOUS-TIME LINEAR TIME-INVARIANT (LTI) SYSTEMS

A system of which both input and output are continuous-time signals and possesses two properties namely linearity and time-invariance is referred to as continuous-time linear time-invariant (LTI) system.

These systems are characterized either by their impulse response $h(t)$ or by their transfer function $H(s)$.

Continuous-time signals are represented mathematically as the superposition of scaled and shifted continuous-time unit impulse functions $\delta(t)$.

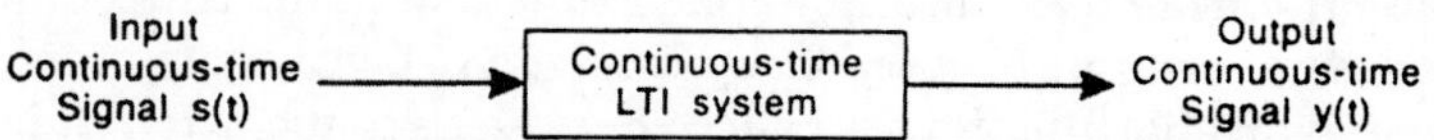

Fig. 2.1 *Continuous-time LTI system.*

Continuous-time input signal $s(t)$ is represented in terms of $\delta(t)$ as

$$s(t) = \int_{\tau=-\infty}^{\infty} s(\tau)\, \delta(t - \tau)\, d\tau \qquad ...(2.1)$$

Continuous-time impulse response $h(t)$ is the output of a continuous-time system when applied input $s(t)$ is equal to unit impulse function $\delta(t)$.

Transfer function of a continuous-time LTI system is the ratio of Laplace transform of output $y(n)$ and Laplace transform of inupt $s(n)$ of a system when the system is initially relaxed. *For an initially relaxed system all the initial conditions are zero.* We will study Laplace transform in the next chapters.

2.2.1 Unit Impulse Response for Continuous-Time LTI Systems

Output of a continuous-time LTI system is called *unit-impulse response* when we apply an input equal to unit-impulse function $\delta(t)$.

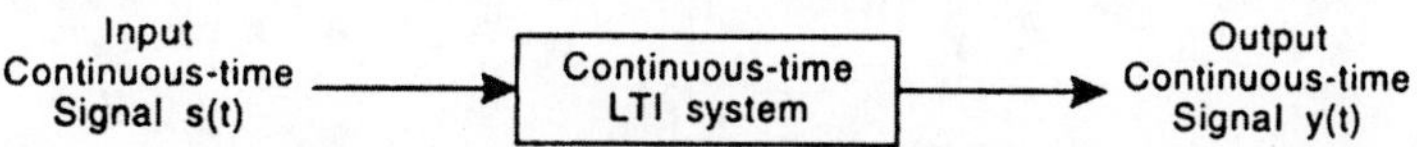

Fig. 2.2 *Continuous-time LTI system.*

If input $s(t) = \delta(t)$ is unit-impulse function then output $y(t) = h(t)$ is unit-impulse response, where unit-impulse function $\delta(t)$ is given by

$$\delta(t) = \begin{cases} 1, & t = 0 \\ 0, & t \neq 0 \end{cases} \qquad ...(2.2)$$

Transfer function of a continuous-time LTI system is equal to Laplace transform of unit-impulse response of a continuous-time LTI system.

Transfer function of a continuous-time LTI system $H(s)$ = Laplace transform of $h(t)$

$$= \text{LT}\,[h(t)] = \int_{-\infty}^{\infty} h(t)\, e^{-st}\, dt \qquad ...(2.3)$$

where s is a complex variable and $s = \sigma + j\omega$, σ real part and ω imaginary part.

Note. We will study Laplace transform in chapter 10.

2.2.2 Convolution Integral for Continuous-Time LTI Systems

Convolution is a mathematical operation applied to two functions of the same independent variables and results in a third function of the same independent variable as used in applied functions. This third function is called the *convolution of two functions*. Convolution of two continuous-variable functions is called *continuous convolution*. It is also called convolution integral. Convolution operation is also useful in certain system analysis where we have functions of the independent-variable frequency.

Convolution integral is referred as superposition integral. It is used to compute the response of a continuous-time LTI system with impulse response $h(t)$ and for any arbitrary input $s(n)$. Impulse response $h(t)$ is the output of system when unit-impulse function $\delta(t)$ is applied as input. Any input signal $s(t)$ can be represented in terms of unit-impulse function (Dirac-Delta function) $\delta(t)$, given by

$$s(t) = \int_{\tau=-\infty}^{\infty} s(\tau)\,\delta(t-\tau)\,d\tau \qquad \text{...(2.4)}$$

$$\text{where } \delta(t) = \text{Unit impulse function } = \begin{cases} 1, & t = 0 \\ 0, & t \neq 0 \end{cases} \qquad \text{...(2.5)}$$

Convolution integral is given as

$$y(t) = s(t) * h(t) = \int_{\tau=-\infty}^{\infty} s(\tau)\,h(t-\tau)\,d\tau$$

$$= h(t) * s(t) = \int_{\tau=-\infty}^{\infty} h(\tau)\,s(t-\tau)\,d\tau \qquad \text{...(2.6)}$$

Here $*$ denotes the convolution operation between input signal $s(t)$ and its impulse response $h(t)$.

Both continuous-time and discrete-time LTI systems are completely characterized by their impulse response $h(t)$ or $h(n)$.

Example 2.1 Find the convolution of the two continuous-time functions

$$s(t) = e^{-t^2}$$

and $$h(t) = 3t^2 \text{ for all } t.$$

Solution. Convolution integral is given by

$$y(t) = \int_{\tau=-\infty}^{\infty} s(\tau)\,h(t-\tau)\,d\tau$$

$$= \int_{\tau=-\infty}^{\infty} e^{-\tau^2}\,[3(t-\tau)^2]\,d\tau$$

$$= \int_{\tau=-\infty}^{\infty} e^{-\tau^2}\, [3t^2 - 6t\tau + 3\tau^2]\, d\tau$$

$$= 3t^2 \int_{\tau=-\infty}^{\infty} e^{-\tau^2}\, d\tau - 6t \int_{\tau=-\infty}^{\infty} \tau\, e^{-\tau^2}\, d\tau + 3 \int_{\tau=-\infty}^{\infty} \tau^2 e^{-\tau^2}\, dt$$

$$= 3t^2 \sqrt{\pi} - 0 + 1.5\sqrt{\pi} = 5.31t^2 + 2.659, \quad \text{for all } t.$$

Example 2.2 Find the convolution of the two continuous-time functions

$$s(t) = 3 \cos 2t \text{ for all } t$$

and
$$h(t) = e^{-|t|} = \begin{cases} e^t, & t < 0 \\ e^{-t}, & t \geq 0 \end{cases}$$

Solution. Convolution integral is given by

$$y(t) = s(t) * h(t) = \int_{\tau=-\infty}^{\infty} s(\tau)\, h(t-\tau)\, d\tau$$

Now
$$s(\tau) = 3 \cos 2\tau \quad \text{for all } \tau$$

and
$$y(t-\tau) = \begin{cases} e^{(t-\tau)}, & t-\tau < 0 \rightarrow \tau > t \\ e^{(\tau-t)}, & t-\tau \geq 0 \rightarrow \tau \leq t \end{cases}$$

Therefore,

$$y(t) = \int_{\tau=-\infty}^{\infty} s(\tau)\, h(t-\tau)\, d\tau$$

$$= \int_{\tau=-\infty}^{\infty} (3 \cos 2\tau)\, e^{(\tau-t)}\, d\tau + \int_{\tau=t}^{\infty} (3 \cos 2\tau)\, e^{(t-\tau)}\, d\tau$$

$$= 3e^{-t} \int_{-\infty}^{t} e^{\tau} \cos 2\tau\, d\tau + 3e^{t} \int_{t}^{\infty} e^{-\tau} \cos 2\tau\, d\tau$$

$$= 3e^{-t}\left[e^{\tau} \frac{(\cos 2\tau + 2 \sin 2\tau}{5} \right]_{-\infty}^{t} + 3e^{t}\left[e^{-\tau} (-\cos 2\tau + 2 \sin 2\tau \right]_{t}^{\infty}$$

$$= \frac{6}{5} \cos 2t \quad \text{for all } t.$$

Example 2.3 Find the convolution integral of input $s(t)$ to an LTI continuous-time system with impulse response $h(t)$

where $\quad s(t) = e^{-At}\, u(t), \; A > 0 \quad$ and $\quad h(t) = u(t).$

Solution. Convolution integral of $s(t)$ and $h(t)$ is given by

$$y(t) = \int_{\tau=-\infty}^{\infty} s(\tau)\, h(t-\tau)\, d\tau$$

$$= \int_{-\infty}^{\infty} e^{-At}\, u(\tau)\, u(t-\tau)\, d\tau$$

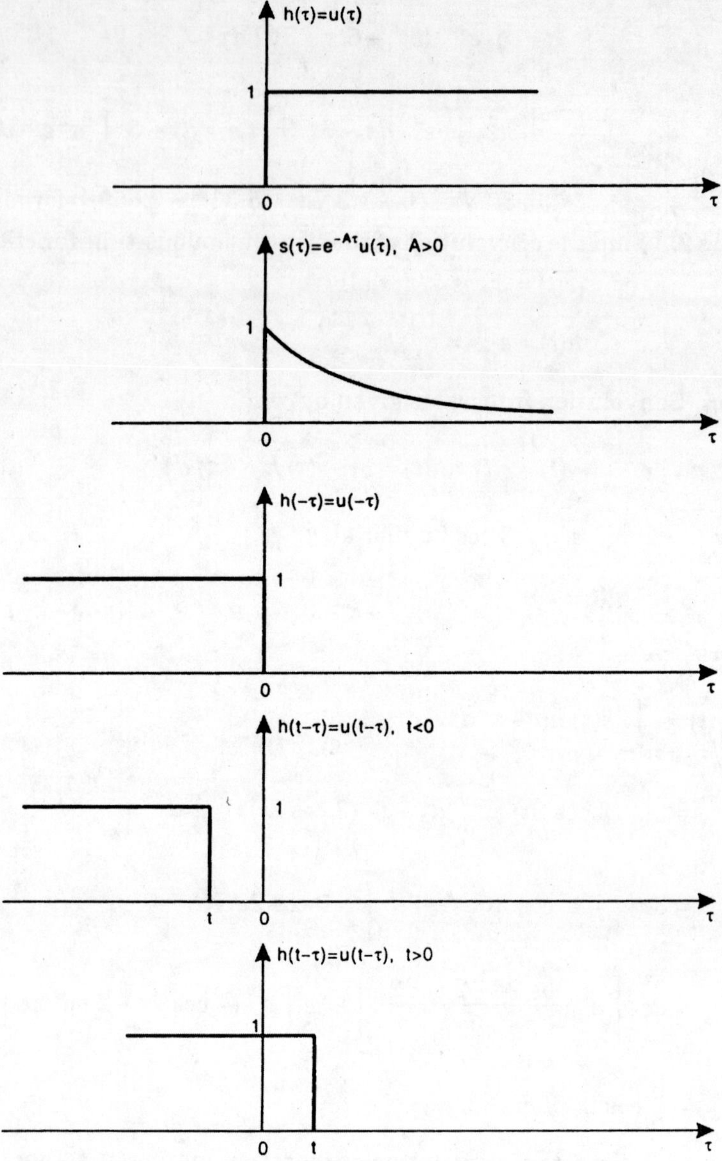

Fig. 2.3 *Graphical illustration for the calculation of convolution integral of $s(n)$ and $h(n)$ given in example 2.3.*

where $$u(\tau) = \begin{cases} 1\,, & \tau > 0 \\ 0\,, & \tau < 0 \end{cases}$$

$$u(t-\tau) = \begin{cases} 1\,, & t-\tau > 0 \rightarrow \tau < t \\ 0\,, & t-\tau < 0 \rightarrow \tau > t \end{cases}$$

$y(t)$ for $t > 0$ will be determined as

$$y(t) = \int_0^t e^{-A\tau}\, d\tau = \left[-\frac{1}{A} e^{-A\tau} \right]_0^t$$

$$= \frac{1}{A}(1 - e^{-At})$$

Thus, for all t, $y(t)$ is

$$y(t) = \frac{1}{A}(1 - e^{-At})\, u(t)$$

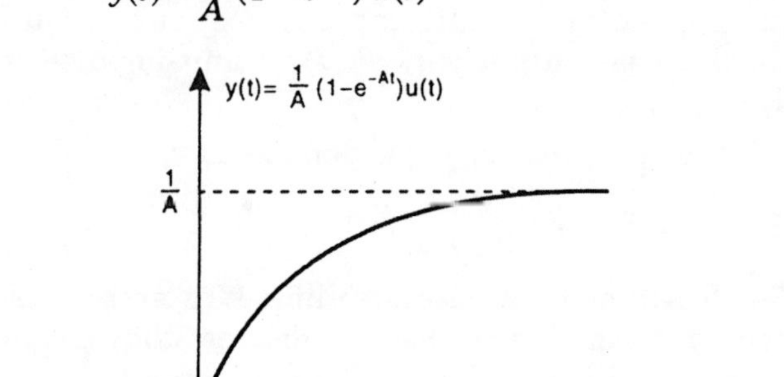

Fig. 2.4 *Resultant of convolution of* $s(n)$ *and* $h(n)$ *given in example* 2.3.

2.3 DISCRETE-TIME LINEAR-TIME INVARIANT (LTI) SYSTEMS

A system of which both input and output are discrete-time signals and posses the two properties of linearity and time-invariance is referred as discrete-time LTI system. These systems are characterised either by their impulse response $h(n)$ or by their transfer function $H(z)$.

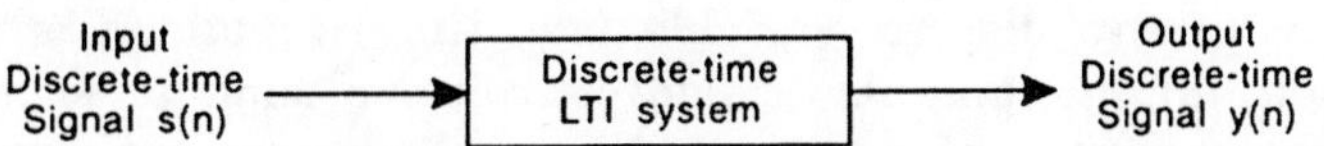

Fig. 2.5 *Discrete-time LTI system.*

Discrete-time signals are represented mathmatically as the superposition of scaled and shifted discrete-time unit-impulse function $\delta(n)$.

$$s(n) = \sum_{k=-\infty}^{\infty} s(k)\, \delta(n-k) \qquad ...(2.7)$$

Discrete-time impulse response $h(n)$ is the output of a discrete-time system when applied input $s(n)$ is equal to discrete-time unit impulse function $\delta(n)$. Transfer function of a discrete-time system is the ratio of z-transform of output $y(n)$ and z-transform of the input $s(n)$of a system when the system is initially relaxed.

2.3.1 Unit Impulse Response for Discrete-Time LTI System

Output of a discrete-time LTI system is called *unit-impulse response* or *unit-sample response* when we apply unit- impulse sequence $\delta(n)$ as input.

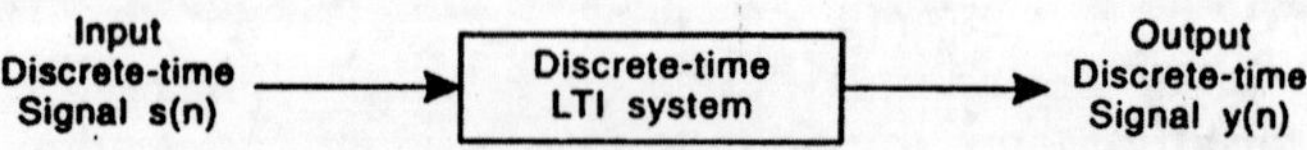

Fig. 2.6 *Discrete-time LTI system.*

If input $s(n) = \delta(n)$ $\Rightarrow$ unit-impluse sequence or unit-sample sequence. In this case output $y(n) = h(n)$ $\Rightarrow$ unit-impulse response or unit-sample response.

where, unit-impulse sequence, $\delta(n)$ is defined as

$$\delta(n) = \begin{cases} 1 \, , & n = 0 \\ 0 \, , & n \neq 0 \end{cases} \qquad \text{...(2.8)}$$

Transfer function of a discrete-time LTI system is equal to z-transform of unit-sample response of a discrete-time LTI system.

Transfer function of discrete-time LTI system

$$H(z) = z\text{-transform of } h(n)$$

$$= \mathcal{Z}\,[h(n)] = \sum_{n=-\infty}^{\infty} h(n)\, z^{-n} \qquad \text{...(2.9)}$$

where z is a complex variable and $z = re^{j\omega}$, r is the magnitude of z and ω is the angle of z, where $h(n)$ is the unit-sample response of discrete-time LTI system.

Note. We will study z-transform in chapter 10.

2.3.2 Convolution Sum for Discrete-Time LTI Systems

Convolution of two discrete-variable (*i.e.,* time) functions is called *discrete convolution*. It is also called convolution sum. It is used to compute the response of a discrete-time LTI system with impulse response $h(n)$ for any arbitrary input $s(n)$.

Any input signal $s(n)$ can be represented in terms of its unit-impulse sequence, $\delta(n)$ which is given by

$$s(n) = \sum_{k=-\infty}^{\infty} s(k)\, \delta(n-k) \qquad \text{...(2.10)}$$

Convolution sum is also called superposition sum and it is given as

$$y(n) = s(n) * h(n) = \sum_{k=-\infty}^{\infty} s(k)\, h(n-k)$$

or $$y(n) = h(n) * s(n) = \sum_{k=-\infty}^{\infty} h(k)\, s(n-k) \qquad ...(2.11)$$

It means both convolution integral and convolution sum are commutative operation.

Example 2.4 Find the convolution sum of two discrete-time sequences $s(n) = e^{-n^2}$ and $h(n) = 3n^2$, for all n.

Solution. Convolution sum of $s(n)$ and $h(n)$ is given as

$$y(n) = s(n) * h(n) = \sum_{k=-\infty}^{\infty} s(k)\, h(n-k)$$

$$= \sum_{k=-\infty}^{\infty} e^{-k^2}\; 3(n-k)^2$$

$$= 3 \sum_{k=-\infty}^{\infty} e^{-k^2}\, [n^2 + k^2 - 2nk]$$

$$= 3n^2 \sum_{k=-\infty}^{\infty} e^{-k^2} + 3 \sum_{k=-\infty}^{\infty} k^2 e^{-k^2} - 6n \sum_{k=-\infty}^{\infty} k e^{-k^2}$$

Example 2.5 Find the convolution sum of two discrete-time signals

$$s(n) = e^{-n^2}, \quad \text{for all } n$$

and $$h(n) = \begin{cases} e^{n}, & n < 0 \\ e^{-n}, & n \geq 0 \end{cases}$$

Solution. Convolution sum of $s(n)$ and $h(n)$ is given by

$$y(n) = s(n) * h(n) = \sum_{k=-\infty}^{\infty} s(k)\, h(n-k)$$

Now $$s(k) = e^{-k^2} \quad \text{for all } k$$

and $$h(n-k) = \begin{cases} e^{n-k}, & n-k < 0 \rightarrow k > n \\ e^{-(n-k)}, & n-k \geq 0 \rightarrow k \leq n \end{cases}$$

Therefore, $$y(n) = \sum_{k=-\infty}^{\infty} s(k)\, h(n-k)$$

$$= \sum_{k=-\infty}^{n} s(k)\, h(n-k) + \sum_{k=n+1}^{\infty} s(k)\, h(n-k)$$

$$= \sum_{k=-\infty}^{n} e^{-k^2} e^{-(n-k)} + \sum_{k=n+1}^{\infty} e^{-k^2} e^{(n-k)}$$

$$= e^{-n} \sum_{k=-\infty}^{n} e^{(k-k^2)} + e^{n} \sum_{k=n+1}^{\infty} e^{-(k+k^2)} \quad \text{for all } k.$$

Example 2.6 Find the convolution sum of two discrete-time signals

$$s(n) = \begin{cases} 0, & n < -5 \\ (1/2)^n, & n \geq -5 \end{cases}$$

and $$h(n) = \begin{cases} 0 \,, & n < 3 \\ (1/3)^n \,, & n \geq 3 \end{cases}$$

Solution. Convolution sum of $s(n)$ and $h(n)$ is given by

$$y(n) = s(n) * h(n) = \sum_{k=-\infty}^{\infty} s(k)\, h(n-k)$$

Now, $$s(k) = \begin{cases} 0 \,, & k < -5 \\ (1/2)^k \,, & k \geq -5 \end{cases}$$

and $$h(n-k) = \begin{cases} 0 \,, & n - k < 3 \rightarrow k > n - 3 \\ (1/2)^{n-k} \,, & n - k \geq 3 \rightarrow k \leq n - 3 \end{cases}$$

Therefore, $$y(n) = \sum_{k=-\infty}^{\infty} s(k)\, h(n-k)$$

$$= \begin{cases} 0, & n < -2 \\ \sum_{k=-5}^{n-3} (1/2)^k\, (1/3)^{n-k}, & n \geq -2 \end{cases}$$

Substituting $l = K + 5$, then for $n \geq -2$

$$y(n) = \sum_{l=0}^{n-8} \left(\frac{1}{2}\right)^{l-5} \left(\frac{1}{3}\right)^{n-l+5}$$

$$= \left(\frac{2}{3}\right)^5 \left(\frac{1}{3}\right)^n \sum_{l=0}^{n-8} \left(\frac{3}{2}\right)^l$$

$$= \left(\frac{2}{3}\right)^5 \left(\frac{1}{3}\right)^n \left[\frac{1 - (3/2)^{n-8+1}}{1 - (3/2)}\right]$$

$$= -2\left(\frac{2}{3}\right)^5 \left(\frac{1}{3}\right)^n \left[1 - \left(\frac{3}{2}\right)^{n-7}\right]$$

$$= -2\left(\frac{2}{3}\right)^5 \left[\left(\frac{1}{3}\right)^n - \left(\frac{2}{3}\right)^7 \left(\frac{1}{2}\right)^n\right]$$

Now $y(n)$ can be written as

$$y(n) = \begin{cases} 0 \,, & n < -2 \\ -64\,(1/3)^{n+5} + 2\,(2/3)^{12}\,(1/2)^n \,, & n \geq -2 \end{cases}$$

Procedure for Computing Convolution Sum of Two Discrete-Time Signals

Convolution sum between $s(n)$ and $h(n)$ involves the following four steps.

Step 1 *Folding*. Folding of a discrete-time signal means to take a mirror image of that signal. Take the mirror image of $h(k)$ about $k = 0$ to obtain $h(-k)$.

Step 2 *Shifting of Folded Sequence*. Shift $h(-k)$ by n_0 to the right (left) if n_0 is positive (negative) to obtain $h(n_0 - k)$.

Step 3 *Multiplication.* Multiply $s(k)$ by $h(n_0 - k)$ to obtain the product sequence $P_{n_0}(k) = s(k)\, h(n_0 - k)$.

Step 4 *Summation.* Sum all the values of the product sequences $P_{n_0}(k)$ to obtain the value of the output at the time $n = n_0$.

Above procedure results in the response of the system at a single time-instant, $n = n_0$.

If we are interested in evaluating the response of the system over all time instants, $-\infty < n < \infty$, then repetition of step 2 to 4 is necessary till the response at all time instants is obtained.

Note : If one sequence has M number of samples or points and other sequence has N number of points then convolution of these sequences will have $M + N - 1$ number of sample points.

Example 2.7 Find the response of the discrete-time system with unit-impulse response

$$h(n) = \{2, \underset{\uparrow}{1}, 3\} \text{ to input } s(n) = \{\underset{\uparrow}{1}, 2, 3\}$$

Solution. Convolution of input $s(n)$ to unit-impulse response $h(n)$ gives response of the system.

Response, $\quad y(n) = \sum_{k=-\infty}^{\infty} s(k)\, h(n-k)$

Given : $\quad s(n) = \{\underset{\uparrow}{1}, 2, 3\}$ and $h(n) = \{2, \underset{\uparrow}{1}, 3\}$

Number of samples in $s(n)$ are M and in $h(n)$ are N threfore number of samples in $y(n)$ will be $M + N - 1$.

Number of samples in $y(n)$ will be

$$M + N - 1 = 3 + 3 - 1 = 5$$

$n = 0, \quad y(0) = \sum_{k=-\infty}^{\infty} s(k)\, h(-k)$

$$s(k) = \underset{\uparrow}{1}, 2, 3$$

$$h(k) = 2, \underset{\uparrow}{1}, 3$$

$$h(-k) = 3, 1, 2$$

$$y(0) = \sum_{k=-\infty}^{\infty} s(k)\, h(-k) = 1 \times 1 + 2 \times 2 = 1 + 4 = 5$$

$n = 1, \quad y(1) = \sum_{k=-\infty}^{\infty} s(k)\, h(1-k)$

$$s(k) = \underset{\uparrow}{1}, 2, 3$$

$h(1-k) = 3, 1, 2$

$$y(1) = \sum_{k=-\infty}^{\infty} s(k)\, h(1-k) = 1 \times 3 + 2 \times 1 + 3 \times 2 = 3 + 2 + 6 = 11$$

$n = 2,$ $y(2) = \sum_{k=-\infty}^{\infty} s(k)\, h(2-k)$

$s(k) = \underset{\uparrow}{1}, 2, 3$

$h(2-k) = 3, 1, 2$

$$y(2) = \sum_{k=-\infty}^{\infty} s(k)\, h(2-k) = 2 \times 3 + 3 \times 1 = 6 + 3 = 9$$

$n = 3,$ $y(3) = \sum_{k=-\infty}^{\infty} s(k)\, h(3-k)$

$s(k) = \underset{\uparrow}{1}, 2, 3$

$h(3-k) = 3, 1, 2$

$$y(3) = \sum_{k=-\infty}^{\infty} s(k)\, h(3-k) = 3 \times 3 = 9$$

$n = 4,$ $y(4) = \sum_{k=-\infty}^{\infty} s(k)\, h(4-k)$

$s(k) = \underset{\uparrow}{1}, 2, 3$

$h(4-k) = 3, 1, 2$

$$y(4) = \sum_{k=-\infty}^{\infty} s(k)\, h(4-k) = 0$$

$y(5) = 0$

.

.

.

and so on

$n = -1,$ $y(-1) = \sum_{k=-\infty}^{\infty} s(k)\, h(-1-k)$

$s(k) = \underset{\uparrow}{1}, 2, 3$

$h(-1-k) = 3, 1, 2$

$$y(-1) = \sum_{k=-\infty}^{\infty} s(k)\, h(-1-k) = 1 \times 2 = 2$$

$$n=-2, \qquad y(-2)=\sum_{k=-\infty}^{\infty} s(k)\,h(-2-k)$$

$$\begin{array}{rl} s(k) = & \quad\; \underset{\uparrow}{1},\ 2,\ 3 \\ h(-2-k) = 3,\ 1,\ 2 & \\ \hline \end{array}$$

$$y(-2)=\sum_{k=-\infty}^{\infty} s(k)\,h(-2-k)=0$$

Similarly

$$y(-3)=0$$

$y(-4)=0$ and so on.

Convolution of $s(n)$ and $h(n)$ is given by

$$y(n)=\{\,y(-1)\,,y(0)\,,y(1)\,,y(2)\,,y(3)\,\}$$

$$=\{\,2\,,\underset{\uparrow}{5}\,,11\,,9\,,9\,\}$$

Example 2.8 Compute the convolution sum of two discrete-time signals

$s(n)=A^n\,u(n)$ and $h(n)=B^n\,u(n)$ for both $A=B$ and $A\neq B$.

Solution. $s(n)$ and $h(n)$ are causal sequences because both these sequences are multiples of $u(n)$, where $u(n)$ is given by

$$u(n)=\begin{cases}1\,, & n\geq 0\\ 0\,, & n<0\end{cases}$$

Convolution sum is defined as

$$y(n)=\sum_{k=-\infty}^{\infty} s(k)\,h(n-k)$$

$$=\sum_{k=-\infty}^{\infty} A^k\,u(k)\,B^{n-k}\,u(n-k)$$

$$\left[\begin{array}{l} u(k)=\begin{cases}1\,, & k\geq 0\\ 0\,, & k<0\end{cases} \quad \text{and} \\ u(n-k)=\begin{cases}1\,, & n-k\geq 0\rightarrow k\leq n\\ 0\,, & n-k<0\rightarrow k>n\end{cases}\end{array}\right]$$

Therefore, $$y(n)=\sum_{k=0}^{n} A^k\,B^{n-k}=B^n\sum_{k=0}^{n}\left(\frac{A}{B}\right)^k$$

For all integer values of k ranging from 0 to n.

Case I : if $A=B$

then $$y(n)=B^n\sum_{k=0}^{n}\left(\frac{A}{A}\right)^k$$

$$= B^n \sum_{k=0}^{n} (1)^k = B^n (n+1), \qquad n = 0, 1, ...$$

Case II : if $A \neq B$

then $$y(n) = B^n \sum_{k=0}^{n} \left(\frac{A}{B}\right)^k$$

$$= B^n \left\{ \frac{1 - \left(\frac{A}{B}\right)^{n+1}}{1 - \frac{A}{B}} \right\}$$

$$= \frac{B^{n+1} - A^{n+1}}{B - A}, \qquad n = 0, 1, ...$$

$$= \frac{1}{(B-A)} \left[B^{n+1} - A^{n+1} \right] u(n)$$

Linear convolution sum of two discrete-time sequences can also be performed by graphical method. Graphical method of linear convolution sum can be understood by an example given below.

Example 2.9 Determine the linear convolution sum of the two discrete-time signals

$$s(n) = A^n\, u(n)\,,\; 0 < A < 1$$

and $$h(n) = u(n).$$

Solution. Linear convolution sum of two discrete-time sequences $s(n)$ and $h(n)$ is given by

$$y(n) = \sum_{k=-\infty}^{\infty} s(k)\, h(n-k)$$

$$= \sum_{k=-\infty}^{\infty} A^k\, u(k)\, u(n-k)$$

$$\left[\begin{array}{l} u(k) = \begin{cases} 1, & k \geq 0 \\ 0, & k < 0 \end{cases} \text{ and} \\ u(n-k) = \begin{cases} 1, & n-k \geq 0 \rightarrow k \leq n \\ 0, & n-k < 0 \rightarrow k > n \end{cases} \end{array} \right]$$

Therefore, $$y(n) = \sum_{k=0}^{n} A^k$$

For $n \geq 0$, $$s(k)\, h(n-k) = \begin{cases} A^k, & 0 \leq k \leq n \\ 0, & \text{otherwise} \end{cases}$$

Computation of linear convolution sum of two discrete-time sequences using graphs is performed as follows :

Sketching of both sequences on graphs.

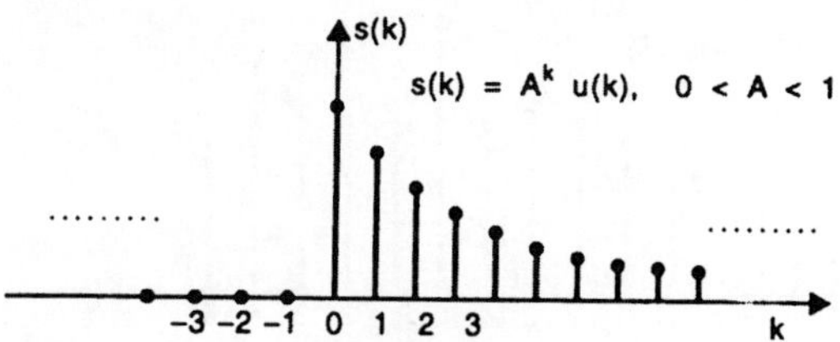

(*a*) *Graphical representation of* $s(k) = A^k u(k)$, $0 < A < 1$.

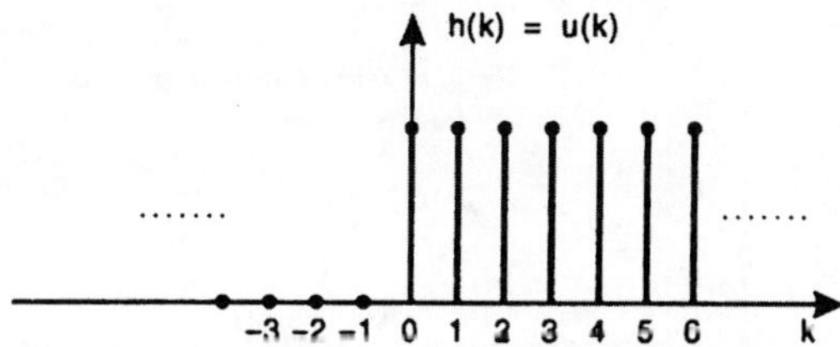

(*b*) Graphical representation of $h(k) = u(k)$.

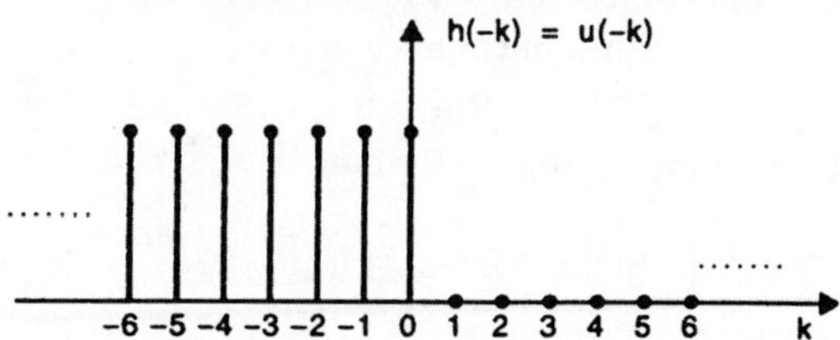

(*c*) *Graphical representation of mirror image of* $h(k) = u(k)$.

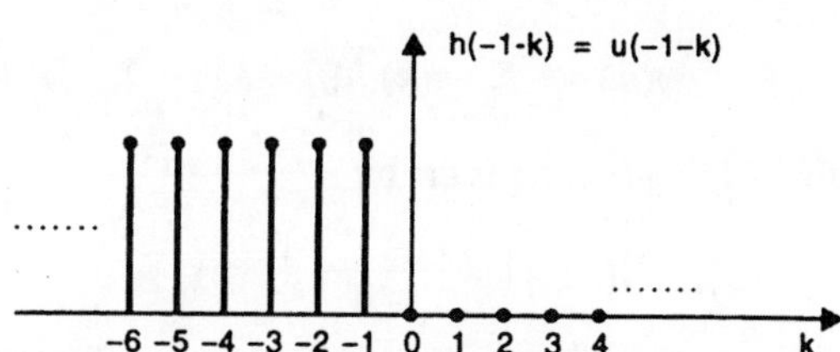

(*d*) Graphical representation of shifted version of mirror image, $h(-k)$ from right to left by one unit of time.

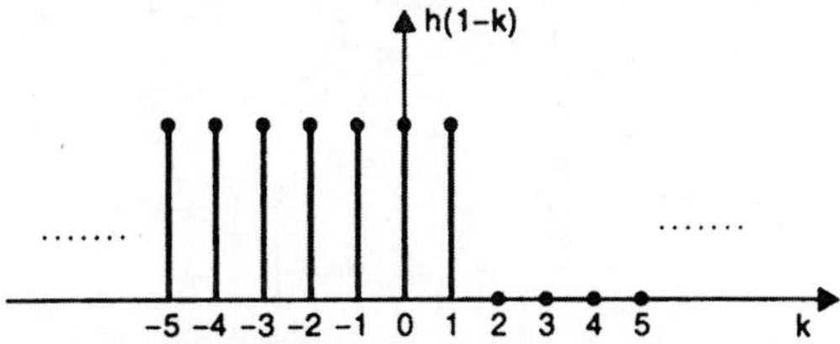

(*e*) Graphical representation of shifted version of mirror image $h(-k)$ from left to right by one unit of time.

Fig. 2.7

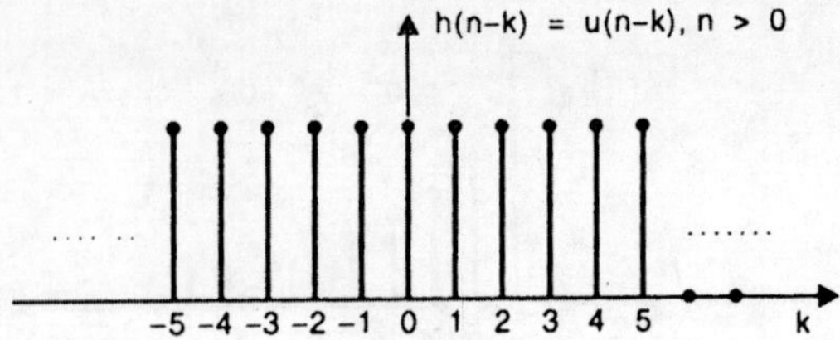

(f) Graphical representation of shifted version of $h(-k)$ from left to right by 'n' unit of time when $n > 0$.

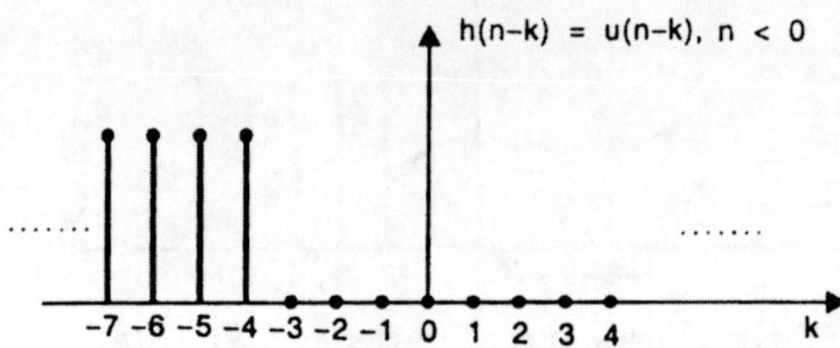

(g) Graphical representation of shifted version of $h(-k)$ from left to right by 'n' unit of time when $n < 0$.

Fig. 2.7

From graphs shown in above figures.

Thus, for $n \geq 0$, $$y(n) = \sum_{k=-\infty}^{\infty} s(k)\,h(n-k)$$

$$= \sum_{k=0}^{n} A^k = \left(\frac{1-A^{n+1}}{1-A}\right)$$

and for $n < 0$, $$y(n) = \sum_{k=-\infty}^{\infty} s(k)\,h(n-k) = \sum_{k=-\infty}^{\infty} (0) = 0$$

Therefore, for all n, $y(n)$ is given by

$$y(n) = \left(\frac{1-A^{n+1}}{1-A}\right)u(n)$$

This is the resultant of convolution of $s(n)$ and $h(n)$ by graphical method. Resultant of convolution, $y(n)$ is graphically represented and it is shown in Fig. 2.7 (*h*).

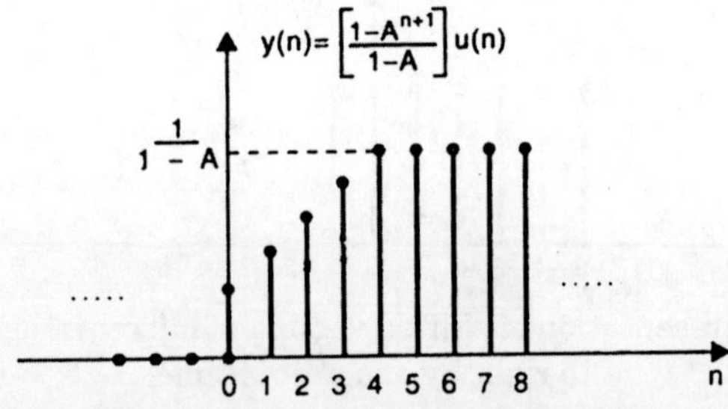

Fig. 2.7 *(h) Graphical representation of $y(n)$.*

2.4 PROPERTIES OF CONTINUOUS-TIME AND DISCRETE-TIME LTI SYSTEMS

We know that continuous-time and discrete-time systems are characterized by their unit-impulse responses. Response of continuous-time LTI system is determined by using convolution integral while the response of discrete-time LTI system is determined by using convolution sum.

Now we will discuss various properties of continuous-time and discrete-time LTI systems.

Convolution integral for continuous-time LTI system is given by

$$y(t) = s(t) * h(t) = \int_{\tau=-\infty}^{\infty} s(\tau)\, h(t-\tau)\, d\tau$$

and convolution sum for discrete-time LTI system is given by

$$y(n) = s(n) * h(n) = \sum_{k=-\infty}^{\infty} s(k)\, h(n-k)$$

Both continuous-time and discrete-time LTI systems are completely characterized by their impulse-responses. These systems are also completely characterized by their transfer function. For continuous-time LTI system, the transfer function is the ratio of Laplace transform of output to Laplace transform of input to the system when system is initially relaxed. It is also defined as Laplace transform of impulse-response of a continuous-time LTI system.

For discrete-time LTI system, transfer function is the ratio of z-transform of output to z-transform of input when initially system is relaxed. It is also defined as z- transform of unit-impulse response of discrete-time LTI system.

Note : We will study Laplace transform and z-transform in chapter 9 and 10, respectively.

In this section, we will study various properties of continuous-time and discrete-time LTI systems in parallel. We will study following properties of LTI systems.

1. Commutative property of LTI systems.
2. Distributive property of LTI systems.
3. Associative property of LTI systems.
4. Static and dynamic LTI systems.
5. Invertibility of LTI systems.
6. Causality of LTI systems.
7. Stability of LTI systems.
8. Unit-step response of LTI systems.

2.4.1 Commutative Property of LTI Systems

Both convolution integral for continuous-time LTI systems and convolution sum for discrete-time LTI systems are commutative.

For continuous-time LTI system, according to commutative property

$$\text{outut} \qquad y(t) = s(t) * h(t) = \int_{\tau=-\infty}^{\infty} s(\tau)\, h(t-\tau)\, d\tau$$

$$= h(t) * s(t) = \int_{\tau=-\infty}^{\infty} h(\tau)\, s(t-\tau)\, d\tau \qquad \text{...(2.13)}$$

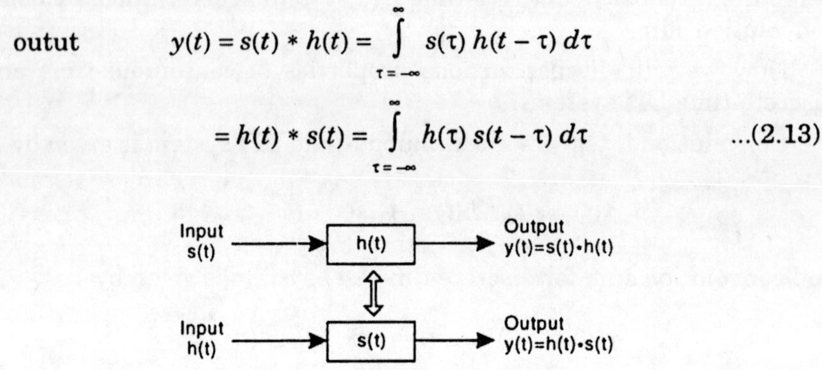

Fig. 2.8 *Continuous-time LTI system showing the commutative property.*

According to commutative property, the output of a continuous-time LTI system with input $s(t)$ and unit impulse response $h(t)$ is identical to the output of a continuous-time LTI system with input $h(t)$ and unit-impulse response $s(t)$.

For discrete-time LTI system, according to commutative property

$$\text{output} \qquad y(n) = s(n) * h(n) = \sum_{k=-\infty}^{\infty} s(k)\, h(n-k)$$

$$= h(n) * s(n) = \sum_{k=-\infty}^{\infty} h(k)\, s(n-k) \qquad \text{...(2.14)}$$

Input
s(n)
h(n)
Output
y(n)=s(n)•h(n)
Input
h(n)
s(n)
Output
y(n)=h(n)•s(n)

Fig. 2.9 *Discrete-time LTI system showing the commutative property.*

According to commutative property, the output of a discrete-time LTI system with input $s(n)$ and unit-impulse response $h(n)$ is identical to the output of a discrete-time LTI system with input $h(n)$ and unit-impulse response $s(n)$.

2.4.2 Distributive Property of LTI Systems

Both convolution integral for continuous-time LTI system and convolution sum for discrete-time LTI systems are distributive.

For continuous-time LTI system, the distributive property is given by

Output $y(t) = s(t) * [h_1(t) + h_2(t)] = s(t) * h_1(t) + s(t) * h_2(t)$ ...(2.15)

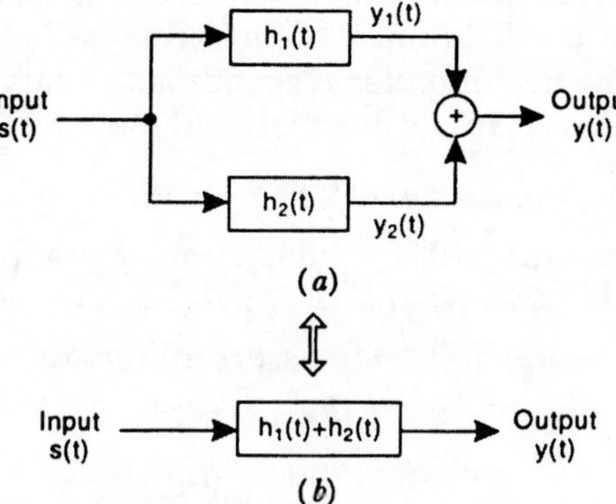

Fig. 2.10 *Illustration of the distributive property of convolution integral for a parallel interconnection of continuous-time LTI systems.*

The two continuous-time LTI systems, with impulse responses $h_1(t)$ and $h_2(t)$, have identical inputs, and outputs are added.

$$y_1(t) = s(t) * h_1(t) \text{ and } y_2(t) = s(t) * h_2(t)$$

From Fig. 2.10 (*a*)

Output $y(t) = y_1(t) + y_2(t) = s(t) * h_1(t) + s(t) * h_2(t)$ ...(2.16)

From Fig. 2.10 (*b*)

Output $y(t) = s(t) * [h_1(t) + h_2(t)]$ ...(2.17)

By comparing Eqs. 2.16 and 2.17, we see that the systems in Fig. 2.10 (*a*) and Fig. 2.10 (*b*) are identical. There is an identical interpretation for discrete-time LTI systems.

For discrete-time LTI system, the distributive property is given by

Output
$$y(n) = s(n) * [h_1(n) + h_2(n)]$$
$$= s(n) * h_1(n) + s(n) * h_2(n) \quad ...(2.18)$$

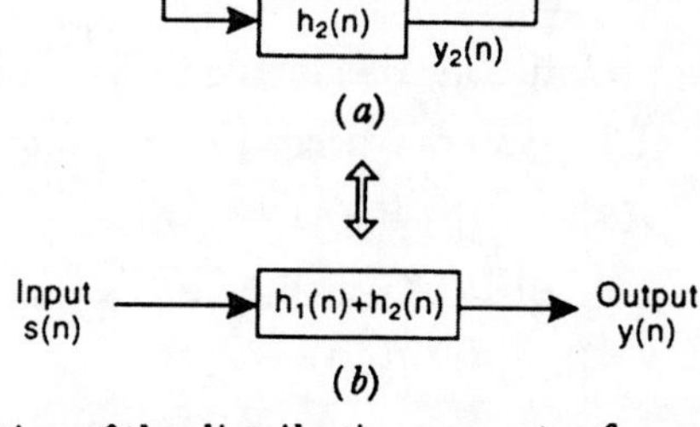

Fig. 2.11 *Illustration of the distributive property of convolution sum for a parallel interconnection of discrete-time LTI systems.*

The distributive property has a useful interpretation in terms of system interconnections.

By distributive property of both convolution integral for continuous-time LTI systems and convolution sum for discrete-time LTI systems, a parallel combination of LTI systems can be replaced by single LTI systems whose unit-impulse response is the sum of the individual unit-impulse responses in the parallel combination.

2.4.3 Associative Property of LTI Systems

Both convolution integral for continuous-time LTI systems and convolution sum for discrete-time LTI systems are associative.

For continuous-time LTI systems, according to associative property.

Output $y(t) = s(t) * [h_1(t) * h_2(t)]$

$$= [s(t) * h_1(t)] * h_2(t) \qquad ...(2.19)$$

or $y(t) = s(t) * h_1(t) * h_2(t)$ ...(2.20)

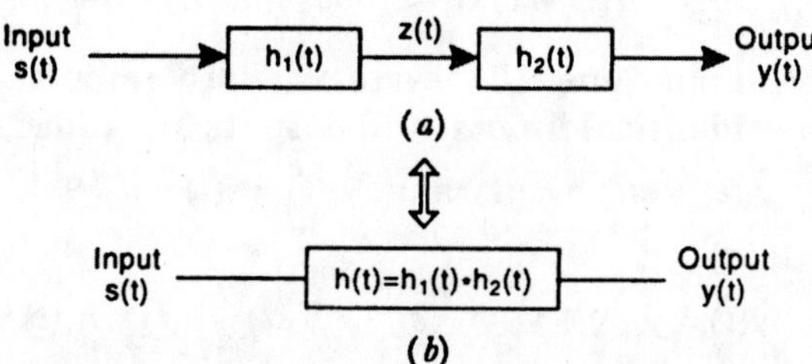

Fig. 2.12 *Illustration of the associative property of convolution integral for a cascade (series) interconnection of continuous-time LTI systems.*

An interpretation of associative property is given below.

In Fig. 2.12 (*a*), $y(t) = z(t) * h_2(t)$

But $z(t) = s(t) * h_1(t)$

Therefore, $y(t) = [s(t) * h_1(t)] * h_2(t)$ ...(2.21)

In Fig. 2.12 (*b*), $y(t) = s(t) * h(t)$

where $h(t) = h_1(t) * h_2(t)$

Therefore, $y(t) = s(t) * [h_1(t) * h_2(t)]$ ...(2.22)

According to the associative property, the cascade interconnection of two systems in Fig. 2.13 (*a*) is equivalent to the single system shown in Fig. 2.13 (*b*).

An analogous interpretation also holds true in discrete-time LTI systems.

For discrete-time LTI systems, according to associative property

Output $y(n) = s(n) * [h_1(n) * h_2(n)]$

$$= [s(n) * h_1(n)] * h_2(n) \qquad ...(2.23)$$

or $y(n) = s(n) * h_1(n) * h_2(n)$ ...(2.24)

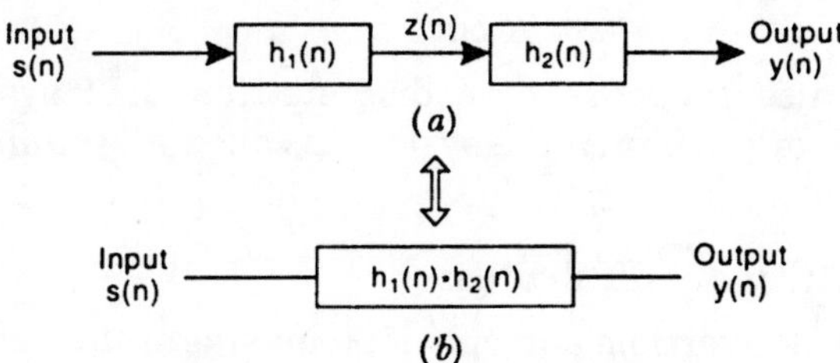

Fig. 2.13 *Illustration of associative property of convolution sum for a cascade (series) interconnection of discrete-time LTI systems.*

An interpretation of associative property is given below.

In Fig. 2.13 (*a*), $y(n) = z(n) * h_2(n)$

But $z(n) = s(n) * h_1(n)$

Therefore, $y(n) = [s(n) * h_1(n)] * h_2(n)$

In Fig. 2.13 (*b*), $y(n) = s(n) * h(n)$

where $h(n) = h_1(n) * h_2(n)$

Therefore, $$y(n) = s(n) * [h_1(n) * h_2(n)] \qquad ...(2.25)$$

Now we can conclude that the impulse response of the cascade (series) of two LTI (either continuous-time or discrete-time) systems is the convolution of their individual impulse responses. Convolution operation is also a commutative operation. So we can determine convolution of two impulse responses in either order.

2.4.4 Static and Dynamic LTI Systems

Static systems are also called *memoryless systems*. A system is called static if its output at any time depends only on the value of the input at the same time.

In particular, a continuous-time LTI system is memoryless (static) if its unit-impluse response $h(t)$ is zero for $t \neq 0$. These memoryless LTI systems are characterized by

$$y(t) = K\, s(t) \qquad ...(2.26)$$

where K is constant.

and its impulse response

$$h(t) = K\, \delta(t) \qquad ...(2.27)$$

If $K = 1$, then these systems are identity systems.

Similarly, a discrete-time LTI system is memoryless (static) if its unit-impulse (unit-sample) response $h(n)$ is zero for $n \neq 0$. Discrete-time memoryless LTI systems are characterized by

$$y(n) = K\, s(n) \qquad ...(2.28)$$

where K is constant.

and its unit-impulse response has the form of

$$h(n) = K\,\delta(n) \qquad \text{...(2.29)}$$

If the impulse response of a discrete-time LTI system is not identically zero for $n \neq 0$ then the system is called a *dynamic system* or *system with memory*.

2.4.5 Invertibility of LTI Systems

A system is called invertible and only if an inverse system exists which, when cascaded (connected in series) with the original system, produces an output equal to the input at first system. If an LTI system is invertible then it has a LTI inverse.

Now we have a continuous-time LTI system with impulse response $h(t)$ and its inverse system with impulse response $h_I(t)$ which results in an output equal to $s(t)$. Cascade interconnection of original continuous-time LTI system with its inverse system is given as identity system.

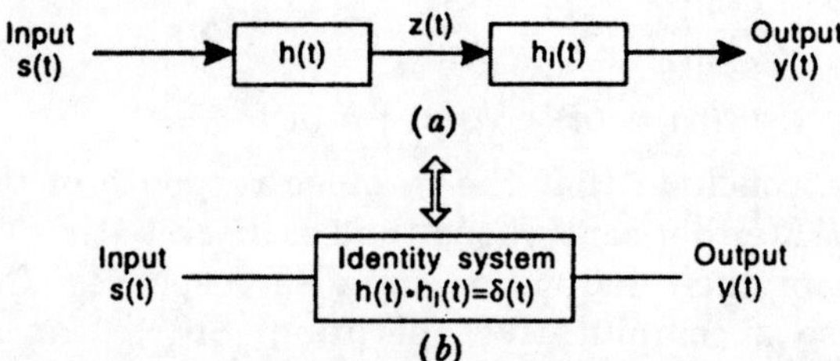

Fig. 2.14 *Illustration of an inverse system for continuous- time LTI systems.*

The overall impulse response of a system with impulse response $h(t)$ cascaded with its inverse system with impulse response $h_I(t)$ is given by

$$h(t) * h_I(t) = \delta(t) \qquad \text{...(2.30)}$$

Similarly, for a discrete-time LTI system, the overall impulse response of a system with impulse response $h(n)$ cascaded with its inverse system with impulse response $h_I(n)$ is given by

$$h(n) * h_I(n) = \delta(n) \qquad \text{...(2.31)}$$

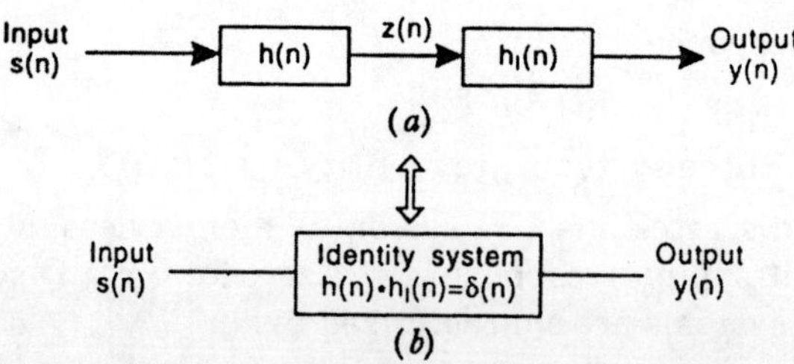

Fig. 2.15 *Illustration of an inverse system for discrete-time LTI systems.*

2.4.6 Causality for LTI Systems

The cutput of a causal system depends only on the present and past values of the input to the system.

A continuous-time LTI system is called *causal system* if its impulse response $h(t)$ is zero for $t < 0$.

For a causal continuous-time LTI system, convolution integral is given by

$$y(t) = s(t) * h(t) = \int_{\tau=-\infty}^{t} s(\tau)\, h(t-\tau)\, d\tau$$

$$= h(t) * s(t) = \int_{\tau=0}^{\infty} h(\tau)\, s(t-\tau)\, d\tau \qquad ...(2.32)$$

A discrete-time LTI system is called causal system if its impulse response $h(n)$ is zero for $n < 0$.

For a causal discrete-time LTI system, convolution sum is given by

$$y(n) = s(n) * h(n) = \sum_{k=-\infty}^{n} s(k)\, h(n-k)$$

$$= h(n) * s(n) = \sum_{k=0}^{\infty} h(k)\, s(n-k) \qquad ...(2.33)$$

Causality of an LTI system is equivalent to its impulse response being a causal signal. Accumulator with impulse response

$$h(n) = u(n) = \begin{cases} 1\,, & n \geq 0. \\ 0\,, & n < 0 \end{cases}$$

is a causal discrete-time LTI system.

A pure time-shift with unit-impulse response $h(t) = \delta(t - t_0)$ is a causal continuous-time LTI system for $t \geq 0$. In this case time-shift is called a *delay*.

A pure time-shift is non-causal continuous-time LTI system for $t < 0$. In this case time-shift is called an *advance*.

2.4.7 Stability for LTI Systems

A stable system is a system which produces bounded output from every bounded input. Now we want to determine conditions under which LTI systems are stable.

Condition of Stability for Continuous-time LTI System. Consider an input $s(t)$ that is bounded in magnitude

$$|s(t)| < M \quad \text{for all values of } t \qquad ...(2.34)$$

Now we apply this input to an continuous-time LTI system with unit-impulse response $h(t)$.

Output of this LTI system is determined by convolution integral and is given by

$$y(t) = \int_{\tau=-\infty}^{\infty} h(\tau)\, s(t-\tau)\, d\tau \qquad \text{...(2.35)}$$

Magnitude of output $y(t)$ is given by

$$|y(t)| = \left| \int_{\tau=-\infty}^{\infty} h(\tau)\, s(t-\tau)\, d\tau \right|$$

$$= \int_{\tau=-\infty}^{\infty} |h(\tau)|\ |s(t-\tau)|\ d\tau \qquad \text{...(2.36)}$$

[Substituting the value $|s(t-\tau)| < M$ for all values of τ and t]

or $$|y(t)| \le \int_{\tau=-\infty}^{\infty} |h(\tau)|\ .\ M\ .\ d\tau$$

$$\le M \int_{\tau=-\infty}^{\infty} |h(\tau)|\ d\tau \quad \text{for all values of } t \qquad \text{...(2.37)}$$

From Eq. 2.37 we can conlcude that if the impulse response $h(t)$ is absolutely integerable then output of a continuous-time LTI system is bounded in magnitude, and hence, the system is bounded-input, bounded-output (BIBO) stable.

A sufficient and necessary condition to stability of a continuous-time LTI system is given by

$$\boldsymbol{S} \equiv \int_{\tau=-\infty}^{\infty} |\boldsymbol{h}(\tau)|\ \boldsymbol{d}\tau < \infty \qquad \text{...(2.38)}$$

Similarly, we can obtain a sufficient and necessary condition to stability of a discrete-time LTI system.

Consider an input $s(n)$ that is bounded in magnitude

$$|s(n)| < M \qquad \text{for all values of } n \qquad \text{...(2.39)}$$

Now we apply this input to a discrete-time LTI system with unit-impulse response $h(n)$. Output of discrete-time LTI system is determined by convolution sum and is given by

$$y(n) = \sum_{k=-\infty}^{\infty} h(k)\, s(n-k) \qquad \text{...(2.40)}$$

Magnitude of $y(n)$ is given by

$$|y(n)| = \left| \sum_{k=-\infty}^{\infty} h(k)\, s(n-k) \right|$$

$$= \sum_{k=-\infty}^{\infty} |h(k)|\ |s(n-k)| \qquad \text{...(2.41)}$$

[Substituting the value of $|s(n-k)| < M$ for all values of k and n]

or $$|y(n)| \leq \sum_{k=-\infty}^{\infty} |h(k)|\, M$$

$$\leq M \sum_{k=-\infty}^{\infty} |h(k)| \qquad \text{...(2.42)}$$

for all values of k.

From Eq. 2.42, we can conclude that if the impulse response $h(n)$ is absolutely summable then output of a discrete-time LTI system is bounded in magnitude, and hence, the system is bounded-input, bounded-output (BIBO) stable.

A sufficient and necessary condition of stability of a discrete-time LTI system is given by

$$S \equiv \sum_{k=-\infty}^{\infty} |h(k)| < \infty \qquad \text{...(2.43)}$$

2.4.8 Unit-Step Response of an LTI System

We have already studied the characterization of LTI system, in terms of their unit-impulse responses. Since impulse responses $h(t)$ or $h(n)$ completely determines the behavior of an LTI system. These impulse responses are used to relate system properties such as stability and causality of LTI systems.

Unit-step response is the output of a LTI system (either continuous-time or discrete-time) for input is equal to unit-step function or sequence.

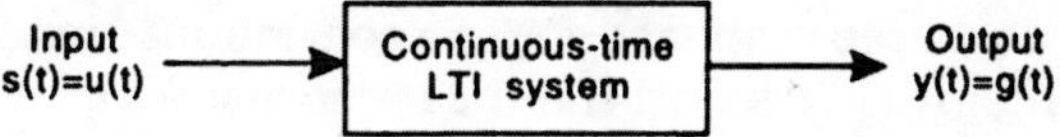

Fig. 2.16 *Continuous-time LTI system. $g(t) \Rightarrow$ unit-step response.*

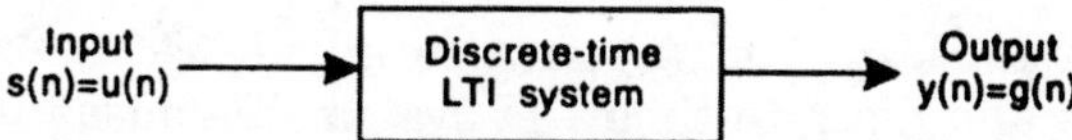

Fig. 2.17 *Discrete-time LTI system. $g(n) \Rightarrow$ unit-step response.*

Unit-step response of continuous-time LTI system is determined by convolution integral of $u(t)$ with unit-impulse response $h(t)$ and is given by

$$g(t) = u(t) * h(t) = h(t) * u(t) \qquad \text{... (2.44)}$$

from commutative property .

Therefore, unit-step response $g(t)$ can be viewed as the response to the input $h(t)$ of a continuous-time LTI system with unit-impulse response $u(t)$.

Unit-step response of discrete-time LTI system is determined by convolution sum of $u(n)$ with unit-impulse response $h(n)$ and is given by

$$g(n) = u(n) * h(n) = h(n) * u(n) \qquad ...(2.45)$$

from commutative property.

Therefore, the unit-step response $g(n)$ can be viewed as the response to the input $h(n)$ of a discrete-time LTI system with unit impulse response $u(n)$.

For example : 1. Unit-impulse response of a accumulator is given by

$$h(n) = u(n)$$

Unit-step response of accumulator is determined by

$$g(n) = u(n) * h(n) = h(n) * u(n)$$

$$= \sum_{k=-\infty}^{\infty} h(k)\, u(n-k) = \sum_{k=-\infty}^{n} h(k) \qquad ...(2.46)$$

For accumulator, unit-impulse response $h(n)$ can be recovered from unit-step response $g(n)$ by using the relation

$$h(n) = g[n] - g[n-1] \qquad ...(2.47)$$

For example : 2. Unit-impulse response of an integrator is given by

$$h(t) = u(t)$$

Unit-step response of integrator is determined by

$$g(t) = u(t) * h(t) = h(t) * u(t)$$

$$= \int_{\tau=-\infty}^{\infty} h(\tau)\, u(t-\tau)\, d\tau = \int_{\tau=-\infty}^{t} h(\tau)\, d\tau \qquad ...(2.48)$$

That is, the unit-step response $g(t)$ of a continuous-time LTI system is the running integral of its unit-impulse response $h(t)$.

For integrator, the unit-impulse response $h(t)$ can be recovered from unit-step response $g(t)$ by using the relation

$$h(t) = g(t) - g(t-1) \qquad ...(2.49)$$

Unit-step responses $g(t)$ or $g(n)$ can also be used to characterize both continuous-time and discrete-time LTI systems because unit-impulse responses are related to unit-step responses.

2.5 DIFFERENTIAL AND DIFFERENCE EQUATIONS OF CAUSAL LTI SYSTEMS

A very important class of continuous-time system is referred to as Linear Time-Invariant (LTI) system for which input-output relation is given by a linear constant-coefficient differential equations. These differential equations arise in the description of a wide variety of systems and physical phenmenon. These differential equations arise in the description of electrical circuits, mechanical systems, kinematics of chemical reactions etc.

We have another class of systems that is a discrete counterpart of continuous-time LTI systems is referred to as discrete-time LTI systems. For discrete-time LTI systems input and output are related through a linear constant-coefficient-difference equations. These equations can be used to describe digital simulation of continuous-time systems which are described by differential equations.

2.5.1 Linear Constant-Coefficient Differential Equations

Continuous-time linear time-invariant (LTI) systems are described by their linear constant-coefficient differential equations. Let us consider a first-order differential equation

$$\frac{dy(t)}{dt} + Ay(t) = s(t) \qquad \text{...(2.50)}$$

where $s(t)$ and $y(t)$ are the input and output of the continuous-time LTI system. A is a constant value.

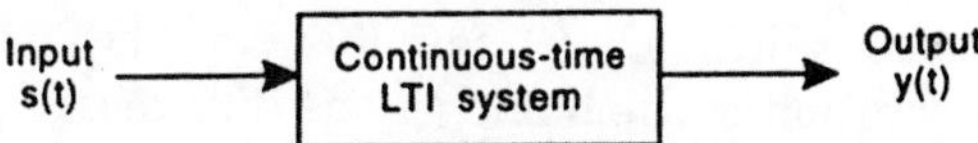

Fig. 2.18 *Continuous-time LTI system.*

Differential equation given by Eq. 2.50 provides an implicit specification of the system. The implicit expressions describe a relationship between the input and the output rather than an explicit expression for the system output as a function of the input. In order to obtain an explicit expression, we must solve the differenial equations.

For the solution of above differential equation (implicit expression) we require more information than that provided by the differential equation alone. For example, to determine the capacitor voltage of an *RC* circuit at the end of a 20 seconds interval, a constant source voltage of 2 volt is applied. Determination of this output capacitor voltage is not possible without knowing some additional information such as initial capacitor voltage.

To solve differential equations, we require one or more auxiliary conditions. By using these auxiliary conditions, we can obtain an explicit expression of the differential equations for the output in terms of the input. In other words, a differential equation given in Eq. 2.50 describes a constraint between the input and the output of a system, but to characterize the system completely, we must also specify auxiliary conditions. Different choices of auxiliary conditions give different relationships between input and output.

Example 2.10 Determine an explicit expression or solve first-order differential equation given as

$$\frac{dy(t)}{dt} + 3\,y(t) = s(t) \qquad ...(1)$$

where $y(t)$ and $s(t)$ are the output and input of the continuous-time LTI system.

Solution. To solve the above differential equation, we assume input signal equal to

$$s(t) = K\,e^{2t}\,u(t) \qquad ...(2)$$

where K is a real number.

The complete solution of differential equation given in Eq. (1) to Eq. (2) consists of the sum of particular solution $y_p(t)$ and homogeneous solution $y_h(t)$.

Complete solution = Particular solution + Homogenous solution

$$y(t) = y_p(t) + y_h(t) \qquad ...(3)$$

where the particular solution $y_p(t)$ satisfies Eq. (1) and homogenous solution $y_h(t)$ is a solution of the homogenous differential equation

$$\frac{d\,y(t)}{dt} + 3\,y(t) = 0 \qquad ...(4)$$

A common method to find the particular solution for an exponential input signal given in Eq. (2) is to look for a so- called force response. Force response is an output signal of the system which has similar form as input signal. Particular solution of differential equation given in Eq. (1) for input $s(t) = K\,e^{2t}$ for $t > 0$ will be of the form

$$y_p(t) = R\,e^{2t}, \quad \text{for } t > 0 \qquad ...(5)$$

where R is a number that we must determine.

For the determination of R, we substitute Eqs. (2) and (5) in Eq. (1) for $t > 0$, we get

$$\frac{dy(t)}{dt} + 3\,y(t) = s(t)$$

or $$\frac{dy(t)}{dt} + 3\,y_p(t) = K\,2^{2t}, \quad \text{for } t > 0$$

or $$\frac{d(R\,e^{2t})}{dt} + 3\,\text{R}e^{2t} = K\,e^{2t}$$

or $$2\,\text{R}e^{2t} + 3\,\text{R}e^{2t} = K\,e^{2t}$$

or $$5\,R = K$$

or $$R = \frac{K}{5} \qquad ...(6)$$

Substituting the value of R in Eq. (5), we get

$$y_p(t) = R\, e^{2t}\ ,\quad t > 0$$

$$= \frac{K}{5} e^{2t}\ ,\quad t > 0 \qquad \text{...(7)}$$

For the determination of homogeneous solution of the differential equation given in Eq. (1), we hypothesize a solution of the form

$$y_h(t) = B\, e^{st} \qquad \text{...(8)}$$

Substituting Eq. (8) in Eq. (4), we get

$$\frac{dy(t)}{dt} + 3\, y(t) = 0$$

or $$\frac{d\,(B\, e^{st})}{dt} + 3\,(B\, e^{st}) = 0$$

or $$B\, s\, e^{st} + 3\, B\, e^{st} = 0$$

or $$B\, e^{st}\,(s + 3) = 0 \qquad \text{...(9)}$$

From Eq. (9), we see that we must take $s = -3$ and that $B\, e^{-3t}$ is a solution to Eq. (4) for any value of B.

$$y_h(t) = B\, e^{-3t} \qquad \text{...(10)}$$

Substituting Eqs. (7) and (10) in Eq. (3), we get

$$y(t) = y_p(t) + y_h(t)$$

$$= \frac{K}{5} e^{2t} + B\, e^{-3t}\ ,\quad t > 0 \qquad \text{...(11)}$$

Now to determine the value of A, we need to specify an auxiliary condition in addition to the differential equation given by Eq. (1). For different choices for auxiliary condition lead to different solution of $y(t)$. Consequently we get different relationships between the input and the output. For causal LTI systems, auxiliary condtions takes the form of the condition of initial rest. For a causal LTI system, if input $s(t) = 0\ ,\ t < t_0$, then output $y(t) = 0$ for $t < t_0$.

From the Eq. (2) if input $s(t) = 0\ ,\ t < 0$, then output $y(t) = 0,\ t < 0$ and thus the condition of initial rest is given by $y(t) = 0\ ,\ t < 0$.

Putting $t = 0$ in Eq. (11), we get

$$y(0) = \frac{K}{5} e^{2\times 0} + B\, e^{-3\times 0}$$

or $$0 = \frac{K}{5} + B$$

or $$B = -\frac{K}{5} \qquad \text{...(12)}$$

Substituting value of B in Eq. (11), we get

$$y(t) = \frac{K}{5} e^{2t} - \frac{K}{5} e^{-3t}, \quad t > 0$$

$$= \frac{K}{5} [e^{2t} - e^{-3t}], \quad t > 0 \quad \text{...(13)}$$

$$y(t) = 0, \quad t < 0, \quad \text{because of the condition of initial rest.} \quad \text{...(14)}$$

Combining Eqs. (13) and (14), we get

$$y(t) = \frac{K}{5} (e^{2t} - e^{-3t})\, u(t) \quad \text{...(15)}$$

From previous example 2.10, we know several very important points concerning linear constant-coefficient differential equations and the systems they represent.

Response to an input $s(t)$ will generally consist of the sum of a particular solution to the differential equation and the homogeneous solution of the differential equation.

The homogeneous solution of the differential equation is solution to the differential equation with input set to zero. The homogeneous solution is also called *natural response of the system*. It is important to note that the condition of initial rest never specify a zero initial condition at a fixed point in time. Condition of initial rest so adjusts a fixed point in time that the response is zero until the input becomes non-zero.

Above example 2.10 also gives some intuition about the condition of initial rest which makes the system [which is described by a linear constant-coefficient differential equation] time-invariant.

The idea of a first-order differential equation can be extended for higher-order differential equations. A general Nth order linear constant-coefficient differential equation is given by

$$\sum_{k=0}^{N} A_k \frac{d^k y(t)}{dt^k} = \sum_{k=0}^{M} B_k \frac{d^k s(t)}{dt^k} \quad \text{...(2.51)}$$

The order of a differential equation means the highest-derivative of the output $y(t)$ in the differential equation. A particular case of above differential equation is determined by putting $N = 0$, we get

$$y(t) = \frac{1}{A_0} \sum_{k=0}^{M} B_k \frac{d^k s(t)}{dt^k} \quad \text{...(2.52)}$$

In this particular case, $y(t)$ is an explicit function of the input $s(t)$ and its derivatives.

For $N \geq 1$, differential equation

$$\sum_{k=0}^{N} A_k \frac{d^k y(t)}{dt^k} = \sum_{k=0}^{M} B_k \frac{d^k s(t)}{dt^k} \quad \text{...(2.53)}$$

specifies an implicit relationship between output and input of a continuous-time LTI system. Analysis of such a type of relationship proceeds just as first-order differential equation. In this case solution $y(t)$ consists of two parts one is particular solution and other is homogeneous solution to differential equation. The homogenous solution of a differential equation is possible by putting

$$\sum_{k=0}^{N} A_k \frac{d^k y(t)}{dt^k} = 0 \qquad ...(2.54)$$

This solution to differential equation is also called natural responses of the system. From condition of initial rest, *i.e.*, if $s(t) = 0$, $t \le t_0$, then $y(t) = 0$, $t \le t_0$. Therefore, the response for $t > t_0$ can be determined from differential Eq. 2.53 with initial conditions

$$y(t_0) = \frac{d\, y(t_0)}{dt} = \frac{d^2 y(t_0)}{dt^2} = \ldots = \frac{d^{N-1} y(t_0)}{dt^{N-1}} = 0 \qquad ...(2.55)$$

Under the condition of initial rest, the system described by Eq. 2.55 is causal LTI system.

In some of the next chapters we will develope some tools such as Laplace transform CTFT etc. for the analysis of continuous-time LTI systems.

2.5.2 Linear Constant-Coefficient Difference Equations

Discrete-time linear time-invariant (LTI) systems are described by their linear constant-coefficient difference equations. Nth order linear constant-coefficient difference equation of a discrete-time LTI system is given by

$$\sum_{k=0}^{N} A_k\, y(n-k) = \sum_{k=0}^{M} B_k\, s(n-k) \qquad ...(2.56)$$

This type of difference equation can also be solved in a manner analogous to that for differential equations of continuous-time LTI systems. The solution of $y(n)$ can be written as the sum of particular solution of difference equation given in Eq. 2.56 and solution to the homogeneous equation.

The homogenous equation is given by

$$\sum_{k=0}^{N} A_k\, y(n-k) = 0 \qquad ...(2.57)$$

The solution to this homogenous equation is called natural responses of the system is described by Eq. 2.56. Condition of initial rest is given by

if $s(n) = 0$ for $n < n_0$, then $y(n) = 0$ for $n < n_0$

With initial rest, the system given by Eq. 2.57 is a causal LTI system.

Now rearranging the Nth order difference equation given in Eq. 2.56

$$\sum_{k=0}^{N} A_k\, y(n-k) = \sum_{k=0}^{M} B_k\, s(n-k)$$

or $$A_0\, y(n) + \sum_{k=1}^{N} A_k\, y(n-k) = \sum_{k=0}^{M} B_k\, s(n-k)$$

or $$y(n) = \frac{1}{A_0}\left[\sum_{k=0}^{M} B_k\, s(n-k) - \sum_{k=1}^{N} A_k\, y(n-k)\right] \qquad ...(2.58)$$

This Eq. 2.58 directly expresses the output of a discrete-time LTI system at time n in terms of previous values of input and output.

In order to determine output $y(n)$, we require $y(n-1)$, $y(n-2)$, ... $y(n-N)$. Therefore, we have given the input for all n and all the auxiliary conditions such as $y(-N)$, $Y(-N+1)$, $y(-N+2)$, $y(-1)$, Eq. 2.56 can be solved for successive values of output $y(n)$.

Both Eqs. 2.56 and 2.58 are called recursive equations because present output $y(n)$ depends upon the input and previous outputs. We can get a special case of Eq. 2.58 by substituting $N = 0$,

$$y(n) = \frac{1}{A_0} \sum_{k=0}^{M} B_k s(n-k) \qquad ...(2.59)$$

It is the convolution sum of B_n and $s(n)$.

This is an explicit relationship between output and input of discrete-time LTI system. Eq. 2.59 is a non-recursive equation because present value of output $y(n)$ does not depend upon previous values of outputs. For solving non-recursive equations, we do not require auxiliary conditions. Eq. 2.59 describes a discrete-time LTI system and impulse response of this system is determined by direct computation as

$$h(n) = \begin{cases} \dfrac{B_n}{A_0}, & 0 \geq n \geq M \\ 0, & \text{otherwise} \end{cases} \qquad ...(2.60)$$

If the impulse response of a discrete-time LTI system has finite duration then the system is called finite impulse response (FIR) system.

Note : We do not require auxiliary conditions for the case $N = 0$ but these conditions are required for the case of $N \leq 1$.

Example 2.11 Determine the current value of output $y(n)$ of a discrete-time LTI system described by

$$y(n) = s(n) + \frac{1}{3} y(n-1) \qquad ...(1)$$

Solution. To solve the above difference equation, we need an initial condition because this difference equation requires previous values of output to determine present value of output. Now we are imposing the condition of initial rest and considering the input

$$s(n) = K\, \delta(n) \qquad ...(2)$$

In this case, since $s(n) = 0$, for $n \leq -1$ and from condition of initial rest $y(n) = 0$, for $n \leq -1$. Now we require one initial condition *i.e.*, $y(-1) = 0$ for solving the above difference equation. We can solve the above difference equation in recursive manner as

$$y(n) = s(n) + \frac{1}{3} y(n-1)$$

$$n = 0, \qquad y(0) = s(0) + \frac{1}{3} y(-1) \qquad [\text{but } s(n) = K\,\delta(n)]$$

$$\text{or} \qquad y(0) = K.1 + \frac{1}{3}\,.\,0 = K$$

$$n = 1, \qquad y(1) = s(1) + \frac{1}{3} y(0) = 0 + \frac{1}{3} K$$

$$n = 2, \qquad y(2) = s(2) + \frac{1}{3} y(1) = 0 + \frac{1}{3} \cdot \frac{1}{3} K = \left(\frac{1}{3}\right)^2 K$$

$$n = 3, \qquad y(3) = s(3) + \frac{1}{3} y(2) = 0 + \frac{1}{3}\left(\frac{1}{3}\right)^2 K = \left(\frac{1}{3}\right)^3 K$$

.

.

.

$$n = n, \qquad y(n) = s(n) + \frac{1}{3} y(n-1) = \left(\frac{1}{2}\right)^n K \qquad ...(3)$$

By setting $K = 1$, then $y(n) = h(n)$ which is the impulse response of above discrete-time LTI system.

$$h(n) = \left(\frac{1}{2}\right)^2 u(n) \qquad ...(4)$$

Note that the causal LTI system given in previous example 2.11 has an impulse response of infinite duration. Such systems are called infinite impulse response (IIR) systems. In most of the cases we will use recursive difference equations to describe and analyse causal LTI systems and also we will make assumption of initial rest. In the next chapters we will develop some powerful tools such as z-transform, DTFT etc. for solving linear constant-coefficient difference equations.

2.6 BASIC BUILDING BLOCKS OR ELEMENTS OF CONTINUOUS-TIME AND DISCRETE-TIME LTI SYSTEMS

Both types of LTI systems can be represented by interconnections of elementary building blocks or elements.

Elementary building blocks for continuous-time systems are

1. Adder

2. Multiplier
3. Differentiator/Integrator.

Adder performs an addition-operation on two or more continuous-time inputs. Multiplier performs multiplication by a coefficient. Differentiator performs differentiation and integrator performs reverse operation of differentiator.

Now we can understand the definition of above building blocks or elements by considering a causal continuous-time LTI system described by a first order differential equation

$$\frac{dy(t)}{dt} + A\,y(t) = B\,s(t) \qquad ...(2.61)$$

This equation can be rewritten as

$$y(t) = -\frac{1}{A}\frac{dy(t)}{dt} + \frac{B}{A}\,s(t) \qquad ...(2.62)$$

Right hand side of Eq. 2.62 has three basic operations viz. addition, multiplication by a coefficient, and differentiation. Therefore, we require three basic network elements such as adder, multiplier and differentiator to represent above differential equation by block diagram interconnections. These elements are shown in Fig. 2.19.

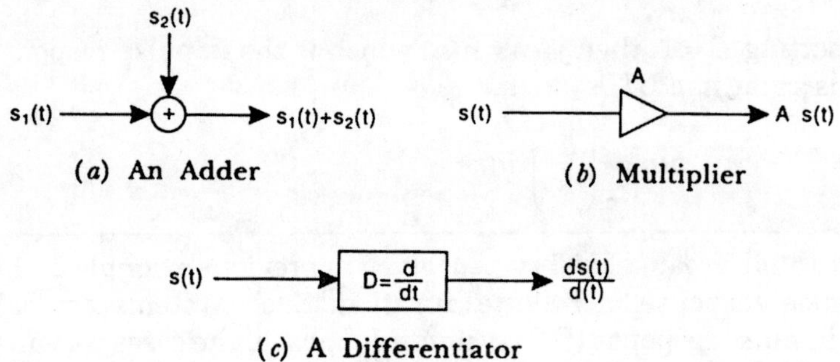

Fig. 2.19 *Basic building blocks or elements of continuous-time systems.*

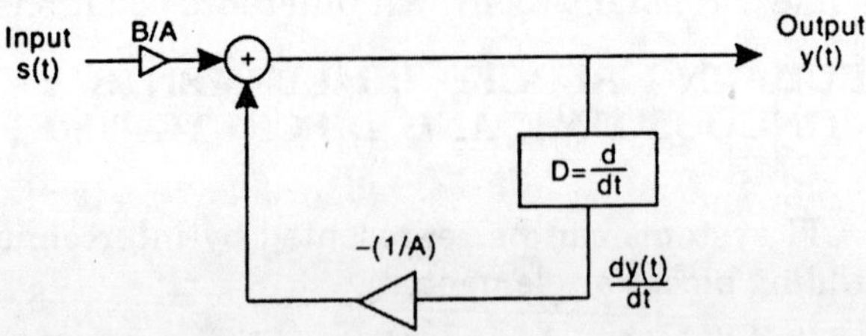

Fig. 2.20 *Block diagram representation for the system described by Eqs.* 2.61 *and* 2.62.

Rewriting Eq. 2.61, we get

$$\frac{dy(t)}{dt} = Bs(t) - Ay(t) \qquad ...(2.63)$$

and integrating Eq. 2.63 from $-\infty$ to t, we get

$$y(t) = \int_{\tau=-\infty}^{t} [B\, s(\tau) - A\, y(\tau)]\, d\tau \qquad ...(2.64)$$

In this form, our system can be implemented using the adder, coefficient multiplier and integrator. We always prefer integrator instead of differentiator because differentiators are both difficult to implement and extremely senstive to errors and noise. Integrators can be readily implemented using operational amplfiers. In continuous-time systems, integrator represents the memory storage element of the system.

Integrating Eq. 2.64 from a finite point in time t_0 results in the expression

$$y(t) = y(t_0) + \int_{\tau=t_0}^{t} [B\, s(\tau) - A\, y(\tau)]\, d\tau \qquad ...(2.65)$$

From Eq. 2.65, it is clear tnat specification of $y(t)$ requires an initial condition, the value of $y(t_0)$. It is the value of $y(t)$ that is stored by integrator at time t_0.

Fig. 2.21 *Block diagram representation of an integrator.*

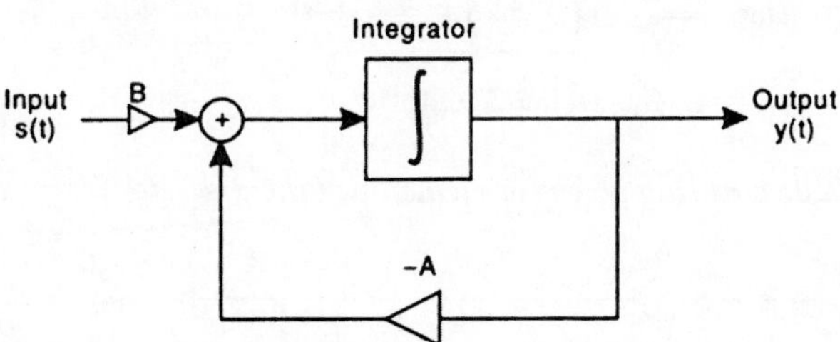

Fig. 2.22 *Block diagram representation for the system described by Eq.* 2.64. *Here we are using integrator instead of differentiator.*

Elementary building blocks for discrete-time systems are

1. Adder
2. Multiplier
3. Delay element.

Adder performs an addition operation on two or more discrete-time inputs and multiplier performs multiplication by a coefficient.

Delay element corresponds to a memory element, as the element must retain the previous value of its input. The initial value of this memory element serves as a necessary initial condition for recursive calculation. For initially at rest condition, the initial value stored in the delay element is zero.

Now we can understand the definition of above building blocks or elements by considering a causal discrete-time LTI system described by first order differential equation

$$y(n) + A\,y(n-1) = B\,s(n) \qquad \text{...(2.66)}$$

This equation can be rewritten as

$$y(n) = -A\,y(n-1) + B\,s(n) \qquad \text{...(2.67)}$$

Right hand side of Eq. 2.67 has three basic operations : addition, multiplication by a coefficient and delay operation. Therefore, we require three basic network elements such as adder, multiplier and delay element to represent above difference equation by block diagram interconnection. The elements are shown in Fig. 2.23.

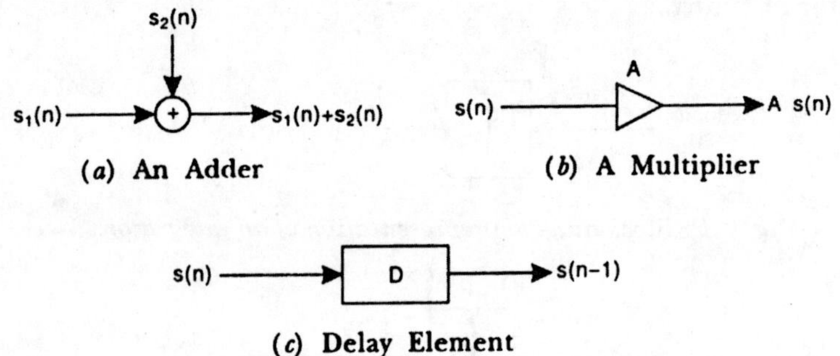

Fig. 2.23 *Basic building blocks or elements of the discrete-time systems.*

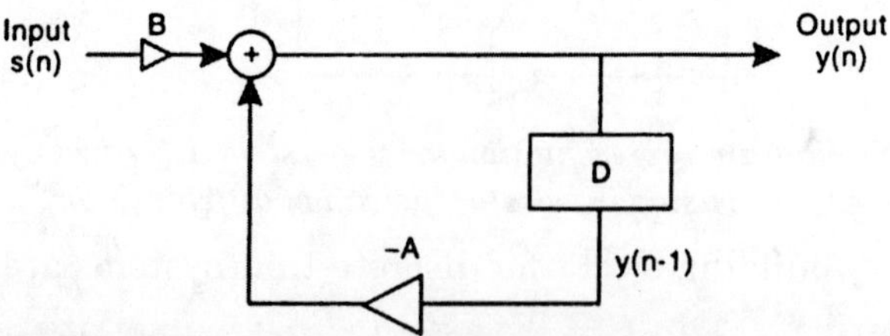

Fig. 2.24 *Block diagram representation for the system described by Eqs.* 2.66 *and* 2.67.

SOLVED EXAMPLES

Example 2.12 Find out the response of a continuous-time system to unit step input given the impulse response

$$h(t) = \frac{t}{RC} e^{-t/RC} u(t)$$

Solution. Response of a continuous-time system is determined by using convolution integral as

$$y(t) = \int_{-\infty}^{\infty} s(\tau)\, h(t-\tau)\, d\tau \qquad ...(1)$$

where $s(t)$ = input; $y(t)$ = output; $h(t)$ = impulse response

Given, input $s(t) = u(t)$, impulse response $h(t) = \frac{1}{RC} e^{-t/RC} u(t)$

Putting values of input $s(t)$ and impulse response $h(t)$ in Eq. (1), we get

$$y(t) = \int_{-\infty}^{\infty} u(\tau) \frac{1}{RC} e^{-(t-\tau)/RC} u(t-\tau)\, d\tau \qquad ...(2)$$

Since we know that $u(\tau) = \begin{cases} 1, & \tau > 0 \\ 0, & \tau < 0 \end{cases}$

and $u(t-\tau) = \begin{cases} 1, & t-\tau > 0 \Rightarrow \tau < t \\ 0, & t-\tau < 0 \Rightarrow \tau > t \end{cases}$

Substituting the values of $u(\tau)$ and $u(t-\tau)$ in Eq. 2, we get

$$y(t) = \frac{1}{RC} \int_0^t e^{-(t-\tau)/RC}\, d\tau = \frac{e^{-t/RC}}{RC} \int_0^t e^{\tau/RC}\, d\tau$$

$$= \frac{e^{-t/RC}}{RC} \left[RC\, e^{\tau/RC}\right]_0^t = \frac{e^{-t/RC}}{RC} \times RC\, [e^{t/RC} - e^{0/RC}]$$

$$= e^{-t/RC} [e^{t/RC} - 1]$$

$$y(t) = 1 - e^{-t/RC} \text{ for } t \geq 0$$

Example 2.13 Find whether the system characterized by its impulse response $h(t) = \frac{1}{RC} e^{-t/RC} u(t)$ is BIBO stable or not.

Solution. We have a given impulse response of a continuous- time system as $h(t) = \frac{1}{RC} e^{-t/RC} u(t)$

From the condition for a system to be BIBO stable

$$S \equiv \int_{-\infty}^{\infty} |h(t)|\, dt < \infty \qquad ...(1)$$

Now we will determine the values of $\int_{-\infty}^{\infty} |h(t)|\, dt$ as

$$\int_{-\infty}^{\infty} |h(t)|\, dt = \int_{-\infty}^{\infty} \left| \frac{1}{RC} e^{-t/RC} u(t) \right| dt$$

$$\left(\text{Since } u(t) = \begin{cases} 1, & t > 0 \\ 0, & t < 0 \end{cases} \right)$$

$$= \int_{0}^{\infty} \frac{1}{RC} e^{-t/RC} (1)\, dt$$

$$= \frac{1}{RC} \Big[(-RC)\, e^{-t/RC} \Big]_0^{\infty} = e^{-0/RC} - e^{-\infty/RC}$$

$$= 1 < \infty$$

or $S \equiv 1$. Hence the system is BIBO stable.

Example 2.14 Find out the convolution of $s(n)$ and $h(n)$. Where

$$s(n) = \delta(n) + 2\delta(n-1) - \delta(n-3)$$

and
$$h(n) = 2\delta(n+1) + 2\delta(n-1)$$

Solution. We know that convolution of $s(n)$ and $h(n)$ is defined as

$$y(n) = \sum_{k=-\infty}^{\infty} h(k)\, s(n-k) \qquad \ldots(1)$$

Signals $s(n)$ and $h(n)$ are shown in Fig. 2.25.

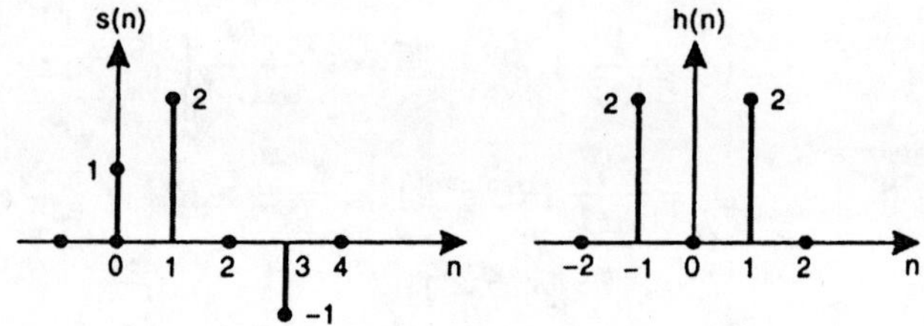

Fig. 2.25

From Fig. 2.25, we can easily see that the above convolution sum reduces to

$$y(n) = h(-1)\, s(n+1) + h(1)\, s(n-1)$$

$$= 2s(n+1) + 2s(n-1)$$

$$= 2[\delta(n+1) + 2\delta(n) - \delta(n-2)$$
$$+ \delta(n-1) + 2\delta(n-2) - \delta(n-4)]$$

$$= 2\,[-\delta(n-4) + \delta(n-2) + \delta(n-1)$$
$$+ 2\delta(n) + \delta(n+1)]$$

Example 2.15 Determine output $y(n)$ if input $s(n) = \delta(n-1)$ for a causal discrete-time LTI system whose input- output relationship is given by

$$y(n) = \frac{1}{4}y(n-1) + s(n)$$

Solution. We have given difference equation of a causal discrete-time LTI system as

$$y(n) = \frac{1}{4}y(n-1) + s(n)$$

Since the system is causal, $y(n) = 0$ for $n < 0$

$n = 0,$ $\quad y(0) = \frac{1}{4}y(0-1) + s(0) = \frac{1}{4}y(-1) + \delta(-1) = \frac{1}{4} \times 0 + 0 = 0$

$n = 1,$ $\quad y(1) = \frac{1}{4}y(1-1) + s(1) = \frac{1}{4}y(0) + \delta(0) = \frac{1}{4} \times 0 + 1 = 1$

$n = 2,$ $\quad y(2) = \frac{1}{4}y(2-1) + s(2) = \frac{1}{4}y(1) + \delta(1) = \frac{1}{4} \times 1 + 0 = \frac{1}{4}$

$n = 3,$ $\quad y(3) = \frac{1}{4}y(3-1) + s(3) = \frac{1}{4}y(2) + \delta(2) = \frac{1}{4} \times \frac{1}{4} + 0 = \frac{1}{16}$

.

.

.

and so on

.

.

$$y(m) = \left(\frac{1}{4}\right)^{m-1}, \text{ where } m \text{ is any arbitrary number.}$$

.

.

.

Therefore $y(n) = \left(\frac{1}{4}\right)^{n-1} u(n-1)$

Example 2.16 Find out the response $y(t)$ of the continuous-time LTI system with impulse response $h(t)$ to the input $s(t)$. Given : input $s(t) = e^{-At}u(t)$ and impulse response $h(t) = e^{-Bt}u(t)$ (*a*) when $A \neq B$, (*b*) when $A = B$

Solution. (*a*) Response $y(t)$ is determined by convolution integral of $s(t)$ and $h(t)$ and is given as

$$y(t) = \int_{-\infty}^{\infty} s(\tau)\, h(t-\tau)\, d\tau$$

or

$$y(t) = \int_{-\infty}^{\infty} e^{-A\tau}\, u(\tau)\, e^{-B(t-\tau)}\, u(t-\tau)\, d\tau \qquad ...(1)$$

$$\left[\text{Since } u(\tau) = \begin{cases} 1, & \tau > 0 \\ 0, & \tau < 0 \end{cases} \text{ and } u(t-\tau) = \begin{cases} 1, & t-\tau > 0 \rightarrow \tau < t \\ 0, & t-\tau < 0 \rightarrow \tau > t \end{cases}\right]$$

Putting above values in Eq. (1), we get

$$y(t) = \int_0^t e^{-A\tau} e^{-B(t-\tau)} d\tau \qquad \text{...(2)}$$

$$= e^{-Bt} \int_0^t e^{-(A-B)\tau} d\tau = e^{-Bt} \left[\frac{e^{-(A-B)\tau}}{-(A-B)}\right]_0^t$$

$$= -\frac{e^{-Bt}}{(A-B)} [e^{-(A-B)t} - e^{-(A-B)\times 0}]$$

$$= \frac{e^{-Bt}}{(A-B)} [1 - e^{-(A-B)t}] = \frac{1}{A-B} [e^{-Bt} - e^{-At}]$$

or $$y(t) = \frac{e^{-Bt} - e^{-At}}{A-B} \text{ for } A \neq B \text{ and } t > 0$$

or $$y(t) = \left(\frac{e^{-Bt} - e^{-At}}{A-B}\right) u(t)$$

(*b*) Putting $A = B$ in Eq. (2), we get

$$y(t) = \int_0^t e^{-A\tau} e^{-A(t-\tau)} d\tau = \int_0^t e^{-At} d\tau$$

$$= e^{-At} [\, t \,]_0^t = t\, e^{-At} u(t) \text{ for } A = B$$

Example 2.17 Examine whether the following systems are stable

(*a*) $y(n) = \cos [s(n)]$ (*b*) $y(n) = s(-n-2)$

Solution. (*a*) $y(n) = \cos [s(n)]$

First of all we determine impulse or unit-sample response $h(n)$ by putting input $s(n) = \delta(n)$.

We are given that $y(n) = \cos [s(n)]$

$n = 0$, $y(0) = \cos [s(0)] = \cos [\delta(0)] = \cos [1]$

$n = 1$, $y(1) = \cos [s(1)] = \cos [\delta(1)] = \cos [0] = 1$

$n = 2$, $y(2) = \cos [s(2)] = \cos [\delta(2)] = \cos [0] = 1$

$\Rightarrow$ $y(n) = 1$, $n > 0$

or $h(n) = 1$, $n > 0$

From condition of stability, we know that

$$S = \sum_{k=-\infty}^{\infty} |h(k)| = \cos(1) + \sum_{k=1}^{\infty} (1) \to \infty$$

Therefore, the system is unstable.

(*b*) $y(n) = s(-n-2)$

$n = 0, \quad y(0) = s(-0-2) = s(-2) = \delta(-2) = 0$

$n = 1, \quad y(1) = s(-1-2) = s(-3) = \delta(-3) = 0$

$n = 2, \quad y(2) = s(-2-2) = s(-4) = \delta(-4) = 0$

$n = 3, \quad y(3) = s(-3-2) = s(-5) = \delta(-5) = 0$

$$y(n) = \delta(-n-2) = 0$$

or $\quad h(n) = \delta(-n-2) = 0$

Now, from condition of stability, we know that

$$S = \sum_{k=-\infty}^{\infty} |h(k)| < \infty$$

$$S = \sum_{k=-\infty}^{\infty} |h(k)| = \sum_{k=-\infty}^{\infty} (0) = 0 < \infty$$

Therefore, the system is stable.

Example 2.18 Find whether the three sample-averager given by the difference equation

$$y(n) = \frac{1}{3}[s(n-1) + s(n) + s(n+1)]$$ is stable or not.

Solution. For checking stability of the system, we determine impulse response of the system by putting input $s(n) = \delta(n)$

For $n = 0, \quad y(0) = \frac{1}{3}[s(0-1) + s(0) + s(0+1)]$

$$= \frac{1}{3}[\delta(-1) + \delta(0) + \delta(1)] = \frac{1}{3}[0+1+0] = \frac{1}{3}$$

For $n = 1, \quad y(1) = \frac{1}{3}[s(1-1) + s(1) + s(1+1)]$

$$= \frac{1}{3}[\delta(0) + \delta(1) + \delta(2)] = \frac{1}{3}[1+0+0] = \frac{1}{3}$$

For $n = 2, \quad y(2) = \frac{1}{3}[s(2-1) + s(2) + s(2+1)]$

$$= \frac{1}{3}[\delta(1) + \delta(2) + \delta(3)] = \frac{1}{3}[0+0+0] = 0$$

For $n = -1$, $y(-1) = \frac{1}{3}[s(-1-1) + s(-1) + s(-1+1)]$

$$= \frac{1}{3}[\delta(-2) + \delta(-1) + \delta(0)] = \frac{1}{3}[0 + 0 + 1] = \frac{1}{3}$$

For $n = -2$, $y(-2) = \frac{1}{3}[s(-2-1) + s(-2) + s(-2+1)]$

$$= \frac{1}{3}[\delta(-3) + \delta(-2) + \delta(-1)] = \frac{1}{3}[0 + 0 + 0] = 0$$

Therefore, $h(n) = \frac{1}{3}$ for $n = -1, 0, 1$

From the condition of stability,

$$S \equiv \sum_{k=-\infty}^{\infty} |h(k)| < \infty$$

$$\sum_{k=-\infty}^{\infty} |h(k)| = \sum_{k=-1}^{1} |h(k)| = h(-1) + h(0) + h(1)$$

$$= \frac{1}{3} + \frac{1}{3} + \frac{1}{3} = 1$$

$$S \equiv 1 < \infty$$

Therefore, the above system is stable.

Example 2.19 Find the impulse or unit-sample response $h(n)$ of a discrete-time system shown in Fig. 2.26. Assume that the system is initially relaxed. Also find the unit step response of the system.

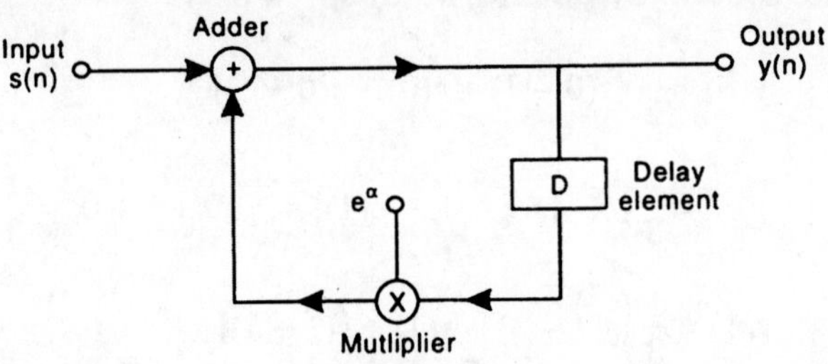

Fig. 2.26

Solution. This discrete-time system can be described by following difference equation

$$y(n) = s(n) + e^{\alpha}\, y(n-1) \qquad ...(1)$$

To determine impulse response, we put input $s(n) = \delta(n)$ in Eq. (1).

For $n = 0$, $\quad y(0) = s(0) + e^{\alpha}\, y(0-1) = \delta(0) + e^{\alpha}\, y(-1)$

[But from the definition of $\delta(n) = \begin{cases} 1, & n = 0 \\ 0, & n \neq 0 \end{cases}$ and as system is initially relaxed, therefore $y(-1) = 0$]

$$y(0) = 1 + e^{\alpha}\,0 = 1$$

$n = 1$, $\quad y(1) = s(1) + e^{\alpha}\,y(0) = \delta(1) + e^{\alpha}\,y(0)$

[Since $\delta(1) = 0$ and for this system $y(0) = 0$]

$$y(1) = 0 + e^{\alpha}\,1 = e^{\alpha}$$

$n = 2$, $\quad y(2) = s(2) + e^{\alpha}\,y(1) = \delta(2) + e^{\alpha}y(1)$

$$= 0 + e^{\alpha}\,e^{\alpha} = e^{2\alpha}$$

$n = n$, $\quad y(n) = e^{n\alpha}$

It is the impulse response of the discrete-time system given in Fig. 2.26.

$$h(n) = e^{n\alpha}u(n)$$

To determine the step response, we put $s(n) = u(n)$ as

$$y(n) = s(n) + e^{\alpha}y(n-1)$$

$n = 0$, $\quad y(0) = s(0) + e^{\alpha}y(0-1)$

[From the defintion of $u(n)$ and as system is initially relaxed, *i.e.*, $y(-1) = 0$]

$$y(0) = u(0) + e^{\alpha}\,y(-1) = 1 + e^{\alpha}(0) = 1$$

$n = 1$, $\quad y(1) = u(1) + e^{\alpha}\,y(1-1) = u(1) + e^{\alpha}\,y(0)$

[Also given that $y(0) = 0$]

$$= 1 + e^{\alpha}\,1 = 1 + e^{\alpha}$$

$n = 2$, $\quad y(2) = s(2) + e^{\alpha}\,y(2-1) = u(2) + e^{\alpha}\,y(1)$

$$= 1 + e^{\alpha}\,(1 + e^{\alpha}) = 1 + e^{\alpha} + e^{2\alpha}$$

$n = n$, $\quad y(n) = 1 + e^{\alpha} + e^{2\alpha} + \ldots + e^{n\alpha} = \sum_{k=0}^{n} e^{k\alpha}$

(It is a geometric progression)

$$= \frac{A[1 - R^{N+1}]}{[1 - R]}$$ where A is the first term of G.P., R is the common ratio

$$= \frac{1[1 - (e^{\alpha})^{N+1}]}{[1 - e^{\alpha}]} = \frac{1 - e^{(n+1)\alpha}}{1 - e^{\alpha}} \qquad \ldots(1)$$

In Eq. (1), the unit-step response of the discrete-time LTI system given in Fig. 2.26.

$$\therefore \qquad g(n) = \left[\frac{1 - e^{(n+1)\alpha}}{1 - e^{\alpha}}\right]$$

Example 2.20 Check the discrete-time system for time-invariance which is described by the following difference equation

$$y(n) = 4ns(n)$$

Solution. The response to a delayed input is

$$y(n - k) = 4n\ s(n - k) \qquad \text{...(1)}$$

The delayed response is

$$y(n - k) = 4(n - k)\ s(n - k) \qquad \text{...(2)}$$

Both responses are not equal

$$y(n\ , k) \neq y(n - k)$$

Therefore, the given discrete-time system $y(n) = 4ns(n)$ is not time-invariant. It is a time-varying system.

Example 2.21 Check the discrete-time system for time-invariance which is described by following difference equation.

$$y(n) = 12\ s(n - 1) + 11\ s(n - 2)$$

Solution. The response to a delayed input is

$$y(n\ , k) = 12\ s(n - k - 1) + 11\ s(n - k - 2) \qquad \text{...(1)}$$

The delayed response is

$$y(n - k) = 12\ s(n - k - 1) + 11\ s(n - k - 1) \qquad \text{...(2)}$$

Both responses are equal, *i.e.*,

$$y(n\ , k) = y(n - k)$$

Therefore, the given discrete-time system is time-invariant.

Example 2.22 Find whether the following systems are causal or not.

(*a*) $y(t) = s(-t)$ (*b*) $y(t) = s(t^2)$

Solution. (*a*) A continuous-time system described by input-output relationship as

$$y(t) = s(-t)$$

The output of the above system at any positive instant of time depends only on the past and present values of the input and not on the future values. For example

$$y(t) = s(-t)$$

Here we will determine output for $t \geq 0$ as

$t = 0$, $\quad y(0) = s(-0) = s(0)$

$t = 1$, $\quad y(1) = s(-1)$

$t = 2\,,\qquad y(2) = s(-2)$

$t = 3\,,\qquad y(3) = s(-3)$

$\vdots$

and so on.

where $s(-1)\,,\, s(-2)\,,\, s(-3)\,,\, \ldots$ are past values of the input. Therefore, the given system is a causal system.

(*b*) A continuous-time system described by input-output relationship as

$$y(t) = s(t^2)$$

Here we determine output $y(t)$ for all times $(-\infty < t < \infty)$ as

$t = -2\,,\qquad y(-2) = s[(-2)^2] = s(4)$

$t = -1\,,\qquad y(-1) = s[(-1)^2] = s(1)$

$t = 0\,,\qquad y(0) = s(0) = s(0)$

$t = 1\,,\qquad y(1) = s[(1)^2] = s(1)$

$t = 2\,,\qquad y(2) = s[(2)^2] = s(4)$

$\vdots$

From the above analysis, we can conclude that output of the system at all time instants depends upon the future values of inputs. Therefore, the given system is non-causal.

Example 2.23 Find the response (or output) of the following discrete-time systems to the input signal

$$s(n) = \begin{cases} |n|\,, & -3 \le n \le 3 \\ 0\,, & \text{otherwise} \end{cases}$$

(*a*) $y(n) = s(n)$

(*b*) $y(n) = s(n-1)$

(*c*) $y(n) = s(n+1)$

(*d*) $y(n) = \dfrac{1}{3}\,[s(n-1) + s(n) + s(n+1)]$

Solution. Input signal

$$s(n) = \begin{cases} |n|\,, & -3 \le n \le 3 \\ 0\,, & \text{otherwise} \end{cases}$$

$$= \{\ldots, 0, 0, 0, 3, 2, 1, \underset{\uparrow}{0}, 1, 2, 3, 0, 0, 0, \ldots\}$$

(*a*) Given $y(n) = s(n)$

The given system is an "identity system", since the output is exactly equal to the input

$$y(n) = s(n) = \{\ldots, 0, 0, 0, 3, 2, 1, \underset{\uparrow}{0}, 1, 2, 3, 0, 0, 0, \ldots\}$$

[Note : Arrow mark represents the zeroth sample ($n = 0$)]

(*b*) Given $y(n) = s(n - 1)$

This system delays the input by one sample. Therefore, the output is given by

$$y(n) = s(n - 1) = \{\ldots, 0, 0, 0, 3, 2, \underset{\uparrow}{1}, 0, 1, 2, 3, 0, 0, 0 \ldots\}$$

(*c*) Given $y(n) = s(n + 1)$

In this case the system advances the input by one sample into the future. For example, the value of the output at time $n = 0$ if $y(0)$ is equal to $s(1)$. The response of this system is given by

$$y(n) = s(n + 1) = \{\ldots, 0, 0, 0, 3, 2, 1, 0, \underset{\uparrow}{1}, 2, 3, 0, 0, 0 \ldots\}$$

(*d*) Given $y(n) = \frac{1}{3}\,[s(n - 1) + s(n) + s(n + 1)]$

This is the input-output relationship or difference equation of the three sample averager. The output of this system at any time is the mean value of the immediate past, present and immediate future samples of the input.

For example, the output at time $n = 0$ is

$$y(0) = \frac{1}{3}\,[s(0 - 1) + s(0) + s(0 + 1)]$$

$$= \frac{1}{3}\,[s(-1) + s(0) + s(1)] = \frac{1}{3}\,[1 + 0 + 1] = \frac{2}{3}$$

Similarly, we can determine the values of $y(-4)$, $y(-3)$, $y(-2)$, $y(-1)$ and $y(1)$, $y(2)$, $y(3)$, $y(4)$, etc.

Repeating previous method for every n, we have the output signal given by

$$y(n) = \left\{\ldots, 0, 0, 1, \frac{5}{3}, 2, 1, \underset{\uparrow}{\frac{2}{3}}, 1, 2, \frac{5}{3}, 1, 0, 0, \ldots\right\}$$

Example 2.24 Check the stability of the second-order discrete- time system given by

$$y(n) = 2r\cos(\omega_0)\, y(n - 1) - r^2\, y(n - 2)$$
$$+ s(n) - r\cos(\omega_0)\, s(n - 1)$$

Solution. Here we are assuming that above system is initially relaxed, *i.e.*, $y(n) = 0$, for $n < 0$

Putting $n = 0, 1, 2, \ldots$ we get

$n = 0,\quad y(0) = 2r\cos(\omega_0)\, y(0-1) - r^2 y(0-2) + s(0) - r\cos(\omega_0)\, s(0-1)$

$$= 2r\cos(\omega_0)\, y(-1) - r^2 y(-2) + s(0) - r\cos(\omega_0)\, s(-1)$$

[Putting input $s(n) = \delta(n)$]

$$= 2r\cos(\omega_0)\, y(-1) - r^2 y(-2) + \delta(0) - r\cos(\omega_0)\, \delta(-1)$$

$$= 2r\cos(\omega_0) \times 0 - r^2 \times 0 + 1 - r\cos(\omega_0) \times 0$$

$$= 1$$

$n = 1,\quad y(1) = 2r\cos(\omega_0)\, y(1-1) - r^2 y(1-2) + s(1) - r\cos(\omega_0)\, s(1-1)$

$$= 2r\cos(\omega_0)\, y(0) - r^2 y(-1) + \delta(1) - r\cos(\omega_0)\, \delta(0)$$

$$= 2r\cos(\omega_0) \times 1 - r^2 \times 0 + 0 - r\cos(\omega_0) \times 1$$

$$= 2r\cos(\omega_0) - r\cos(\omega_0) = r\cos(\omega_0)$$

$n = 2,\quad y(2) = 2r\cos(\omega_0)\, y(2-1) - r^2 y(2-2) + s(2) - r\cos(\omega_0)\, s(2-1)$

$$= 2r\cos(\omega_0)\, y(1) - r^2 y(0) + \delta(2) - r\cos(\omega_0)\, \delta(1)$$

$$= 2r\cos(\omega_0) \times r\cos(\omega_0) - r^2 \times 1 + 0 - r\cos(\omega_0) \times 0$$

$$= 2[r\cos(\omega_0)]^2 - r^2 = 2r^2\cos^2(\omega_0) - r^2$$

$$= r^2\,[2\cos^2(\omega_0) - 1] = r^2\cos(2\omega_0)$$

$n = 3,\quad y(3) = 2r\cos(\omega_0)\, y(3-1) - r^2 y(3-2) + s(3) - r\cos(\omega_0)\, s(3-1)$

$$= 2r\cos(\omega_0)\, y(2) - r^2 y(1) + \delta(3) - r\cos(\omega_0)\, \delta(2)$$

$$= 2r\cos(\omega_0) \times \{r^2\cos(2\omega_0)\} - r^2 \times r\cos(\omega_0) + 0 - r\cos(\omega_0) \times 0$$

$$= 2r\cos(\omega_0)\,\{r^2\cos(2\omega_0)\} - r^3\cos(\omega_0)$$

$$= 2r^3\cos(\omega_0)\,\{2\cos^2(\omega_0) - 1\} - r^3\cos(\omega_0)$$

$$= 4r^3\cos^3(\omega_0) - 2r^3\cos(\omega_0) - r^3\cos(\omega_0)$$

$$= 4r^3\cos^3(\omega_0) - 3r^3\cos(\omega_0)$$

$$= r^3\,[4\cos^3(\omega_0) - 3\cos(\omega_0)]$$

$$= r^3\cos(3\omega_0)$$

$n = n,\quad y(n) = r^n\cos(n\omega_0)$ for $n \geq 0$

This output is known as impulse response of the second-order discrete-time system, *i.e.*,

$$h(n) = r^n \cos(n\omega_0)\, u(n)$$

From condition of stability, we know that

$$S \equiv \sum_{k=-\infty}^{\infty} |h(k)| < \infty$$

Now, we determine

$$\sum_{k=-\infty}^{\infty} |h(k)| = \sum_{k=-\infty}^{\infty} \left| r^n \cos(n\omega_0)\, u(n) \right|$$

(From the definition of $u(n)$)

$$= \sum_{k=0}^{\infty} \left| r^n \cos(n\omega_0) \right| = \sum_{k=0}^{\infty} |r^n|$$

(Because $|\cos(n\omega_0)| = 1$)

$$= |r^0| + |r^1| + |r^2| + ... \infty$$

This is a geometric progression of infinite number of terms.

$$= \frac{\text{First term}}{1 - [\text{common ratio}]^2} = \frac{1}{1-|r|} < \infty$$

for $|r| < 1$

or $$S \equiv \frac{1}{1-|r|} < \infty, \quad \text{for } |r| < 1$$

Hence, the discrete-time system is stable for $|r| < 1$.

Example 2.25 Find whether the first-order discrete-time system is stable or not. Its difference equation is given by

$$y(n) = Ay(n-1) + \delta(n)$$

Solution. For checking stability of this system, we first find out its impulse response by putting input

$$s(n) = \delta(n) = \begin{cases} 1, & n = 0 \\ 0, & n \neq 0 \end{cases}$$

We also assume that the system is initially relaxed, *i.e.*, $y(n) = 0$, $n < 0$.

Putting $n = 0, 1, 2, ...$ in the difference equation of the system, we get

$$y(n) = Ay(n-1) + s(n)$$

$n = 0$, $$y(0) = Ay(0-1) + s(0) = Ay(-1) + \delta(0)$$
$$= A \times 0 + 1 = 1$$

$n = 1,\qquad y(1) = Ay(1-1) + s(1) = Ay(0) + \delta(1)$

$\qquad\qquad = A \times 1 + 0 = A$

$n = 2,\qquad y(2) = Ay(2-1) + s(2) = Ay(1) + \delta(2)$

$\qquad\qquad = A \times A + 0 = A^2$

--------- --

--------- --

$n = n,\qquad y(n) = A^n\ ,\ \ n \geq 0$

or $\qquad y(n) = A^n\ u(n)$

This special output of the system is called impulse response of the first-order discrete-time system and is given by

$$h(n) = A^n u(n)$$

From the condition of the stability, we know that

$$S \equiv \sum_{k=-\infty}^{\infty} |h(k)| = \sum_{k=-\infty}^{\infty} |A^k\ u(k)|$$

(From the definition $u(n)$)

$$= \sum_{k=0}^{\infty} |A^k| = |A^0| + |A^1| + |A^2| + \ldots\ \infty$$

[This is a geometric progression of infinite number of terms]

$$= \frac{\text{First term}}{1-(\text{common ratio})^2} = \frac{1}{1-|A|}$$

$$S \equiv \frac{1}{1-|A|} < \infty \text{ for } |A| < 1$$

This system is stable for $|A| < 1$.

Example 2.26 Determine $y(n)$ if input $s(n) = n + 2$ for $0 \leq n \leq 3$

$$h(n) = A^n\ u(n) \text{ for all } n.$$

Solution. Convolution sum of $s(n)$ and $h(n)$ is given by

$$y(n) = s(n) * h(n) = \sum_{k=-\infty}^{\infty} s(k)\ h(n-k) \qquad \ldots(1)$$

Given $\qquad s(n) = n + 2$ for $0 \leq n \leq 3$ and $h(n) = A^n u(n)$ for all n

Substituting these values in Eq. (1), we get

$$y(n) = \sum_{k=-\infty}^{\infty} (k+2)\ A^{n-k}\ u(n-k)$$

$$\left[\text{But we know that } u(n-k) = \begin{cases} 1\ , & n-k \geq 0 \rightarrow k \leq n \\ 0\ , & n-k < 0 \rightarrow k > n \end{cases}\right]$$

$$y(n) = \sum_{0}^{n} (k+2)\ A^{n-k} = (0+2)\ A^{n-0} + (1+2)\ A^{n-1}$$

$$+ (2+2)\ A^{n-2} + \ldots + (n+2)\ A^{n-n}$$

$$= 2A^n + 3A^{n-1} + 4A^{n-2} + \ldots + (n+2)A^0$$

or $$y(n) = 2A^n + 3A^{n-1} + 4A^{n-2} + \ldots + (n+2)$$

Example 2.27 Consider a causal and stable discrete-time LTI system whose input $s(n)$ and output $y(n)$ are related through the second-order difference equation

$$y(n) = \frac{1}{6}y(n-1) + \frac{1}{6}y(n-2) + s(n)$$

Determine the impulse response $h(n)$ for the system.

Solution. Given difference equation

$$y(n) = \frac{1}{6}y(n-1) + \frac{1}{6}y(n-2) + s(n)$$

or $$y(n) - \frac{1}{6}y(n-1) - \frac{1}{6}y(n-2) = s(n) \quad \ldots(1)$$

For impulse response the particular solution $y_p(n) = 0$

So the complete solution of Eq. (1), will be given by

$$y(n) = y_p(n) + y_h(n) = 0 + y_h(n) = y_h(n)$$

Let $y_h(n) = \lambda^n$

Substituitng $y(n) = y_h(n) = \lambda^n$ in Eq. (1), we get

$$\lambda^n - \frac{1}{6}\lambda^{n-1} - \frac{1}{6}\lambda^{n-2} = 0$$

($\because$ $s(n) = 0$ for homogeneous solution)

or $$\lambda^2 - \frac{\lambda}{6} - \frac{1}{6} = 0$$

or $$6\lambda^2 - \lambda - 1 = 0 \quad \ldots(2)$$

This is the characteristic equation of the homogeneous solution.

The roots of the characteristic equations are

$$\lambda = \frac{1}{2} \text{ and } -\frac{1}{3}$$

or $$\lambda_1 = \frac{1}{2} \text{ and } \lambda_2 = -\frac{1}{3}$$

The solution $y_h(n) = C_1(\lambda_1)^n + C_2(\lambda_2)^n$

$$= C_1\left(\frac{1}{2}\right)^n + C_2\left(-\frac{1}{3}\right)^n \quad \ldots(3)$$

For impulse response

$$s(n) = \delta(n) = \begin{cases} 1, & n = 0 \\ 0, & n \neq 0 \end{cases}$$

or $\quad s(0) = 1$

$$s(n \neq 0) = 0$$

Substituting the above relations in Eq. (1), we get

For $n = 0$ $\quad y(0) = \frac{1}{6}y(0-1) + \frac{1}{6}y(0-2) + s(0)$

$$= \frac{1}{6}y(-1) + \frac{1}{6}y(-2) + s(0)$$

[Here we assume that the system is initially relaxed, *i.e.*, $y(-1) = y(-2) = ... = 0$]

$$y(0) = \frac{1}{6} \times 0 + \frac{1}{6} \times 0 + 1 = 1$$

$n = 1,$ $\quad y(1) = \frac{1}{6}y(1-1) + \frac{1}{6}y(1-2) + s(1)$

$$= \frac{1}{6}y(0) + \frac{1}{6}y(-1) + s(1)$$

$$= \frac{1}{6} \times 1 + \frac{1}{6} \times 0 + 0 = \frac{1}{6}$$

From Eq. (3),

$$y(n) = y_h(n) = C_1\left(\frac{1}{2}\right)^n + C_2\left(-\frac{1}{3}\right)^n$$

$n = 0,$ $\quad y(0) = C_1\left(\frac{1}{2}\right)^0 + C_2\left(-\frac{1}{3}\right)^0$

or $\quad y(0) = C_1 + C_2$

$n = 1,$ $\quad y(1) = C_1\left(\frac{1}{2}\right)^1 + C_2\left(-\frac{1}{3}\right)^1$

$$y(0) = C_1 + C_2$$

or $\quad C_1 + C_2 = 1 \quad$ ...(4)

$$y(1) = \frac{1}{2}C_1 - \frac{1}{3}C_2$$

or $\quad 3C_1 - 2C_2 = 6y(1)$

or $\quad 3C_1 - 2C_2 = 6 \times \frac{1}{6}$

or $\quad 3C_1 - 2C_2 = 1 \quad$ ...(5)

Solving Eqs. (4) and (5), we get

$$C_1 = \frac{3}{5} \quad \text{and} \quad C_2 = \frac{2}{5}$$

Putting the value of C_1 and C_2 in Eq. (3), we get

$$y(n) = y_h(n) = C_1\left(\frac{1}{2}\right)^n + C_2\left(-\frac{1}{3}\right)^n$$

$$= \frac{3}{5}\left(\frac{1}{2}\right)^n + \frac{2}{5}\left(-\frac{1}{3}\right)^n$$

This is the complete solution for Eq. (1).

Example 2.28 Find the step response of the discrete-time system described by following difference equation

$$y(n) = \frac{1}{6}y(n-1) + \frac{1}{6}y(n-2) + s(n)$$

This is a causal as well as stable system.

Solution. We know that complete solution of any difference equation is the sum of homogeneous solution and particular solution.

$$y(n) = y_h(n) + y_p(n) \qquad ...(1)$$

The particular solution for the step input is of the form $y_p(n) = K$.

Given the difference equation of the system

$$y(n) = \frac{1}{6}y(n-1) + \frac{1}{6}y(n-2) + s(n) \qquad ...(2)$$

For $n > 2$, substituting $y_p(n) = K$ and $s(n) = 1$ in Eq. (2), we get

$$y(n) = \frac{1}{6}y(n-1) + \frac{1}{6}y(n-2) + s(n)$$

or $$K = \frac{1}{6}K + \frac{1}{6}K + 1$$

or $$K - \frac{1}{3}K = 1$$

or $$\frac{2}{3}K = 1 \text{ or } K = \frac{3}{2}$$

i.e., $$y_p(n) = \frac{3}{2} \qquad ...(3)$$

Let homogeneous solution be

$$y_h(n) = \lambda^n \qquad ...(4)$$

Substituting Eq. (4) in Eq. (2), we get

$$y(n) = \frac{1}{6}y(n-1) + \frac{1}{6}y(n-2) + s(n)$$

$$\lambda^n = \frac{1}{6}\lambda^{n-1} + \frac{1}{6}\lambda^{n-2} + 0$$

(For homogeneous solution $s(n) = 0$)

$$6\lambda^2 - \lambda - 1 = 0 \qquad ...(5)$$

The roots of Eq. (5) are

$$\lambda = \frac{1}{2} \text{ and } -\frac{1}{3}$$

or $$\lambda_1 = \frac{1}{2} \text{ and } \lambda_2 = -\frac{1}{3}$$

The homogeneous solution is given by

$$y_h(n) = C_1\,\lambda_1^n + C_2\,\lambda_2^n$$

or $$y_h(n) = C_1\left(\frac{1}{2}\right)^n + C_2\left(-\frac{1}{3}\right)^n \qquad ...(6)$$

Substituting Eqs. (3) and (6) in Eq. (1), we get

$$y(n) = y_h(n) + y_p(n)$$

or $$y(n) = C_1\left(\frac{1}{2}\right)^n + C_2\left(-\frac{1}{3}\right)^n + \frac{3}{2} \qquad ...(7)$$

From Eq. (7),

$n = 0,$ $$y(0) = C_1\left(\frac{1}{2}\right)^0 + C_2\left(-\frac{1}{3}\right)^0 + \frac{3}{2}$$

or $$y(0) = C_1 + C_2 + \frac{3}{2} \qquad ...(8)$$

$n = 1,$ $$y(1) = C_1\left(\frac{1}{2}\right)^1 + C_2\left(-\frac{1}{3}\right)^1 + \frac{3}{2}$$

or $$y(1) = \frac{1}{2}C_1 - \frac{1}{3}C_2 + \frac{3}{2} \qquad ...(9)$$

From the difference equation of the discrete-time system, assuming for causal system $y(-1) = y(-2) = 0$, we get

$$y(n) = \frac{1}{6}y(n-1) + \frac{1}{6}y(n-2) + s(n)$$

$n = 0,$ $$y(0) = \frac{1}{6}y(0-1) + \frac{1}{6}y(0-2) + s(0)$$

$$= \frac{1}{6}y(-1) + \frac{1}{6}y(-2) + u(0) = \frac{1}{6}\times 0 + \frac{1}{6}\times 0 + 1$$

or $$y(0) = 1 \qquad ...(10)$$

$n = 1,$ $$y(1) = \frac{1}{6}y(1-1) + \frac{1}{6}y(1-2) + s(1)$$

$$= \frac{1}{6}y(0) + \frac{1}{6}y(-1) + u(1) = \frac{1}{6}\times 1 + \frac{1}{6}\times 0 + 1$$

$$= \frac{1}{6} + 0 + 1 = \frac{7}{6}$$

or $$y(1) = \frac{7}{6} \qquad ...(11)$$

Substituting Eqs. (10) and (11) in Eqs. (8) and (9), respectively, we get

$$C_1 + C_2 + \frac{3}{2} = 1$$

or $$C_1 + C_2 = 1 - \frac{3}{2}$$

or $$C_1 + C_2 = -\frac{1}{2} \qquad ...(12)$$

$$\frac{1}{2}C_1 - \frac{1}{3}C_2 + \frac{3}{2} = \frac{7}{6}$$

or $$\frac{3C_1 - 2C_2}{6} = \frac{7}{6} - \frac{3}{2}$$

or $$\frac{3C_1 - 2C_2}{6} = \frac{7-9}{6}$$

or $$3C_1 - 2C_2 = -\frac{2}{6}$$

or $$3C_1 - 2C_2 = -\frac{1}{3} \qquad ...(13)$$

On solving Eqs. (12) and (13), we get

$$C_1 + C_2 = -\frac{1}{2}$$

$$3C_1 - 2C_2 = -\frac{1}{3}$$

$$2C_1 + 2C_2 = -1$$

$$3C_1 - 2C_2 = -2$$

$$5C_1 = -3$$

or $$C_1 = -\frac{3}{5}$$

and $$C_2 = \frac{1}{10}$$

Substituting the values of C_1 and C_2 in Eq. 7, we get

$$y(n) = C_1\left(\frac{1}{2}\right)^n + C_2\left(-\frac{1}{3}\right)^n + \frac{3}{2}$$

or $$y(n) = -\frac{3}{5}\left(\frac{1}{2}\right)^n + \frac{1}{10}\left(-\frac{1}{3}\right)^n + \frac{3}{2} \quad \text{for } n \geq 0$$

or $$y(n) = -\frac{3}{5}\left(\frac{1}{2}\right)^n u(n) + \frac{1}{10}\left(-\frac{1}{3}\right)^n u(n) + \frac{3}{2}u(n)$$

Example 2.29 Find out the following convolution integral

(*a*) $s_1(t) * s_2(t)$ (*b*) $s_1(t) * s_3(t)$

for the following continuous-time signals

$$s_1(t) = 1.5\ \delta(t + 1.5) + 1.5\ \delta(t - 1.5)$$

$$s_2(t) = 2\ \delta(t - 0.5) + 4\ \delta(t - 1)$$

and $$s_3(t) = \begin{cases} t\ , & 0 \le t < 2 \\ 0\ , & \text{elsewhere} \end{cases}$$

Solution. (*a*) Convolution integral of $s_1(t)$ and $s_2(t)$ is given by

$$\begin{aligned} y_1(t) &= s_1(t) * s_2(t) \\ &= [1.5\ \delta(t + 1.5) + 1.5\ \delta(t - 1.5)] \\ &\qquad * [2\ \delta(t - 0.5) + 4\ \delta(t - 1)] \\ &= 1.5\ \delta(t + 1.5) * 2\ \delta(t - 0.5) + 1.5\ \delta(t - 1.5) \\ &\qquad * 2\ \delta(t - 0.5) + 1.5\ \delta(t + 1.5) * 4\ \delta(t - 1) \\ &\qquad + 1.5\ \delta(t - 1.5) * 4\ \delta(t - 1) \end{aligned}$$

or $$y_1(t) = 3\ \delta(t + 1) + 3\ \delta(t - 2) + 6\ \delta(t + 0.5) + 6\ \delta(t - 2.5)$$

(*b*) Convolution integral of $s_1(t)$ and $s_3(t)$ is given by

$$\begin{aligned} y_2(t) &= s_1(t) * s_3(t) \\ &= [1.5\ \delta(t + 1.5) + 1.5\ \delta(t - 1.5)] * s_3(t) \\ &= 1.5\ \delta(t + 1.5) * s_3(t) + 1.5\ \delta(t - 1.5) * s_3(t) \\ &= 1.5\ s_3(t + 1.5) + 1.5\ s_3(t - 1.5) \end{aligned}$$

or $$y_2(t) = \begin{cases} 1.5(t + 1.5)\ , & -1.5 \le t < 0.5 \\ 1.5(t - 1.5)\ , & 1.5 \le t < 3.5 \\ 0\ , & \text{elsewhere} \end{cases}$$

Example 2.30 Determine the unit impulse response of the causal discrete-time system described by the difference equation

$$y(n) - \frac{y(n-2)}{n} = 2\ s(n)$$

Solution. Impulse response is the output of the system when input $s(n) = \delta(n) = \begin{cases} 1\ , & n \ge 0 \\ 0\ , & n < 0 \end{cases}$, *i.e.*, $y(n) = h(n)$ for $s(n) = \delta(n)$

Since the system is causal, $h(-2) = h(-1) = 0$. Hence from the equation

$$h(n) - \frac{h(n-2)}{n} = 2\ \delta(n)$$

or $$h(n) = \frac{h(n-2)}{n} + 2\ \delta(n) \qquad ...(1)$$

$n = 0,$ $$h(0) = \frac{h(-2)}{0} + 2\ \delta(0) = 0 + 2 \times 1 = 2$$

$$n = 1, \qquad h(1) = \frac{h(1-2)}{1} + 2 \times \delta(1)$$

$$= h(-1) + 2\,\delta(1) = 0 + 2 \times 0 = 0$$

$$n = 2, \qquad h(2) = \frac{h(2-2)}{2} + 2\,\delta(2)$$

$$= \frac{h(0)}{2} + 2\,\delta(2) = \frac{2}{2} + 2 \times 0 = 1$$

$$n = 3, \qquad h(3) = \frac{h(3-2)}{3} + 2\,\delta(3)$$

$$= \frac{h(1)}{3} + 2\,\delta(3) = \frac{0}{3} + 2 \times 0 = 0$$

$$n = 4, \qquad h(4) = \frac{h(4-2)}{4} + 2\,\delta(4)$$

$$= \frac{h(2)}{4} + 2\,\delta(4) = \frac{1}{4} + 2 \times 0 = \frac{1}{4}$$

$$\vdots$$

Note that for $n > 0$, the difference equation is

$$h(n) = \frac{h(n-2)}{n}$$

When n is odd, $n - 2$ is also odd and is the immediately preceding odd number. Since $h(1) = 0$, we have $h(3) = 0$, and since $h(3) = 0$, $h(5)$ is also zero, and so on. Hence $h(n) = 0$ for n odd.

For n even, the pattern is as follows

$$h(2) = \frac{2}{2}$$

$$h(4) = \frac{2}{2 \times 4}$$

$$h(6) = \frac{2}{2 \times 4 \times 6}$$

$$\vdots$$

and so on.

Hence, in general, for n even,

$$h(n) = \frac{2}{2 \times 4 \times 6 \times \ldots \times n}$$

$$= \frac{2}{\left(2 \times 2 \times 2 \times \ldots \times \frac{n}{2} \text{ times}\right) \times 1 \times 2 \times 3 \times \times \frac{n}{2}}$$

$$= \frac{2}{2^{n/2}\left(\frac{n}{2}\right)!}$$

Hence the final result is

$$h(n) = \begin{cases} \dfrac{2}{2^{n/2}\,(n/2)\,!}, & n \text{ even} \\ 0\,, \text{ odd} \end{cases}$$

Example 2.31 Find convolution integral of $s_1(t) = 3 \cos 2t$, for all t, and $s_2(t) = e^{-|t|}$ for all t.

Solution. The convolution integral of $s_1(t)$ and $s_2(t)$ is given by

$$s(t) = s_1(t) * s_2(t) = s_2(t) * s_1(t)$$

$$= \int_{-\infty}^{\infty} s_2(\tau)\, s_1(t-\tau)\, d\tau$$

$$= \int_{-\infty}^{\infty} e^{-|\tau|}\, 3 \cos [2(t-\tau)]\, d\tau$$

$$= 3 \int_{-\infty}^{\infty} e^{-|\tau|} \left\{ \frac{e^{j(2t-2\tau)} + e^{-j(2t-2\tau)}}{2} \right\} d\tau$$

$$= \frac{3}{2}\left[\int_{-\infty}^{\infty} e^{j(2t-2\tau)-|\tau|}\, d\tau + \int_{-\infty}^{\infty} e^{-j(2t-2\tau)-|\tau|}\, d\tau \right]$$

$$s(t) = \frac{3}{2}\,[I_1 + I_2] \qquad \text{...(1)}$$

The first integral I_1 can be broken up into two parts as follows

$$I_1 = \int_{-\infty}^{\infty} e^{j(2t-2\tau)-|\tau|}\, d\tau$$

$$= \int_{-\infty}^{0} e^{j(2t-2\tau)+\tau}\, d\tau + \int_{0}^{\infty} e^{j(2t-2\tau)-\tau}\, d\tau$$

$$= \left[\frac{e^{j(2t-2\tau)+\tau}}{-2j+1} \right]_{-\infty}^{0} + \left[\frac{e^{j(2t-2\tau)-\tau}}{-2j-1} \right]_{0}^{\infty}$$

$$= \frac{e^{j2t}}{1-2j} - \frac{e^{j2t}}{-2j-1} = e^{j2t}\left[\frac{1}{1-2j} + \frac{1}{1+2j} \right]$$

$$= e^{j2t}\, \frac{2}{1+4} = \frac{2}{5}\, e^{j2t}$$

or $$I_1 = \frac{2}{5}\, e^{j2t} \qquad \text{...(2)}$$

In a similar manner, we can determine I_2 as

$$I_2 = \int_{-\infty}^{\infty} e^{-j(2t-2\tau)-|\tau|}\, d\tau$$

$$= \int_{-\infty}^{0} e^{-j(2t-2\tau)+\tau} d\tau + \int_{0}^{\infty} e^{-j(2t-2\tau)-\tau} d\tau$$

$$= \left[\frac{e^{-j(2t-2\tau)+\tau}}{2j+1}\right]_{-\infty}^{0} + \left[\frac{e^{-j(2t-2\tau)-\tau}}{2j-1}\right]_{0}^{\infty}$$

$$= \frac{e^{-j2t}}{1+2j} - \frac{e^{-j2t}}{2j-1}$$

or $$I_2 = \frac{2}{5} e^{-j2t} \quad ...(3)$$

Hence $$s(t) = \frac{3}{2}[I_1 + I_2] = \frac{3}{2}\left[\frac{2}{5}e^{j2t} + \frac{2}{5}e^{-j2t}\right]$$

$$= \frac{6}{5}\left[\frac{e^{j2t} + e^{-j2t}}{2}\right]$$

or $$s(t) = \frac{6}{5} \cos 2t$$

REVIEW QUESTIONS

1. Define linearity property of a system with examples.
2. Define time-invariance property of the system with examples.
3. What is linear time-invariant (LTI) system ? Give example.
4. What is superposition condition or theorem ?
5. Discuss the following systems
   (*a*) Continuous-time LTI system
   (*b*) Discrete-time LTI system
6. What is impulse response ?
7. Define following :
   (*a*) Impulse response of a continuous-time LTI system
   (*b*) Impulse or unit sample response of a discrete-time system
8. What is physical significance of impulse response of a system.
9. What do you mean by convolution ?
10. Explain the following with examples
    (*a*) Convolution integral for continuous-time signals or continuous convolution
    (*b*) Convolution sum for discrete-time signals or discrete convolution
11. What do you mean by the following ?
    (*a*) Superposition integral (*b*) Superposition sum
12. Give the procedure for computing convolution sum or superposition sum of two discrete-time signals.
13. What is folding ? Explain with an example.

14. Why is folding of a signal required in convolution of two signals ?
15. Discuss various properties of continuous and discrete-time LTI systems in brief.
16. Discuss the following properties of continuous-time and discrete- time LTI system with examples
    (*a*) Commutative property (*b*) Distributive property
    (*c*) Associative property
17. Distinguish dynamic LTI systems from static LTI systems. Give example of each system.
18. Describe
    (*a*) Invertible system (*b*) Inverse system
19. Discuss causal and non-causal LTI systems. Give examples of each system.
20. What is the physical significance of causality of LTI systems.
21. Distinguish between systems with memory from memoryless systems. Give example of each system.
22. Give the condition of causality of continuous-time and discrete- time LTI systems in terms of impulse responses.
23. Give the condition of BIBO stability of continuous-time and discrete-time LTI systems in terms of impulse response.
24. Discuss unit-step responses of continuous-time and discrete-time LTI systems.
25. Give the relationship between unit-step responses and impulse responses of continuous-time and discrete-time LTI systems.
26. Give the physical significance of unit-step responses in signals and systems analysis.
27. What do you mean by differential equations which describe continuous-time systems ? Give an example.
28. Discuss difference equations which describe discrete-time systems. Give an example.
29. What are basic building blocks or elements of a system ?
30. Write short notes on following
    (*a*) Basic building blocks or elements of continuous-time systems
        (*i*) Adder (*ii*) Multiplier
        (*iii*) Differentiator/Integrator
    (*b*) Basic building blocks or elements of discrete-time systems
        (*i*) Adder (*ii*) Multiplier
        (*iii*) Delay elements.

NUMERICAL PROBLEMS

1. Find out the convolution sum of $s(n)$ and $h(n)$ as shown in Fig. 2.27.

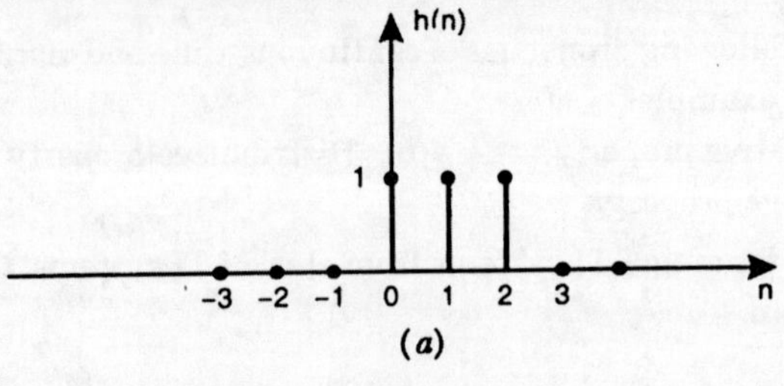

(a)

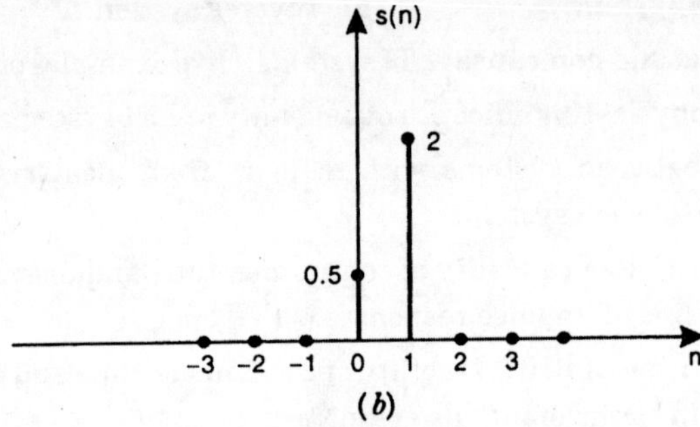

(b)

Fig. 2.27

Ans.

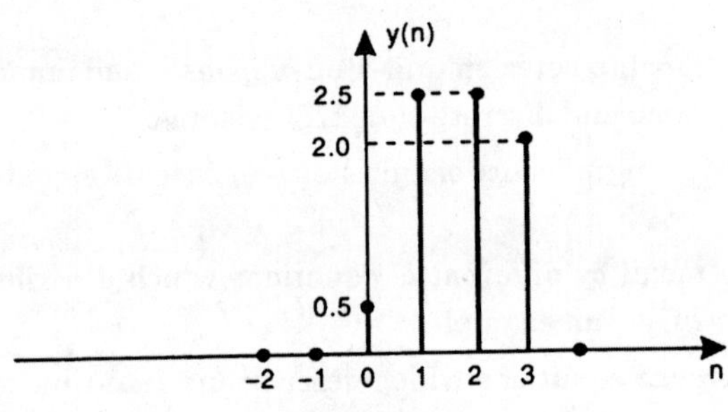

Fig. 2.28

2. Find the discrete-time convolution of

$$s(n) = \left(\frac{1}{2}\right)^n u(n)$$

$$h(n) = u(n) \qquad \textbf{Ans. } y(n) = 2\left[1 - \left(\frac{1}{2}\right)^{n+1}\right] u(n)$$

3. Find the convolution sum of $s(n)$ and $h(n)$ given as

$$s(n) = \{1, 2, 3, 4\}$$
$$\uparrow$$

$$h(n) = \{1, 1, 1, 1\}$$
$$\uparrow$$

4. Find the response of the discrete-time system with impulse response

$$h(n) = \{\underset{\uparrow}{2}, 1, 3\} \text{ to input}$$

$$s(n) = \{\underset{\uparrow}{1}, 2, 3\}$$

5. Compute the convolution sum of two discrete-time signals

$$s(n) = \left(\frac{1}{2}\right)^n u(n) \text{ and } h(n) = \left(\frac{1}{3}\right)^n u(n)$$

6. Find the convolution sum of following discrete-time sequences

$$s(n) = \begin{cases} 1, & 0 \le n \le 4 \\ 0, & \text{otherwise} \end{cases} \text{ and}$$

$$h(n) = \begin{cases} \alpha^n, & 0 \le n \le 6 \\ 0, & \text{otherwise} \end{cases}$$

$$\textbf{Ans. } y(n) = \begin{cases} 0, & n < 0 \\ \dfrac{1-\alpha^{n+1}}{1-\alpha}, & 0 \le n \le 4 \\ \dfrac{\alpha^{n-4}-\alpha^{n+1}}{1-\alpha}, & 4 < n \le 6 \\ \dfrac{\alpha^{n-4}-\alpha^{7}}{1-\alpha}, & 6 < n \le 10 \\ 0, & 10 < n \end{cases}$$

7. Determine the convolution integral of input $s(t)$ to impulse response of continuous-time LTI system. Given

$$s(t) = e^{(-1/2)t}\, u(t) \text{ and } h(t) = u(t)$$

Ans. $y(t) = 2(1 - e^{(-1/2)t})\, u(t)$

8. Find the convolution integral of following two signals

$$s(t) = \begin{cases} 1, & 0 < t < T_0 \\ 0, & \text{otherwise} \end{cases}$$

$$h(t) = \begin{cases} 1, & 0 < t < T_0 \\ 0, & \text{otherwise} \end{cases}$$

Ans.

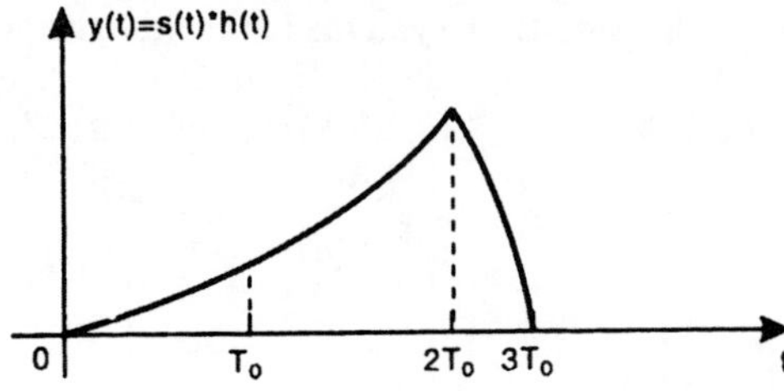

Fig. 2.29

9. Find the convolution integral of the following signals

$$s(t) = e^{2t}\, u(-t)$$
$$h(t) = u(t-3)$$

Ans.

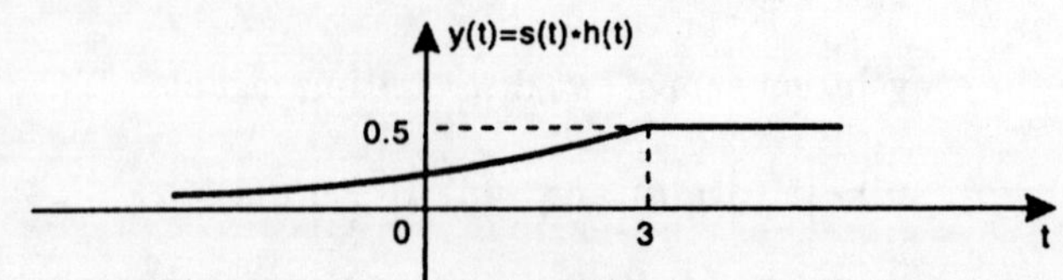

Fig. 2.30

10. Find the convolution sum of following two discrete-time sequences

$$s(n) = \left[\frac{1}{2}\right]^n u(n) + [2]^n u(-n)$$

and $$h(n) = u(n)$$

11. Given $s(n) = \delta(n) + 2\delta(n-1) - \delta(n-3)$ and $h(n) = 2\,\delta(n+1) + 2\,\delta(n-1)$

Compute and plot each of the following convolutions

(a) $y_1(n) = s(n+2) * h(n)$ (b) $y_2(n) = s(n) * h(n+2)$

Ans. (a) $y_1(n) = y(n+2)$ (b) $y_2(n) = y(n+2)$ where $y(n) = s(n) * h(n)$

12. Compute and plot the convolution sum of $s(n)$ and $h(n)$ where

$s(n) = \left(\frac{1}{3}\right)^{-n} u(-n-1)$ and $h(n)$ **Ans.** $y(n) = \begin{cases} (3^n/2)\,, & n < 0 \\ (1/2)\,, & n \geq 0 \end{cases}$

13. Compute and plot the convolution integral of $s(t)$ and $h(t)$ where

$$s(t) = \begin{cases} t+1\,, & 0 \leq t \leq 1 \\ 2-t\,, & 1 < t \leq 2 \\ 0\,, & \text{elsewhere} \end{cases}$$

$$h(t) = \delta(t+2) + 2\,\delta(t+1)$$

Ans. $y(t) = \begin{cases} t+3\,, & -2 < t \leq -1 \\ t+4\,, & -1 < t \leq 0 \\ 2-2t\,, & 0 < t \leq 1 \\ 0\,, & \text{otherwise} \end{cases}$

14. Check the following discrete-time systems for causality and stability.

(a) $h(n) = \left(\frac{1}{5}\right)^n u(n)$ (b) $h(n) = (0.8)^n\, u(n+2)$

(c) $h(n) = \left(\frac{1}{2}\right)^n u(-n)$ (d) $h(n) = \left(-\frac{1}{2}\right)^n u(n) + (1.01)^n\, u(n-1)$

(e) $h(n) = n\left(\frac{1}{3}\right)^n u(n-1)$

Ans. (a) causal, stable; (b) non-causal, stable; (c) anticausal, unstable; (d) causal, unstable; (e) causal, stable

15. Check the following continuous-time systems for causality and stability

(*a*) $h(t) = e^{-4t}\, u(t-2)$ (*b*) $h(t) = e^{-2t}\, u(t+50)$

(*c*) $h(t) = t\, e^{-t}\, u(t)$ (*d*) $h(t) = e^{-6|t|}$

Ans. (*a*) causal, stable; (*b*) non causal, stable; (*c*) causal, stable; (*d*) non causal, stable

16. The difference equation of a discrete-time system is given by

$$y(n) = -2\, y(n-1) + s(n) + 2s(n-2)$$

This system is initially at rest. Determine the response of above system to the input shown in Fig. 2.31 by solving the difference equation recursively.

Ans. $y(n) = -110\, (-2)^{n-5}$, for $n \geq 5$

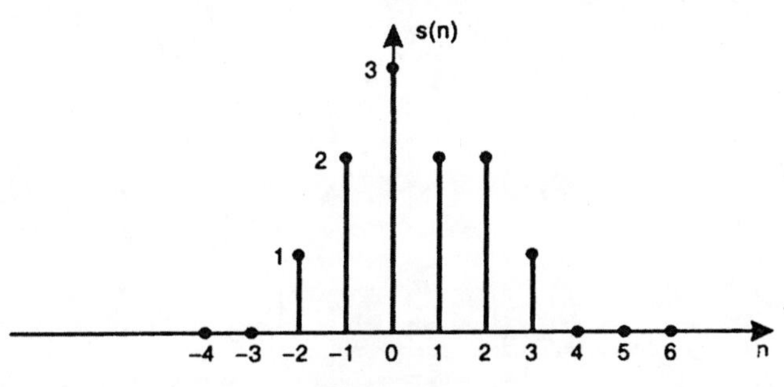

Fig. 2.31

17. The differential equation of a first-order continuous-time system is given by

$$\frac{dy(t)}{dt} + 2y(t) = s(t)$$

This system is initially at rest. Determine the system output $y(t)$ when input $s(t) = e^{3t}\, u(t)$.

Ans. $y(t) = \left[\frac{e^{3t}}{5} - \frac{1}{5} e^{-2t}\right] u(t)$

18. (*a*) Determine the system output $y_1(t)$ when the input is $s_1(t) = A\, e^{2t} u(t)$.

(*b*) Determine the system output $y_2(t)$ when the input is

$$s_2(t) = Ae^{2(t-T_0)}\, u(t-T_0)$$

Show that $y_2(t) = y_1(t - T_0)$

Given the differential equation of the continuous-time system as

$$\frac{dy(t)}{dt} + 2\, y(t) = s(t)$$

This system is initially at rest.

Ans. (*a*) $y_1(t) = \frac{A}{4}\, [e^{2t} - e^{-2t}]\, u(t)$ (*b*) $y_2(t) = Ae^{-2t} + \frac{A}{4} e^{2(t-T_0)}$, for $t > T_0$

19. Draw block diagram representations for causal continuous-time LTI system given by

$$y(t) = -\left(\frac{1}{2}\right)\frac{d\,y(t)}{dt} + 4\,s(t)$$

20. Draw block diagram representations for causal discrete-time LTI system given by

$$y(n) = \frac{1}{3}\,y(n-1) + s(n-1)$$

3

Fourier Series of Continuous-Time Signals or Continuous-Time Fourier Series (CTFS)

3.1 INTRODUCTION

In this chapter we will study about the Fourier series representation of continuous-time periodic signals. A signal $s(t)$ is said to be periodic of period T_s if $s(t) = s(t + T_s)$ for all t.

Fourier series is used to represent any arbitrary signal into weighted sum of harmonically related complex exponentials that share a common period with the signal being represented. Fourier series representation is used for both continuous-time and discrete-time periodic signals.

The French mathematician J.B.J. Fourier (1768-1830) showed that any arbitrary periodic signals or functions can be represented by an infinite series of sinusoids of harmoncially related frequencies. Periodic signals are represented by Fourier series but non-periodic signals are represented by Fourier Transforms. We will study Fourier series for discrete-time periodic signals in chapter 4.

We will disucss the Continuous-Time Fourier Transforms (CTFT) and Discrete-Time Fourier Transform (DTFT) in chatpers 5 and 6 respectively.

Fourier series can be represented either in the form of infinite trignometric series or infinite exponential series. Fourier series consists of dc terms as well as ac terms of all harmonics.

3.2 RESPONSE OF LTI SYSTEMS TO COMPLEX EXPONENTIALS

It is advantageous in the study of LTI systems to represent signals as a linear combination of basic signals due to the following advantages.

1. These basic signals can be used to construct a broad and useful class of signals.

2. The response of an LTI system to each basic signal should be simple enough in structure to provide us with a convenient representation for the response of the system to any signal that is constructed as a linear combination of the basic signals.

Above two properties are achieved in Fourier analysis by using a set of complex-exponential signals in continuous and discrete-time. These exponential signals are of the form e^{st} in continuous-time and z^n in discrete-time, respectively, where s and z are complex numbers.

The importance of complex exponential functions in the study of LTI systems stems from the fact that the response of an LTI system to a complex exponential input is the same complex exponential with only a change in amplitude.

For continuous-time sytsems

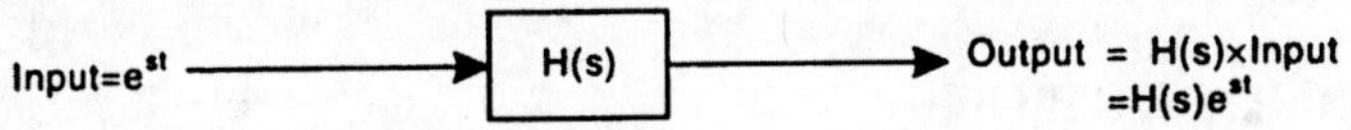

Input=e^{st} → H(s) → Output = H(s)×Input =$H(s)e^{st}$

where $H(s)$ is a function of complex variable s and is called the transfer function of the continuous-time system.

For discrete-time systems

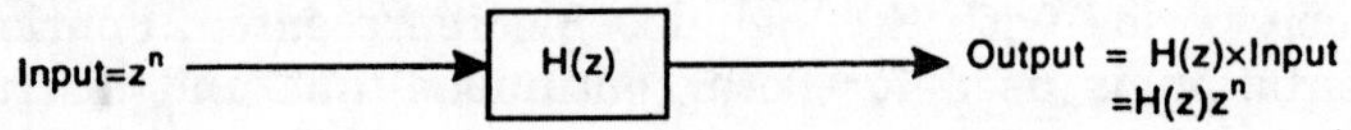

Input=z^n → H(z) → Output = H(z)×Input =$H(z)z^n$

where $H(z)$ is a function of complex variable z and is called the transfer function of the discrete-time system.

A signal for which the system output is a constant times the input is called an Eigen function of the system and the amplitude multiplication factor is referred to as the system's eigen value.

Now we will show that complex exponentials are indeed eigen functions of LTI systems. Let us consider a continuous-time LTI system. Its impulse response is denoted by $h(t)$. The output $y(t)$ for the system to input $s(t)$ is determined by using convolution integral.

$$y(t) = \int_{-\infty}^{\infty} h(\tau)\, s(t-\tau)\, d\tau \qquad \text{...(3.1)}$$

Substituting input $s(t) = e^{st}$ in Eq. 3.1, we get

$$y(t) = \int_{-\infty}^{\infty} h(\tau)\, e^{s(t-\tau)}\, d\tau$$

or

$$y(t) = e^{st} \int_{-\infty}^{\infty} h(\tau)\, e^{-s\tau}\, d\tau \qquad \text{...(3.2)}$$

$$y(t) = H(s)e^{st} \qquad ...(3.3)$$

where $H(s)$ is a complex constant whose value depends on s. It is related to the system impluse response $h(t)$ by

$$H(s) = \int_{-\infty}^{\infty} h(\tau)\, e^{-s\tau}\, d\tau \qquad ...(3.4)$$

Hence, we have proved that complex exponentials are eigen functions of LTI systems. The constant $H(s)$ for a specific value of s is then the eigen-value associated with eigen function e^{st}.

In a similar manner, we can also show that complex exponential sequences are eigen functions of discrete-time LTI systems.

Let us consider a discrete-time LTI system. It has an impulse response $h(n)$.

The output $y(n)$ for the system to input $s(n)$ is determined by using convolution sum.

$$y(n) = \sum_{k=-\infty}^{\infty} h(k)\, s(n-k)$$

$$= \sum_{k=-\infty}^{\infty} h(k)\, z^{n-k} = z^n \sum_{k=-\infty}^{\infty} h(k)\, z^{-k} \qquad ...(3.5)$$

From this expression, we see that if the input $s(n) = z^n$(complex exponential) then, assuming that the summation on the right hand side of Eq. 3.5 converges and the output is a complex exponential (z^n) multiplied by a constant. This constant depends on the value of z. That is

$$y(n) = H(z)\, z^n \qquad ...(3.6)$$

where $$H(z) = \sum_{k=-\infty}^{\infty} h(k)\, z^{-k} \qquad ...(3.7)$$

Consequently, we have seen that complex exponentials are eigen functions of discrete-time LTI systems. The constant $H(z)$ for a specified value of complex variable z is the eigen value associated with the eigen function z^n.

For the analysis of LTI systems, the usefulness of decomposing more general signals in terms of eigen function such as e^{st} can be seen from the following example.

We have a signal $s(t)$ which is a linear combination of four complex exponential functions.

$$s(t) = A_1\, e^{s_1 t} + A_2\, e^{s_2 t} + A_3\, e^{s_3 t} + A_4\, e^{s_4 t}$$

$$= s_1(t) + s_2(t) + s_3(t) + s_4(t) \qquad ...(3.8)$$

From the property of eigen functions, response to each part of the above input separately is

$$s_1(t) = A_1 e^{s_1 t} \xrightarrow{\text{Response}} H(s_1) A_1 e^{s_1 t} = y_1(t)$$

$$s_2(t) = A_2 e^{s_2 t} \xrightarrow{\text{Response}} H(s_2) A_2 e^{s_2 t} = y_2(t)$$

$$s_3(t) = A_3 e^{s_3 t} \xrightarrow{\text{Response}} H(s_3) A_3 e^{s_3 t} = y_3(t)$$

$$s_4(t) = A_4 e^{s_4 t} \xrightarrow{\text{Response}} H(s_4) A_4 e^{s_4 t} = y_4(t)$$

and from the superposition property the response of the sum of eigen functions of a signal is the sum of responses of each part (eigen functions), so that

$$y(t) = y_1(t) + y_2(t) + y_3(t) + y_4(t)$$
$$= A_1 H(s_1) e^{s_1 t} + A_2 H(s_2) e^{s_2 t} + A_3 H(s_3) e^{s_3 t} + A_4 H(s_4) e^{s_4 t} \quad ...(3.9)$$

If the input to a continuous-time LTI system is represented as a linear combination of complex exponential functions, that is, if input

$$s(t) = \sum_k A_k e^{s_k t} \quad ...(3.10)$$

then the output to this system will be given by

$$y(t) = \sum_k A_k H(s_k) e^{s_k t} \quad ...(3.11)$$

In an exactly similar manner, if the input to a discrete-time LTI system is represented as a linear combination of complex exponential function of the form z^n, that is, if input

$$s(n) = \sum_k A_k z_k^n \quad ...(3.12)$$

then the output to this sytem is given by

$$y(n) = \sum_k A_k H(z_k) z_k^n \quad ...(3.13)$$

Output of an LTI system to the input that is represented as a linear combination of complex exponential functions can also be represetned as a linear combination of the same complex exponential functions. Each coefficient in this representation of the output of the LTI system is equal to the product of the corresponding coefficient of the input and system's eigen value associated with Eigen function. In this representation coefficient is denoted by A_k. For continuous-time LTI system, system's eigen value associated with the eigen function $e^{s_k t}$ is denoted by $H(s_k)$. For discrete-time LTI system, system's eigen value associated with the eigen function z_k^n is denoted by $H(z_k)$.

Example 3.1 A continuous-time LTI system is described by input-output relationship given as

$$y(t) = s(t - 5)$$

Determine the output to this system if input is the complex exponential signal $s(t) = e^{j2t}$.

Solution. We have a given input-output relationship for a continuous-time LTI system

$$y(t) = s(t-5) \qquad ...(1)$$

Substituting $s(t) = e^{j2t}$ in above Eq. (1), we get

$$y(t) = e^{j2(t-5)} = e^{-j10}e^{j2t} \qquad ...(2)$$

Eq. (2) is a form of equation $y(t) = H(s)e^{st}$, since e^{j2t} is an eigen function. The associated eigen value is $H(j2) = e^{-j10}$.

The impulse response of the system will be determined by putting input $s(t) = \delta(t)$ in Eq. (1)

$$h(t) = \delta(t-5) \qquad ...(3)$$

But $$H(s) = \int_{-\infty}^{\infty} h(\tau)\, e^{-s\tau}\, d\tau$$

Substituting the value $h(t)$

$$H(s) = \int_{-\infty}^{\infty} \delta(\tau - 5)\, e^{-s\tau}\, d\tau = e^{-5s} \qquad ...(4)$$

Substituting $s = j2$ in Eq. (4), we get

$$H(j2) = e^{-5j2} = e^{-j10} \qquad ...(5)$$

Example 3.2 A continuous-time LTI system is described by input-output relationship given by

$$y(t) = s(t-3) \qquad ...(1)$$

Determine the output of the system if input

$$s(t) = \cos(5t) + \cos(8t) \qquad ...(2)$$

Solution. Substituting the value of $s(t)$ in Eq. (1), we get

$$y(t) = \cos[5(t-3)] + \cos[8(t-3)] \qquad ...(3)$$

Now we expand $s(t)$ using Euler's relation

$$s(t) = \cos(5t) + \cos(8t)$$

$$= \frac{e^{j5t} + e^{-j5t}}{2} + \frac{e^{j8t} + e^{-j8t}}{2}$$

$$= \frac{1}{2}e^{j5t} + \frac{1}{2}e^{-j5t} + \frac{1}{2}e^{j8t} + \frac{1}{2}e^{-j8t} \qquad ...(4)$$

An input that is linear combination of complex exponential functions is given as

$$s(t) = A_1e^{s_1t} + A_2e^{s_2t} + A_3e^{s_3t} + A_4e^{s_4t} \qquad ...(5)$$

Output of LTI systems can be also represented as a linear combination of exponential functions and given as

$$y(t) = H(s_1)\,A_1e^{s_1t} + H(s_2)\,A_2e^{s_2t} + H(s_3)\,A_3e^{s_3t} + H(s_4)\,A_4e^{s_4t} \qquad ...(6)$$

Comparing Eqs. (4) and (5), we get

$$A_1 = A_2 = A_3 = A_4 = \frac{1}{2}$$

$$s_1 = j5\,,\ s_2 = -j5\,,\ s_3 = j8 \text{ and } s_4 = -j8$$

Substituting above values in Eq. (4), we get

$$y(t) = H(j5)\frac{1}{2}e^{j5t} + H(-j5)\frac{1}{2}e^{-j5t} + H(j8)\frac{1}{2}e^{j8t} + H(-j8)\frac{1}{2}e^{-j8t} \quad ...(7)$$

The impulse response of the system will be determined by putting $s(t) = \delta(t)$ in Eq. (1) which gives us Eq. (8), we get

$$h(t) = \delta(t-3) \quad ...(8)$$

We know that

$$H(s) = \int_{-\infty}^{\infty} h(\tau)\, e^{-s\tau}\, d\tau \quad ...(9)$$

Substituting Eq. (8) in Eq. (9), we get

$$H(s) = \int_{-\infty}^{\infty} h(\tau)\, e^{-s\tau}\, d\tau$$

$$= \int_{-\infty}^{\infty} \delta(\tau - 3)\, e^{-s\tau}\, d\tau = e^{-3s}$$

or $$H(s) = e^{-3s} \quad ...(10)$$

Substituting $s = j5\,,\ -j5\,, j8$ and $-j8$ in Eq. (10), we get

$$H(j5) = e^{-3(j5)} = e^{-j15}$$

$$H(-j5) = e^{-3(-j5)} = e^{j15}$$

$$H(j8) = e^{-3(j8)} = e^{-j24}$$

and $$H(-j8) = e^{-3(-j8)} = e^{j24}$$

Substituting above values in Eq. 7, we get

$$y(t) = \frac{1}{2}e^{-j15}\,e^{j5t} + \frac{1}{2}e^{j15}\,e^{-15t} + \frac{1}{2}e^{-j24}\,e^{j8t} + \frac{1}{2}e^{j24}\,e^{-j8t}$$

$$= \frac{e^{j[5(t-3)]} + e^{-j[5(t-3)]}}{2} + \frac{e^{j[8(t-3)]} + e^{-j[8(t-3)]}}{2}$$

$$= \cos\,[5(t-3)] + \cos\,[8(t-3)]$$

In this example, the multiplication of each periodic exponential component of input $s(t)$ by corresponding eigen value effectively causes the input component to shift in time by 3.

The equation $\quad y(t) = A_1H(s_1)\,e^{s_1t} + A_2H(s_2)\,e^{s_2t} + ...$

not only allows us to calculate the responses of more complex LTI systems, but also provides the basis for frequency-domain representation and analysis of LTI systems.

3.3 FOURIER SERIES REPRESENTATION OF CONTINUOUS-TIME PERIODIC SIGNALS

Fourier series is a linear combination of harmonically related complex exponential functions. Fourier series is used to represent both continuous-time and discrete-time periodic signals into linear combination of harmonically related complex exponentials. A continuous-time signal is periodic if and only if it satisfies the condition $s(t) = s(t + T_0)$ for all time t. Where T_0 is the fundamental period and $\omega_0 = 2\pi/T_0$ is the fundamental frequency of the periodic signal.

We are already familiar with two important basic periodic signals. These basic peridoc signals are sinusoidal signal $s(t) = \cos \omega_0 t$ and complex exponential signal $s(t) = e^{j\omega_0 t}$. Both these signals are periodic with fundamental frequency ω_0 and fundamental period $T_0 = 2\pi/\omega_0$.

A set of harmonically related complex exponentials associated with periodic complex exponential signal $s(t) = e^{j\omega_0 t}$ is given by

$$f_k(t) = e^{jk\omega_0 t} = e^{jk(2\pi/T_0)t}\ , \quad k = 0\ , \pm 1\ , \pm 2\ , \ldots \qquad \ldots.(3.14)$$

Each of these signals has a fundamental frequency that is a multiple of ω_0 and therefore, each is periodic with period T_0. Thus, the general form of linear combination of harmonically related complex exponentials is given by

$$s(t) = \sum_{k=-\infty}^{\infty} A_k\, e^{jk\omega_0 t} = \sum_{k=-\infty}^{\infty} A_k\, e^{jk(2\pi/T_0)t} \qquad \ldots(3.15)$$

This linear combination of harmonically related complex exponentials is also periodic with fundamental period T_0. In Eq. 3.15, the term $k = 0$ is a constant. The terms for $k = 1$ and $k = -1$ both have fundamental frequency equal to ω_0 and are collectively called fundamental components. These are also called first harmonic components. The terms for $k = 2$ and $k = -2$ both have fundamental frequency equal to $2\omega_0$ and these are called second harmonic components. More generally, the components $k = N$ and $k = -N$ are called the Nth harmonic components.

The representation of a periodic signal in the form of Eq. 3.15 is referred to as the Fourier series representation.

Example 3.3 Given a continuous-time periodic signal $s(t) = \sum_{k=-3}^{3} A_k\, e^{jk2\pi t}$ with fundamental frequency 2π.

where, coefficients are $A_0 = 1\ ,\ A_1 = A_{-1} = \frac{1}{4}, A_2 = A_{-2} = \frac{1}{2}, A_3 = A_{-3} = \frac{1}{3}$

Show graphically how $s(t)$ is built up from its harmonic components.

Solution. Given $\quad s(t) = \sum_{k=-3}^{3} A_k e^{jk2\pi t}$

$$= A_{-3}\, e^{-j3.2\pi t} + A_{-2}\, e^{-j2.2\pi t} + A_{-1}\, e^{-j1.2\pi t} + A_0\, e^{-j0.2\pi t}$$

$$+ A_1\, e^{j1.2\pi t} + A_2\, e^{j2.2\pi t} + A_3\, e^{j3.2\pi t}$$

$$= \frac{1}{3} e^{-j6\pi t} + \frac{1}{2} e^{-j4\pi t} + \frac{1}{4} e^{-j2\pi t} + \frac{1}{4} e^{j2\pi t}$$

$$+ \frac{1}{2} e^{j4\pi t} + \frac{1}{3} e^{j6\pi t}$$

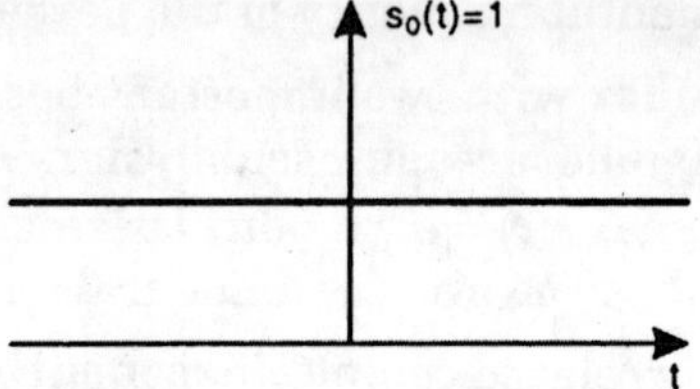

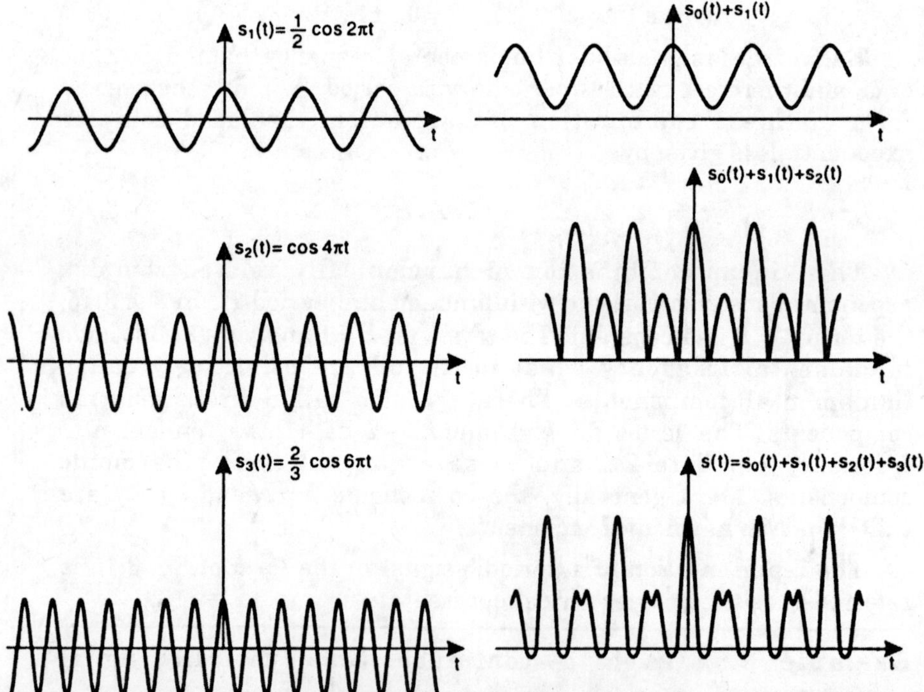

Fig. 3.1 *Construction of a signal given in example 3.3 from its harmonic components.*

$$= 1 + \frac{1}{4}[e^{j2\pi t} + e^{-j2\pi t}] + \frac{1}{2}[e^{j4\pi t} + e^{-j4\pi t}] + \frac{1}{3}[e^{j6\pi t} + e^{-j6\pi t}]$$

or $$s(t) = 1 + \frac{1}{2}\cos 2\pi t + \cos 4\pi t + \frac{2}{3}\cos 6\pi t$$

$$= s_0(t) + s_1(t) + s_2(t) + s_3(t)$$

Fig. 3.1 illustrates construction of a signal from its harmonic components.

Equation $s(t) = 1 + \frac{1}{2}\cos 2\pi t + \cos 4\pi t + \frac{2}{3}\cos 6\pi t$ is an example of an alternative form of the Fourier series of real continuous-time periodic signals.

If signal $s(t)$ is real then $s^*(t) = s(t)$, where $s^*(t)$ is the complex conjugate of $s(t)$.

$$s(t) = s^*(t) \qquad ...(3.16)$$

or $$s(t) = \sum_{k=-\infty}^{\infty} A_k^* e^{-jk\omega_0 t}$$

Replacing k by $-k$ in the above summation, we get

$$s(t) = \sum_{k=-\infty}^{\infty} A_{-k}^* e^{jk\omega_0 t} \qquad ...(3.17)$$

But we know that

$$s(t) = \sum_{k=-\infty}^{\infty} A_k e^{jk\omega_0 t} \qquad ...(3.18)$$

Comparing Eqs. 3.17 and 3.18, we get

$$A_k = A_{-k}^* \qquad ...(3.19)$$

or equivalently $$A_k^* = A_{-k} \qquad ...(3.20)$$

If A_k's are real then $A_k = A_{-k}$.

Derivation of Alternative Forms of the Fourier Series. We know that one form of Fourier series is a linear combination of harmonically related complex exponentials, it is expressed by

$$s(t) = \sum_{k=-\infty}^{\infty} A_k e^{jk\omega_0 t} \qquad ...(3.21)$$

Rearranging the summation in Eq. 3.21 as

$$S(t) = A_0 + \sum_{k=1}^{\infty} [A_k e^{jk\omega_0 t} + A_{-k} e^{-jk\omega_0 t}] \qquad ...(3.22)$$

Putting $A_{-k} = A_k^*$ in Eq. 3.22, we get

$$s(t) = A_0 + \sum_{k=1}^{\infty} [A_k e^{jk\omega_0 t} + A_k^* e^{-jk\omega_0 t}] \qquad ...(3.23)$$

Since the two terms inside the summation of Eq. 3.23 are complex conjugates of each other, this equation can be written as

$$s(t) = A_0 + \sum_{k=1}^{\infty} 2\text{Re}\left\{ A_k e^{jk\omega_0 t} \right\} \qquad ...(3.24)$$

If coefficient A_k is expressed in polar form as

$$A_k = \alpha_k e^{j\theta_k} \qquad ...(3.25)$$

then Eq. 3.24 becomes

$$s(t) = A_0 + \sum_{k=1}^{\infty} 2\text{Re}\left\{ \alpha_k e^{j\theta_k} . e^{jk\omega_0 t} \right\}$$

$$= A_0 + \sum_{k=1}^{\infty} 2\text{Re}\left\{ \alpha_k e^{j(k\omega_0 t + \theta_k)} \right\}$$

$$= A_0 + \sum_{k=1}^{\infty} 2\text{Re}\left\{ \alpha_k [\cos k\omega_0 t + j \sin k\omega_0 t] \right\}$$

or

$$s(t) = A_0 + 2 \sum_{k=1}^{\infty} \alpha_k \cos (k\omega_0 t + \theta_k) \qquad ...(3.26)$$

We generally come across equations in the form of Eq. 3.26 in the Fourier series of real continuous-time periodic signals.

Another form of Fourier series is obtained by writing coefficient A_k in rectangular form as

$$A_k = B_k + jC_k \qquad ...(3.27)$$

Both B_k and C_k are real.

Substituting Eq. 3.27 in Eq. 3.24, we get

$$s(t) = A_0 + \sum_{k=1}^{\infty} 2\text{Re}\left\{ A_k e^{jk\omega_0 t} \right\}$$

$$= A_0 + 2 \sum_{k=1}^{\infty} \text{Re}\left\{ (B_k + j C_k) e^{jk\omega_0 t} \right\}$$

$$= A_0 + 2 \sum_{k=1}^{\infty} \text{Re}\left\{ (B_k + j C_k) (\cos k\omega_0 t + j \sin k\omega_0 t) \right\}$$

$$= A_0 + 2 \sum_{k=1}^{\infty} \text{Re}\left\{ B_k \cos k\omega_0 t + j B_k \sin k\omega_0 t + j C_k \cos k\omega_0 t + j^2 C_k \sin k\omega_0 t \right\}$$

$$= A_0 + 2 \sum_{k=1}^{\infty} \text{Re}\left\{ (B_k \cos k\omega_0 t - C_k \sin k\omega_0 t) + j(B_k \sin k\omega_0 t + C_k \cos k\omega_0 t) \right\}$$

or $$s(t) = A_0 + 2 \sum_{k=1}^{\infty} (B_k \cos k\omega_0 t - C_k \sin k\omega_0 t) \quad ...(3.28)$$

If coefficients A_k's are real then $A_k = \alpha_k = B_k$ and therefore, both forms of Fourier series reduce to the same form.

Thus, for real periodic functions, the Fourier series in terms of complex exponentials is mathematically equivalent to either of the two forms given in Eqs. 3.26 and 3.28 that use trignometric functions. Although the two trignometric Fourier series given in Eqs. 3.26 and 3.28 are common forms, but the complex exponential form of equation $s(t) = \sum_{k=-\infty}^{\infty} A_k e^{jk\omega_0 t}$ is particularly convenient for our purpose.

3.4 DETERMINATION OF THE FOURIER SERIES REPRESENTATION OF CONTINUOUS-TIME PERIODIC SIGNALS

In this section, we will study a procedure for determining the coefficients A_k. Any continuous-time periodic signal can be represented with the series given by

$$s(t) = \sum_{k=-\infty}^{\infty} A_k e^{jk\omega_0 t} \quad ...(3.29)$$

Multiplying both sides of Eq. 3.29 by $e^{-jn\omega_0 t}$, we get

$$s(t)\, e^{-jn\omega_0 t} = \sum_{k=-\infty}^{\infty} A_k e^{jk\omega_0 t} e^{-jn\omega_0 t}$$

Now integrating both sides of above Eq. from 0 to T_0, we get

$$\int_0^{T_0} s(t)\, e^{-jn\omega_0 t}\, dt = \int_0^{T_0} \sum_{k=-\infty}^{\infty} A_k e^{jk\omega_0 t} e^{-jn\omega_0 t}$$

Here, T_0 is the fundamental period of $s(t)$. Interchaning the order of integration and summation gives

$$\int_0^{T_0} s(t)\, e^{-jn\omega_0 t}\, dt = \sum_{k=-\infty}^{\infty} A_k \left[\int_0^{T_0} e^{j(k-n)\omega_0 t}\, dt \right] \quad ...(3.30)$$

To evaluate $\int_0^{T_0} e^{j(k-n)\omega_0 t}\, dt$ we proceed as follows

$$\int_0^{T_0} e^{j(k-n)\omega_0 t}\, dt = \int_0^{T_0} [\cos (k-n)\, \omega_0 t + j \sin (k-n)\, \omega_0 t]\, dt$$

$$= \int_0^{T_0} \cos (k-n)\, \omega_0 t\, dt + j \int_0^{T_0} \sin (k-n)\, \omega_0 t\, dt \quad ..(3.31)$$

For $k \neq n$, $\cos (k-n)\, \omega_0 t$ and $\sin (k-n)\, \omega_0 t$ are periodic sinusoids with fundamental period $\frac{T_0}{|k-n|}$. Therefore, in Eq. 3.31, we integrate over an interval (*i.e.*, fundamental period T_0 that is an integral number) of periods of these signals. But the integration is equivalent to measuring the total area under the functions over the interval.

Now we have

$$\int_0^{T_0} e^{j(k-n)\omega_0 t}\, dt = \begin{cases} T_0, & k = n \\ 0, & k \neq n \end{cases} \qquad ...(3.32)$$

Substituting Eq. 3.32 in Eq. 3.30, we get

$$\int_0^{T_0} s(t)\, e^{-jn\omega_0 t}\, dt = \sum_{k=-\infty}^{\infty} A_k \left[\int_0^{T_0} e^{j(k-n)\omega_0 t}\, dt \right]$$

or $$\int_0^{T_0} s(t)\, e^{-jn\omega_0 t}\, dt = A_n T_0$$

or $$A_n = \frac{1}{T_0} \int_0^{T_0} s(t)\, e^{-jn\omega_0 t} \qquad ...(3.33)$$

This equation provides for determining the coefficients.

Note that in determining Eq. 3.31, we used interval of integration equal to T_0 which is an integral number of periods cos $(k-n)\,\omega_0 t$ and sin $(k-n)\,\omega_0 t$. Therefore, we will obtain the same result if we integrate over an interval of length T_0. It is represented as

$$\int_{T_0} e^{j(k-n)\omega_0 t}\, dt = \begin{cases} T_0, & k = n \\ 0, & k \neq n \end{cases} \qquad ...(3.34)$$

Consequently, Eq. 3.33 can be written as

$$A_n = \frac{1}{T_0} \int_{T_0} s(t)\, e^{-jn\omega_0 t}\, dt \qquad ...(3.35)$$

Now we can summarize the Fourier series representation of a continuous-time periodic signal and their Fourier series coefficients A_k as

$$s(t) = \sum_{k=-\infty}^{\infty} A_k\, e^{jk\omega_0 t} \qquad ...(3.36)$$

where $$A_k = \frac{1}{T_0} \int_{T_0} s(t)\, e^{-jk\omega_0 t}\, dt \qquad ...(3.37)$$

where ω_0 = fundamental frequency = $\frac{2\pi}{T_0}$; T_0 = fundamental period

Eq. 3.36 is referred to as the Synthesis Equation and Eq. 3.37 as the Analysis Equation. The set of coefficients $\{A_k\}$ is called the Fourier series coefficients or the spectral coefficients of signal $s(t)$. These complex Fourier series coefficients A_k are the measure of a portion of the signal $s(t)$ that is at each harmonic of the fundamental component.

The coefficient A_0 is the direct current (dc) or constant component of $s(t)$ and it is given as

$$A_0 = \frac{1}{T} \int_T s(t)\, dt \qquad ...(3.38)$$

A_0 is simply the average value of $s(t)$ over one period (T_0).

Now we will understand continuous-time fourier series by means of a few numerical examples.

Example 3.4 Determine the Fourier series coefficients of a continuous-time periodic signal $s(t) = \cos \omega_0 t$, where ω_0 is the fundamental frequency.

Solution. One approach to determine the Fourier series coefficients for the signal, $s(t) = \cos \omega_0 t$ is to use the equation given below.

$$A_k = \frac{1}{T_0} \int_{T_0} s(t)\, e^{-jk\omega_0 t}\, dt \qquad ...(1)$$

But this is a simple case so we can easily expand the sinusoidal signal as a linear combination of complex exponential functions and Fourier series coefficients A_k's are easily identified by inspection.

We know that

$$s(t) = \cos \omega_0 t = \frac{e^{j\omega_0 t} + e^{-j\omega_0 t}}{2} \qquad ...(2)$$

(This is the Euler's relation).

But from equation of Fourier series representation given as

$$s(t) = \sum_{k=-\infty}^{\infty} A_k e^{jk\omega_0 t} = \ldots + A_{-1}\, e^{-j\omega_0 t} + A_0 + A_1\, e^{j\omega_0 t} + \ldots \qquad ...(3)$$

On comparing Eqs. 2 and 3, we get Fourier series coefficients A_k's

$$A_1 = \frac{1}{2}$$

$$A_{-1} = \frac{1}{2}$$

$$A_k = 0\,,\ k \neq +1 \text{ or } -1$$

Example 3.5 Determine the Fourier series coefficients A_k's of a continuous-time periodic signal

$$s(t) = 1 + \sin \omega_0 t + 2 \cos \omega_0 t + \cos\left(2\omega_0 t + \frac{\pi}{4}\right) \qquad ...(1)$$

where ω_0 is the fundamental frequency.

Solution. To determine the Fourier series coefficients, we first expand the above signal $s(t)$ as

$$s(t) = 1 + \sin \omega_0 t + 2 \cos \omega_0 t + \cos\left(2\omega_0 t + \frac{\pi}{4}\right)$$

$$= 1 + \left[\frac{e^{j\omega_0 t} - e^{-j\omega_0 t}}{2j}\right] + 2\left[\frac{e^{j\omega t} + e^{j\omega_0 t}}{2}\right]$$

$$+ \left[\frac{e^{j(2\omega_0 t + \pi/4)\,t} + e^{-j(2\omega_0 t + \pi/4)\,t}}{2}\right]$$

$$= 1 + \left(1 + \frac{1}{2j}\right)e^{j\omega_0 t} + \left(1 - \frac{1}{2j}\right)e^{-j\omega_0 t} + \left(\frac{1}{2}\,e^{j(\pi/4)}\right)e^{j2\omega_0 t}$$

$$+ \left(\frac{1}{2}\,e^{-j(\pi/4)}\right)e^{-j2\omega_0 t} \qquad ...(2)$$

But we know from the equation of Fourier series representation of a continuous-time periodic signal that

$$s(t) = \sum_{k=-\infty}^{\infty} A_k\, e^{jk\omega_0 t}$$

$$= ... + A_{-2}\, e^{-j2\omega_0 t} + A_{-1}\, e^{-j\omega_0 t} + A_0 + A_1\, e^{j\omega_0 t} + A_2\, e^{j2\omega_0 t} + ... \qquad ...(3)$$

On comparing Eqs. 2 and 3, we get Fourier series coefficients A_k's as

$$A_0 = 1$$

$$A_1 = \left(1 + \frac{1}{2j}\right) = 1 - \frac{1}{2}j$$

$$A_{-1} = \left(1 - \frac{1}{2j}\right) = 1 + \frac{1}{2}j$$

$$A_2 = \frac{1}{2}\,e^{j(\pi/4)} = \frac{1}{2}\left(\cos\frac{\pi}{4} + j\sin\frac{\pi}{4}\right) = \frac{\sqrt{2}}{4}(1 + j)$$

$$A_{-2} = \frac{1}{2}\,e^{-j(\pi/4)} = \frac{1}{2}\left(\cos\frac{\pi}{4} - j\sin\frac{\pi}{4}\right) = \frac{\sqrt{2}}{4}(1 - j)$$

The Fourier series coefficients are complex coefficients so they can be written as $A_k = \text{Re}\,[A_k] + j\text{Im}\,[A_k]$.

The magnitude of Fourier series coefficients A_k's are determined by

$$|A_k| = \sqrt{\{\text{Re}\,[A_k]\}^2 + \{\text{Im}[A_k]\}^2}$$

and the phase of A_k's are determined by

$$\angle A_k = \tan^{-1}\left\{\frac{\text{Im}\,[A_k]}{\text{Re}\,[A_k]}\right\}$$

Determination of magnitude of Fourier series coefficients.

$$|A_0| = 1$$

$$|A_{-1}| = \sqrt{(1)^2 + (1/2)^2} = \sqrt{\frac{5}{4}} = \frac{\sqrt{5}}{2} = 1.118$$

$$|A_1| = \sqrt{(1)^2 + (1/2)^2} = \sqrt{\frac{5}{4}} = \frac{\sqrt{5}}{2} = 1.118$$

$$|A_{-2}| = \sqrt{(\sqrt{2}/4)^2 + (\sqrt{2}/4)^2} = \sqrt{\frac{2}{16} + \frac{2}{16}} = \sqrt{\frac{4}{16}} = \frac{1}{2} = 0.5$$

$$|A_2| = \sqrt{(\sqrt{2}/4)^2 + (-\sqrt{2}/4)^2} = \sqrt{\frac{2}{16} + \frac{2}{16}} = \sqrt{\frac{4}{16}} = \frac{1}{2} = 0.5$$

Determination of phase of Fourier series coefficients.

$$\angle A_k = \tan^{-1}\left\{\frac{\text{Im}\,[A_k]}{\text{Re}\,[A_k]}\right\}$$

$$\angle A_0 = \tan^{-1}\left\{\frac{\text{Im}\,[A_k]}{\text{Re}\,[A_k]}\right\} = \tan^{-1}\left\{\frac{0}{1}\right\} = \tan^{-1}\{0\} = 0^\circ$$

$$\angle A_{-1} = \tan^{-1}\left\{\frac{1/2}{1}\right\} = \tan^{-1}\{0.5\} = 26.565^\circ$$

$$\angle A_1 = \tan^{-1}\left\{\frac{-1/2}{1}\right\} = -\tan^{-1}\{0.5\} = -26.565^\circ$$

$$\angle A_{-2} = \tan^{-1}\left\{\frac{\sqrt{2}/4}{-\sqrt{2}/4}\right\} = -\tan^{-1}\{1\} = -45^\circ$$

$$\angle A_2 = \tan^{-1}\left\{\frac{\sqrt{2}/4}{\sqrt{2}/4}\right\} = \tan^{-1}\{1\} = 45^\circ$$

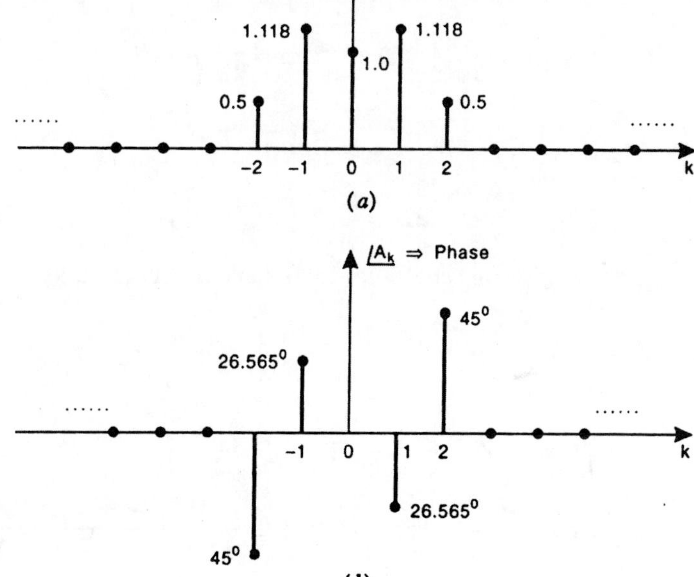

Fig. 3.2 *Plots of magnitude and phase of the Fourier series coefficients of the signal* $s(t) = 1 + \sin \omega_0 t + 2 \cos \omega_0 t + \cos(2\omega_0 t + \pi/4)$

Example 3.6 Determine the Fourier series coefficients A_k's of a continuous-time periodic signal

$$s(t) = \begin{cases} 1, -T_1 \le t \le T_1 \\ 0, \ T_1 < |t| < \dfrac{T_0}{2} \end{cases} \quad ...(1)$$

This is a periodic square wave which is defined over one period and sketched in Fig. 3.3, where T_0 is the fundamental period and $\omega_0 = 2\pi/T_0$ is fundamental frequency.

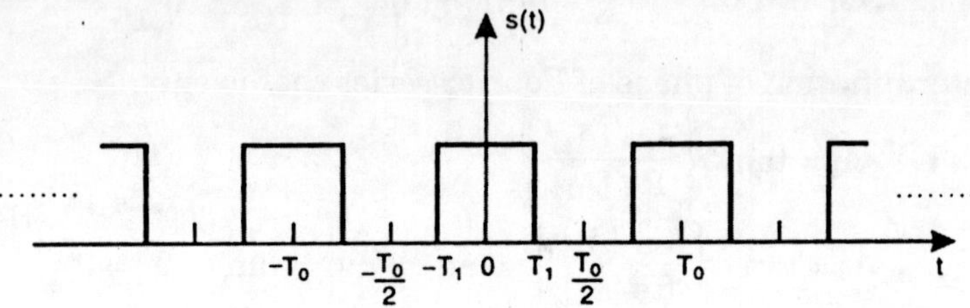

Fig. 3.3 *Periodic square wave.*

Solution. To determine the fourier series coefficients A_k's for the signal $s(t)$, we use equation

$$A_k = \frac{1}{T_0} \int_{T_0} s(t)\, e^{-jk\omega_0 t}\, dt \quad ...(2)$$

This waveform is symmetric about $t = 0$, so it is convenient to choose $-T_0/2 \le t \le T_0/2$ as the interval over which the integration is performed.

Using these limits of integration and substituting $k = 0$ in Eq. 2, we get

$$A_0 = \frac{1}{T_0} \int_{-T_0/2}^{T_0/2} s(t)\, e^{-j.0\omega_0 t}\, dt = \frac{1}{T_0} \int_{-T_1}^{T_1} 1.dt$$

$$= \frac{1}{T_0} [\, T\,]_{-T_1}^{T_1} = \frac{1}{T_0} [T_1 + T_1]$$

$$= \frac{1}{T_0} \,.\, 2T_1 = \frac{2T_1}{T_0} \quad ...(3)$$

For $k \ne 0$, Fourier series coefficients A_k's are determined as

$$A_k = \frac{1}{T_0} \int_{-T_0/2}^{T_0/2} s(t)\, e^{-jk\omega t}\, dt = \frac{1}{T_0} \int_{-T_1}^{T_1} 1\,.\, e^{-jk\omega t}\, dt$$

$$= \frac{1}{T_0} \int_{-T_1}^{T_1} e^{-jk\omega t}\, dt = \frac{1}{T_0\,.\,(-jk\omega_0)} [e^{-jk\omega t}]_{-T_1}^{T_1}$$

$$= -\frac{1}{jk\omega_0 T_0} [e^{-jk\omega T_1} - e^{jk\omega_0 T_1}] = \frac{2}{k\omega_0 T_0} \left[\frac{e^{jk\omega_0 T_1} - e^{-jk\omega_0 T_1}}{2j} \right]$$

$$= \frac{2 \sin (k\omega_0 T_1)}{k\omega_0 T_0} \qquad \left(\text{But we know that } \omega_0 = \frac{2\pi}{T_0}\right)$$

$$= \frac{2 \sin (k\omega_0 T_1)}{k2\pi}$$

$$= \frac{\sin (k\omega_0 T_1)}{k\pi} \quad \text{for } k \neq 0 \qquad \text{...(4)}$$

Fourier series coefficients A_k's (only for $k \neq 0$) for $T_0 = 4T_1$, $s(t)$ is a square wave that is unity for one half of the period and zero for the other half of the period. In this case $\omega_0 T_1 = \pi/2$, and from Eq. 4, we get

$$A_k = \frac{\sin (k\,\pi/2)}{k\pi}, \quad k \neq 0 \qquad \text{...(5)}$$

Substituting $T_0 = 4T_1$ in Eq. 3, we get

$$A_0 = \frac{2T_1}{T_0} = \frac{2T_1}{4T_1} = \frac{1}{2} \qquad \text{...(6)}$$

Now from Eq. 5, $A_k = 0$ for k even and non-zero.

For odd values of k, Fourier series coefficients A_k's are determined as

$k = 1$, $\quad A_1 = \frac{\sin (1\,\pi/2)}{1.\pi} = \frac{\sin \pi 2}{\pi} = \frac{1}{\pi}$

$k = -1$, $\quad A_{-1} = \frac{\sin (-1\,\pi/2)}{(-1)\,\pi} = \frac{-\sin \pi/2}{-\pi} = \frac{1}{\pi}$

$$A_1 = A_{-1} = \frac{1}{\pi}$$

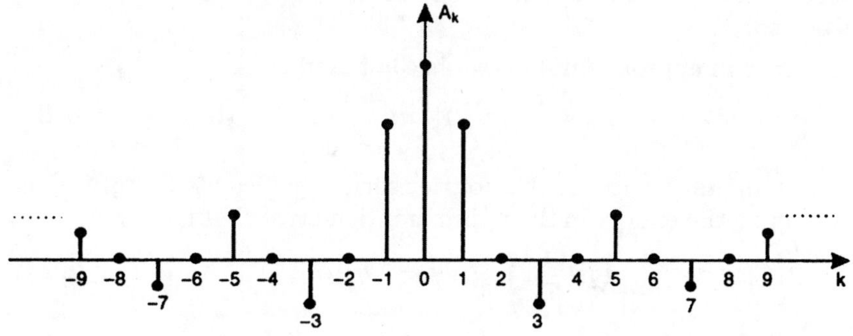

Fig. 3.4 *Plot of scaled Fourier series coefficients for the continuous-time periodic square wave with T_1 fixed and fundamental period $T_0 = 4T_1$.*

Similarly, $A_3 = A_{-3} = -\frac{1}{3\pi}$

$$A_5 = A_{-5} = \frac{1}{5\pi}$$

$$\vdots$$

3.5 CONVERGENCE OF FOURIER SERIES

Euler and Lagrange were satisfied with the results of Fourier series representation of sinusoidal periodic signals but they objected to results of Fourier series representation of square wave periodic signal since square wave is discontinuous while each of its harmonic component is continuous. Fourier, on the other hand, considered the same signals and maintained that the Fourier series representation of square wave signal is valid. In fact, Fourier maintained that any periodic signal can be represented by Fourier series. But it is not true for all periodic signals and Fourier series can be used to represent an extremely large class of periodic signals including square wave signals.

Now we will see the validity of Fourier series representation. Consider a periodic signal $s(t)$ which is a linear combination of a finite number of harmonically related complex exponentials. It is given as

$$s_N(t) = \sum_{k=-N}^{N} A_k\, e^{jk\omega_0 t} \qquad ...(3.40)$$

Approximation error is given as

$$e_N(t) = s(t) - s_N(t) \qquad ...(3.41)$$

A quantitative measure of the size of the approximation error is determined by computing the energy in the approximation error over one period.

Energy in approximation error is computed as

$$E_N = \int_{T_0} |e_N(t)|^2\, dt \qquad ...(3.42)$$

A particular choice for the Fourier series coefficients in Eq. 3.40 is to minimize the energy in the approximation error that is

$$A_k = \frac{1}{T_0} \int_{T_0} s(t)\, e^{-jk\omega_0 t}\, dt \qquad ...(3.43)$$

Fourier series coefficients A_k's are determined by the same equation for both infinite and finite Fourier series. A finite Fourier series is obtained by truncating the infinite Fourier series to the desired number of terms. As N increases, new terms in Fourier series are added and the energy in approximation E_N decreases. If, in fact, signal $s(t)$ has a Fourier series representation, then the limit of E_N as $N \to \infty$ is zero.

One class of periodic signals that can be represented through the Fourier series are those signals which have finite amount of energy over a single period.

These signals satisfy the condition given as

$$\int_{T_0} |s(t)|^2 \, dt < \infty \qquad ...(3.44)$$

When this condition is satisfied, we are guaranteed that the fourier series coefficients A_k's obtained from equation $A_k = \frac{1}{T_0} \int_{T_0} s(t)\, e^{-jk\omega_0 t}$ are finite.

Let $s_N(t)$ be the approximation to the signal $s(t)$. Now $s_N(t)$ is obtained as

$$s_N(t) = \sum_{k=-N}^{N} A_k \, e^{jk\omega_0 t} \qquad ...(3.45)$$

Here we are guaranteed that the energy E_N in the approximatioon error $e_N(t)$ converges to zero as we add more and more terms. It means if $N \to \infty$ then $E_N \to 0$.

If approximation error is defined as

$$e(t) = s(t) - \sum_{k=-\infty}^{\infty} A_k \, e^{jk\omega_0 t} \qquad ...(3.46)$$

then $$E_N = \int_{T_0} |e(t)|^2 \, dt = 0 \qquad ...(3.47)$$

This type of convergence is guaranteed when signal $s(t)$ has finite energy over a single period and it is quite useful. In the case of Eq. 3.47 which states that there is zero energy in the difference between original signal $s(t)$ and its Fourier series represenation. Since physical systems respond to signal energy, from this perspective signal $s(t)$ and its Fourier series representation are indistinguishable. Most of the periodic signals that have finite energy over a single period are represented by Fourier series.

Dirichlet Conditions. An alternative set of conditions developed by P.L. Dirichlet are satisfied by all of the signals with which we will be concerned. These conditions guarantee that a signal $s(t)$ equals its Fourier series representation (except at isolated values of t for which $s(t)$ is discontinuous). At these values, the infinite Fourier series of equation

$$s(t) = \sum_{k=-\infty}^{\infty} A_k \, e^{jk\omega_0 t}$$

converges to the average of the values on either side of the discontinuity. These conditions are called *Dirichlet conditions*. There are three Dirichlet conditions given as

1. Signal $s(t)$ must be absolutely integrable over any period. It is mathematically given as

$$\int_{T_0} |s(t)|\, dt < \infty \qquad \text{...(3.48)}$$

If $\int_{T_0} |s(t)|\, dt < \infty$ then $|A_k| < \infty$.

Signal $s(t) = \frac{1}{t}$, $0 < t \le 1$ is a periodic signal when period 1 violates the first Dirichlet condition.

2. There should be a finite number of maxima and minima during any single period of the signal. Signal $s(t) = \sin(2\pi/t)$, $0 < t \le 1$ satisfies first Dirichlet condition but it does not satisfy condition 2. This signal has infinite number of maxima and minima in a single period.
3. There should be a finite number of discontinuities in any finite interval of time.

Signals that do not satisfy the Dirichlet conditions are generally pathdological in nature and consequently do not arise in practical situations. Therefore, convergence of Fourier series is not important in such cases. For a periodic signal that has no discontinuities, the Fourier series representation converges and equals the original signal at every value of time t. For a periodic signal with finite number of discontinuities in each period, the Fourier series representation equals the original signal everywhere except at the isolated points of discontinuity, at which the fourier series converges to the average value of the signal on either side of the discontinuity. In the case of zero energy in difference between the original signal $s(t)$ and its Fourier series representation, these two signals are identical for all practical purposes. These two signals differ only at isolated points. The integrals of both signals over any interval are identical. Due to this reason, the two signals behave identically under convolution operation and consequently are identical from the point of analysis of linear time-invariant (LTI) systems.

3.6 PROPERTIES OF CONTINUOUS-TIME FOURIER SERIES (CTFS)

Fourier series representations possess a number of important properties which are very useful for developing conceptual insights into such representations. These properties can also help reducing in the complexity of the determination of Fourier series of many signals.

Here it will be convenient to use a short hand notion to indicate the relationship between a periodic signal $s(t)$ and its Fourier series coefficients A_k's. We have a continuous-time periodic signal $s(t)$ with period T_0 and fundamental frequency $\omega_0 = 2\pi/T_0$.

Continuous-time periodic signal $s(t)$ and its Fourier series coefficients A_k's are denoted by

$$s(t) \xrightarrow{\text{CTFS}} A_k$$

It signifies the pairing of continuous-time periodic signal $s(t)$ and its Fourier series coefficients A_k's. Now we will discuss the following properties of Continuous-time Fourier series (CTFS).

1. Linearity
2. Time Shifting
3. Time Reversal
4. Time Scaling
5. Multiplication
6. Conjugation
7. Parseval's relation for continuous-time periodic signals.

We will discuss each of these property one by one in the following sections.

3.6.1 Linearity

Let $s_1(t)$ and $s_2(t)$ be two continuous-time periodic signals with period T_0 and let them have Fourier series coefficients denoted by A_k and B_k, respectively. That is

$$s_1(t) \xleftrightarrow{\text{CTFS}} A_k$$

$$s_2(t) \xleftrightarrow{\text{CTFS}} B_k$$

Since $s_1(t)$ and $s_2(t)$ have the same period T_0. Linear combination of $s_1(t)$ and $s_2(t)$ will also be periodic with period T_0.

The Fourier series coefficients C_k of the linear combination of $s_1(t)$ and $s_2(t)$, $s_3(t) = \alpha\, s_1(t) + \beta\, s_2(t)$, are given by the same linear combination of the Fourier series coefficients of $s_1(t)$ and $s_2(t)$. That is

$$s_3(t) = \alpha\, s_1(t) + \beta\, s_2(t) \xleftrightarrow{\text{CTFS}} C_k = \alpha A_k + \beta B_k$$

3.6.2 Time Shifting

By time-shifting of a periodic signal, its period is preserved. The Fourier series coefficients B_k of the resulting signal $s_2(t) = s_1(t - t_0)$ can be expressed as

$$B_k = \frac{1}{T_0} \int_{T_0} s(t - t_0)\, e^{-jk\omega_0 t}\, dt \qquad ...(3.48)$$

Subsituting $\tau = t - t_0$ in Eq. (3.48), we get

$$\frac{1}{T_0}\int_{T_0} s(t-t_0)\, e^{-jk\omega_0 t} = \frac{1}{T_0}\int_{T_0} s(\tau)\, e^{-jk\omega_0(\tau+t_0)}\, dt$$

$$= \frac{e^{-jk\omega_0 t_0}}{T_0}\int_{T_0} s(\tau)\, e^{-jk\omega_0\tau}\, d\tau$$

$$= \frac{e^{-jk\omega_0 t_0}}{T_0} A_k \qquad ...(3.49)$$

where A_k is the kth Fourier series coefficient of $s_1(t)$.

If $$s_1(t) \xleftrightarrow{\text{CTFS}} A_k$$

then $$s(t-t_0) \xleftrightarrow{\text{CTFS}} e^{-jk\omega_0 t_0} A_k$$

Magnitude of Fourier series coefficients remain unaltered when periodic signal is shifted in time.

3.6.3 Time Reversal

Folding of a signal is called *time reversal*. The period T_0 of a periodic signal $s(t)$ also remains unaltered when the signal undergoes time reversal.

Consider a continuous-time periodic signal $s_1(t)$ its mirror image (folded) form will be given as

$$s_2(t) = s_1(-t)$$

If $$s_1(t) \xleftrightarrow{\text{CTFS}} A_k$$

then $$s_1(-t) \xleftrightarrow{\text{CTFS}} A_{-k}$$

In other words, time reversal (folding) applied to a continuous-time periodic signal results in a time reversal of the corresponding sequence of Fourier series coefficients $\{A_k\}$.

If signal $s_1(t)$ is even, that is

$$s_1(-t) = s_1(t)$$

then its Fourier series coefficients are also even, *i.e.*,

$$A_{-k} = A_k$$

Similarly, if signal $s_1(t)$ is odd, *i.e.*,

$$s_1(-t) = -\, s_1(t)$$

then its Fourier series coefficients are also odd, *i.e.*, $A_{-k} = -A_k$

3.6.4 Time Scaling

Time scaling is an operation by which period of an underlying signal changes. If $s(t)$ is a periodic signal with period T_0 then $s(\alpha t)$ is periodic with period T_0/α, where α is a positive real number. Since the time scaling operation applies directly to each of the harmonic components of signal $s(t)$, we can easily conclude that the Fourier series coefficients for each of these harmonic components remain the same.

If signal $s(t)$ has the Fourier series representation as

$$s(t) = \sum_{k=-\infty}^{\infty} A_k e^{jk\omega_0 t} \quad ...(3.50)$$

then Fourier series representation for the time scaled version of $s(t)$, *i.e.*, $s(\alpha t)$ will be given by

$$s(\alpha t) = \sum_{k=-\infty}^{\infty} A_k e^{jk(\alpha\,\omega_0)\,t} \quad ...(3.51)$$

We can conclude that in time scaling operation, Fourier series coefficients have not changed but the Fourier series representation has changed because of the change in the fundamental frequency.

3.6.5 Multiplication

We have two periodic signals $s_1(t)$ and $s_2(t)$ with period T_0.

$$s_1(t) \xleftrightarrow{\text{CTFS}} A_k$$

and $$s_2(t) \xleftrightarrow{\text{CTFS}} B_k$$

Since the product of $s_1(t)$ and $s_2(t)$ is also periodic with period T_0. We can expand $s_1(t)\, s_2(t)$ in a Fourier series with Fourier series coefficients C_k expressed in terms of coefficients $s_1(t)$ and $s_2(t)$.

The result will be

$$s_1(t)\, s_2(t) \xleftrightarrow{\text{CTFS}} C_k = \sum_{l=-\infty}^{\infty} A_l B_{k-l} \quad ...(3.52)$$

We observe that the summation on the right hand side of Eq. 3.52 may be interpreted as the discrete-time convolution of the sequence representing the Fourier series coefficients of $s_1(t)$ and the sequence representing the Fourier series coefficients of $s_2(t)$.

3.6.6 Conjugation

Taking the complex conjugate of a periodic signal $s(t)$ has the effect of complex conjugation and time reversal (folding) on the corresponding Fourier series coefficients.

That is, if $$s(t) \xleftrightarrow{\text{CTFS}} A_k$$

then $$s^*(t) \xleftrightarrow{\text{CTFS}} A_{-k}^*$$

Some of the important consequences of complex conjugation property of the signal $s(t)$ are

For real signal

If $$s(t) = s^*(t)$$

then, Fourier series coefficients A_k will be conjugate symmetric, *i.e.*,

$$A_{-k} = A_k^* \qquad ...(3.53)$$

If $s(t)$ is real then A_0 is real and

$$|A_k| = |A_{-k}| \qquad ...(3.54)$$

If signal $s(t)$ is real and even, then so are its fourier series coefficients.

Similarly, if $s(t)$ is real and odd, then its Fourier series coefficients are purely imaginary and odd. Thus, for example, $A_0 = 0$ if $s(t)$ is real and odd.

3.6.7 Parseval's Relation for Continuous-Time Periodic Signals

Parseval's relation for continuous-time periodic signals is given as

$$\frac{1}{T_0}\int_{T_0} |s(t)|^2\, dt = \sum_{k=-\infty}^{\infty} |A_k|^2 \qquad ...(3.55)$$

where, A_k's are the Fourier series coefficients of signal $s(t)$, and T_0 is the period of the signal.

Note that left hand side of Eq. 3.55 is the average power in one period of the periodic signal $s(t)$. Average power is the energy per unit time in one period of the periodic signal $s(t)$.

or $$\frac{1}{T_0}\int_{T_0} |A_k e^{jk\omega_0 t}|^2\, dt = \frac{1}{T_0}\int_{T_0} |A_k|^2\, dt = |A_k|^2 \qquad ...(3.56)$$

where, $|A_k|^2$ is the average power in the kth harmonic of the signal $s(t)$. Thus, Parseval's relation states that the total average power ina periodic signal is equal to the sum of the average powers in all of its harmonic components.

3.7 TABULATION OF PROPERTIES OF CONTINUOUS-TIME FOURIER SERIES (CTFS)

All the important properties of CTFS are given in Table 3.1.

Table 3.1 Properties of CTFS

Name of property	*Continuous-time periodic signals*	*Fourier series coefficients*
	$s_1(t)$, $s_2(t)$ both are periodic with period T_0	A_k B_k
Linearity	$\alpha s_1(t) + \beta s_2(t)$	$\alpha A_k + \beta B_k$
Time shifting	$s(t - t_0)$	$A_k e^{-jk\omega_0 t_0}$

Contd...

Name of property	*Continuous-time periodic signals*	*Fourier series coefficients*
Frequency shifting	$e^{jM\omega_0 t}\, s(t)$	A_{k-M}
Compelx conjugation	$s^*(t)$	A^*_{-k}
Time reversal	$s(-t)$	A_{-k}
Time scaling	$s(\alpha t)$, $\alpha > 0$ [periodic with period T_0/α]	A_k
Periodic convolution	$\int_{T_0} s_1(\tau)\, s_{(t} - \tau)\, d\tau$	$T_0\, A_k\, B_k$
Multiplication	$s_1(t)\, s_2(t)$	$\sum_{l=-\infty}^{\infty} A_l\, B_{k-l}$
Differentiation	$\frac{ds(t)}{dt}$	$jk\omega_0 A_k$
Integration	$\int_{-\infty}^{t} s(t)\, dt$ (Finite valued and periodic only if $A_0 = 0$)	$\left(\frac{1}{jk\omega_0}\right) A_k$
Conjugate symmetry for real signals	$s(t)$ is real	$\begin{cases} A_k = A^*_{-k} \\ \mathrm{Re}\{A_k\} = \mathrm{Re}\{A_{-k}\} \\ \mathrm{Im}\{A_k\} = -\,\mathrm{Im}\{A_{-k}\} \\ \lvert A_k \rvert = \lvert A_{-k} \rvert \\ \angle A_k = -\angle A_{-k} \end{cases}$
Real and even signals	$s(t)$ is real and even	A_k is real and even
Real and odd signals	$s(t)$ is real and odd	A_k is purely imaginary and odd
Even-odd decomposition of real signals	$s_e(t) = \mathrm{Ev}\{s(t)\}$ $s_0(t) = \mathrm{Od}\{s(t)\}$ Here $s(t)$ is real	$\mathrm{Re}\{A_k\}$ $j\,\mathrm{Im}\{A_k\}$

Parseval's relation for periodic signals

$$\frac{1}{T_0}\int_{T_0} \lvert s(t)\rvert^2\, dt = \sum_{k=-\infty}^{\infty} \lvert A_k \rvert^2$$

3.8 TRIGNOMETRIC FOURIER SERIES

If any continuous-time signal or function is periodic and satisfies the Dirichlet's conditions then the periodic continuos- time signal $s(t)$ can be expanded into an infinite trignometric Fourier series as

$$s(t) = A_0 + A_1 \cos \omega t + A_2 \cos 2\omega t + \dots + A_n \cos n\omega t + \dots$$
$$+ B_1 \sin \omega t + B_2 \sin 2\omega t + \dots + B_n \sin n\omega t + \dots$$

or $$s(t) = A_0 + \sum_{n=1}^{\infty} (A_n \cos n\omega t + B_n \sin n\omega t) \quad ...(3.57)$$

The Fourier series coefficients A_n and B_n are given by

$$A_n = \frac{2}{T_0} \int_0^{T_0} s(t) \cos n\omega t \, dt \quad ...(3.58)$$

$$A_0 = \frac{1}{T_0} \int_0^{T_0} s(t) \, dt \quad ...(3.59)$$

and $$B_n = \frac{2}{T_0} \int_0^{T_0} s(t) \sin n\omega t \, dt \quad ...(3.60)$$

3.8.1 Evaluation of Fourier Series Coefficients

The evaluation of Fourier series coefficients A_n and B_n is accomplished by using simple integral equations. These integral equations are desired from the orthogonality property of the set of functions involved, namely, cos $n\omega t$ and sin $n\omega t$ with integer values of m and n. These functions are orthogonal over time period T_0.

We have observed that

$$\int_0^{T_0} \sin m\omega t = 0, \text{ for all } m \quad ...(3.61)$$

and $$\int_0^{T_0} \cos n\omega t = 0, \text{ for } n \neq 0 \quad ...(3.62)$$

Since the average value of a sinusoid (sine and cosine) over m or n, complete cycles in the period T_0 is zero.

We also observed that these cross-product terms are also zero for the stated orthogonal relationship of m and n

$$\int_0^{T_0} \sin m\omega t \cos n\omega t \, dt = 0, \text{ for } m \text{ and } n \quad ...(3.63)$$

$$\int_0^{T_0} \sin m\omega t \sin n\omega t \, dt = 0, \; m \neq n \quad ...(3.64)$$

and $$\int_0^{T_0} \cos m\omega t \cos n\omega t \, dt = 0 \,, \quad m \neq n \qquad ...(3.65)$$

The integrals becomes non-zero when $m = n$; thus

$$\int_0^{T_0} \sin m\omega t \sin m\omega t \, dt = \frac{T_0}{2} \,, \text{ for all } m$$

or $$\int_0^{T_0} \sin^2 m\omega t \, dt = T_0 \,, \text{ for all } m \qquad ...(3.66)$$

and $$\int_0^{T_0} \cos n\omega t \cos n\omega t \, dt = \frac{T_0}{2} \,, \text{ for all } n$$

or $$\int_0^{T_0} \cos^2 n\omega t \, dt = \frac{T_0}{2} \,, \text{ for all } n \qquad ...(3.67)$$

Fourier series coefficients can be determined by application of above relations.

By integrating Eq. $s(t) = A_0 + \sum_{n=1}^{\infty} (A_n \cos n\omega t + B_n \sin n\omega t)$ we get the Fourier series coefficient A_0 as

$$\int_0^{T_0} s(t) \, dt = \int_0^{T_0} \left[A_0 + \sum_{n=1}^{\infty} (A_n \cos n\omega t + B_n \sin n\omega t) \right] dt$$

$$= A_0 \int_0^{T_0} dt + \int_0^{T_0} s_1(t) \, dt. \qquad ..(3.68)$$

where $$s_1(t) = A_0 + \sum_{n=1}^{\infty} (A_n \cos n\omega t + B_n \sin n\omega t) \qquad ...(3.69)$$

Eq. 3.68 can be written as

$$\int_0^{T_0} s(t) \, dt = A_0 \, [\, t \,]_0^{T_0} + \int_0^{T_0} s_1(t) \, dt$$

$$= A_0 T_0 + \int_0^{T_0} s_1(t) \, dt \qquad ...(3.70)$$

The first term of R.H.S. of Eq. 3.70 has the value of $A_0 T_0$ while every term in the infinte summation of $s_1(t)$ when integrated from 0 to T_0, has zero value as given by Eqs. 3.61 and 3.62.

Then from Eq. 3.70, we get

$$A_0 = \frac{1}{T_0} \int_0^{T_0} s(t) \, dt \qquad ...(3.71)$$

If $T_0 = 2\pi$, then

$$A_0 = \frac{1}{2\pi} \int_0^{2\pi} s(t)\, dt \qquad ...(3.72)$$

where T_0 is the time-period of the continuous-time periodic signal. In order to determine A_n, we multiply Eq. 3.57 by cos $m\omega t$ and integrating between the limits 0 to T_0, we get

$$\int_0^{T_0} s(t) \cos m\omega t\, dt = \int_0^{T_0} A_0 \cos m\omega t\, dt + \int_0^{T_0} \cos m\omega t \sum_{n=1}^{\infty} A_n \cos n\omega t\, dt$$

$$+ \int_0^{T_0} \cos m\omega t + \sum_{n=1}^{\infty} B_n \sin n\omega t\, dt. \qquad ..(3.73)$$

Now, by using following relations

$$\int_0^{T_0} \cos n\omega t = 0\,, \text{ for } n \neq 0$$

$$\int_0^{T_0} \sin m\omega t \cos n\omega t\, dt = 0\,, \text{ for all } m \text{ and } n$$

$$\int_0^{T_0} \cos m\omega t \cos n\omega t\, dt = 0\,, \text{ for } m \neq n$$

and $\int_0^{T_0} \cos^2 n\omega t\, dt = \frac{T_0}{2}$, for all n in Eq. 3.73, we obtain

$$A_0 \int_0^{T_0} \cos m\omega t\, dt = 0$$

$$\int_0^{T_0} \cos m\omega t \sum_{n=1}^{\infty} B_n \sin n\omega t\, dt = 0$$

$$\int_0^{T_0} \cos m\omega t \sum_{\substack{n=1 \\ n \neq m}}^{\infty} A_n \cos n\omega t\, dt = 0$$

and $\int_0^{T_0} A_m \cos m\omega t \cos m\omega t\, dt = \left(\frac{T_0}{2}\right) A_m$

Therefore, for $n = m$

$$A_n = \frac{2}{T_0} \int_0^{T_0} s(t) \cos n\omega t\, dt \quad \text{for all integer values of } n. \qquad ...(3.74)$$

Similarly, in order to determine B_n, we multiply Eq. 3.57 by sin $m\ \omega t$ and integrating between the limits 0 to T_0, we get

$$\int_0^{T_0} s(t) \sin m\omega t \, dt = \int_0^{T_0} A_0 \sin m\omega t \, dt$$

$$+ \int_0^{T_0} \sin m\omega t \sum_{n=1}^{\infty} A_n \cos n\omega t \, dt$$

$$+ \int_0^{T_0} \sin m\omega t \sum_{n=1}^{\infty} B_n \sin n\omega t \, dt \qquad ...(3.75)$$

Now by using following relations

$$\int_0^{T_0} \sin m\omega t = 0 , \text{ for all } m$$

$$\int_0^{T_0} \sin m\omega t \cos n\omega t \, dt = 0 , \text{ for all } m \text{ and } n$$

$$\int_0^{T_0} \sin m\omega t \sin n\omega t \, dt = 0 , \text{ for } m \neq n$$

and $\int_0^{T_0} \sin^2 m\omega t \, dt = \frac{T_0}{2}$, for all m, in Eq. 3.75, we get

$$A_0 \int_0^{T_0} \sin m\omega t \, dt = 0$$

$$\int_0^{T_0} \sin m\omega t \sum_{n=1}^{\infty} A_n \cos n\omega t \, dt = 0$$

$$\int_0^{T_0} \sin m\omega t \sum_{\substack{n=1 \\ n \neq m}}^{\infty} B_n \sin n\omega t \, dt = 0$$

and $\int_0^{T_0} B_m \sin m\omega t \sin m\omega t \, dt = \left(\frac{T_0}{2} \right) B_n$

Therefore, for $n = m$

$$B_n = \frac{2}{T_0} \int_0^{T_0} s(t) \sin n\omega t \, dt \qquad ...(3.76)$$

Hence, any periodic signal or function $s(t)$ can be represented by the Fourier series as

$$s(t) = A_C + \sum_{n=1}^{\infty} (A_n \cos n\omega t + B_n \sin n\omega t)$$

The sum of the harmonically related ac components A_n and B_n together with the dc component A_0 is equal to the continuous-time signal

$s(t)$. Signal $s(t)$ is the input excitation for driving a network as shown in Fig. 3.5 (*a*). The multi-generator representation of signal $s(t)$ in the network is shown in Fig. 3.5 (*b*). The total response of the network shown in Fig. 3.5 (*b*) may be determined by superposition.

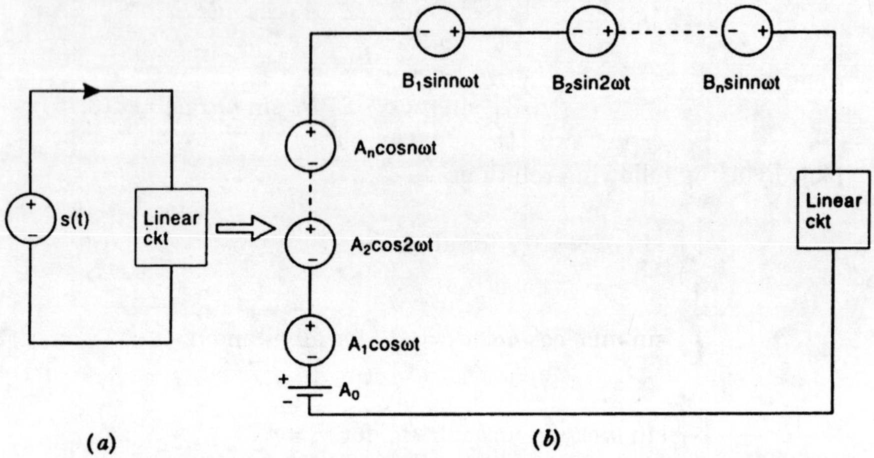

Fig. 3.5 *Illustration of how continuous-time signal s(t) can be related by Fourier series as 'n' sinusoidal voltage sources in series of harmonically related frequencies nω where 'n' is an integer.*

A more compact representation of Fourier series is obtained by combining the sine and cosine terms of the equation

$$s(t) = A_0 + \sum_{n=1}^{\infty} (A_n \cos n\omega t + B_n \sin n\omega t)$$

either in the sine or cosine form as follows :

Now let us consider the nth harmonic term

$$s_n(t) = A_n \cos n\omega t + B_n \sin n\omega t$$

Divide and multiply R.H.S. of the above equation by $\sqrt{A_n^2 + B_n^2}$, we get

$$s_n(t) = \sqrt{A_n^2 + B_n^2}\left[\frac{A_n}{\sqrt{A_n^2 + B_n^2}} \cos n\omega t + \frac{B_n}{\sqrt{A_n^2 + B_n^2}} \sin n\omega t\right]$$

$$= \sqrt{A_n^2 + B_n^2}\left[\cos \theta_n \cos n\omega t + \sin \theta_n \sin n\omega t\right]$$

or $$s_n(t) = C_n \cos (n\omega t - \theta_n) \qquad ...(3.77)$$

where, $C_n = \sqrt{A_n^2 + B_n^2}$ = amplitude of the nth harmonic of the $s(t)$

and $\theta_n = \tan^{-1}\left\{\dfrac{B_n}{A_n}\right\}$ = phase of the nth harmonic of the $s(t)$

Putting $A_0 = C_0$ and $s_n(t)$ in Eq. 3.57, we get

$$s(t) = A_0 + \sum_{n=1}^{\infty} (A_n \cos n\omega t + B_n \sin n\omega t)$$

$$= A_0 + \sum_{n=1}^{\infty} s_n(t)$$

or $$s(t) = C_0 + \sum_{n=1}^{\infty} C_n \cos(n\omega t - \theta_n) \qquad ...(3.78)$$

Eq. 3.78 represents the Fourier series of Eq. 3.57 in the cosine form.

In a similar way, the Fourier series of Eq. 3.57 can be represented in the sine form

$$s(t) = C_0 + \sum_{n=1}^{\infty} C_n \sin(n\omega t + \phi_n) \qquad ...(3.79)$$

where $\phi_n = \tan^{-1}\left\{\frac{A_n}{B_n}\right\}$

We oberved that if a Fourier series is to be constructed in the form of Eq. 3.79, then the set of numbers C_n and θ_n or ϕ_n contains all the needed information. The plot of C_n as function of n or $n\omega$ is known as the Amplitude Spectrum and the plot of θ_n as a function of n or $n\omega$ is known as Phase Spectrum.

3.8.1 Waveform Symmetry

Symmetry saves considerable amount of labour in the Fourier series analysis. In waveform synthesis we can see that

Sum of even functions is an even function and sum of odd functions is an odd function. However, if an odd function is added to an even function, the resultant is neither even nor odd. The new function $s(t)$ may be said to have an even and an odd part

$$s(t) = s_E(t) + s_O(t) \qquad ...(3.80)$$

where $s_E(t) = \text{Ev}\{s(t)\}$

and $s_O(t) = \text{Od}\{s(t)\}$

Putting $t = -t$ in Eq. 3.80, we get

$$s(-t) = s_E(-t) + s_O(-t)$$

or $$s(-t) = s_E(t) - s_O(t) \qquad ...(3.81)$$

From Eqs. 3.80 and 3.81, we obtain

$$s_E(t) = \frac{1}{2}[s(t) + s(-t)]$$

and $$s_O(t) = \frac{1}{2}[s(t) - s(-t)]$$

These relations can be used in the equations of Fourier cofficients.

Now to compare $s(t)$ and $s(-t)$, we choose the interval from $-\frac{T_0}{2}$ and

$\frac{T_0}{2}$ instead of 0 to T_0. Fourier series coefficients A_n and B_n can be rewritten as

$$A_n = \frac{2}{T_0} \int_{-T_0/2}^{T_0/2} s(t) \cos n\omega t \, dt$$

$$= \frac{2}{T_0} \int_{-T_0/2}^{T_0/2} [s_E(t) + s_O(t)] \cos n\omega t \, dt$$

[Since $s(t) = s_E(t) + s_O(t)$]

$$= \frac{2}{T_0} \int_{-T_0/2}^{T_0/2} s_E(t) \cos n\omega t \, dt + \frac{2}{T_0} \int_{-T_0/2}^{T_0/2} s_O(t) \cos n\omega t \, dt$$

Similarly, $$B_n = \frac{2}{T_0} \int_{-T_0/2}^{T_0/2} s(t) \sin n\omega t \, dt$$

$$= \frac{2}{T_0} \int_{-T_0/2}^{T_0/2} [s_E(t) + s_O(t)] \sin n\omega t \, dt$$

$$= \frac{2}{T_0} \int_{-T_0/2}^{T_0/2} s_E(t) \sin n\omega t \, dt + \frac{2}{T_0} \int_{-T_0/2}^{T_0/2} s_O(t) \sin n\omega t \, dt$$

Again, we know that

$$s_{E1}(t) \times s_{O1}(t) = s_O(t)$$

$$s_{E1}(t) \times s_{E2}(t) = s_E(t)$$

$$s_{O1}(t) \times s_{O2}(t) = s_E(t)$$

and $$\int_{-t_0}^{t_0} s_E(t) \, dt = 2 \int_{0}^{t_0} s_E(t) \, dt$$

$$\int_{-t_0}^{t_0} s_O(t) \, dt = 0$$

Even Functions or Signals. A signal or function $s(t)$ is said to be even if $s(t) = s(-t)$. In other words, if one signal and its mirror image are equal then it is called an *even signal*.

Examples of even signals or functions are t^2, $\cos t$, $\cosh t$, $\tan (t^2)$, $1 + 2t^2 + 4t^4$ etc. Also we know that even nature is preserved on addition of a constant. The sum of even functions remains even. The cosine function is an even function as it can be expressed as a sum of a constant and even functions.

$$\cos t = 1 - \frac{t^2}{2} + \frac{t^2}{4} - \frac{t^6}{6} + \ldots$$

Even functions exhibit mirror symmetry about the vertical axis, the negative half being the mirror image of the positive half.

In even functions, the Fourier series cofficients, B_n are absent.

Equation $A_n = \frac{2}{T_0}\int_0^{T_0} s(t)\cos n\omega t\,dt$ can be rewritten by using the limits of integration from $-\frac{T_0}{2}$ to $+\frac{T_0}{2}$ and representing the coefficients as a sum of two parts.

$$A_n = \frac{2}{T_0}\int_{-T_0/2}^{T_0/2} s(t)\cos n\omega t\,dt$$

$$= \frac{2}{T_0}\int_{-T_0/2}^{0} s(t)\cos n\omega t\,dt + \frac{2}{T_0}\int_{0}^{T_0/2} s(t)\cos n\omega t\,dt$$

Substituting $t = -\tau$ in first integral and $t = \tau$ in the second integral, we get

$$A_n = \frac{2}{T_0}\int_{T_0/2}^{0} s(-\tau)\cos n\omega(-\tau)\,(-d\tau) + \frac{2}{T_0}\int_{0}^{T_0/2} s(\tau)\cos n\omega\tau\,d\tau$$

$$= \frac{2}{T_0}\int_{0}^{T_0/2} s(-\tau)\cos n\omega\tau\,d\tau + \frac{2}{T_0}\int_{0}^{T_0/2} s(\tau)\cos n\omega\tau\,d\tau$$

$$= \frac{2}{T_0}\int_{0}^{T_0/2} [s(\tau) + s(-\tau)]\cos n\omega\tau\,d\tau$$

$$\left[\text{Since } \cos n\omega(-\tau) = \cos n\omega\tau \text{ and } -\int_{T_0/2}^{0} = \int_{0}^{T_0/2}\right]$$

$$A_n = \frac{4}{T_0}\int_{0}^{T_0/2} s(\tau)\cos n\omega\tau\,d\tau$$

[Since for even functions, $s(-t) = s(t)$]

or $$A_n = \frac{4}{T_0}\int_{0}^{T_0/2} s(t)\cos n\omega t\,dt \qquad \text{...(3.82)}$$

Similarly, we can find out B_n as

$$B_n = \frac{2}{T_0}\int_{0}^{T_0/2} [s(t) + s(-t)]\sin n\omega t\,dt$$

[Since for odd functions, $s(-t) = -\,s(t)$]

or $B_n = 0$ ...(3.83)

For even functions,

$$A_n = \frac{4}{T_0} \int_0^{T_0/2} s(t) \cos n\omega t \, dt$$

and $B_n = 0$

If the average value of the function is zero, the coefficient A_0 is also absent.

Now, we can conclude that an even function exhibits mirror symmetry about the vertical axis and its Fourier series does not have sine terms ($B_n = 0$).

Odd Functions or Signals. A function or signal is termed as odd if $s(t) = -s(-t)$. Examples of odd functions are t, t^3, $\sin t$, $t \cos 10t$ etc. Also, we know that the sum of odd functions is odd.

$$\sin t = t - \frac{t^3}{\angle 3} + \frac{t^5}{\angle 5} - \frac{t^7}{\angle 7} + \ldots$$

The addition of a constant in an odd function removes the odd nature of the function $s(t)$ since $s(t) \neq -s(-t)$.

For odd functions, $A_n = 0$ for all n

and $$B_n = \frac{4}{T_0} \int_0^{T_0/2} s(t) \sin n\omega t \, dt$$

Now we can conclude that an odd function does not exhibit mirror symmetry about the vertical axis and its Fourier series does not have cosine terms ($A_n = 0$). Its Fourier series expression contains only the sine terms.

3.8.3 Half-Wave Symmetry

A periodic signal or function is said to have half-wave symmetry if

$$s(t) = -s\left(t \pm \frac{T_0}{2}\right)$$

or in terms of angle $s(\theta) = -s(\theta \pm \pi)$

Half-wave symmetric functions have only odd harmonics.

$$A_n = \frac{2}{2\pi} \int_{-\pi}^{\pi} s(\theta) \cos n\theta \, d\theta$$

$$= \frac{1}{\pi} \left[\int_{-\pi}^{0} s(\theta) \cos n\theta \, d\theta + \int_{0}^{\pi} s(\theta) \cos n\theta \, d\theta \right]$$

or $$A_n = \frac{1}{\pi} [I_1 + I_2]$$ ...(3.84)

Now, by changing the available θ to $\theta + \pi$ in the first integral,

$$I_1 = \int_{-\pi}^{0} s(\theta) \cos n\theta \, d\theta$$

$$= \int_{0}^{\pi} s(\theta + \pi) \cos n(\theta + \pi) \, d\theta$$

$$= \int_{0}^{\pi} s(\theta + \pi) [\cos n\theta \cos n\pi - \sin n\theta \sin n\pi] \, d\theta$$

[Using half-wave symmetry property $s(\theta) = - s(\theta \pm \pi)$]

$$= \int_{0}^{\pi} - s(\theta) [\cos n\theta \cos n\pi - \sin n\theta \times 0] \, d\theta$$

$$= - \int_{0}^{\pi} s(\theta) \cos n\theta \cos n\pi \, d\theta$$

$$= - \cos n\pi \int_{0}^{\pi} s(\theta) \cos n\theta \, d\theta$$

or
$$I_1 = - \cos n\pi \int_{0}^{\pi} s(\theta) \cos n\theta \, d\theta \qquad ...(3.85)$$

Substituting Eq. 3.85 in Eq. 3.84, we get

$$A_n = \frac{1}{\pi} [I_1 + I_2]$$

$$= \frac{1}{\pi} \left[- \cos n\pi \int_{0}^{\pi} s(\theta) \cos n\theta \, d\theta + \int_{0}^{\pi} s(\theta) \cos n\theta \, d\theta \right]$$

or
$$A_n = \left(\frac{1 - \cos n\pi}{\pi} \right) \int_{0}^{\pi} s(\theta) \cos n\theta \, d\theta \qquad ...(3.86)$$

Hence, $A_n = 0$, for n even ...(3.87)

(Since $\cos n\pi = 1$)

and
$$A_n = \frac{2}{\pi} \int_{0}^{\pi} s(\theta) \cos n\theta \, d\theta \, , \text{ for } n \text{ odd}$$

(Since $\cos n\pi = -1$)

or
$$A_n = \frac{4}{T_0} \int_{0}^{T_0/2} s(t) \cos n\omega t \, dt \, , \; n \text{ odd} \qquad ...(3.88)$$

In a similar manner, we can also show that

$$B_n = 0 \, , \text{ for } n \text{ even} \qquad ...(3.89)$$

$$B_n = \frac{2}{\pi}\int_0^{\pi} s(\theta)\sin n\theta\, d\theta\,, \text{ for } n \text{ odd}$$

or $$B_n = \frac{4}{T_0}\int_0^{T_0/2} s(t)\sin n\omega t\, dt\,, \text{ for } n \text{ odd} \qquad ...(3.90)$$

Now, we can conclude that waveforms having half-wave symmetry have both A_n and B_n coefficients, *i.e.*, odd harmonics of sine and cosine terms are present unless the function is even or odd as well.

3.9 EXPONENTIAL FORM OF FOURIER SERIES

We have already studied the trignometric Fourier series of continuous-time periodic functions or signals. It is given as

$$s(t) = A_0 + \sum_{n=1}^{\infty} [A_n \cos n\omega t + B_n \sin n\omega t]$$

We are also familiar with other compact form of Fourier series. These compact forms of Fourier series are

$$s(t) = C_0 + \sum_{n=1}^{\infty} C_n \cos(n\omega t - \theta_n)$$

and $$s(t) = C_0 + \sum_{n=1}^{\infty} C_n \sin(n\omega t + \phi_n)$$

where $C_0 = A_0$

$$C_n = \sqrt{A_n^2 + B_n^2}$$

$$\theta_n = \tan^{-1}\left\{\frac{B_n}{A_n}\right\}$$

$$\phi_n = \tan^{-1}\left\{\frac{A_n}{B_n}\right\}$$

These Fourier series are associated only with the positive frequency $[0 < \omega < \infty]$, *i.e.*, one side of the frequency spectrum. Now we can obtain a convenient and elegant representation of Fourier series which is obtained by expressing the sine and cosine terms as exponential functions with complex constants.

By using Euler's relation, we get

$$\cos n\omega t = \frac{e^{jn\omega t} + e^{-jn\omega t}}{2}$$

$$\sin n\omega t = \frac{e^{jn\omega t} - e^{-jn\omega t}}{2j}$$

Substituting values of cos $n\omega t$ and sin $n\omega t$ in following equation, we get

$$s(t) = A_0 + \sum_{n=1}^{\infty} [A_n \cos n\omega t + B_n \sin n\omega t]$$

$$= A_0 + \sum_{n=1}^{\infty} \left\{ A_n \left[\frac{e^{jn\omega t} + e^{-jn\omega t}}{2} \right] + B_n \left[\frac{e^{jn\omega t} - e^{-jn\omega t}}{2j} \right] \right\}$$

$$= A_0 + \sum_{n=1}^{in} \left[\left(\frac{A_n - jB_n}{2} \right) e^{jn\omega t} + \left(\frac{A_n + jB_n}{2} \right) e^{-jn\omega t} \right] \qquad ...(3.91)$$

Now we define

$$\left. \begin{aligned} \tilde{C}_n &= \frac{A_n - jB_n}{2} \\ \tilde{C}_{-n} &= \frac{A_n + jB_n}{2} \\ \tilde{C}_0 &= A_0 \end{aligned} \right\} \qquad ...(3.92)$$

Then, $$s(t) = \tilde{C}_0 + \sum_{n=1}^{\infty} \left[\tilde{C}_n e^{jn\omega t} + \tilde{C}_{-n} e^{-jn\omega t} \right]$$

$$= \sum_{n=0}^{\infty} \tilde{C}_n e^{jn\omega t} + \sum_{n=1}^{\infty} \tilde{C}_n e^{-jn\omega t}$$

[Putting $n = -n$ in second summation of above equation]

$$= \sum_{n=0}^{\infty} \tilde{C}_n e^{jn\omega t} + \sum_{n=-1}^{-\infty} \tilde{C}_n e^{jn\omega t}$$

$$= \sum_{n=0}^{\infty} \tilde{C}_n e^{jn\omega t} + \sum_{n=-\infty}^{-1} \tilde{C}_n e^{jn\omega t}$$

$$= \sum_{n=-\infty}^{\infty} \tilde{C}_n e^{jn\omega t}$$

or $$s(t) = \sum_{n=-\infty}^{\infty} \tilde{C}_n e^{jn\omega t} \qquad ...(3.93)$$

Eq. 3.93 is the exponential or complex form of Fourier Series.

Substituting the expressions of A_n and B_n in the expression of C_n in equation

$$\tilde{C}_n = \frac{A_n - jB_n}{2}$$

$$= \frac{1}{2} \left[\frac{2}{T_0} \int_{-T_0/2}^{T_0/2} s(t) \cos n\omega t \, dt \right] - j \frac{2}{T_0} \int_{-T_0/2}^{T_0/2} s(t) \sin n\omega t \, dt$$

$$= \frac{1}{T_0} \int_{-T_0/2}^{T_0/2} s(t) \, [\cos n\omega t - j \sin n\omega t] \, dt$$

$$= \frac{1}{T_0} \int_{-T_0/2}^{T_0/2} s(t)\, e^{-jn\omega t}\, dt$$

or
$$\tilde{C}_n = \frac{1}{T_0} \int_{-T_0/2}^{T_0/2} s(t)\, e^{-jn\omega t}\, dt \qquad ...(3.94)$$

where n ranges over all integral values from $-\infty$ to ∞.

Eqs. 3.93 ar 3.94 are required in the representation of a periodic functions in the exponential form.

The exponential form of the Fourier series has a number of advantages over the trignometric form. In exponential form of Fourier series only one integral has to be calculated instead of three integrals in the trigometric form, for the calculation of the Fourier series coefficients. Moreover, single integral required in exponential Fourier series is a simple integration. The effect of waveform symmetry is also present in the exponential form of Fourier series representation.

For an even function, sine terms are absent, *i.e.*, $B_n = 0$, hence the $\tilde{C}_n$ coefficients are all real, with $\tilde{C}_n = \tilde{C}_{-n} = \frac{A_n}{2}$.

For an odd function, the cosine terms are zero, *i.e.*, $A_n = 0$, hence $\tilde{C}_n$ coefficients are imaginary, with $\tilde{C}_n = -\frac{jB_n}{2}$ and $\tilde{C}_{-n} = \frac{jB_n}{2}$

Eq. 3.92 relates complex Fourier series coefficients $\tilde{C}_n$ and $\tilde{C}_{-n}$ and the real Fourier series coefficients A_0, A_n and B_n.

This relationship can be seen more clearly from phasor representation shown in Fig. 3.6. At $t = 0$ for a pair of general terms $[\tilde{C}_n e^{jn\omega t} + \tilde{C}_{-n} e^{-jn\omega t}]$ since $\tilde{C}_n$ and $\tilde{C}_{-n}$ are complex conjugates. These phasors represent the pair of terms which rotate with equal and opposite angular frequencies. The sum of these two terms will always be real and will vary sinusoidally at the rate of $n\omega$ radians/sec.

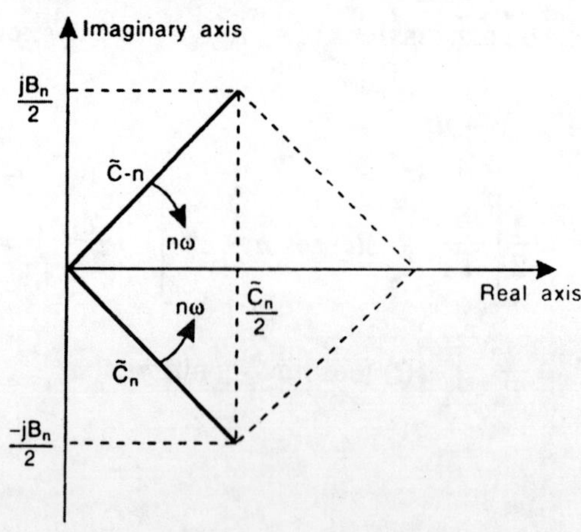

Fig. 3.6 *Phasor representation of* $\tilde{C}_n e^{jn\omega t} + \tilde{C}_{-n} e^{-jn\omega t}$.

SOLVED PROBLEMS

Example 3.7 Find the trignometric Fourier series for the continuous-time saw-tooth wave shown in Fig. 3.7 (*a*).

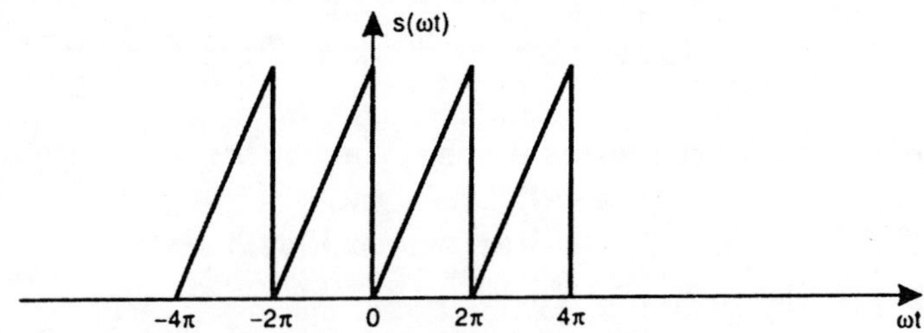

Fig. 3.7 (*a*) *Continuous-time saw-tooth wave.*

Solution. The given waveform is periodic and continuous for $0 < \omega t < 2\pi$. This waveform also satisfies Dirichlet's conditions. The analytical form of this waveform is given by

$$s(\omega t) = \frac{1}{2\pi}\,\omega t\,, \quad \text{for } 0 < \omega t < 2\pi \qquad \text{...(1)}$$

The trignometric Fourier series representation of any signal $s(\omega t)$ is given by

$$s(\omega t) = A_0 + \sum_{n=1}^{\infty} [A_n \cos n\omega t + B_n \sin n\omega t] \qquad \text{...(2)}$$

Now we will determine trignometric Fourier series coefficients A_0 , A_n and B_n

$$A_0 = \text{Average value of signal } s(\omega t)$$

$$= \frac{1}{2\pi}\int_0^{2\pi} s(\omega t)\, d(\omega t) = \frac{1}{2\pi}\int_{\omega t=0}^{\omega t=2\pi}\left(\frac{\omega t}{2\pi}\right) d(\omega t)$$

$$= \frac{1}{(2\pi)^2}\left[\frac{(\omega t)^2}{2}\right]_{\omega t=0}^{\omega t=2\pi} = \frac{1}{(2\pi)^2}\left[\frac{(2\pi)^2}{2}\right]$$

$$= \frac{1}{2} = 0.5$$

or $$A_0 = 0.5$$

$$A_n = \frac{2}{2\pi}\int_{\omega t=0}^{\omega t=2\pi} s(\omega t)\cos n\omega t\, d(\omega t)$$

$$= \frac{1}{\pi}\int_0^{2\pi} \frac{\omega t}{2\pi}\cos n\omega t\, d(\omega t)$$

$$= \frac{1}{2\pi^2} \int_0^{2\pi} \omega t \cos n\omega t \, d(\omega t)$$

$$= \frac{1}{2\pi^2} \left[\frac{\omega t}{n} \sin n\omega t + \frac{\cos n(\omega t)}{n^2} \right]_{\omega t = 0}^{\omega t = 2\pi}$$

or $\qquad A_n = 0$

It means that all the cosine terms of the Fourier series are zero

$$B_n = \frac{2}{2\pi} \int_{\omega t = 0}^{\omega t = 2\pi} s(\omega t) \sin n\omega t \, d(\omega t)$$

$$= \frac{1}{\pi} \int_0^{2\pi} \frac{\omega t}{2\pi} \sin n\omega t \, d(\omega t)$$

$$= \frac{1}{2\pi^2} \int_0^{2\pi} \omega t \sin n\omega t \, d(\omega t)$$

$$= \frac{1}{2\pi^2} \left[-\frac{\omega t}{n} \cos n\omega t + \frac{1}{n^2} \sin n\omega t \right]_0^{2\pi}$$

$$= -\frac{1}{n\pi}$$

or $\qquad B_n = -\dfrac{1}{n\pi}$

Substituting the vlaues of A_0 , A_n and B_n in Eq. (2) we get

$$s(\omega t) = A_0 + \sum_{n=1}^{\infty} [A_n \cos n\omega t + B_n \sin n\omega t]$$

$$= 0.5 + \sum_{n=1}^{\infty} (0) \cos n\omega t + \left(-\frac{1}{n\pi} \right) \sin n\omega t$$

$$= 0.5 + \sum_{n=1}^{\infty} \frac{\sin n\omega t}{-n\pi}$$

or $$s(\omega t) = 0.5 - \sum_{n=1}^{\infty} \frac{\sin n\omega t}{n\pi} \qquad ...(3)$$

This is the required trignometric Fourier series representation of a continuous-time saw-tooth wave shown in Fig. 3.7 (*a*).

For plotting amplitude spectrum, we determine

$$C_0 = A_0 = 0.5$$

$$C_n = \sqrt{A_n^2 + B_n^2} = \sqrt{(0)^2 + \left[-\frac{1}{n\pi} \right]^2}$$

or $$C_n = -\frac{1}{n\pi}$$

The amplitude spectrum is shown in Fig. 3.7 (*b*).

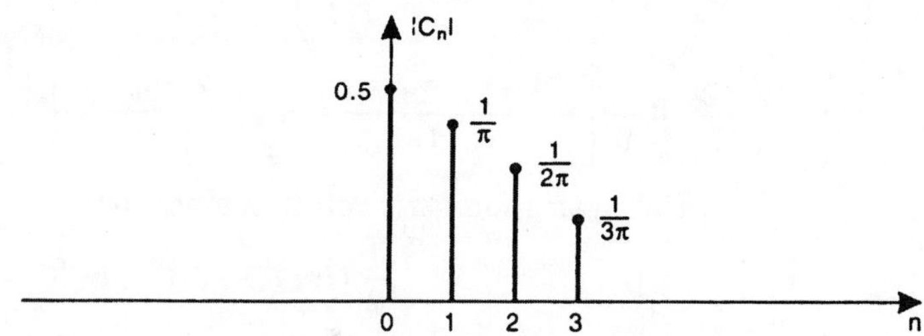

Fig. 3.7 (*b*) *Amplitude (or magnitude) spectrum of continuous-time Fourier series shown in Fig. 3.7 (a).*

Example 3.8 Find the trignometric Fourier series representation for the rectified sine wave shown in Fig. 3.8.

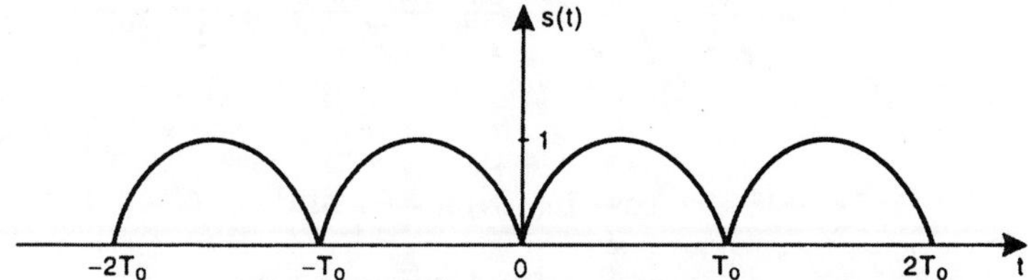

Fig. 3.8 *Rectified continuous-time sine wave.*

Solution. The analytical form of the function or signal $s(t)$ is given by

$$s(t) = \sin\left(\frac{\pi}{T_0}\right)t\ ,\ \text{for } 0 < t < T_0 \qquad ...(1)$$

Trignometric Fourier series representation of signal $s(t)$ is given by

$$s(t) = A_0 + \sum_{n=1}^{\infty} [A_n \cos n\omega t + B_n \sin n\omega t] \qquad ...(2)$$

$$A_0 = \frac{1}{T_0}\int_0^{T_0} s(t)\,dt = \frac{1}{T_0}\int_0^{T_0} \sin\left(\frac{\pi}{T_0}\right)t\,dt$$

$$= \frac{1}{\pi}\left[-\cos\frac{\pi}{T_0}t\right]_0^{T_0} = \frac{2}{\pi}$$

or $$A_0 = \frac{2}{\pi} \qquad ...(3)$$

$$A_n = \frac{2}{T_0}\int_0^{T_0} s(t)\cos n\omega t\,dt$$

$$= \frac{2}{T_0} \int_0^{T_0} \sin\left(\frac{\pi}{T_0}\right) t \cos n\omega t \, dt \qquad \left(\text{Since } \omega = \frac{2\pi}{T_0}\right)$$

$$= \frac{2}{T_0} \int_0^{T_0} \sin\left(\frac{\pi t}{T_0}\right) \cos\left(\frac{2n\pi t}{T_0}\right) dt$$

$$= \frac{2}{T_0} \int_0^{T_0} \frac{1}{2}\left[\sin \frac{(1+2n)\pi t}{T_0} + \sin \frac{(1-2n)\pi t}{T_0}\right] dt$$

$\{$Using trignometric relation $\sin C \cos D = \frac{1}{2}[\sin(C+D) + \sin(C-D)]\}$

$$= \frac{1}{T_0}\left\{\frac{\cos\left[\frac{(1+2n)\pi t}{T_0}\right]}{\left[\frac{(1+2n)\pi}{T_0}\right]} - \frac{\cos\left[\frac{(1-2n)\pi t}{T_0}\right]}{\left[\frac{(1-2n)\pi}{T_0}\right]}\right\}$$

$$= -\frac{\cos(1+2n)\pi}{(1+2n)\pi} - \frac{\cos(1-2n)\pi}{(1-2n)\pi} + \frac{1}{(1+2n)\pi} + \frac{1}{(1-2n)\pi}$$

[Since $\cos(1+2n)\pi = -1$ and $\cos(1-2n)\pi = -1$]

$$A_n = \frac{1}{\pi}\left[\frac{2}{1+2n} + \frac{2}{1-2n}\right] = \frac{4}{\pi(1-4n^2)}, \quad n = 1, 2, \ldots \qquad \ldots(4)$$

$$B_n = \frac{2}{T_0} \int_0^{T_0} s(t) \sin n\omega t \, dt$$

$$= \frac{2}{T_0} \int_0^{T_0} \sin\left(\frac{\pi}{10} t\right) \sin\left(\frac{2n\pi t}{T_0}\right) dt \qquad \left(\text{Since } \omega = \frac{2\pi}{T_0}\right)$$

$\{$Using trignometric relation $\sin C \sin D = \frac{1}{2}[\cos(C-D) - \cos(C+D)]\}$

$$= \frac{2}{T_0} \int_0^{T_0} \frac{1}{2}\left[\cos \frac{(1-2n)\pi t}{T_0} - \cos \frac{(1+2n)\pi t}{T_0}\right] dt$$

$$= 0$$

or $\quad B_n = 0 \qquad \ldots(5)$

It means all the sine terms are absent. Substituting A_0, A_n and B_n in Eq. (2), we get

$$s(t) = A_0 + \sum_{n=1}^{\infty} [A_n \cos n\omega t + B_n \sin n\omega t]$$

$$= \frac{2}{\pi} + \sum_{n=1}^{\infty} \left[\frac{4}{\pi(1-4n^2)} \cos n\omega t + (0) \sin n\omega t \right]$$

$$= \frac{2}{\pi} + \sum_{n=1}^{\infty} \frac{4}{\pi(1-4n^2)} \cos n\omega t$$

$$= \frac{2}{\pi} + \frac{4}{\pi} \sum_{n=1}^{\infty} \frac{\cos n\omega t}{(1-4n^2)}, \quad \omega = \frac{2\pi}{T_0}$$

Example 3.9 Find the trignometric Fourier series representation of the continuous-time square-wave voltage signal shown in Fig. 3.9 (*a*).

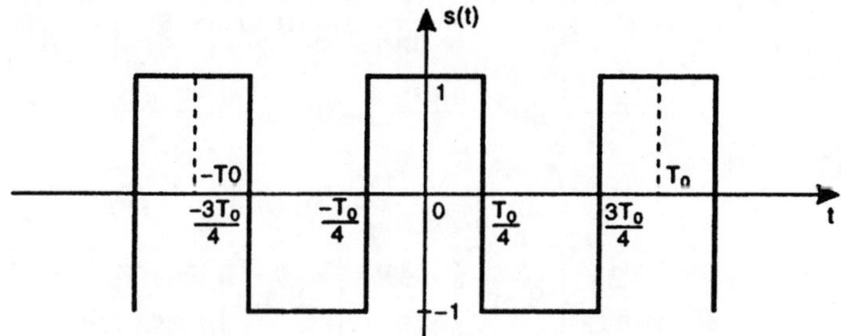

Fig. 3.9 (*a*) *Continuous-time square-wave signal.*

Solution. The analytical form of the signal $s(t)$ is given by

$$s(t) = \begin{cases} 1, & \text{for } 0 < t < \frac{T_0}{4} \\ -1, & \text{for } \frac{T_0}{4} < t < \frac{3T_0}{4} \\ 1, & \text{for } \frac{3T_0}{4} < t < T_0 \end{cases} \quad ...(1)$$

By inspection of the square wave, the average value over one period is zero, *i.e.*,

$$A_0 = 0$$

We obtained A_0 without using formula $A_0 = \frac{1}{T_0} \int_0^{T_0} s(t)\, dt$.

$$A_n = \frac{2}{T_0} \int_0^{T_0} s(t) \cos n\omega t \, dt$$

$$= \frac{2}{T_0} \left[\int_0^{T_0/4} (1) \cos n\omega t \, dt + \int_{T_0/4}^{3T_0/4} (-1) \cos n\omega t \, dt + \int_{3T_0/4}^{T_0} (1) \cos n\omega t \, dt \right]$$

$$= \frac{2}{T_0}\left\{\left[\frac{\sin n\omega t}{n\omega}\right]_0^{T_0/4} - \left[\frac{\sin n\omega t}{n\omega}\right]_{T_0/4}^{3T_0/4} + \left[\frac{\sin n\omega t}{n\omega}\right]_{3T_0/4}^{T_0}\right\}$$

$$= \frac{2}{T_0}\left\{\frac{1}{n\pi}\left[\sin\left(\frac{n\omega T_0}{4}\right) - \sin n\omega(0)\right]\right.$$

$$- \frac{1}{n\pi}\left[\sin\left(\frac{n\omega 3T_0}{4}\right) - \sin\left(\frac{n\omega T_0}{4}\right)\right]$$

$$\left. + \frac{1}{n\pi}\left[\sin(n\omega T_0) - \sin\left(\frac{n\omega 3T_0}{4}\right)\right]\right\}$$

$$= \frac{2}{n\pi T_0}\left\{\sin\left(\frac{n\omega T_0}{4}\right) - 0 - \sin\left(\frac{3n\omega T_0}{4}\right)\right.$$

$$\left. + \sin\left(\frac{n\omega T_0}{4}\right) + \sin(n\omega T_0) - \sin\left(\frac{3n\omega T_0}{4}\right)\right\}$$

$$= \frac{2}{n\pi T_0}\left\{2\sin\left(\frac{n\omega T_0}{4}\right) - 2\sin\left(\frac{3n\omega T_0}{4}\right)\right.$$

$$\left. + \sin(n\omega T_0)\right\}$$

$$= \frac{2}{n\pi T_0}\left\{2\sin\left(\frac{n\omega T_0}{4}\right) - 2\sin\left(\frac{3n\omega T_0}{4}\right) + 0\right\}$$

[Since $\sin(n\omega T_0) = 0$, for $\omega T_0 = 2\pi$]

or $$A_n = \frac{4}{n\pi T_0}\left[\sin\left(\frac{n\omega T_0}{4}\right) - \sin\left(\frac{3n\omega T_0}{4}\right)\right] \quad ..(2)$$

For all integer values of n,

$$A_n = \frac{4}{n\pi}, \text{ for } n = 1, 5, 9, 13, \ldots$$

$$= -\frac{4}{n\pi}, \text{ for } n = 3, 7, 11, 15, \ldots$$

$$= 0, \text{ for } n = 2, 4, 6, 8, \ldots$$

Again, $B_n = 0$, for all n

Substituting values of A_0, A_n and B_n in formula for trignometric Fourier series given as

$$s(t) = A_0 + \sum_{n=1}^{\infty}[A_n \cos n\omega t + B_n \sin n\omega t]$$

$$= 0 + \sum_{n=1}^{\infty}[A_n \cos n\omega t + 0 \times \sin n\omega t]$$

$$= \sum_{n=1}^{\infty} A_n \cos n\omega t$$

$$= A_1 \cos 1\omega t + A_2 \cos 2\omega t + A_3 \cos 3\omega t + A_4 \cos 4\omega t + A_5 \cos 5\omega t + A_6 \cos 6\omega t + \ldots$$

$$= \frac{4}{\pi} \cos \omega t + 0 \times \cos 2\omega t + \left(-\frac{4}{3\pi} \right) \cos 3\omega t + 0 \times \cos 4\omega t + \frac{4}{5\pi} \cos 5\omega t + 0 \times \cos 6\omega t + \ldots$$

$$= \frac{4}{\pi} \cos \omega t - \frac{4}{3\pi} \cos 3\omega t + \frac{4}{5\pi} \cos 5\omega t + \ldots$$

$$s(t) = \frac{4}{\pi} \left(\cos \omega t - \frac{1}{3} \cos 3\omega t + \frac{1}{5} \cos 5\omega t - \frac{1}{7} \cos 7\omega t + \ldots \right)$$

The amplitude and phase spectrum is shown in Fig. 3.9 (*b*) and Fig. 3.9 (*c*), respectively.

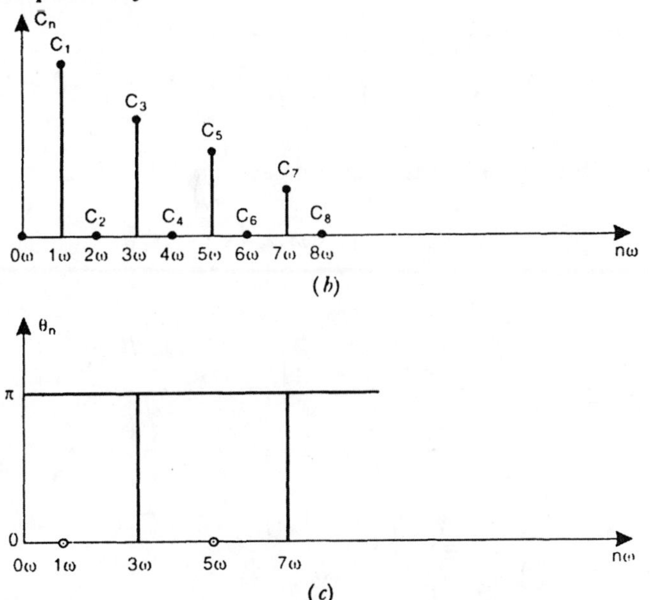

(*c*)

(*b*) *Amplitude spectrum or magnitude spectrum of a continuous-time periodic square-wave signal.* (*b*) *Phase spectrum of a continuous-time periodic square wave signal.*

Fig. 3.9

Example 3.10 Determine the trignometric Fourier series represenation of the periodic rectangular waveform shown in Fig. 3.10 (*a*).

Solution. The rectangular waveform $s(t)$ is even. Hence, only cosine terms will be present, *i.e.*, $B_n = 0$.

The analytical form of the function $s(t)$ is given by

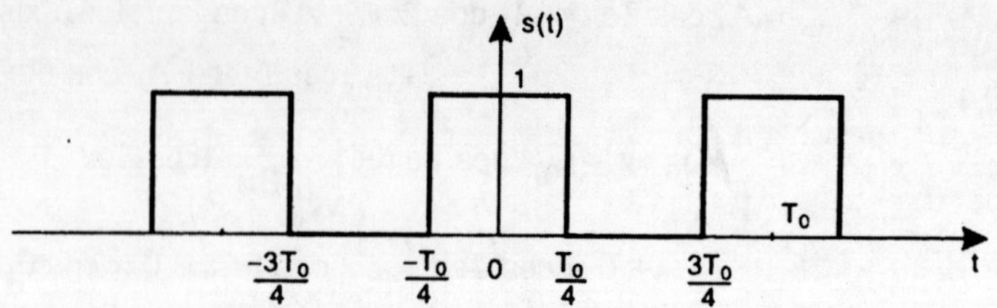

Fig. 3.10 *(a) Periodic rectangular waveform.*

$$s(t) = \begin{cases} 0, & \text{for } -\frac{T_0}{2} < t < -\frac{T_0}{4} \\ 1, & \text{for } -\frac{T_0}{4} < t < \frac{T_0}{4} \\ 0, & \text{for } \frac{T_0}{4} < t < \frac{T_0}{2} \end{cases} \quad \text{...(1)}$$

Now we will determine values of A_0 and A_n.

$$A_0 = \frac{1}{T_0} \int_{-T_0/2}^{T_0/2} s(t)\, dt \quad \text{...(2)}$$

Substituting Eq. 1 in Eq. 2, we get

$$A_0 = \frac{1}{T_0}\left[\int_{-T_0/2}^{T_0/4} s(t)\, dt + \int_{-T_0/4}^{T_0/4} s(t)\, dt + \int_{T_0/4}^{T_0/2} s(t)\, dt\right]$$

$$= \frac{1}{T_0}\left[\int_{-T_0/2}^{-T_0/4} (0)\, dt + \int_{-T_0/4}^{T_0/4} (1)\, dt + \int_{T_0/4}^{T_0/2} (0)\, dt\right]$$

$$= \frac{1}{T_0} \int_{-T_0/4}^{T_0/4} dt = \frac{1}{T_0} [\,t\,]_{-T_0/4}^{T_0/4} = \frac{1}{T_0}\left[\frac{T_0}{4} - \left(-\frac{T_0}{4}\right)\right]$$

$$= \frac{1}{T_0}\left[\frac{2T_0}{4}\right] = \frac{1}{2} = 0.5$$

or $$A_0 = 0.5 \quad \text{...(3)}$$

$$A_n = \frac{2}{T_0} \int_{-T_0/2}^{T_0/2} s(t) \cos n\omega t\, dt$$

$$= \frac{2}{T_0}\left[\int_{-T_0/2}^{T_0/4} s(t) \cos n\omega t\, dt + \int_{-T_0/4}^{T_0/4} s(t) \cos n\omega t\, dt + \int_{T_0/4}^{T_0/2} s(t) \cos n\omega t\, dt\right]$$

[Substituting Eq. 1]

$$= \frac{2}{T_0} \int_{-T_0/4}^{T_0/4} (1) \cos n\omega t \, dt = \frac{2}{T_0} \left[\frac{\sin n\omega t}{n\omega} \right]_{-T_0/4}^{T_0/4}$$

$$= \frac{2}{n\pi T_0} \left[\sin \frac{n\omega T_0}{4} - \sin n\omega \left(- \frac{T_0}{4} \right) \right]$$

$$= \frac{2}{n\pi T_0} \left[\sin \frac{n\omega T_0}{4} - \left(- \sin \frac{n\omega T_0}{4} \right) \right]$$

$$= \frac{2}{n\pi T_0} 2 \sin \frac{n\omega T_0}{4}$$

or
$$A_n = \frac{4}{n\pi T_0} \sin \frac{n\omega T_0}{4} \qquad \text{...(4)}$$

From Eq. 4, we get

$$A_n = \begin{cases} 0, & \text{for } n = 2, 4, 6, \ldots \\ \dfrac{2}{n\pi}, & \text{for } n = 1, 5, 9, \ldots \\ -\dfrac{2}{n\pi}, & \text{for } n = 3, 7, 11, \ldots \end{cases} \qquad \text{...(5)}$$

Substituting the values of A_0, A_n and B_n in the expression of trignometric Fourier series representation given as

$$s(t) = A_0 + \sum_{n=1}^{\infty} [A_n \cos n\omega t + B_n \sin n\omega t]$$

$$= 0.5 + \sum_{n=1}^{\infty} [A_n \cos n\omega t + 0 \times \sin n\omega t]$$

$$= 0.5 + \sum_{n=1}^{\infty} A_n \cos n\omega t$$

$$= 0.5 + A_1 \cos (1) \omega t + A_2 \cos 2\omega t + A_3 \cos 3\omega t + A_4 \cos 4\omega t + A_5 \cos 5\omega t + A_6 \cos 6\omega t + A_7 \cos 7\omega t + \ldots$$

$$= 0.5 + \frac{2}{\pi} \left[\cos \omega t - \frac{1}{3} \cos 3\omega t + \frac{1}{5} \cos 5\omega t + \frac{1}{7} \cos 7\omega t + \ldots \right]$$

or
$$s(t) = \frac{1}{2} + \frac{2}{\pi} \sum_{n=1}^{\infty} (-1)^{n-1} \frac{1}{(2n-1)} \cos (2n-1) \omega t \qquad \text{...(6)}$$

Spectrum of rectangular square wave is shown in Fig. 3.10 (*b*).

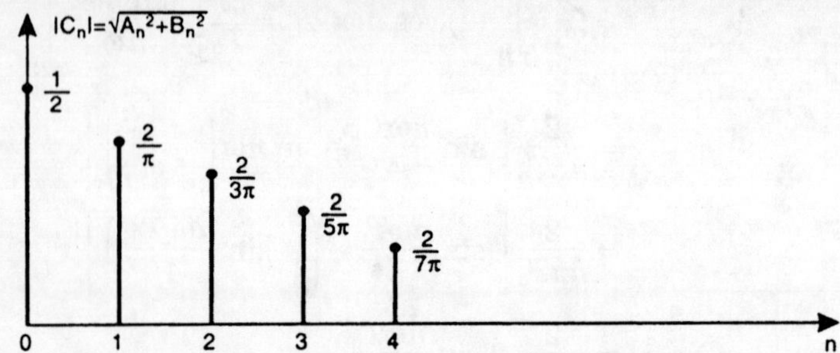

Fig. 3.10 (*b*) *Amplitude spectrum of periodic rectangular waveform.*

Example 3.11 Find the trignometric Fourier series representation of the triangular waveform shown in Fig. 3.11.

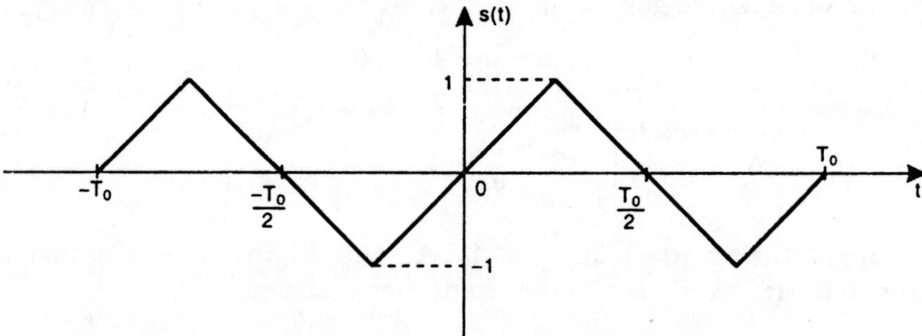

Fig. 3.11 *Continuous-time rectangular waveform.*

Solution. This is an odd function because $s(t) = -s(-t)$ and also having half wave symmetry as

$$s(t \pm 2) = -s(t).$$

The Fourier series expansion will contain only the odd harmonics of sine terms, *i.e.*, $A_n = 0$ and $A_0 = 0$. The analytical form of the function or signal is given by

$$s(t) = \begin{cases} \dfrac{4t}{T_0}, & \text{for } 0 < t < \dfrac{T_0}{4} \\ -\dfrac{4t}{T_0} + 2, & \text{for } \dfrac{T_0}{4} < t < \dfrac{3T_0}{4} \end{cases} \quad \text{...(1)}$$

The Fourier series coefficient B_n can be determined as

$$B_n = \frac{4}{T_0} \int_0^{T_0/2} s(t) \sin n\omega t \, dt$$

$$= \frac{4}{T_0} \left[\int_0^{T_0/4} s(t) \sin n\omega t \, dt + \int_{T_0/4}^{T_0/2} s(t) \sin n\omega t \, dt \right]$$

$$= \frac{4}{T_0}\left[\int_0^{T_0/4}\left(\frac{4}{T_0}t\right)\sin n\omega t\,dt + \int_{T_0/4}^{T_0/2}\left(-\frac{4}{T_0}t+2\right)\sin n\omega t\,dt\right]$$

$$= \frac{16}{T_0^2}\int_0^{T_0/4} t\sin n\omega t\,dt - \frac{16}{T_0^2}\int_{T_0/4}^{T_0/2} t\sin n\omega t\,dt + 2\int_{T_0/4}^{T_0/2} \sin n\omega t\,dt$$

Example 3.12 Find the exponential Fourier series representation for the square wave shown in Fig. 3.12.

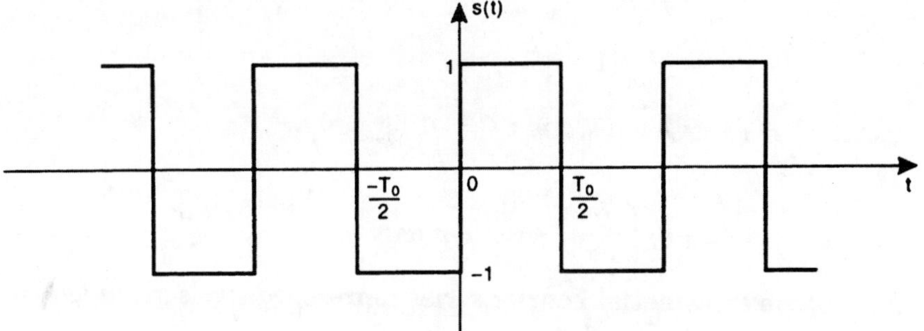

Fig. 3.12 *Continuous-time square wave.*

Solution. The analytical form of the function $s(t)$ is given by

$$s(t) = \begin{cases} 1, & \text{for } 0 < t < \frac{T_0}{2} \\ -1, & \text{for } \frac{T_0}{2} < t < T_0 \end{cases} \quad ...(1)$$

Since function $s(t)$ is an odd function, the coefficients, C_n will be purely imaginary.

We know that

$$\tilde{C}_n = \frac{1}{T_0}\int_{-T_0/2}^{T_0/2} s(t)\, e^{-jn\omega t}\,dt, \text{ for } -\infty < n < \infty \quad ...(2)$$

Substituting Eq. 1 in Eq. 2, we get

$$\tilde{C}_n = \frac{1}{T_0}\int_{-T_0/2}^{0} (-1)\, e^{-jn\omega t}\,dt + \frac{1}{T_0}\int_0^{T_0/2} (1)\, e^{-jn\omega t}\,dt$$

$$= -\frac{1}{T_0}\left[\frac{e^{-jn\omega t}}{-jn\omega}\right]_{-T_0/2}^{0} + \frac{1}{T_0}\left[\frac{e^{-jn\omega t}}{-jn\omega}\right]_0^{T_0/2}$$

$$= \frac{1}{-jn\omega T_0}\left\{-e^{-jn\omega(0)} + e^{-jn\omega(-T_0/2)} + e^{-jn\omega(T_0/2)} - e^{-jn\omega(0)}\right\}$$

$$= \frac{1}{-jn\omega T_0}\left\{-1 + e^{j(n\omega T_0)/2} + e^{-j(n\omega T_0)/2} - 1\right\}$$

$$\left(\text{Since } \omega = \frac{2\pi}{T_0} \text{ or } \omega T_0 = 2\pi\right)$$

$$= \frac{1}{-jn2\pi}\left[-2 + e^{j(n2\pi)/2} + e^{-j(n2\pi)/2}\right]$$

$$= \frac{1}{j2n\pi}\left[2 - e^{jn\pi} - e^{-jn\pi}\right]$$

or

$$\tilde{C}_n = \frac{1}{jn\pi}\left[1 - e^{jn\pi}\right]$$

Here, $e^{jn\pi} = +1$ for even n and $e^{jn\pi} = -1$ for odd n.

Therefore,

$$\tilde{C}_n = -j\left(\frac{2}{n\pi}\right) \text{ for odd } n \text{ only} \qquad ...(3)$$

Hence, the exponential Fourier series representation is given as

$$s(t) = \sum_{n=-\infty}^{\infty} \tilde{C}_n\, e^{jn\omega t} \qquad ...(4)$$

Substituting Eq. (3) in Eq. (4), we get

$$s(t) = \sum_{\substack{n=-\infty \\ n = \text{odd only}}}^{\infty} \left(\frac{-j2}{n\pi}\right) e^{jn\omega t}$$

$$= \ldots + \left(\frac{-j2}{-3\pi}\right) e^{j(-3)\omega t} + \left(\frac{-j2}{-1\pi}\right) e^{j(-1)\omega t} + \left(\frac{-j2}{1\pi}\right) e^{j(1)\omega t} + \left(\frac{-j2}{3\pi}\right) e^{j(3)\omega t} + \ldots$$

$$= \ldots + \frac{j2}{3\pi} e^{-j3\omega t} + \frac{j2}{\pi} e^{-j\omega t} - \frac{j2}{\pi} e^{j\omega t} - \frac{j2}{3\pi} e^{j3\omega t} + \ldots$$

Example 3.13 Find out the exponential Fourier series representation for the periodic waveform shown in Fig. 3.13 (*a*).

Solution. The analytical form of the function $s(t)$ is given by

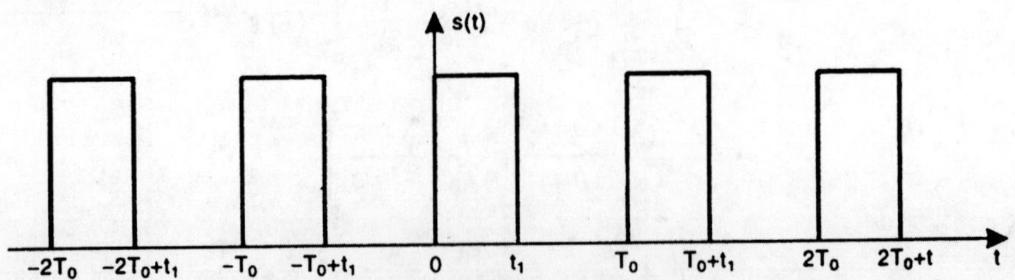

Fig. 3.13 (*a*) *Periodic waveform.*

$$s(t) = \begin{cases} 1\,, & \text{for } 0 < t < t_1 \\ 0\,, & \text{for } t_1 < t < \dfrac{T_0}{2} \end{cases} \qquad \text{...(1)}$$

C_n can be determined as

$$C_n = \frac{1}{T_0} \int_0^{T_0} s(t)\, e^{-jn\omega t}\, dt \qquad \left(\text{Since } \omega = \frac{2\pi}{T_0}\right)$$

$$= \frac{1}{T_0} \int_0^{T_0} s(t)\, e^{-jn(2\pi/T_0)t}\, dt$$

$$= \frac{1}{T_0}\left[\int_0^{t_1} s(t)\, e^{-(j2n\pi t)/T_0}\, dt + \int_{t_1}^{T_0} s(t)\, e^{-(j2n\pi t)/T_0}\, dt\right]$$

$$= \frac{1}{T_0}\left[\int_0^{t_1} (1)\, e^{-(j2n\pi t)/T_0}\, dt + \int_{t_1}^{T_0} (0)\, e^{-(j2n\pi t)/T_0}\, dt\right]$$

$$= \frac{1}{T_0} \int_0^{t_1} e^{-(j2n\pi t)/T_0}\, dt = \frac{1}{T_0}\left[\frac{e^{-(j2n\pi t)/T_0}}{-\dfrac{j2\pi n}{T_0}}\right]_0^{t_1}$$

$$= \frac{1}{T_0} \times \frac{T_0}{-j2\pi n}\left[e^{-(j2\pi n t_1)/T_0} - e^{-(j2\pi n \times 0)/T_0}\right] = \frac{1}{-j2\pi n}\left[e^{-(j2\pi n t_1)/T_0} - e^{0}\right]$$

$$= \frac{1}{-j2\pi n}\left[e^{-(j2\pi n t_1)/T_0} - 1\right] = \frac{1}{j2\pi n}\left[1 - e^{-j2\pi n t/T_0}\right]$$

$$= \frac{1}{jn2\pi} e^{-(jn\pi t_1)/T_0}\left[e^{(jn\pi t_1)/T_0} - e^{-(jn\pi t_1)/T_0}\right]$$

$$= \frac{1}{n\pi} e^{-(jn\pi t_1)/T_0}\left[\frac{e^{(jn\pi t_1)/T_0} - e^{-(jn\pi t_1)/T_0}}{2j}\right]$$

or

$$C_n = \frac{1}{n\pi} e^{-(jn\pi t_1)/T_0}\left[\sin \frac{n\pi t_1}{T_0}\right] \qquad \text{...(2)}$$

Exponential Fourier series representation is expressed as

$$s(t) = \sum_{n=-\infty}^{\infty} C_n\, e^{jn\omega t} \qquad \text{...(3)}$$

Substituting Eq. 2 in Eq. 3, we get

$$s(t) = \sum_{n=-\infty}^{\infty}\left[\frac{1}{n\pi} e^{-(jn\pi t_1)/T_0} \sin \frac{n\pi t_1}{T_0}\, e^{jn\omega t}\right]$$

$$\left(\text{Putting } \omega = \frac{2\pi}{T_0}\right)$$

$$= \sum_{n=-\infty}^{\infty}\left[\frac{1}{n\pi} e^{-(jn\pi t_1)/T_0} \sin \frac{n\pi t_1}{T_0}\, e^{(jn2\pi t)/T_0}\right]$$

$$= \sum_{n=-\infty}^{\infty} \left[\frac{1}{n\pi} e^{jn\pi/T_0(2t - t_1)} \sin \frac{n\pi t_1}{T_0} \right]$$

or $$s(t) = \sum_{n=-\infty}^{\infty} \left[\frac{1}{n\pi} e^{jn2\pi/T_0[t - (t_1/2)]} \sin \frac{n\pi t_1}{T_0} \right] \quad \text{...(4)}$$

The magnitude spectrum is a plot of magnitude

$$|C_n| = 2\,|\tilde{C}_n|$$

$$= \frac{2}{n\pi} \left| \sin\left(\frac{n\pi t_1}{T_0}\right) \right|, \text{ for } n = 1, 2, 3 \ldots$$

versus $n\omega$. It is shown in Fig. 3.13 (*b*).

For $n = 0$, we have to apply L'Hospital's rule (otherwise it will be indeterminate).

$$C_0 = \lim_{n \to 0} \frac{1}{n\pi} \sin\left(\frac{n\pi t_1}{T_0}\right)$$

$$= \lim_{n \to 0} \frac{\dfrac{d}{dn}\left[\sin\left(\dfrac{n\pi t_1}{T_0}\right)\right]}{\dfrac{d}{dn}[n\pi]}$$

$$= \lim_{n \to 0} \left[\frac{\cos\left(\dfrac{n\pi t_1}{T_0}\right) \times \dfrac{\pi t_1}{T_0}}{\pi} \right] = \frac{t_1}{T_0}$$

or $$C_0 = \frac{t_1}{T_0}$$

The amplitude spectrum is plotted for $\frac{t_1}{T_0} = \frac{1}{6}$. The values of coefficient C_n with n is given in Table 3.2.

Table 3.2

n	0	1	2	3	4	5	6	7	8	9
$\tilde{C}_n$	$\frac{1}{6}$	$\frac{1}{\pi}$	$\frac{\sqrt{3}}{2\pi}$	$\frac{2}{3\pi}$	$\frac{\sqrt{3}}{-4\pi}$	$-\frac{1}{5\pi}$	0	$-\frac{1}{7\pi}$	$\frac{\sqrt{3}}{-8\pi}$	$\frac{2}{-9\pi}$

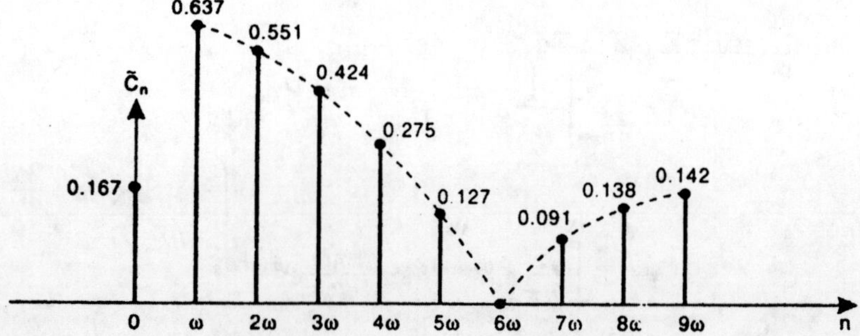

Fig. 3.13 (*b*) *Amplitude spectrum of exponential Fourier series of a function*

Example 3.14 Find the trignometric and exponential Fourier series for the continuous-time waveform shown in Fig. 3.14.

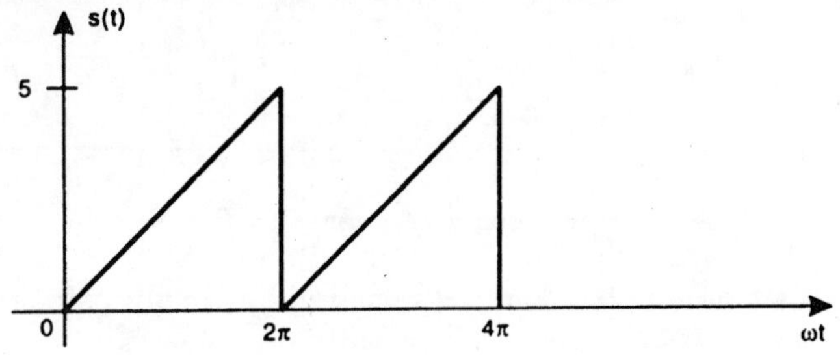

Fig. 3.14

Solution. As the waveform is periodic with period 2π in ωt and continuous for $0 < \omega t < 2\pi$, with discontinuities at $\omega t = n\,(2\pi)$, where $n = 0, 1, 2, \ldots$, the Derichlet's conditions are satisfied.

First of all, we will find signal $s(t)$ for the given waveform of region $0 < \omega t < 2\pi$.

The equation of straight line is given by

$$\frac{y - y_1}{x - x_1} = \frac{y_1 - y_2}{x_1 - x_2} \qquad \ldots(1)$$

Substituting $y = s(t)$, $x = \omega t$ and $(x_1, y_1) = (0, 0)$ and $(x_2, y_2) = (2\pi, 5)$ in Eq. 1, we obtain

$$\frac{s(t) - 0}{\omega t - 0} = \frac{(0 - 5)}{0 - 2\pi}$$

Therefore, $$s(t) = \left(\frac{5}{2\pi}\right)\omega t$$

Determination of Fourier Series Coefficients for Trignometric Fourier Series Representation :

D.C.part, $$A_0 = \frac{2}{T_0}\int_{-T_0/2}^{T_0/2} s(t)\,dt = \frac{2}{2\pi}\int_0^{2\pi} \frac{5}{2\pi}(\omega t)\,d(\omega t)$$

$$= \frac{1}{\pi} \times \frac{5}{2\pi}\int_0^{2\pi} (\omega t)\,d(\omega t) = \frac{5}{2\pi^2}\left[\frac{(\omega t)^2}{2}\right]_0^{2\pi}$$

$$= \frac{5}{(2\pi)^2}[(2\pi)^2 - (0)^2]$$

or $$A_0 = 5 \qquad \ldots(2)$$

$$A_n = \frac{2}{T_0}\int_{-T_0/2}^{T_0/2} s(t)\cos n\omega t\,dt$$

$$= \frac{2}{2\pi} \int_0^{2\pi} \left(\frac{5}{2\pi} \right) \omega t \cos n\omega t \, d(\omega t)$$

$$= \frac{5}{2\pi^2} \left[\frac{\omega t}{n} \sin n\omega t + \frac{1}{n^2} \cos n\omega t \right]_0^{2\pi}$$

or $$A_n = \frac{5}{2\pi^2 n^2} [\cos n2\pi - \cos (0) = 0 \quad ...(3)$$

Hence, the trignometric Fourier series will have no cosine terms.

$$B_n = \frac{2}{T_0} \int_{-T_0/2}^{T_0/2} s(t) \sin n\omega t \, dt$$

$$= \frac{2}{2\pi} \int_0^{2\pi} \left(\frac{5}{2\pi} \right) \omega t \sin n\omega t \, d(\omega t)$$

$$= \frac{5}{2\pi^2} \left[-\frac{\omega_0 t}{n} \cos n\omega t + \frac{1}{n^2} \sin n\omega t \right]_0^{2\pi}$$

or $$B_n = -\frac{5}{n\pi} \quad ...(4)$$

Substituting the values of A_0, A_n and B_n from Eqs. 2, 3 and 4, respectively in the epxression of trignometric Fourier series representation.

$$s(t) = A_0 + \sum_{n=1}^{\infty} [A_n \cos n\omega t + B_n \sin n\omega t]$$

$$s(t) = 5 + \sum_{n=1}^{\infty} \left[(0) \cos n\omega t + \left(\frac{-5}{n\pi} \right) \sin n\omega t \right]$$

$$= 5 - \frac{5}{\pi} \sum_{n=1}^{\infty} \frac{\sin n\omega t}{n}$$

$$= 5 - \frac{5}{\pi} \sin \omega t - \frac{5}{2\pi} \sin 2\omega t - \frac{5}{3\pi} \sin 3\omega t$$

Determination of Exponential Fourier Series

Here $$C_0 = \left| \frac{A_0}{2} \right| = \frac{5}{2}$$

Exponential Fourier series representation of signal $s(t)$ is expressed as

$$s(t) = \sum_{n=-\infty}^{\infty} \tilde{C}_n e^{jn\omega t} \quad ...(5)$$

Now we determine $\tilde{C}_n$ as

$$\tilde{C}_n = \frac{1}{T_0}\int_0^{T_0} s(t)\, e^{-jn\omega t}\, dt$$

$$= \frac{1}{2\pi}\int_0^{2\pi}\left[\frac{5}{2\pi}\right]\omega t\, e^{-jn\omega t}\, dt$$

$$= \frac{5}{(2\pi)^2}\left[\frac{e^{-jn\omega t}}{(-jn)^2}(-jn\omega t - 1)\right]_0^{2\pi}$$

or $$\tilde{C}_n = \frac{j5}{2\pi n} \qquad ...(6)$$

Substituting Eq. 6 in Eq. 5, we obtain

$$s(t) = \sum_{n=-\infty}^{\infty} \frac{j5}{2\pi n}\, e^{jn\omega t} = \frac{j5}{2\pi}\sum_{n=-\infty}^{\infty}\frac{e^{jn\omega t}}{n}$$

or $$s(t) = \ldots - \frac{j5}{4\pi}e^{-j2\omega t} - \frac{j5}{2\pi}e^{-j\omega t} + \frac{5}{2\pi}e^{j\omega t} + \frac{j5}{4\pi}e^{-j2\omega t} + \ldots$$

The trignometric Fourier series coefficients A_n and B_n can be evaluated as

$$A_n = 2\ \mathrm{Re}\ \{\tilde{C}_n\} = 2\ \{\tilde{C}_n\} = 0$$

and $$B_n = -2\ \mathrm{Im}\ \{\tilde{C}_n\} = \frac{-5}{n\pi}$$

Hence, $$s(t) = \frac{5}{2} - \frac{5}{\pi}\sin \omega t - \frac{5}{2\pi}\sin 2\omega t - \frac{5}{3\pi}\sin 3\omega t - \ldots$$

This result is the same as that of the trignometric Fourier series.

REVIEW QUESTIONS

1. **What is Fourier series ?**
2. **Why do we need Fourier seires representation ?**
3. **Give the two forms of Fourier series.**
4. **What is periodic signal ? Give two examples of periodic signals.**
5. **What do you mean by d.c term in Fourier series representation ?**
6. **What are a.c. terms of all harmonics in Fourier series representation ?**
7. **What is Fourier analysis ?**
8. **Define the following :**
   (*a*) **Eigen value and Eigen function of continuous-time LTI system**
   (*b*) **Eigen value and Eigen function of discrete-time LTI system**
9. **Define the following :**
   (*a*) **Transfer function of continuous-time LTI system $H(s)$**
   (*b*) **Transfer function of discrete-time LTI system $H(z)$**

10. What are s and z in transfer function of continuous-time and discrete-time systems, respectively ?

11. What is usefulness of decomposing more general signals in terms of Eigen functions in the analysis of LTI systems ?

12. Define fundamental frequency and fundamental period of a periodic signal.

13. Give the general form of linear combination of harmonically related complex exponentials.

14. Discuss the determination of the Fourier series representation of continuous-time periodic signals.

15. What are Fourier series cofficients ?

16. What are Derichlet conditions ?

17. Discuss various properties of continuous-time Fourier series (CTFS) in brief.

18. Write Parseval's relation for CTFS.

19. Tabulate various properties of CTFS.

NUMERICAL PROBLEMS

1. A continuous-time periodic signal $s(t)$ is real valued and has a fundamental period $T_0 = 8$. The non-zero Fourier series coefficients for $s(t)$ are

$$A_1 = A_{-1} = 2\ ,\ A_3 = A^*_{-3} = 4j.$$

Express $s(t)$ in the form of

$$s(t) = \sum_{k=0}^{\infty} A_k \cos(\omega_k t + \phi_k)$$

Ans. $s(t) = 4\cos\left(\frac{\pi}{4}t\right) + 8\cos\left(\frac{3\pi}{4}t + \frac{\pi}{2}\right)$

2. Determinme the fundamental frequency ω_0 and continuous-time Fourier series coefficients A_k for the continuous-time signal

$$s(t) = 2 + \cos\left(\frac{2\pi}{3}t\right) + 4\sin\left(\frac{5\pi}{3}t\right).$$

Ans. $\omega_0 = \frac{\pi}{3}$ and $A_0 = 2\ , A_2 = A_{-2} = \frac{1}{2}$ and $A_5 = A^*_{-5} = -2j$

3. Determine the Fourier series coefficients A_k by using continuous-time Fourier series (CTFS) analysis equation

$$A_k = \frac{1}{T_0}\int_{T_0} s(t)\, e^{-jk\omega_0 t}\, dt$$

for the continuous-time periodic signal

$$s(t) = \begin{cases} 1.5, & 0 \le t < 1 \\ -1.5, & 1 \le t < 2 \end{cases}$$

Given fundamental frequency $\omega_0 = \pi$.

Ans. $A_0 = 0$, for $k = 0$; $A_k = \frac{3}{k\pi} e^{-jk(\pi/2)} \sin\left(\frac{k\pi}{2}\right)$, for $k \neq 0$

4. Specify two different continuous-time signals which satisfy following conditions. We have given following information about a continuous-time signal $s(t)$

(a) $s(t)$ is real and odd

(b) $s(t)$ is periodic with period $T_0 = 2$ and has Fourier series coefficient A_k

(c) $A_k = 0$ for $|k| > 1$

(d) $\frac{1}{2} \int_0^2 |s(t)|^2 dt = 1$ **Ans.** $s_1(t) = -\sqrt{2} \sin(\pi t)$; $s_2(t) = \sqrt{2} \sin(\pi t)$

5. We have given following three continuous-time signals with a fundamental frequency $\omega_0 = 4\pi$

$$s_1(t) = \cos(4\pi t)$$

$$s_2(t) = \sin(4\pi t)$$

$$s_3(t) = s_1(t)\, s_2(t) = \cos(4\pi t) \sin(4\pi t)$$

(a) Determine the Fourier series coefficients of $s_1(t)$

(b) Determine the Fourier series coefficients of $s_2(t)$

(c) Determine the Fourier series coefficients of $s_3(t)$ using multiplication property of CTFS and results of part (a) and (b)

(d) Determine the Fourier series coefficients of $s_3(t)$ through direct expansion of $s_3(t)$ in trignometric form

(e) Compare results of part (c) and (d)

Ans. (a) $A_1 = A_{-1} = \frac{1}{2}$ (b) $B_1 = B^*_{-1} = \frac{1}{2j}$ (c) $C_2 = C^*_{-2} = \frac{1}{4j}$ (d) $C_2 = C^*_{-2} = \frac{1}{4j}$

(e) Results obtained from part (c) and (d) are same

6. Obtain the trignometric Fourier series representation of the continuous-time periodic square wave signal symmetrical with respect to the vertical axis at time $t = 0$. This waveform is shown in Fig. 3.15.

Ans. $A_0 = 0$; $A_n = \frac{4A}{n\pi} \sin\left(\frac{n\pi}{2}\right)$

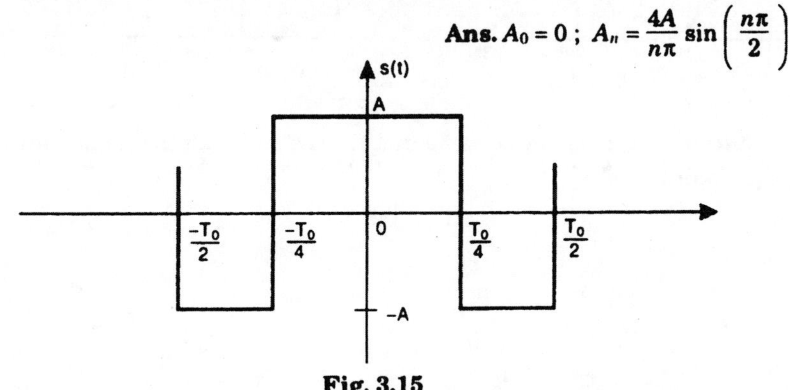

Fig. 3.15

7. Find the trignometric Fourier series for the half-wave rectified sine wave shown in Fig. 3.16.

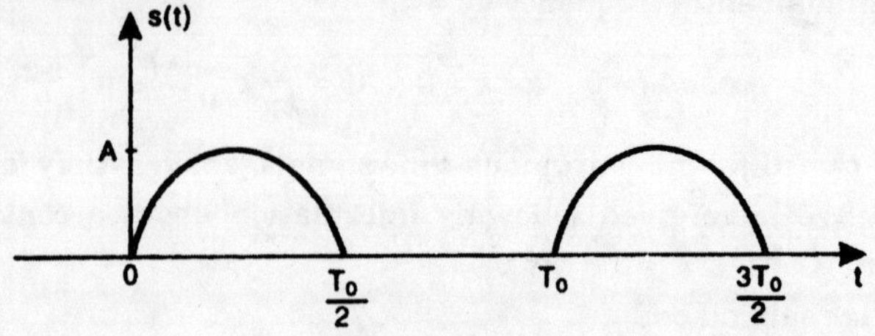

Fig. 3.16

Ans. $A_0 = \frac{2A}{\pi}$; $A_n = \frac{2A}{\pi(1-n^2)}$, for n even

$= 0$, for n odd

$B_n = 0$

8. Obtain the trignometric Fourier series for the full wave rectified sine wave as shown in Fig. 3.17.

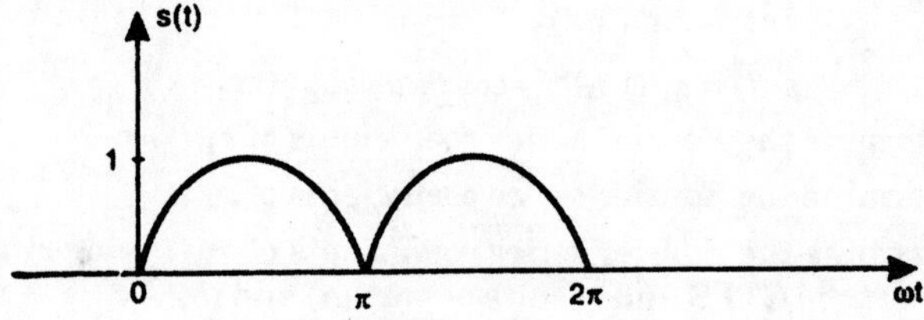

Fig. 3.17

$$\textbf{Ans. } s(t) = \frac{2}{\pi} - \frac{4}{\pi} \sum_{n=1}^{\infty} \frac{\cos 2nt}{4n^2 - 1}$$

9. Find the exponential Fourier series for the waveforms shown in Fig. 3.18.

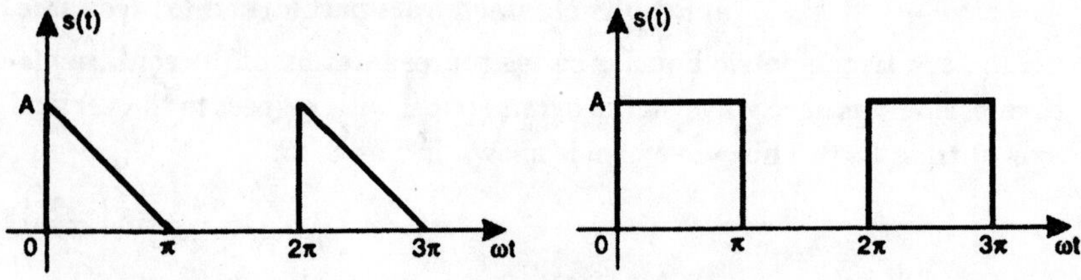

Fig. 3.18

10. Determine the complex exponential Fourier series for the continuous-time periodic signal

$$s(t) = A\,|\sin \omega t\,|$$

4

Discrete-Time Fourier Series (DTFS)

4.1 INTRODUCTION

In this section, we will study about the Fourier series representation of discrete-time periodic signals or sequences. Discrete-time Fourier series (DTFS) has some important similarities as well as differences with continuous-time Fourier series (CTFS). In particular, the Fourier series representation of a discrete-time periodic signal is a finite series, unlike the infinite series representation required for continuous-time periodic signals. Discrete-time Fourier series has no mathematical issues of convergence.

4.2 FOURIER SERIES REPRESENTATION OF DISCRETE-TIME PERIODIC SIGNALS

Fourier series is a linear combination of harmonically related complex exponential functions. Fourier series is used to represent both continuous-time and discrete-time periodic signals into linear combination of harmonically related complex exponentials. A discrete-time signal or sequence is periodic with period N_0 if

$$s(n) = s(n + N_0) \quad \text{...(4.1)}$$

The fundamental period is the smallest positive integer N_0 for which Eq. 4.1 is satisfied and $\omega_0 = 2\pi/N_0$ is the fundamental frequency. For example, the complex exponential $e^{j\omega_0 n}$ is periodic with period N_0. A set of all discrete-time complex exponential signals which are periodic with period N_0 are given by

$$F_k(n) = e^{jk\omega_0 n} \quad \text{for } k = 0, \pm 1, \pm 2, \ldots \quad \text{...(4.2)}$$

where $\omega_0 = \dfrac{2\pi}{N_0}$

All these signals have fundamental frequencies that are multiples of $2\pi/N_0$ and thus these are harmonically related.

There are only N_0 distinct signals in the set of Eq. 4.2. Thus, discrete-time complex exponentials which differ in frequency by a multiple of 2π are identical.

Specifically, $F_0(n) = F_{N_0}(n)$

$$F_1(n) = F_{N_0+1}(n)$$

$$F_2(n) = F_{N_0+2}(n)$$

$$\vdots$$

and in general $F_k(n) = F_{k+rN_0}(n)$ ...(4.3)

From Eq. 4.3, if k is changed by an integer multiple of N_0, an identical sequence is generated.

Now we can represent more general periodic discrete-time signals or sequences in terms of linear combinations of the sequences $F_k(n)$ given by Eq. 4.2. This linear combination of $F_k(n)$ is written as

$$s(n) = \sum_k A_k F_k(n) = \sum_k A_k e^{jk\omega_0 n} . \quad ..(4.4)$$

Since the sequence $F_k(n)$ is distinct only over a range of N_0 successive values of k. Therefore, summation in Eq. 4.4 includes terms over this range. Summation of Eq. 4.4 varies over a range of N_0 successive integers and expressing this equation in the limit $k = \langle N \rangle$ as

$$s(n) = \sum_{k=\langle N \rangle} A_k F_k(n) = \sum_{k=\langle N \rangle} A_k e^{jk\omega_0 n} \quad ...(4.5)$$

For example, k can be chosen in any of the combinations such as

$$k = 0, 1, \ldots, N_0 - 1, \text{ or}$$

$$k = 1, 2, \ldots, N_0, \text{ or}$$

$$k = 2, 3, \ldots, N_0 + 1, \text{ or}$$

and so on.

Eq. 4.5 is referred to as the discrete-time Fourier series and A_k's are Fourier series coefficients.

4.3 FOURIER SERIES REPRESENTATION OF A DISCRETE-TIME PERIODIC SIGNAL

Let us consider a discrete-time signal or sequence $s(n)$ that is periodic with fundamental period N_0. In this section we want to determine whether a representation of $s(n)$ in the form of equation $s(n) = \sum_{k=\langle N_0 \rangle} A_k e^{jk\omega_0 n}$ exists. If this representation exists then we determine the value of Fourier series coefficients A_k for the discrete-time periodic signal $s(n)$.

$$s(n) = \sum_{k=\langle N_0 \rangle} A_k e^{jk\omega_0 n}$$

$$\left.\begin{aligned}
n=0\,,\ s(0) &= \sum_{k=\langle N_0\rangle} A_k\, e^{jk\omega_0.0} = \sum_{k=\langle N_0\rangle} A_k \\
n=1\,,\ s(1) &= \sum_{k=\langle N_0\rangle} A_k\, e^{jk\omega_0.1} = \sum_{k=\langle N_0\rangle} A_k\, e^{jk\omega_0} \\
n=2\,,\ s(2) &= \sum_{k=\langle N_0\rangle} A_k\, e^{jk\omega_0.2} = \sum_{k=\langle N_0\rangle} A_k\, e^{jk2\omega_0} \\
n=N_0-1\,,\ s(N_0-1) &= \sum_{k=\langle N_0\rangle} A_k\, e^{jk\omega_0(N_0-1)} = \sum_{k=\langle N_0\rangle} A_k\, e^{jk(N_0-1)\,\omega}
\end{aligned}\right\} \quad \text{...(4.6)}$$

Thus, Eq. 4.6 reresents a set of N_0 linear equations for N_0 unknown Fourier series coefficients A_k as k ranges over a set of N_0 successive integers.

This set of equations is linearly independent and consequently can be solved to obtain the Fourier series coefficients A_k in terms of given values of the sequence $s(n)$. We have a parallel method as continuous-time case to determine Fourier series coefficients A_k of a discrete-time periodic sequence $s(n)$. It is possible to obtain a closed-form expression for A_k in terms of the values of the sequence $s(n)$.

The basis for this result is the fact that

$$\sum_{n=\langle N_0\rangle} s(n)\, e^{jk\omega_0 n} = \sum_{n=\langle N_0\rangle} s(n)\, e^{jk\,(2\pi/N_0)n}$$

$$= \begin{cases} N_0\,, & k=0\,,\pm N_0\,,\pm 2N_0\,,\ldots \\ 0\,, & \text{otherwise} \end{cases} \quad \text{...(4.7)}$$

This equation states that the sum over one period of the values of a periodic complex exponential function is zero, unless the complex exponential function is of a constant value.

Now let us consider a Fourier series representation as

$$s(n) = \sum_{k=\langle N_0\rangle} A_k\, e^{jk(2\pi/N_0)n}$$

Multiplying both sides of above equation by $e^{-jl(2\pi/N_0)\,n}$ and adding over N_0 terms, we get

$$\sum_{n=\langle N_0\rangle} s(n)\, e^{-jl(2\pi/N_0)n} = \sum_{n=\langle N_0\rangle}\ \sum_{k=\langle N_0\rangle} A_k\, e^{jk(2\pi/N_0)n}\ e^{-jl(2\pi/N_0)n}$$

$$= \sum_{n=\langle N_0\rangle}\ \sum_{k=\langle N_0\rangle} A_k\, e^{j(k-l)\,(2\pi/N_0)n} \quad \text{...(4.8)}$$

By interchanging the order of summation on the right hand side of Eq. 4.8, we get

$$\sum_{n=\langle N_0\rangle} s(n)\, e^{-jl(2\pi/N_0)n} = \sum_{n=\langle N_0\rangle} A_k\ \sum_{k=\langle N_0\rangle} e^{j(k-l)\,(2\pi/N_0)n} \quad \text{...(4.9)}$$

From the identity of Eq. 4.7, we have

$$\sum_{n=\langle N_0\rangle} e^{jk(2\pi/N_0)n} = \begin{cases} N_0\,, & \text{if } k=0\,,\pm N_0\,,\pm 2N_0\,,\ldots \\ 0\,, & \text{otherwise} \end{cases}$$

Similarly $$\sum_{n=\langle N_0\rangle} e^{j(k-l)\,(2\pi/N_0)n} = \begin{cases} N_0\,, & \text{if } k=l \\ 0\,, & \text{if } k\neq l \end{cases} \quad \text{...(4.10)}$$

Substituting Eq. 4.10 in Eq. 4.9, we get

$$\sum_{n=\langle N_0\rangle} s(n)\, e^{-jl(2\pi/N_0)n} = \sum_{k=\langle N_0\rangle} A_k . N_0$$

or
$$\sum_{n=\langle N_0\rangle} s(n)\, e^{-jl(2\pi/N_0)n} = A_l . N_0$$

or
$$A_l = \frac{1}{N_0} \sum_{n=\langle N_0\rangle} s(n)\, e^{-jl(2\pi/N_0)n} \qquad ...(4.11)$$

This Eq. 4.11 gives a close-form expression for obtaining the Fourier series coefficients A_k. Now discrete-time Fourier series (DTFS) pair is given as

$$s(n) = \sum_{k=\langle N_0\rangle} A_k\, e^{jk\omega_0 n} \qquad ...(4.12)$$

where
$$A_k = \frac{1}{N_0} \sum_{n=\langle N_0\rangle} s(n)\, e^{-jk\omega_0 n} \qquad ...(4.13)$$

where $\omega_0 = \dfrac{2\pi}{N_0}$

Equation 4.12 is called the *synthesis equation* and Eq. 4.13 is called the *analysis equation*. As in continuous-time periodic signals, the discrete-time Fourier series coefficients A_k are often called the spectral coefficients of signal $s(n)$. These Fourier series coefficients A_k specify a decomposition of signal $s(n)$ into a sum of N_0 harmonically related complex exponential functions.

From equation $s(n) = \sum_{k=\langle N_0\rangle} A_k F_k(n)$, for k ranges from 0 to $N - 1$, we get

$$s(n) = A_0 F_0(n) + A_1 F_1(n) + A_2 F_2(n) + \ldots + A_{N_0-1} F_{N_0-1}(n) \qquad ...(4.14)$$

Similarly, if k ranges from 1 to N_0, then

$$s(n) = A_1 F_1(n) + A_2 F_2(n) + A_3 F_3(n) + \ldots + A_{N_0} F_{N_0}(n) \qquad ...(4.15)$$

But we know that $F_0(n) = F_{N_0}(n)$ and comparing Eqs. 4.14 and 4.15, we get, $A_0 = A_{N_0}$.

Now we can conclude for k ranges over any set of N_0 consecutive integers as

$$A_k = A_k + N_0 \qquad ...(4.16)$$

If we consider more than N_0 sequential values of k then the values A_k repeats periodically with period N_0. In particular, since there are only N_0 distinct complex exponential functions that are periodic with period N_0, the DTFS representation is a finite series with N_0 terms.

Now we can understand discrete-time Fourier series (DTFS) representation of discrete-time periodic sequence $s(n)$ by the following numerical examples.

Example 4.1 Determine the Fourier series coefficients A_k for a discrete-time periodic signal given as $s(n) = \cos \omega_0 n$ where $\omega_0 = 2\pi/N_0$.

Solution. Given a signal

$$s(n) = \cos \omega_0 n \qquad \text{...(1)}$$

is a periodic signal with fundamental period T_0 and $\omega_0 = 2\pi/N_0$ is the fundamental frequency.

Determination of Fourier series for signal $s(n) = \cos \omega_0 n$

Expanding the signal into two complex exponential functions, we get

$$s(n) = \cos \omega_0 n = \frac{e^{j\omega_0 n} + e^{-j\omega_0 n}}{2} = \frac{1}{2} e^{j\omega_0 n} + \frac{1}{2} e^{-j\omega_0 n} \qquad \text{...(2)}$$

But we know that

$$s(n) = \sum_{k = \langle N_0 \rangle} A_k e^{jk\omega_0 n} = \dots + A_{-1} e^{j(-1)\omega_0 n} + A_0 e^{j(0)\omega_0 n}$$

$$+ A_1 e^{j(1)\omega_0 n} + \dots \qquad \text{...(3)}$$

Comparing Eqs. 2 and 3, we can see by inspection that Fourier series coefficients A_k's are

$$A_{-1} = \frac{1}{2}, \; A_0 = 0, \; A_1 = \frac{1}{2} \qquad \text{...(4)}$$

and remaining coefficients over the interval of summation are zero. These coefficients A_k repeat with period N_0 and thus

$$A_{N_0+1} = \frac{1}{2} \text{ and } A_{N_0-1} = \frac{1}{2}$$

Fourier series coefficients A_k with $N_0 = 5$ are shown in Fig. 4.1.

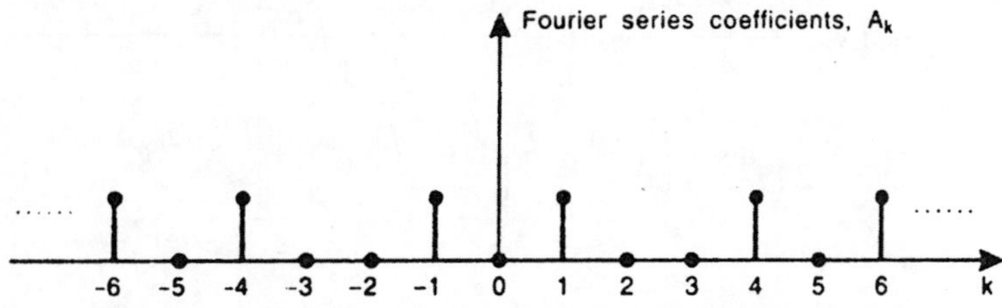

Fig. 4.1 *Fourier series coefficients for* $s(n) = \sin(2\pi/5)\, n$.

Now considering the case when $\frac{\omega_0}{2\pi} = \frac{M}{N_0}$, where $\frac{M}{N_0}$ is the ratio of integers which have no common factor.

Now again expanding $s(n)$ as a sum of two complex exponentials as

$$s(n) = \cos \omega_0 n = \cos\left(\frac{2\pi M}{N_0}\right) n$$

$$= \frac{e^{jM(2\pi/N_0)n} + e^{-jM(2\pi/N_0)n}}{2} = \frac{1}{2} e^{jM(2\pi/N_0)n} + \frac{1}{2} e^{-jM(2\pi/N_0)n} \qquad \text{...(5)}$$

Comparing Eqs. 3 and 5, we see by inspection that Fourier series coefficients A_k's are

$$A_M = \frac{1}{2} \text{ and } A_{-M} = \frac{1}{2}$$

and the remaining coefficients over one period of length N_0 are zero. Fourier series coefficients A_k with $M = 3$ and $N_0 = 5$ are shown in Fig. 4.2.

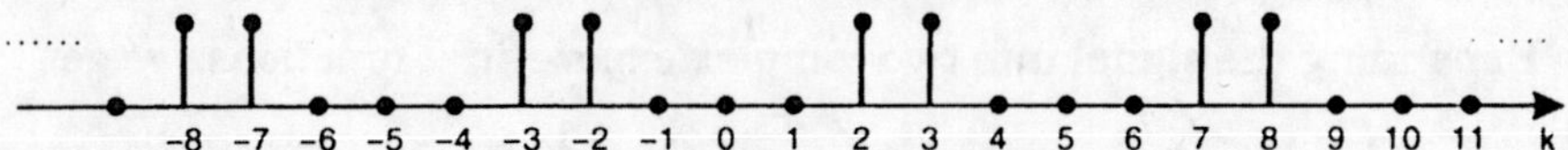

Fig. 4.2 *Fourier series coefficients A_k for $s(n) = \sin M\,(2\pi/N_0)n$, where $M = 3$ and $N_0 = 5$.*

Note, however, that over any period of length 5 there are only two non-zero Fourier series coefficients, and therefore, there are only two non-zero terms in the synthesis equation.

Example 4.2 Determine the Fourier series coefficients A_k for a discrete-time periodic signal

$$s(n) = 1 + \sin \omega_0 n + 3 \cos \omega_0 n + \cos\left(2\omega_0 n + \frac{\pi}{2}\right)$$

where notations carry their usual meaning.

Solution. For determining Fourier series coefficients A_k we expand the above signal in terms of complex exponential functions as

$$s(n) = 1 + \sin \omega_0 n + 3 \cos \omega_0 n + \cos\left(2\omega_0 + \frac{\pi}{2}\right)$$

$$= 1 + \frac{e^{j\omega_0 n} - e^{-j\omega_0 n}}{2} + 3\left[\frac{e^{j\omega_0 n} + e^{-j\omega_0 n}}{2}\right] + \frac{e^{j[2\omega_0 n + (\pi/2)]} + e^{-j[2\omega_0 n + (\pi/2)]}}{2}$$

$$= 1 + \frac{1}{2}\left(3 + \frac{1}{j}\right)e^{j\omega_0 n} + \frac{1}{2}\left(3 - \frac{1}{j}\right)e^{-j\omega_0 n} + \frac{1}{2}e^{j(\pi/2)}\,e^{j2\omega_0 n}$$

$$+ \frac{1}{2}e^{-j(\pi/2)}\,e^{-j2\omega_0 n}$$

$$= 1 + \frac{1}{2}(3 - j)\,e^{j\omega_0 n} + \frac{1}{2}(3 + j)\,e^{-j\omega_0 n} + \frac{1}{2}e^{j(\pi/2)}\,e^{j2\omega_0 n}$$

$$+ \frac{1}{2}e^{-j(\pi/2)}\,e^{-j2\omega_0 n} \qquad \ldots(1)$$

But we know that

$$s(n) = \sum_{k = \langle N_0 \rangle} A_k\, e^{jk\omega_0 n}$$

$$= \ldots + A_{-2}\,e^{j(-2)\omega_0 n} + A_{-1}\,e^{j(-1)\omega_0 n} + A_0\,e^{j(0)\omega_0 n} + A_1\,e^{j(1)\omega_0 n}$$

$$+ A_2\,e^{j(2)\omega_0 n} + \ldots \qquad \ldots(2)$$

Comparing Eqs. 1 and 2 we can see by inspection that Fourier series coefficients A_k are

$$A_0 = 1$$

$$A_1 = \frac{1}{2}(3 - j) = \frac{3}{2} - \frac{1}{2}j$$

$$A_{-1} = \frac{1}{2}(3 + j) = \frac{3}{2} + \frac{1}{2}j$$

$$A_2 = \frac{1}{2}e^{j(\pi/2)} = \frac{1}{2}\left(\cos\frac{\pi}{2} + j\sin\frac{\pi}{2}\right) = \frac{1}{2}(0 + j.1) = \frac{1}{2}j$$

$$A_{-2} = \frac{1}{2}e^{-j(\pi/2)} = \frac{1}{2}\left(\cos\frac{\pi}{2} - j\sin\frac{\pi}{2}\right) = \frac{1}{2}(0 - j.1) = -\frac{1}{2}j$$

and the remaining Fourier series coefficients of A_k are zero. Again the Fourier series coefficients A_k are periodic with period N_0.

Magnitude and phase of the Fourier series coefficients are shown in Fig. 4.3.

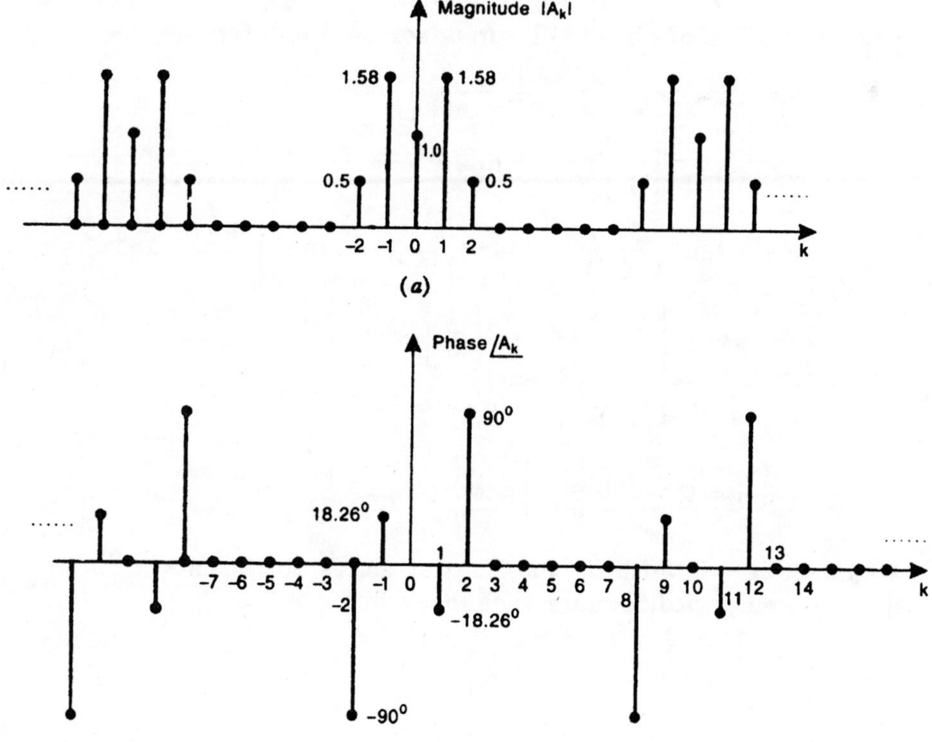

Fig. 4.3 *Plots of (a) magntiude (b) phase of Fourier series coefficients A_k of signal s(n).*

Determination of magnitude of Fourier series coefficients A_k

Fourier series coefficients A_k are complex coefficients and are represented as

$$A_k = \text{Re}\{A_k\} + j\,\text{Im}\{A_k\} \tag{3}$$

Magnitude of A_k, $|A_k| = \sqrt{[Re\{A_k\}]^2 + [Im\{A_k\}]^2}$

$$|A_0| = 1$$

$$|A_1| = \sqrt{(3/2)^2 + (-1/2)^2} = \sqrt{\frac{9}{4} + \frac{1}{4}} = \sqrt{\frac{10}{4}} = \frac{3.16}{2}$$

$$|A_{-1}| = \sqrt{(3/2)^2 + (1/2)^2} = \sqrt{\frac{9}{4} + \frac{1}{4}} = \sqrt{\frac{10}{4}} = 1.58$$

$$|A_2| = \sqrt{(0)^2 + (1/2)^2} = \sqrt{0 + \frac{1}{4}} = \frac{1}{2}$$

$$|A_{-2}| = \sqrt{(0)^2 + (-1/2)^2} = \sqrt{0 + \frac{1}{4}} = \frac{1}{2}$$

Determination of phase of Fourier series coefficients A_k

$$\text{Phase of } A_k = \angle A_k = \tan^{-1}\left\{\frac{\text{Im}\{A_k\}}{\text{Re}\{A_k\}}\right\}$$

$$\angle A_0 = \tan^{-1}\left[\frac{0}{1}\right] = \tan^{-1}(0) = 0^\circ$$

$$\angle A_1 = \tan^{-1}\left[\frac{(-1/2)}{(3/2)}\right] = \tan^{-1}\left[-\frac{1}{3}\right] = -\tan^{-1}\left[\frac{1}{3}\right] = -18.26^\circ$$

$$\angle A_{-1} = \tan^{-1}\left[\frac{(1/2)}{(3/2)}\right] = \tan^{-1}\left[\frac{1}{3}\right] = +18.26^\circ$$

$$\angle A_2 = \tan^{-1}\left[\frac{1/2}{0}\right] = \tan^{-1}(\infty) = \frac{\pi}{2} = 90^\circ$$

$$\angle{-A_2} = \tan^{-1}\left[\frac{(-1/2)}{0}\right] = \tan^{-1}(-\infty) = -\frac{\pi}{2} = -90^\circ$$

Example 4.3 Determine the Fourier series coefficients A_k of a discrete-time periodic square wave shown in Fig. 4.4.

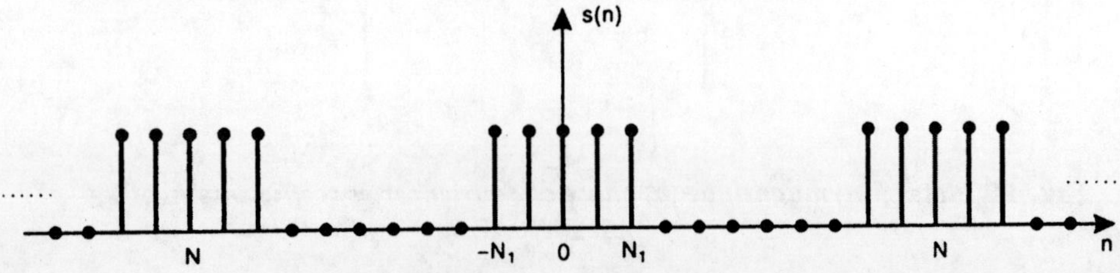

Fig. 4.4 *Discrete-time periodic squarewave.*

Solution. We can determine the discrete-time Fourier series (DTFS) for a discrete-time periodic signal by using the expression

$$A_k = \frac{1}{N_0} \sum_{n=\langle N_0 \rangle} s(n)\, e^{-jk\omega_0 n} \qquad ...(1)$$

Because signal $s(n) = \begin{cases} 1, & -N_1 \le n \le N_1 \\ 0, & \text{otherwise} \end{cases}$

Substituting this value in Eq. 1, we get

$$A_k = \frac{1}{N_0} \sum_{n=-N_1}^{N_1} 1 \,.\, e^{-jk\omega_0 n} = \frac{1}{N_0} \sum_{n=-N_1}^{N_1} e^{-jk\omega_0 n} \qquad ...(2)$$

Now, substituting $m = n + N_1$ in Eq. 2, we get

$$A_k = \frac{1}{N_0} \sum_{m=0}^{2N_1} e^{jk\omega_0(m-N_1)} = \frac{1}{N_0} \sum_{m=0}^{2N_1} e^{-jk\omega_0 n}\, e^{jk\omega_0 N_1}$$

$$= \frac{1}{N_0} e^{jk\omega_0 N_1} \sum_{m=0}^{2N_1} e^{-jk\omega_0 n} \qquad ...(3)$$

The summation of Eq. 3 consists of the sum of first $2N_1 + 1$ terms in a geometric progression (GR). By using the formula given in Eq. 4

$$S_N = \text{Sum of } N \text{ terms of GP}$$

$$= a\,\frac{(1 - r^N)}{(1 - r)} \qquad ...(4)$$

where a = first term of GP, r = common ratio of GP, N = total number of terms in GP.

Given : $N = 2N_1 + 1$

$a = 1$

$r = e^{-jk\omega_0}$

Substituting these values in Eq. 4, we get

$$\sum_{m=0}^{2N_1} e^{-jk\omega_0 m} = \frac{1\,[1 - (e^{-jk\omega_0 n})^{2N_1+1}]}{[1 - e^{-jk\omega_0}]}$$

$$= \left[\frac{1 - e^{-jk\omega_0(2N_1+1)}}{1 - e^{-jk\omega_0}}\right] \qquad ...(5)$$

Substituting Eq. 5 in Eq. 2, we get

$$A_k = \frac{1}{N_0} e^{jk\omega_0 N_1} \sum_{m=0}^{2N_1} e^{-jk\omega_0 m}$$

$$= \frac{1}{N_0} e^{jk\omega_0 N_1} \left[\frac{1 - e^{-jk\omega_0(2N_1+1)}}{1 - e^{-jk\omega_0}}\right]$$

$$= \frac{1}{N_0} \frac{e^{-(jk\omega_0)/2}}{e^{-(jk\omega_0)/2}} \left[\frac{e^{jk\omega_0(N_1+1/2)} - e^{-jk\omega_0(N_1+1/2)}}{e^{(jk\omega_0)/2} - e^{-(jk\omega_0)/2}}\right]$$

$$= \frac{1}{N_0} \frac{\sin\left[k\omega_0\left(N_1 + \frac{1}{2}\right)\right]}{\sin\left(\frac{k\omega_0}{2}\right)}, \quad k \neq 0, \pm N_0, \pm 2N_0, \ldots \quad \ldots(6)$$

where T_0 is the fundamental period.

Substituting $\omega_0 = \frac{2\pi}{N_0}$ in Eq. 6, we get

$$A_k = \frac{1}{N_0} \frac{\sin\left[k\frac{2\pi}{N_0}\left(N_1 + \frac{1}{2}\right)\right]}{\sin\left(\frac{k}{2}\frac{2\pi}{N_0}\right)}$$

or
$$A_k = \frac{1}{N_0} \frac{\sin\left[2\pi k \frac{(N_1 + 1/2)}{N_0}\right]}{\sin\left(\frac{k\pi}{N_0}\right)}, \quad k = 0, \pm N_0, \pm 2N_0, \ldots \quad \ldots(7)$$

Hence, $$A_k = \frac{2N_1 + 1}{N_0}, \quad k = 0, \pm N_0, \pm 2N_0, \ldots \quad \ldots(8)$$

The Fourier series coefficients A_k for $2N_1 + 1 = 5$ are sketched in Fig. 4.5 for $N_0 = 10$.

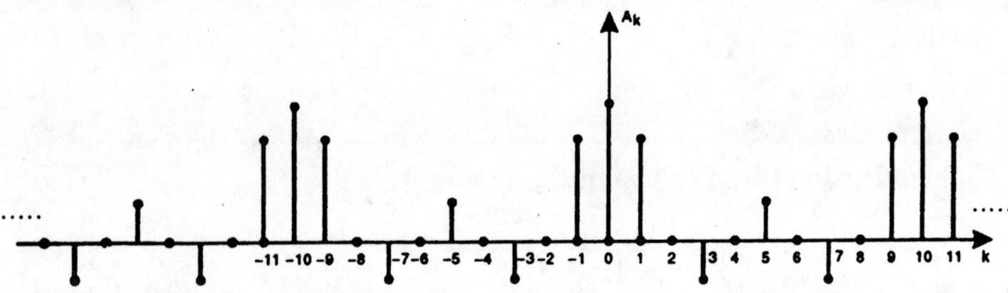

Fig. 4.5 *Fourier series coefficients A_k for discrete-time periodic square wave. Plots for $2N_1 + 1 = 5$ and $N_0 = 10$.*

4.4 PROPERTIES OF DISCRETE-TIME FOURIER SERIES (DTFS)

In this section we will study different properties of the discrete-time Fourier series (DTFS). There are strong similarities between the properties of discrete-time Fourier series (DTFS) and continuous-time Fourier series (CTFS). The similarity between these two Fourier series can be readily seen by comparing properties of DTFS with CTFS. The

derivation of many of the properties of DTFS is very similar to those of the corresponding properties of CTFS.

Here we will also use a short hand notation to indicate the relationship between a discrete-time periodic signal $s(n)$ and its Fourier series coefficients A_k. Fourier series representation of $s(n)$ and its Fourier series coefficients are denoted by

$$s(n) \xleftrightarrow{\text{CTFS}} A_k$$

Here we will study only a few properties of DTFS. These are :

1. Multiplication of two discrete-time periodic signals.
2. First difference of a discrete-time periodic signal.
3. Parseval's relation for a discrete-time periodic signal.

Other properties of DTFS are very similar to the properties of CTFS.

4.4.1 Multiplication of Two Discrete-Time Periodic Signals

The multiplication property of Fourier series representation is one example of property which reflects the difference between DTFS and CTFS. The multiplication of two continuous-time periodic signals of period T_0 results in a periodic signal with period T_0 whose sequence of Fourier series coefficient $\{A_k\}$ is the convolution of the sequences of Fourier series coefficients of the two signals being multiplied.

For discrete-time periodic signals,

$$s_1(n) \xleftrightarrow{\text{DTFS}} A_k$$

and

$$s_2(n) \xleftrightarrow{\text{DTFS}} B_k$$

Both signals $s_1(n)$ and $s_2(n)$ are periodic with period N_0. The product $s_1(n)\, s_2(n)$ is also periodic with period N_0 and its Fourier series coefficients C_k are given by

$$s_1(n)\, s_2(n) \xleftrightarrow{\text{DTFS}} C_k = \sum_{l=\langle N_0 \rangle} A_l\, B_{k-l} \qquad \text{...(4.17)}$$

Eq. 4.17 is equivalent to the definition of convolution, except that the summations variable is restricted to an interval of N consecutive samples. This type of operation is called a periodic convolution between two periodic sequences of Fourier series coefficients, $\{A_k\}$ and $\{B_k\}$.

The usual form of convolution sum is sometimes referred to as an aperiodic convolution.

4.4.2 First Difference of a Discrete-Time Periodic Signal

First-difference of a discrete-time signal is an operation parallel to the differentiation property of the CTFS. The first difference operation of a discrete-time periodic signal is defined as $s(n) - s(n-1)$.

If $s(n)$ is a discrete-time periodic signal with period N_0 and its shifted version $s(n-1)$ is also a periodic signal with same period N_0.

If $$s(n) \xleftrightarrow{\text{DTFS}} A_k$$

then $$s(n) - s(n-1) \xleftrightarrow{\text{DTFS}} (1 - e^{-jk\omega_0})\, A_k \qquad \text{.. (4.18)}$$

First difference is easily obtained by applying the time shifting and linearity properties of DTFS given in Table 4.1. This property is commonly used in a situation where determination of Fourier series coefficients is easier for first difference than the original sequence.

4.4.3 Parseval's Relation for Discrete-Time Periodic Signals

Parseval's relation for discrete-time periodic signals is given by

$$\frac{1}{N_0} \sum_{m=\langle N_0 \rangle} |s(m)|^2 = \sum_{k=\langle N_0 \rangle} |A_k|^2 \qquad \text{...(4.19)}$$

where A_k are the Fourier series coefficients of $s(n)$ and N_0 is period. In continuous-time periodic signal, left hand side of the Parseval's relation is the average power in one period for the periodic signal $s(n)$.

Parseval's relation states that the average power in a periodic signal is equal to the sum of average powers in all of its harmonic components.

4.5 TABULATION OF PROPERTIES OF DTFS

Properties of discrete-time Fourier series (DTFS) are summarized in Table 4.1.

Table 4.1 Summary of Properties of DTFS

Property	*Periodic signal*	*Fourier series coefficients or sequence*
	$\left.\begin{matrix} s_1(n) \\ s_2(n) \\ s(n) \end{matrix}\right\}$ period with N_0 $\omega_0 = 2\pi/N_0 =$ fundamental frequency	$\left.\begin{matrix} A_k \\ B_k \\ C_k \end{matrix}\right\}$ Period with N_0
Linearity	$\alpha\, s_1(n) + \beta\, s_2(n)$	$\alpha A_k + \beta B_k$
Time shifting	$s(n - n_0)$	$C_k\, e^{-jk\omega_0 n_0}$
Frequency shifting	$e^{jM\omega_0 n}\, s(n)$	C_{k-M}
Conjugation	$s^*(n)$	C^*_{-k}
Time reversal	$s(-n)$	C_{-k}

(Contd..)

Table 4.1 (Contd...)

Property	*Periodic signal*	*Fourier series coefficients or sequence*
Time scaling	$s_m(n) = \begin{cases} s(n/m), & \text{if } n \text{ is multiple of } m \\ 0, & \text{if } n \text{ is not multiple of } m \end{cases}$ → It is periodic with period mN_0	$\frac{1}{m} C_k$ → It is viewed as a periodic with period mN_0
Periodic convolution	$\sum_{l=\langle N_0\rangle} s_1(l)\, s_2(n-l)$	$N_0 A_k B_k$
Multiplication	$s_1(n)\, s_2(n)$	$\sum_{l=\langle N_0\rangle} A_l B_{k-l}$
First difference	$s(n) - s(n-1)$	$(1 - e^{-jk\omega_0})\, C_k$
Running sum	$\sum_{k=-\infty}^{\infty} s(k)$ It will be finite valued and periodic only if $C_0 = 0$	$\left(\frac{1}{1 - e^{-jk\omega_0}}\right) C_k$
Conjugate symmetry for real periodic signals	$s(n)$ is a real signal	$\begin{cases} C_k = C_{-k}^* \\ \text{Re}\{C_k\} = \text{Re}\{C_{-k}\} \\ \text{Im}\{C_k\} = -\text{Im}\{C_{-k}\} \\ \lvert C_k \rvert = \lvert C_{-k} \rvert \\ \angle C_k = -\angle C_{-k} \end{cases}$
Real and even signals	$s(n)$ real and even	C_k real and even
Real and odd signals	$s(n)$ real and odd	C_k purely imaginary and odd
Even and odd decomposition of real signals	$s_e(n) = Ev\{s(n)\}$, $s(n)$ is real $s_0(n) = Od\{s(n)\}$ $s(n)$ is real	$\text{Re}\{C_k\}$ $j\text{Im}\{C_k\}$
Parseval's relation for discrete-time periodic signals $\frac{1}{N_0} \sum_{n=\langle N_0\rangle} \lvert s(n)\rvert^2 = \sum_{k=\langle N_0\rangle} \lvert C_k \rvert^2$		

4.6 FOURIER SERIES (BOTH CTFS AND DTFS) AND LTI SYSTEMS

We have seen that Fourier series representation can be used to construct any continuous-time or discrete-time periodic signal. The response of an LTI system to a linear combination of complex exponential functions takes a simple form.

In continuous-time LTI system, if the input is $s(t) = e^{st}$ then the output in this system is given by

$$H(s) = \int_{-\infty}^{\infty} h(\tau)\, e^{-s\tau}\, d\tau \qquad ...(4.20)$$

where $h(t)$ is the impulse response of the continuous-time LTI system.

Similarly, in discrete-time LTI system, if the input is $s(n) = z^n$ then the output in this system is given by

$$H(z) = \sum_{k=-\infty}^{\infty} h(k)\, z^{-k} \qquad ...(4.21)$$

where $h(k)$ is the unit-sample (impulse) response of the discrete-time LTI system. Both s and z are complex numbers and $H(s)$ and $H(z)$ are called systems functions of corresponding systems.

For continuous-time LTI system, a specific case in which real part of complex number s, *i.e.*, $\text{Re}\{s\}$ is equal to zero, so that

$$s = \text{Re}\{s\} + j\,\text{Im}\{s\} = \sigma + j\omega = 0 + j\omega = j\omega$$

and consequently Eq. 4.20 becomes

$$H(j\omega) = \int_{-\infty}^{\infty} h(t)\, e^{-j\omega t}\, dt \qquad ...(4.22)$$

Hence $H(j\omega)$ is viewed as a function of ω and it is referred to as frequency response of the continuous-time LTI system.

Similarly, for discrete-time LTI system, a specific case in which complex number $|z| = 1$, so that $z = e^{j\omega}$ and $z^n = e^{j\omega n}$ and consequently Eq. 4.21 becomes

$$H(e^{j\omega}) = \sum_{n=-\infty}^{\infty} h(n)\, e^{-j\omega n} \qquad ...(4.23)$$

Hence $H(e^{j\omega})$ is viewed as a function of ω and it is referred to as frequency response of the discrete-time LTI system.

It is simple to express the response of an LTI system to a complex exponential signal of the form $e^{j\omega t}$ for continuous-time and $e^{j\omega n}$ for discrete-time LTI system in terms of the frequency response.

Now we first consider the continuous-time LTI system case with a periodic signal $s(t)$. Its Fourier series representation is given by

$$s(t) = \sum_{k=-\infty}^{\infty} A_k\, e^{jk\omega_0 t} \qquad ...(4.24)$$

We are applying this signal as input to an LTI system with impulse response $h(t)$.

Since each of the complex exponential functions in Eq. 4.24 is an eigen function of the system as in following equations.

Input, $s(t) = \sum_{k} A_k e^{s_k t}$

and output $y(t) = \sum_{k} A_k e^{s_k t} H(s_k)$

with $s_k = jk\omega_0$

The output of the system is given by

$$y(t) = \sum_{k=-\infty}^{\infty} A_k e^{jk\omega_0 t} H(jk\omega_0) \qquad ...(4.25)$$

Thus, the output $y(t)$ is also periodic with same fundamental frequency as input $s(t)$. Therefore, if $\{A_k\}$ is the set of Fourier series coefficients for input signal $s(t)$ then $\{A_k H(jk\omega_0)\}$ is the set of Fourier series coefficients for the output $y(t)$.

Now let us consider the discrete-time LTI system case with a periodic signal $s(n)$. Its Fourier series representation is given by

$$s(n) = \sum_{k=\langle N_0 \rangle} A_k e^{jk\omega_0 n} \qquad ...(4.26)$$

and we apply the above signal $s(n)$ as input to an LTI system with impulse response $h(n)$.

Since each of the complex exponential functions in Eq. 4.26 is an eigen function of the system as in following equations

Input $s(n) = \sum_{k=\langle N_0 \rangle} A_k z_k^n$

and output $y(n) = \sum_{k=\langle N_0 \rangle} A_k z_k^n H(z_k)$

with $z_k = e^{jk\omega_0}$

The output of above system is given by

$$y(n) = \sum_{k=\langle N_0 \rangle} A_k e^{jk\omega_0 n} H(e^{jk\omega_0}) \qquad ...(4.27)$$

Thus, the output $y(n)$ is also periodic with same fundamental frequency as input $s(n)$. Therefore, if $\{A_k\}$ is the set of Fourier series coefficients for input signal $s(n)$ then $\{A_k H(e^{jk\omega_0})\}$ is the set of Fourier series coefficients for the output $y(n)$.

Example 4.4 Determine the Fourier series coefficients of the output $y(t)$ for a continuous-time periodic signal

$$s(t) = \sum_{k=-3}^{3} A_k e^{jk\omega_0} \qquad ...(1)$$

where coefficients $A_0 = 1$, $A_1 = A_{-1} = \frac{1}{4}$, $A_2 = A_{-2} = \frac{1}{2}$ and $A_3 = A_{-3} = \frac{1}{3}$ to an LTI system with impulse response $h(t) = e^{-t}\, u(t)$.

Solution. To calculate the Fourier series coefficients of the output $y(t)$, we first determine the frequency response by using the formula given below :

$$H(j\omega) = \int_{-\infty}^{\infty} h(\tau)\, e^{-j\omega\tau}\, d\tau = \int_{-\infty}^{\infty} e^{-\tau}\, u(\tau)\, e^{-j\omega\tau}\, d\tau \quad ...(2)$$

But $$u(\tau) = \begin{cases} 1\,, & \tau > 0 \\ 0\,, & \tau < 0 \end{cases}$$

or $$= \int_{0}^{\infty} e^{-\tau}\, e^{-j\omega\tau}\, d\tau$$

$$= \int_{0}^{\infty} e^{-(1+j\omega)\tau}\, d\tau = \left[\frac{e^{-(1+j\omega)\tau}}{-(1+j\omega)}\right]_{0}^{\infty} = \frac{e^{-(1+j\omega)\infty} - e^{-(1+j\omega)0}}{-(1+j\omega)}$$

or $$H(j\omega) = \frac{1}{(1+j\omega)} \quad ...(3)$$

Therefore, using the following equation given as

$$y(t) = \sum_{k=-\infty}^{\infty} A_k\, H(jk\omega_0)\, e^{jk\omega_0 t}$$

Output $$y(t) = \sum_{k=-3}^{3} B_k\, e^{jk\omega_0 t} = \sum_{k=-3}^{3} B_k\, e^{jk\omega_0 t} \quad ...(4)$$

where, $$B_k = A_k\, H(jk\omega_0) \quad ...(5)$$

The values of Fourier series coefficients A_k are given and $H(jk\omega_0) = \dfrac{1}{1+jk\omega_0}$ from Eq. 3, we get

$$\left.\begin{aligned}
k = 0\,,\ & B_0 = A_0\, H(j.0\omega_0) = 1 \cdot \frac{1}{1+j.0\omega_0} = 1 \\
k = 1\,,\ & B_1 = A_1\, H(j.1\omega_0) = \frac{1}{4} \cdot \left(\frac{1}{1+j.1\omega_0}\right) = \frac{1}{4}\left(\frac{1}{1+j\omega_0}\right) \\
k = -1\,,\ & B_{-1} = A_{-1}\, H(j(-1)\omega_0) = \frac{1}{4}\left(\frac{1}{1+j(-1)\omega_0}\right) = \frac{1}{4}\left(\frac{1}{1-j\omega_0}\right) \\
k = 2\,,\ & B_2 = A_2\, H(j.2\omega_0) = \frac{1}{2}\left(\frac{1}{1+j.2\omega_0}\right) = \frac{1}{2}\left(\frac{1}{1+j.2\omega_0}\right) \\
k = -2\,,\ & B_{-2} = A_{-2}\, H(j(-2)\omega_0) = \frac{1}{2}\left(\frac{1}{1+j(-2)\omega_0}\right) = \frac{1}{2}\left(\frac{1}{1-j2\omega_0}\right) \\
k = 3\,,\ & B_3 = A_3\, H(j.3\omega_0) = \frac{1}{3}\left(\frac{1}{1+j.3\omega_0}\right) = \frac{1}{3}\left(\frac{1}{1+j.3\omega_0}\right) \\
k = -3\,,\ & B_{-3} = A_{-3}\, H(j(-3)\omega_0) = \frac{1}{3}\left(\frac{1}{1+j(-3)\omega_0}\right) = \frac{1}{3}\left(\frac{1}{1-j3\omega_0}\right)
\end{aligned}\right\} \quad ...(6)$$

Note that output $y(t)$ must be a real valued signal because it is the convolution integral of input $s(t)$ and impulse response $h(t)$ and both $s(t)$ and $h(t)$ are real. It can be verified by examining Eq. 6 and observing that $B_k^* = B_{-k}$. Therefore, output $y(t)$ can also be expressed by any two forms given as

$$y(t) = B_0 + 2 \sum_{k=1}^{\infty} C_k \cos (k\omega_0 t + \theta_k)$$

$$= 1 + 2 \sum_{k=1}^{3} C_k \cos (k\omega_0 t + \theta_k) \qquad ...(7)$$

where $B_k = C_k e^{j\theta_k}$ in polar form and $B_0 = C_0$

or $$y(t) = B_0 + 2 \sum_{k=1}^{\infty} [D_k \cos k\omega_0 t - E_k \sin k\omega_0 t]$$

$$= 1 + 2 \sum_{k=1}^{3} [D_k \cos k\omega_0 t - E_k \sin k\omega_0 t] \qquad ...(8)$$

where $B_k = D_k + j E_k$ in rectangular form.

These coefficients $[C_k, \theta_k, D_k$ and $E_k]$ are directly determined from Eq. 6, for example

$$C_1 = |B_1| = \frac{1}{4\sqrt{1+\omega_0^2}} = \frac{1}{4\sqrt{1+4\pi^2}}$$

$$\theta_1 = \angle B_1 = \tan^{-1}\left(\frac{2\pi}{1}\right) = \tan^{-1}(2\pi) \qquad ...(9)$$

$$D_k = \text{Re}\{B_1\} = \frac{1}{4(1+4\pi^2)}$$

and $$E_k = \text{Im}\{B_1\} = -\frac{\pi}{2(1+4\pi^2)}$$

Here, we have assumed that fundamental frequency $\omega = 2\pi$.

Example 4.5 Determine the Fourier series representation for the output to a discrete-time LTI system with unit-sample (impulse) response

$$h(n) = \left(\frac{1}{2}\right)^n u(n) \quad \text{and input } s(n) = \cos \omega_0 n$$

Solution. Input $s(n)$ can be expressed in Fourier series form as

$$s(n) = \cos \omega_0 n = \frac{e^{j\omega_0 n} + e^{-j\omega_0 n}}{2}$$

$$= \frac{1}{2} e^{j\omega_0 n} + \frac{1}{2} e^{-j\omega_0 n} \qquad ...(1)$$

Frequency response

$$H(e^{j\omega}) = \sum_{n=-\infty}^{\infty} h(n)\, e^{-j\omega n}$$

$$= \sum_{n=-\infty}^{\infty} \left(\frac{1}{2}\right)^n u(n)\, e^{-j\omega n} \qquad \left[\text{But, } u(n) = \begin{cases} 1, & n \geq 0 \\ 0, & n < 0 \end{cases}\right]$$

$$= \sum_{n=0}^{\infty} \left(\frac{1}{2}\right)^n e^{-j\omega n} = \sum_{n=0}^{\infty} \left(\frac{1}{2} e^{-j\omega}\right)^n$$

[This is a geometric progression, with first term equal to 1 and common ratio is equal to $1/2\, e^{-j\omega}$]

or $$H(e^{j\omega}) = \frac{\text{First term}}{1 - \text{common ratio}} = \frac{1}{1 - \frac{1}{2} e^{-j\omega}} \qquad \text{...(2)}$$

Fourier series for output,

$$y(n) = \sum_{k=\langle N_0 \rangle} A_k\, H(e^{jk\omega_0})\, e^{jk\omega_0 n}$$

$$= A_{-1}\, H(e^{j(-1)\omega_0})\, e^{j(-1)\omega_0 n} + A_1\, H(e^{j(1)\omega_0})\, e^{j(1)\omega_0 n}$$

or $$y(n) = A_{-1}\, H(e^{-j\omega_0})\, e^{-j\omega_0 n} + A_1\, H(e^{j\omega_0})\, e^{j\omega_0 n} \qquad \text{...(3)}$$

Fourier series coefficients A_k can be obtained from the expression of

$$s(n) = \cos \omega_0 n = \frac{1}{2} e^{j\omega_0 n} + \frac{1}{2} e^{-j\omega_0 n}$$

$$\left. \begin{aligned} A_1 &= \frac{1}{2} \\ A_{-1} &= \frac{1}{2} \end{aligned} \right\} \qquad \text{...(4)}$$

We have determined

$$\left. \begin{aligned} H(e^{j\omega}) &= \frac{1}{1 - \frac{1}{2} e^{-j\omega}} \\ \text{Similarly, } H(e^{-j\omega}) &= \frac{1}{1 - \frac{1}{2} e^{j\omega}} \end{aligned} \right\} \qquad \text{...(5)}$$

Substituting Eqs. 4 and 5 in Eq. 3, we get

$$y(n) = \frac{1}{2} \left(\frac{1}{1 - \frac{1}{2} e^{-j\omega}} \right) e^{j\omega_0 n} + \frac{1}{2} \left(\frac{1}{1 - \frac{1}{2} e^{j\omega}} \right) e^{-j\omega_0 n} \qquad \text{...(6)}$$

Putting $\dfrac{1}{1 - \frac{1}{2} e^{-j\omega}} = \gamma e^{j\theta}$ and $\dfrac{1}{1 - \frac{1}{2} e^{j\omega}} = \gamma e^{-j\theta}$ in Eq. 6, we get

$$y(n) = \frac{1}{2}\,[\gamma e^{j\theta}]\,e^{j\omega_0 n} + \frac{1}{2}\,[\gamma e^{-j\theta}]\,e^{-j\omega_0 n}$$

$$= \frac{\gamma}{2}\,[e^{j(\theta+\omega_0 n)} + e^{-j(\theta+\omega_0 n)}] = \gamma\left[\frac{e^{j(\theta+\omega_0 n)} + e^{-j(\theta+\omega_0 n)}}{2}\right]$$

or $$y(n) = \gamma \cos(\omega_0 n + \theta) \qquad \text{...(7)}$$

$$= \gamma \cos\left(\frac{2\pi n}{N_0} + \theta\right) \qquad \text{where } \omega_0 = \frac{2\pi}{N_0}$$

If the fundamental period $N_0 = 4$, then

$$\gamma e^{j\theta} = \frac{1}{1 - \frac{1}{2}e^{-j\omega_0}} = \frac{1}{1 - \frac{1}{2}e^{-j(2\pi/N_0)}} = \frac{1}{1 - \frac{1}{2}e^{-j(2\pi/4)}}$$

$$= \frac{1}{1 - \frac{1}{2}e^{-j(\pi/2)}} = \frac{1}{1 + j\frac{1}{2}} = \frac{1}{\sqrt{1 + (1/2)^2}}\,e^{j[-\tan^{-1}(1/2)]}$$

It means, $$\gamma = \frac{1}{\sqrt{1 + (1/2)^2}} \qquad \text{...(8)}$$

and $$\theta = -\tan^{-1}\left(\frac{1}{2}\right) \qquad \text{...(9)}$$

Substituting $N_0 = 4$, $\gamma = \frac{2}{\sqrt{5}}$ and $\theta = -\tan^{-1}\left(\frac{1}{2}\right)$ these values in Eq. 7, we get

$$y(n) = \frac{1}{\sqrt{1 + (1/2)^2}} \cos\left[\frac{\pi n}{2} - \tan^{-1}\left(\frac{1}{2}\right)\right]$$

$$= \frac{2}{\sqrt{5}} \cos\left(\frac{\pi n}{2} - 0.147\pi\right)$$

Example 4.6 Find the Fourier series coefficients C_k of the discrete-time sequence $s(n)$ shown in Fig. 4.6 (*a*). This sequence has a fundamental period of 5.

Solution. We observed that sequence $s(n)$ may be viewed as the sum of the square wave $s_1(n)$ and the d.c. sequence $s_2(n)$. The sequences $s_1(n)$ and $s_2(n)$ are shown in Fig. 4.6 (*b*) and Fig. 4.6 (*c*), respectively.

We are denoting Fourier series coefficients of sequence $s_1(n)$ by A_k and of $s_2(n)$ by B_k. Here we are using property of linearity and we have

$$C_k = A_k + B_k \qquad \text{...(1)}$$

where C_k are the Fourier series coefficients of the periodic sequence $s(n)$.

Fourier series coefficients for $s_1(n)$ is determined as

$$A_k = \frac{1}{N} \sum_{n=-N_1}^{N_1} s_1(n)\,e^{-jk\omega n} \qquad \text{...(2)}$$

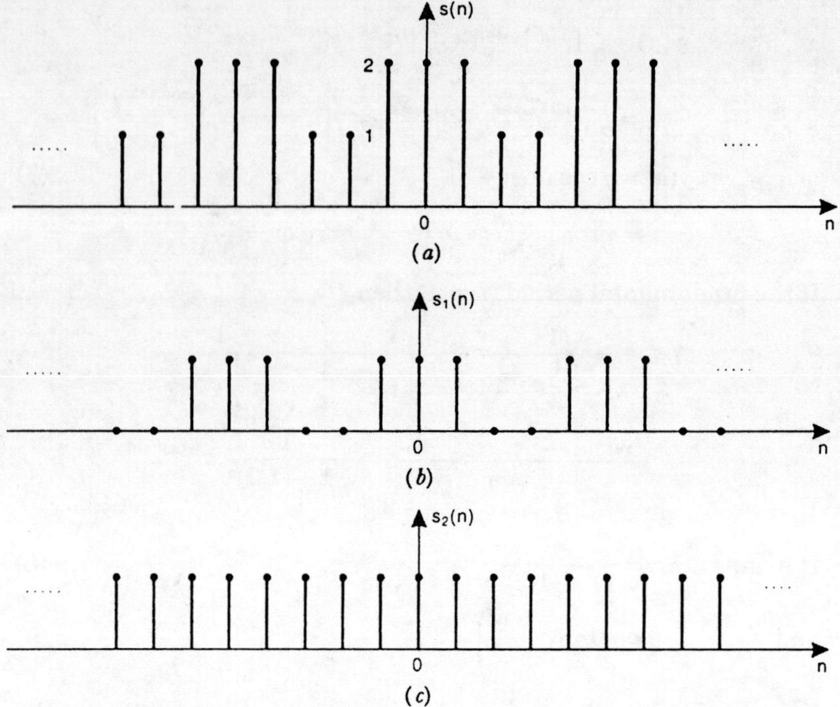

Fig. 4.6 *(a) Periodic sequence $s(n)$ (b) sequence $s_1(n)$ (c) sequence $s_2(n)$.*

Substituting $m = n + N_1$, in Eq. 2, we obtain

$$A_k = \frac{1}{N} \sum_{m=0}^{2N_1} e^{-jk\omega(m - N_1)}$$

$$= \frac{1}{N} e^{jk\omega N_1} \sum_{m=0}^{2N_1} e^{-jk\omega m} \qquad \text{...(3)}$$

The summation of Eq. 3 consists of the sum of the first $2N_1 + 1$ terms in the geometric series, it can be evaluated as

$$A_k = \frac{1}{N} e^{jk\omega N_1} \left[\frac{1 - e^{-jk\omega(2N_1 + 1)}}{1 - e^{-jk\omega}} \right]$$

$$= \frac{1}{N} \frac{e^{-(jk\omega)/2} \left[e^{jk\omega(N_1 + 1/2)} - e^{-jk\omega(N_1 + 1/2)}\right]}{e^{-(jk\omega)/2} \left[e^{+(jk\omega)/2} - e^{-(jk\omega)/2}\right]}$$

or

$$A_k = \frac{1}{N} \frac{\sin\left[\omega k \left(N_1 + \frac{1}{2}\right)\right]}{\sin\left[\frac{\omega k}{2}\right]}, \; k \neq 0, \pm N, \pm 2N, \ldots \qquad \text{...(4)}$$

and

$$A_k = \frac{2N_1 + 1}{N}, \; k = 0, \pm N, \pm 2N, \ldots \qquad \text{...(5)}$$

Substituting $N_1 = 1$ and $N = 5$ in Eqs. 4 and 5, we obtain

$$A_k = \frac{1}{5}\frac{\sin\left[\frac{2\pi}{5}k\left(1+\frac{1}{2}\right)\right]}{\sin\left[\frac{2\pi}{2\times 5}k\right]}$$

or
$$A_k = \frac{1}{5}\frac{\sin\left(\frac{3}{5}\pi k\right)}{\sin\left(\frac{\pi k}{5}\right)}, \text{ for } k \neq \pm 0, \pm 5, \pm 10, \ldots \quad \ldots(6)$$

and
$$A_k = \frac{2N1+1}{N} = \frac{2\times 1+1}{5} = \frac{3}{5}$$

$$\text{for } k = 0, \pm 5, \pm 10, \ldots \quad \ldots(7)$$

Eqs. 6 and 7 can be written as

$$A_k = \begin{cases} 1/5\,\dfrac{\sin[(3\pi k)/5]}{\sin\,[(\pi k)/5]}, & \text{for } k \neq \pm 0, \pm 5, \pm 10, \ldots \\ 3/5, & \text{for } k = 0, \pm 5, \pm 10, \ldots \end{cases} \quad \ldots(8)$$

The sequence $s_2(n)$ has only d.c. value. Its Fourier series coefficients are determined by

$$B_0 = \frac{1}{N}\sum_{n=N_1}^{N_1} s_2(n)\,e^{-jk\omega n} = \frac{1}{5}\sum_{n=0}^{4} s_2(n)\,e^{-j(0)\omega n}$$

$$= \frac{1}{5}\sum_{n=0}^{4} s_2(n)\,e^{0} = \frac{1}{5}\sum_{n=0}^{4} s_2(n)$$

$$= \frac{1}{5}[s_2(0) + s_2(1) + s_2(2) + s_2(3) + s_2(4)]$$

$$= \frac{1}{5}[1+1+1+1+1] = \frac{5}{5} = 1$$

or $\quad B_0 = 1 \quad \ldots(9)$

Since the discrete-time Fourier series coefficients are periodic, it follows that $B_k = 1$ whenever k is an integer multiple of 5. The remaining Fourier series coefficients of $s_2(n)$ must be zero, because $s_2(n)$ contains only a d.c. component. Putting the values of A_n and B_n in Eq. 1, we obtain

$$C_k = A_k + B_k = \begin{cases} 1/5\,\dfrac{\sin[(3\pi k)/5]}{\sin\,[(\pi k)/5]}, & \text{for } k \neq 0, \pm 5, \pm 10, \pm 15, \ldots \\ \dfrac{3}{5} + 1 = \dfrac{8}{5}, & \text{for } k = 0, \pm 5, \pm 10, \pm 15, \ldots \end{cases}$$

REVIEW QUESTIONS

1. Distinguish Discrete-time Fourier series (DTFS) from Continuous-time Fourier series (CTFS).
2. Discuss Fourier series representation of discrete-time periodic signals.
3. Give the more general form of periodic discrete-time signals or sequences in terms of linear combination of the exponential sequences.
4. Give the synthesis equation and analysis equation for discrete- time signal and Fourier series coefficients.
5. Discuss determination of magnitude and phase of Fourier series coefficient. Give one example.
6. Discuss similarities and differences between properties of CTFS and DTFS.
7. Discuss various properties of DTFS in brief.
8. Tabulate various properties of DTFS.
9. Distinguish an aperiodic convolution from periodic convolution.
10. What do you mean by first difference of a discrete-time periodic signal ?
11. Write Parseval's relation for discrete-time periodic signals.

NUMERICAL PROBLEMS

1. A discrete-time periodic signal $s(n)$ is real valued and has fundamental period $N_0 = 5$. The non-zero Fourier series coefficients for $s(n)$ are

$$A_0 = 1\,,\ A_2 = A_{-2}^{*} = e^{j(\pi/4)}$$
$$A_4 = A_{-4}^{*} = 2e^{j(\pi/3)}$$

Express $s(n)$ in the form of $s(n) = A_0 + \sum_{k=1}^{\infty} A_k \sin(\omega_k n + \phi_k)$

Ans. $s(n) = 1 + 2\sin\left(\frac{4\pi}{5}n + \frac{3\pi}{4}\right) + 4\sin\left(\frac{8\pi}{5}n + \frac{5\pi}{6}\right)$

2. Determine the numerical values of one period of the Fourier series coefficients of periodic signal

$$s(n) = \sum_{m=-\infty}^{\infty} \{4\delta(n-4m) + 8\delta(n-4m-1)\}$$

Ans. $A_0 = 3\,,\ A_1 = 1 - 2j\,,\ A_2 = -1$ and $A_3 = 1 + 2j$

3. Frequency response of a continuous-time LTI system is given by

$$H(j\omega) = \int_{-\infty}^{\infty} h(t)\,e^{-j\omega t}\,dt = \frac{\sin(4\omega)}{\omega}$$

Determine the system output $y(t)$ for input $s(t) = \begin{cases} 1\,, & 0 \le t < 4 \\ -1\,, & 4 \le t < 8 \end{cases}$. Given $T_0 = 8$.

Ans. $y(t) = 0$

4. Determine the values of $H(e^{j(k\pi/2)})$ for $k = 0, 1, 2$, and 3 for discrete-time LTI system with frequency response $H(j\omega)$.

Given, input $s(n) = \sum_{k=-\infty}^{\infty} \delta(n-4k)$

and output $y(n) = \cos\left(\frac{5\pi}{2}n + \frac{\pi}{4}\right)$

Ans. $H(e^{j0}) = H(e^{j\pi}) = 0$, $H(e^{j\pi/2}) = 2e^{j\pi/4}$, $H(e^{j3\pi/2}) = 2e^{-j\pi/4}$

5. We have given three discrete-time signals with fundamental period of 6 :

$$s_1(n) = 1 + \cos\left(\frac{2\pi}{6}n\right)$$

$$s_2(n) = \sin\left(\frac{2\pi}{6}n + \frac{\pi}{4}\right)$$

$$s_3(n) = s_2(n)\, s_1(n)$$

Determine the following :

(*a*) Fourier series coefficient of $s_1(n)$

(*b*) Fourier series coefficient of $s_2(n)$

(*c*) Fourier series coefficient of $s_3(n)$ using multiplication property of DTFS.

(*d*) Fourier series coefficient of $s_3(n)$ through direct evaluation.

(*e*) Compare results of part (*c*) and (*d*).

Ans. (*a*) $A_0 = 1$, $A_1 = A_{-1} = \frac{1}{2}$ (*b*) $B_1 = B_{-1}^{*} = \frac{1}{2}e^{-j\pi/4}$

(*c*) $C_0 = \frac{1}{2}\cos\left(\frac{\pi}{4}\right)$, $C_1 = C_{-1}^{*} = \frac{1}{2}e^{-j\pi/4}$, $C_2 = C_{-2}^{*} = \frac{1}{4}e^{-j\pi/4}$

(*d*) $C_0 = \frac{1}{2}\cos\left(\frac{\pi}{4}\right)$, $C_1 = C_{-1}^{*} = \frac{1}{2}e^{-j\pi/4}$, $C_2 = C_{-2}^{*} = \frac{1}{4}e^{-j\pi/4}$

(*e*) Results obtained from part (*c*) and (*d*) are same.

6. Differential equation of a causal continuous-time LTI system is given by

$$\frac{d}{dt}y(t) + 4\,y(t) = s(t)$$

Determine the Fourier series representation of the output $y(t)$ for the following inputs

(*a*) $s(t) = \cos 2\pi t$ (*b*) $s(t) = \sin 4\pi t + \cos\left(6\pi t + \frac{\pi}{4}\right)$

Ans. (*a*) $B_1 = \frac{1}{2\,(4+j2\pi)}$, $B_{-1} = \frac{1}{2\,(4-j2\pi)}$ (*b*) $B_2 = \frac{1}{2j\,(4+j4\pi)}$,

$B_{-2} = \frac{1}{-2j\,(4-j4\pi)}$, $B_3 = \frac{e^{j\pi/4}}{2\,(4+j6\pi)}$, $B_{-3} = \frac{e^{-j\pi/4}}{2\,(4-j6\pi)}$

7. The difference equation of a causal discrete-time LTI system is given by

$$y(n) = \frac{1}{4} y(n-1) + s(n)$$

Determine the Fourier series representation of the output $y(n)$ for each of the following inputs

(a) $s(n) = \sin\left(\frac{3\pi}{4} n\right)$ (b) $s(n) = \cos\left(\frac{\pi}{4} n\right) + 2\cos\left(\frac{\pi}{2} n\right)$

Ans. (a) $B_3 = \dfrac{1}{2j\left[1 - \frac{1}{4} e^{-j3\pi/4}\right]}$, $B_{-3} = \dfrac{-1}{2j\left[1 - \frac{1}{4} e^{j3\pi/4}\right]}$

(b) $B_1 = \dfrac{1}{2\left[1 - \frac{1}{4} e^{-j\pi/4}\right]}$, $B_{-1} = \dfrac{-1}{2\left[1 - \frac{1}{4} e^{j\pi/4}\right]}$,

$B_2 = \dfrac{1}{\left[1 - \frac{1}{4} e^{-j\pi/2}\right]}$, $B_{-2} = \dfrac{-1}{\left[1 - \frac{1}{4} e^{j\pi/2}\right]}$

5

Continuous-Time Fourier Transform (CTFT)

5.1 COMPARISON BETWEEN VARIOUS TRANSFORMS AND LOGARITHMS

In various transforms (Transformation Tools) such as Laplace transform, z-transform, continuous-time Fourier transform (CTFT), discrete-time Fourier transform (DTFT), etc. We use one transformation (conversion) formula which converts signals from one domain to another corresponding domain depending upon a particular transform, so that domain is easy and suitable for computations we can also extract more information in that domain which is necessary for signal processing.

We use an inverse transform formula to reconvert that signal into previous domain.

These transformations serve the following two purposes :

(*a*) to extract more information from the transformed domain for the purpose of signal processing.

(*b*) for the purpose of simplication of computations in the transformed domain.

Logarithms are used only for simplification and making computations easy. These are not used for extracting any information from a signal which is usually done in transformation techniques.

Transform techniques are used for the analysis of signals and linear-time invariant (LTI) systems.

5.2 INTRODUCTION TO FOURIER TRANSFORM

We have already studied representation of continuous-time and discrete-time periodic signals as linear combinations of complex exponentials in chapters 3 and 4 respectively. This representation is called Fourier series representation for periodic signals (continuous-time/discrete-time).

In this and the next chapter 6, we extend these concepts and apply them to non-periodic signals. In this chapter we will study representation of continuous-time periodic signals such as linear combinations of complex exponentials. A large class of signals, including all signals with finite energy, can also be represented through a linear combination of complex exponentials. For periodic signals, the complex exponential building blocks are harmonically related. But for non-periodic signals, the complex exponential building blocks are infinitesimally close in frequency and the representation in terms of a linear combination takes the form of an integral rather than a sum. The resulting spectrum of coefficients in this representation is called the Fourier Transform. Synthesis of a signal from these coefficients is called the inverse Fourier transform. Fourier transform of a continuous-time periodic signal is called *continuous-time Fourier transform* (CTFT) and Fourier transform of a discrete-time periodic signal is termed as discrete-time Fourier transform (DTFT). In this chapter we will study continuous-time Fourier transform (CTFT), discrete-time Fourier transform (DTFT) will be discussed in chapter 6.

The development of representation for continuous-time periodic signal is one of Fourier's most important contributions. CTFT is the original work done by J.B.J. Fourier. In particular, Fourier reasoned that any non-periodic signal can be viewed as a periodic signal with an infinite time period ($T_0 \to \infty$).

In the Fourier series representation of a periodic signal, as the fundamental period 'T_0' increases the fundamental frequency $\omega_0 = 2\pi/T_0$ decreases and the harmonically related components become closer in frequency. As the fundamental period 'T_0' become infinite, the frequency components form a continuum and the Fourier series sum becomes a Fourier integral.

5.3 DEVELOPMENT OF CONTINUOUS-TIME FOURIER TRANSFORM (CTFT)

Continuous-time Fourier series (CTFS) is used to represent continuous-time periodic signals through a linear combination of complex exponential function ($e^{j\omega t}$) whereas continuous-time Fourier transform (CTFT) is used to represent continuous-time non-periodic signals thorugh a linear combination of complex exponential functions.

Fourier transform representation of a continuous-time non-periodic signal can be developed by revisiting the Fourier series representation for the continuous-time periodic square wave.

This square wave can be represented over one period as

$$s(t) = \begin{cases} 1, & |t| < T_1 \\ 0, & T_1 < |t| < T_0/2 \end{cases} \qquad \text{...(5.1)}$$

This square wave periodically repeats with T_0.

Fourier series coefficients A_k for the square wave will be given by

$$A_k = \frac{2 \sin (k\omega_0 T_1)}{k\omega_0 T_0} \quad ...(5.2)$$

where $\omega_0 = \dfrac{2\pi}{T_0}$ is the fundamental frequency of the square wave.

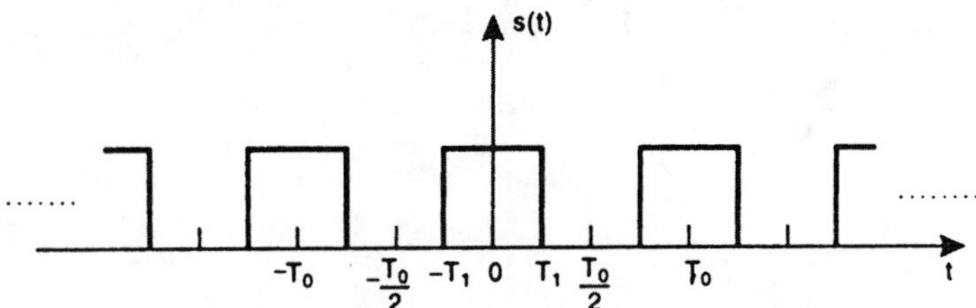

Fig. 5.1 *Continuous-time periodic signal (square wave).*

Alternatively, Eq. 5.2 can be interpreted as samples of an envelope function as

$$T_0 A_k = \frac{2 \sin k\omega_0 T_1}{k\omega_0}$$

$$= \left. \frac{2 \sin \omega T_1}{\omega} \right|_{\omega = k\omega_0} \quad ...(5.3)$$

where $\omega = k\omega_0$ is a continuous frequency variable and the function $\dfrac{2 \sin \omega T_1}{\omega}$ is the envelople of $T_0 A_k$.

Coefficients A_k are the equally spaced samples of this envelople. The envelope ($T_0 A_k$) is independent of period T_0 for fixed value of T_1. Fourier series coefficients for the periodic square wave as samples of the envelope of $T_0 A_k$ are shown in Fig. 5.2.

From Fig. 5.2, we see that as the period T_0 increases or as the fundamental frequency ω_0 increases the envelope is sampled with closer and closer spacing.

As T_0 becomes arbitrarily large the original periodic square wave approaches a rectangular pulse. The multiplication of Fourier series coefficients and fundamental period T_0 *i.e.* ($T_0 A_k$) becomes more and more closely spaced samples of the envelope. In a sense, the set of Fourier series coefficients approaches the envelope function as the period T_0 tends to infinity.

The example of periodic square wave shows the basic idea behind Fourier's development of a representation for a non-periodic signal. Specifically, we can say that a non-periodic signal is a limited form of a periodic signal as in arbitrarily large period. Now we can examine the limiting behavior of the Fourier series representation for a non-periodic signal.

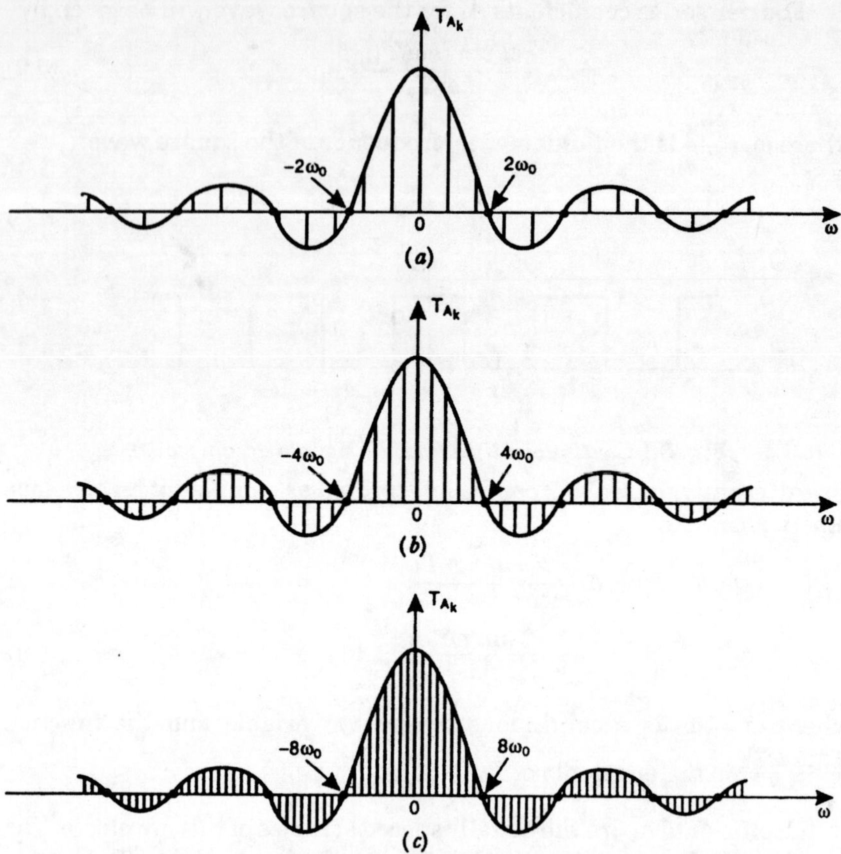

Fig. 5.2 *The Fourier series coefficients and their envelope for the continuous-time periodic square wave shown in Fig.* 5.1. *(a)* $T_0 = 4T_1$ *(b)* $T_0 = 8T_1$ *(c)* $T_0 = 16T_1$.

In particular, consider a non-periodic signal $s(t)$ that is of finite duration. A non-periodic signal $s(t)$ is shown in Fig 5.3 (*a*) and given as

$$s(t) = \begin{cases} \neq 0\,, & -T_1 \geq t \leq T_1 \\ = 0\,, & |t| > T_1 \end{cases} \quad \text{...(5.4)}$$

where T_1 is a some finite time.

We can construct a periodic signal $\tilde{s}(t)$ for this non-periodic signal $s(t)$. Here $s(t)$ is for one period. It is shown in Fig. 5.3 (*b*). For larger period, $\tilde{s}(t)$ is identical to $s(t)$ over longer interval. In other words we can say that as $T \to \infty$, $\tilde{s}(t)$ is equal to $s(t)$ for any finite value of time t.

Now let us examine the effect of this on the Fourier series representation of $\tilde{s}(t)$. Fourier series coefficients A_k of periodic signal $\tilde{s}(t)$ is determined by

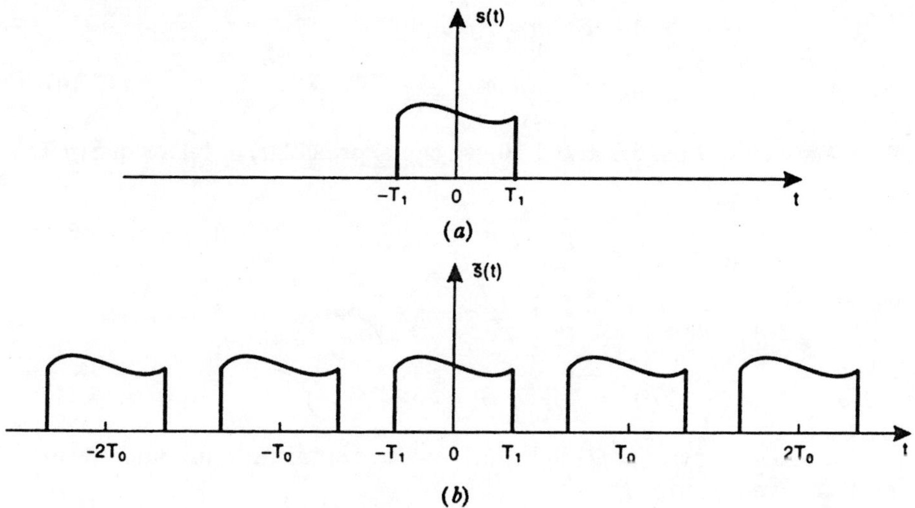

Fig. 5.3 (*a*) *Non-periodic signal s(t),* (*b*) *period signal* $\tilde{s}(t)$. *It is constructed from s(t) over one period.*

$$A_k = \frac{1}{T_0} \int_{-T_0/2}^{T_0/2} \tilde{s}(t) \; e^{-jk\omega_0 t} \; dt \qquad \text{...(5.5)}$$

and
$$\tilde{s}(t) = \sum_{k=-\infty}^{\infty} A_k \, e^{jk\omega_0 t} \qquad \text{...(5.6)}$$

where $\omega_0 = 2\pi/T_0$, is fundamental frequency of the periodic signal $\tilde{s}(t)$.

Since $\tilde{s}(t) = s(t)$ for $|t| < \frac{T_0}{2}$ and also since $s(t) = 0$ outside this interval.

Now, we can rewrite Eq. 5.5 as

$$A_k = \frac{1}{T_0} \int_{-T_0/2}^{T_0/2} s(t) \, e^{-jk\omega_0 t} \; dt$$

$$= \frac{1}{T_0} \int_{-\infty}^{\infty} s(t) \, e^{-jk\omega_0 t} \; dt \qquad \text{...(5.7)}$$

From Eq. 5.7, we obtain

$$T_0 A_k = \int_{-\infty}^{\infty} s(t) \, e^{-jk\omega_0 t} \; dt \qquad \text{...(5.8)}$$

Therefore, we can define the envelope of $T_0 A_k$ as

$$S(j\omega) = \int_{-\infty}^{\infty} s(t) \, e^{-j\omega t} \; dt \qquad \text{...(5.9)}$$

where $\omega = k\omega_0$

Fourier series coefficients A_k,

$$T_0 A_k = S(j\omega) = S(jk\omega_0)$$

or
$$A_k = \frac{1}{T_0} S(jk\omega_0) \qquad ...(5.10)$$

Comparing Eqs. 5.6 and 5.10, we can express $\tilde{s}(t)$ in terms of $S(j\omega)$ as

$$\tilde{s}(t) = \sum_{k=-\infty}^{\infty} A_k e^{jk\omega_0 t}$$

$$= \sum_{k=-\infty}^{\infty} \left[\frac{1}{T_0} S(jk\omega_0) \right] e^{jk\omega_0 t}$$

or
$$\tilde{s}(t) = \frac{1}{T_0} \sum_{k=-\infty}^{\infty} S(jk\omega_0) e^{jk\omega_0 t} \qquad ...(5.11)$$

Substituting $\omega_0 = 2\pi/T_0$ in Eq. 5.11, we can obtain an equivalent form of the above equation.

$$\tilde{s}(t) = \frac{1}{T_0} \sum_{k=-\infty}^{\infty} S(jk\omega_0) e^{jk\omega_0 t}$$

$$= \frac{\omega_0}{2\pi} \sum_{k=-\infty}^{\infty} S(jk\omega_0) e^{jk\omega_0 t}$$

or
$$\tilde{s}(t) = \frac{1}{2\pi} \sum_{k=-\infty}^{\infty} S(jk\omega_0) e^{jk\omega_0 t} \omega_0 \qquad ...(5.12)$$

As period $T_0 \to \infty$, $\tilde{s}(t)$ approaches $s(t)$ and consequently, the limiting form of Eq. 5.12 becomes representation of signal $s(t)$. Furthermore $\omega_0 \to 0$ as $T \to \infty$, and the right hand side of Eq. 5.12 *passes to an integral*. This can be seen by considering the graphical interpretation of the Eq. 5.12. It is shown in Fig. 5.4. Each term in the summation on the right hand side of Eq. 5.12 is the area of a rectangle of height $S(jk\omega_0) e^{jk\omega_0 t}$ and width ω_0. Here t is assumed constant.

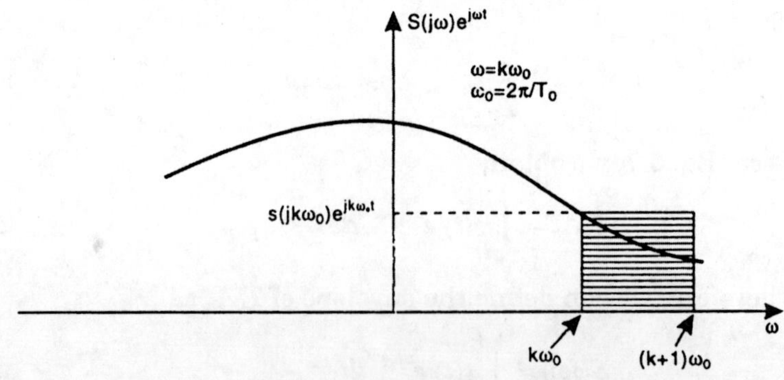

Fig. 5.4 *Graphical interpretation of Eq.* 5.12.

As fundamental frequency ω_0 tends to zero, the summation in Eq. 5.12 converges to an integral of $S(j\omega)\, e^{j\omega t}$

Therefore, using the fact that $\tilde{s}(t) \rightarrow s(t)$ as $T \rightarrow \infty$ and Eq. 5.9 becomes

$$S(j\omega) = \int_{-\infty}^{\infty} s(t)\, e^{-j\omega t}\, dt \qquad ...(5.13)$$

and Eq. 5.12 becomes

$$s(t) = \frac{1}{2\pi} \int_{-\infty}^{\infty} S(j\omega)\, e^{j\omega t}\, d\omega \qquad ...(5.14)$$

Eqs. 5.13 and 5.14 are referred to as a Fourier transform pair. The function $S(j\omega)$ is called the Fourier transform of continuous-time non-periodic signal $s(t)$. It is also called Fourier integral. Eq. 5.14 is referred to as synthesis or inverse Fourier transform equation. $s(t)$ is the inverse Fourier transform of $S(j\omega)$. For periodic signals, complex exponentials have amplitude $\{A_k\}$ and occur at discrete set of harmonically related frequencies $k\omega_0$, $k = 0, \pm 1, \pm 2 ...$

The complex exponentials occur at a continuum of frequencies in a non-periodic signal.

The Fourier transform $S(j\omega)$ of a non-periodic signal $s(t)$ is called the spectrum of $s(t)$. This spectrum provides us information needed for describing $s(t)$ as a linear combination (integral) of sinusoidal signals at different frequencies.

5.4 CONVERGENCE OF CONTINUOUS-TIME FOURIER TRANSFORMS (CTFT)

In driving a CTFT pair, we have assumed that signal $s(t)$ was an arbitrary signal with finite duration. Eqs. 5.13 and 5.14 remain valid for an extremely broad class of signals of infinite duration. In fact, the derivation of CTFT suggests that we require a set of conditions like those required for convergence of Fourier series. Inverse CTFT of $S(j\omega)$ is denoted by $\hat{s}(t)$ and is given by

$$\hat{s}(t) = \frac{1}{2\pi} \int_{-\infty}^{\infty} S(j\omega)\, e^{j\omega t}\, d\omega \qquad ...(5.15)$$

$\hat{s}(t)$ is a valid representation of the original signal $s(t)$, original signal $s(t)$ should have finite energy. It means that the original signal $s(t)$ is square integrable, that is

$$\int_{-\infty}^{\infty} |s(t)|^2\, dt < \infty \qquad ...(5.15)$$

then we are guaranteed that CTFT of $s(t)$, $S(j\omega)$ is finite.

The error between $s(t)$ and $\hat{s}(t)$ is denoted by $e(t)$ and is given by

$$e(t) = \hat{s}(t) - s(t) \quad \text{...(5.16)}$$

If a signal $s(t)$ is square integrable then it must satisfy Eq. 5.17 given as

$$\int_{-\infty}^{\infty} |e(t)|^2 \, dt = 0 \quad \text{...(5.17)}$$

Thus, in a manner similar to that for periodic signals, if $s(t)$ has finite energy, then, although original signal $s(t)$ and its Fourier transform representation $\hat{s}(t)$ may differ significantly at individual values of t, there is no energy in their differences, *i.e.*, $e(t) = \hat{s}(t) - s(t)$.

There is also an alternative set of conditions for non-periodic signals like the ones we had for periodic signals. These set of conditions are sufficient to ensure that $\hat{s}(t) = s(t)$ for any t except at the discontinuity. At the discontinuity, $\hat{s}(t)$ is equal to the average of the values on either side of the discontinuity. These conditions are called *Dirichlet conditions*. There are three Dirichlet conditions for CTFT and are given as :

1. Signal $s(t)$ be absolutely integrable, that is

$$\int_{-\infty}^{\infty} |s(t)| < \infty \quad \text{...(5.18)}$$

2. Signal $s(t)$ should have a finite number of maxima and minima within any finite interval.
3. Signal $s(t)$ should have a finite number of discontinuities within any finite interval.

Therefore, absolutely integrable signals which are continuous or the ones which have a finite number of discontinuities have Fourier transforms.

Example 5.1 Determine the CTFT of a continuous-time signal given as

$$s(t) = e^{-At} \, u(t) \, , \; A > 0 \quad \text{...(1)}$$

where A is a scalar constant.

Solution. $\text{CTFT}\{ s(t) \} = S(j\omega) = \int_{-\infty}^{\infty} s(t) \, e^{-j\omega t} \, dt$

$$= \int_{-\infty}^{\infty} e^{-At} \, u(t) \, e^{-j\omega t} \, dt$$

$$= \int_{-\infty}^{\infty} e^{-(A + j\omega)t} \, u(t) \, dt$$

But, $\quad u(t) = \begin{cases} 1, & t > 0 \\ 0, & t < 0 \end{cases}$

or $$S(j\omega) = \int_0^\infty e^{-(A+j\omega)t}\,dt = \left[-\frac{1}{(A+j\omega)}e^{-(A+j\omega)t}\right]_0^\infty$$

$$= \frac{1}{(A+j\omega)},\ A > 0$$

or $$S(j\omega) = \frac{1}{(A+j\omega)},\ A > 0$$

Since this Fourier transform $S(j\omega)$ is complex valued, to plot $S(j\omega)$ as a function of frequency ω, we express $S(j\omega)$ in terms of its magnitude and phase.

Magnitude of $S(j\omega) = |S(j\omega)| = \dfrac{1}{\sqrt{A^2+\omega^2}}$

and Phase of $S(j\omega) = \angle S(j\omega) = -\tan^{-1}(\omega/A)$

Magnitude and phase of $S(j\omega)$ are sketched in Fig. 5.5.

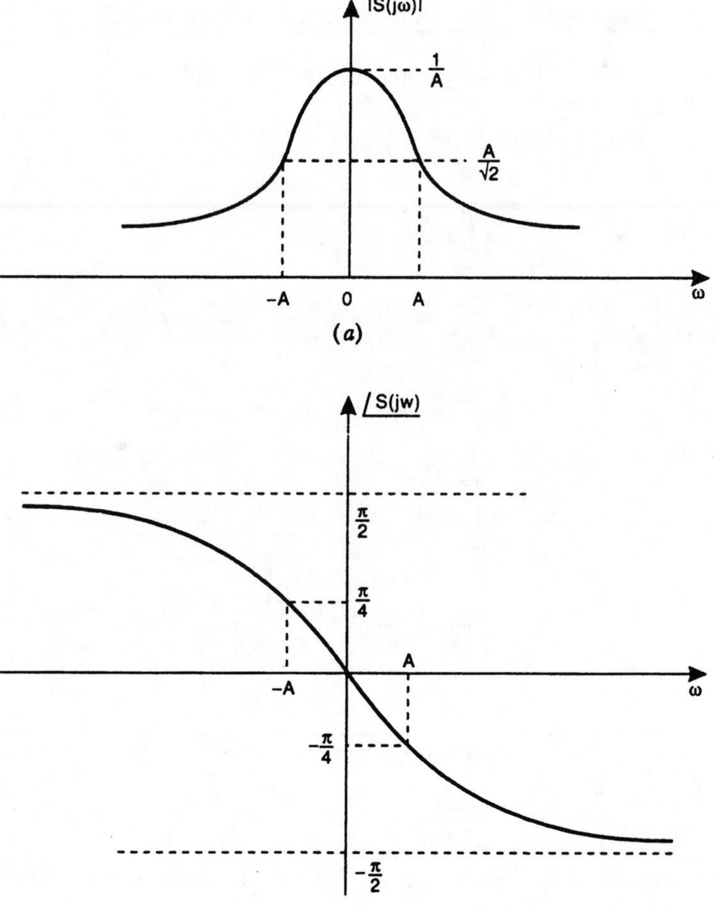

Fig. 5.5 *CTFT of the signal $s(t) = e^{-At}u(t)$ for $A > 0$. (a) Magnitude spectrum of $S(j\omega)$ (b) Phase spectrum of $S(j\omega)$.*

Note **: If A is complex, then the signal $s(t)$ is absolutely integrable as long as $\text{Re}\{A\} > 0$. In this case, we get that same result for $S(j\omega)$. That is**

$$S(j\omega) = \frac{1}{(A + j\omega)}, \quad \text{Re}\{A\} > 0.$$

Example 5.2 Determine the CTFT of a continuous-time signal

$$s(t) = e^{-A|t|}, \quad A > 0$$

Solution. Signal $s(t) = e^{-A(t)}$ is sketched in Fig. 5.6.

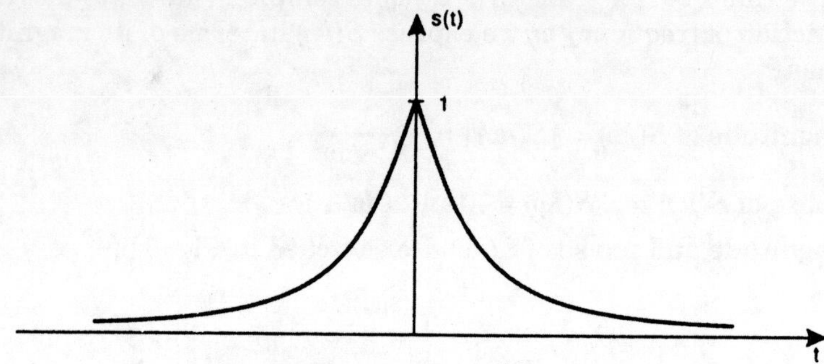

Fig. 5.6 *Signal* $s(t) = e^{-A|t|}$, $A > 0$.

$$\text{CTFT}\{s(t)\} = S(j\omega) = \int_{-\infty}^{\infty} s(t)\, e^{-j\omega t}\, dt$$

$$= \int_{-\infty}^{\infty} e^{-A|t|}\, e^{-j\omega t}\, dt$$

$$= \int_{-\infty}^{0} e^{-A(-t)}\, e^{-j\omega t}\, dt + \int_{0}^{\infty} e^{-A(t)}\, e^{-j\omega t}\, dt$$

$$= \int_{-\infty}^{0} e^{At}\, e^{-j\omega t}\, dt + \int_{0}^{\infty} e^{-At}\, e^{-j\omega t}\, dt$$

$$= \int_{-\infty}^{0} e^{(A-j\omega)t}\, dt + \int_{0}^{\infty} e^{-(A+j\omega)t}\, dt$$

$$= \left[\frac{1}{(A-j\omega)}\, e^{(A-j\omega)t}\right]_{-\infty}^{0} + \left[\frac{1}{-(A+j\omega)}\, e^{-(A+j\omega)t}\right]_{0}^{\infty}$$

$$= \frac{1}{A-j\omega} + \frac{1}{A+j\omega} = \frac{2A}{A^2+\omega^2}$$

$$S(j\omega) = \frac{2A}{A^2+\omega^2}$$

In this case, CTFT $S(j\omega)$ is real. We can sketch $S(j\omega)$ directly. CTFT $S(j\omega)$ is sketched in Fig. 5.7.

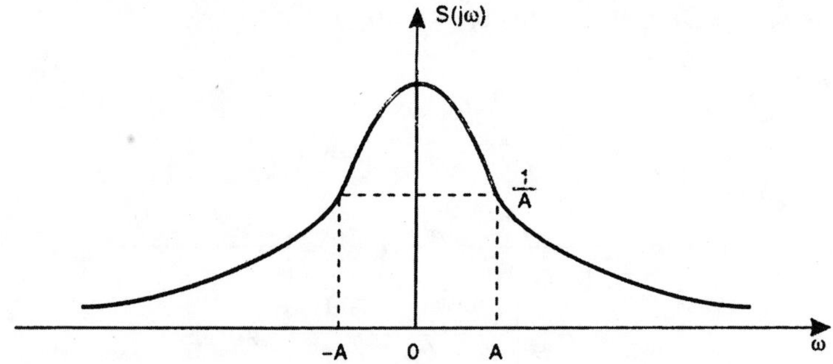

Fig. 5.7 *CTFT of* $s(n) = e^{-A|t|}$, $A > 0$.

Example 5.3 Determine the CTFT of the unit impulse function $s(t) = \delta(t)$.

Solution. $\text{CTFT}\{s(t)\} = S(j\omega) = \int_{-\infty}^{\infty} s(t)\, e^{-j\omega t}\, dt$

$$= \int_{-\infty}^{\infty} \delta(t)\, e^{-j\omega t}\, dt$$

But $$\delta(t) = \begin{cases} 1, & t = 0 \\ 0, & t \neq 0 \end{cases}$$

$$S(j\omega) = \int_{-\infty}^{\infty} \delta(t)\, e^{-j\omega t}\, dt = 1$$

That is, the unit impulse function or Dirac Delta function $\delta(t)$ has a CTFT consisting of equal contribution at all frequencies.

Example 5.4 Determine the CTFT of the rectangular pulse signal

$$s(t) = \begin{cases} 1, & |t| < T_1 \\ 0, & |t| > T_1 \end{cases}$$

This signal is shown in Fig. 5.8.

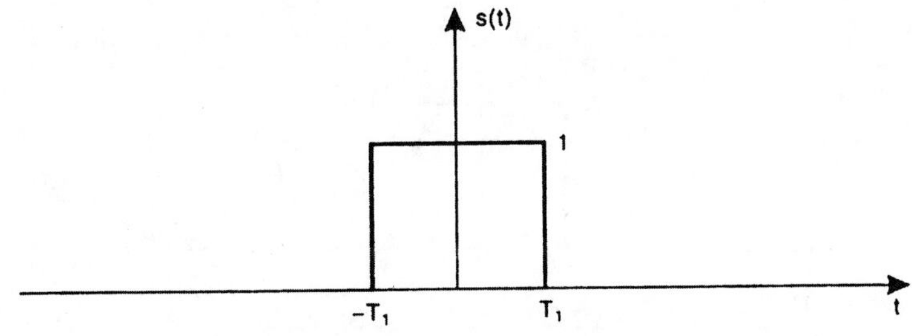

Fig. 5.8 *Rectangular pulse signal.*

Solution. CTFT $\{ s(t) \} = S(j\omega) = \int_{-\infty}^{\infty} s(t)\, e^{-j\omega t}\, dt$

$$= \int_{-T_1}^{T_1} 1.e^{-j\omega t}\, dt = \left[\frac{1}{-j\omega} e^{-j\omega t} \right]_{-T_1}^{T_1}$$

$$= \frac{1}{-j\omega} [e^{-j\omega T_1} - e^{-j\omega(-T_1)}] = \frac{e^{j\omega T_1} - e^{-j\omega T_1}}{j\omega}$$

$$= \frac{2}{\omega} \left[\frac{e^{j\omega T_1} - e^{-j\omega T_1}}{2j} \right] = \frac{2}{\omega} \sin \omega T_1$$

$$S(j\omega) = \frac{2}{\omega} \sin \omega T_1$$

is sketched in Fig. 5.9.

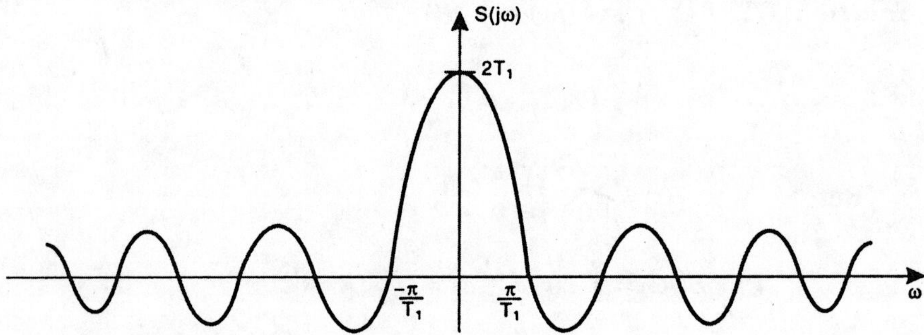

Fig. 5.9 *CTFT of a rectangular pulse signal in shown in Fig.* 5.8.

Example 5.5 Determine the continuous-time signal $s(t)$ whose CTFT is given by

$$S(j\omega) = \begin{cases} 1, & |\omega| < W \\ 0, & |\omega| > W \end{cases}$$

and it is shown in Fig. 5.10.

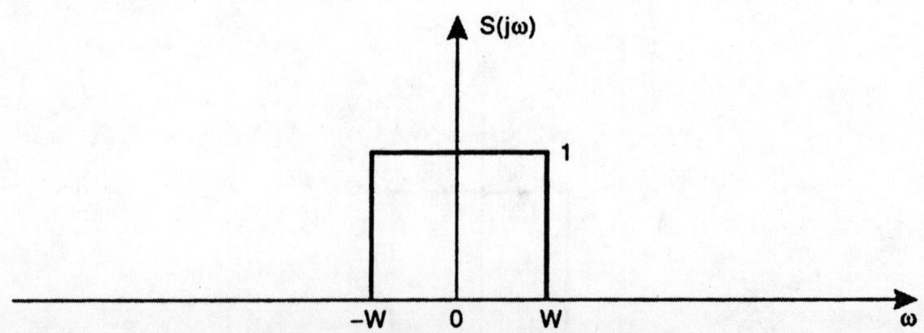

Fig. 5.10 *CTFT of* $s(n)$, $S(j\omega)$.

Solution. The continuous-time signal $s(t)$ can be determined by using CTFT synthesis equation.

$$s(t) = \frac{1}{2\pi} \int_{-\infty}^{\infty} S(j\omega)\, e^{j\omega t}\, d\omega$$

$$= \frac{1}{2\pi} \int_{-W}^{W} 1.e^{j\omega t}\, d\omega = \frac{1}{2\pi} \left[\frac{1}{jt} e^{j\omega t} \right]_{-W}^{W}$$

$$= \frac{1}{\pi t} \left[\frac{e^{jWt} - e^{-jWt}}{2j} \right] = \frac{1}{\pi t} \sin Wt$$

Therefore, $s(n) = \dfrac{\sin Wt}{\pi t}$ and signal $s(n) = \dfrac{\sin Wt}{\pi t}$ is illustrated in Fig. 5.11.

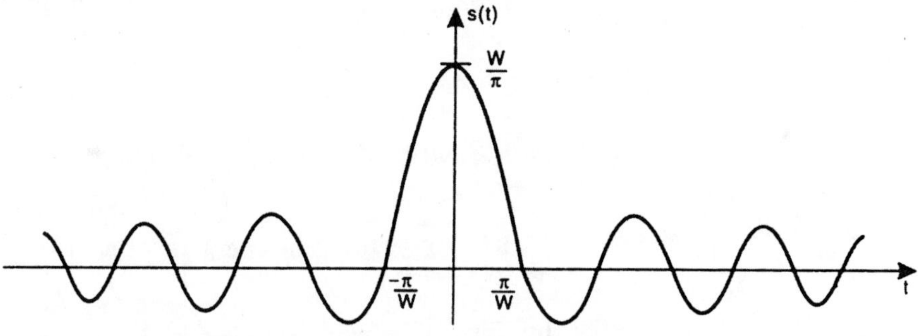

Fig. 5.11 *Signal* $s(t) = \dfrac{\sin Wt}{\pi t}$

5.5 CTFT OF CONTINUOUS-TIME PERIODIC SIGNALS

In the section (5.3), we discussed Fourier Transform representation of non-periodic signals. Here we will discuss Fourier Transform representation of periodic signals. Thus both periodic and non-periodic signals are discussed within a unified context. Now, we can construct the Fourier Transform of a periodic signal directly from its Fourier Series representation. The resulting Fourier Transform consists of a train of impulses in the frequency domain. The area of these impulses is directly proportional to the Fourier series coefficients A_k. This representation can be very useful to us.

Let us consider a signal $s(t)$ with Fourier transform $S(j\omega)$. Here $S(j\omega)$ is a single impulse of area 2π at $\omega = \omega_0$. It can be written as

$$S(j\omega) = 2\pi\, \delta(\omega - \omega_0) \qquad ...(5.19)$$

To determine the signal $s(t)$ for this Fourier transform $S(j\omega) = 2\pi\, \delta(\omega - \omega_0)$, we use inverse Fourier transform relation.

$$s(t) = \frac{1}{2\pi} \int_{-\infty}^{\infty} S(j\omega)\, e^{j\omega t}\, d\omega$$

$$= \frac{1}{2\pi} \int_{-\infty}^{\infty} 2\pi\, \delta(\omega - \omega_0)\, e^{j\omega t}\, d\omega$$

$$= \frac{2\pi}{2\pi} \int_{-\infty}^{\infty} \delta(\omega - \omega_0)\, e^{j\omega t}\, d\omega = e^{j\omega_0 t} \qquad ...(5.20)$$

In a more general sense, if $S(j\omega)$ is of the form of a linear combination of impulses equally spaced in frequency. It can be written as

$$S(j\omega) = \sum_{k=-\infty}^{\infty} 2\pi A_k\, \delta(\omega - k\omega_0) \qquad ...(5.21)$$

Form Fourier Transform synthesis equation

$$s(t) = \frac{1}{2\pi} \int_{-\infty}^{\infty} S(j\omega)\, e^{j\omega t}\, d\omega \qquad ...(5.22)$$

Substituting Eq. 5.21 in Eq. 5.22, we get

$$s(t) = \frac{1}{2\pi} \int_{-\infty}^{\infty} S(j\omega)\, e^{j\omega t}\, d\omega$$

$$= \frac{1}{2\pi} \int_{-\infty}^{\infty} \left[\sum_{k=-\infty}^{\infty} 2\pi A_k\, \delta(\omega - k\omega_0) \right] e^{j\omega t}\, d\omega$$

$$= \frac{2\pi}{2\pi} \sum_{k=-\infty}^{\infty} A_k \left[\int_{-\infty}^{\infty} \delta(\omega - k\omega_0)\, e^{j\omega t}\, d\omega \right]$$

or

$$s(t) = \sum_{k=-\infty}^{\infty} A_k\, e^{jk\omega_0 t} \qquad ...(5.23)$$

We see that Eq. 5.23 corresponds exactly to the Fourier Series representation of a periodic signal.

Thus, the Fourier transform of a periodic signal with Fourier series coefficients $\{A_k\}$ can be interpreted as a train of impulses occuring at harmonically related frequencies. The area of the impulse at kth harmonic frequency $k\omega_0$ for $S(j\omega)$ is 2π times the kth Fourier series coefficients A_k.

Example 5.6 Determine the CTFT of the square wave signal given in Fig. 5.12.

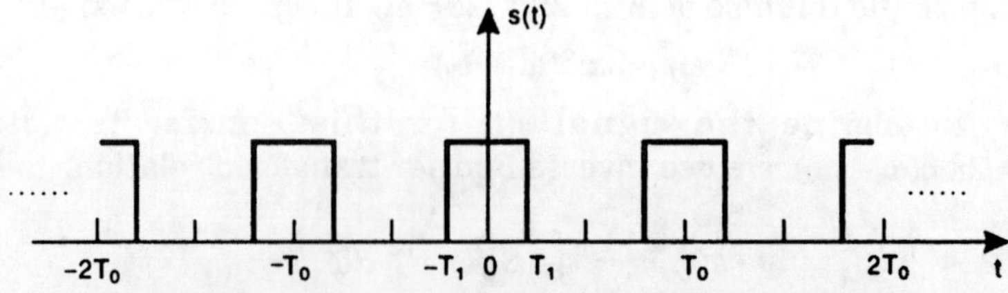

Fig. 5.12 *A continuous-time periodic square wave signal.*

Solution. This signal is periodic with fundamental period T_0 and fundamental frequency $\omega_0 = 2\pi/T_0$.

To determine the CTFT of this periodic signal, we first determine Fourier series coefficients A_k. This signal can be defined over one period as

$$s(t) = \begin{cases} 1, & |t| < T_1 \\ 0, & T_1 < |t| < T_0/2 \end{cases} \quad ...(1)$$

Because signal $s(t)$ is symmetric about time $t = 0$ and it is convenient to choose $-\frac{T_0}{2} \leq |t| \leq \frac{T_0}{2}$ interval over which the integration is to be performed, although we can take any interval of length T_0. Fourier series coefficiens A_k are determined as

$$A_k = \frac{1}{T_0} \int_{T_0} s(t)\, e^{-jk\omega_0 t}\, dt$$

$$= \frac{1}{T_0} \int_{-T_1}^{T_1} 1.e^{-jk\omega_0 t}\, dt$$

$$= \frac{1}{T_0} \left[\frac{1}{-jk\omega_0} e^{-jk\omega_0 t} \right]_{-T_1}^{T_1}$$

$$= \frac{1}{T_0} \frac{1}{-jk\omega_0} [e^{-jk\omega_0 T_1} - e^{jk\omega_0 T_1}]$$

$$= \frac{2}{k\omega_0 T_0} \left(\frac{e^{jk\omega_0 T_1} - e^{-jk\omega_0 T_1}}{2j} \right)$$

$$= \frac{2}{k \frac{2\pi}{T_0} T_0} \sin(k\omega_0 T_1)$$

or $$A_k = \frac{\sin(k\omega_0 T_1)}{k\pi}, \quad k \neq 0 \quad ...(2)$$

Now we can determine CTFT for this signal in terms of Fourier series coefficients using following formula.

CTFT {Continuous-Time Periodic Signal $s(t)$}

$$= S(j\omega) = \sum_{k=-\infty}^{\infty} 2\pi A_k\, \delta(\omega - k\omega_0) \quad ...(3)$$

Substituting values of A_k from Eq. 2 into Eq. 3, we get

$$S(j\omega) = \sum_{k=-\infty}^{\infty} 2\pi A_k\, \delta(\omega - k\omega_0)$$

$$= \sum_{k=-\infty}^{\infty} 2\pi \left[\frac{\sin k\omega_0 T_1}{k\pi} \right] \delta(\omega - k\omega_0)$$

or $$S(j\omega) = \sum_{k=-\infty}^{\infty} \frac{2 \sin k\omega_0 T_1}{k} \delta(\omega - k\omega_0) \quad ...(4)$$

Fig. 5.13 illustrates the CTFT of symmetric continuous-time periodic square wave signal form $T_0 = 4T_1$.

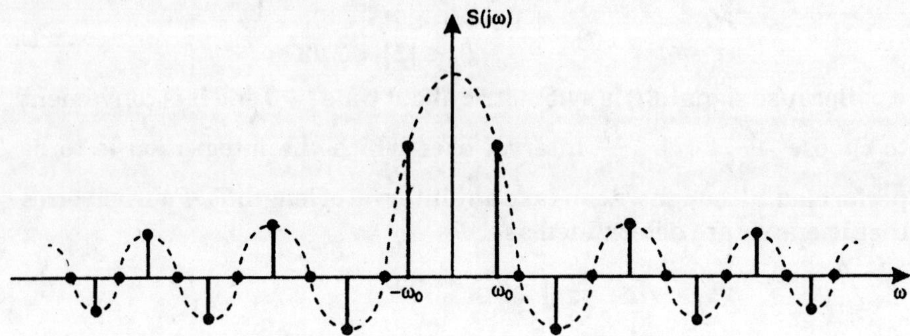

Fig. 5.13 *CTFT of a symmetric continuous-time periodic square wave signal.*

Now Eq. 4 becomes

$$S(j\omega) = \sum_{k=-\infty}^{\infty} \frac{2 \sin\left(k \frac{2\pi}{4T_1} T_1\right)}{k} \delta(\omega - k\omega_0)$$

$$= \sum_{k=-\infty}^{\infty} \frac{2 \sin \frac{k\pi}{2}}{k} \delta(\omega - k\omega_0)$$

Example 5.7 Detemine the CTFT of a periodic continuous-time sinusoidal signal $s(t) = \sin \omega_0 t$

Solution. To determine the CTFT of a periodic continuous- time signal $s(t)$, we determine Fourier series coefficients A_k for this periodic signal as

$$s(t) = \sin \omega_0 t = \frac{e^{j\omega_0 t} - e^{-j\omega_0 t}}{2j}$$

$$= \frac{1}{2j} e^{j\omega_0 t} - \frac{1}{2j} e^{-j\omega_0 t} \quad ...(1)$$

But, Fourier series representation for a periodic signal $s(t)$ is given by

$$s(t) = \sum_{k=-\infty}^{\infty} A_k e^{jk\omega_0 t}$$

$$= ... + A_{-1} e^{j(-1)\omega_0 t} + A_0 e^{j(0)\omega_0 t} + A_1 e^{j(1)\omega_0 t} + ... \quad ...(2)$$

Now, comparing Eqs. 1 and 2, we get

$$A_1 = \frac{1}{2j} \text{ and } A_{-1} = -\frac{1}{2j} \quad ...(3)$$

and Fourier series coefficients $A_k = 0$ for all the values of k other than $k = -1$ and $k = 1$.

CTFT {Continuous-time periodic signal $s(t)$}

$$= S(j\omega) = \sum_{k=-\infty}^{\infty} 2\pi A_k\, \delta(\omega - k\omega_0) \qquad \text{...(4)}$$

Substituting Eq. 3 in Eq. 4, we get

$$\begin{aligned} S(j\omega) &= \sum_{k=-\infty}^{\infty} 2\pi A_k\, \delta(\omega - k\omega_0) \\ &= \ldots + 2\pi A_{-1}\, \delta[\omega - (-1)\,\omega_0] + 2\pi A_0\, \delta[\omega - (0)\,\omega_0] \\ &\qquad + 2\pi A_1\, \delta[\omega - (1)\,\omega_0] + \ldots \\ &= \ldots + 2\pi A_{-1}\, \delta[\omega + \omega_0] + 2\pi A_0\, \delta[\omega] \\ &\qquad + 2\pi A_1\, \delta[\omega - \omega_0] + \ldots \\ &= \ldots + 2\pi\left(-\frac{1}{2j}\right)\delta(\omega + \omega_0) + 2\pi\,(0)\,\delta(\omega) \\ &\qquad + 2\pi\left(\frac{1}{2j}\right)\delta(\omega - \omega_0) + \ldots \\ &= -\frac{\pi}{j}\,\delta(\omega + \omega_0) + \frac{\pi}{j}\,\delta(\omega - \omega_0) \end{aligned}$$

or $$S(j\omega) = \pi j\,[\delta(\omega + \omega_0) - \delta(\omega - \omega_0)\,] \qquad \text{...(5)}$$

CTFT of $s(t) = \sin \omega_0 t$ is shown in Fig. 5.14.

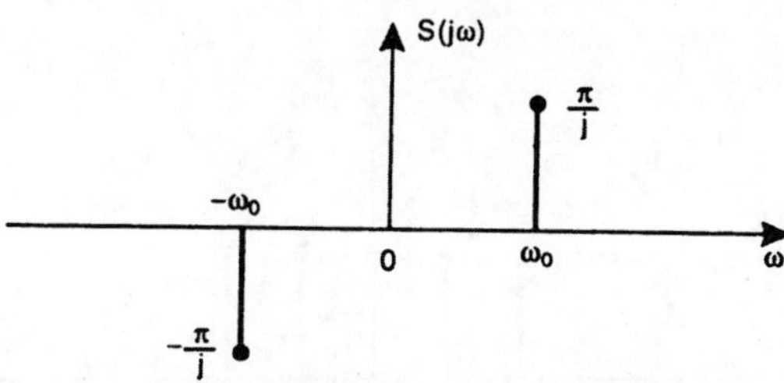

Fig. 5.14 *CTFT of a continuous-time periodic sinusoidal signal $s(t) = \sin \omega_0 t$.*

Example 5.8 Determine the CTFT for the periodic impulse train given by

$$s(t) = \sum_{k=-\infty}^{\infty} \delta(t - kT_0) \qquad \text{...(1)}$$

This impulse train is periodic with period T_0.

Solution. To determine CTFT of a periodic signal $s(t)$ we determine Fourier series coefficients A_k. We use equation

$$A_k = \frac{1}{T_0} \int_{T_0} s(t)\, e^{-jk\omega_0 t}\, dt$$

to determine the Fourier series coefficients in the interval of integration $-\frac{T_0}{2} \le t \le \frac{T_0}{2}$.

$$A_k = \frac{1}{T_0} \int_{-T_0/2}^{T_0/2} s(t)\, e^{-jk\omega_0 t}\, dt = \frac{1}{T_0} \qquad \text{... (2)}$$

In other words, all the Fourier series coefficients of the impulse train are identical. Fourier series coefficients A_k of the periodic impulse train has the same value, *i.e.*, $1/T_0$.

CTFT {Continuous-time periodic signal $s(t)$}

$$= S(j\omega) = \sum_{k=-\infty}^{\infty} 2\pi A_k\, \delta(\omega - k\omega_0) \qquad \text{...(3)}$$

Substituting the vlaue of Fourier series coefficients A_k from Eq. 2 into Eq. 3, we get

$$S(j\omega) = \sum_{k=-\infty}^{\infty} \frac{2\pi}{T_0} A_k\, \delta(\omega - k\omega_0)$$

$$= \frac{2\pi}{T_0} \sum_{k=-\infty}^{\infty} \delta\left(\omega - k\frac{2\pi}{T_0}\right) \qquad \text{...(4)}$$

Thus, the CTFT of a periodic impulse train in the time-domain with fundamental period T_0 is a periodic impulse train in frequency domain with period $\omega_0 = 2\pi/T_0$. It again shows that there is an inverse relationship between time and frequency-domains. The periodic impulse train $s(t)$ and its CTFT are shown in Fig. 5.15 and Fig. 5.16, respectively.

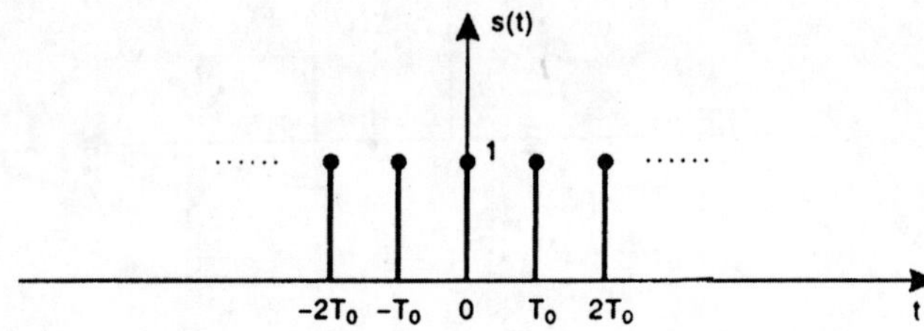

Fig. 5.15 *Periodic impulse train signal.*

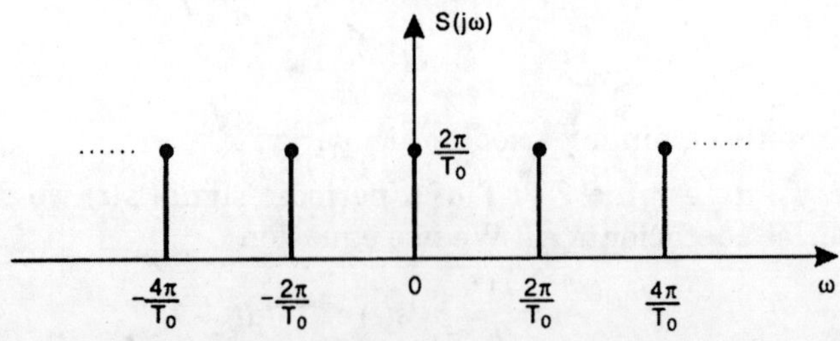

Fig. 5.16 *CTFT of periodic impulse train signal.*

5.6 PROPERTIES OF CTFT

In this section, we will study a number of properties of the Fourier transform of continuous-time non-periodic signal. As in the case with Fourier series representation of periodic signals, the properties of CTFT provide us the relationship between the time-domain and frequency-domain descriptions of a signal. In addition, many of the properties of CTFT are often useful in reducing the complexity of evaluation of Fourier transforms or inverse Fourier transforms.

Furthermore, we know that there is a close relationship between the Fourier series and Fourier transform representation of a periodic signal. By using this relationship, we can translate many of the properties of CTFT into properties of Fourier series.

In this book, we will use a shorthand notation to indicate the pairing of a signal and its Fourier transform. A signal $s(t)$ and its Fourier transform $S(j\omega)$ are related by the Fourier transform analysis and synthesis equations given by

$$S(j\omega) = \int_{-\infty}^{\infty} s(t)\, e^{-j\omega t}\, dt \qquad ...(5.24)$$

and

$$s(t) = \frac{1}{2\pi} \int_{-\infty}^{\infty} S(j\omega)\, e^{j\omega t}\, d\omega \qquad ...(5.25)$$

We refer signal $s(t)$ and its Fourier transform $S(j\omega)$ as a Fourier transform pair with the following notation

$$s(t) \xrightarrow{\text{CTFT}} S(j\omega)$$

or

$$S(j\omega) = \text{CTFT}\left\{ s(n) \right\}$$

In this section, we will study following properties of CTFT

1. Linearity
2. Time shifting
3. Conjugation
4. Conjugate symmetry
5. Differentiation and integration
6. Time scaling
7. Frequency scaling
8. Duality
9. Parseval's relation for CTFT
10. Convolution property
11. Multiplication property.

Now we will discuss each property one by one in subsequent sections.

5.6.1 Linearity

CTFT is also a linear transformation tool which is used to transform a signal from time-domain to frequency-domain. CTFT also satisfies the principle of superposition.

Let us consider two continuous-time signals $s_1(t)$ and $s_2(t)$.

If $$s_1(t) \xleftrightarrow{\text{CTFT}} S_1(j\omega)$$

and $$s_2(t) \xleftrightarrow{\text{CTFT}} S_2(j\omega)$$

then from linearity of CTFT, we have

$$A_1 s_1(t) + A_2 s_2(t) \xleftrightarrow{\text{CTFT}} A_1 S_1(j\omega) + A_2 S_2(j\omega) \quad ...(5.26)$$

where A_1 and A_2 are constants.

Also the property of linearity of CTFT is easily extended to a linear combination of an arbitrary number of signals. Therefore, we can write as

$$A_1 s_1(t) + A_2 s_2(t) + A_3 s_3(t) + \ldots \xleftrightarrow{\text{CTFT}} A_1 S_1(j\omega) + A_2 S_2(j\omega) + A_3 S_3(j\omega) + \ldots \quad ...(5.27)$$

Example 5.9 Determine the CTFT of the signal shown in Fig. 5.17.

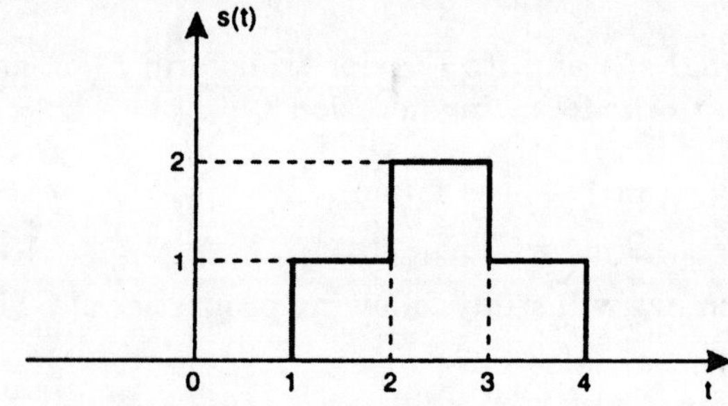

Fig. 5.17

Solution. Above signal $s(t)$ can be expressed as

$$s(t) = \begin{cases} 1, & 1 \le t \le 2 \\ 2, & 2 \le t \le 3 \\ 1, & 3 \le t \le 4 \end{cases} \quad ...(1)$$

$$\text{CTFT}\{s(t)\} = S(j\omega) = \int_{-\infty}^{\infty} s(t)\, e^{-j\omega t}\, dt$$

$$= \int_1^2 1e^{-j\omega t}\, dt + \int_2^3 2e^{-j\omega t}\, dt + \int_3^4 1e^{-j\omega t}\, dt$$

$$= \left[\frac{e^{-j\omega t}}{-j\omega}\right]_1^2 + \left[\frac{2e^{-j\omega t}}{-j\omega}\right]_2^3 + \left[\frac{e^{-j\omega t}}{-j\omega}\right]_3^4$$

$$= \frac{1}{(-j\omega)}[e^{-j\omega 2} - e^{-j\omega 1} + 2e^{-j\omega 3} - 2e^{-j\omega 2} + e^{-j\omega 4} - e^{-j\omega 3}]$$

$$= \frac{1}{(-j\omega)}[e^{-j2\omega} - e^{-j\omega} + 2e^{-j3\omega} - 2e^{-j2\omega} + e^{-j4\omega} - e^{-j3\omega}]$$

$$= \frac{1}{(-j\omega)}[-e^{-j\omega} - e^{-j2\omega} + e^{-j3\omega} + e^{-j4\omega}]$$

$$= \frac{e^{-(j5\omega)/2}}{\omega j}[e^{j3/2\omega} - e^{-j3/2\omega} + e^{j1/2\omega} - e^{-j1/2\omega}]$$

$$= \frac{e^{-(j5\omega)/2}}{\omega}\left[\frac{e^{j3/2\omega} - e^{-j3/2\omega}}{2j} + \frac{e^{j1/2\omega} - e^{-j1/2\omega}}{2j}\right]$$

$$= \frac{e^{-(j5\omega)/2}}{\omega}\left[2\sin\frac{3\omega}{2} + 2\sin\frac{1}{2}\omega\right]$$

$$= \frac{2e^{-(j5\omega)/2}}{\omega}\left[\sin\left(\frac{\omega}{2}\right) + \sin\left(\frac{3\omega}{2}\right)\right]$$

Example 5.10 Determine the CTFT of the unit step function $s(t) = u(t)$.

Solution. To determine the CTFT of a unit step function, we will use the knowledge that

$$g(t) = \delta(t) \xleftrightarrow{\text{CTFT}} G(j\omega) = 1 \qquad ...(1)$$

So we can represent $s(t)$ in terms of $g(t)$ as

$$s(t) = \int_{-\infty}^{t} g(\tau)\, d\tau \qquad ...(2)$$

Taking CTFT of both sides of Eq. 2, we get

$$\text{CTFT}\{s(t)\} = \text{CTFT}\left\{\int_{-\infty}^{t} g(\tau)\, d\tau\right\}$$

or
$$S(j\omega) = \frac{G(j\omega)}{j\omega} + \pi\, G(0)\, \delta(\omega) \qquad ...(3)$$

Eq. 3 is called integration property of CTFT.

Since $g(t) = \delta(t)$ then its CTFT will be

$G(\omega) = 1$

Substituting $G(\omega) = 1$ in Eq. 3, we obtain

$$S(j\omega) = \frac{1}{j\omega} + \pi\, 1\, \delta(\omega)$$

or
$$S(j\omega) = \frac{1}{j\omega} + \pi\, \delta(\omega)$$

5.6.2 Time Shifting

Here we will study the effect of shifting the time-variable t on the CTFT of a signal $s(t)$.

if $$s(t) \xleftrightarrow{\text{CTFT}} S(j\omega)$$

then $$s(t-t_0) \xleftrightarrow{\text{CTFT}} e^{-j\omega t_0}\, S(j\omega) \qquad ...(5.28)$$

To prove this property, let us consider the CTFT synthesis equation given as

$$s(t) = \frac{1}{2\pi}\int_{-\infty}^{\infty} S(j\omega)\, e^{j\omega t}\, d\omega$$

Substituting $t = (t - t_0)$ in above equation, we get

$$s(t-t_0) = \frac{1}{2\pi}\int_{-\infty}^{\infty} S(j\omega)\, e^{j\omega(t-t_0)}\, d\omega$$

$$= \frac{1}{2\pi}\int_{-\infty}^{\infty} S(j\omega)\, e^{j\omega t}\, e^{-j\omega t_0}\, d\omega$$

or $$s(t-t_0) = \frac{1}{2\pi}\int_{-\infty}^{\infty} [S(j\omega)\, e^{-j\omega t_0}]\, e^{j\omega t}\, d\omega \qquad ...(5.29)$$

Therefore, we conclude from Eq. 5.29 that

$$\text{CTFT}\{s(t-t_0)\} = e^{-j\omega t_0}\, S(j\omega) \qquad ...(5.30)$$

The magnitude of CTFT of a signal does not change in time-shifting of a signal. CTFT of a signal $s(t)$ can be expressed in polar form as

$$\text{CTFT}\{s(t)\} = S(j\omega) = |S(j\omega)|\, e^{j\angle S(j\omega)}$$

then $$\text{CTFT}\{s(t-t_0)\} = e^{-j\omega t_0}\, S(j\omega) = |S(j\omega)|\, e^{j[\angle S(j\omega) - \omega t_0]} \qquad ...(5.31)$$

Thus, the effect of time-shift on a signal is to introduce into its Fourier transform a phase shift, namely $(-\omega t_0)$. Here phase shift is a linear function of frequency ω.

5.6.3 Conjugation and Conjugate Symmetry

Complex conjugate of complex signal $s(t)$ is obtained by changing the sign of imaginary part of the signal $s(t)$.

Signal $s(t)$ can be written in rectangular form as

$$s(t) = s_R(t) + j\, s_I(t)$$

where $s_R(t)$ and $s_I(t)$ are the real and imaginary parts of the complex signal $s(t)$.

Complex conjugate of $s(t)$ will be expressed as

$$s^*(t) = s_R(t) - j\, s_I(t)$$

The conjugation property states that

if $$s(t) \xleftrightarrow{\text{CTFT}} S(j\omega)$$

then $$s^*(t) \xleftrightarrow{\text{CTFT}} S^*(-j\omega)$$

where $S^*(j\omega)$ is the complex conjugate of CTFT of $s(t)$, *i.e.*, $S(j\omega)$.

$S(j\omega)$ can be expressed in rectangular form as $S(j\omega) = S_R(j\omega) + j\,S_I(j\omega)$ and its complex conjugate can be expressed as

$$S^*(j\omega) = S_R(j\omega) - j\,S_I(j\omega)$$

Therefore, $S^*(-j\omega)$ can be written as

$$S^*(-j\omega) = S_R(-j\omega) - S_I(-j\omega)$$

To prove this property, we use the CTFT analysis equation given as

$$S(j\omega) = \text{CTFT}\left\{ s(t) \right\} = \int_{-\infty}^{\infty} s(t)\, e^{-j\omega t}\, dt$$

The complex conjugate of $S(j\omega)$ is determined as

$$S^*(j\omega) = \int_{-\infty}^{\infty} s^*(t)\, [e^{-j\omega t}\,]^* dt$$

$$= \int_{-\infty}^{\infty} s^*(t)\, e^{j\omega t}\, dt$$

Replacing ω by –ω, we get

$$S^*(-j\omega) = \int_{-\infty}^{\infty} s^*(t)\, e^{-j\omega t}\, dt \qquad ...(5.32)$$

The right hand side of Eq. 5.32 is the CTFT analysis equation for $s^*(t)$

Hence $$\text{CTFT}\left\{ s^*(t) \right\} = S^*(-j\omega) \qquad ...(5.33)$$

Now let us consider a special case that is when $s(t)$ is a real signal then its CTFT, $S(j\omega)$ will have conjugate symmetry. It is given as

$$S(-j\omega) = S^*(j\omega) \;\text{ for } s(t) \text{ is real} \qquad ...(5.34)$$

If $s(t)$ is real so that $s^*(t) = s(t)$, then from Eq. 5.32, we get

$$S^*(-j\omega) = \int_{-\infty}^{\infty} s^*(t)\, e^{-j\omega t}\, dt$$

$$= \int_{-\infty}^{\infty} s(t)\, e^{-j\omega t}\, dt = S(j\omega) \qquad ...(5.35)$$

Replacing ω with –ω, we get

$$S(-j\omega) = S^*(j\omega)$$

If we express $S(j\omega)$ in rectangular form as

$$S(j\omega) = S_R(j\omega) + j\, S_I(j\omega)$$

then if $s(t)$ is real,

$$S_R(j\omega) = S_R(-j\omega)$$

and $$S_I(j\omega) = -\, S_I(-j\omega)$$

That is, the real part of the CTFT of $s(t)$ is an even function of frequency and the imaginary part is an odd function of frequency.

In a similar manner, we can express $S(j\omega)$ in polar form as

$$S(j\omega) = |S(j\omega)|\, e^{j\angle S(j\omega)}$$

That is, the magnitude of $S(j\omega)$, *i.e.*, $|S(j\omega)|$ is an even function of ω and phase of $S(j\omega)$ *i.e.*, $\angle S(j\omega)$ is an odd function of ω. Thus, when computing or displaying the Fourier transform of a real valued signal, the real and imaginary parts (or magnitude and phase) of the Fourier transform need only be specified for positive frequencies. The values for negative frequencies can be determined directly from the values for positive frequencies ($\omega > 0$).

If signal $s(t)$ is both real and even, then $S(j\omega)$ will also be real and even.

From Eq. 5.34

$$S(-j\omega) = \int_{-\infty}^{\infty} s(t)\, e^{j\omega t}\, dt$$

Substituting $t = -\tau$ in above equation, we get

$$S(-j\omega) = \int_{\tau=-\infty}^{-\infty} s(-\tau)\, e^{-j\omega\tau}\,(-d\tau)$$

$$= -\int_{\tau=\infty}^{-\infty} s(-\tau)\, e^{-j\omega\tau}\, d\tau$$

$$= \int_{\tau=\infty}^{-\infty} s(-\tau)\, e^{-j\omega\tau}\, d\tau \qquad ...(5.36)$$

Since $s(-\tau) = s(\tau)$, we get

$$S(-j\omega) = \int_{\tau=-\infty}^{\infty} s(\tau)\, e^{-j\omega\tau}\, d\tau = S(j\omega) \qquad ...(5.37)$$

Thus, $S(j\omega)$ is an even function of frequency ω.

In a similar manner, it can be shown that if a signal $s(t)$ is real and odd function of time, such that $s(t) = -\, s(t)$, then $S(j\omega)$ is purely imaginary and odd function of frequency ω.

We already know that a real function $s(t)$ can always be represented in terms of the sum of even function $s_E(t) = \mathrm{Ev}\{s(t)\}$ and odd function

$s_O(t) = \text{Od}\{s(t)\}$. It can be represented as

$$s(t) = s_E(t) + s_O(t)$$

From the property of linearity of the CTFT

$$\text{CTFT}\{s(t)\} = \text{CTFT}\{s_E(t)\} + \text{CTFT}\{s_O(t)\}$$

We have already discussed that CTFT of $s_E(t)$ is a real function and CTFT of $s_O(t)$ is purely imaginary. Thus, we can conclude that, with $s(t)$ as a real function,

$$s(t) \xleftrightarrow{\text{CTFT}} S(j\omega)$$

$$\text{Ev}\{s(t)\} \xleftrightarrow{\text{CTFT}} \text{Re}\{S(j\omega)\}$$

$$\text{Od}\{s(t)\} \xleftrightarrow{\text{CTFT}} j\,\text{Im}\{S(j\omega)\}$$

5.6.4 Differentiation and Integration of Signal $s(t)$

We will study the Fourier transform of differentiation and integration of a signal $s(t)$. Let $s(t)$ be a signal with CTFT $S(j\omega)$.

By differentiation the CTFT synthesis equation given as

$$s(t) = \frac{1}{2\pi}\int_{-\infty}^{\infty} S(j\omega)\, e^{j\omega t}\, d\omega$$

$$\frac{d\,s(t)}{dt} = \frac{d}{dt}\left[\frac{1}{2\pi}\int_{-\infty}^{\infty} S(j\omega)\, e^{j\omega t}\, d\omega\right]$$

$$= \frac{1}{2\pi}\int_{-\infty}^{\infty} S(j\omega)\left[\frac{d}{dt} e^{j\omega t}\right] d\omega$$

$$= \frac{1}{2\pi}\int_{-\infty}^{\infty} S(j\omega)\,[j\omega e^{j\omega t}]\, d\omega$$

or $$\frac{d\,s(t)}{dt} = \frac{1}{2\pi}\int_{-\infty}^{\infty} [j\omega\, S(j\omega)]\, e^{j\omega t}\, d\omega$$

Therefore, $$\frac{d\,s(t)}{dt} \xleftrightarrow{\text{CTFT}} j\omega\, S(j\omega) \qquad ...(5.38)$$

This is a very important property because it replaces the operation of differentiation in the time-domain with that of multiplication by $j\omega$ in the frequency domain. This property of CTFT is extremely useful for the analysis of continuous-time LTI systems described by differential equations.

We know that integration is the inverse operation of differentiation. So we can say that integration should involve division by $j\omega$ in frequency domain. In integration we add one more term that represnets an average or a dc value that results from integration.

The precise relationship can be written as

$$\int_{-\infty}^{t} s(\tau)\, d\tau \xleftrightarrow{\text{CTFT}} \frac{1}{j\omega} S(j\omega) + \underbrace{\pi\, S(0)\, \delta(\omega)}_{\text{dc term}} \qquad ...(5.39)$$

5.6.5 Time and Frequency Scaling

Alteration of time variable t in a signal is called *time scaling*. Similarly, alteration of frequency variable ω in a signal is referred as frequency scaling.

If $$s(t) \xleftrightarrow{\text{CTFT}} S(j\omega)$$

then $$S(At) \xleftrightarrow{\text{CTFT}} \frac{1}{|A|} S\left(\frac{j\omega}{A}\right) \qquad ...(5.40)$$

where A is a real constant.

This property can be proved directly from the definition of the Fourier transform given as

$$S(j\omega) = \text{CTFT}\{ s(t) \} = \int_{-\infty}^{\infty} s(t)\, e^{-j\omega t}\, dt$$

We will determine the Fourier transform for time-scaled version of $s(t)$ *i.e.,* $S(At)$.

$$\text{CTFT}\{ S(At) \} = \int_{-\infty}^{\infty} S(At)\, e^{-j\omega t}\, dt \qquad ...(5.41)$$

Substituting $\tau = At$ in Eq. 5.41, we obtain

$$\text{CTFT}\{ S(At) \} = \int_{-\infty}^{\infty} s(\tau)\, e^{-j\omega(\tau/A)} \frac{d\tau}{A}$$

$$\text{CTFT}\{ S(At) \} = \frac{1}{A} \int_{-\infty}^{\infty} s(\tau)\, e^{-j(\omega/A)\tau}\, d\tau \qquad \text{for } A > 0$$

$$= \frac{1}{A} S\left(\frac{j\omega}{A}\right) \qquad \text{for } A > 0 \qquad ...(5.42)$$

and $$\text{CTFT}\{ S(At) \} = -\frac{1}{A} \int_{-\infty}^{\infty} s(\tau)\, e^{-j(\omega/A)\tau}\, d\tau \qquad \text{for } A < 0$$

$$= -\frac{1}{A} S\left(\frac{j\omega}{A}\right) \qquad \text{for } A < 0 \qquad ...(5.43)$$

By combining Eq. 5.42 and Eq. 5.43, Eq. 5.40 can be written as

$$\text{CTFT}\{ S(At) \} = \frac{1}{|A|} S\left(\frac{j\omega}{A}\right)$$

or it can be written as

$$S(At) \xleftrightarrow{\text{CTFT}} \frac{1}{|A|} S\left(\frac{j\omega}{A}\right)$$

Thus, besides the amplitude multiplication factor $1/|A|$, a linear time scaling by a factor of A corresponds to a linear frequency scaling by a factor of $1/A$, and vice-versa.

For a special case, putting $A = -1$ in Eq. 5.40, we get

$$s(-t) \xleftrightarrow{\text{CTFT}} S(-j\omega) \quad \text{...(5.44)}$$

That is, folding or reversing a signal in time also reverses its Fourier transform.

A common example which shows the effect on frequency content that results when an audio tape is recorded at one speed and played back at a different speed. If the playback speed is higher than the recording speed. Due to higher play back speed than recording speed, time will compress and corresponding frequency of the spectrum will expand. The audible effect is that the play back frequencies are higher. Conversely, for slower play back speed than recording speed, the time will expand and the corresponding frequency of the spectrum will compress. The audible effect is that the play back frequencies are lower.

The inverse relationship between time and frequency domains is of great importance in a variety of signals and systems context. This inverse relationship is encountered in filtering and filter design.

5.6.6 Duality

By comparing the Fourier transform and inverse Fourier transform relations given as

$$S(j\omega) = \int_{-\infty}^{\infty} s(t)\, e^{-j\omega t}\, dt \quad \text{...(5.45)}$$

and

$$s(t) = \frac{1}{2\pi} \int_{-\infty}^{\infty} S(j\omega)\, e^{j\omega t}\, d\omega \quad \text{...(5.46)}$$

We observe that these equations are similar but not identical in form. This symmetry leads to a property of the Fourier transform that is called *duality*. Specifically, because of the symmetry between Eqs. 5.45 and 5.46, for any Fourier transform pair, there is a dual pair with the time and frequency variables interchanged.

Example 5.11 Determine the CTFT $S(j\omega)$ for the signal $s(t) = \dfrac{2}{1+t^2}$ using duality property of CTFT.

Solution. We have already studied that CTFT of signal $s(t) = e^{-2|t|}$ is

$S(j\omega) = \frac{2}{1+\omega^2}$ that is,

$$s(t) = e^{-2|t|} \xleftrightarrow{\text{CTFT}} S(j\omega) = \frac{2}{1+\omega^2} \qquad ...(1)$$

Now, using CTFT synthesis equation

$$s(t) = \frac{1}{2\pi}\int_{-\infty}^{\infty} S(j\omega)\, e^{j\omega t}\, d\omega$$

$$e^{-2|t|} = \frac{1}{2\pi}\int_{-\infty}^{\infty} \left(\frac{2}{1+\omega^2}\right) e^{j\omega t}\, d\omega$$

Replacing t by $-t$, we get

$$2\pi\, e^{-2|t|} = \int_{-\infty}^{\infty} \left(\frac{2}{1+\omega^2}\right) e^{-j\omega t}\, d\omega$$

Interchanging the names of variables t and ω, we find that

$$2\pi\, e^{-2|\omega|} = \int_{-\infty}^{\infty} \left(\frac{2}{1+t^2}\right) e^{-j\omega t}\, dt \qquad ...(2)$$

The R.H.S. of Eq. 2 is the CTFT analysis equation for signal $s(t) = \frac{2}{1+t^2}$ and thus we conclude that

$$\text{CTFT}\left\{\frac{2}{1+t^2}\right\} = 2\pi\, e^{-2|\omega|}$$

5.6.7 Parseval's Relation for CTFT

Parseval's relation says that the total energy of a signal may be determined either by computing the energy per unit time and integrating over all time or by computing the energy per unit frequency and integrating over all frequencies.

Parseval's relation for CTFT is given as

$$\int_{-\infty}^{\infty} |s(t)|^2\, dt = \frac{1}{2\pi}\int_{-\infty}^{\infty} |S(j\omega)|^2\, d\omega \qquad ...(5.47)$$

This relation can be derived by direct application of the Fourier transform.

$$\text{L.H.S. of Eq. 5.47} = \int_{-\infty}^{\infty} |s(t)|^2\, dt = \int_{-\infty}^{\infty} s(t)\, s^*(t)\, dt \qquad ...(5.48)$$

Substituting $s^*(t) = \frac{1}{2\pi}\int_{-\infty}^{\infty} S^*(j\omega)\, e^{-j\omega t}\, d\omega$ in Eq. 5.47, we get

$$\int_{-\infty}^{\infty} |s(t)|^2\, dt = \int_{-\infty}^{\infty} s(t)\, s^*(t)\, dt$$

$$= \int_{-\infty}^{\infty} s(t) \left[\frac{1}{2\pi} \int_{-\infty}^{\infty} S^*(j\omega)\, e^{-j\omega t}\, d\omega \right] dt$$

Reversing the order of integration gives

$$= \frac{1}{2\pi} \int_{-\infty}^{\infty} S^*(j\omega) \left[\int_{-\infty}^{\infty} s(t)\, e^{-j\omega t}\, dt \right] d\omega$$

But we know that

$$S(j\omega) = \int_{-\infty}^{\infty} s(t)\, e^{-j\omega t}\, dt = \frac{1}{2\pi} \int_{-\infty}^{\infty} S^*(j\omega)\, S(j\omega)\, d\omega$$

$$= \text{L.H.S. of Eq. 5.48}$$

This becomes the Parseval's relation for CTFT

$$\int_{-\infty}^{\infty} |s(t)|^2\, dt = \frac{1}{2\pi} \int_{-\infty}^{\infty} |S(j\omega)|^2\, d\omega$$

The term $|S(j\omega)|^2$ is often called the energy- density spectrum of the signal $s(t)$. Parseval's relation and other CTFT properties are useful in determining some time-domain characteristics of a signal directly from the Fourier transform.

5.6.8 Convolution Property of CTFT

If a periodic signal is represented in a Fourier series then the response of an LTI system to this input can also be represented by Fourier series.

In other words, if a periodic signal is represented as a linear combination of harmonically related complex exponential functions then the response of an LTI system to this input can also be represented in the same fashion. This is because complex exponential functions are eigen functions of LTI systems. It means that the Fourier series coefficients of the output of an LTI system is equal to product of input and the frequency response of the system determined at the corresponding harmonic frequencies.

Fourier transform makes the convolution of two signals into the product of their Fourier transforms.

Let us consider two continuous-time signals $s_1(t)$ and $s_2(t)$ and their Fourier transforms are $S_1(j\omega)$ and $S_2(j\omega)$ respectively.

$$s_1(t) \xleftrightarrow{\text{CTFT}} S_1(j\omega)$$

and

$$s_2(t) \xleftrightarrow{\text{CTFT}} S_2(j\omega)$$

Therefore, from the property of convolution of CTFT

$$s_1(t) * s_2(t) \xleftrightarrow{\text{CTFT}} S_1(j\omega)\, S_2(j\omega) \quad ...(5.49)$$

Now, we have a continuous-time LTI system with input $s(t)$ and output $y(t)$. The impulse response of the system is $h(t)$. The output $y(t)$ of this system is determined by convolving input $s(t)$ and its impulse response $h(t)$.

$$y(t) = s(t) * h(t) \quad ...(5.50)$$

Now, by using the convolution property of CTFT, we get

$$y(t) = s(t) * h(t) \xleftrightarrow{\text{CTFT}} Y(j\omega) = S(j\omega)\, H(j\omega) \quad ...(5.51)$$

where $S(j\omega)$, $H(j\omega)$ and $Y(j\omega)$ are the Fourier transforms of $s(t)$, $h(t)$ and $y(t)$ respectively.

Now let us prove the convolution property for CTFT.

From CTFT synthesis equation

$$s(t) = \frac{1}{2\pi} \int_{-\infty}^{\infty} S(j\omega)\, e^{j\omega t}\, d\omega$$

or $$s(t) = \underset{\omega \to 0}{\text{Lim}} \frac{1}{2\pi} \sum_{k=-\infty}^{\infty} S(jk\omega_0)\, e^{jk\omega_0 t}\, \omega_0 \quad ...(5.52)$$

The response of an LTI system with impulse response $h(t)$ to a complex exponential $e^{jk\omega_0 t}$ is $H(jk\omega_0)\, e^{jk\omega_0 t}$.

where $$H(jk\omega_0) = \int_{-\infty}^{\infty} h(t)\, e^{-jk\omega_0 t}\, dt \quad ...(5.53)$$

Frequency response $H(j\omega)$ is recognised as the CTFT of the system impulse response $h(t)$. In other words, the CTFT of the impulse response evaluated at $\omega = k\omega_0$ is the complex scaling factor that the LTI system applies to the eigen function $e^{jk\omega_0 t}$. From superposition principle, we obtain

$$\frac{1}{2\pi} \sum_{k=-\infty}^{\infty} S(jk\omega_0)\, e^{jk\omega_0 t}\, \omega_0 \longrightarrow \frac{1}{2\pi} \sum_{k=-\infty}^{\infty} S(jk\omega_0)\, H(jk\omega_0)\, e^{jk\omega_0 t}\, \omega_0$$

Thus, from Eq. 5.52, the response of LTI system to input $s(t)$ is

$$y(t) = \underset{\omega_0 \to 0}{\text{Lim}} \frac{1}{2\pi} \sum_{k=-\infty}^{\infty} S(jk\omega_0)\, H(jk\omega_0)\, e^{jk\omega_0 t}\, \omega_0$$

$$= \frac{1}{2\pi} \int_{k=-\infty}^{\infty} S(j\omega_0)\, H(j\omega)\, e^{j\omega t}\, d\omega \quad ...(5.54)$$

Since CTFT synthesis equation for output $y(t)$ is given by

$$y(t) = \frac{1}{2\pi} \int_{k=-\infty}^{\infty} Y(j\omega)\, e^{j\omega t}\, d\omega \quad ...(5.55)$$

Comparing Eqs. 5.54 and 5.55, we get

$$Y(j\omega) = S(j\omega)\, H(j\omega) \qquad \text{...(5.56)}$$

But, convolution integral is also used to determine the output from input and impulse response of an LTI system.

$$y(t) = s(t) * h(t) = \int_{-\infty}^{\infty} s(\tau)\, h(t-\tau)\, d\tau \qquad \text{...(5.57)}$$

Taking CTFT of Eq. 5.57, we get

$$Y(j\omega) = \text{CTFT}\{\, y(t)\,\} = \int_{-\infty}^{\infty} \left[\int_{-\infty}^{\infty} s(\tau)\, h(t-\tau)\, d\tau \right] e^{-j\omega t}\, dt \qquad \text{...(5.58)}$$

Interchanging the order of integration and assuming that $s(\tau)$ does not depend on t, we get

$$Y(j\omega) = \int_{-\infty}^{\infty} s(\tau) \left[\int_{-\infty}^{\infty} h(t-\tau)\, e^{-j\omega t}\, dt \right] d\tau \qquad \text{...(5.59)}$$

But using time-shifting property

$$h(t-\tau) \xleftrightarrow{\text{CTFT}} e^{-j\omega\tau}\, H(j\omega)$$

in the bracketed term in $e^{-j\omega\tau}\, H(j\omega)$. Substituting the value of bracketed term in Eq. 5.59, we get

$$Y(j\omega) = \int_{-\infty}^{\infty} s(\tau)\, e^{-j\omega\tau}\, H(j\omega)\, d\tau$$

$$= H(j\omega) \int_{-\infty}^{\infty} s(\tau)\, e^{-j\omega\tau}\, d\tau$$

$$= H(j\omega)\, S(j\omega) \qquad \text{...(5.60)}$$

That is, $\quad y(t) = s(t) * h(t) \xleftrightarrow{\text{CTFT}} Y(j\omega) = S(j\omega)\, H(j\omega)$

Hence, proved.

$H(j\omega)$ is the Fourier transform of impulse response $h(t)$. $H(j\omega)$ is also called frequency response and captures the change in complex amplitude of the Fourier transform of the input at each frequency ω. The frequency response $H(j\omega)$ plays an important role in the analysis of LTI systems. Both impulse response $h(t)$ and its frequency response $H(j\omega)$ completely characterize an LTI system. But many of the properties of LTI systems can be conveniently interpreted in terms of frequency response $H(j\omega)$. We already know that the impulse response of cascading two LTI systems is the convolution of the impulse responses of individual LTI systems. The overall impulse response does not depend on the order in which the systems are cascaded.

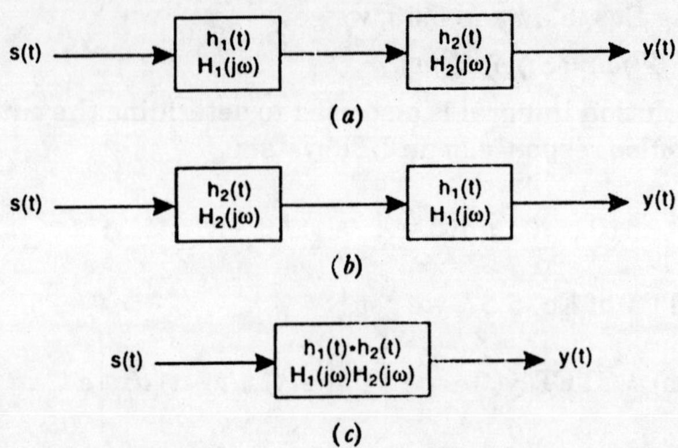

Fig. 5.18 *Illustration of three equivalent LTI systems. Each block represents an LTI systems. Here $h_1(t)$ and $h_2(t)$ are the impulse responses. $H_1(j\omega)$ and $H_2(j\omega)$ are frequency responses of systems 1 and 2, respectively.*

As shown in Fig. 5.18, since the impulse response of the cascade of two LTI systems is the convolution of individual impulse responses.

$$h(t) = h_1(t) * h_2(t) = \int_{-\infty}^{\infty} h_1(\tau)\, h_2(t-\tau)\, d\tau \qquad ...(5.61)$$

But convolution property implies that the overall frequency response of cascading two LTI systems is the product of individual frequency responses.

$$H(j\omega) = H_1(j\omega)\, H_2(j\omega) \qquad ...(5.62)$$

Convergence of the Fourier transform is guaranteed only under certain conditions and the frequency response $H(j\omega)$ cannot be defined for every LTI system. If the LTI system is stable then its impulse response $h(t)$ is absolutely integrable.

i.e.,
$$\int_{-\infty}^{\infty} |h(t)|\, dt < \infty \qquad ...(5.63)$$

Eq. 5.63 is one of the three Dirichlet conditions that together guarantee the existance of the CTFT of $h(t)$, *i.e.,* $H(j\omega)$. Here we are assuming that $h(t)$ satisfies the other two Dirichlet conditions. In chapter 9, we will study generalization of CTFT, the Laplace transform.

Example 5.12 Determine the output of a continuous-time LTI system with impuse response

$$h(t) = \delta(t - t_0)$$

Solution. The frequency response of this system is the CTFT of the impulse response $h(t)$

$$H(j\omega) = \text{CTFT}\{\, h(t)\,\} = \int_{-\infty}^{\infty} h(t)\, e^{-j\omega t}\, dt$$

$$= \int_{-\infty}^{\infty} \delta(t - t_0)\, e^{-j\omega t}\, dt \qquad ...(1)$$

Since $\delta(t - t_0) = \begin{cases} 1, & t - t_0 = 0 \text{ or } t = t_0 \\ 0, & t - t_0 \neq 0 \text{ or } t \neq t_0 \end{cases}$ ...(2)

From Eqs. 1 and 2, we get

$$H(j\omega) = \int_{-\infty}^{\infty} \delta(t - t_0)\, e^{-j\omega t}\, dt$$

or $H(j\omega) = e^{-j\omega t_0}$ ...(3)

Thus, for any input $s(t)$ with CTFT $S(j\omega)$, the CTFT of the output $y(t)$ is determined as

$$Y(j\omega) = H(j\omega)\, S(j\omega) \qquad ...(4)$$

Substituting Eq. 3 in Eq. 4, we get

$$Y(j\omega) = e^{-j\omega t_0}\, S(j\omega) \qquad ...(5)$$

Now, taking inverse CTFT of Eq. 5, we get

$$y(t) = \text{Inverse CTFT}\{\, Y(j\omega) \,\}$$
$$= \text{Inverse CTFT}\{\, e^{-j\omega t_0}\, S(j\omega) \,\}$$

or $y(t) = s(t - t_0)$

(From time shifting property of CTFT)

Specifically, an LTI system for which the impulse response $h(t) = \delta(t - t_0)$ applies a time-shift of constant 't_0' to the input.

Example 5.13 Determine the frequency response of a differentiator

$$y(t) = \frac{d\, s(t)}{dt}$$

Solution. From differentiation property of CTFT

$$Y(j\omega) = j\omega\, S(j\omega) \qquad ...(1)$$

But we know from convolution property of CTFT

$$y(t) = h(t) * s(t) \xleftrightarrow{\text{CTFT}} Y(j\omega) = H(j\omega)\, S(j\omega)$$

or $Y(j\omega) = H(j\omega)\, S(j\omega)$ ...(2)

Now, by comparing Eqs. 1 and 2, we get

$$H(j\omega) = \text{Frequency response of differentiator}$$
$$= j\omega$$

Example 5.14 Determine the CTFT of the integrator described by

$$y(t) = \int_{-\infty}^{t} s(\tau)\, d\tau$$

and its impulse response $h(t) = u(t)$.

Solution. Frequency response of an integrator

$$H(j\omega) = \text{CTFT}\{\, h(t) \,\} = \text{CTFT}\{\, u(t) \,\}$$

$$= \frac{1}{j\omega} + \pi\, \delta(\omega) \qquad \text{...(1)}$$

From convolution property of CTFT,

$$y(t) = h(t) * s(t) \xleftrightarrow{\text{CTFT}} Y(j\omega) = H(j\omega)\, S(j\omega)$$

or $$Y(j\omega) = H(j\omega)\, S(j\omega) \qquad \text{...(2)}$$

Putting the value of $H(j\omega)$ from Eq. 1 into Eq. 2, we obtain

$$Y(j\omega) = H(j\omega)\, S(j\omega) = \left[\frac{1}{j\omega} + \pi\, \delta(\omega)\right] S(j\omega)$$

$$= \frac{S(j\omega)}{j\omega} + \pi\, \delta(\omega)\, S(j\omega)$$

or $$Y(j\omega) = \frac{1}{j\omega}\, S(j\omega) + \pi\, S(0)\, \delta(\omega)$$

Example 5.15 Determine the impulse response $h(t)$ of a low pass filter whose frequency response $H(j\omega)$ is given by

$$H(j\omega) = \begin{cases} 1, & |\omega| < \omega_c \\ 0, & |\omega| > \omega_c \end{cases}$$

and it is shown in Fig. 5.19.

where ω_c is the cutoff frequency of the low pass filter.

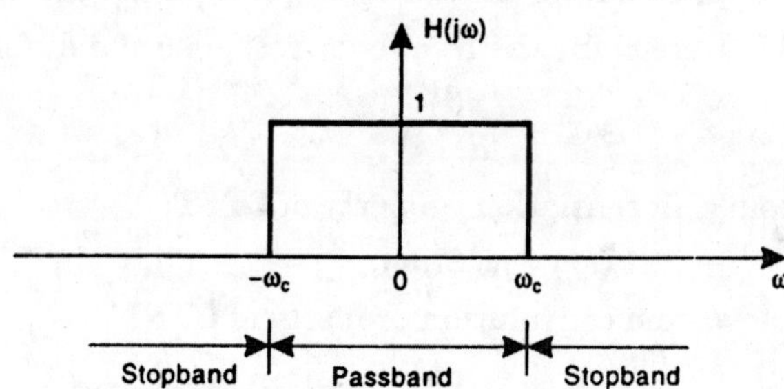

Fig. 5.19 *Frequency response of ideal low pass filter.*

Solution. Impulse response $h(t)$ is determined by

$$h(t) = \text{Inverse CTFT}\{\, H(j\omega) \,\}$$

$$= \frac{1}{2\pi} \int_{-\infty}^{\infty} H(j\omega)\, e^{j\omega t}\, d\omega = \frac{1}{2\pi} \int_{-\omega_c}^{\omega_c} 1 e^{j\omega t}\, d\omega$$

$$= \frac{1}{2\pi} \left[\frac{e^{j\omega t}}{jt}\right]_{-\omega_c}^{\omega_c} = \frac{1}{\pi t} \left[\frac{e^{j\omega_c t} - e^{-j\omega_c t}}{2j}\right]$$

or $$h(t) = \frac{1}{\pi t} \sin \omega_c t \qquad ...(1)$$

The impulse response of a low pass filter $h(t) = \dfrac{\sin \omega_c t}{\pi t}$ is plotted in Fig. 5.20.

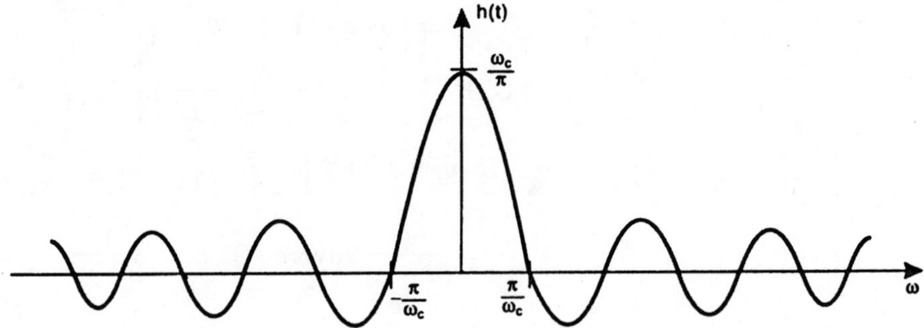

Fig. 5.20 *Impulse response of an ideal low pass filter.*

Example 5.16 Determine the output $y(t)$ of a continuous-time LTI system with impulse response

$$h(t) = e^{-At}\, u(t)\,, \quad A > 0$$

to input $$s(t) = e^{-Bt}\, u(t)\,, \quad B > 0$$

Solution. The output of an LTI can be determined by using convolution of $h(t)$ and $s(t)$ as

$$y(t) = h(t) * s(t) \qquad ...(1)$$

But from convolution property of CTFT,

$$y(t) = h(t) * s(t) \xleftrightarrow{\text{CTFT}} Y(j\omega) = H(j\omega)\, S(j\omega)$$

or $$Y(j\omega) = H(j\omega)\, S(j\omega) \qquad ...(2)$$

Now to determine the CTFT of $h(t)$ and $s(t)$, we proceed as follows

$$H(j\omega) = \text{CTFT}\{\, h(t)\,\} = \text{CTFT}\{\, e^{-At}\, u(t)\,\}$$

$$= \frac{1}{A + j\omega} \qquad ...(3)$$

$$S(j\omega) = \text{CTFT}\{\, s(t)\,\} = \text{CTFT}\{\, e^{-Bt}\, u(t)\,\}$$

$$= \frac{1}{B + j\omega} \qquad ...(4)$$

Substituting Eqs. 3 and 4 in Eq. 2, we get

$$Y(j\omega) = H(j\omega)\, S(j\omega) = \left(\frac{1}{A + j\omega}\right)\left(\frac{1}{B + j\omega}\right) \qquad ...(5)$$

Now output $y(t)$ can be determined by taking inverse CTFT of Eq. 5, *i.e.*,

$$y(t) = \text{Inverse CTFT}\{Y(j\omega)\}$$

$$= \text{Inverse CTFT}\left\{\frac{1}{(A+j\omega)(B+j\omega)}\right\}$$

(By using partial fraction expansion method)

$$= \text{Inverse CTFT}\left\{\frac{\left(\frac{1}{B-A}\right)}{A+j\omega} - \frac{\left(\frac{1}{B-A}\right)}{B+j\omega}\right\}$$

$$= \left(\frac{1}{B-A}\right)\text{Inverse CTFT}\left\{\frac{1}{A+j\omega}\right\}$$

$$-\frac{1}{(B-A)}\text{Inverse CTFT}\left\{\frac{1}{B+j\omega}\right\}$$

$$= \left(\frac{1}{B-A}\right)e^{-At}\,u(t) - \left(\frac{1}{B-A}\right)e^{-Bt}\,u(t)$$

or
$$y(t) = \left(\frac{1}{B-A}\right)[e^{-At} - e^{-Bt}]\,u(t)$$

When $B = A$, the partial fraction expansion is not valid.

We have following formula

$$\left[\begin{array}{c} e^{-At}\,u(t) \xleftrightarrow{\text{CTFT}} \dfrac{1}{A+j\omega} \\ \text{and } e^{-Bt}\,u(t) \xleftrightarrow{\text{CTFT}} \dfrac{1}{B+j\omega} \end{array}\right]$$

Example 5.17 Determine the response of an ideal low pass filter to an input signal $s(t)$ that has the form of sinc function

$$s(t) = \text{sinc function} = \text{sinc } t = \frac{\sin \omega_i t}{\pi t} \qquad \text{...(1)}$$

Solution. We already know that the impulse response of the ideal low pass filter is of similar form and is given by

$$h(t) = \frac{\sin \omega_c t}{\pi t} \qquad \text{...(2)}$$

Therefore, the output of the filer $y(t)$ is the convolution of $s(t)$ and $h(t)$. Both $s(t)$ and $h(t)$ are sinc functions. Also, we will see that the resultant of convolution of two sinc functions will also give a sinc function.

From the convolution propperty of the CTFT

$$y(t) = h(t) * s(t) \xleftrightarrow{\text{CTFT}} Y(j\omega) = H(j\omega)\,S(j\omega)$$

or
$$Y(j\omega) = H(j\omega)\,S(j\omega) \qquad \text{...(3)}$$

where $$S(j\omega) = \begin{cases} 1, & |\omega| \le \omega_i \\ 0, & \text{otherwise} \end{cases}$$

and $$H(j\omega) = \begin{cases} 1, & |\omega| \le \omega_c \\ 0, & \text{otherwise} \end{cases}$$

Therefore, $$Y(j\omega) = H(j\omega)\, S(j\omega) = \begin{cases} 1, & |\omega| \le \omega_0 \\ 0, & \text{ctherwise} \end{cases} \qquad ...(4)$$

where ω_0 is the smaller of the two numbers ω_i and ω_c.

Now taking inverse CTFT of Eq. 4, we get

$$y(t) = \text{Inverse CTFT}\{\, Y(j\omega) \,\}$$

$$= \begin{cases} \dfrac{\sin \omega_c t}{\pi t}, & \text{if } \omega_c \le \omega_i \\[2mm] \dfrac{\sin \omega_i t}{\pi t}, & \text{if } \omega_i \le \omega_c \end{cases}$$

That is, $y(t)$ depending upon whether ω_c is smaller or ω_i, the output is equal to either $s(t)$ or $h(t)$.

5.6.9 Multiplication Property of CTFT

We know from convolution property that convolution in the time-domain corresponds to multiplication in frequency-domain. This is because of a duality that exists between the time and frequency-domains. Due to dual property between time and frequency domains, the multiplication in the time-domain corresponds to convolution in the frequency domain. This can be stated as

$$r(t) = p(t)\, q(t) \xleftrightarrow{\text{CTFT}} R(j\omega) = \frac{1}{2\pi}\,[P(j\omega) * Q(j\omega)] \qquad ...(5.64)$$

This property can be derived by exploiting the duality property along with the convolution property or by directly using CTFT. Multiplication of one signal by another is called amplitude modulation. For this reason, Eq. 5.64 is sometimes called the modulation property. This property has several important applications such as in the study of amplitude modulation in communication engineering.

Example 5.18 Determine the CTFT of $r(t) = s(t)\, p(t)$ where $p(t) = \cos \omega_o t$ and CTFT of $s(t)$ is shown in Fig. 5.21.

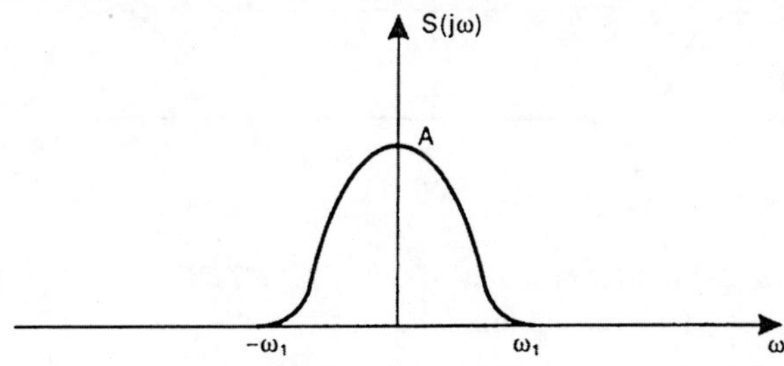

Fig. 5.21 *CTFT of* $s(t)$, $S(j\omega)$.

Solution. We know from multiplication property of CTFT which is given by

$$r(t) = s(t)\,p(t) \xleftrightarrow{\text{CTFT}} R(j\omega) = \frac{1}{2\pi}\,[S(j\omega) * P(j\omega)] \quad \text{...(1)}$$

Determination of CTFT of $p(t) = \cos \omega_0 t$

$$P(j\omega) = \text{CTFT}\{p(t)\} = \text{CTFT}\{\cos \omega_0 t\}$$

$$= \text{CTFT}\left\{\frac{e^{j\omega_0 t} + e^{-j\omega_0 t}}{2}\right\}$$

or

$$P(j\omega) = \pi\,\delta(\omega - \omega_0) + \pi\,\delta(\omega + \omega_0) \quad \text{...(2)}$$

From Eq. 1

$$R(j\omega) = \frac{1}{2\pi}\,[S(j\omega) * P(j\omega)] \quad \text{...(3)}$$

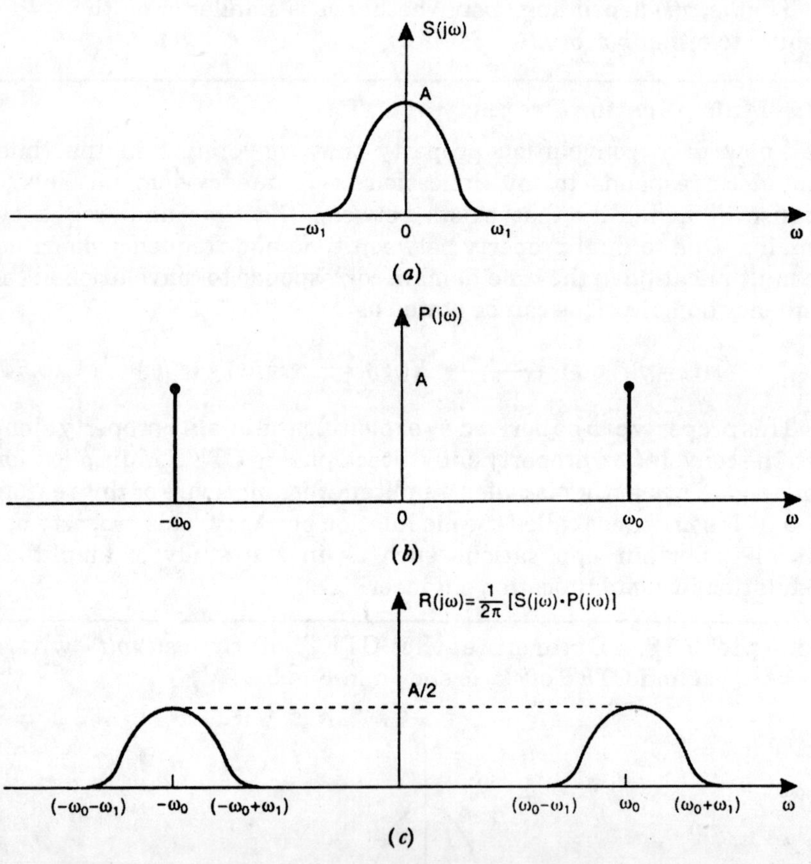

Fig. 5.22 *Multiplication property of CTFT (a) CTFT of $s(t)$ (b) CTFT of $p(t) = \cos \omega_0 t$ (c) CTFT of $r(t) = s(t)\,p(t)$.*

From Eq. 3,

$$R(j\omega) = \frac{1}{2\pi}\,[S(j\omega) * P(j\omega)\,]$$

$$= \frac{1}{2\pi}\left\{S(j\omega) * [\pi\,\delta(\omega - \omega_0) + \pi\,\delta(\omega + \omega_0)]\right\}$$

$$= \frac{1}{2\pi}\,[S(j\omega) * \delta(\omega - \omega_0) + S(j\omega) * \delta(\omega + \omega_0)\,]$$

or $$R(j\omega) = \frac{1}{2}\,[S(j(\omega - \omega_0)) + s(j(\omega + \omega_0))\,] \qquad ...(4)$$

CTFT of $r(t) = s(t)\,p(t)$, *i.e.*, $R(j\omega)$ is given Eq. 4 and shown in Fig. 5.22 (*c*). Here we have assumed that $\omega_0 > \omega_1$, so that the two non-zero portions of $R(j\omega)$ do not overlap. So we can say that the spectrum of $r(t)$ consists of the sum of two shifted and scaled versions of $S(j\omega)$.

From Eq. 4 and Fig. 5.22 (*c*), we see that all of the information in the signal $s(t)$ is preserved when we multiply this signal by sinusoidal signal $[p(t) = \cos\omega_0 t]$. But in multiplication of $s(t)\,p(t)$, the information has been shifted to higher frequencies. This fact is used for amplitude modulation (AM) systems in communication engineering.

5.7 TABULATION FOR CTFT PROPERTIES

The various properties of continuous-time Fourier transform (CTFT) are summarized in Table 5.1.

Table 5.1 Summary of CTFT Properties

CTFT propety	*Non-periodic signal*	*Continuous-time Fourier transform (CTFT)*
	$s(t)$ $s_1(t)$ $s_2(t)$	$S(j\omega)$ $S_1(j\omega)$ $S_2(j\omega)$
Linearity	$A_1\,s_1(t) + A_2\,s_2(t)$ where A_1 and A_2 constants	$A_1\,S_1(j\omega) + A_2\,S_2(j\omega)$
Time shifting	$s(t - t_0)$ t_0 = constant time	$e^{-j\omega t_0}\,S(j\omega)$
Frequency shifting	$e^{j\omega_0 t}\,s(t)$	$S[j(\omega - \omega_0)]$
Complex conjugation	$s^*(t)$	$S^*(-j\omega)$
Time reversal or time folding	$s(-t)$	$S(-j\omega)$
Time and frequency scaling	$s(At)$ where A is a constant	$\frac{1}{\lvert A\rvert}S\left(\frac{j\omega}{A}\right)$

Contd...

CTFT propety	*Non-periodic signal*	*Continuous-time Fourier transform (CTFT)*
Convolution of two continuous-time signals	$s_1(t) * s_2(t)$	$S_1(j\omega)\, S_2(j\omega)$
Multiplication of two continuous-time signals	$s_1(t)\, s_2(t)$	$\frac{1}{2\pi} S_1(j\omega) * S_2(j\omega)$
Differentiation of a signal in time	$\frac{d\, s(t)}{dt}$	$j\omega\, S(j\omega)$
Integration of a signal in time	$\int_{-\infty}^{\infty} s(\tau)\, d\tau$	$\frac{1}{j\omega} S(j\omega) + \pi\, S(0)\, \delta(\omega)$
Differentiation of CTFT of a signal in frequency	$t\, s(t)$	$j \frac{d\, S(j\omega)}{d\omega}$
Complex conjugate symmetry for real signals	$s(t)$ is a real signal	$\begin{cases} S(j\omega) = S^*(-j\omega) \\ \text{Re}\{S(j\omega)\} = \text{Re}\{S(-j\omega)\} \\ \text{Im}\{S(j\omega)\} = -\text{Im}\{S(-j\omega)\} \\ \lvert S(j\omega)\rvert = \lvert S(-j\omega)\rvert \\ \angle S(j\omega) = -\angle S(-j\omega) \end{cases}$
Symmetry for real and even signals	$s(t)$ real and even	$S(j\omega)$ real and even
Symmetry for real and odd signals	$s(t)$ real and odd	$S(j\omega)$ purely imaginary and odd
Decomposition of a real signal in even and odd parts	$S_E(t) = \text{Ev}\{s(t)\}$ $S_O(t) = \text{Od}\{s(t)\}$ Here signal $s(t)$ is real	$\text{Re}\{S(j\omega)\} = S_R(j\omega)$ $j\,\text{Im}\{S(j\omega)\} = S_I(j\omega)$ Here CTFT of $s(t)$ will be complex, $S(j\omega) = S_R(j\omega) + j\, S_I(j\omega)$

Parseval's relation for a continuous-time non-periodic signals

$$\int_{-\infty}^{\infty} |s(t)|^2\, dt = \frac{1}{2\pi} \int_{-\infty}^{\infty} |S(j\omega)|^2\, d\omega$$

5.8 TABULATION FOR BASIC CTFT PAIRS

The basic continuous-time Fourier transform (CTFT) pairs are summarized in Table 5.2.

Table 5.2 The Summary of Basic CTFT Pairs

Continuous-time signal $s(t)$	*Continuous-time Fourier transform (CTFT),* $S(j\omega)$
$\sum_{k=-\infty}^{\infty} A_k e^{jk\omega_0 t}$ where $\omega_0 = \frac{2\pi}{T_0}$ = fundamental frequency T_0-fundamental period of the periodic signal	$2\pi \sum_{k=-\infty}^{\infty} A_k \delta(\omega - k\omega_0)$ where A_k = constant
Exponential function = $e^{j\omega_0 t}$	$2\pi\, \delta(\omega - \omega_0)$
Cosine function = $\cos \omega_0 t$	$\pi\, [\delta(\omega - \omega_0) + \delta(\omega + \omega_0)]$
Sine function = $\sin \omega_0 t$	$\pi j\, [\delta(\omega + \omega_0) - \delta(\omega - \omega_0)]$
$s(t) = 1$	$2\pi\, \delta(\omega)$
Periodic square wave $s(t) = \begin{cases} 1, & \text{for } \lvert t \rvert < T_1 \\ 0, & \text{for } T_1 < \lvert t \rvert \le T_0/2 \end{cases}$ and $s(t + T_0) = s(t)$	$S(j\omega) = \sum_{k=-\infty}^{\infty} \frac{2 \sin k\omega_0 T_1}{k} \delta(\omega - k\omega_0)$
$\sum_{n=-\infty}^{\infty} \delta(t - nT_0)$	$\frac{2\pi}{T_0} \sum_{k=-\infty}^{\infty} \delta\left(\omega - \frac{2\pi k}{T_0}\right)$
$s(t) = \begin{cases} 1, & \lvert t \rvert < T_1 \\ 0, & \lvert t \rvert > T_1 \end{cases}$	$\frac{2 \sin \omega T_1}{\omega}$
$s(t) = \frac{\sin Wt}{\pi t}$	$S(j\omega) = \begin{cases} 1, & \lvert\omega\rvert < W \\ 0, & \lvert\omega\rvert > W \end{cases}$
Dirac delta function or impulse function, $\delta(t)$	1
Unit step function, $u(t)$	$\frac{1}{j\omega} + \pi\, \delta(\omega)$
Shifted impulse function $\delta(t - t_0)$	$e^{-j\omega t_0}$
$e^{-At} u(t)$, for $A > 0$ where A is a real constant	$\frac{1}{A + j\omega}$
$t\, e^{-At} u(t)$, for $A > 0$ where A is a real constant	$\frac{1}{(A + j\omega)^2}$
$\frac{t^{n-1}}{(n-1)!} e^{-At} u(t)$, for $A > 0$ where A is a real constant	$\frac{1}{(A + j\omega)^n}$

5.9 CONTINUOUS-TIME LTI SYSTEMS DESCRIBED BY LINEAR CONSTANT-COEFFICIENT DIFFERENTIAL EQUATIONS

We already know the continuous-time LTI systems for which the input-output relationship is described by a linear constant- coefficient differential equation of the form

$$\sum_{k=0}^{N} A_k \frac{d^k y(t)}{dt^k} = \sum_{k=0}^{M} B_k \frac{d^k s(t)}{dt^k} \quad \text{...(5.65)}$$

In this section, we will determine the frequency response of such an LTI system. In this section, we will assume that the system is stable, so that its frequency response exists.

We have two closely related methods for the determination of the frequency response $H(j\omega)$ for an LTI system described by linear constant-coefficient differential equation given by Eq. 5.65. First method relies on the fact that complex exponential signals are eigen functions of LTI systems. Specifically, if signal $s(t) = e^{j\omega t}$, then the output must be $y(t) = H(j\omega)\, e^{j\omega t}$. Substituting the values of $s(t)$ and $y(t)$ in the Eq. 5.65 and performing some algebraical manipulations, we can then solve for frequency response $H(j\omega)$.

$$\sum_{k=0}^{N} A_k \frac{d^k y(t)}{dt^k} = \sum_{k=0}^{M} B_k \frac{d^k s(t)}{dt^k}$$

or $$\sum_{k=0}^{N} A_k \frac{d^k [H(j\omega)\, e^{j\omega t}]}{dt^k} = \sum_{k=0}^{M} B_k \frac{d^k [e^{j\omega t}]}{dt^k}$$

or $$H(j\omega) \sum_{k=0}^{N} A_k \frac{d^k [e^{j\omega t}]}{dt^k} = \sum_{k=0}^{M} B_k \frac{d^k [e^{j\omega t}]}{dt^k}$$

or $$H(j\omega) = \frac{\sum_{k=0}^{M} B_k \frac{d^k [e^{j\omega t}]}{dt^k}}{\sum_{k=0}^{N} A_k \frac{d^k [e^{j\omega t}]}{dt^k}}$$

or $$H(j\omega) = \frac{\sum_{k=0}^{M} B_k (j\omega)^k}{\sum_{k=0}^{N} A_k (j\omega)^k} \quad \text{...(5.66)}$$

Alternative Method. Now we will discuss the second method of determination of frequency response of an LTI system described Eq. 5.65. In this method, we make use of the differentiation property of the CTFT. From the convolution property of the CTFT given as

$$y(t) = s(t) * h(t) \xleftrightarrow{\text{CTFT}} Y(j\omega) = S(j\omega)\, H(j\omega) \quad \text{...(5.67)}$$

or $$Y(j\omega) = S(j\omega)\, H(\omega)$$

or $$H(j\omega) = \frac{Y(j\omega)}{S(j\omega)} \qquad ...(5.68)$$

where $S(j\omega)$, $Y(j\omega)$ and $H(j\omega)$ are the CTFTs of the input $s(t)$, output $y(t)$ and impulse response $h(t)$, respectively.

Now applying CTFT to both sides of Eq. 5.65, we get

$$\sum_{k=0}^{N} A_k \frac{d^k y(t)}{dt^k} = \sum_{k=0}^{M} B_k \frac{d^k s(t)}{dt^k} \qquad ...(5.69)$$

$$\text{CTFT}\left\{ \sum_{k=0}^{N} A_k \frac{d^k y(t)}{dt^k} \right\} = \text{CTFT}\left\{ \sum_{k=0}^{M} B_k \frac{d^k s(t)}{dt^k} \right\} \qquad ...(5.70)$$

From the property of linearity, Eq. 5.70 becomes

$$\sum_{k=0}^{N} A_k \,\text{CTFT}\left\{ \frac{d^k y(t)}{dt^k} \right\} = \sum_{k=0}^{M} B_k \,\text{CTFT}\left\{ \frac{d^k s(t)}{dt^k} \right\} \qquad ...(5.71)$$

Now using differentiation property given as

$$\left.\begin{aligned} &\frac{d^k}{dt^k} s(t) \xleftrightarrow{\text{CTFT}} (j\omega)^k S(j\omega) \\ \text{and } &\frac{d^k}{dt^k} y(t) \xleftrightarrow{\text{CTFT}} (j\omega)^k Y(j\omega) \end{aligned}\right\} \qquad ...(5.72)$$

in Eq. 5.71, we get

$$\sum_{k=0}^{N} A_k \,(j\omega)^k\, Y(j\omega) = \sum_{k=0}^{M} B_k \,(j\omega)^k\, S(j\omega)$$

or $$Y(j\omega) \sum_{k=0}^{N} A_k (j\omega)^k = S(j\omega) \sum_{k=0}^{M} B_k (j\omega)^k$$

or $$\frac{Y(j\omega)}{S(j\omega)} = \frac{\sum_{k=0}^{M} B_k (j\omega)^k}{\sum_{k=0}^{N} A_k (j\omega)^k} \qquad ...(5.73)$$

But from convolution property, we have

$$Y(j\omega) = S(j\omega)\, H(j\omega)$$

or $$H(j\omega) = \frac{Y(j\omega)}{S(j\omega)} \qquad ...(5.74)$$

Therefore, from Eqs. 5.73 and 5.74, we get

$$H(j\omega) = \frac{Y(j\omega)}{S(j\omega)} = \frac{\sum_{k=0}^{M} B_k (j\omega)^k}{\sum_{k=0}^{N} A_k (j\omega)^k} \qquad ...(5.75)$$

From Eq. 5.75, we have observed that the frequency response $H(j\omega)$ is a rational function. In other words, we can say that $H(j\omega)$ is a ratio of polynomials in $(j\omega)$. The coefficients of the numerator polynomial are the same as those that appeared on the RHS of Eq 5.65 and coefficients of the denominator polynomial are the same coefficients as those on the LHS of the Eq. 5.65.

Hence the frequency response $H(j\omega)$ for an LTI system characterized by its linear constant-coefficient differential equation of the form of Eq. 5.65 can be written directly by inspection.

Example 5.19 Determine the frequency response $H(j\omega)$ and impulse response $h(t)$ for a stable continuous-time LTI system characterized by the linear constant-coefficients differential equation given as

$$\frac{d\,y(t)}{dt} + A.y(t) = s(t) \qquad ...(1)$$

where A = constant coefficient and $A > 0$.

Solution. We know that frequency response is

$$H(j\omega) = \frac{Y(j\omega)}{S(j\omega)} = \frac{\sum_{k=0}^{M} B_k\,(j\omega)^k}{\sum_{k=0}^{N} A_k\,(j\omega)^k}$$

In the above differential equation $M = 0$ and $N = 1$

$$H(j\omega) = \frac{B_0\,(j\omega)^0}{\sum_{k=0}^{1} A_k\,(j\omega)^k}$$

$$= \frac{B_0}{A_0\,(j\omega)^0 + A_1\,(j\omega)^1}$$

or $$H(j\omega) = \frac{1}{A + j\omega} \qquad ...(2)$$

Taking the inverse CTFT of Eq. 2, we get

$$h(t) = \text{Inverse CTFT}\left\{H(j\omega)\right\} = \text{Inverse CTFT}\left\{\frac{1}{A + j\omega}\right\}$$

$$= e^{-At}\,u(t)$$

or Impluse response, $h(t) = e^{-At}\,u(t)$

Example 5.20 Determine the frequency response $H(j\omega)$ and impulse response $h(t)$ for a stable continuous-time LTI system characterized by the linear constant-coefficient differential equation given as

$$\frac{d^2\,y(t)}{dt^2} + 4\,\frac{d\,y(t)}{dt} + 3\,y(t) = \frac{d\,s(t)}{dt} + 2\,s(t) \qquad ...(1)$$

Solution. Frequency response of an LTI system is given by

$$H(j\omega) = \frac{Y(j\omega)}{S(j\omega)} = \frac{\sum_{k=0}^{M} B_k (j\omega)^k}{\sum_{k=0}^{N} A_k (j\omega)^k} \qquad ...(2)$$

From Eq. 1, we obtain $M = 1$ and $N = 2$, then frequency response will be

$$H(j\omega) = \frac{\sum_{k=0}^{1} B_k (j\omega)^k}{\sum_{k=0}^{2} A_k (j\omega)^k} = \frac{B_0 (j\omega)^0 + B_1 (j\omega)^1}{A_0 (j\omega)^0 + A_1 (j\omega)^1 + A_2 (j\omega)^2}$$

$$= \frac{2 \times 1 + 1 \times (j\omega)}{3 \times 1 + 4 \times (j\omega) + 1 \times (j\omega)^2} = \frac{2 + j\omega}{3 + 4j\omega + (j\omega)^2}$$

$$= \frac{2 + j\omega}{3 + 4j\omega + j^2\omega^2} = \frac{j\omega + 2}{j^2\omega^2 + 4j\omega + 3}$$

Hence, frequency response $H(j\omega) = \dfrac{j\omega + 2}{j^2\omega^2 + 4j\omega + 3}$ ...(3)

Now we will determine impulse response $h(t)$ by taking inverse CTFT of $H(j\omega)$ as

$$h(t) = \text{Inverse CTFT } [H(j\omega)]$$

$$= \text{Inverse CTFT}\left[\frac{j\omega + 2}{j^2\omega^2 + 4j\omega + 3}\right] \qquad ...(4)$$

or $$h(t) = \text{Inverse CTFT}\left[\frac{(j\omega + 2)}{(j\omega + 1)(j\omega + 3)}\right] \qquad ...(5)$$

Now, using the partial fraction expansion, we find that

$$\frac{(j\omega + 2)}{(j\omega + 1)(j\omega + 3)} = \frac{\alpha_1}{(j\omega + 1)} + \frac{\alpha_2}{(j\omega + 3)} \qquad ...(6)$$

Determination of values of α_1 and α_2.

or $j\omega + 1 = 0$, $j\omega = -1$ then, $\alpha_1 = \dfrac{j\omega + 2}{j\omega + 3} = \dfrac{-1 + 2}{-1 + 3} = \dfrac{1}{2}$

or $j\omega + 3 = 0$, $j\omega = -3$ then, $\alpha_2 = \dfrac{j\omega + 2}{j\omega + 1} = \dfrac{-3 + 2}{-3 + 1} = \dfrac{1}{2}$

Substituting the values of α_1 and α_2 in Eq. 6, we get

$$\frac{j\omega + 2}{(j\omega + 1)(j\omega + 3)} = \frac{1/2}{(j\omega + 1)} + \frac{1/2}{(j\omega + 3)} \qquad ...(7)$$

Substituting Eq. 7 in Eq. 5, we get

$$h(t) = \text{Inverse CTFT}\left[\frac{1/2}{(j\omega + 1)} + \frac{1/2}{(j\omega + 3)}\right]$$

$$= \text{Inverse CTFT}\left[\frac{1/2}{j\omega + 1}\right] + \text{Inverse CTFT}\left[\frac{1/2}{j\omega + 3}\right]$$

or $$h(t) = \frac{1}{2}e^{-t}\,u(t) + \frac{1}{2}e^{-3t}\,u(t) \qquad ...(8)$$

This is the impulse response of the LTI system characterized by differential equation given in Eq. 1.

Example 5.21 Find the output response of the low-pass RC network for an input signal $s(t) = e^{-t/\tau}$ where $\tau = RC$. The low-pass RC network is shown in Fig. 5.23 (a).

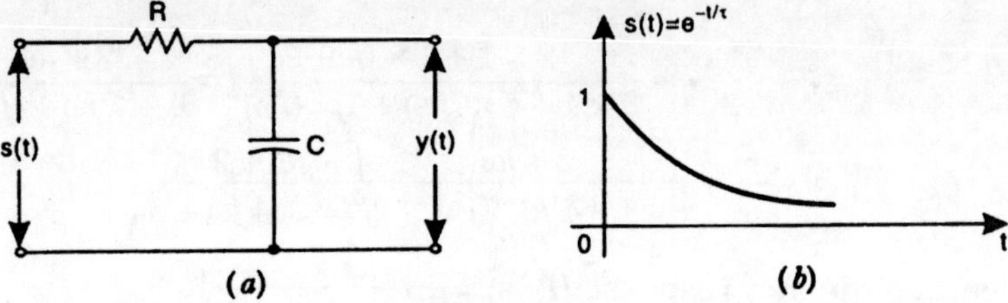

Fig. 5.23

Solution. Input $s(t)$ is shown in Fig. 5.23 (b).

Output response can be determined by using convolution theorem for CTFT as

$$y(t) = s(t) * h(t) \xleftrightarrow{\text{CTFT}} Y(j\omega) = S(j\omega)\,H(j\omega)$$

$$S(j\omega) = \text{CTFT}\,\{s(t)\} = \text{CTFT}\,\{e^{-t/\tau}\}$$

$$= \int_0^\infty e^{-t/\tau}\,e^{-j\omega t}\,dt = \int_0^\infty e^{-[(1/\tau)+j\omega]t}\,dt$$

$$= \frac{1}{j\omega + \frac{1}{\tau}} \qquad (\text{Since } \tau = RC)$$

or $$S(j\omega) = \frac{1}{j\omega + \frac{1}{RC}} \qquad ...(1)$$

Similarly, the frequency response of the network is given by

$$H(j\omega) = \frac{Y(j\omega)}{S(j\omega)} \qquad ...(2)$$

From the network shown in Fig. 5.23 (a), frequency response

$$H(j\omega) = \frac{\frac{1}{j\omega C}}{\left(R + \frac{1}{j\omega C}\right)} = \frac{1}{(j\omega RC + 1)}$$

or $$H(j\omega) = \frac{1}{RC}\frac{1}{\left(j\omega + \frac{1}{RC}\right)} \qquad ...(3)$$

Substituting Eqs. 1 and 3 in Eq. 2, we get

$$H(j\omega) = \frac{Y(j\omega)}{S(j\omega)}$$

$$\frac{1}{RC\left(j\omega + \frac{1}{RC}\right)} = \frac{Y(j\omega)}{\frac{1}{\left(j\omega + \frac{1}{RC}\right)}}$$

or $$Y(j\omega) = \frac{1}{RC\left(j\omega + \frac{1}{RC}\right)^2} \qquad ...(4)$$

$$y(t) = \text{Inverse CTFT } \{y(j\omega)\}$$

$$= \text{Inverse CTFT}\left\{\frac{1}{RC\left(j\omega + \frac{1}{RC}\right)^2}\right\}$$

or $$y(t) = \frac{1}{RC}\, t\, e^{-t/RC}\, u(t)$$

Example 5.22 Find the CTFT of $s(t) \cos \omega_0 t$ by using frequency shifting property of CTFT.

Solution. The CTFT or spectrum of $s(t) \cos \omega_0 t$ is

$$\text{CTFT } \{s(t) \cos \omega_0 t\} = \text{CTFT}\left\{s(t)\left[\frac{e^{j\omega_0 t} + e^{-j\omega_0 t}}{2}\right]\right\}$$

$$= \frac{1}{2}\text{CTFT } \{s(t)\, e^{j\omega_0 t}\} + \frac{1}{2}\text{CTFT } \{s(t)\, e^{-j\omega_0 t}\}$$

$$= \frac{1}{2}\{S(j\omega - j\omega_0) + S(j\omega + j\omega_0)\}$$

$$\left[\text{Since CTFT } \{s(t)\, e^{j\omega_0 t}\} = S(j\omega - j\omega_0)\right]$$

Thus, due to product of $s(t)$ and $\cos \omega_0 t$, the original spectrum is shifted in such a way that half the original spectrum is centered about ω_0 and other half is centered about $-\omega_0$.

Example 5.23 A continuous-time signal has the following CTFT

$$S(j\omega) = \frac{1}{\omega^2 + 1}\, e^{-[2\omega^2/(\omega^2 + 1)]}$$

Using the various properties of CTFT, write the CTFT of the following

(*a*) $s(2t)$ (*b*) $s(t-2)\, e^{jt}$

(c) $4\dfrac{d}{dt}s(t)$ (d) $\int_{-\infty}^{t} s(\tau)\,d\tau$

Also name the property you will use.

Solution. Given $S(j\omega) = \dfrac{1}{\omega^2+1} e^{-[2\omega^2/(\omega^2+1)]}$

(*a*) To find CTFT of $s(2t)$, we use scaling property of CTFT

$$s(At) \xleftrightarrow{\text{CTFT}} \frac{1}{|A|} S\left(\frac{j\omega}{A}\right) \quad \text{...(1)}$$

where A is scalar constant.

$$s(2t) \xleftrightarrow{\text{CTFT}} \frac{1}{|2|} S\left(\frac{j\omega}{2}\right)$$

$$\xleftrightarrow{\text{CTFT}} \frac{1}{2}\frac{1}{(\omega/2)^2+1} e^{-[\{2(\omega/2)^2\}/\{(\omega/2)^2+1\}]}$$

$$\xleftrightarrow{\text{CTFT}} \frac{1}{2}\frac{(2)^2}{\omega^2+(2)^2} e^{-[2\omega^2/\{\omega^2+(2)^2\}]}$$

$$\xleftrightarrow{\text{CTFT}} \frac{2}{\omega^2+4} e^{-[2\omega^2/(\omega^2+4)]}$$

(*b*) To find CTFT of $s(t-2)\,e^{jt}$, we use time shifting property of CTFT as

$$s(t-t_0) \xleftrightarrow{\text{CTFT}} e^{-j\omega t_0} S(j\omega) \qquad \text{(Putting } t_0 = 2)$$

$$s(t-2) \xleftrightarrow{\text{CTFT}} e^{-j\omega 2} S(j\omega)$$

$$\xleftrightarrow{\text{CTFT}} e^{-j2\omega} S(j\omega)$$

Therefore, $$s(t-2) \xleftrightarrow{\text{CTFT}} e^{-j2\omega}\left[\frac{1}{\omega^2+1} e^{-[2\omega^2/(\omega^2+1)]}\right]$$

or $$s(t-2) \xleftrightarrow{\text{CTFT}} \frac{1}{(\omega^2+1)} e^{-2\omega^2/(\omega^2+1)}\, e^{-j2\omega} \quad \text{...(2)}$$

Using the frequency shifting property

$$s(t)\,e^{j\omega_0 t} \xleftrightarrow{\text{CTFT}} S(j\omega - j\omega_0), \text{ we get}$$

$$s(t-2)\,e^{jt} \xleftrightarrow{\text{CTFT}} S(j\omega - j1)$$

or $$s(t-2)\,e^{jt} \xleftrightarrow{\text{CTFT}} S(j\omega - j)$$

or $$s(t-2)\,e^{jt} \xleftrightarrow{\text{CTFT}} S\{j(\omega-1)\} \quad \text{...(3)}$$

From Eqs. 2 and 3, we get

$$s(t-2)\,e^{jt} \xleftrightarrow{\text{CTFT}} \frac{1}{[(\omega-1)^2+1]}\, e^{-[\{2(\omega-1)^2\}/\{(\omega-1)^2+1\}]}\, e^{-j2(\omega-1)}$$

(c) To find CTFT of $4\,\frac{d\,s(t)}{dt}$, we use the differentiation property as

$$\frac{d\,s(t)}{dt} \xleftrightarrow{\text{CTFT}} j\omega\, S(j\omega)$$

as
$$4\,\frac{d\,s(t)}{dt} \xleftrightarrow{\text{CTFT}} 4j\omega\, S(j\omega)$$

$$\xleftrightarrow{\text{CTFT}} j4\omega\,\frac{1}{(\omega^2+1)}\, e^{-[2\omega^2/(\omega^2+1)]}$$

(d) To find CTFT of $\int_{-\infty}^{t} s(\tau)\,d\tau$, we use integration property of CTFT

as

$$\int_{-\infty}^{t} s(\tau)\,d\tau \xleftrightarrow{\text{CTFT}} \frac{S(j\omega)}{j\omega} + \pi\, S(0)\,\delta(\omega)$$

or
$$\int_{-\infty}^{t} s(\tau)\,d\tau \xleftrightarrow{\text{CTFT}} \frac{1}{j\omega}\,\frac{1}{(\omega^2+1)}\, e^{-[2\omega^2/(\omega^2+1)]} + \pi\,\delta(\omega)$$

[Since $S(0)=1$]

Example 5.24 Determine the CTFT of the Gaussian pulse $s(t)=e^{-A^2t^2}$, where A is scalar constant.

Solution. $S(j\omega) = \text{CTFT of } s(t) = \int_{-\infty}^{\infty} s(t)\, e^{-j\omega t}\, dt$

$$= \int_{-\infty}^{\infty} e^{-A^2t^2}\, e^{-j\omega t}\, dt = \int_{-\infty}^{\infty} e^{-(A^2t^2+j\omega t)}\, dt$$

Substituting $A^2t^2 + j\omega t = \left(At + \frac{j\omega}{2A}\right)^2 + \frac{\omega^2}{4A^2}$, we get

$$S(j\omega) = \int_{-\infty}^{\infty} e^{-[\{At+(j\omega/2A)\}^2 + \omega^2/4A^2]}\, dt$$

$$= e^{-\omega^2/4A^2} \int_{-\infty}^{\infty} e^{-[At+(j\omega/2A)]^2}\, dt$$

Putting $B = At + \frac{j\omega}{2A}$, $dB = A\,dt$ and so $dt = \frac{dB}{A}$

$$S(j\omega) = e^{-\omega^2/4A^2}\, 2 \int_0^\infty e^{-B^2} \frac{dB}{A} \qquad ...(1)$$

$$\left[\text{Here } \int_0^\infty e^{-B^2}\, dB = \frac{\sqrt{\pi}}{2}\right]$$

$$S(j\omega) = e^{-\omega^2/4A^2} \frac{2}{A} \frac{\sqrt{\pi}}{2}$$

$$= \frac{\sqrt{\pi}}{A} e^{-\omega^2/4A^2} = \frac{\sqrt{\pi}}{A} e^{-[\omega/2A]^2}$$

or $$S(j\omega) = e^{-[\omega/2A]^2}$$

or $$\text{CTFT } \{e^{-A^2t^2}\} = \frac{\sqrt{\pi}}{A} e^{-[\omega/2A]^2}$$

Example 5.25 Find the CTFT of the signal $s(t) = e^{-A|t|}\, sgn(t)$.

Solution. Given $s(t) = e^{-A|t|}\, sgn(t)$

$$= \begin{cases} -e^{At}, & \text{for } t < 0 \\ 0, & \text{for } t = 0 \\ +e^{-At}, & \text{for } t > 0 \end{cases}$$

Therefore, $S(j\omega) = \text{CTFT } \{s(t)\}$

$$= -\int_{-\infty}^{0} e^{At} e^{-j\omega t}\, dt + \int_0^\infty e^{-At} e^{-j\omega t}\, dt$$

$$= -\int_{-\infty}^{0} e^{(A-j\omega)t}\, dt + \int_0^\infty e^{-(A+j\omega)t}\, dt$$

$$= -\left[\frac{e^{(A-j\omega)t}}{(A-j\omega)}\right]_{-\infty}^{0} + \left[\frac{e^{-(A+j\omega)t}}{-(A+j\omega)}\right]_0^\infty$$

$$= \left[\frac{1}{A-j\omega}\right] + \left[\frac{1}{A+j\omega}\right] = \frac{-j2\omega}{A^2+\omega^2}$$

Hence, $\text{CTFT } \{e^{-A|t|}\, sgn(t)\} = \dfrac{-j2\omega}{A^2+\omega^2}$

It is a pure imaginary function of ω.

Since $$S(-j\omega) = \frac{-j2(-\omega)}{A^2+(-\omega)^2} = \frac{j2\omega}{A^2+\omega^2} = -F(j\omega)$$

It is also an odd function of ω. Hence, this is proved that if $s(t)$ is real and odd, then $S(j\omega)$ is pure imaginary and odd.

The phase of $S(j\omega)$ is $90°$ for all $\omega < 0$ and $-90°$ for all $\omega > 0$. The amplitude of $S(j\omega)$ is $|S(j\omega)| = \dfrac{2|\omega|}{A^2+\omega^2}$.

Example 5.26 Find the magnitude and phase spectrum of the pulse shown in Fig. 5.24 (*a*).

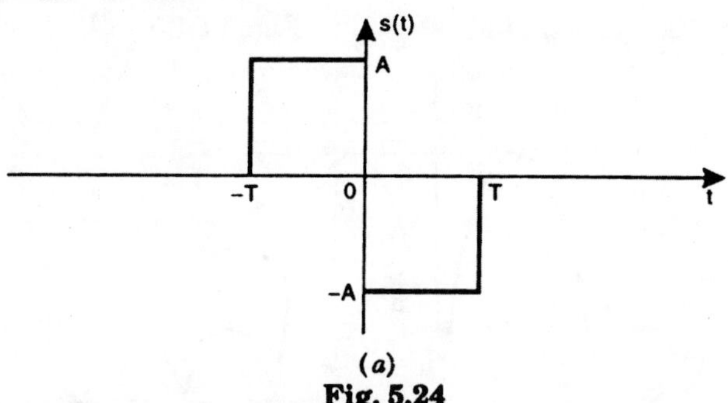

(*a*)

Fig. 5.24

Solution.
$$s(t) = \begin{cases} A\,, & \text{for } -T \le t \le 0 \\ -A\,, & \text{for } 0 \le t \le T_0 \\ 0\,, & \text{otherwise} \end{cases}$$

$$S(j\omega) = \text{CTFT}\,\{s(t)\} = \int_{-\infty}^{\infty} s(t)\, e^{-j\omega t}$$

$$= \int_{-\infty}^{0} s(t)\, e^{-j\omega t}\, dt + \int_{0}^{\infty} s(t)\, e^{-j\omega t}\, dt$$

$$= \int_{-T}^{0} s(t)\, e^{-j\omega t}\, dt + \int_{0}^{T} s(t)\, e^{-j\omega t}\, dt$$

$$= \int_{-T}^{0} A\, e^{-j\omega t}\, dt + \int_{0}^{T} (-A)\, e^{-j\omega t}\, dt$$

Changing t by $-t$ in the first integral, we get

$$S(j\omega) = \int_{0}^{T} A\, e^{j\omega t}\, dt - \int_{0}^{T} A\, e^{-j\omega t}\, dt$$

$$= A \int_{0}^{T} (e^{j\omega t} - e^{-j\omega t})\, dt = j2A \int_{0}^{T} \sin \omega t\, dt$$

$$= j2A \left[\frac{-\cos \omega t}{+\omega} \right]_0^T$$

or $$S(j\omega) = \frac{j2A}{\omega}\,[1 - \cos \omega T] \qquad ...(1)$$

Therefore, the magnitude of

$$S(j\omega) = |S(j\omega)| = \frac{2A}{\omega}\,[1 - \cos \omega T]$$

Therefore, the magnitude is zero when $(1 - \cos \omega T) = 0$

i.e., $\omega T = 2n\pi$ or $\omega = \dfrac{n2\pi}{T}$

The magnitude spectrum is shown in Fig. 5.24 (*b*).

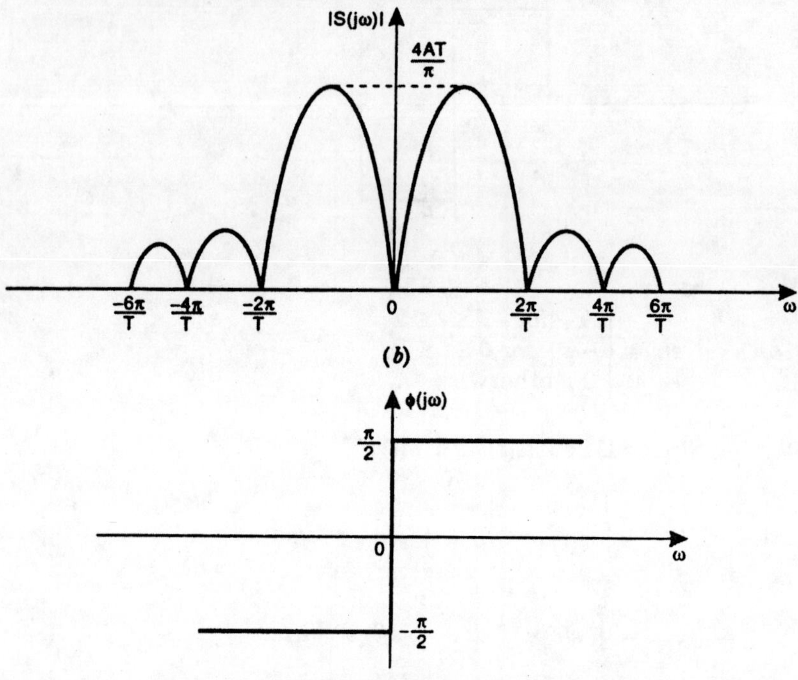

Fig. 5.24 (*b*) *Magnitude spectrum* (*b*) *Phase spectrum*

Phase $\quad \phi(j\omega) = \tan^{-1}\left\{\dfrac{\text{Im}\,[F(j\omega)]}{\text{Re}\,[F(j\omega)]}\right\}$

$$= \tan^{-1}\left\{\frac{\dfrac{2A}{\omega\,(1-\cos\omega T)}}{0}\right\}$$

$$= \tan^{-1}\{+\infty\} = +\frac{\pi}{2} \text{ for } \omega \text{ positive}$$

$$= \tan^{-1}\{-\infty\} = -\frac{\pi}{2} \text{ for } \omega \text{ negative}$$

The phase spectrum is shown in Fig. 5.24 (*c*).

Example 5.27 Determine the CTFT of the signal $s(t)$ shown in Fig. 5.25.

Solution. The equation of the line *OP* is given by

$$\frac{y-y_1}{x-x_1} = \frac{y_1-y_2}{x_1-x_2}$$

Hence, $\quad \dfrac{s(t)-0}{t-0} = \dfrac{0-A}{0-T}$

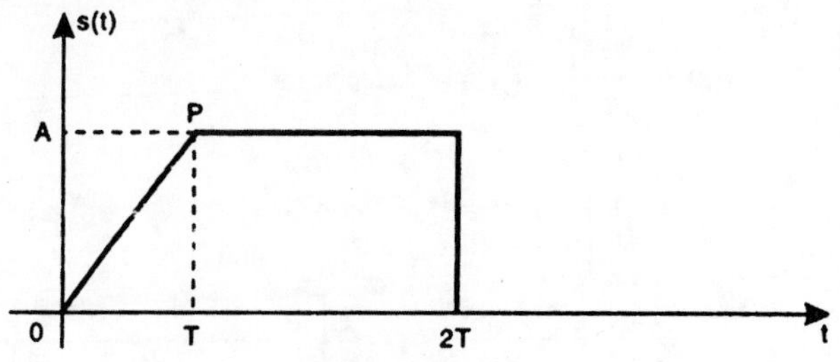

Fig. 5.25

or $$s(t) = \begin{cases} \frac{A}{T}t\,, & \text{for } 0 < t < T \\ A\,, & \text{for } T < t < 2T \end{cases} \qquad ...(1)$$

$$S(j\omega) = \text{CTFT}\{s(t)\}$$

$$= \int_0^T \frac{A}{T} t\, e^{-j\omega t}\, dt + \int_T^{2T} A\, e^{-j\omega t}\, dt$$

$$= -j\frac{2A}{T\omega^2} e^{-j\omega T/2} \sin\frac{\omega T}{2} + j\frac{A}{\omega} e^{-j2\omega T}$$

or $$S(j\omega) = \frac{A}{j\omega}\left[e^{-j\omega T/2} \operatorname{sinc}\frac{\omega T}{2} - e^{-j2\omega T} \right]$$

Example 5.28 The magnitude $|S(j\omega)|$ and the phase $\phi(j\omega)$ of the CTFT of a signal $s(t)$ are shown in Fig. 5.26 (*a*) and (*b*) respectively.

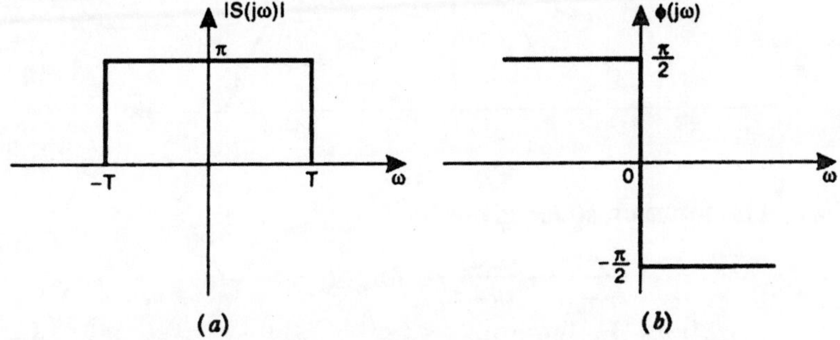

Fig. 5.26 (*a*) *Magnitude spectrum* (*b*) *Phase spectrum.*

Solution. The magnitude $|S(j\omega)| = \pi$, for $-T \le \omega \le T$

$$\phi(j\omega) = \angle S(j\omega) = \frac{\pi}{2}, \text{ for } \omega < 0$$

$$= -\frac{\pi}{2}, \text{ for } \omega > 0$$

$$= 0, \text{ for } \omega = 0$$

For the limits $-T \le \omega \le 0$, $S(j\omega) = \pi\, e^{j\pi/2}$

For the limits $0 \le \omega \le T$, $S(j\omega) = \pi\, e^{-j\pi/2}$

$$s(t) = \frac{1}{2\pi} \int_{-\infty}^{\infty} S(j\omega)\, e^{j\omega t}\, d\omega$$

$$= \frac{1}{2\pi}\left[\int_{-T}^{0} \pi\, e^{j\pi/2} e^{j\omega t}\, d\omega + \int_{0}^{T} \pi\, e^{-j\pi/2}\, e^{j\omega t}\, d\omega \right]$$

$$= \frac{1}{t}\left[\frac{e^{j\pi/2} - e^{-j\pi/2}}{2j} \right] - \frac{1}{t}\left[\frac{e^{j[(\pi/2) - Tt]} - e^{-j[(\pi/2) - Tt]}}{2j} \right]$$

$$= \frac{1}{t}\left[\sin\left(\frac{\pi}{2}\right) - \sin\left(\frac{\pi}{2} - Tt\right) \right]$$

$$= \frac{1}{t}\,[1 - \cos Tt\,] = \frac{2\sin^2\left(\frac{Tt}{2}\right)}{t}$$

or $$s(t) = \frac{T^2 t}{2}\,\text{sinc}^2\left(\frac{Tt}{2}\right)$$

Example 5.29 Find the CTFT of the trapezoidal pulse shown in Fig. 5.27.

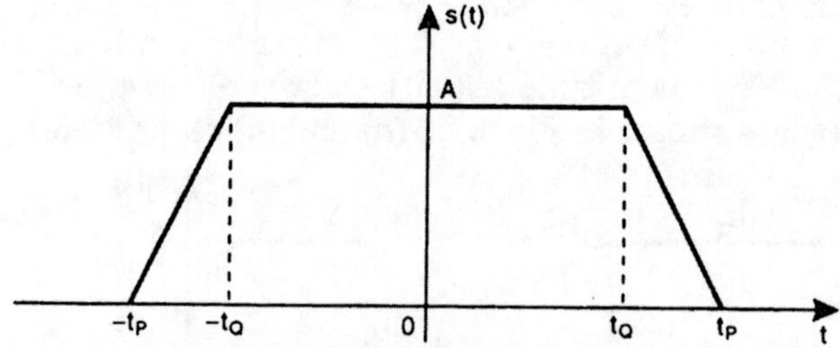

Fig. 5.27

Solution. The function $s(t)$ is given by

$$s(t) = \begin{cases} \dfrac{At}{t_P - t_Q} + \dfrac{At_P}{t_P - t_Q}, & \text{for } -t_P < t < -t_Q \\ A, & \text{for } -t_Q < t < t_Q \\ \dfrac{At_P}{t_P - t_Q} - \dfrac{At}{t_P - t_Q}, & \text{for } t_Q < t < t_P \end{cases} \quad ...(1)$$

The CTFT of a function $s(t)$ is determined by

$$S(j\omega) = \text{CTFT}\,\{s(t)\} = \int_{-\infty}^{\infty} s(t)\, e^{-j\omega t}\, dt$$

$$= \int_{-t_P}^{-t_Q} \left[\frac{At}{t_P - t_Q} + \frac{At_P}{t_P - t_Q} \right] e^{-j\omega t}\, dt + \int_{-t_Q}^{t_Q} [A]\, e^{-j\omega t}\, dt$$

$$+ \int_{t_Q}^{t_P} \left[\frac{At_P}{t_P - t_Q} - \frac{At}{t_P - t_Q} \right] e^{-j\omega t}\, dt$$

$$= \frac{A}{t_P - t_Q}\left[\int_{-t_P}^{-t_Q} t\, e^{-j\omega t}\, dt - \int_{t_Q}^{t_P} t\, e^{-j\omega t}\, dt\right]$$

$$+ \frac{At_P}{t_P - t_Q}\left[\int_{-t_P}^{-t_Q} e^{-j\omega t}\, dt + \int_{t_Q}^{t_P} e^{-j\omega t}\, dt\right] + A\int_{-t_Q}^{t_Q} e^{-j\omega t}\, dt$$

or $$S(j\omega) = \frac{2A}{\omega^2 (t_P - t_Q)}[\cos \omega t_Q - \cos \omega t_P]$$

This is the CTFT of the trapezoidal pulse shown in Fig. 5.27.

Example 5.30 Determine the CTFT of the signal shown in Fig. 5.28.

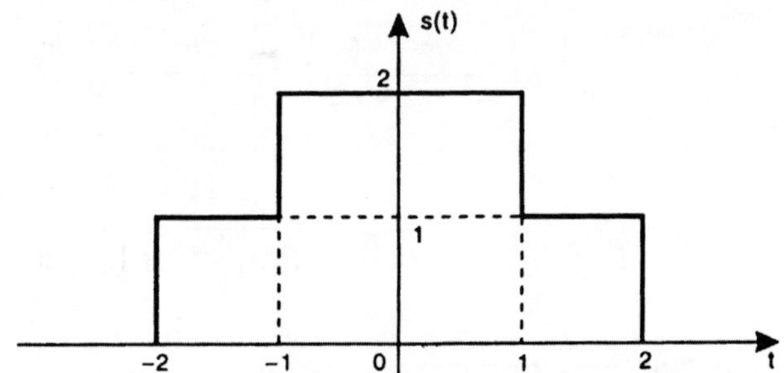

Fig. 5.28

Solution. Above signal can be represented analytically as

$$s(t) = \begin{cases} 2, & \text{for } -1 < t < 1 \\ 1, & \text{for } -2 < t < 1 \\ 1, & \text{for } 1 < t < 2 \end{cases} \qquad \text{...(1)}$$

$$S(j\omega) = \text{CTFT}\{s(t)\} = \int_{-\infty}^{\infty} s(t)\, e^{-j\omega t}\, dt$$

$$= \int_{-1}^{1} (2)\, e^{-j\omega t}\, dt + \int_{-2}^{-1} (1)\, e^{-j\omega t}\, dt + \int_{1}^{2} (1)\, e^{-j\omega t}\, dt$$

$$= \frac{1}{-j\omega}\left\{ [2e^{-j\omega t}]_{-1}^{1} + [e^{-j\omega t}]_{-2}^{-1} + [e^{-j\omega t}]_{1}^{2} \right\}$$

$$= \frac{1}{-j\omega}\{ 2(e^{-j\omega} - e^{j\omega}) + (e^{j\omega} - e^{2j\omega}) + (e^{-j2\omega} - e^{-j\omega}) \}$$

$$= \frac{1}{\omega}\left\{ 4\left[\frac{e^{j\omega} - e^{-j\omega}}{2j}\right] + 2\left[\frac{e^{j2\omega} - e^{-j2\omega}}{2j}\right] - 2\left[\frac{e^{j\omega} - e^{-j\omega}}{2j}\right]\right\}$$

$$= \frac{4}{\omega}\left\{ \sin \omega + \frac{\sin 2\omega}{2} - \frac{\sin \omega}{2} \right\}$$

$$= \frac{2 \sin \omega}{\omega} + \frac{2 \sin 2\omega}{\omega}$$

or $$S(j\omega) = 2 \text{ sinc } \omega + 4 \text{ sinc } 2\omega$$

Example 5.31 Determine the CTFT of the signal $s(t) = t \cos At$

Solution. CTFT of the signal $s(t)$ is given by

$$S(j\omega) = \text{CTFT}\{s(t)\} = \int_0^{\infty} s(t)\, e^{-j\omega t}\, dt$$

$$= \int_0^{\infty} t \cos At = \int_0^{\infty} t\left[\frac{e^{jAt} + e^{-jAt}}{2}\right] e^{-j\omega t}\, dt$$

$$= \frac{1}{2}\int_0^{\infty} [t\, e^{-j(-A+\omega)t} + t\, e^{-j(A+\omega)t}]\, dt$$

$$= \frac{1}{2}\left[t\left\{\frac{e^{-j(-A+\omega)t}}{-j(-A+\omega)}\right\} - \frac{e^{-j(-A+\omega)t}}{[-j(-A+\omega)]^2}\right]_0^{\infty}$$

$$+ \frac{1}{2}\left[t\left\{\frac{e^{-j(A+\omega)t}}{-j(A+\omega)}\right\} - \frac{e^{-j(A+\omega)t}}{[-j(A+\omega)]^2}\right]_0^{\infty}$$

$$S(j\omega) = \frac{-(\omega^2 + A^2)}{(\omega^2 - A^2)^2}$$

Example 5.32 Determine the CTFT of the sinusoidal pulse shown in Fig. 5.29.

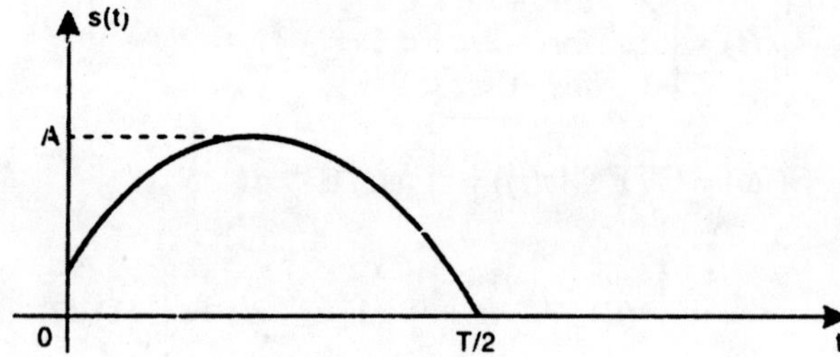

Fig. 5.29

Solution. The CTFT of signal $s(t)$ is defined as

$$S(j\omega) = \text{CTFT}\{s(t)\}$$

$$= \int_{-\infty}^{\infty} s(t)\, e^{-j\omega t}\, dt \qquad \text{...(1)}$$

The signal $s(t)$ shown in Fig. 5.29 is anlaytically given by

$$s(t) = \begin{cases} A \sin \omega_0 t\,, & \text{for } 0 < t \le \dfrac{T}{2} \\ 0\,, & \text{otherwise} \end{cases} \qquad \text{...(2)}$$

Substituting Eq. 2 in Eq. 1, we get

$$S(j\omega) = \int_0^{T/2} A \sin \omega_0 t \, e^{-j\omega t} \, dt$$

$$= A \int_0^{T/2} \left[\frac{e^{j\omega_0 t} - e^{-j\omega_0 t}}{2j} \right] e^{-j\omega t} \, dt$$

$$= \frac{A}{2j} \int_0^{T/2} e^{j(\omega_0 - \omega)t} \, dt - \frac{A}{2j} \int_0^{T/2} e^{-j(\omega_0 + \omega)t} \, dt$$

or $$S(j\omega) = \frac{A}{2j} \left[\frac{e^{j(\omega_0 - \omega)t}}{j(\omega_0 - \omega)} - \frac{e^{-j(\omega_0 + \omega)t}}{-j(\omega_0 + \omega)} \right]_0^{T/2}$$

On solving above equation, we get

$$S(j\omega) = \frac{A\omega_0}{\omega_0^2 - \omega^2} \, [e^{j\omega T4} + e^{-j\omega T/4}] \, e^{-j\omega T/4}$$

$$= \left(\frac{2A\omega_0}{\omega_0^2 - \omega^2} \cos \frac{\omega T}{4} \right) e^{-j\omega T/4}$$

REVIEW QUESTIONS

1. Compare transforms with logarithms.
2. What do you mean by Fourier anlysis ? What is Fourier transform ?
3. Distinguish Fourier transform from Fourier series.
4. Write a short note on development of continuous-time Fourier transform (CTFT).
5. What do you mean by convergence of continuous-time Fourier transform (CTFT) ?
6. Write three Derichlet's conditions for CTFT.
7. What do you mean by magnitude and phase of CTFT ?
8. Is it possible to determine CTFT of continuous-time periodic signal?
9. Discuss CTFT of continuous-time periodic signal.
10. Write CTFT analysis and synthesis equation.
11. Discuss properties of CTFT in brief.
12. Tabulate various properties of CTFT.
13. Tabulate various basic CTFT pairs.
14. Discuss differential equations of continuous-time LTI system.
15. Discuss two methods of determination of frequency response of continuous-time LTI system.
16. Discuss linearity and shifting property of CTFT with example.
17. What do you mean by conjugation ? Give example.

18. What is conjugate symmetry ? Give conjugate symmetry properties of CTFT in brief.

19. Discuss differentiation and integration properties for CTFT.

20. What do you mean by time scaling and frequency scaling ? Also discuss both properties for CTFT.

21. What is duality ? Give one example.

22. Give an expression for Parseval's relation for CTFT.

23. Discuss convolution property of CTFT. Also give physical significance of convolution property of CTFT.

24. State and explain multiplication property of CTFT.

25. Distinguish CTFT from CTFS.

NUMERICAL PROBLEMS

1. Find the output response of the low-pass RC network due to an input $s(t) = t\, e^{-t/RC}$ by convolution.

Ans. $y(t) = \frac{1}{RC} \frac{t^2 e^{-t/RC}}{2} u(t)$

2. Find the output signal $y(t)$ when $B \neq A$ and $B = A$ for a continuous-time system whose impulse response $h(t) = e^{-Bt} u(t)$, $B > 0$, for input $s(t) = e^{-At} u(t)$, $A > 0$.

Ans. $y(t) = \frac{1}{(B - A)} [e^{-At} u(t) - e^{-Btu(t)]}$ for $A \neq B$

$y(t) = te^{-At} u(t)$ for $A = B$

3. Find the amplitude and phase spectrum of the time-shifted impulse function $s(t) = 10\, \delta(t - 2)$.

4. Determine the CTFT of the rectangular pulse or gate function shown in Fig. 5.30.

Ans. $S(j\omega) = T \operatorname{sinc}\left(\frac{\omega T}{2}\right)$

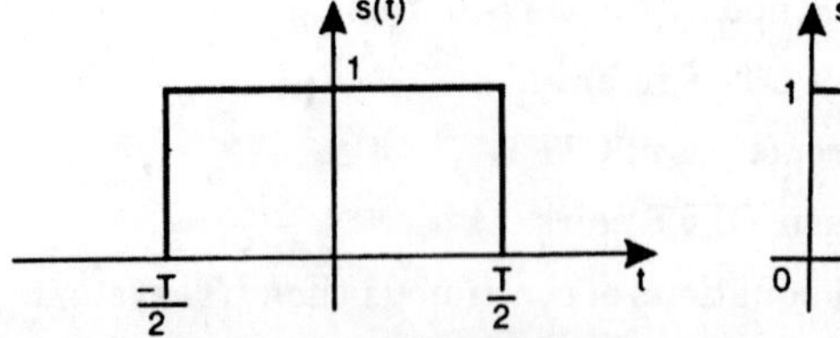

Fig. 5.30 *Rectangular function.* **Fig. 5.31**

5. Determine CTFT of the rectangular pulse shown in Fig. 5.31.

Ans. $S(j\omega) = T\, e^{-j\omega T/2} \times \operatorname{sinc}\left(\frac{\omega T}{2}\right)$

6. Find CTFT of exponential pulse shown in Fig. 5.32.

Ans. $S(j\omega) = \dfrac{1}{A + j\omega}$

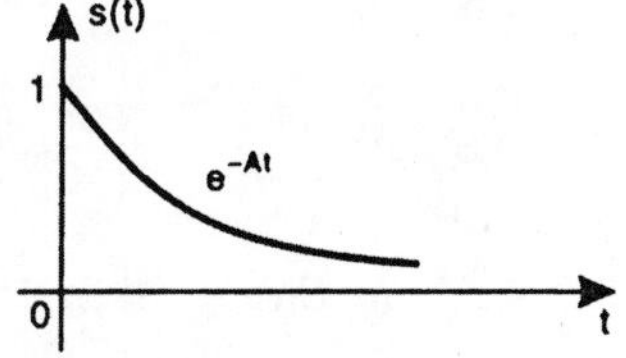

Fig. 5.32

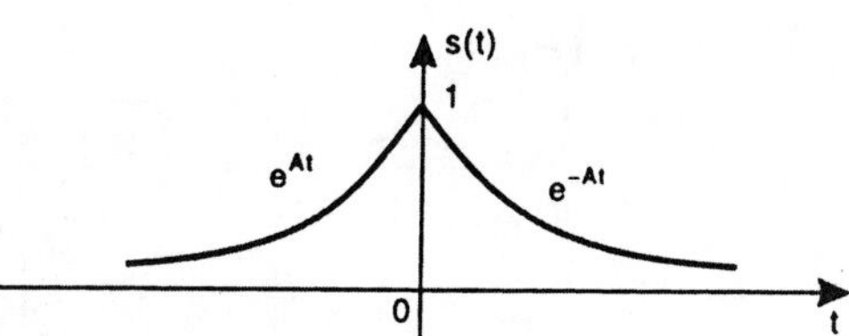

Fig. 5.33 *Double exponential pulse.*

7. Find the CTFT of double exponential pulse shown in Fig. 5.33.

Ans. $S(j\omega) = \dfrac{2A}{A^2 + \omega^2}$

8. Find the CTFT of triangular pulse shown in Fig. 5.34.

Ans. $S(j\omega) = \dfrac{AT}{2} \operatorname{sinc}^2\left(\dfrac{\omega T}{4}\right)$

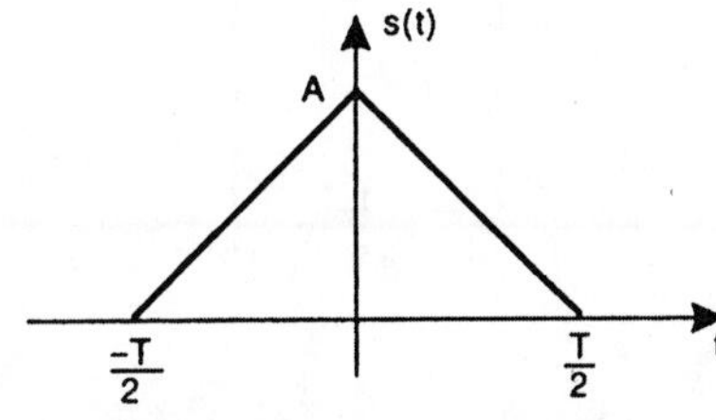

Fig. 5.34 *Triangular pulse.*

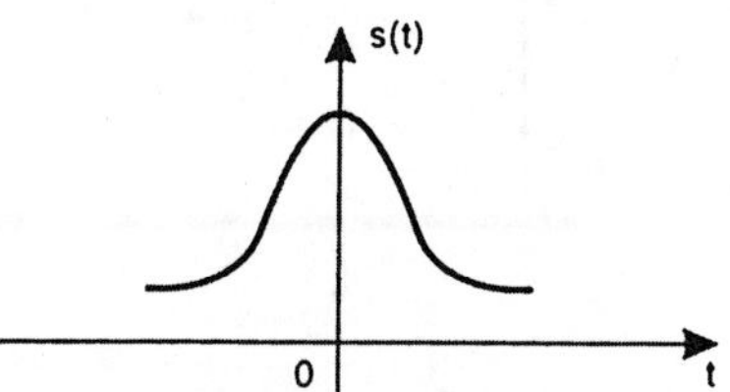

Fig. 5.35 *Gaussian pulse.*

9. Find the CTFT of the Gaussian pulse shown in Fig. 5.35.

10. Find the CTFT of the following functions

(*a*) Impulse function $\delta(t)$ (*b*) Unit step function $u(t)$

(*c*) Sinusoidal function (*d*) Signum functions

Ans. (*a*) $S(j\omega) = 1$ (*b*) $S(j\omega) = \dfrac{1}{j\omega} + \pi\,\delta(\omega)$ (*c*) $S(j\omega) = \text{CTFT}\{\cos \omega_0 t\}$

$= \pi\,[\delta(\omega - \omega_0) + \delta(\omega + \omega_0)]$; $S(j\omega) = \text{CTFT}\{\sin \omega_0 t\}$

$= j\,\pi\,[\delta(\omega + \omega_0) - \delta(\omega - \omega_0)]$ (*d*) $S(j\omega) = \dfrac{2}{j\omega}$

11. Find the CTFT of the function $s(t) = e^{-At} \cos Bt$ where A and B are constants.

Ans. $S(j\omega) = \dfrac{A + j\omega}{(A + j\omega)^2 + B^2}$

12. Find the CTFT of $s(t) = \sin(\omega_C t + \theta)$

Ans. $S(j\omega) = \dfrac{\omega_C \cos\theta + j\omega \sin\theta}{\omega_C^2 - \omega^2}$

13. Find the CTFT of $s(t) = e^{jAt^2}$ and hence find the CTFT of the following functions

(a) $s(t) = \cos At^2$ (b) $s(t) = \sin At^2$

(c) $s(t) = e^{-B^2t^2}$ (d) $s(t) = e^{-Bt^2}$

Ans. CTFT $\{e^{jAt^2}\} = \sqrt{\frac{\pi}{A}}\, e^{j[(\pi/4) - (\omega^2/4A)]}$ (a) $S(j\omega) = \sqrt{\frac{\pi}{A}} \cos\left(\frac{\pi}{4} - \frac{\omega^2}{4A}\right)$

(b) $S(j\omega) = \sqrt{\frac{\pi}{A}} \sin\left(\frac{\pi}{4} - \frac{\omega^2}{4A}\right)$ (c) $S(j\omega) = \sqrt{\frac{\pi}{B}}\, e^{-(\omega/2B)^2}$

(d) $S(j\omega) = \sqrt{\frac{\pi}{B}}\, e^{-\omega^2/4B}$

14. Find the CTFT of the double gate function shown in Fig. 5.36 using time-shifting property.

Ans. $S(j\omega) = 2T \cos(\omega T_1)\, \text{sinc}\left(\frac{\omega T}{2}\right)$

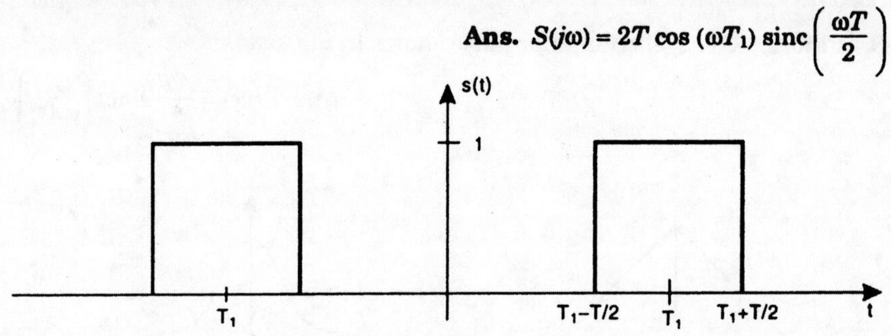

Fig. 5.36

6

The Discrete-Time Fourier Transform (DTFT)

6.1 INTRODUCTION TO DISCRETE-TIME FOURIER TRANSFORM (DTFT)

In chapter 5, we have already studied about the Fourier transform of continuous-time signals. It is also called continuous-time Fourier transform (CTFT). In this chapter, we will discuss about the Fourier transform of discrete-time signals, *i.e.,* discrete-time Fourier transform (DTFT). There are many similarities in the analysis of continuous-time and discrete-time signals using Fourier series and there are also important differences between continuous-time Fourier series (CTFS) and discrete-time Fourier series (DTFS). For example, the Fourier series representation of a discrete-time periodic signal is finite series but the Fourier series representation of continuous-time periodic signal is infinite series. Also we will see in this chapter, there are corresponding differences between continuous-time Fourier transform (CTFT) and discrete-time Fourier transform (DTFT).

Here we will extend the Fourier series description of discrete-time periodic signals in order to develop a Fourier transform representation for discrete-time non-periodic signals.

6.2 DEVELOPMENT OF THE DISCRETE-TIME FOURIER TRANSFORM (DTFT)

Discrete-time Fourier transform (DTFT) is a transformation tool that transforms discrete-time signal from time-domain to frequency-domain. In this section, we will develop an expression for discrete-time Fourier transform (DTFT) for discrete-time signals. In chapter 5, we have already learnt that the Fourier series coefficient for a continuous-time periodic square wave can be viewed as samples of an envelope function. As the fundamental period of the continuous-time periodic square wave increases, its samples become more and more finely spaced. Here a non-periodic signal $s(t)$ is used to construct a periodic signal $\tilde{s}(t)$ that equals to $s(t)$ over one fundamental period. As the period approaches infinitely, $\tilde{s}(t)$ will be equal to $s(t)$ over larger and larger intervals of time and the Fourier series representation for $\tilde{s}(t)$ converges to the Fourier transform representation for $s(t)$. In this section, we will develop an analogous procedure for deriving an expression for discrete-time Fourier transform (DTFT) for discrete-time non-periodic signals.

Consider a general discrete-time sequence $s(n)$ of finite duration. That is,

$$\begin{aligned} s(n) &\neq 0, \text{ for } -N_1 \leq n \leq N_2 \\ &= 0, \text{ otherwise} \end{aligned} \quad ...(6.1)$$

From this non-periodic signal, we can construct a periodic signal or sequence $\tilde{s}(n)$ for which $s(n)$ is of one period. A discrete-time non-periodic signal $s(n)$ is shown in Fig. 6.1 (*a*) and a periodic signal $\tilde{s}(n)$ which is constructed from $s(n)$ for which $s(n)$ is of one fundamental period is shown in Fig. 6.1 (*b*).

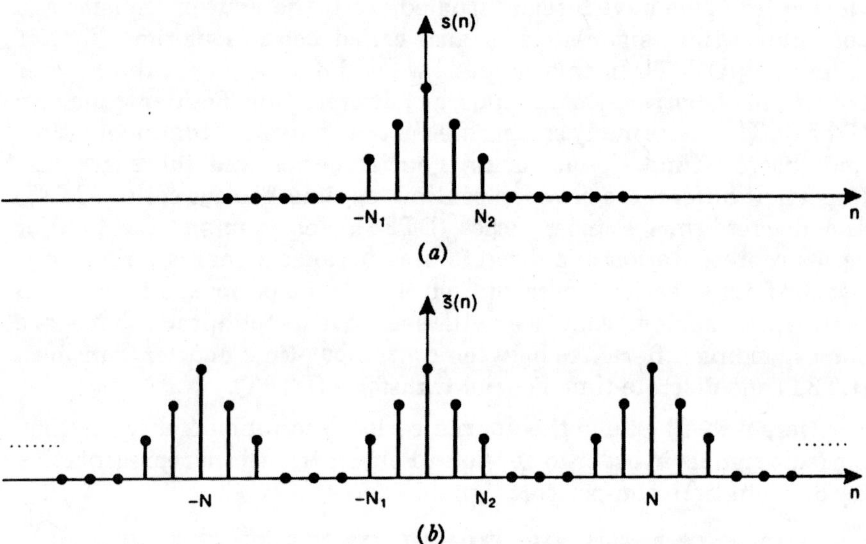

Fig. 6.1 (*a*) *Finite-duration discrete-time signal s(n)* (*b*) *Periodic signal $\tilde{s}(n)$ constructed to be equal to s(n) over one period.*

If fundamental period N_0 approaches infinity then $\tilde{s}(n) = s(n)$ for any finite value of n.

Fourier series representation of $\tilde{s}(n)$ can be expressed as

$$\tilde{s}(n) = \sum_{k=\langle N_0 \rangle} A_k e^{jk\omega_0 n} \quad ...(6.2)$$

where A_k is the Fourier series coefficient and $\omega_0 = 2\pi/N_0$ is the fundamental frequency. N_0 is the fundamental period.

Fourier series coefficients A_k are given by

$$A_k = \frac{1}{N_0} \sum_{n=\langle N_0 \rangle} \tilde{s}(n)\; e^{-jk\omega_0 n} \qquad ...(6.3)$$

Since $s(n) = \tilde{s}(n)$ over one period that includes the interval $-N_1 \le n \le N_2$. Therefore replacing $\tilde{s}(n)$ by $s(n)$ in Eq. 6.3, we get

$$A_k = \frac{1}{N_0} \sum_{n=N_1}^{N_2} s(n)\, e^{-jk\omega_0 n} \qquad ...(6.4)$$

But $s(n)$ is zero outside the interval $-N_1 \le n \le N_2$.

Defining the function

$$S(e^{j\omega}) = \sum_{n=-\infty}^{\infty} s(n)\, e^{-j\omega n} \qquad ...(6.5)$$

We see that the Fourier series coefficients A_k are directly proportional to samples of $S(j\omega)$

$$A_k = \frac{1}{N_0} S(e^{jk\omega_0}) \qquad ...(6.6)$$

Now combining Eqs. 6.2 and 6.6, we get

$$\tilde{s}(n) = \sum_{k=\langle N_0 \rangle} A_k\, e^{jk\omega_0 n} = \sum_{k=\langle N_0 \rangle} \frac{1}{N_0} S(e^{jk\omega_0})\, e^{jk\omega_0 n} \qquad ...(6.7)$$

Since $\omega_0 = \frac{2\pi}{N_0}$ or $\frac{1}{N_0} = \frac{\omega_0}{2\pi}$, Eq. 6.7 can be rewritten as

$$\tilde{s}(n) = \sum_{k=\langle N_0 \rangle} \frac{\omega_0}{2\pi} S(e^{jk\omega_0})\, e^{jk\omega_0 n}$$

$$= \frac{1}{2\pi} \sum_{k=\langle N_0 \rangle} S(e^{jk\omega_0})\, e^{jk\omega_0 n}\, \omega \qquad ...(6.8)$$

As fundamental period N_0 increases, fundamental frequency ω_0 decreases. As N_0 approaches infinity, the Eq. 6.8 passes to an integral.

Eq. 6.8 can be graphically interpreted in Fig. 6.2.

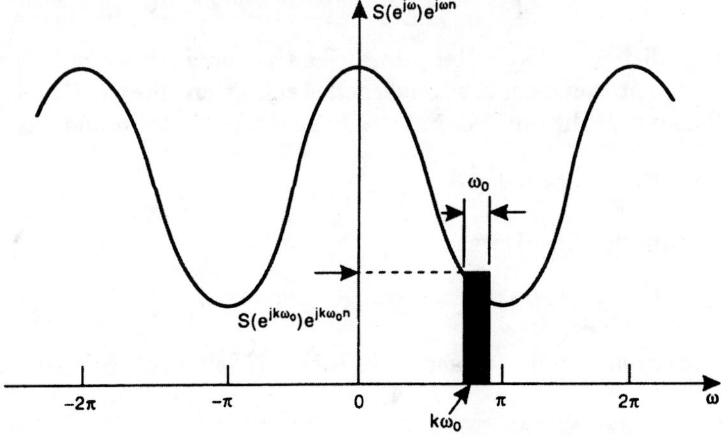

Fig. 6.2 *Graphical representation of equation* $\tilde{s}(n) = \frac{1}{2\pi} \sum_{k=\langle N_0 \rangle} S(e^{jk\omega_0})\, e^{jk\omega_0 n}\, \omega_0$

From Eq. 6.5, $S(e^{j\omega})$ is periodic in ω with period 2π and therefore $e^{j\omega n}$ is also periodic in ω with same period. Thus the, product $S(e^{j\omega})\, e^{j\omega n}$ will also be periodic in ω with period 2π. Each term in the summation in Eq. 6.8 represents the area of rectangle of height $S(e^{jk\omega_0})\, e^{j\omega_0 n}$ and width ω_0. As fundamental frequency ω_0 approaches zero, the summation in Eq. 6.8 becomes an integral. The summation is carried out over N_0 at consecutive intervals of width ω_0 but the total interval of integration will always be of width 2π.

Therefore, as $N_0 \to \omega_0$ or $\omega_0 \to 0$, $\tilde{s}(n) = s(n)$, the Eq. 6.8 becomes

$$s(n) = \frac{1}{2\pi} \int_{2\pi} S(e^{j\omega})\, e^{j\omega n}\, d\omega \qquad \text{...(6.9)}$$

where
$$S(e^{j\omega}) = \sum_{n=-\infty}^{\infty} s(n)\, e^{-j\omega n} \qquad \text{...(6.10)}$$

The function $S(e^{j\omega})$ is called the *Discrete- time Fourier Transform* (DTFT). Eq. 6.9 is called DTFT synthesis equation. Eq. 6.10 is referred to as DTFT analysis equation. These equations show how a non-periodic sequence can be thought of as a linear combination of complex exponential functions. The discrete-time Fourier transform (DTFT) has many similarities with the continuous-time Fourier transform (CTFT) such as linearity, convolution property etc. The major differences between the DTFT and CTFT are :

(*i*) DTFT is periodic in ω with period 2π but CTFT is not periodic.

(*ii*) DTFT has finite interval of integration in the synthesis equation but CTFT has infinite interval of integration in the synthesis equation.

6.3 CONVERGENCE OF THE DTFT

Here we will discuss about the convergence of the infinite summation in the DTFT analysis equation given as

$$S(e^{j\omega}) = \sum_{n=-\infty}^{\infty} s(n)\, e^{-j\omega n} \qquad \text{...(6.11)}$$

The conditions on $s(n)$ that guarantee the convergence of this sum are direct counterparts of the convergence conditions for the CTFT. Eq. 6.11 will converge either if signal $s(n)$ is absolutely summable, *i.e.*,

$$\sum_{n=-\infty}^{\infty} |s(n)| < \infty \qquad \text{...(6.12)}$$

or, if the sequence has finite energy, *i.e.*,

$$\sum_{n=-\infty}^{\infty} |s(n)|^2 < \infty \qquad \text{...(6.13)}$$

But, there are no issues associated with DTFT synthesis equation given by

$$s(n) = \frac{1}{2\pi} \int_{2\pi} S(e^{j\omega})\, e^{j\omega n}\, d\omega \qquad ...(6.14)$$

as the integral in this synthesis equation is over a finite interval of integration. This is a similar situation as for the discrete-time Fourier series (DTFS) synthesis equation which involves a finite sum. Consequently, there is no issue of convergence associated with DTFT synthesis equation.

Example 6.1 Determine DTFT of a discret-time signal $s(t) = A^n u(n)$, $|A| < 1$.

Solution. DTFT of $s(n)$ is given by

$$S(e^{j\omega}) = \text{DTFT}\{s(n)\} = \sum_{n=-\infty}^{\infty} s(n)\, e^{-j\omega n}$$

$$= \sum_{n=-\infty}^{\infty} A^n\, u(n)\, e^{-j\omega n} \qquad ...(1)$$

But $u(n)$ is the unit-step sequence and it is defined as

$$u(n) = \begin{cases} 1\,, & n \geq 0 \\ 0\,, & n < 0 \end{cases} \qquad ...(2)$$

Substituting Eq. 2 in Eq. 1, we obtain

$$S(e^{j\omega}) = \sum_{n=0}^{\infty} A^n\, 1.e^{-j\omega n} = \sum_{n=0}^{\infty} (A\, e^{-j\omega})^n$$

This is a geometric progression of infinite number of terms.

$$= \frac{1}{1 - A\, e^{-j\omega}}$$

or $$S(e^{j\omega}) = \frac{1}{1 - A\, e^{-j\omega}}$$

Note that DTFT of any discrete-time sequence $s(n)$ is periodic function in ω with period 2π.

Example 6.2 Determine the DTFT of the discrete-time signal $s(n) = A^{|n|}$, $|A| < 1$.

Solution. DTFT of the discrete-time signal is determined;

$$S(e^{j\omega}) = \text{DTFT}\{s(n)\} = \sum_{n=-\infty}^{\infty} s(n)\, e^{-j\omega n}$$

$$= \sum_{n=-\infty}^{\infty} A^{|n|}\, e^{-j\omega n} = \underbrace{\sum_{n=0}^{\infty} A^n\, e^{-j\omega n}}_{\text{I}} + \underbrace{\sum_{n=-\infty}^{-1} A^{-n}\, e^{-j\omega n}}_{\text{II}}$$

Substituting $n = -m$ in the second summation, we obtain

$$S(e^{j\omega}) = \sum_{n=0}^{\infty} A^n\, e^{-j\omega n} + \sum_{m=1}^{\infty} A^m\, e^{j\omega m} \qquad ...(1)$$

Both of the summations given in Eq. 1 are infinite geometric progressions that we can evaluate in closed form, producing

$$S(e^{j\omega}) = \sum_{n=0}^{\infty} (Ae^{-j\omega})^n + \sum_{m=1}^{in} (Ae^{j\omega})^m$$

$$= \frac{1}{1 - Ae^{-j\omega}} + \frac{Ae^{j\omega}}{1 - Ae^{j\omega}}$$

$$= \frac{1 - Ae^{j\omega} + Ae^{j\omega}\,(1 - Ae^{-j\omega})}{(1 - Ae^{-j\omega})\,(1 - Ae^{j\omega})}$$

$$= \frac{1 - Ae^{j\omega} + Ae^{j\omega} - A^2}{1 - Ae^{j\omega} - Ae^{-j\omega} + A^2}$$

$$= \frac{1 - A^2}{1 - 2A\left(\dfrac{e^{j\omega} + e^{-j\omega}}{2}\right) + A^2}$$

or
$$S(e^{j\omega}) = \frac{1 - A^2}{1 - 2A\cos\omega + A^2} \qquad \text{...(2)}$$

The discrete-time signal $s(n)$ is shown in Fig. 6.3 and its DTFT is shown in Fig. 6.4.

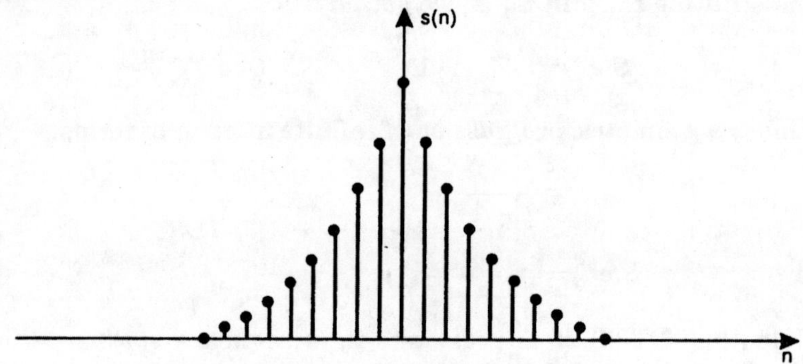

Fig. 6.3 *Discrete-time signal* $s(n) = A^{|n|}$, $|A| < 1$.

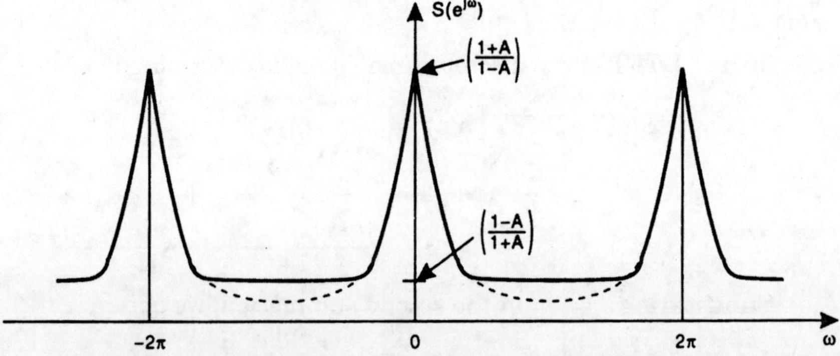

Fig. 6.4 *DTFT of the discrete-time signal* $s(n) = A^{|n|}$, $|A| < 1$.

Here $S(e^{j\omega}) = \dfrac{1 - A^2}{1 - 2A\cos\omega + A^2}$ is a real valued function of ω.

Example 6.3 Determine the DTFT of the discrete-time rectangular pulse given as

$$s(n) = \begin{cases} 1, & |n| \le N_1 \\ 0, & |n| > N_1 \end{cases}$$

Solution. This rectangular pulse is shown in Fig. 6.5.

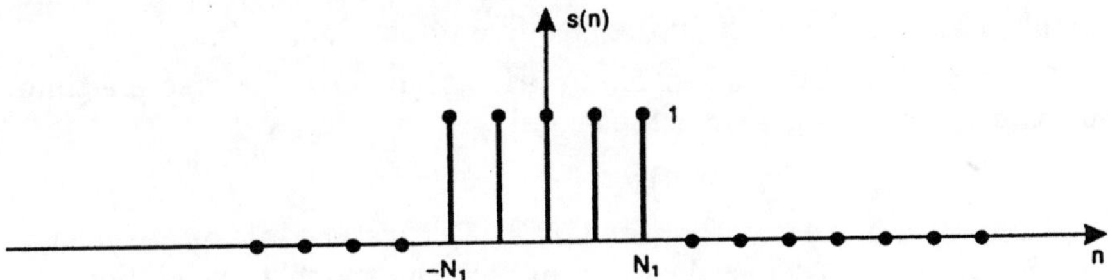

Fig. 6.5 *Discrete time rectangular pulse*

$$s(n) = \begin{cases} 1, & |n| \le N_1 \\ 0, & |n| > N_1 \end{cases}$$

DTFT of this rectangular pulse is determined as

$$S(e^{j\omega}) = \text{DTFT}\{s(n)\} = \sum_{n=-\infty}^{\infty} s(n)\, e^{-j\omega n}$$

$$= \sum_{n=-N_1}^{N_1} 1 e^{-j\omega n} = \sum_{n=-N_1}^{N_1} e^{-j\omega n} \qquad \text{...(1)}$$

This is a geometric series or progression of finite number of terms. Eq. 1 can be written as

$$S(e^{j\omega}) = \sum_{n=-N_1}^{N_1} e^{-j\omega n} = \frac{\sin\omega\,[N_1 + (1/2)]}{\sin(\omega/2)}$$

The discrete-time Fourier transform of rectangular pulse is shown in Fig. 6.6.

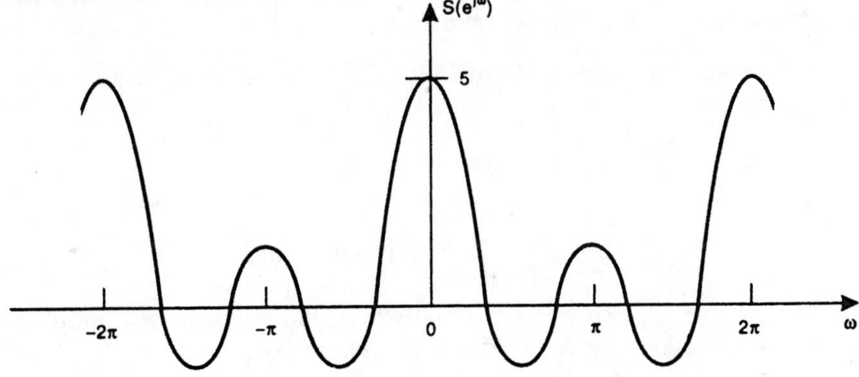

Fig. 6.6 *DTFT of rectangular pulse shown in Fig.* 6.5 *for* $N_1 = 2$.

DTFT of a discrete-time rectangular pulse is periodic with period 2π but the CTFT of the continuous-time rectangular pulse is not periodic.

6.4 FOURIER TRANSFORM OF DISCRETE-TIME PERIODIC SIGNALS

In this section, we will study about the Fourier transform for discrete-time periodic signals. Fourier transform for discrete- time periodic signals is interpreted as the Fourier transform of a periodic signal as an impulse train in the frequency domain.

To derive an expression for Fourier transform for discrete-time periodic signals, we consider the signal

$$s(n) = e^{j\omega n} \quad ...(6.15)$$

In chatper 5, we saw that the CTFT of $e^{j\omega_0 t}$ can be interpreted as an impulse at $\omega = \omega_0$. Therefore, we expect the same type of Fourier transform that results for the discrete-time signal of Eq. 6.15. However, the DTFT must be periodic in ω with period 2π. Fourier transform of $s(n) = e^{j\omega n}$ should have impulses at ω_0, $\omega_0 \pm 2\pi$, $\omega_0 \pm 4\pi$ and so on. Fourier transform of $s(n)$ is the impulse train. It is given by Eq. 6.16 and illustrated in Fig. 6.7.

$$S(e^{j\omega}) = \sum_{m=-\infty}^{\infty} 2\pi\, \delta(\omega - \omega_0 - 2\pi m) \quad ...(6.16)$$

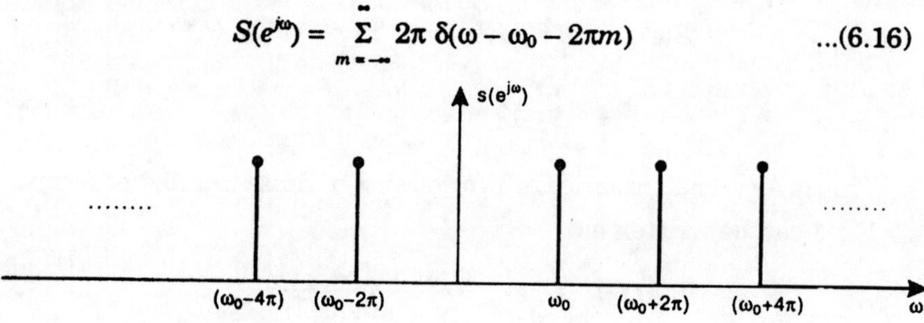

Fig. 6.7 *Illustration of Fourier transform of signal* $s(n) = e^{j\omega n}$.

For checking the validity of expression given in Eq. 6.16, we will determine the inverse Fourier transform.

Substituting the Eq. 6.16 into the DTFT synthesis equation given as

$$\frac{1}{2\pi}\int_{2\pi} S(e^{j\omega})\, e^{j\omega n}\, d\omega$$

$$= \frac{1}{2\pi}\int_{2\pi}\left[\sum_{m=-\infty}^{\infty} 2\pi\, \delta(\omega - \omega_0 - 2\pi m)\right] e^{j\omega n}\, d\omega \quad ...(6.17)$$

We know that any interval of length 2π includes exactly one impulse in the summation given by Eq. 6.16. Therefore, if the interval of integration is so chosen that it includes the impulse located at $\omega_0 + 2\pi l$, then Eq. 6.17 becomes

$$\frac{1}{2\pi}\int_{2\pi} S(e^{j\omega})\, e^{j\omega n}\, d\omega = e^{j(\omega_0 + 2\pi l)n} = e^{j\omega n} \qquad ...(6.18)$$

Fourier series representation for a discrete-time periodic signal $s(n)$ is given by

$$s(n) = \sum_{k=\langle N_0\rangle} A_k\, e^{jk\omega_0 n} \qquad ...(6.19)$$

and the DTFT is given by

$$S(e^{j\omega}) = \sum_{k=-\infty}^{\infty} 2\pi A_k\, \delta\left(\omega - \frac{2\pi k}{N_0}\right) \qquad ...(6.20)$$

From the above analysis we conclude that the Fourier transform of a discrete-time periodic signal can be directly constructed from its Fourier series coefficients A_k.

Example 6.4 Determine the DTFT of the discrete-time periodic signal

$$s(n) = \cos \omega_0 n$$

with fundamental frequency $\omega_0 = 2\pi/5$.

Solution. Using Euler's relation, signal $s(n) = \cos \omega_0 n$ can written as

$$s(n) = \cos \omega_0 n = \frac{e^{j\omega_0 n} + e^{-j\omega_0 n}}{2} \qquad ...(1)$$

$$= \frac{1}{2} e^{j\omega_0 n} + \frac{1}{2} e^{-j\omega_0 n}$$

But we know that DTFT of a period signal $s(n) = e^{j\omega_0 n}$ is given by

$$S(e^{j\omega}) = \text{DTFT}\{s(n)\} = \text{DTFT}\{e^{j\omega_0 n}\}$$

$$= \sum_{m=-\infty}^{\infty} 2\pi\, \delta(\omega - \omega_0 - 2\pi m) \qquad ...(2)$$

Similarly, DTFT of Eq. 1 can be determined as

$$S(e^{j\omega}) = \text{DTFT}\{s(n)\} = \text{DTFT}\left\{\frac{1}{2} e^{j\omega_0 n} + \frac{1}{2} e^{-j\omega_0 n}\right\}$$

$$= \frac{1}{2}\text{DTFT}\{e^{j\omega_0 n}\} + \frac{1}{2}\text{DTFT}\{e^{-j\omega_0 n}\}$$

$$= \frac{1}{2}\sum_{m=-\infty}^{\infty} 2\pi\, \delta(\omega - \omega_0 - 2\pi m)$$

$$+ \frac{1}{2}\sum_{m=-\infty}^{\infty} 2\pi\, \delta(\omega + \omega_0 - 2\pi m)$$

$$\left(\text{Substituting } \omega_0 = \frac{2\pi}{5}\right)$$

or
$$S(e^{j\omega}) = \sum_{m=-\infty}^{\infty} \pi\,\delta\left(\omega - \frac{2\pi}{5} - 2\pi m\right) + \sum_{m=-\infty}^{\infty} \pi\,\delta\left(\omega + \frac{2\pi}{3} - 2\pi m\right) \quad ...(3)$$

$$S(e^{j\omega}) = \sum_{m=-\infty}^{\infty} \pi\,\delta\left(\omega - \frac{2\pi}{5} - 2\pi m\right) + \sum_{m=-\infty}^{\infty} \pi\,\delta\left(\omega + \frac{2\pi}{5} - 2\pi m\right)$$

$$= \left\{ ... + \pi\,\delta\left(\omega - \frac{2\pi}{5} + 2\pi\right) + \pi\,\delta\left(\omega - \frac{2\pi}{5}\right) + \pi\,\delta\left(\omega - \frac{2\pi}{5} - 2\pi\right) + ... \right\}$$

$$+ \left\{ ... + \pi\,\delta\left(\omega + \frac{2\pi}{5} + 2\pi\right) + \pi\,\delta\left(\omega + \frac{2\pi}{3}\right) + \pi\,\delta\left(\omega + \frac{2\pi}{5} - 2\pi\right) + ... \right\}$$

or
$$S(e^{j\omega}) = \pi\,\delta\left(\omega - \frac{2\pi}{5}\right) + \pi\,\delta\left(\omega + \frac{2\pi}{3}\right), \quad -\pi \le \omega \le \pi \quad ...(4)$$

Here $S(e^{j\omega})$ repeats periodically with a period of 2π. DTFT of $s(n) = \cos \omega_0 n$ is shown in Fig. 6.8.

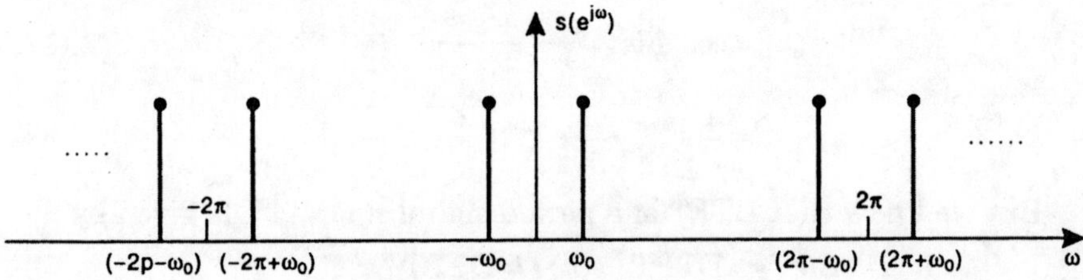

Fig. 6.8 *DTFT of* $s(n) = \cos \omega_0 n$.

Example 6.5 Determine the DTFT of the discrete-time periodic impulse train given by

$$s(n) = \sum_{k=-\infty}^{\infty} \delta(n - kN_0)$$

Solution. Discrete-time periodic impulse train $s(n) = \sum_{k=-\infty}^{\infty} \delta(n - kN_0)$ is sketched in Fig. 6.9.

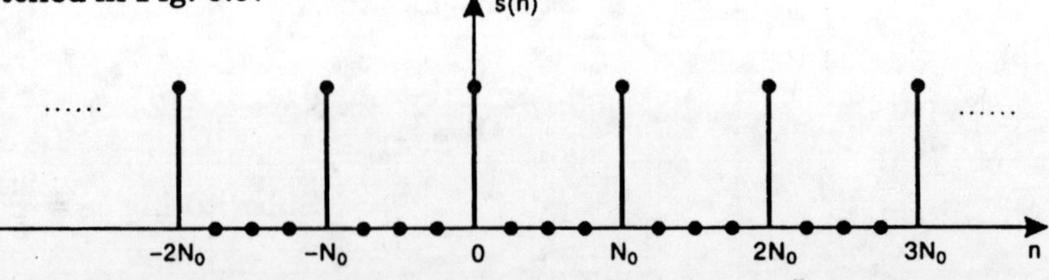

Fig. 6.9 *Discrete-time periodic impulse train* $s(n) = \sum_{k=-\infty}^{\infty} \delta(n - kN_0)$.

Now, we can calculate the Fourier series coefficients for this discrete-time periodic impulse train $s(n) = \sum_{k=-\infty}^{\infty} \delta(n - kN_0)$ directly from equation given below

$$A_k = \frac{1}{N_0} \sum_{n=<N_0>} s(n)\, e^{-jk\omega_0 n} \qquad ...(1)$$

Substituting $s(n) = \sum_{k=-\infty}^{\infty} \delta(n - kN_0)$ and choosing the interval of summation as $0 \le n \le N_0 - 1$ in Eq. 1, we obtain

$$A_k = \frac{1}{N_0} \sum_{n=0}^{N_0-1} \sum_{k=-\infty}^{\infty} \delta(n - kN_0)\, e^{-jk\omega_0 n} = \frac{1}{N_0} \qquad ...(2)$$

DTFT of $s(n)$ is determined as

$$S(e^{j\omega}) = \text{DTFT}\{\, s(n) \,\} = \text{DTFT}\left\{ \sum_{k=-\infty}^{\infty} \delta(n - kN_0) \right\}$$

$$= \sum_{k=-\infty}^{\infty} 2\pi A_k\, \delta\left(\omega - \frac{2\pi k}{N_0}\right) \qquad ...(3)$$

Substituting Eq. 2 in Eq. 3, we obtain

$$S(e^{j\omega}) = \frac{2\pi}{N_0} \sum_{k=-\infty}^{\infty} \delta\left(\omega - \frac{2\pi k}{N_0}\right) \qquad ...(4)$$

DTFT of discrete-time periodic impulse train $s(n) = \sum_{k=-\infty}^{\infty} \delta(n - kN_0)$ is shown in Fig. 6.10.

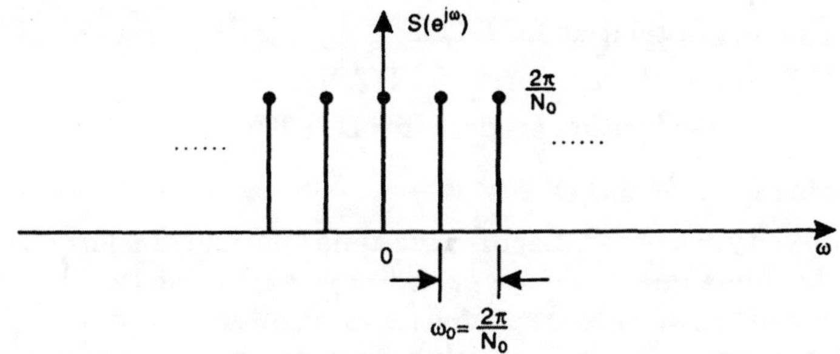

Fig. 6.10 *DTFT of discrete-time periodic impulse train*

$$s(n) = \sum_{k=-\infty}^{\infty} \delta(n - kN_0).$$

6.5 PROPERTIES OF THE DTFT

In this section, we will study about the various properties of Fourier transform of discrete-time signals. Fourier transform of discrete-time signals is also referred to as discrete-time Fourier transform (DTFT). These properties are often useful in reducing the complexity in evaluation of the Fourier transform and inverse Fourier transform. Also, we will see some of the similarities and differences between CTFT and DTFT. The derivation of DTFT properties is essentially identical to its continuous-time counterpart, *i.e.,* CTFT. Also, because of the close relationship between the Fourier series and Fourier transform, many of the DTFT properties translate directly into corresponding DTFS properties.

Here we will use one notation similar to that used for CTFT to indicate the pairing of a signal and its Fourier transform. That is

$$S(e^{j\omega}) = \text{DTFT}\{s(n)\}$$

$$s(n) = \text{Inverse DTFT}\{S(e^{j\omega})\}$$

or $$s(n) \xleftrightarrow{\text{DTFT}} S(e^{j\omega})$$

In this section, we will discuss following properties of the DTFT :

1. Periodicity of the DTFT
2. Linearity of the DTFT
3. Time shifting and frequency shifting
4. Complex conjugation and conjugate symmetry
5. Differencing and accumulation in time-domain
6. Time reversal of a discrete-time sequence
7. Time-expansion
8. Differentiation in frequency-domain
9. Parseval's relation for DTFT
10. The convolution property for DTFT
11. The multiplication property for DTFT.

6.5.1 Periodicity of the DTFT

We have already discussed that discrete-time Fourier transform (DTFT) of a discrete-time signal is always periodic in ω with period 2π. A periodic function is a function which repeates its value after some fixed value of independent variable. This fixed value of independent variable is called the period of the periodic function.

The DTFT of $s(n)$ is always periodic in ω with period, *i.e.,*

$$s(n) \xleftrightarrow{\text{DTFT}} S(e^{j\omega})$$

$$S(e^{j(\omega+2\pi)}) = S(e^{j\omega}) \qquad \text{...(6.21)}$$

This is in contrast of CTFT, which in general is not a periodic function.

6.5.2 Linearity of DTFT

The DTFT is also a linear transformation tool as CTFT.

$$s_1(n) \xleftrightarrow{\text{DTFT}} S_1(e^{j\omega})$$

$$s_2(n) \xleftrightarrow{\text{DTFT}} S_2(e^{j\omega})$$

where $s_1(n)$ and $s_2(n)$ are discrete-time sequences whose DTFTs are $S_1(e^{j\omega})$ and $S_2(e^{j\omega})$, respectively.

From the property of linearity, it is true for above two discrete-time sequences $s_1(n)$ and $s_2(n)$ given as

$$A_1\, s_1(n) + A_2\, s_2(n) \xleftrightarrow{\text{DTFT}} A_1\, S_1(e^{j\omega}) + A_2\, S_2(e^{j\omega}) \qquad \text{...(6.22)}$$

where A_1 and A_2 are constants.

6.5.3 Time-Shifting and Frequency Shifting

Here we will first study the effect of time shifting of a discrete-time sequence $s(n)$ on its DTFT $S(e^{j\omega})$ and then the effect of frequency shifting.

If $$s(n) \xleftrightarrow{\text{DTFT}} S(e^{j\omega})$$

then $$s(n - n_0) \xleftrightarrow{\text{DTFT}} e^{-j\omega n_0}\, S(e^{j\omega}) \qquad \text{...(6.23)}$$

where $s(n - n_0)$ is the time-shifted version of $s(n)$.

Now will see the effect of frequency shifting.

If $$s(n) \xleftrightarrow{\text{DTFT}} S(e^{j\omega})$$

then $$e^{j\omega_0 n}\, s(n) \xleftrightarrow{\text{DTFT}} S(e^{j(\omega - \omega_0)}) \qquad \text{...(6.24)}$$

where $S(e^{j(\omega - \omega_0)})$ is the frequency-shifted version of $S(e^{j\omega})$.

6.5.4 Complex Conjugation and Conjugate Symmetry

We can obtain complex conjugation of a complex discrete-time signal $s(n)$ by reversing the sign of the imaginary part of the complex signal $s(n)$.

Let $s(n)$ be the complex discrete-time signal, $s(n) = s_R(n) + j\, s_I(n)$ and its DTFT is also complex, *i.e.*,

$$S(e^{j\omega}) = S_R(e^{j\omega}) + j\, S_I(e^{j\omega})$$

where $s_R(n)$ and $s_I(n)$ are real and imaginary parts of $s(n)$, and $S_R(e^{j\omega})$ and $S_I(e^{j\omega})$ are also real and imaginary parts of $S(e^{j\omega})$, respectively.

Now complex conjugation of $s(n)$ is given by

$$s^*(n) = s_R(n) - j\, s_I(n)$$

and complex conjugation of $S(e^{j\omega})$ is given by

$$S^*(e^{j\omega}) = S_R(e^{j\omega}) - j\, S_I(e^{j\omega})$$

If $$s(n) \xleftrightarrow{\text{DTFT}} S(e^{j\omega})$$

then $$s^*(n) \xleftrightarrow{\text{DTFT}} S^*(e^{-j\omega}) \qquad ...(6.25)$$

Also, if signal $s(n)$ is a real-valued function, then its DTFT $S(e^{j\omega})$ will be conjugate symmetric. That is

$$S(e^{j\omega}) = S^*(e^{-j\omega}) \quad \text{for } s(n) \text{ real} \qquad ...(6.26)$$

From Eq. 6.26, we can say that Re $\{S(e^{j\omega})\}$ is an even function of ω and Im $\{S(e^{j\omega})\}$ is an odd function of ω. Similarly, the magnitude of $S(e^{j\omega})$ is an even function and the phase angle $\angle S(e^{j\omega})$ is an odd function.

Also, we can decompose sequence $s(n)$ into even and odd parts. Furthermore,

$$\text{Ev}\{s(n)\} = s_E(n) \xleftrightarrow{\text{DTFT}} \text{Re}\{S(e^{j\omega})\} = S_R(e^{j\omega})$$

and $$\text{Od}\{s(n)\} = s_O(n) \xleftrightarrow{\text{DTFT}} j\,\text{Im}\{S(e^{j\omega})\} = j\,S_I(e^{j\omega})$$

where Ev $\{s(n)\}$ and Od $\{s(n)\}$ are the even and odd parts, respectively of $s(n)$. Specifically, if $s(n)$ is real and even, its DTFT is also real and even.

6.5.5 Differencing and Accumulation in time

Accumulation is the discrete-time counterpart of integration. Integration is used for continuous-time signals. Inverse of accumulation is referred to as first differencing.

Let $s(n)$ be a discrete-time signal with DTFT $S(e^{j\omega})$. The first differencing of signal $s(n)$ is given by $s(n) - s(n-1)$. DTFT of first differencing of signal $s(n)$ can be determined by using the properties of linearity and time-shifting as

$$s(n) - s(n-1) \xleftrightarrow{\text{DTFT}} S(e^{j\omega}) - e^{-j\omega} S(e^{j\omega})$$
$$= (1 - e^{-j\omega})\, S(e^{j\omega}) \qquad ...(6.27)$$

The accumulation of signal $s(n)$ is given by

$$y(n) = \sum_{m=-\infty}^{n} s(m) \qquad ...(6.28)$$

Eq. 6.28 can also be expressed as

$$y(n) - y(n-1) = s(n) \qquad ...(6.29)$$

Thus, we can conclude that DTFT of $y(n)$ should be related to the DTFT of $s(n)$ by division by $(1 - e^{-j\omega})$. This is not perfectly correct but the precise relation is given by

$$\sum_{m=-\infty}^{n} s(m) \xleftrightarrow{\text{DTFT}} \frac{1}{1-e^{-j\omega}} S(e^{j\omega}) + \pi\, S(e^{j0}) \sum_{k=-\infty}^{\infty} \delta(\omega - 2\pi k) \qquad ...(6.30)$$

The impulse train on the RHS of Eq. 6.30 shows the dc or average value that can result from summation.

Example 6.6 Determine the DTFT of the unit step function $s(n) = u(n)$ using the accumulation property of DTFT.

Solution. We know that

$$g(n) = \delta(n) \xleftrightarrow{\text{DTFT}} G(e^{j\omega}) = 1 \quad ...(1)$$

Also we have studied that the unit-step function $u(n)$ is the running sum of the unit-impulse function $\delta(n)$. This realtion is given as

$$s(n) = \sum_{m=-\infty}^{\infty} g(m) \quad ...(2)$$

Now, taking the DTFT of both sides of Eq. 2 and using accumulation property of DTFT, we obtain

$$S(e^{j\omega}) = \text{DTFT}\left\{ \sum_{m=-\infty}^{\infty} g(m) \right\}$$

or
$$S(e^{j\omega}) = \frac{1}{(1-e^{-j\omega})} G(e^{j\omega}) + \pi\, G(e^{j0}) \sum_{k=-\infty}^{\infty} \delta(\omega - 2\pi k) \quad ...(3)$$

But from Eq. 1, we have

$$G(e^{j\omega}) = 1 \quad \text{and} \quad G(e^{j.0}) = 1$$

Substituting these values in Eq. 3, we obtain

$$S(e^{j\omega}) = \frac{1}{(1-e^{-j\omega})} + \pi \sum_{k=-\infty}^{\infty} \delta(\omega - 2\pi k) \quad ...(4)$$

6.5.6 Time Reversal of a Discrete-Time Signal

Taking mirror image or folding of a discrete-time sequence is called time reversal. Consider a discrete-time signal or sequence $s(n)$ whose mirror image is given by $s(-n)$.

If
$$s(n) \xleftrightarrow{\text{DTFT}} S(e^{j\omega})$$

then
$$s(-n) \xleftrightarrow{\text{DTFT}} S(e^{-j\omega}) \quad ...(6.31)$$

Now we can prove this property as

$$y(n) = s(-n) \quad ...(6.32)$$

Taking DTFT of both sides of Eq. 6.32, we obtain

$$Y(e^{j\omega}) = \text{DTFT}\,\{y(n)\} = \sum_{n=-\infty}^{\infty} s(-n)\, e^{-j\omega n} \quad ...(6.33)$$

Now substituting $m = -n$ in Eq. 6.33, we obtain

$$Y(e^{j\omega}) = \sum_{m=-\infty}^{\infty} s(m)\, e^{-j\omega(-m)} = S(e^{j\omega}) \quad ...(6.34)$$

That is,

$$s(-n) \xleftrightarrow{\text{DTFT}} S(e^{-j\omega}) \qquad \text{...(6.35)}$$

Hence, it is to be proved.

6.5.7 Time Expansion of a Discrete-Time Signal

Time expansion of a discrete-time signal is equivalent to multiplication of independent variable, *i.e.,* time of a signal by an integer scalar quantity A, which is greater than unity.

We have already derived time expansion property of CTFT in chapter 5 which is given by

$$s(At) \xleftrightarrow{\text{CTFT}} \frac{1}{|A|} S\left(\frac{j\omega}{A}\right) \qquad \text{...(6.36)}$$

We cannot slow down the signal by choosing $A < 1$. On the otherhand, if we let A be an integer other than ± 1. For example $A = 2$ then $s(2n)$, we cannot merely speed up the original signal. That is, since n can take on only integer values the signal $s(2n)$ consists of the even samples of original signal $s(n)$ alone.

Time-expanded signal is given by

$$s_{(k)}(n) = \begin{cases} s(n/k), & \text{if } n \text{ is a multiple of } k \\ 0, & \text{if } n \text{ is not a multiple of } k \end{cases} \qquad \text{...(6.37)}$$

where k is a positive integer.

Form $k = 2$, the sequence $s_{(k)}(n)$ is obtained from original sequence $s(n)$ by placing $k - 1$ zeros between successive values of the original sequence. Now we can say that $s_{(k)}(n)$ is a slowed-down version of $s(n)$.

One sequence $s(n)$ and its slowed down version $s_{(k)}(n)$ for $k = 2$ are shown in Fig. 6.11.

Since $s_{(k)}(n)$ equals zero unless n is multiple of k.

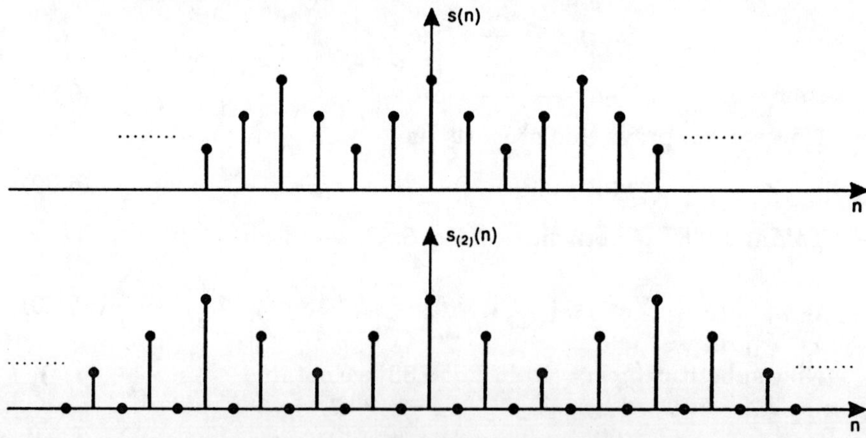

Fig. 6.11 *The signal $s_{(2)}(n)$ obtained from $s(n)$ by inserting one zero between successive values of the original signal $s(n)$.*

For $n = mk$, DTFT of $s_{(k)}(n)$ is given by

$$S_{(k)}(e^{j\omega}) = \text{DTFT}\left\{ s_{(k)}(n) \right\} = \sum_{n=-\infty}^{\infty} s_{(k)}(n)\, e^{-j\omega n}$$

$$= \sum_{m=-\infty}^{\infty} s_{(k)}(mk)\, e^{-j\omega mk}$$

Furthermore, since $s_{(k)}(mk) = s(m)$, we find that

$$S_{(k)}(e^{j\omega}) = \sum_{m=-\infty}^{\infty} s(m)\, e^{-j(k\omega)m} = S(e^{jk\omega})$$

That is $\quad s_{(k)}(n) \xleftrightarrow{\text{DTFT}} S(e^{jk\omega})$ ...(6.38)

As the discrete-time signal is spread out and slowed down in time by taking $k > 1$, its DTFT is compressed. For example, since $S(e^{j\omega})$ is periodic in ω with period 2π and $S(e^{jk\omega})$ is also periodic in ω with period $2\pi/k$.

Now we can conclude that there is inverse relationship between time and frequency domains. In other words, we can say that as value of k increases, $s_{(k)}(n)$ spreads out while its discrete-time Fourier transform (DTFT) is compressed.

Example 6.7 Determime the DTFT of the discrete-time sequence $s(n)$ shown in Fig. 6.12 using time-expansion property of DTFT.

Solution. This sequence $s(n)$ can be related to the simpler sequence

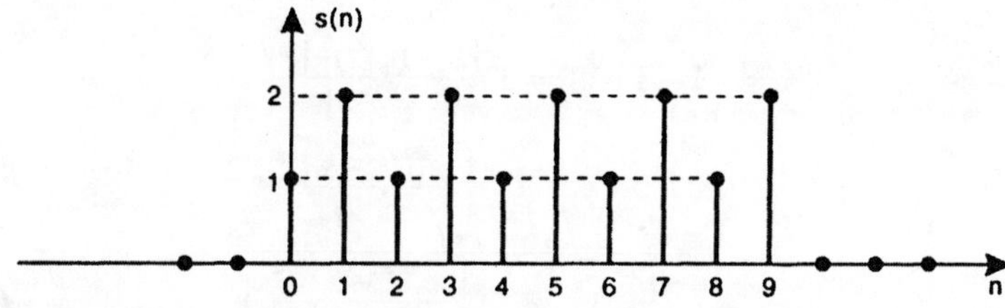

Fig. 6.12 *Discrete-time sequence s(n).*

$g(n)$ shown in Fig. 6.13 (*a*).

The discrete-time sequence $s(n)$ is related to the discrete-time sequence $g(n)$ as

$$s(n) = g_{(2)}(n) + 2\, y_{(2)}(n-1) \qquad \text{...(1)}$$

where

$$g_2(n) = \begin{cases} g(n/2)\,, & \text{if } n \text{ is even} \\ 0\,, & \text{if } n \text{ is odd} \end{cases} \qquad \text{...(2)}$$

and $g_{(2)}(n-1)$ is the shifted version of $g_{(2)}(n)$ by one unit to the right. The signals $g_{(2)}(n)$ and $2\, g_{(2)}(n-1)$ are depicted in Fig. 6.13 (*b*) and Fig. 6.13 (*c*) respectively.

Here we relate given sequence $s(n)$ in terms of a simpler sequence

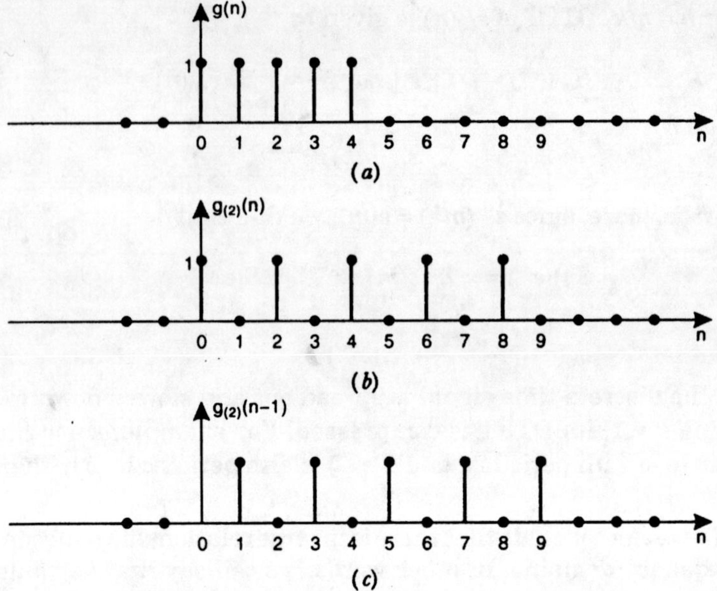

Fig. 6.13 (*a*) *Simpler sequence g(n).* (*b*) *Time- expanded version of sequence g(n).* (*c*) *Time-shifted version of* $g_{(2)}(n)$ *by one unit to the right.*

$g(n)$ which is a discrete-time rectangular pulse.

$$\text{DTFT of } g(n) = G(e^{j\omega}) = \sum_{n=-\infty}^{\infty} g(n)\, e^{-j\omega n}$$

$$= \sum_{n=0}^{4} 1.e^{-j\omega n} = 1\,\frac{[1-(e^{-j\omega})^5]}{[1-e^{-j\omega}]}$$

$$= \frac{1-e^{-j5\omega}}{1-e^{-j\omega}} = \frac{e^{-j5\omega/2}\left[\dfrac{e^{j5\omega/2}-e^{-j5\omega/2}}{2j}\right]}{e^{-j\omega/2}\left[\dfrac{e^{j\omega/2}-e^{-j\omega/2}}{2j}\right]}$$

or $$G(e^{j\omega}) = e^{-j2\omega}\left\{\frac{\sin\left(\dfrac{5\omega}{2}\right)}{\sin\left(\dfrac{\omega}{2}\right)}\right\} \qquad ...(3)$$

$$g(n) \xleftrightarrow{\text{DTFT}} G(e^{j\omega}) = e^{-j2\omega}\left\{\frac{\sin\left(\dfrac{5\omega}{2}\right)}{\sin\left(\dfrac{\omega}{2}\right)}\right\}$$

then from time-expansion property

$$g_{(k)}(n) \xleftrightarrow{\text{DTFT}} G(e^{jk\omega}) = e^{-j2k\omega}\left\{\frac{\sin\left(\frac{5k\omega}{2}\right)}{\sin\left(\frac{k\omega}{2}\right)}\right\}$$

For $k = 2$, $g_{(2)}(n) \xleftrightarrow{\text{DTFT}} G(e^{j2\omega}) = e^{-j2\times 2\omega}\left\{\frac{\sin\left(\frac{5\times 2\omega}{2}\right)}{\sin\left(\frac{2\omega}{2}\right)}\right\}$

$$= e^{-j4\omega}\left\{\frac{\sin(5\omega)}{\sin(\omega)}\right\} \qquad ...(4)$$

Now using the linearity and time-shifting properties, we get

$$2\,g_{(2)}(n) \xleftrightarrow{\text{DTFT}} e^{-j\omega}\left\{2\,G(e^{j2\omega})\right\}$$

$$= 2\,e^{-j\omega}\left\{e^{-j4\omega}\left[\frac{\sin(5\omega)}{\sin(\omega)}\right]\right\}$$

$$= 2\,e^{-j5\omega}\left[\frac{\sin(5\omega)}{\sin(\omega)}\right] \qquad ...(5)$$

Combining Eqs. 4 and 5, we have

$$s(n) = g_{(2)}(n) + 2\,g_{(2)}(n-1) \xleftrightarrow{\text{DTFT}} e^{-j4\omega}\left[\frac{\sin(5\omega)}{\sin(\omega)}\right] + 2\,e^{-j5\omega}\left[\frac{\sin(5\omega)}{\sin(\omega)}\right]$$

$$= e^{-j4\omega}(1 - 2e^{-j\omega})\left[\frac{\sin(5\omega)}{\sin(\omega)}\right]$$

6.5.8 Differentiation in Frequency-Domain

Here we will study the differentiation of DTFT of $s(n)$, $S(e^{j\omega})$ with respect to ω. Let

$$s(n) \xleftrightarrow{\text{DTFT}} S(e^{j\omega})$$

Using the DTFT analysis equation $S(e^{j\omega}) = \sum_{n=-\infty}^{\infty} s(n)\,e^{-j\omega n}$ and differentiating both sides with respect to ω, we obtain

$$\frac{d\,S(e^{j\omega})}{d\omega} = \frac{d}{d\omega}\left[\sum_{n=-\infty}^{\infty} s(n)\,e^{-j\omega n}\right]$$

$$= \sum_{n=-\infty}^{\infty} s(n)\,[-jne^{-j\omega n}]$$

$$= -j \sum_{n=-\infty}^{\infty} n\, s(n)\, e^{-j\omega n}$$

or $$\frac{j\, d\, S(e^{j\omega})}{d\omega} = -j^2 \sum_{n=-\infty}^{\infty} n\, s(n)\, e^{-j\omega n}$$

or $$\frac{j\, d\, S(e^{j\omega})}{d\omega} = \sum_{n=-\infty}^{\infty} (n\, s(n))\, e^{-j\omega n} = \text{DTFT}\{n\, s(n)\}$$

or $$n\, s(n) \xleftrightarrow{\text{DTFT}} \frac{j\, d\, S(e^{j\omega})}{d\omega} \qquad \text{...(6.39)}$$

6.5.9 Parseval's Relation for DTFT

Parseval's relation for DTFT states that the total energy in a discrete-time signal $s(n)$ may be determined either by computing the energy per unit time $|s(n)|^2$ and summing over all time or by computing the energy per unit frequency $\frac{|S(e^{j\omega})|^2}{2\pi}$ and integrating over a full 2π interval of distinct discrete-time frequencies. In analog with the continuous-time signal, $|S(e^{j\omega})|$ is called the energy-density spectrum of the signal $s(n)$. Parseval's relation for DTFT is given as

$$\sum_{n=-\infty}^{\infty} |s(n)|^2 = \frac{1}{2\pi} \int_{2\pi} |S(e^{j\omega})|^2\, d\omega \qquad \text{...(6.40)}$$

6.5.10 The Convolution Property

In chapter 5, we have discussed the importance of the CTFT with respect to its effect on the convolution operation and its application in dealing with continuous-time LTI systems. Here we will discuss the importance of the DTFT with regard to its effect on the convolution operation and analysis of discrete-time LTI systems. By using DTFT, convolution of two discrete-time signals $s_1(n)$ and $s_2(n)$ is converted into multiplication of DTFT of individual discrete-time signals.

If $$s_1(n) \xleftrightarrow{\text{DTFT}} S_1(e^{j\omega})$$

and $$s_2(n) \xleftrightarrow{\text{DTFT}} S_2(e^{j\omega})$$

then from convolution property of DTFT

$$s(n) = s_1(n) * s_2(n) \xleftrightarrow{\text{DTFT}} S(e^{j\omega}) = S_1(e^{j\omega})\, S_2(e^{j\omega}) \qquad \text{...(6.41)}$$

Now we are going to apply DTFT for representing and analyzing discrete-time LTI systems. Specifically, if $s(n)$, $h(n)$ and $y(n)$ are the input, impulse response and output, respectively of a discrete-time LTI system.

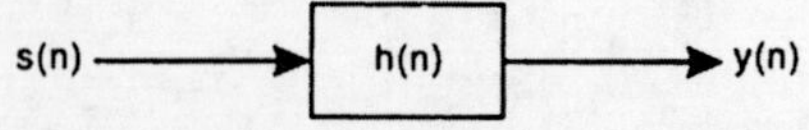

Fig. 6.14 *Discrete-time LTI system.*

Output $y(n)$ is determined by convolving $s(n)$ and $h(n)$ given as

$$y(n) = s(n) * h(n) \quad ...(6.42)$$

By using DTFT, Eq. 6.42 can be expressed as

$$Y(e^{j\omega}) = S(e^{j\omega})\, H(e^{j\omega}) \quad ...(6.43)$$

where $S(e^{j\omega})$, $H(e^{j\omega})$ and $Y(e^{j\omega})$ are the DTFTs of $s(n)$, $h(n)$ and $y(n)$ respectively.

Now, combining Eqs. 6.42 and 6.43, we get

$$y(n) = s(n) * h(n) \xleftrightarrow{\text{DTFT}} Y(e^{j\omega}) = S(e^{j\omega})\, H(e^{j\omega}) \quad ...(6.44)$$

where $H(e^{j\omega})$ is the discrete-time Fourier transform (DTFT) of the impulse response $h(n)$ of the discrete-time LTI system. It is also called frequency response of discrete-time LTI system.

Example 6.8 Determine the frequency response of a discrete-time LTI system with impulse response $h(n) = \delta(n - n_0)$. Also determine output for this system.

Solution. The frequency response of a discrete-time LTI system is equal to the DTFT of the impulse response $h(n)$ of the system. The frequency response is determined as

$$H(e^{j\omega}) = \text{DTFT}\left\{h(n)\right\} = \sum_{n=-\infty}^{\infty} h(n)\, e^{-j\omega n}$$

$$= \sum_{n=-\infty}^{\infty} \delta(n - n_0)\, e^{-j\omega n} = e^{-j\omega n_0} \quad ...(1)$$

We know from convolution property of DTFT

$$y(n) = h(n) * s(n) \xleftrightarrow{\text{DTFT}} Y(e^{j\omega}) = H(e^{j\omega})\, S(e^{j\omega})$$

or

$$Y(e^{j\omega}) = H(e^{j\omega})\, S(e^{j\omega}) \quad ...(2)$$

Substituting Eq. 1 in Eq. 2, we obtain

$$Y(e^{j\omega}) = e^{-j\omega n_0}\, S(e^{j\omega}) \quad ...(3)$$

Output $y(n)$ of above discrete-time LTI system is determined by taking the inverse DTFT of Eq. 3. Taking inverse DTFT, we get

$$y(n) = \text{Inverse DTFT}\left\{Y(e^{j\omega})\right\}$$

$$= \text{Inverse DTFT}\left\{e^{-j\omega n_0}\, S(e^{j\omega})\right\} = s(n - n_0) \quad ...(4)$$

Note in this example output $y(n)$ is equal to shifted version of input $s(n)$ by a constant time n_0. The frequency response $H(e^{j\omega}) = e^{-j\omega n_0}$ is purely time-shifted and has unity magnitude at all frequencies. Its phase characteristics are equal to $-\omega n_0$, *i.e.*, it is linear with frequency.

Example 6.9 Determine the impulse response $h(n)$ of a discrete-time ideal low pass filter whose frequency response $H(e^{j\omega})$ is shown in Fig. 6.15.

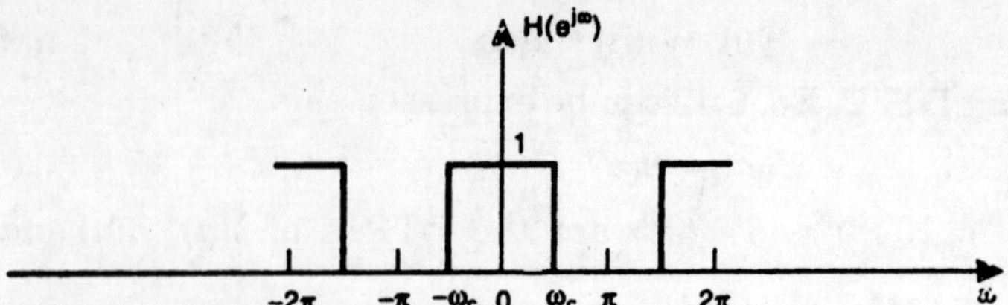

Fig. 6.15 *Frequency response of a discrete-time ideal low pass filter.*

Solution. Impulse response $h(n)$ of discrete-time ideal low pass filter is equal to inverse DTFT of the frequency response $H(e^{j\omega})$.

$$h(n) = \text{Inverse DTFT}\left\{ H(e^{j\omega}) \right\}$$

$$= \frac{1}{2\pi} \int_{2\pi} H(e^{j\omega})\, e^{j\omega n}\, d\omega \qquad \text{...(1)}$$

In particular, using $-\pi \le \omega \le \pi$ as the interval of integration in Eq. 1, we obtain

$$h(n) = \frac{1}{2\pi} \int_{-\pi}^{\pi} H(e^{j\omega})\, e^{j\omega n}\, d\omega$$

$$= \frac{1}{2\pi} \int_{-\omega_c}^{\omega_c} 1.e^{j\omega n}\, d\omega = \frac{1}{2\pi} \left[\frac{1}{jn} e^{j\omega n} \right]_{-\omega_c}^{\omega_c}$$

$$= \frac{1}{n\pi} \left[\frac{e^{j\omega_c n} - e^{-j\omega_c n}}{2j} \right] = \frac{1}{n\pi} \sin \omega_c n$$

or $$h(n) = \frac{\sin \omega_c n}{\pi n} \qquad \text{...(2)}$$

This impulse response $h(n)$ is shown in Fig. 6.16.

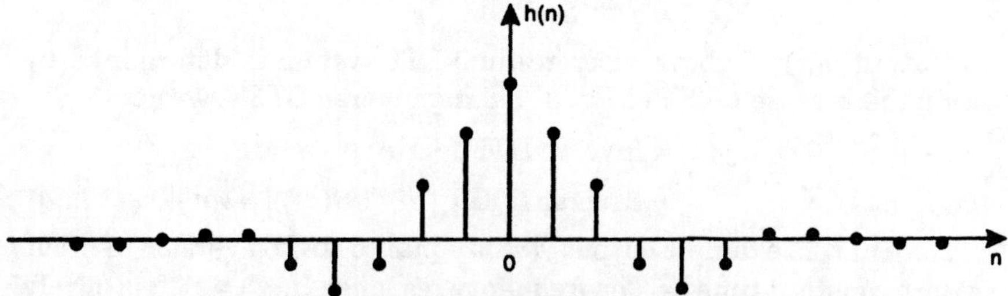

Fig. 6.16 *Impulse response of a discrete-time ideal low pass filter.*

Example 6.10 Determine the output $y(n)$ of a discrete-time LTI system with impulse response $h(n) = A^n\, u(n)$ with $|A| < 1$ to an input $s(n) = B^n\, u(n)$ with $|B| < 1$.

Solution. Output $y(n)$ is determined by using convolution property of DTFT as

$$y(n) = h(n) * s(n) \xleftrightarrow{\text{DTFT}} Y(e^{j\omega}) = H(e^{j\omega})\, S(e^{j\omega})$$

or $$Y(e^{j\omega}) = H(e^{j\omega})\, S(e^{j\omega}) \qquad ...(1)$$

Now we will determine $H(e^{j\omega})$ and $S(e^{j\omega})$ as

$$H(e^{j\omega}) = \text{DTFT}\{\, h(n) \,\} = \sum_{n=-\infty}^{\infty} h(n)\, e^{-j\omega n}$$

$$= \sum_{n=-\infty}^{\infty} A^n\, u(n)\, e^{-j\omega n}$$

But $u(n)$ is defined as $u(n) = \begin{cases} 1\,, & n \geq 0 \\ 0\,, & n < 0 \end{cases}$

$$= \sum_{n=0}^{\infty} A^n\, e^{-j\omega n} = \sum_{n=0}^{\infty} (A\, e^{-j\omega})^n$$

or $$H(e^{j\omega}) = \frac{1}{1 - A\, e^{-j\omega}} \qquad ...(2)$$

Similarly, we can determine $S(e^{j\omega})$ as

$$S(e^{j\omega}) = \text{DTFT}\{\, s(n) \,\} = \sum_{n=-\infty}^{\infty} s(n)\, e^{-j\omega n}$$

$$= \sum_{n=-\infty}^{\infty} B^n\, u(n)\, e^{-j\omega n} = \sum_{n=0}^{\infty} B^n\, e^{-j\omega n}$$

$$= \sum_{n=0}^{\infty} (B\, e^{-j\omega})^n = \frac{1}{1 - B\, e^{-j\omega}}$$

or $$S(e^{j\omega}) = \frac{1}{1 - B\, e^{-j\omega}} \qquad ...(3)$$

Substituting Eqs. 2 and 3 in Eq. 1, we get

$$Y(e^{j\omega}) = H(e^{j\omega})\, S(e^{j\omega})$$

$$= \left(\frac{1}{1 - A\, e^{-j\omega}}\right)\left(\frac{1}{1 - B\, e^{-j\omega}}\right) = \frac{1}{(1 - A\, e^{-j\omega})\,(1 - B\, e^{-j\omega})}$$

(using partial fraction expansion method)

$$= \frac{\alpha_1}{1 - A\, e^{-j\omega}} + \frac{\alpha_2}{1 - B\, e^{-j\omega}} \qquad ...(4)$$

Determination of values of α_1 and α_2 :

$1 - A\, e^{-j\omega} = 0$ or $e^{-j\omega} = \frac{1}{A}$,

$$\alpha_1 = \frac{1}{(1 - B\, e^{-j\omega})} = \frac{1}{\left(1 - B\,\frac{1}{A}\right)} = \frac{A}{(A - B)}$$

$1 - B\,e^{-j\omega} = 0$ or $e^{-j\omega} = \frac{1}{B}$,

$$\alpha_2 = \frac{1}{(1 - A\,e^{-j\omega})} = \frac{1}{\left(1 - A\frac{1}{B}\right)} = \frac{B}{(B-A)} = \frac{-B}{(A-B)}$$

Substituting the values of α_1 and α_2 in Eq. 4, we get

$$Y(e^{j\omega}) = \frac{\left(\frac{A}{A-B}\right)}{1 - A\,e^{-j\omega}} + \frac{\left(-\frac{B}{A-B}\right)}{1 - B\,e^{-j\omega}} \qquad ...(5)$$

For determining output $y(n)$, we take inverse DTFT of both sides of Eq. 5, we get

$$y(n) = \text{Inverse DTFT}\left\{Y(e^{j\omega})\right\}$$

$$= \text{Inverse DTFT}\left\{\frac{\left(\frac{A}{A-B}\right)}{1 - A\,e^{-j\omega}} + \frac{\left(-\frac{B}{A-B}\right)}{1 - B\,e^{-j\omega}}\right\}$$

$$= \left(\frac{A}{A-B}\right)\text{Inverse DTFT}\left\{\frac{1}{1 - A\,e^{-j\omega}}\right\}$$

$$- \left(\frac{B}{A-B}\right)\text{Inverse DTFT}\left\{\frac{1}{1 - B\,e^{-j\omega}}\right\}$$

$$= \left(\frac{A}{A-B}\right)A^n\,u(n) - \left(\frac{B}{A-B}\right)B^n\,u(n)$$

or
$$y(n) = \left(\frac{1}{A-B}\right)\left\{A^{n+1}\,u(n) - B^{n+1}\,u(n)\right\} \qquad ...(6)$$

Note that for $\alpha_1 = \alpha_2$, the partial fraction expansion of Eq. 4 is not valid.

6.5.11 The Multiplication Property

The multiplication property of DTFT is used in the sampling and communication theory.

Consider $s(n)$ is a sequence which is a product of two sequences $s_1(n)$ and $s_2(n)$.

$$s(n) = s_1(n)\,s_2(n) \qquad ...(6.45)$$

Let
$$s_1(n) \xleftrightarrow{\text{DTFT}} S_1(e^{j\omega})$$

$$s_2(n) \xleftrightarrow{\text{DTFT}} S_2(e^{j\omega})$$

and
$$s(n) \xleftrightarrow{\text{DTFT}} S(e^{j\omega})$$

Taking the DTFT of both sides of Eq. 6.45, we get

$$S(e^{j\omega}) = \text{DTFT}\{s(n)\}$$

$$= \sum_{n=-\infty}^{\infty} s(n)\, e^{-j\omega n} = \sum_{n=-\infty}^{\infty} s_1(n)\, s_2(n)\, e^{-j\omega n} \quad \text{...(6.46)}$$

Since from inverse DTFT

$$s_1(n) = \frac{1}{2\pi} \int_{2\pi} S_1(e^{j\theta})\, e^{j\theta n}\, d\theta \quad \text{...(6.47)}$$

Substituting Eq. 6.47 in Eq. 6.46, we obtain

$$S(e^{j\omega}) = \sum_{n=-\infty}^{\infty} s_1(n)\, s_2(n)\, e^{-j\omega n}$$

$$= \sum_{n=-\infty}^{\infty} s_2(n) \left\{ \frac{1}{2\pi} \int_{2\pi} S_1(e^{j\theta})\, e^{j\theta n}\, d\theta \right\} e^{-j\omega n} \quad \text{...(6.48)}$$

Interchanging the order of summation and integration in Eq. 6.48, we get

$$S(e^{j\omega}) = \frac{1}{2\pi} \int_{2\pi} S_1(e^{j\theta}) \left[\sum_{n=-\infty}^{\infty} s_2(n)\, e^{-j(\omega-\theta)n} \right] d\theta \quad \text{...(6.49)}$$

But $$S_2(e^{j(\omega-\theta)}) = \sum_{n=-\infty}^{\infty} s_2(n)\, e^{-j(\omega-\theta)n} \quad \text{...(6.50)}$$

Substituting the Eq. 6.50 in Eq. 6.49, we get

$$S(e^{j\omega}) = \frac{1}{2\pi} \int_{2\pi} S_1(e^{j\theta})\, S_2(e^{j(\omega-\theta)})\, d\theta \quad \text{...(6.51)}$$

Eq. 6.51 corresponds to a periodic convolution of $S_1(e^{j\omega})$ and $S_2(e^{j\omega})$. Integral in Eq. 6.51 can be evaluated over an interval of length 2π. The usual form of convolution is often referred to as non-periodic convolution as the integral ranges from $-\infty$ to ∞.

Example 6.11 Determine the DTFT of the multiplication of two discrete-time signals given as $s(n) = s_1(n)\, s_2(n)$.

where $s_1(n) = \dfrac{\sin\left(\dfrac{3\pi n}{4}\right)}{\pi n}$ and $s_2(n) = \dfrac{\sin\left(\dfrac{\pi n}{2}\right)}{\pi n}$

Solution. From the multiplication property of DTFT, we know that DTFT of $s(n)$, *i.e.*, $S(e^{j\omega})$ is the periodic convolution of $S_1(e^{j\omega})$ and $S_2(e^{j\omega})$. Here integration is done over any interval of length 2π. In this periodic convolution, we have chosen the interval $-\pi < \theta < \pi$. The multiplication property of DTFT is given as

$$s(n) = s_1(n)\, s_2(n) \xleftrightarrow{\text{DTFT}} S(e^{j\omega}) = \frac{1}{2\pi} \int_{2\pi} S_1(e^{j\theta})\, S_2(e^{j(\omega-\theta)})\, d\theta$$

or $$S(e^{j\omega}) = \frac{1}{2\pi} \int_{2\pi} S_1(e^{j\theta})\, S_2(e^{j(\omega-\theta)})\, d\theta \qquad ...(1)$$

Eq. 1 represents a periodic convolution, it resembles non- periodic convolution except for that the integration is limited to the interval $-\pi < \theta \leq \pi$.

Eq. 1 can be converted into an ordinary convolution by defining

$$\hat{S}_1(e^{j\omega}) = \begin{cases} S_1(e^{j\omega}), & \text{for } -\pi < \omega \leq \pi \\ 0, & \text{otherwise} \end{cases} \qquad ...(2)$$

Substituting Eq. 2 in Eq. 1, we get

$$S(e^{j\omega}) = \frac{1}{2\pi} \int_{-\pi}^{\pi} \hat{S}_1(e^{j\theta})\, S_2(e^{j(\omega-\theta)})\, d\theta$$

$$= \frac{1}{2\pi} \int_{-\infty}^{\infty} \hat{S}_1(e^{j\theta})\, S_2(e^{j(\omega-\theta)})\, d\theta \qquad ...(3)$$

Thus, $S(e^{j\omega})$ is $1/2\pi$ the non-periodic convolution of the rectangular pulse $\hat{S}_1(e^{j\omega})$ and periodic square wave $S_2(e^{j\omega})$. Both $\hat{S}_1(e^{j\omega})$ and $S_2(e^{j\omega})$ are shown in Fig. 6.17. The result of this convolution is the DTFT of $s(n)$, *i.e.*, $S(e^{j\omega})$ shown in Fig. 6.17 (*c*).

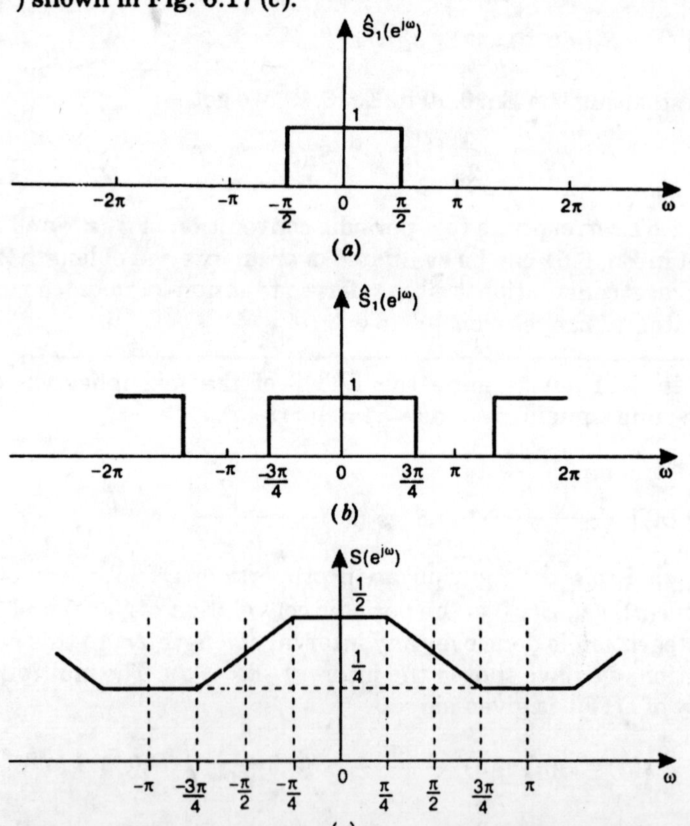

Fig. 6.17 (*a*) $\hat{S}_1(e^{j\omega})$ *shows one period of* $S_1(e^{j\omega})$. (*b*) *DTFT of* $s_2(n)$, *i.e* , $S_2(e^{j\omega})$. (*c*)*Resultant of the periodic convolution of* $\hat{S}_1(e^{j\omega})$ *and* $S_2(e^{j\omega})$.

6.6 TABULATION OF PROPERTIES OF DTFT

A number of important properties of the DTFT are summarised in Table 6.1.

Table 6.1 Summary of Important Properties of DTFT

DTFT property	*Non-periodic signal*	*DTFT*
	$s(n)$ $s_1(n)$ $s_2(n)$	$S(e^{j\omega})$, $S_1(e^{j\omega})$ and $S_2(e^{j\omega})$ These are periodic with period 2π
Linearity	$A_1 s_1(n) + A_2 s_2(n)$	$A_1 S_1(e^{j\omega}) + A_2 S_2(e^{j\omega})$ where A_1 and A_2 are scalar constants
Time-shifting	$s(n - n_0)$	$e^{-j\omega n_0} S(e^{j\omega})$ where n_0 is constant time
Frequency shifting	$e^{j\omega_0 n} s(n)$	$S(e^{j(\omega - \omega_0)})$
Complex conjugation	$s^*(n)$	$S^*(e^{-j\omega})$
Time reversal	$s(-n)$	$S(e^{-j\omega})$
Time expansion	$s_{(k)}(n) = \begin{cases} s(n/k), & \text{if } n \text{ is multiple of } k \\ 0, & \text{if } n \text{ is not multiple of } k \end{cases}$	$S(e^{jk\omega})$
Convolution of two discrete-time signal	$s_1(n) * s_2(n)$	$S_1(e^{j\omega}) S_2(e^{j\omega})$
Multiplication of two discrete-time signal	$s_1(n)\, s_2(n)$	$\frac{1}{2\pi}\int_{2\pi} S_1(e^{j\theta}) S_2(e^{j(\omega-\theta)})\, d\theta$
Difference of a discrete-time signal in time	$s(n) - s(n-1)$	$(1 - e^{-j\omega}) S(e^{j\omega})$
Accumulation of a discrete-time signal	$\sum_{k=-\infty}^{n} s(k)$	$\frac{1}{1-e^{-j\omega}} S(e^{j\omega}) + \pi S(e^{j0}) \times \sum_{k=-\infty}^{\infty} \delta(\omega - 2\pi k)$
Differentiation in frequency	$n\, s(n)$	$\frac{j\, d\, S(e^{j\omega})}{d\omega}$
Conjugate symmetry for real signals	$s(n)$ real	$\begin{cases} S(j\omega) = S^*(e^{-j\omega}) \\ \text{Re}\{S(e^{j\omega})\} = \text{Re}\{S(e^{-j\omega})\} \\ \text{Im}\{S(e^{j\omega})\} = -\text{Im}\{S(e^{-j\omega})\} \\ \lvert S(e^{j\omega})\rvert = \lvert S(e^{-j\omega})\rvert \\ \angle S(e^{j\omega}) = -\angle S(e^{-j\omega}) \end{cases}$

(Contd...)

DTFT property	*Non-periodic signal*	*DTFT*
Symmetry for real and even signals	$s(n)$ real and even	$S(e^{j\omega})$ real and even
Symmetry for real and odd signals	$s(n)$ real and odd	$S(e^{j\omega})$ purely imaginary and odd
Decomposition of a real signal into even and odd parts.	$s_E(n) = \text{Ev}\{s(n)\}$ $s_O(n) = \text{Od}\{s(n)\}$ Here $s(n)$ is real signal	$\text{Re}\{S(e^{j\omega})\}$ $j\,\text{Im}\{S(e^{j\omega})\}$
Parseval's relation for non-periodic signals $\sum_{n=-\infty}^{\infty} \lvert s(n)\rvert^2 = \frac{1}{2\pi}\int_{2\pi} \lvert S(e^{j\omega})\rvert^2\, d\omega$		

6.7 TABULATION OF DTFT PAIRS

A summary of the basic discrete-time Fourier transform (DTFT) pairs is shown in Table 6.2.

Table 6.2 Summary of Basic DTFT Pairs

Discrete-time signal $s(n)$	*Discrete-time Fourier transform (DTFT)* $S(e^{j\omega})$
Exponential function $e^{j\omega_0 n}$	$2\pi \sum_{m=-\infty}^{\infty} \delta(\omega - \omega_0 - 2\pi m)$ where $\omega_0 = \frac{2\pi m}{N_0}$
Cosine function $\cos \omega_0 n = \frac{e^{j\omega_0 n} + e^{-j\omega_0 n}}{2}$	$\frac{2\pi}{2}\left[\sum_{m=-\infty}^{\infty} \delta(\omega - \omega_0 - 2\pi m) + \sum_{m=-\infty}^{\infty} \delta(\omega + \omega_0 - 2\pi m)\right]$
Sine function $\sin \omega_0 n = \frac{e^{j\omega_0 n} - e^{-j\omega_0 n}}{2j}$	$\frac{2\pi}{2j}\left[\sum_{m=-\infty}^{\infty} \delta(\omega - \omega_0 - 2\pi m) + \sum_{m=-\infty}^{\infty} \delta(\omega + \omega_0 - 2\pi m)\right]$
$s(n) = 1 = e^{j\omega_0 n}\Big\rvert_{\omega_0 = 0}$	$2\pi \sum_{m=-\infty}^{\infty} \delta(\omega - 0 - 2\pi m)$ $= 2\pi \sum_{m=-\infty}^{\infty} \delta(\omega - 2\pi m)$
$s(n) = \sum_{k=\langle N_0\rangle} A_k e^{jk\omega_0 n}$	$S(e^{j\omega}) = 2\pi \sum_{k=-\infty}^{\infty} A_k\, \delta\left(\omega - \frac{2\pi k}{N_0}\right)$

(Contd...)

Discrete-time signal $s(n)$	*Discrete-time Fourier transform (DTFT)* $S(e^{j\omega})$
Periodic square wave $s(n) = \begin{cases} 1, & \lvert n \rvert \le N_1 \\ 0, & N_1 < \lvert n \rvert \le N_0/2 \end{cases}$ and for periodic signal $s(n + N_0) = s(n)$	$S(e^{j\omega}) = 2\pi \sum_{k=-\infty}^{\infty} A_k\, \delta\left(\omega - \frac{2\pi k}{N_0}\right)$ where, $A_k = \begin{cases} \frac{\sin[(2\pi k/N_0)(N_1 + 1/2)]}{N_0 \sin[2\pi k/2N_0]}, \\ k \ne 0, \pm N_0, \pm 2N_0, \ldots \\ (2N_1 + 1)/N_0, \\ k = 0, \pm N_0, \pm 2N_0, \ldots \end{cases}$
$\sum_{k=-\infty}^{\infty} \delta(n - kN_0)$	$\frac{2\pi}{N_0} \sum_{k=-\infty}^{\infty} \delta\left(\omega - \frac{2\pi k}{N_0}\right)$
$s(n) = A^n u(n),\ \lvert A \rvert < 1$	$S(e^{j\omega}) = \frac{1}{1 - Ae^{-j\omega}}$
$s(n) = \begin{cases} 1, & \lvert n \rvert \le N_1 \\ 0, & \lvert n \rvert > N_1 \end{cases}$	$S(e^{j\omega}) = \frac{\sin\left[\omega\left(N_1 + \frac{1}{2}\right)\right]}{\sin\left(\frac{\omega}{2}\right)}$
$s(n) = \frac{\sin Wn}{\pi n} = \frac{W}{\pi}\left(\frac{\sin Wn}{Wn}\right) = \frac{W}{\pi} \operatorname{sinc}\left(\frac{Wn}{\pi}\right)$ for $0 < W < \pi$	$S(e^{j\omega}) = \begin{cases} 1, & 0 \le \lvert \omega \rvert \le W \\ 0, & W < \lvert \omega \rvert \le \pi \end{cases}$ where $S(e^{j\omega})$ is periodic with period 2π
Impulse function $\delta(n)$	1
Unit step function $u(n)$	$\frac{1}{1 - e^{-j\omega}} + \sum_{k=-\infty}^{\infty} \pi\, \delta(\omega - 2\pi k)$
Shifted version of $\delta(n)$, $\delta(n - n_0)$, where n_0 is constant time	$e^{-j\omega n_0}$
$(n + 1) A^n u(n),\ \lvert A \rvert < 1$	$\frac{1}{(1 - Ae^{-j\omega})^2}$
$\frac{(n + m - 1)!}{n!\,(m - 1)!} A^n u(n)$ for $\lvert A \rvert < 1$	$\frac{1}{(1 - Ae^{-j\omega})^m}$

6.8 DUALITY

There is a duality between the CTFT analysis equation and CTFT synthesis equation. Both equations are given as

CTFT analysis equation,

$$S(j\omega) = \int_{-\infty}^{\infty} s(t)\, e^{-j\omega t}\, dt \qquad ...(6.52)$$

and CTFT synthesis equation

$$s(n) = \frac{1}{2\pi} \int_{-\infty}^{\infty} S(j\omega)\, e^{j\omega t}\, d\omega \qquad ...(6.53)$$

But there is no corresponding duality between the DTFT analysis equation and DTFT synthesis equation. These equations are given as

DTFT analysis equation

$$S(e^{j\omega}) = \sum_{n=-\infty}^{\infty} s(n)\, e^{-j\omega n} \qquad ...(6.54)$$

and DTFT synthesis equation

$$s(n) = \frac{1}{2\pi} \int_{2\pi} S(e^{j\omega})\, e^{j\omega n}\, d\omega \qquad ...(6.55)$$

However, there is a duality between the DTFS equations. These equations are given as

DTFS analysis equation,

$$A_k = \frac{1}{N_0} \sum_{n=<N_0>} s(n)\, e^{-jk\omega_0 n} \qquad ...(6.56)$$

and DTFS synthesis equation,

$$s(n) = \sum_{k=<N_0>} A_k\, e^{jk\omega_0 n} \qquad ...(6.57)$$

where A_k is the Fourier series coefficients of a discrete-time periodic signal $s(n)$.

Also, there is a duality relationship between the DTFT and the CTFS.

6.8.1 Duality in the DTFS

Since the Fourier series coefficients A_k of the periodic signal $s(n)$ are themselves a periodic sequence, we can expand the sequence A_k in a Fourier series. The duality property for discrete-time Fourier series (DTFS) implies that the Fourier series coefficieints for the periodic sequence A_k are the values of $(1/N_0)\, s(-n)$.

In other words, the duality property for DTFS implies that the Fourier series coefficients for the periodic sequence A_k is proportional to the values of the original sequence which is reversed or folded in time.

To prove the duality property for DTFS, we consider two periodic sequences with period N_0. These two sequences are related through the summation given as

$$s_1(m) = \frac{1}{N_0} \sum_{l=\langle N_0\rangle} s_2(l)\, e^{-jl\omega_0 m} \qquad \text{...(6.58)}$$

where $\omega = \dfrac{2\pi}{N_0}$

If we let $m = k$ and $l = n$, Eq. (6.55) becomes

$$s_1(k) = \sum_{n=\langle N_0\rangle} s_2(n)\, e^{-jn\omega_0 t}$$

$$= \sum_{n=\langle N_0\rangle} s_2(n)\, e^{-jk\omega_0 n} \qquad \text{...(6.59)}$$

Comparing Eq. (6.59) with DTFS analysis equation given as

$$A_k = \frac{1}{N_0} \sum_{n=\langle N_0\rangle} s(n)\, e^{-jk\omega_0 n} \qquad \text{...(6.60)}$$

We see that the sequence $s_1(k)$ corresponds to the Fourier series coefficients of the signal $s_2(n)$.

Eq. 6.60 can be represented as

$$s(n) \xleftrightarrow{\text{DTFT}} A_k$$

Similarly, Eq. 6.59 can be represented as

$$s_2(n) \xleftrightarrow{\text{DTFT}} s_1(k) \qquad \text{...(6.61)}$$

Alternatively, if we let $m = n$ and $l = -k$, Eq. 6.58 becomes

$$s_1(n) = \frac{1}{N_0} \sum_{k=\langle N_0\rangle} s_2(-k)\, e^{jk\omega_0 n}$$

$$= \sum_{k=\langle N_0\rangle} \frac{1}{N_0} s_2(-k)\, e^{jk\omega n} \qquad \text{...(6.62)}$$

Now, comparing Eq. 6.62 with DTFS synthesis equation given as

$$s(n) = \sum_{k=\langle N_0\rangle} A_k\, e^{jk\omega_0 n} \qquad \text{...(6.63)}$$

We obtain, $$s_1(n) \xleftrightarrow{\text{DTFT}} \frac{1}{N_0} s_2(-k) \qquad \text{...(6.64)}$$

Now, we can conclude that $\dfrac{1}{N_0} s_2(-k)$ corresponds to the sequence of the Fourier series coefficients of discrete-time signal $s_1(n)$.

Thus it is to prove.

As in continuous-time, this duality implies that every property of the DTFS has a dual charater.

For example, we have two DTFS pairs given as

$$s(n - n_0) \xleftrightarrow{\text{DTFS}} A_k\, e^{-jk\omega_0 n_0} \quad ...(6.65)$$

and

$$e^{jm\omega_0 n}\, s(n) \xleftrightarrow{\text{DTFS}} A_{k-m} \quad ...(6.66)$$

Both DTFS pairs are dual.

Duality is often useful is reducing the complexity of the calculations involved in determining Fourier series representations.

Example 6.12 Verify the duality of the following periodic signal with a period of $N_0 = 9$

$$s(n) = \begin{cases} 1/9\, \dfrac{\sin (5\pi n/9)}{\sin (\pi n/9}, & n \neq \text{multiple of } 9 \\ 5/9, & n = \text{multiple of } 9 \end{cases} \quad ...(1)$$

Solution. We have already studied that the Fourier series coefficients of a discrete-time square (rectangular) wave is given as

$$A_k = \begin{cases} 1/N_0 \dfrac{\sin\left[2\pi k\left(N_1 + \dfrac{1}{2}\right)/N_0\right]}{\sin\left(\dfrac{\pi k}{N_0}\right)}, & k \neq 0, \pm N_0, \pm 2N_0, \ldots \\ \dfrac{2N_1+1}{N_0}, & k = 0, \pm N_0, \pm 2N_0, \ldots \end{cases} \quad ...(2)$$

Duality, then suggests that the Fourier series coefficients for $s(n)$ must be in the form of a rectangular square wave. The rectangular square wave with period $N_0 = 9$ is such that

$$g(n) = \begin{cases} 1, & |n| \leq 2 \\ 0, & 2 < |n| \leq 4 \end{cases} \quad ...(3)$$

The Fourier series coefficients B_k for discrete-time square wave $g(n)$ is determined as

$$B_k = \begin{cases} \dfrac{1}{9} \dfrac{\sin\left(\dfrac{5\pi n}{9}\right)}{\sin\left(\dfrac{\pi n}{9}\right)}, & \text{for } k \neq \text{multiple of } 9 \\ \dfrac{5}{9}, & \text{for } k = \text{multiple of } 9 \end{cases} \quad ...(4)$$

The discrete-time Fourier series (DTFS) analysis equation for sequence $g(n)$ is given as

$$B_k = \frac{1}{N_0} \sum_{n=\langle N_0 \rangle} g(n)\, e^{-jk\omega_0 n}$$

$$= \frac{1}{9} \sum_{n=-2}^{2} (1)\, e^{-jk(2\pi/9)n}$$

$$= \frac{1}{9} \sum_{n=-2}^{2} e^{-(jk2\pi n)/9} \qquad ...(5)$$

Interchanging the names of the variables k and n and putting $B_n = s(n)$, we obtain

$$B_n = \frac{1}{9} \sum_{k=-2}^{2} e^{-(jn2\pi k)/9}$$

or

$$s(n) = \frac{1}{9} \sum_{k=-2}^{2} e^{-(j2\pi nk)/9} \qquad ...(6)$$

Also letting $k = -k'$ in the sum of RHS of Eq. 6, we get

$$s(n) = \frac{1}{9} \sum_{k'=-2}^{2} e^{(j2\pi nk')/9} \qquad ...(7)$$

or

$$s(n) = \sum_{k'=-2}^{2} \frac{1}{9} e^{(j2\pi nk')/9} \qquad ...(8)$$

Finally we can say that RHS of Eq. 8 has the form of DTFS synthesis equation for $s(n)$.

$$s(n) = \sum_{k=\langle N_0 \rangle} A_k e^{jk\omega_0 n} \qquad ...(9)$$

We thus conclude that the Fourier series coefficients of sequence $s(n)$ is given by

$$A_k = \begin{cases} \frac{1}{9}, & |k| \leq 2 \\ 0, & 2 < |k| \leq 4 \end{cases} \qquad ...(10)$$

These Fourier series coefficients A_k, of course, are periodic with period $N_0 = 9$.

6.8.2 Duality between the DTFT and CTFT

We have already studied duality in discrete-time Fourier series (DTFS). In this section, we will study the property of duality between the discrete-time Fourier transform (DTFT) and continuous-time Fourier series (CTFS). Here we will compare CTFS analysis and synthesis equations with DTFT analysis and synthesis equations. These equations are given as follows :

CTFS synthesis equation,

$$s(t) = \sum_{k=-\infty}^{\infty} A_k e^{jk\omega_0 t} \qquad ...(6.67)$$

and CTFS analysis equation,

$$A_k = \frac{1}{T_0} \int_{T_0} s(t) e^{-jk\omega_0 t} dt \qquad ...(6.68)$$

DTFT synthesis equation,

$$s(n) = \frac{1}{2\pi} \int_{2\pi} S(e^{j\omega})\, e^{j\omega n}\, d\omega \qquad ...(6.69)$$

DTFT analysis equation,

$$S(e^{j\omega}) = \sum_{n=-\infty}^{\infty} s(n)\, e^{-j\omega n} \qquad ...(6.70)$$

Eqs. 6.69 and 6.68 are very similar. Similarly, Eqs. 6.70 and 6.67 are also very similar. Now we can interpret Eqs. 6.69 and 6.70 as a Fourier series representation of the periodic frequency response $S(e^{j\omega})$. In particular, since $S(e^{j\omega})$ is a periodic function of ω with period 2π and It has a Fourier series representation as a weighted sum of harmonically related periodic exponential functions of ω, all of which have the common period of 2π.

DTFT of $s(n)$, $S(e^{j\omega})$ can be represented in a Fourier seires as a weighted sum of the exponential functions $e^{j\omega n}$, $n = 0, \pm 1, \pm 2, \ldots$. From equation $S(e^{j\omega}) = \sum_{n=-\infty}^{\infty} s(n) = e^{-j\omega n}$, we see that the nth Fourier series coefficients in this expansion (that is, the Fourier series coefficient multiplying $s^{j\omega n}$ is $s(-n)$). Since the fundamental period of $S(e^{j\omega})$ is 2π, equation $s(n) = \frac{1}{2\pi} \int_{2\pi} S(e^{j\omega})\, e^{j\omega n}\, d\omega$ can be interpreted as the Fourier series analysis equation for the Fourier series coefficient $s(n)$ (That is, for the Fourier series coefficient multiplying $e^{-j\omega n}$ in the expression for $S(e^{j\omega})$ in equation $S(e^{j\omega}) = \sum_{n=-\infty}^{\infty} s(n)\, e^{-j\omega n}$.

Example 6.13 Determine the DTFT of the sequence $s(n) = \frac{\sin(\pi n/2)}{\pi n}$.

Solution. To determine the DTFT of the sequence $s(n) = \frac{\sin(\pi n/2)}{\pi n}$, we will exploit the duality between the DTFT synthesis equation and the CTFS analysis equation.

To use duality, we must first identify a continuous-time signal $g(t)$ with period $T_0 = 2\pi$ and Fourier series coefficients $A_k = s(k)$. If $g(t)$ is a periodic square wave with period 2π, *i.e.*,

$$g(t) = \begin{cases} 1, & |t| \le T_1 \\ 0, & T_1 < |t| \le \pi \end{cases} \qquad ...(1)$$

then the Fourier series coefficients of $g(t)$ are

$$A_k = \frac{\sin(kT_1)}{k\pi} \qquad ...(2)$$

If we take $T_1 = \frac{\pi}{2}$, then

$$A_k = \frac{\sin\left(k\frac{\pi}{2}\right)}{k\pi} = s(k) \qquad ...(3)$$

In this case, CTFS analysis equation for $g(t)$ is

$$\frac{\sin\left(\frac{\pi k}{2}\right)}{\pi k} = \frac{1}{2\pi}\int_{-\pi}^{\pi} g(t)\, e^{-jkt}$$

$$= \frac{1}{2\pi}\int_{-\pi}^{\pi} (1)\, e^{-jkt} \qquad ...(4)$$

Renaming k to n and t as ω in Eq. 4, we get

$$\frac{\sin\left(\frac{\pi n}{2}\right)}{\pi n} = \frac{1}{2\pi}\int_{-\pi/2}^{\pi/2} (1)\, e^{-jn\omega}\, d\omega \qquad ...(5)$$

Replacing n by $-n$ on both sides of Eq. 5, we get

$$\frac{\sin\left(\frac{\pi n}{2}\right)}{\pi n} = \frac{1}{2\pi}\int_{-\pi/2}^{\pi/2} (1)\, e^{jn\omega}\, d\omega \qquad ...(6)$$

The RHS of Eq. 6 has the form of the DTFT synthesis equation for sequence $s(n)$, where

$$S(e^{j\omega}) = \begin{cases} 1, & |\omega| \le \pi/2 \\ 0, & \pi/2 < |\omega| \le \pi \end{cases}$$

6.8.3 Summary of Fourier Series (CTFS and DTFS) and Fourier Transform (CTFT and DTFT) Expressions

Now we will summarize the Fourier series and Fourier transform both in time-domain and frequency-domain. The summary of Fourier series and Fourier transform expressions is given in Table 6.3.

6.9 DISCRETE-TIME LTI SYSTEMS CHARACTERIZED BY LINEAR CONSTANT-COEFFICIENT DIFFERENCE EQUATIONS

The general form of a linear constant-coefficient difference equation for a discrete-time LTI system is given as

$$\sum_{k=0}^{N} A_k\, y(n-k) = \sum_{k=0}^{M} B_k\, s(n-k) \qquad ...(6.71)$$

This difference equation is generally called Nth order difference equation.

Table 6.3 Summary of Fourier Series and Fourier Transform Expressions

	Time-domain	*Frequency domain*		*Time-domain*	*Frequency-domain*
Continuou-time Fourier series (CTFS)	CTFS synthesis equation, $s(t) = \sum_{k=-\infty}^{\infty} A_k e^{jk\omega_0 t}$ where A_k is the Fourier series coefficients and $s(t)$ is the continuous-time, periodic in time signal. $\omega_0 = \frac{2\pi}{T_0}$, fundamental frequency	CTFS analysis, equation, $A_k = \frac{1}{T_0} \int_{T_0} s(t)\, e^{-jk\omega_0 t}$ where A_k is the discrete-frequency, non-periodic in frequency function. Where $\omega_0 = \frac{2\pi}{T_0}$, fundamental frequency, T_0 = fundamental period	Discrete-time Fourier series (DTFS) ←Duality→	DTFS syntehsis, equation $s(n) = \sum_{k=\langle N_0 \rangle} A_k e^{jk\omega_0 n}$ where A_k is the Fourier series coefficients and $s(n)$ is the discrete-time, periodic in time signal $\omega_0 = \frac{2\pi}{N_0}$ is the discrete fundamental frequency and N_0 is the fundamental period	DTFS analysis equation $A_k = \frac{1}{N_0} \sum_{k=\langle N_0 \rangle} s(n)\, e^{-jk\omega_0 n}$ where A_k is the Fourier series coefficients and it is a discrete-frequency, periodic in frequency sequence.
Continuous-time Fourier transform (CTFT)	CTFT synthesis equation, $s(t) = \frac{1}{2\pi} \int_{-\infty}^{\infty} S(j\omega)\, e^{j\omega t}\, d\omega$ where $s(t)$ is the continuous-time, non-periodic in time signal	CTFT analysis equation, $S(j\omega) = \int_{-\infty}^{\infty} s(t)\, e^{-j\omega t}\, dt$ where $S(j\omega)$ is the continuous frequency, non-periodic in frequency function ←Duality→	Discrete-time Fourier transform (DTFT)	DTFT synthesis equation, $s(n) = \frac{1}{2\pi} \int_{2\pi} S(e^{j\omega})\, e^{j\omega n}\, d\omega$ where $s(n)$ is a discrete-time, non-periodic in time signal	DTFT analysis equation, $S(e^{j\omega}) = \sum_{n=-\infty}^{\infty} s(n)\, e^{-j\omega n}$ where $S(e^{j\omega})$ is a continuous frequency, periodic in frequency function.

Here, A_k and B_k are the constant coefficients. $s(n)$ and $y(n)$ are the inputs and outputs of the discrete-time LTI system, respectively.

Here, we will determine the frequency response $H(e^{j\omega})$ for an LTI system described by Eq. 6.71. In determining the frequency response $H(e^{j\omega})$ of a discrete-time LTI system, we will take advantage of several important properties of the DTFT.

There are two related methods for determining the frequency response $H(e^{j\omega})$ of the discrete-time LTI system.

In the first-method, we explicity use the fact that complex exponential functions are eigen functions of LTI systems. Specifically, if $s(n) = e^{j\omega n}$ is the input to a discrete-time LTI system then the output of this system is given by the multiplication of $H(e^{j\omega})$ and $e^{j\omega n}$.

Now, substituting $s(n) = e^{j\omega n}$ and $y(n) = H(e^{j\omega})\, e^{j\omega n}$ in Eq. 6.71, we get

$$\sum_{k=0}^{N} A_k\, y(n-k) = \sum_{k=0}^{M} B_k\, s(n-k)$$

or $$\sum_{k=0}^{N} A_k\, H(e^{j\omega})\, e^{j\omega(n-k)} = \sum_{k=0}^{M} B_k\, e^{j\omega(n-k)}$$

or $$H(e^{j\omega}) \sum_{k=0}^{N} A_k\, e^{j\omega n}\, e^{-j\omega k} = \sum_{k=0}^{M} B_k\, e^{j\omega n}\, e^{-j\omega k}$$

or $$H(e^{j\omega}) = \frac{\sum_{k=0}^{M} B_k\, e^{-j\omega k}}{\sum_{k=0}^{N} A_k\, e^{-j\omega k}} \qquad ...(6.72)$$

In the second-method, we make use of the convolution, linearity and time-shifting properties of DTFT. Let $S(e^{j\omega})$, $Y(e^{j\omega})$ and $H(e^{j\omega})$ be the DTFT of the input $s(n)$, output $y(n)$ and impulse response $h(n)$ respectively. Fig. 6.18 shows a discrete-time LTI system.

Fig. 6.18 *Discrete-time LTI system.*

From the convolution property of the DTFT,

$$y(n) = s(n) * h(n) \xleftrightarrow{\text{DTFT}} Y(e^{j\omega}) = S(e^{j\omega})\, H(e^{j\omega}) \qquad ...(6.73)$$

$$Y(e^{j\omega}) = S(e^{j\omega})\, H(e^{j\omega})$$

or $$H(e^{j\omega}) = \frac{Y(e^{j\omega})}{S(e^{j\omega})} \qquad ...(6.74)$$

Taking the DTFT of both sides of Eq. 6.71 and using the properties of linearity and time-shifting of DTFT, we get

$$\sum_{k=0}^{N} A_k\, y(n-k) = \sum_{k=0}^{M} B_k\, s(n-k) \qquad ...(6.75)$$

$$\text{DTFT}\left\{\sum_{k=0}^{N} A_k\, y(n-k)\right\} = \text{DTFT}\left\{\sum_{k=0}^{M} B_k\, s(n-k)\right\}$$

or $$\sum_{k=0}^{N} A_k\, \text{DTFT}\,\{y(n-k)\} = \sum_{k=0}^{M} B_k\, \text{DTFT}\,\{s(n-k)\}$$

or $$\sum_{k=0}^{N} A_k\, e^{-jk\omega}\, Y(e^{j\omega}) = \sum_{k=0}^{M} B_k\, e^{-jk\omega}\, S(e^{j\omega})$$

or $$Y(e^{j\omega}) \sum_{k=0}^{N} A_k\, e^{-jk\omega} = S(e^{j\omega}) \sum_{k=0}^{M} B_k\, e^{-jk\omega}$$

or $$\frac{Y(e^{j\omega})}{S(e^{j\omega})} = \frac{\sum_{k=0}^{M} B_k\, e^{-jk\omega}}{\sum_{k=0}^{N} A_k\, e^{-jk\omega}} \quad \text{...(6.76)}$$

From Eqs. 6.74 and 6.76, we obtain

$$H(e^{j\omega}) = \frac{Y(e^{j\omega})}{S(e^{j\omega})} = \frac{\sum_{k=0}^{M} B_k\, e^{-jk\omega}}{\sum_{k=0}^{N} A_k\, e^{-jk\omega}} \quad \text{...(6.77)}$$

We see from Eq. 6.77 that frequency response $H(e^{j\omega})$ is a ratio of two polynomials in the variable $e^{-j\omega}$. The coefficient of the numerator polynomial are the same as the ones that appear on the RHS of Eq. 6.71 and the coefficients of the denominator polynomial are the same as the ones that appear on the LHS of Eq. 6.71. Therefore, frequency response of an LTI system described by Eq. 6.71 can be written by inspection.

Example 6.14 Determine the frequency response and impulse response of a causal discrete-time LTI system that is characterized by the difference equation given as

$$y(n) - A\, y(n-1) = s(n) \quad \text{...(1)}$$

with $|A| < 1$

Solution. To determine frequency response, we take DTFT of both sides of Eq. 1, we obtain

$$\text{DTFT}\,\{y(n) - A\, y(n-1)\} = \text{DTFT}\,\{s(n)\}$$

or $$Y(e^{j\omega}) - A\, e^{-j\omega}\, Y(e^{j\omega}) = S(e^{j\omega})$$

or $$Y(e^{j\omega})\,[1 - A\, e^{-j\omega}] = S(e^{j\omega})$$

or $$\frac{Y(e^{j\omega})}{S(e^{j\omega})} = \frac{1}{1 - A\, e^{-j\omega}} \quad \text{...(2)}$$

But we know that frequency response $H(e^{j\omega})$ is given by

$$H(e^{j\omega}) = \frac{Y(e^{j\omega})}{S(e^{j\omega})} \quad \text{...(3)}$$

From Eqs. 2 and 3, we obtain

$$H(e^{j\omega}) = \frac{1}{1 - A\,e^{-j\omega}} \qquad ...(4)$$

Now we will determine impulse response $h(n)$ as

$$h(n) = \text{Inverse DTFT } [H(e^{j\omega})]$$

$$= \text{Inverse DTFT}\left[\frac{1}{1 - A\,e^{-j\omega}}\right] = A^n\,u(n)$$

Thus, the impulse of the LTI system is

$$h(n) = A^n\,u(n) \qquad ...(5)$$

Example 6.15 Determine the frequency response $H(e^{j\omega})$ and impulse response $h(n)$ of a causal discrete-time LTI system that is characterized by the difference equation given as

$$y(n) - \frac{3}{4}y(n-1) + \frac{1}{8}y(n-2) = 2\,s(n) \qquad ...(1)$$

Solution. To determine frequency response, taking DTFT of both sides of Eq. 1, we get

$$\text{DTFT}\left\{y(n) - \frac{3}{4}y(n-1) + \frac{1}{8}y(n-2)\right\} = \text{DTFT}\{2\,s(n)\}$$

or $$Y(e^{j\omega}) - \frac{3}{4}e^{-j\omega}\,Y(e^{j\omega}) + \frac{1}{8}e^{-j2\omega}\,Y(e^{j\omega}) = 2\,S(e^{j\omega})$$

or $$Y(e^{j\omega})\left[1 - \frac{3}{4}e^{-j\omega} + \frac{1}{8}e^{-j2\omega}\right] = 2\,S(e^{j\omega})$$

or $$\frac{Y(e^{j\omega})}{S(e^{j\omega})} = \frac{2}{1 - \frac{3}{4}e^{-j\omega} + \frac{1}{8}e^{-j2\omega}} \qquad ...(2)$$

But frequency response is given by

$$H(e^{j\omega}) = \frac{Y(e^{j\omega})}{S(e^{j\omega})} \qquad ...(3)$$

From Eqs. 2 and 3, we get frequency response,

$$H(e^{j\omega}) = \frac{2}{1 - \frac{3}{4}e^{-j\omega} + \frac{1}{8}e^{-j2\omega}} \qquad ...(4)$$

Now, we will determine impulse response $h(n)$ by taking inverse DTFT of Eq. 4 as

$$h(n) = \text{Inverse DTFT } \{H(e^{j\omega})\}$$

$$= \text{Inverse DTFT}\left\{\frac{2}{1 - \frac{3}{4}e^{-j\omega} + \frac{1}{8}e^{-j2\omega}}\right\} \qquad ...(5)$$

By using partial fraction expansion method :

$$\frac{2}{1-\frac{3}{4}e^{-j\omega}+\frac{1}{8}e^{-j2\omega}}=\frac{2}{\left(1-\frac{1}{2}e^{-j\omega}\right)\left(1-\frac{1}{4}e^{-j\omega}\right)}$$

$$=\frac{\alpha_1}{\left(1-\frac{1}{2}e^{-j\omega}\right)}+\frac{\alpha_2}{\left(1-\frac{1}{4}e^{-j\omega}\right)} \qquad ...(6)$$

Determination of α_1 and α_2 :

$1-\frac{1}{2}e^{-j\omega}=0$ or $e^{-j\omega}=2$,

$$\alpha_1=\frac{2}{\left(1-\frac{1}{4}e^{-j\omega}\right)}=\frac{2}{\left(1-\frac{1}{4}2\right)}=4$$

$1-\frac{1}{4}e^{-j\omega}=0$ or $e^{-j\omega}=4$,

$$\alpha_2=\frac{2}{\left(1-\frac{1}{2}e^{-j\omega}\right)}=\frac{2}{\left(1-\frac{1}{2}4\right)}=-2$$

Putting the value of α_1 and α_2 in Eq. 6, we get

$$\frac{2}{1-\frac{3}{4}e^{-j\omega}+\frac{1}{8}e^{-j2\omega}}=\frac{4}{\left(1-\frac{1}{2}e^{-j\omega}\right)}-\frac{2}{\left(1-\frac{1}{4}e^{-j\omega}\right)} \qquad ...(7)$$

Substituting Eq. 7 in Eq. 5, we obtain

$$h(n)=\text{Inverse DTFT}\left\{\frac{4}{\left(1-\frac{1}{2}e^{-j\omega}\right)}-\frac{2}{\left(1-\frac{1}{4}e^{-j\omega}\right)}\right\}$$

$$=\text{Inverse DTFT}\left\{\frac{4}{\left(1-\frac{1}{2}e^{-j\omega}\right)}\right\}$$

$$-\text{Inverse DTFT}\left\{\frac{2}{\left(1-\frac{1}{4}e^{-j\omega}\right)}\right\}$$

or $$h(n)=4\left(\frac{1}{2}\right)^n u(n)-2\left(\frac{1}{4}\right)^n u(n) \qquad ...(8)$$

This is the impulse response of given system.

Example 6.16 Determine the output $y(n)$ of a causal discrete- time LTI system that is characterized by the difference equation

$$y(n)-\frac{3}{4}y(n-1)+\frac{1}{8}y(n-2)=2\,s(n) \qquad ...(1)$$

for input $s(n) = \left(\frac{1}{4}\right)^n u(n)$ ...(2)

Solution. From Eqs. 1 and 2, we get

$$y(n) - \frac{3}{4}y(n-1) + \frac{1}{8}y(n-2) = 2\,s(n)$$

or
$$y(n) - \frac{3}{4}y(n-1) + \frac{1}{8}y(n-2) = 2\left(\frac{1}{4}\right)^n u(n) \quad ...(3)$$

Now taking the DTFT of both sides of Eq. 3, we obtain

$$\text{DTFT}\left\{y(n) - \frac{3}{4}y(n-1) + \frac{1}{8}y(n-2)\right\} = \text{DTFT}\left\{2\left(\frac{1}{4}\right)^n u(n)\right\}$$

or
$$\text{DTFT}\,\{y(n)\} - \frac{3}{4}\,\text{DTFT}\,\{y(n-1)\} + \frac{1}{8}\,\text{DTFT}\,\{y(n-2)\}$$

$$= 2\,\text{DTFT}\left\{\left(\frac{1}{4}\right)^n u(n)\right\}$$

or
$$Y(e^{j\omega}) - \frac{3}{4}e^{-j\omega}\,Y(e^{j\omega}) + \frac{1}{8}e^{-j2\omega}\,Y(e^{j\omega}) = \frac{2}{\left(1 - \frac{1}{4}e^{-j\omega}\right)}$$

or
$$Y(e^{j\omega})\left[1 - \frac{3}{4}e^{-j\omega} + \frac{1}{8}e^{-j2\omega}\right] = \frac{2}{\left(1 - \frac{1}{4}e^{-j\omega}\right)}$$

or
$$Y(e^{j\omega}) = \frac{2}{\left(1 - \frac{3}{4}e^{-j\omega} + \frac{1}{8}e^{-j2\omega}\right)\left(1 - \frac{1}{4}e^{-j\omega}\right)}$$

$$= \frac{2}{\left(1 - \frac{1}{2}e^{-j\omega}\right)\left(1 - \frac{1}{4}e^{-j\omega}\right)\left(1 - \frac{1}{4}e^{-j\omega}\right)}$$

$$= \frac{2}{\left(1 - \frac{1}{2}e^{-j\omega}\right)\left(1 - \frac{1}{4}e^{-j\omega}\right)^2} \quad ...(4)$$

By using partial fraction expansion method

$$Y(e^{j\omega}) = \frac{2}{\left(1 - \frac{1}{2}e^{-j\omega}\right)\left(1 - \frac{1}{4}e^{-j\omega}\right)^2}$$

$$= \frac{\alpha_1}{\left(1 - \frac{1}{2}e^{-j\omega}\right)} + \frac{\alpha_2}{\left(1 - \frac{1}{4}e^{-j\omega}\right)} + \frac{\alpha_3}{\left(1 - \frac{1}{4}e^{-j\omega}\right)^2}$$

$$= \frac{8}{\left(1 - \frac{1}{2}e^{-j\omega}\right)} + \frac{(-4)}{\left(1 - \frac{1}{4}e^{-j\omega}\right)} + \frac{(-2)}{\left(1 - \frac{1}{4}e^{-j\omega}\right)^2} \quad ...(5)$$

To determine $y(n)$, we take inverse DTFT of Eq. 5, we get

$$y(n) = \text{Inverse DTFT } \{Y(e^{j\omega})\}$$

$$= \text{Inverse DTFT} \left\{ \frac{2}{\left(1 - \frac{1}{2}e^{-j\omega}\right)\left(1 - \frac{1}{4}e^{-j\omega}\right)^2} \right\}$$

$$= \text{Inverse DTFT} \left\{ \frac{8}{\left(1 - \frac{1}{2}e^{-j\omega}\right)} - \frac{4}{\left(1 - \frac{1}{4}e^{-j\omega}\right)} - \frac{2}{\left(1 - \frac{1}{4}e^{-j\omega}\right)^2} \right\}$$

$$= 8 \times \text{Inverse DTFT} \left\{ \frac{1}{\left(1 - \frac{1}{2}e^{-j\omega}\right)} \right\} - 4 \times \text{Inverse DTFT} \left(\frac{1}{\left(1 - \frac{1}{4}e^{-j\omega}\right)} \right\} - 2 \times \text{Inverse DTFT} \left(\frac{1}{\left(1 - \frac{1}{4}e^{-j\omega}\right)^2} \right\}$$

or $$y(n) = 8\left(\frac{1}{2}\right)^n u(n) - 4\left(\frac{1}{4}\right)^n u(n) - 2\,(n+1)\left(\frac{1}{4}\right)^n u(n) \quad ...(6)$$

This is the output of the system for a given input.

REVIEW QUESTIONS

1. What is Discrete-time Fourier Transform (DTFT) ?
2. Distinguish between CTFT and DTFT.
3. Distinguish between Discrete-time Fourier Transform (DTFT) and Discrete- time Fourier Sequence (DTFS).
4. Write a short note on development of Discrete-time Fourier Transform (DTFT).
5. Are there any issues associated with DTFT ? Discuss.
6. Give DTFT analysis and synthesis equations.
7. Discuss Fourier transform for discrete-time signals.
8. Discuss various properties of DTFT in brief.

9. Tabluate various properties of DTFT.

10. Tabulate various basic DTFT pairs.

11. What is periodicity ? Discuss property of periodicity for DTFT.

12. What do you mean by linearity property of DTFT ?

13. Discuss properties of time-shifting and frequency-shifting of DTFT.

14. State and explain differencing and accumulation properties of DTFT.

15. Distinguish time-reversal from time expansion.

16. Give the expression of Parseval's relation of DTFT.

17. Discuss following properties of DTFT :

(*a*) The convolution property of DTFT.

(*b*) The multiplication property of DTFT.

18. Discuss the following :

(*a*) Duality in the DTFS analysis and synthesis equations.

(*b*) Duality between the DTFT and the CTFT.

NUMERICAL PROBLEMS

1. Determine the DTFT of following signals :

(*a*) $s(n) = u(n-2) - u(n-6)$ (*b*) $s(n) = \left(\frac{1}{3}\right)^{|n|} u(-n-2)$

(c) $s(n) = \left(\frac{1}{2}\right)^{|n|} \cos\left[\frac{\pi}{8}(n-1)\right]$ (*d*) $s(n) = \sin\left(\frac{\pi}{2}n\right) + \cos(n)$

(*e*) $s(n) = \left(\dfrac{\sin\frac{\pi n}{5}}{\pi n}\right)\cos\left(\frac{7\pi}{2}n\right)$

Ans. (*a*) $S(e^{j\omega}) = e^{-2j\omega} + e^{-3j\omega} + e^{-4j\omega} + e^{-5j\omega}$ (*b*) $S(e^{j\omega}) = \dfrac{e^{2j\omega}}{9\left(1-\frac{1}{3}e^{j\omega}\right)}$

(c) $S(e^{j\omega}) = \dfrac{1}{2j}\left[\dfrac{e^{j\pi/4}\, e^{j\omega}}{1-\left(\frac{1}{2}\right)e^{j\pi/8}\, e^{j\omega}} + \dfrac{e^{-j\pi/4}\, e^{j\omega}}{1-\left(\frac{1}{2}\right)e^{-j\pi/8}\, e^{j\omega}}\right]$

(*d*) $S(e^{j\omega}) = \dfrac{\pi}{j}\left[\delta\left(\omega-\frac{\pi}{2}\right) - \delta\left(\omega+\frac{\pi}{2}\right)\right] + \pi\,[\delta(\omega-1)+\delta(\omega+1)]$

in $0 \le |\omega| \le \pi$ (*e*) $S(e^{j\omega}) = \begin{cases} 1, & 3\pi/10 < |\omega| \le 7\pi/10 \\ 0, & \text{elsewhere} \end{cases}$

2. Find the inverse DTFT of the following DTFTs :

(*a*) $S(e^{j\omega}) = 1 + 3e^{-j\omega} + 2e^{-j2\omega} - 4e^{-j3\omega} + e^{-j10\omega}$

(*b*) $S(e^{j\omega}) = \cos^2\omega + \sin^2 3\omega$

(c) $S(e^{j\omega}) = \begin{cases} 1, & \pi/4 \le |\omega| \le 3\pi/4 \\ 0, & 3\pi/4 \le |\omega| \le \pi,\ 0 \le |\omega| < \pi/4 \end{cases}$

(d) $S(e^{j\omega}) = \sum_{k=-\infty}^{\infty} (-1)^k \delta\left(\omega - \frac{\pi k}{2}\right)$

Ans. (a) $s(n) = \delta(n) + 3\delta(n-1) + 2\delta(n-2) - 4\delta(n-3) + \delta(n-10)$

(b) $s(n) = \delta(n) + \frac{1}{4}\delta(n-2) + \frac{1}{4}\delta(n+2) - \frac{1}{4}\delta(n-3) - \frac{1}{4}\delta(n+3)$

(c) $s(n) = 1 - e^{j\pi n/2} + e^{j\pi n} - e^{j3\pi n/2}$ (d) $s(n) = \frac{1}{\pi n}\left[\sin\left(\frac{3\pi n}{4}\right) - \sin\left(\frac{\pi n}{4}\right)\right]$

3. Find the response of a discrete-time LTI system with impulse response $h(n) = \left(\frac{1}{2}\right)^n u(n)$ for input $s(n) = \left(\frac{3}{4}\right)^n u(n)$

Use DTFT analysis and synthesis equations.

Ans. $y(n) = 3\left(\frac{3}{4}\right)^n u(n) - 2\left(\frac{1}{2}\right)^n u(n)$

4. The difference equation of a causal discrete-time LTI system is given as

$$y(n) = -\frac{1}{2}y(n-1) + s(n)$$

(a) **Find frequency response $H(e^{j\omega})$ for the system**

(b) **Find output response to the input given as**

$$s(n) = \left(\frac{1}{2}\right)^n u(n)$$

Ans. (a) $H(e^{j\omega}) = \frac{1}{1 + \frac{1}{2}e^{-j\omega}}$ (b) $y(n) = \frac{1}{2}\left(\frac{1}{2}\right)^n u(n) + \frac{1}{2}\left(-\frac{1}{2}\right)^n u(n)$

5. Two discrete-time LTI systems are cascaded. These two systems have frequency responses

$$H_1(e^{j\omega}) = \frac{2 - e^{-j\omega}}{1 + \frac{1}{2}e^{-j\omega}}$$

$$H_2(e^{j\omega}) = \frac{1}{1 - \frac{1}{2}e^{-j\omega} + \frac{1}{4}e^{-j2\omega}}$$

(a) **Find the difference equation describing the overall system**

(b) **Find the impulse response of the over all system**

Ans. (a) $y(n) + \frac{1}{8}y(n-3) = 2s(n) - s(n-1)$

(b) $h(n) = \frac{4}{3}\left(-\frac{1}{2}\right)^n u(n) + \frac{1+j\sqrt{3}}{3}\left(\frac{1}{2}e^{j/120}\right)^n u(n)$

$+ \frac{1-j\sqrt{3}}{3}\left(\frac{1}{2}e^{-j/120}\right)^n u(n)$

6. **Determine the impulse response of the inverse system and the difference equation that characterize the inverse system. Following difference equations for causal discrete-time LTI systems are given :**

(a) $y(n) = s(n) - \frac{1}{4} s(n-1)$ (b) $y(n) = -\frac{1}{2} y(n-1) + s(n)$

Ans. (a) $y(n) - \frac{1}{4} y(n-1) = s(n)$ (b) $y(n) = s(n) + \frac{1}{2} s(n-1)$

7. **Find a linear constant-coefficient difference equation relating the input and output. We have given impulse response for the system**

$$h(n) = \left(\frac{1}{2}\right)^n u(n) + \frac{1}{2}\left(\frac{1}{4}\right)^n u(n)$$

Ans. $y(n) = \frac{3}{4} y(n-1) - \frac{1}{8} y(n-2) + \frac{3}{2} s(n) - \frac{1}{2} s(n-1)$

8. **A block diagram implementation of causal discrete-time LTI system is depicted in Fig. 6.19.**

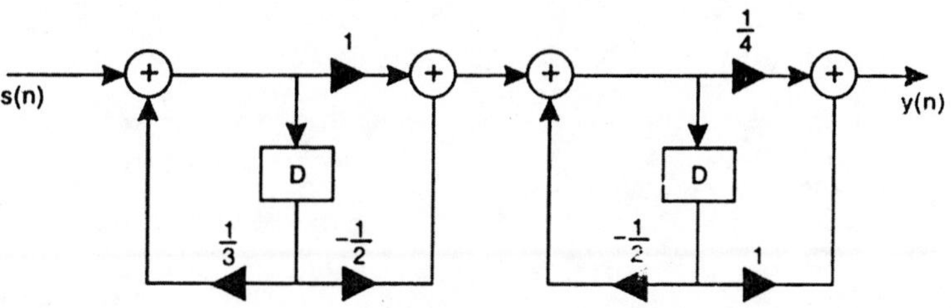

Fig. 6.19

(a) **Find a difference equation relating input $s(n)$ and output $y(n)$**

(b) **Determine frequency response of the system**

(c) **Determine the system's impulse response.**

Ans. (a) $y(n) = \frac{1}{4} y(n-2) + \frac{1}{4} s(n) + \frac{7}{8} s(n-1) - \frac{1}{2} s(n-2)$

$$(b)\ H(e^{j\omega}) = \frac{\frac{1}{4} + \frac{7}{8} e^{-j\omega} - \frac{1}{2} e^{2j\omega}}{1 - \frac{1}{4} e^{-2j\omega}}$$

$$(c)\ h(n) = 2\,\delta(n) - \frac{21}{16}\left(-\frac{1}{2}\right)^n u(n) + \frac{7}{16}\left(\frac{1}{2}\right)^n u(n)$$

9. **We have given the fact that**

$$A^n \xleftrightarrow{\text{DTFT}} \frac{1 - A^2}{1 - 2A\cos\omega + A^2}, \quad |A| < 1$$

Use property of duality to determine the Fourier series coefficients of the following continuous-time signal with period $T = 1$

$$s(t) = \frac{1}{5 - 4\cos(2\pi t)}$$

Ans. Fourier series coefficients of the signal are $\frac{1}{[5 - 4\cos(2\pi t)]}$

10. A discrete-time LTI system with impulse response $h_1(n) = (1/3)^n u(n)$ is connected in parallel with another causal discrete-time LTI system with impulse response $h_2(n)$. The resulting parallel interconnection has the frequency response

$$H(e^{j\omega}) = \frac{-12 + 5e^{-j\omega}}{12 - 7e^{-j\omega} + e^{-j2\omega}}$$

Find impulse response $h_2(n)$. **Ans.** $h_2(n) = -2\left(\frac{1}{4}\right)^n u(n)$

7

The Discrete Fourier Transform (DFT)

7.1 INTRODUCTION

Frequency-domain analysis of discrete-time signals is usually and conveniently performed on a Digital Signal Processor. This digital signal processor may be a general purpose digital computer or a specially designed digital hardware. Fourier transform of discrete-time signal $s(n)$ is called *Discrete-time Fourier Transform* (DTFT) and denoted by $S(e^{j\omega})$ or $S(\omega)$. DTFT $S(\omega)$ is a continuous function of frequency (ω). Therefore, this representation is not a computationally convenient representation of $\{s(n)\}$. We represent a sequence by samples of its continuous spectrum. This frequency-domain representation of a signal is called *Discrete Fourier Transform.* It is a very powerful tool for frequency-domain analysis of discrete- time signals.

The DFT is itself a sequence rather than a function of a continuous variable and it corresponds to equally spaced frequency samples of the Discrete-Time Fourier Transform (DTFT) of a signal. DFT plays a central role in the implementation of various DSP algorithms. Fourier series representation of the periodic sequence corresponds to the Discrete Fourier Transform (DFT) of the finite length sequence. DFT is used for transforming discrete-time sequence $s(n)$ of finite length into discrete-frequency sequence $S(K)$ of finite length. By DFT, discrete-time sequence $s(n)$ in transformed into corresponding discrete-frequency sequence $S(K)$.

Continuous-Time Fourier Transform (CTFT) of a continuous-time signal $s(t)$ is given by

$$S(j\omega) = \int_{-\infty}^{\infty} s(t)\, e^{-j\omega t}\, d\omega \qquad ...(7.1)$$

This is an analysis equation of CTFT.

It is also called Fourier integral. Inverse CTFT is given by the following synthesis equation

$$s(t) = \frac{1}{2\pi} \int_{-\infty}^{\infty} S(j\omega)\ e^{j\omega t}\ d\omega \qquad ...(7.2)$$

Continuous-time Fourier transform (CTFT) is used for non-periodic continuous-time signals. It produces a continuous spectrum of $s(t)$.

Similarly, Discrete-Time Fourier Transform (DTFT) of a discrete-time signal $s(n)$ is given by

$$S(w) \text{ or } S(e^{j\omega}) = \sum_{n=-\infty}^{\infty} s(n)\, e^{-j\omega n} \qquad ...(7.3)$$

This is called analysis equation of DTFT of a discrete-time sequence.

Inverse DTFT is given by the following synthesis equation

$$s(n) = \frac{1}{2\pi} \int_{2\pi} S(e^{j\omega})\, e^{j\omega n}\ d\omega \qquad ...(7.4)$$

It also produces a continuous spectrum of $s(n)$.

Discrete Fourier Transform (DFT) of a discrete-time sequence $s(n)$ is obtained by performing the sampling operation in both the time and frequency domains. But Discrete-Time Fourier Transform (DTFT) of a discrete-time sequence $s(n)$ is obtained by performing the sampling operation in time-domain only.

DFT of a discrete-time sequence $s(n)$ is related to DTFT of same sequence by

$$S(K) = S(\omega)\ \Big|_{\omega = 2\pi K/N}$$

where $S(K)$ is the DFT of $s(n)$ and is given by

$$S(K) = \sum_{n=-\infty}^{\infty} s(n)\ e^{-j2\pi Kn/N}$$

and $S(\omega)$ is the DTFT of the sequence $s(n)$ and is given by

$$S(\omega) = \sum_{n=-\infty}^{\infty} s(n)\, e^{-j\omega n}$$

DTFT gives a continuous spectrum but DFT gives a discrete spectrum.

7.2 DEFINITION OF DFT

DFT is used for transforming a discrete-time sequence, $s(n)$ of a finite length of N points into a discrete-frequency sequence of the same finite length as $s(n)$.

DFT is defined as

$$S(K) = \text{DFT}\ [\ s(n)\ ] = \sum_{n=0}^{N-1} s(n)\, e^{-(j2\pi K/N)\, n} \qquad ...(7.5)$$

where $K = 0, 1, \ldots N-1$.

and *inverse DFT (IDFT) is defined as*

$$s(n) = \text{IDFT}\,[\,S(K)\,] = \sum_{k=0}^{N-1} S(K)\, e^{(j2\pi K/N)\,n} \qquad ...(7.6)$$

where $n = 0, 1, ..., N-1$

$S(K)$ is the DFT of the signal $s(n)$ and is given by

$$S(K) = S_R(K) + j\,S_I(K)$$

Magnitude or amplitude of $S(k)$ is given by

$$|S(K)| = \sqrt{S_R^2(K) + S_I^2(K)}$$

Phase of $S(K)$ is given by

$$\angle S(K) = \tan^{-1}\left[\frac{S_I(K)}{S_R(K)}\right]$$

where $S_R(K)$ is the real part of the $S(K)$ and $S_I(K)$ is the imaginary part of $S(K)$.

Spectrum drawn between amplitude of $S(K)$ and discrete-frequency variable (K) is called amplitude or magnitude spectrum of DFT of $s(n)$, *i.e.*, $S(K)$. Spectrum drawn between phsae of $S(K)$ and discrete frequency variable (K) is called phase spectrum of $S(K)$. Both spectrums are shown by the example given below.

Example 7.1 Compute the 8-point DFT of the signal

$$s(n) = \{\underset{\uparrow}{1}, 1, 1, 1, 1, 1, 0, 0\}$$

Also sketch its magnitude and phase.

Solution. DFT of a signal $s(n)$ is given by

$$S(K) = \sum_{n=0}^{N-1} s(n)\, W_N^{nK} = \sum_{n=0}^{7} s(n)\, W_N^{nK}$$

where $W_N = e^{-j2\pi/N} = e^{-j2\pi/8} = e^{-j\pi/4}$

$$= s(0)\,W_N^{0K} + s(1)\,W_N^{1K} + s(2)\,W_N^{2K} + s(3)\,W_N^{3K} + s(4)\,W_N^{4K} + s(5)\,W_N^{5K} + s(6)\,W_N^{6K} + s(7)\,W_N^{7K}$$

$$= 1 + W_N^{K} + W_N^{2K} + W_N^{3K} + W_N^{4K} + W_N^{5K}$$

$$= 1 + e^{-j\pi K/4} + e^{-j2\pi K/4} + e^{-j3\pi K/4} + e^{-j4\pi K/4} + e^{-j5\pi K/4}$$

$K = 0$, $S(0) = 6$

$K = 1$, $S(1) = -0.7071 - j1.7071$

$K = 2$, $S(2) = 1 - j$

$K = 3$, $S(3) = 0.7071 + j0.2929$

$K = 4$, $S(4) = 0$

$K = 5$, $S(5) = 0.7071 - j0.2929$

$K = 6\,,\ \ S(6) = 1 + j$

$K = 7\,,\ \ S(7) = -\,0.7071 + j1.7071$

DFT of $s(n)$ is given by

$$S(K) = (\,S(0)\,,\,S(1)\,,\,S(2)\,,\,S(3)\,,\,S(4)\,,\,S(5)\,,\,S(6)\,,\,S(7)\,\}$$
$$= \{\underset{\uparrow}{6}, -0.7071 - j1.7071,\ 1 - j,\ 0.7071 + j0.2929,\ 0,\ 0.7071 - j0.2029,\ 1 + j,\ -\,0.7071 + j1.7071\}$$

$$\text{Amplitude} = \sqrt{(\text{Real part})^2 + (\text{Imaginary part})^2}$$

$$\text{Phase} = \tan^{-1}\left(\frac{\text{Imaginary part}}{\text{Real part}}\right)$$

Amplitude of $S(K)$ is given by

$$|\,S(K)\,| = \{6,\ 1.8478,\ 1.4142,\ 0.7654,\ 0,\ 0.7654,\ 1.4142,\ 1.8478\}$$

Phase of $s(K)$ is given by

$$\underline{/S(K)} = \{0, -\,1.9635, -\,0.785,\ 0.3927,\ 0, -\,0.3927,\ 0.785,\ 1.9635\}$$

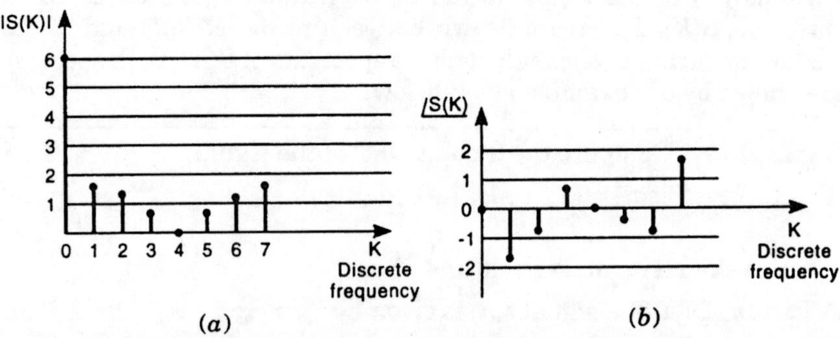

Fig. 7.1 *(a) Amplitude spectrum (b) phase spectrum.*

Example 7.2 Determine the N-point DFT of a finite-duration sequence given by

$$s(n) = \begin{cases} 1, & 0 \le n \le L - 1 \\ 0 & \text{otherwise} \end{cases}$$

Solution. DFT of sequence $s(n)$ is given by

$$S(K) = \text{DFT}\ [\ s(n)\ ] = \sum_{n=0}^{N-1} s(n)\ e^{-(j2\pi K/N)\,n}$$

Now
$$s(n) = \begin{cases} 1, & 0 \le n \le L - 1 \\ 0, & \text{otherwise} \end{cases}$$

$$S(K) = \sum_{n=0}^{L-1} 1\,.\,e^{(-\,j2\pi K/N)\,n}$$
$$= e^{-(j2\pi K/N)\,0} + e^{-(j2\pi K/N)\,1} + \ldots + e^{-(j2\pi K/N)\,(L-1)}$$
$$= 1 + e^{-j2\pi K/N} + \ldots + e^{-(j2\pi K/N)\,(L-1)}$$

This is a geometric progression of the form.

$$= \frac{A\,[1 - R^{M+1}]}{[1 - R]} \qquad ...(1)$$

where A is the First term of G.P.

R is the Common ratio of G.P.

Putting the values of A, R and M in Eq. (1), we get

$$S(K) = \frac{1\left[1 - \left(e^{-j2\pi K/N}\right)^{L-1+1}\right]}{\left[1 - e^{-j2\pi K/N}\right]} = \frac{1 - e^{-j2\pi KL/N}}{1 - e^{-j2\pi K/N}}, \quad \text{where } K = 0, 1, ..., N-1$$

or $$S(K) = \frac{1 - e^{-j2\pi KL/N}}{1 - e^{-j2\pi K/N}} \quad \text{where } K = 0, 1, ..., N-1$$

This is the N-point DFT of the sequence $s(n)$.

Example 7.3 Find the DFT of the sequence

$$s(n) = \begin{cases} 1, & 2 \le n \le 6 \\ 0, & \text{for } n = 0, 1, 7, 8, 9 \end{cases}$$

Given that $N = 10$

Solution. DFT of $s(n)$ is given by

$$S(K) = \sum_{n=0}^{N-1} s(n)\, e^{-(j2\pi k/N)\,n}$$

Let $e^{j(2\pi/N)} = W$, then

$$S(K) = \sum_{n=0}^{N-1} s(n)\, W^{-nK} = \sum_{n=2}^{6} (1)\, W^{-nK} \quad \text{where } W = e^{j(2\pi/N)}$$

$$= \sum_{n=0}^{6} W^{-nK} - \sum_{n=0}^{1} W^{-nK}$$

$$= \frac{1 - (W^{-K})^{6+1}}{1 - W^{-K}} - \frac{1 - (W^{-K})^{1+1}}{1 - W^{-K}}$$

$$= \frac{1 - W^{-7K}}{1 - W^{-K}} - \frac{1 - W^{-2K}}{1 - W^{-K}}$$

or $$S(K) = \frac{1 - W^{-7K} - 1 + W^{-2K}}{1 - W^{-K}} = \frac{W^{-2K} - W^{-7K}}{1 - W^{-K}} \qquad ...(1)$$

Substituting $W = e^{j2\pi/N} = e^{j2\pi/10} = e^{j\pi/5}$ in Eqn. (1), we get

$$S(K) = \frac{e^{-j(\pi K/5)\,2} - e^{-j(\pi K/5)\,7}}{1 - e^{-j(\pi K/2)\,(1)}} = e^{-j(4\pi K/5)} \frac{\sin\left(\dfrac{\pi K}{2}\right)}{\sin\left(\dfrac{\pi K}{10}\right)}$$

7.3 THE DFT AS A LINEAR TRANSFORMATION

DFT and IDFT both are linear transformations on sequences $s(n)$ and $S(K)$ respectively. Where $s(n)$ and $S(K)$ are discrete-time sequence and discrete-frequency sequence, respectively.

DFT and Inverse DFT (IDFT) satisfy the principle of superposition. DFT is a linear transformation tool just like z-transform. Property of linearity for z-transform is already explained in Chapter 3. Same property also applies for both DFT and its inverse DFT.

The DFT as a linear transformation tool is explained and studied by using matrices. DFT and IDFT can be computed by using matrices. Computation of N-point DFT requires $N \times N$ complex multiplications and $N(N-1)$ complex additions.

N-point vector $\mathbf{S_N}$ of frequency samples is given by

$$\mathbf{S_N} = \begin{bmatrix} S(0) \\ S(1) \\ \vdots \\ S(N-1) \end{bmatrix} \qquad \text{...(7.7)}$$

N-point vector $\mathbf{s_N}$ of signal sequence $s(n)$ is given by

$$\mathbf{s_N} = \begin{bmatrix} S(0) \\ S(1) \\ \vdots \\ S(n-1) \end{bmatrix} \qquad \text{...(7.8)}$$

$$\text{DFT,} \qquad S(K) = \sum_{n=0}^{N-1} s(n)\, e^{-(j2\pi K/N)\,n} = \sum_{n=0}^{N-1} s(n)\, W_N^{Kn}$$

where $K = 0, 1, ..., N-1$

$$\text{IDFT,} \qquad s(n) = \frac{1}{N} \sum_{K=0}^{N-1} S(K)\, e^{(j2\pi K/N)\,n} = \frac{1}{N} \sum_{K=0}^{N-1} S(K)\, W_N^{-Kn}$$

where $n = 0, 1, \ldots, N-1$

where $W_N = e^{-j2\pi/N}$ is an Nth root of unity.

N-point DFT in matrix form is defined as

$$\mathbf{S_N} = \mathbf{W_N} \cdot \mathbf{s_N} \qquad \text{...(7.9)}$$

where $\mathbf{W_N}$ is the matrix of the linear transformation and is a $N \times N$ symmetric matrix.

$\mathbf{W_N}$ is given by

$$\mathbf{W_N} = \begin{bmatrix} 1 & 1 & 1 & \cdots & 1 \\ 1 & W_N & W_N^2 & \cdots & W_N^{N-1} \\ \vdots & W_N^2 & W_N^4 & \cdots & W_N^{2(N-1)} \\ \vdots & \vdots & \vdots & & \vdots \\ 1 & W_N^{N-1} & W_N^{2(N-1)} & \cdots & W_N^{(N-1)(N-1)} \end{bmatrix} \qquad \text{...(7.10)}$$

N-point IDFT in matrix form is defined as

$$s_N = W_N^{-1} S_N = \frac{1}{N} W_N^{*} S_N \qquad \text{...(7.11)}$$

where W_N^* denotes the complex conjugate of matrix W_N and W_N^{-1} is the inverse of W_N.

The relationship between W_N and its complex conjugate will be

$$W_N \cdot W_N^* = N \cdot I_N \qquad \text{...(7.12)}$$

where I_N is an $N \times N$ identity matrix.

Example 7.4 Determine the DFT of the 4-point discrete-time sequence $s(n) = \{1, 2, 3, 4\}$ using DFT transformation matrix. Also determine IDFT from DFT.

Solution. Here we require 4×4 DFT transformation matrix because $s(n)$ has 4-points.

$$W_4 = \begin{bmatrix} W_4^0 & W_4^0 & W_4^0 & W_4^0 \\ W_4^0 & W_4^1 & W_4^2 & W_4^3 \\ W_4^0 & W_4^2 & W_4^4 & W_4^6 \\ W_4^0 & W_4^3 & W_4^6 & W_4^9 \end{bmatrix} \qquad \text{...(1)}$$

where $W_4^0 = \left[e^{-j2\pi/4}\right]^0 = 1$

Now we will use the property of periodicity and symmetry of W_4.

The property of periodicity for W_N is given by

$$W_N^{K+N} = W_N^K$$

From this property W_N is repeated after a period N.

The property of symmetry for W_N is given by

$$W_N^{K+N/2} = -W_N^K$$

From this property W_N is repeated with inverse sign after a half period $\frac{N}{2}$.

where $W_N = e^{-j2\pi/N}$ is called *phase rotation factor or twiddle factor* for DFT.

From the property of periodicity

$$W_4^4 = W_4^{(0+4)} = W_4^0$$

$$W_4^6 = W_4^{(2+4)} = W_4^2 \qquad \text{because } N = 4$$

$$W_4^9 = W_4^{(1+2\times 4)} = W_4^1$$

Substituting these values in Eq. 1, we get

$$W_4 = \begin{bmatrix} 1 & 1 & 1 & 1 \\ 1 & W_4^1 & W_4^2 & W_4^3 \\ 1 & W_4^2 & W_4^0 & W_4^2 \\ 1 & W_4^3 & W_4^2 & W_4^1 \end{bmatrix} \quad ...(2)$$

$$W_4^1 = \left(e^{-j2\pi/4}\right) = -j$$

$$W_4^2 = \left(e^{-j2\pi/4}\right)^2 = -1$$

$$W_4^3 = \left(e^{-j2\pi/4}\right)^3 = j$$

Putting above values in Eq. 2, we get

$$W_4 = \begin{bmatrix} 1 & 1 & 1 & 1 \\ 1 & -j & -1 & j \\ 1 & -1 & 1 & -1 \\ 1 & j & -1 & -j \end{bmatrix}$$

Then $S_4 = W_4 s_4 = \begin{bmatrix} 1 & 1 & 1 & 1 \\ 1 & -j & -1 & j \\ 1 & -1 & 1 & -1 \\ 1 & j & -1 & -j \end{bmatrix} \begin{bmatrix} 1 \\ 2 \\ 3 \\ 4 \end{bmatrix}$

$$= \begin{bmatrix} 1+2+3+4 \\ 1-2j-3+4j \\ 1-2+3-4 \\ 1+2j-3-4j \end{bmatrix} = \begin{bmatrix} 10 \\ -2+2j \\ -2 \\ -2-2j \end{bmatrix}$$

Inverse DFT of S_4 is determined by

$s_N = \dfrac{1}{N} W_N^* S_N$ **where W_N^* is the complex conjugate W_N and $N = 4$.**

$$s_4 = \frac{1}{4} W_4^* S_4 = \frac{1}{4} \begin{bmatrix} 1 & 1 & 1 & 1 \\ 1 & j & -1 & -j \\ 1 & -1 & 1 & -1 \\ 1 & -j & -1 & j \end{bmatrix} \begin{bmatrix} 10 \\ -2+2j \\ -2 \\ -2-2j \end{bmatrix}$$

$$= \frac{1}{4} \begin{bmatrix} 10-2+2j-2-2-2j \\ 10-2j+2j^2+2+2j+2j^2 \\ 10+2-2j-2+2+2j \\ 10+2j-2j^2+2-2j-2j^2 \end{bmatrix}$$

$$= \frac{1}{4} \begin{bmatrix} 4 \\ 8 \\ 12 \\ 16 \end{bmatrix} = \begin{bmatrix} 1 \\ 2 \\ 3 \\ 4 \end{bmatrix}$$

Note. W_N^* is determined from W_N by changing the sign of j in W_N.

7.4 PROPERTIES OF THE DFT

Most of the properties of the DFT and z-transform have some similarity because they have some relationship with each other. DFT also differs in some properties such as circular convolution property. This property is explained in detail in Art. 7.6.

Circular convolution is the special property of DFT of a discrete-time sequence. There are several properties of the DFT that play an important role in practical techniques for processing signals. These properties are given as follows :

1. Periodicity.
2. Linearity.
3. Symmetry properties.
4. Circular convolution of two sequences.
5. Time reserval of a sequence.
6. Circular time-shift of a sequence.
7. Circular frequency-shift of a sequence.
8. Circular correlation of two sequences.
9. Multiplication of two sequences.
10. Parseral's theorem.

These properties are discussed one by one in subsequent sections.

7.4.1 Periodicity

If a discrete-time signal is periodic then its DFT will also be periodic. If a signal or sequence repeats its waveform after N number of samples then it is called a periodic signal or sequence and N is called the period of signal.

If $s(n)$ is a discrete-time signal and $S(K)$ is the N-point DFT of $s(n)$. Then

$$s(n+N) = s(n) \text{ for all values of } n \qquad ...(7.13)$$

This is true for periodic signals only.

$$S(K+N) = S(K) \text{ for all values of } K \qquad ...(7.14)$$

DFT of $s(n)$ is also periodic.

This property is used in some applications of DSP such as in FFT algorithms used for linear filtering, power spectrum estimation etc.

7.4.2 Linearity

DFT is also a linear transform just like z-transform. It satisfies the principle of superposition.

If $\qquad s_1(n) \xleftrightarrow{N.\text{pt-DFT}} S_1(K)$

$$s_2(n) \xleftrightarrow{N.\text{pt-DFT}} S_2(K)$$

Then for any real-valued or complex-valued constants A_1 and A_2.

$$A_1 s_1(n) + A_2 s_2(n) \xleftrightarrow{N.\text{p-DFT}} A_1 S_1(K) + A_2 S_2(K)$$

7.4.3 Symmetry Properties of the DFT

If a signal or sequence repeates its waveform in a negative direction after $N/2$ number of samples, then it is called symmetric sequence or signal. Symmetry properties of the DFT is studied in two cases :

Case I. Signal $s(n)$ and its DFT $S(K)$ both are complex and periodic sequences.

These properties are given in Table 7.1.

We know that for periodic signals

$$s(n) = s(N-n) \quad \text{and} \quad S(K) = S(N-K)$$

It concludes that if a discrete-time sequence $s(n)$ is periodic then its DFT $S(K)$ is also periodic in the same period.

Table 7.1 Symmetry properties of the DFT for $s(n)$ and $S(K)$ both are periodic and complex sequences

N-point discrete-time sequence $s(n)$, $0 \le n \le N-1$	*N-point DFT* $S(K)$, $0 \le K \le N-1$
Complex conjugate of $s(n) \Rightarrow s^*(n)$	$S^*(N-K)$
$s^*(N-n)$	$S^*(K)$
Real part of complex sequence $\Rightarrow s_R(n)$	Conjugate even part of $S(K)$ $\Rightarrow S_{ce}(K) = \frac{1}{2}[S(K) + S^*(N-K)]$
Imaginary part of complex sequence $\Rightarrow j\,s_I(n)$	Conjugate odd part of $S(K)$ $\Rightarrow S_{co}(K) = \frac{1}{2}[S(K) - S^*(N-K)]$
Conjugate even part of $s(n)$ $\Rightarrow s_{ce}(n) = \frac{1}{2}[s(n) + s^*(N-n)]$	$S_R(K)$
Conjugate odd part of $s(n)$ $\Rightarrow s_{co}(n) = \frac{1}{2}[s(n) - s^*(N-nm)]$	$jS_I(K)$
where $s^*(n)$ is complex conjugate of $s(n)$ and $S^*(K)$ is complex conjugate of $S(K)$	

Case II. Now in this case $s(n)$ is a real and periodic sequence but its DFT is complex and periodic sequence of N samples.

Now $S(K)$ satisfies the following symmetry conditions.

$$S(K) = S^*(N-K) \quad \text{...(7.15)}$$

Real part of $S(K)$, $S_R(K) = S_R(N-K)$ ...(7.16)

Imaginary part of $S(K)$, $S_I(K) = -S_I(N-K)$ ...(7.17)

Both $S_R(K)$ and $S_R(N-K)$ are even functions and $S_I(K)$ and $S_I(N-K)$ are odd functions.

Amplitude of $S(K)$, $|S(K)| = |S(N-K)|$ ...(7.18)

Phase of $S(K)$, $\angle S(K) = -\angle S(N-K)$ ...(7.19)

Both $|S(K)|$ and $|S(N-K)|$ are even.

and $S(K)$ and $S(N-K)$ are odd.

Conjugate even part of

$$s(n) = \frac{1}{2}[s(n) + s(N-n)] \xleftrightarrow{N.\text{pt - DFT}} S_R(K)$$

Conjugate odd part of

$$s(n) = \frac{1}{2}[s(n) - s(N-n)] \xleftrightarrow{N.\text{pt - DFT}} j\,S_I(K)$$

where $S_R(K)$ is the real part of $S(K)$
and $S_I(K)$ is the imaginary part of $S(K)$

Note. Complex discrete-time sequence is represented as

$$s(n) = s_R(n) + j\,s_I(n)$$

and complex discrete-fourier transform of $s(n)$ is represented as

$$S(K) = S_R(K) + j\,S_I(K)$$

$S(K)$ is also represented in polar form as

$$S(K) = |S(K)|\,.\,e^{j\,S(K)}$$

where $|S(K)|$ is the amplitude of $S(K)$
$\angle S(K)$ is the phase of $S(K)$

7.4.4 Circular Convolution

This is the special property of the DFT. This property is not satisfied by z-transform.

Circular convolution is defined as

$$s_3(m) = \sum_{n=0}^{N-1} s_1(n)\, s_2\,[(m-n)]_N \quad \text{...(7.20)}$$

where $m = 0, 1, ..., N-1$

where $s_2[(m-n)]_N$ is circular shift of sequence $s_2(n)$

and $s_1(n)$ and $s_2(n)$ are two discrete-time sequences of finite length of N-point.

Circular convolution of two-discrete-time sequences $s_1(n)$ and $s_2(n)$ is denoted by

$$s_3(n) = s_1(n) \; \textcircled{N} \; s_2(n) = s_2(n) \; \textcircled{N} \; s_1(n) \qquad ...(7.21)$$

Here Ⓝ denotes N-pt. circular convolution of two sequences. Circular convolution like linear convolution is a commutative operation.

Multiplication of the DFTs of the two sequences is equivalent to the circular convolution of the two sequences in the time-domain.

The basic difference between circular convolution and linear convolution is that, in circular convolution, the folding and shifting operations are performed in a circular fashion by computing the index of one of the sequences *modulo N*. Modulo N operation is not used in linear convolution. This operation is used for circular shifting of sequence $s(n)$. It can be explained with the help of an example given below:

If $\qquad s_1(n) \xleftrightarrow{N.\text{pt - DFT}} S_1(K)$

and $\qquad s_2(n) \xleftrightarrow{N.\text{pt - DFT}} S_2(K)$

then from the special property of the DFT

$$s_1(n) \; \textcircled{N} \; s_2(n) = \sum_{n=0}^{N-1} s_1(n)\, s_2[(m-n)]_N \xleftrightarrow{N.\text{pt - DFT}} S_1(K) \,.\, S_2(K)$$

Example for Modulo N-operation

In general, the circular shift of the sequence $s(n)$ can be represented as the Index modulo N.

$$s'(n) = s(n - K \text{, modulo } N) = s[(n-K)]_N$$

if $K = 2$ and $N = 4$, then

$$s'(n) = s[(n-2)]_4$$

$n = 0\,,\quad s'(0) = s[(0-2)]_4 = s(2)$

$n = 1\,,\quad s'(1) = s[(1-2)]_4 = s[(-1)]_4 = s(3)$

$n = 2 = \quad s'(2) = s[(2-2)]_4 = s[(0)]_4 = s(0)$

$n = 3\,,\quad s'(3) = s[(3-2)]_4 = s[(1)]_4 = s(1)$

Steps in Computation of Circular Convolution

There are four steps in computation of circular convolution just as those involved in computation of ordinary linear convolution. These four steps are :

1. Folding one of two sequences.
2. Shifting the folded sequence.
3. Multiplying the two sequences for obtaining the product sequence.

4. Summing the values of the product sequences.

This is illustrated by example 7.5.

Example 7.5 Compute the circular convolution of the two discrete- time sequences $s_1(n) = \{1, 2, 1, 2\}$ and $s_2(n) = \{3, 2, 1, 4\}$.

Solution. Circular convolution is defined as

$$s_3(m) = \sum_{n=0}^{N-1} s_1(n)\, s_2[(m-n)]_N$$

where $s_1(n) = \{s_1(0), s_1(1), s_1(2), s_1(3)\} = \{1, 2, 1, 2\}$

and $s_2(n) = \{s_2(0), s_2(1), s_2(2), s_2(3)\} = \{3, 2, 1, 4\}$

Computation of circular convolution using graph is shown in Fig. 7.2.

Circular convolution is defined as

$$s_3(m) = \sum_{n=0}^{N-1} s_1(n)\, s_2[(m-n)]_N$$

From Fig. 7.2, we can compute following :

$$m = 0, \quad s_3(0) = \sum_{n=0}^{3} s_1(n)\, s_2[(-n)]_4 = 3 + 8 + 1 + 4 = 16$$

$$m = 1, \quad s_3(1) = \sum_{n=0}^{3} s_1(n)\, s_2[(1-n)]_4 = 2 + 6 + 4 + 2 = 14$$

$$m = 2, \quad s_3(2) = \sum_{n=0}^{3} s_1(n)\, s_2[(2-n)]_4 = 2 + 4 + 3 + 8 = 17$$

$$m = 3, \quad s_3(3) = \sum_{n=0}^{3} s_1(n)\, s_2[(3-n)]_4 = 4 + 2 + 2 + 6 = 14$$

Now circularly convoluted sequence

$$s_3(n) = \{s_3(0), s_3(1), s_3(2), s_3(3)\} = \{16, 14, 17, 14\}$$

Example 7.6 Compute the $s_3(n)$ corresponding to circular convolution of the sequences $s_1(n)$ and $s_2(n)$ by using DFT and IDFT approach.

Given $s_1(n) = \{2, 1, 2, 1\}$ and $s_2(n) = \{1, 2, 3, 4\}$

Solution. DFT of $s_1(n)$ is given by

$$S_1(K) = \sum_{n=0}^{3} s_1(n)\, e^{-j2\pi Kn/4}, \qquad K = 0, 1, 2, 3$$

$$= s_1(0)\, e^{-j2\pi K.0/4} + s_1(1)\, e^{-j2\pi K.1/4} + s_1(2)\, e^{-j2\pi K.2/4} + s_1(3)\, e^{-j2\pi K.3/4}$$

$$= 2(1) + 1.e^{-j2\pi K/4} + 2.e^{-j4\pi K/4} + 1.e^{-j6\pi K/4}$$

$$= 2 + e^{-j\pi K/2} + 2e^{-j\pi K} + e^{-j3\pi K/2} \quad \text{where } K = 0, 1, 2, 3$$

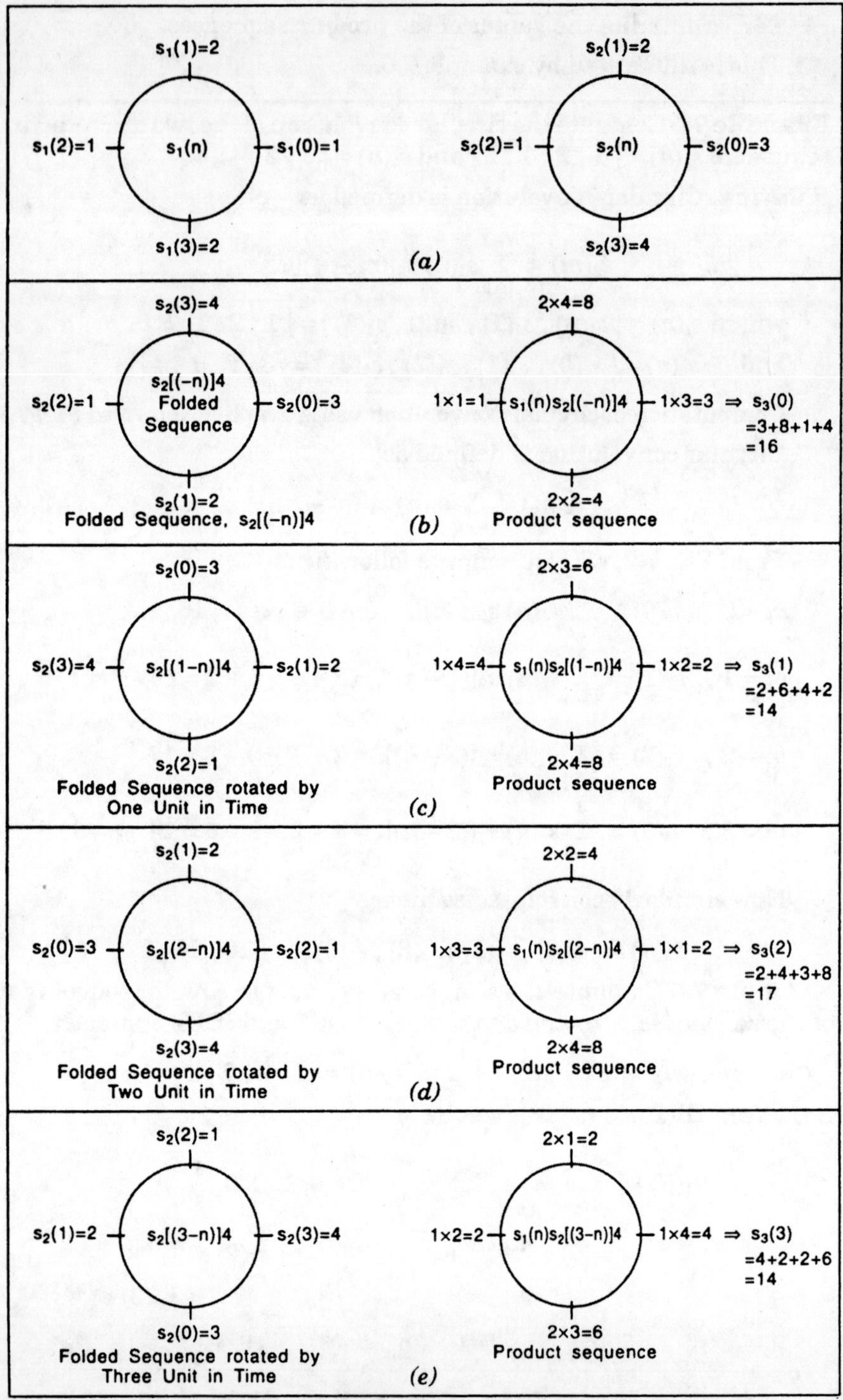

Fig. 7.2 *Circular convolution of the two sequences $s_1(n)$ and $s_2(n)$ given in problem* 7.5.

Thus, $K=0\,,\ S_1(0)=2+e^{-j\pi\,.\,0/2}+2e^{-j\pi\,.\,0}+e^{-j3\pi\,.\,0/2}$

$=2+1+2+1=6$

$K=1\,,\ S_1(1)=2+e^{-j\pi\,.\,1/2}+2e^{-j\pi\,.\,1}+e^{-j3\pi\,.\,1/2}=0$

$K=2\,,\ S_1(2)=2+e^{-j\pi\,.\,2/2}+e^{-j\pi\,.\,2}+e^{-j3\pi\,.\,2/2}=2$

$K=3\,,\ S_1(3)=2+e^{-j\pi\,.\,3/2}+2e^{-j\pi\,.\,3}+e^{-j3\pi\,.\,3/2}=0$

$$S_1(K)=\{S_1(0)\,,S_1(1)\,,S_1(2)\,,S_1(3)\}=\{6,0,2,0\}$$

DFT of $s_2(n)$ is given by

$$S_2(K)=\sum_{n=0}^{3} s_2(n)\,e^{-j2\pi Kn/4},\qquad K=0,1,2,3$$

$$=s_2(0)\,e^{-j2\pi K.0/4}+s_2(1)\,e^{-j2\pi K.1/4}+s_2(2)\,e^{-j2\pi K.2/4}+s_2(3)\,e^{-j2\pi K.3/4}$$

$$=1e^0+2e^{-j2\pi K/4}+3e^{-j4\pi K/4}+4e^{-j6\pi K/4}$$

$$=1+2e^{-j\pi K/2}+3e^{-j\pi K}+4e^{-j3\pi K/2}\quad \text{where } K=0,1,2,3$$

$K=0\,,\ S_2(0)=1+2e^{-j\pi.0/2}+3e^{-j\pi.0}+4e^{-j3\pi.0/2}$

$=1+2+3+4=10$

$K=1\,,\ S_2(1)=1+2e^{-j\pi.1/2}+3e^{-j\pi.1}+4e^{-j3\pi.1/2}=-2+j2$

$K=2\,,\ S_2(2)=1+2e^{-j\pi.2/2}+3e^{-j\pi.2}+4e^{-j3\pi.2/2}=-2$

$K=3\,,\ S_2(3)=1+2e^{-j\pi.3/2}+3e^{-j\pi.3}+4e^{-j3\pi.3/2}=-2-j2$

$$S_2(K)=\{S_2(0)\,,S_1(1)\,,S_2(2)\,,S_2(3)\}$$

$$=\{10\,,-2+j2\,,-2\,,-2-j2\}$$

But from the important property of the DFT that circular convolution of two discrete-time sequences is equivalent to multiplication of DFTs of individual sequences.

$$S_3(K)=S_1(K)\,.\,S_2(K)$$

$$=\{6\,,0\,,2\,,0\}\{10\,,-2+j2\,,-2\,,-2-j2\}$$

$$=\{60,0,-4,0\}$$

$$S_3(K)=\{60\,,0\,,-4\,,0\}$$

Taking inverse DFT of $S_3(K)$ for computing $s_3(n)$.

$$s_3(n)=\text{IDFT}\,[\,S_3(K)\,]$$

$$=\frac{1}{4}\sum_{K=0}^{3} S_3(K)\,e^{j2\pi Kn/4},\qquad n,0,1,2,3$$

$$=\frac{1}{4}\,[\,S_3(0)\,e^{(j2\pi 0.n)/4}+S_3(1)\,e^{(j2\pi 1.n)/4}+S_3(2)\,e^{(j2\pi 2.n)/4}+S_3(3)\,e^{(j2\pi 3.n)/4}\,]$$

$$= \frac{1}{4}\left[60e^{0} + 0e^{j2\pi n/4} + (-4)\, e^{j4\pi n/4} + 0.e^{j6\pi n/4}\right]$$

$$= \frac{1}{4}\left[60 - 4e^{j\pi n}\right] = 15 - e^{j\pi n}$$

or $\quad s_3(n) = 15 - e^{j\pi n}$

$n = 0\,,\ s_3(0) = 15 - e^{j\pi.0} = 15 - 1 = 14$

$n = 1\,,\ s_3(1) = 15 - e^{j\pi 1} = 16$

$n = 2\,,\ s_3(2) = 15 - e^{j\pi 2} = 14$

$n = 3\,,\ s_3(3) = 15 - e^{j\pi 3} = 16$

$$s_3(n) = \{\, s_3(0)\,,\, s_3(1)\,,\, s_3(2)\,,\, s_3(3)\,\} = \{14,\ 16,\ 14,\ 16\}$$

This is the result of output sequence from circular convolution of two sequences by DFT method.

7.4.5 Time Reversal of a Sequence

Reversing the N-point discrete-time sequence is equivalent to reversing the Discrete Fourier Transform (DFT) values.

If $\qquad s(n) \xleftrightarrow{N.\text{pt}-\text{DFT}} S(K)$

Now we are reversing the sequence $s(n)$ in time

$$s[(-n)]_N = s(N-m) \xleftrightarrow{N.\text{pt}-\text{DFT}} S[(-K)]_N = S(N-K)$$

7.4.6 Circular Time-Shift of a Sequence

If $\qquad s(n) \xleftrightarrow{N.\text{pt}-\text{DFT}} S(K)$

then, $\quad s[(n-m)]_N \xleftrightarrow{N.\text{pt}-\text{DFT}} e^{-j2\pi Km/N}\, S(K)$

where $s[(n-m)]_N$ is a circularly time shifted version of $s(n)$.

7.4.7 Circular Frequency-Shift of a Sequence

If $\qquad s(n) \xleftrightarrow{N.\text{pt}-\text{DFT}} S(K)$

then, $\quad e^{j2\pi nm/N}\, s(n) \xleftrightarrow{N.\text{pt}-\text{DFT}} S[(K-m)]_N$

where $S[(K-m)]_N$ is a circularly frequency shifted version of $S(K)$.

7.4.8 Circular Correlation of Two Sequences

Let us consider two complex valued discrete-time sequences $s_1(n)$ and $s_2(n)$.

If $\qquad s_1(n) \xleftrightarrow{N.\text{pt}-\text{DFT}} S_1(K)$

and $\qquad s_2(n) \xleftrightarrow{N.\text{pt}-\text{DFT}} S_2(K)$

Then,

$$R_{s_1s_2(m)} = \sum_{n=0}^{N-1} s_1(n)\, s_2^*[(n-m)]_N \xleftrightarrow{N.\text{pt} - \text{DFT}} S_1(K)\, S_2^*(K)$$

where $s_2^*[(n-m)]_N$ is the complex conjugate of $s_2[(n-m)]_N$

and $S_2^*(K)$ is the complex conjugate of $S_2(K)$

$R_{s_1s_2(m)}$ is called circular correlation sequence.

7.4.9 Multiplication of Two Sequences

If $s_1(n) \xleftrightarrow{N.\text{pt} - \text{DFT}} S_1(K)$

and $s_2(n) \xleftrightarrow{N.\text{pt} - \text{DFT}} S_2(K)$

then, $s_1(n)\,.\,s_2(n) \xleftrightarrow{N.\text{pt} - \text{DFT}} \frac{1}{N} S_1(K) \circledN S_2(K)$

Ⓝ denotes circular convolution.

This property is dual of circular convolution and defined as

$$s_1(n) \circledN s_2(n) \xleftrightarrow{N.\text{pt} - \text{DFT}} S_1(K)\,.\,S_2(K)$$

7.4.10 Parseval's Theorem

For complex valued sequences $s_1(n)$ and $s_2(n)$

If $s_1(n) \xleftrightarrow{N.\text{pt} - \text{DFT}} S_1(K)$

and $s_2(n) \xleftrightarrow{N.\text{pt} - \text{DFT}} S_2(K)$

Then, $\sum_{n=0}^{N-1} s_1(n)\, s_2^*(n) = \frac{1}{N} \sum_{K=0}^{N-1} S_1(K)\, S_2^*(K)$ [This expression is a general form of parseval's theorem.]

If $s_1(n) = s_2(n)$ then $S_1(K) = S_2(K) = S(K)$

Then the above theorem will be

$$\sum_{n=0}^{N-1} s(n)\, s^*(n) \xleftrightarrow{N.\text{pt} - \text{DFT}} \sum_{K=0}^{N-1} S(K)\, S^*(K)$$

But we know that $s(n)\,.\,s^*(n) = |s(n)|^2$

and $S(K)\,.\,S^*(K) = |S(K)|^2$

Putting these values in above expression, we get

$$\sum_{n=0}^{N-1} |s(n)|^2 = \sum_{K=0}^{N-1} |S(K)|^2 \qquad ...(7.22)$$

When $s_1(n) = s_2(n)$ and $S_1(K) = S_2(K)$ it is called special form of Parseval's theorem

7.5 TABULATION OF PROPERTIES OF DFT

All the properties of the DFT is given in Table 7.2.

Table 7.2

Properties of DFT	*Time-domain*	*Frequency-domain*
Notation	$s_1(n)$, $s_2(n)$ Two discrete-time sequences	$S_1(K)$ and $S_2(K)$ Two discrete frequency sequences
Periodicity	$s_1(n) = s_1(N+n)$ and $s_2(n) = s_2(N+n)$	$S_1(K) = S_1(N+K)$ and $S_2(K) = S_2(N+K)$
Linearity	$A_1s_1(n) + A_2s_2(n)$	$A_1S_1(K) + A_2S_2(K)$
Time-Reversal (Folding)	$s_1(N-n)$ and $s_2(N-n)$	$S_1(N-K)$ and $S_2(N-K)$
Circular time shift	$s_1[(n-m)]_N$ and $s_2[(n-m)]_N$	$e^{-j2\pi Km/N}\, S_1(K)$ and $e^{-j2\pi Km/N}\, S_2(K)$
Circular Frequency Shift	$e^{j2\pi nm/N}\, s_1(n)$ and $e^{j2\pi nm/N}\, s_2(n)$	$S_1[(K-m)]_N$ and $S_2[(K-m)]_N$
Complex conjugate	$s_1^*(n)$ and $s_2^*(n)$	$S_1^*(N-K)$ and $S_2^*(N-K)$
Circular convolution	$s_1(n) \circledN s_2(n)$	$S_1(K) \,.\, S_2(K)$
Circular correlation	$s_1(N) \circledN s_2^*(-n)$	$S_1(K) \,.\, S_2^*(K)$
Multiplication of two sequences	$s_1(n) \,.\, s_2(n)$	$\frac{1}{N} S_1(K) \circledN S_2(K)$
Parseval's theorem	$\sum_{n=0}^{N-1} s_1(n)\, s_2^*(n)$	$\frac{1}{N} \sum_{K=0}^{N-1} S(K)\, S^*(K)$

7.6 RELATIONSHIP BETWEEN DFT AND z-TRANSFORM

Now we will develop the relationship between DFT and z-transform. z-transform of a discrete-time sequence of finite duration is defined as

$$S(z) = \sum_{n=-\infty}^{\infty} s(n)\, z^{-n} \quad ...(7.23)$$

Consider a finite-duration sequence $s(n)$, $0 \le n \le N-1$

Now Eqn. 7.23 reduces to

$$S(z) = \sum_{n=0}^{N-1} s(n)\, z^{-n} \quad ...(7.24)$$

DFT is given by

$$S(K) = \sum_{n=0}^{N-1} s(n)\, e^{-(j2\pi K/N)\, n} \quad ...(7.25)$$

Substituting $z = e^{j2\pi K/N}$ in Eq. (7.24), we get

$$S(z)\Big|_{z=e^{j2\pi K/n}} = \sum_{n=0}^{N-1} s(n)\, e^{-(j2\pi K/N)\,n} \qquad \text{...(7.26)}$$

Comparing Eqs. (7.26) and (7.25), we get

$$S(K) = S(z)\Big|_{z=e^{+j2\pi K/N}} \qquad \text{...(7.27)}$$

Now z-transform is uniquely determined by its N-point DFT. $S(z)$ can be expressed as a function of the DFT, $S(K)$ as follows :

$$S(z) = \sum_{n=0}^{N-1} s(n)\, z^{-n} \qquad \text{...(7.24)}$$

But inverse DFT is given by

$$s(n) = \frac{1}{N} \sum_{K=0}^{N-1} S(K)\, e^{j2\pi Kn/N} \qquad \text{...(7.28)}$$

Substituting the value of $s(n)$ in above Eq. (7.24), we get

$$S(z) = \sum_{n=0}^{N-1} \left[\frac{1}{N} \sum_{K=0}^{N-1} S(K)\, e^{j2\pi Kn/N} \right] z^{-n}$$

Rearranging the summations, we get

$$= \frac{1}{N} \sum_{K=0}^{N-1} S(K) \sum_{n=0}^{N-1} \left(e^{j2\pi K/N} z^{-1} \right)^{n}$$

$$= \frac{1}{N} \sum_{K=0}^{N-1} S(K) \frac{\left[1 - (e^{j2\pi K/N} z^{-1})^{N-1+1} \right]}{\left[1 - e^{j2\pi K/N} z^{-1} \right]}$$

$$= \frac{1}{N} \sum_{K=0}^{N-1} S(K) \left[\frac{1 - e^{j2\pi KN/N} z^{-N}}{1 - e^{j2\pi K/N} z^{-1}} \right]$$

$$= \frac{1}{N} \sum_{K=0}^{N-1} S(K) \left[\frac{1 - e^{j2\pi K} z^{-N}}{1 - e^{j2\pi K/N} z^{-1}} \right]$$

But $e^{j2\pi K} = 1$

or
$$S(z) = \frac{1 - z^{-N}}{N} \sum_{K=0}^{N-1} \frac{S(K)}{1 - e^{j2\pi K/N} z^{-1}} \qquad \text{...(7.29)}$$

Eq. (7.29) is a very important result and gives the relationship between z-transform and DFT of a finite-duration discrete-time sequence.

7.7 LINEAR CONVOLUTION USING DFT

The product of two DFTs is equivalent to the circular convolution of the corresponding time-domain sequences. Here we have no use of circular convolution because our objective is to determine the output of a linear filter to a given input sequence. The purpose of linear convolution is linear filtering the input of a linear system.

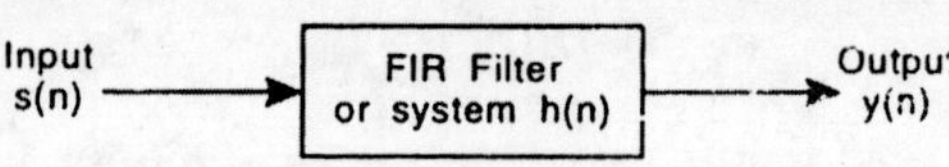

Fig. 7.3 *FIR filter.*

The output of a FIR filter is given by linear convoltuion of $s(n)$ and $y(n)$ as

$$y(n) = \sum_{k=-\infty}^{\infty} h(k)\, s(n-k)$$

Here k is a discrete-time index.

Here we are considering a finite-duration input sequence $s(n)$ of length L and unit-sample response of FIR filter $h(n)$ of length M. Output of this FIR filter will be given by

$$y(n) = \sum_{k=0}^{M-1} h(k)\, s(n-k) \qquad \text{...(7.30)}$$

The duration of $y(n)$ will be $N \geq L + M - 1$.

The frequency-domain equivalent of Eq. (7.30) will be

$$Y(\omega) = S(\omega)\, H(\omega) \qquad \text{...(7.31)}$$

DFT, $\quad Y(K) = Y(\omega)\big|_{\omega = 2\pi K/N} = S(\omega)\, H(\omega)\big|_{\omega = 2\pi K/N}$

$$K = 0, 1, ..., N-1$$

DFT of size $N \geq L + M - 1$ is required to represent $y(n)$ in the frequency domain

$$Y(K) = S(K)\, H(K) \qquad \text{...(7.32)}$$

$$K = 0, 1, 2, ..., N-1$$

Here K is a discrete frequency index.

where $S(K)$ is the DFT of $s(n)$ and represented by

$$S(K) = \sum_{n=0}^{N-1} s(n)\, e^{-(j2\pi K/N)\, n}, \qquad K = 0, 1, 2, ..., N-1$$

and $H(K)$ is the DFT of $h(n)$ and represented by

$$H(K) = \sum_{n=0}^{N-1} h(n)\, e^{-(j2\pi K/N)\, n}, \qquad K = 0, 1, 2, ..., N-1$$

Since $s(n)$ and $h(n)$ have a duration less than N. If DFT is to be used for performing linear convolution then we are required to increase the length of $s(n)$ and $h(n)$ to N by padding these sequences with zero.

After padding these sequencs with zero, the N-point circular convolution of $s(n)$ with $h(n)$ must be equivalent to linear convolution of $s(n)$ with $h(n)$.

Thus we see that with zero as padding, the DFT can be used to perform linear filtering.

Example 7.7 Compute linear convolution of $s(n)$ and $h(n)$ using DFT method

Given $s(n) = \begin{cases} 1\,, n=0 \\ \frac{1}{2}\,,\ n=1 \\ 0\,,\ \text{otherwise} \end{cases}$ and $h(n) = \begin{cases} \frac{1}{2}\,, n=0 \\ 1\,,\ n=1 \\ 0\,,\ \text{otherwise} \end{cases}$

Solution. Sample points of $s(n)$, $N_s = 2$

Sample points of $h(n)$, $N_h = 2$

Sample points of $y(n) = s(n) * h(n)$, $N_y = N_s + N_h - 1$

$= 2 + 2 - 1 = 3$

To avoid time-aliasing, we convert 2-sample sequences into 3-sample sequence by padding with zero. Fig. 7.4 illustrates the graphical representation of sequences $s(n)$ and $h(n)$.

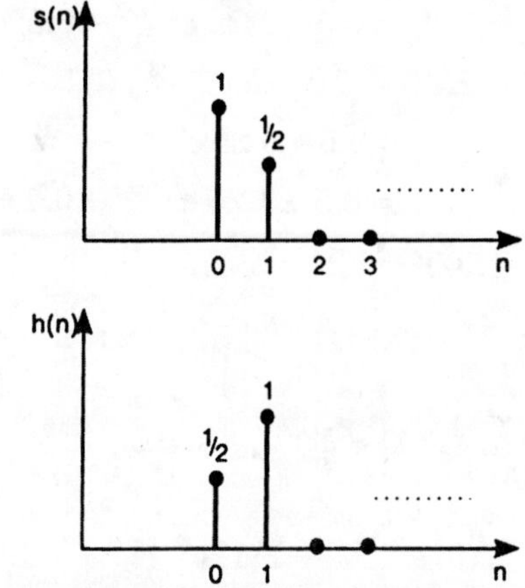

Fig. 7.4 *Graphical representation of s(n) and h(n).*

Three point DFT sequences are given by

$$S(K) = \sum_{n=0}^{2} s(n)\ e^{-(j2\pi K/3)\,n}\ ,\ K = 0\,,1\,,2$$

$$= s(0)\,e^{-j2\pi K.0/3} + s(1)\,e^{-j2\pi K.1/3} + s(2)\,e^{-j2\pi K.2/3}$$

$$= 1\,.\,1 + \frac{1}{2}\,e^{-j2\pi K/3} + 0\,.\,e^{-j4\pi K/3} = 1 + 0.5\,e^{-j2\pi K/3}$$

$K = 0$, $S(0) = 1 + 0.5\,e^{-(j2\pi/3)\,0} = 1 + 0.5 = 1.5$

$K = 1$, $S(1) = 1 + 0.5\,e^{-(j2\pi/3)\,1} = 1 + 0.5\,e^{-j2\pi/3}$

$K = 2$, $S(2) = 1 + 0.5\,e^{-(j2\pi/3)\,2} = 1 + 0.5\,e^{-j4\pi/3}$

$$H(K) = \sum_{n=0}^{2} h(n)\, e^{-(j2\pi K/3)\, n}, \quad K = 0, 1, 2$$

$$= h(0)\, e^{-(j2\pi K/3)\, 0} + h(1)\, e^{-(j2\pi K/3)\, 1} + h(2)\, e^{-(j2\pi K/3)\, 2}$$

$$= \frac{1}{2} e^{0} + 1 \,.\, e^{-j2\pi K/3} + 0 \,.\, e^{-j4\pi K/3} = 0.5 + e^{-j2\pi K/3}$$

$K = 0, \;\; H(0) = 0.5 + e^{-(j2\pi/3)\, 0} = 0.5 + 1 = 1.5$

$K = 1, \;\; H(1) = 0.5\, e^{-(j2\pi K/3)\, 1} = 0.5 + e^{-j2\pi/3}$

$K = 2, \;\; H(2) = 0.5\, e^{-(j2\pi K/3)\, 2} = 0.5 + e^{-j4\pi/3}$

Since a time-domain convolution is equivalent to a frequency-domain multiplication

$$Y(K) = S(K) \,.\, H(K), \quad K = 0, 1, 2$$

$K = 0, \;\; Y(0) = S(0) \,.\, H(0) = 1.5 \times 1.5 = 2.25$

$K = 1, \;\; Y(1) = S(1) \,.\, H(1) = (1 + 0.5\, e^{-j2\pi/3}) \,.\, (0.5 + e^{-j2\pi/3})$

$$= 0.5 + 0.25\, e^{-j2\pi/3} + e^{-j2\pi/3} + 0.5\, e^{-j4\pi/3}$$

$$= 0.5 + 1.25\, e^{-j2\pi/3} + 0.5\, e^{-j4\pi/3}$$

$K = 2, \;\; Y(2) = S(2) \,.\, H(2) = (1 + 0.5\, e^{-j4\pi/3})\, (0.5 + e^{-j4\pi/3})$

$$= 0.5 + 0.25\, e^{-j4\pi/3} + e^{-j4\pi/3} + 0.5\, e^{-j8\pi/3}$$

$$= 0.5 + 1.25\, e^{-j4\pi/3} + 0.5\, e^{-j8\pi/3}$$

The inverse DFT is given by

$$y(n) = \frac{1}{N} \sum_{K=0}^{N-1} Y(K)\, e^{(j2\pi K/N)\, n}, \quad 0 \le n \le N - 1 \qquad (N = 3)$$

$$y(n) = \frac{1}{3} \sum_{K=0}^{2} Y(K)\, e^{j2\pi Kn/3}, \quad n = 0, 1, 2$$

$$= \frac{1}{3}\left[Y(0)\, e^{j2\pi 0.n/3} + Y(1)\, e^{j2\pi 1.n/3} + Y(2)\, e^{j2\pi 2.n/3} \right]$$

$$= \frac{1}{3} [\, 2.25 + (0.5 + 1.25\, e^{-j2\pi/3} + 0.5\, e^{-j4\pi/3})\, e^{j2\pi n/3}$$

$$+ (0.5 + 1.25\, e^{-j4\pi/3} + 0.5\, e^{-j8\pi/3})\, e^{j4\pi n/3}\,]$$

$n = 0, \; y(0) = \frac{1}{3} [2.25 + (0.25 + 1.25\, e^{-j2\pi/3} + 0.5\, e^{-j4\pi/3})\, e^{j2\pi\,.0/3}$

$$+ (0.5 + 1.25\, e^{-j4\pi/3} + 0.5\, e^{-j8\pi/3})\, e^{j4\pi\,.\,0/3}\,]$$

$$= \frac{1}{3} [2.25 + 0.5 + 1.25\, e^{-j2\pi/3} + 0.5\, e^{-j4\pi/3} + 0.5$$

$$+ 1.25\, e^{-j4\pi/3} + 0.5\, e^{-j8\pi/3}\,]$$

$$= \frac{1}{3}\,[3.25 + 1.25\,e^{-j2\pi/3} + 1.75\,e^{-j4\pi/3} + 0.5\,e^{-j8\pi/3}\,]$$

$$= 0.5$$

In similar manner, we can compute

$$y(1) = 1.25 \text{ at } n = 1 \text{ and } y(2) = 0.50 \text{ at } n = 2$$

$$\text{Hence } \; y(n) = \begin{cases} 0.50\,, & n = 0 \\ 1.25\,, & n = 1 \\ 0.50\,, & n = 2 \\ 0\,, & \text{elsewhere} \end{cases}$$

Fig. 7.5 illustrates the graphical representation of resultant of convolution of $s(n)$ and $h(n)$.

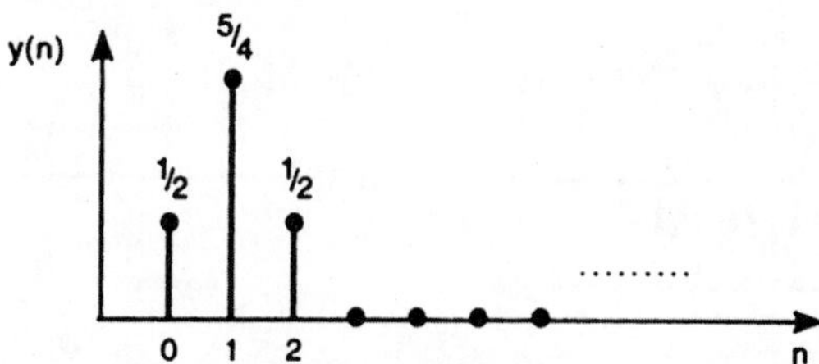

Fig. 7.5 *Graphical representation of resultant of convolution of* $s(n)$ *and* $h(n)$.

7.8 PITFALLS IN USING DFT

The DFT is primarily concerned with the analysis and processing of discrete-time periodic signals and it is an approximation of the CTFT (Continuous-Time Fourier Transform). Fast Fourier Transform (FFT) is a computationally efficient and speedier method of performing DFT. FFT can be used directly for numerical spectral analysis of sampled versions of continuous-time signals. Most of the problems of using DFT arise due to misunderstanding of what this approximation involves.

Important characteristics of DFT are :

1. Continuous-Time Fourier Transform (CTFT) of a periodic function is also periodic.
2. Both DFT and IDFT operations are done over a finite number of samples of continuous-time signal.
3. Both DFT and IDFT produces discrete samples which are equally periodic with period N.
4. DFT of convolution between two discrete-time sequence is equal to the product of their individual DFTs.

Fig. 7.6 (a) ilustrates the sinusoidal continuous-time signal $s(t) = \cos\, 2\pi f_0 t$ and its fourier transform $S(f)$. $S(f)$ consists of two impulses at $f = \pm f_0$ because $s(t)$ has a single frequency.

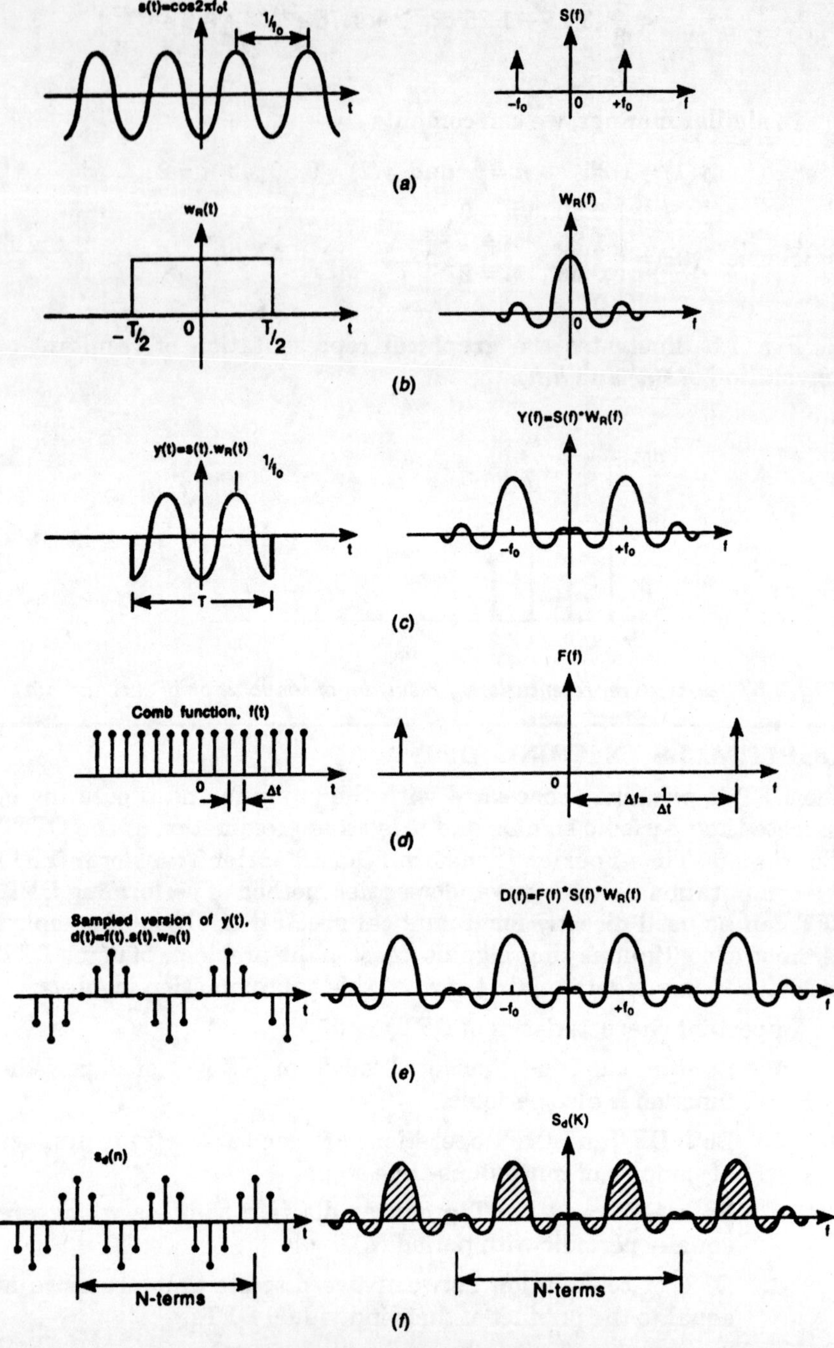

Fig. 7.6 *Discrete fourier transform (DFT) of discrete-time sequence $s(n)$, $S(K)$ viewed as a corruptive estimate of continuous-time fourier transform (CTFT) of continuous-time signal $s(t)$.*

Fig. 7.6 (*b*) shows the rectangular window function $w_R(t)$ and its Fourier transform $W_R(f)$. The spectrum of window function extends from $-\infty$ to ∞. Here the window function $w_R(t)$ is a time-limited function.

Fig. 7.6 (*c*) respresents the multiplication of $s(t)$ and $w_R(t)$, *i.e.*, $y(t)$ and its Fourier transform $Y(f) = S(f) * W_R(f)$. This product results in "truncation" of continuous-time signal $s(t)$. The product in time-domain corresponds to convolution in the frequency domain. Star (*) denotes the convolution operation.

Fig. 7.6 (*d*) illustrates a comb function $f(t)$ and its fourier transform $F(f)$. Comb function $f(t)$ is a periodic train of impulses with a time-interval of Δt between the impulses. Fourier transform of comb function, $F(f)$ is the corresponding Fourier spectra which are again a set of periodic impulses separated by $f_s = \dfrac{1}{\Delta t}$. It is the sampling frequency due to Nyquist.

Fig. 7.6 (*e*) illustrates the sampled version of product $y(t) = s(t)\, w_R(t)$ and it is denoted $d(t)$. Sampling is done by multiplying $y(t)$ with comb function $f(t)$ then $d(t) = y(t)\, f(t) = s(t)\, w_R(t)\, f(t)$ and corresponding Fourier transfrm

$$D(f) = Y(f) * F(f) = S(f) * W_R(f) * F(f)$$

This figure shows that the sampled truncated cosine wave $s(t)$ produces a continuous spectra from $-\infty$ to ∞.

Fig. 7.6 (*f*) shows the discrete-time sequence $s_d(n)$ and its discrete Fourier transform (DFT). DFT of sequence $s(n)$ is given by

$$S_d(K) = \sum_{n=0}^{N-1} s_d(n)\, e^{-j2\pi nK/N}$$

Note. The dotted pattern representss DFT of $s_d(n)$, *i.e.*, $S_d(K)$.

$S_d(K)$ are the sample values of $S(f)$ where $S(f)$ is the Fourier transform of continuous-time signal $s(t)$.

7.8.1 Problems of Pitfalls in using DFT

For convenience we have taken cosine sinusoidal signal $s(t)$ of single frequency f_0. We expect an arbitrary waveform.

In this article we are going to discuss the problems which arise in using DFT and their remedies.

1. Truncation error.
2. Aliasing error.
3. Spectral leakage.
4. Improper use of DFT.
5. Picket-Fence effect.

Now these problems will be discussed one by one in detail.

1. Truncation Error

This error is caused by using the rectangular window function $w_R(t)$. But a time-limited signal has an infinite frequency spectra and spectrum of $w_R(t)$ is spread into frequenices other than $\pm f_0$.

The resultant $y(t)$ which is product of $s(t)$ and $w_R(t)$ gives a truncated version in time-domain. Fourier transform of $y(t)$, *i.e.*, $Y(f)$ is the spectral function which spread into frequencies other than only those due to $W_R(f)$. These are not useful and it is called truncation error.

2. Aliasing Error

We have already studied aliasing error in Chapter 2. To avoid aliasing effect, sampling rate should be greater than or equal to Nyquist rate. Nyquist rate is equal to twice the highest- frequency component of the signal. However, due to truncation of signal $s(t)$, the frequencies spread into the frequency-domain for larger than the allowable range. Even after sampling a signal higher than the Nyquist rate, the aliasing error cannot be entirely eliminated.

3. Spectral Leakage

We have one more effect other than trunction and aliasing error. It is called spectral leakage. Leakage of energy takes place when the original signal $s(t)$ is truncated and sampled. It means energy in the original signal frequency f_0 will now leaks into other frequencies. This leakage is due to truncating original signal $s(t)$. Due to this leakage we get undesirable modification of the total spectrum. The spectral leakage effect cannot always be isolated from the aliasing effect because leakage leads to aliasing if the higher frequency component of the composite spectrum shifts beyond the folding frequency $f_s/2$.

4. Improper Use of DFT

This problem is shown in Fig. 7.7 (*a*). In this figure the waveforms are continuous-time in shape. For correct use of DFT separate the waveform of $s(t)$ with duration T_1 wihch is equal to impulse duration T. Here $y(t) = s(t) * h(t)$, where * denotes convolution operation between $s(t)$ and $h(t)$ and results in $y(t)$. We have no signal between T and $T + T_1$. If this gap is not provided during computation by Fast Fourier Transform (FFT), the aliasing effect will arise and its other associated problems will also occur.

5. Picket-Fence Effect

This problem is caused from the inability of the DFT to observe the spectrum as a continuous function $y(t)$ because DFT computation is limited to integer multiples of the fundamental frequency f_0, where f_0 is given by $f/N = f_s/N$.

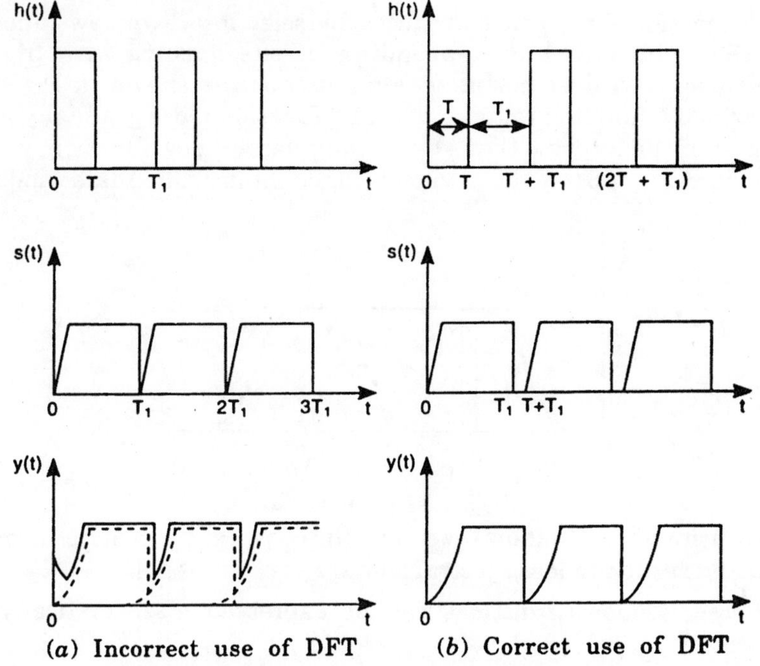

Fig. 7.7 *Correct and incorrect use of DFT.*

The observation of spectrum with the discrete fourier transform is equivalent to looking at it through a sort of "Fence". It means we observe the exact waveform of $y(t)$ only at discrete points of signals.

7.8.2 Remedies to Problems of Pitfalls in using DFT

Now we shall discuss very briefly the remedies to problems to pitfalls in using discrete Fourier transform.

1. Remedy to truncation error.
2. Remedy to aliasing error.
3. Remedy to spectral leakage.
4. Remedy to improper use of DFT.
5. Remedy to picket-fence effect.

Now we shall discuss each and every remedy one by one as follows.

1. Remedy to Trucation Error

The truncation is shown in Fig. 7.6 (*b*). Here we have used a rectangular window function $w_R(t)$. If we want to perform DFT using FFT on a digital computer, we would be required to perform on a finite piece of the signal $s(t)$ because we have limitations in computer storage memory. Thus truncation is usually necessary (if signal $s(t)$ is not naturally limited) and it results in smearing effect. This effect is shown in Fig. 7.6 (*c*).

Truncation error can be reduced by selecting a window function $w(t)$ that minimizes the spreading of the spectra into higher frequencies. A raised cosine window function is shown in Fig. 7.8. This window function reduces the time function more gradually near the ends of the interval than the rectangular window function, which is shown in Fig. 7.6 (*b*). These window functions are called data windows.

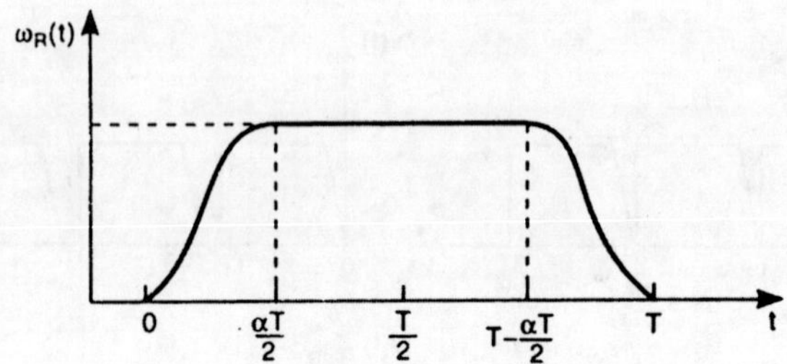

Fig. 7.8 *Data window.*

By using these window functions, the convergence of series is more rapid and results in less spectral leakage.

These windows functions can be expressed mathematically as follows :

$$w_R(t) = 0.50 + 0.50 \cos \frac{2\pi t}{\alpha T}, \qquad 0 < t \le \frac{\alpha T}{2}$$

$$= 1, \qquad \frac{\alpha T}{2} < t < T - \frac{\alpha T}{2}$$

$$= 0.50 + 0.50 \cos \frac{2\pi(T-t)}{\alpha T}, \quad T - \frac{\alpha T}{2} < t < T$$

Typical values of α lie between 0.10 and 0.20.

2. Remedy to Aliasing Error

Aliasing error can be reduced by choosing a sampling rate higher than Nyquist rate to avoid any spectral overlap. There is need of prior knowledge of the spectrum of the signal so we can choose appropriate sampling rate. It we have no prior knowledge of the signal then we must filter the signal before sampling using low pass filter. Prefiltering is done in order to ensure about the highest frequency component so as to decide the appropriate sampling rate. By doing so we can reduce aliasing error.

3. Remedy to Spectral Leakage

This leakage of energy into other frequencies can be minimised by selecting sampling frequency in such a way that the zero and peak samples of the truncated signal $s(t)$ are always present. Longer bandwidth which is some integral multiple of the bandwidth of the signal will reduce spectral leakage. This is shown in Fig. 7.6 (*e*). Spectral leakage can also be reduced by choosing the window function for which the first and the last set of samples are weighted by a raised cosine wave, which is shown in Fig. 7.8.

4. Remedy to Improper Use of DFT

Improper use of DFT can be taken care of by padding some extra zero samples to the set of non-zero samples, so that the N- point DFT produces the correct result.

5. Remedy to Picket-Fence Effect

Picket-fencing can be reduced by varying the number of points N in a given time period T and adding zeros at the end of the original record, while maintaining the original record intact. This method artificially alters the period and this new period changes the locations of spectral lines without changing the continuous form of the original spectrum. By shifting the sample points we can easily observe those spectral components which were originally hidden from view.

Example 7.8 Find out the Discrete Fourier Transform (DFT) of the sequence $s(n)$ illustrated in figure below for $N = 6$.

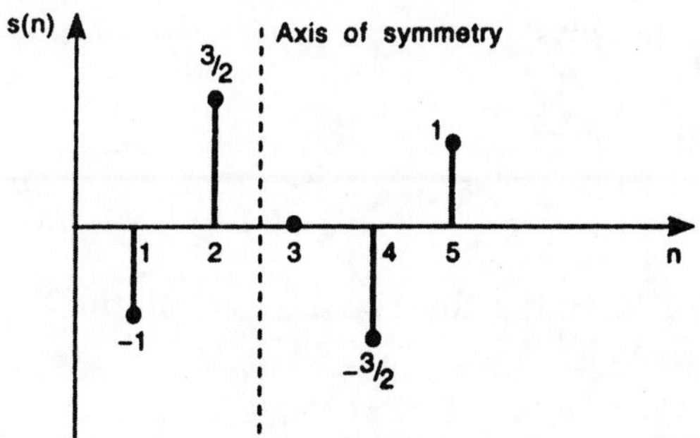

Fig. 7.9 *Graphical representation of sequence $s(n)$ with $N = 6$.*

Solution. For the above sequence $s(n)$

$$s(0) = s\left(\frac{N}{2}\right) = s(3) = 0$$

$$s(1) = -1$$

$$s(2) = -s(4) = \frac{3}{2}$$

$$s(5) = -s(1) = 1$$

Using formula of DFT given by

$$S(K) = \sum_{n=0}^{5} s(n)\, W_N^{nK}, \qquad \text{where } W_N = e^{-j2\pi/N}$$

$$= \sum_{n=0}^{5} s(n)\, e^{-j2\pi nK/N}$$

$$= \sum_{n=0}^{5} s(n)\, e^{-j2\pi nK/6} = \sum_{n=0}^{5} s(n)\, e^{-j\pi nK/3}$$

$$= s(0)\, e^{-j\pi.0K/3} + s(1)\, e^{-j\pi.1K/3} + s(2)\, e^{j\pi.2K/3} + s(3)\, e^{-j\pi.3K/3} + s(4)\, e^{-j\pi.4K/3} + s(5)\, e^{-j\pi.5K/3}$$

$$= 0 \times 1 + (-1)\, e^{-j\pi K/3} + \frac{3}{2} e^{-j\pi 2K/3} + 0\, e^{-j\pi 3K/3} + \left(-\frac{3}{2}\right) e^{-j\pi 4K/3} + (1)\, e^{-j\pi 5K/3}$$

$$= -\, e^{-j\pi K/3} + \frac{3}{2} e^{-j\pi 2K/3} - \frac{3}{2} e^{-j\pi 4K/3} + e^{-j\pi 5K/3}$$

$$S(K) = -\, e^{-j\pi K/3} + \frac{3}{2} e^{-j\pi 2K/3} - \frac{3}{2} e^{-j\pi 4K/3} + e^{-j\pi 5K/3}$$

$K = 0,\quad S(0) = 0$

$K = 1,\quad S(1) = -j\dfrac{\sqrt{3}}{2}$

$K = 2,\quad S(2) = j\dfrac{5}{2}\sqrt{3}$

$K = 3,\quad S(3) = 0$

$K = 4,\quad S(4) = -j\dfrac{5}{2}\sqrt{3}$

$K = 5,\quad S(5) = +j\dfrac{\sqrt{3}}{2}$

$$S(K) = \left\{ 0, -j\frac{\sqrt{3}}{2}, j\frac{5}{2}\sqrt{3}, 0, -j\frac{5}{2}\sqrt{3}, j\frac{\sqrt{3}}{2} \right\}$$

Example 7.9 Determine the circular convolution of two sequences $s_1(n)$ and $s_2(n)$ given by

$$s_1(n) = s_2(n) = 1, \text{ for } 0 \le n \le N-1$$
$$= 0, \text{ otherwise.}$$

Solution. Since both the sequences $s_1(n)$ and $s_2(n)$ are of unity value.

DFTs of these sequences will be

$$S_1(K) = S_2(K) = \sum_{n=0}^{N-1} 1 \,.\, e^{-j2\pi nK/N}$$

Above computation is applied for $N = 4$

$$S_1(K) = S_2(K) = \sum_{n=0}^{3} 1\, e^{-j2\pi nK/4}$$

$$= 1 + e^{-j\pi 1K/2} + e^{-j\pi K} + e^{-j\pi 3K/2}$$

$K = 0,\quad S_1(0) = S_2(0) = 4$

$K = 1,\quad S_1(1) = S_2(1) = 0$

$K = 2,\quad S_1(2) = S_2(2) = 0$

$K = 3,\quad S_1(3) = S_2(n) = 0$, etc.

That is, $S_1(K) = S_2(K) = N$, for $K = 0$

$= 0$, otherwise

Resultant of circular convolution of two sequences

$$s_1(n) = s_2(n) = 1, \ 0 \le n \le N-1$$
$$= 0, \text{ otherwise}$$

is given by

$$S_3(K) = S_1(K) \,.\, S_2(K) = N \,.\, N = N^2, \text{ for } K = 0$$

Inverse DFT

$$s_3(n) = \frac{1}{N} \sum_{K=0}^{N-1} S_3(K)\, e^{j2\pi nK/N}$$
$$= \frac{1}{N} \,.\, N^2 = N \ \text{ for } \ 0 \le n \le N-1$$

This is the result of circular convolution.

Example 7.10 Find the symmetry in each case for the periodic sequence $s(n)$ as shown in Fig. 7.10.

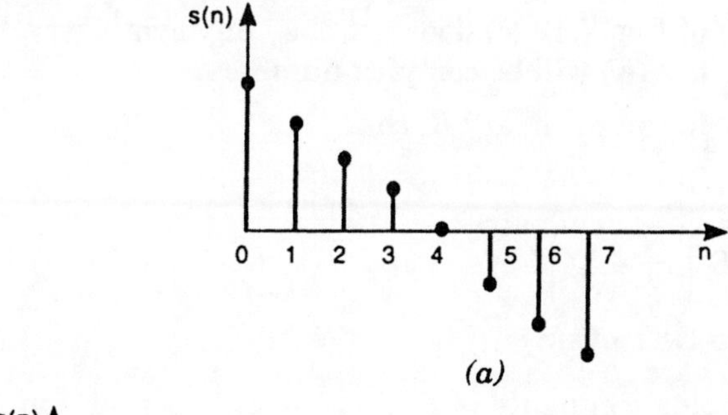

(a)

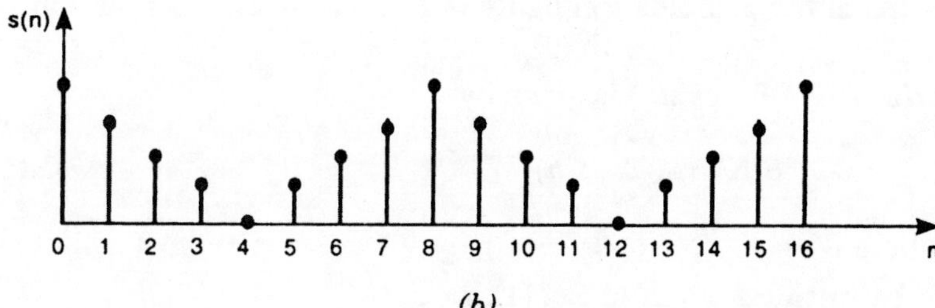

(b)

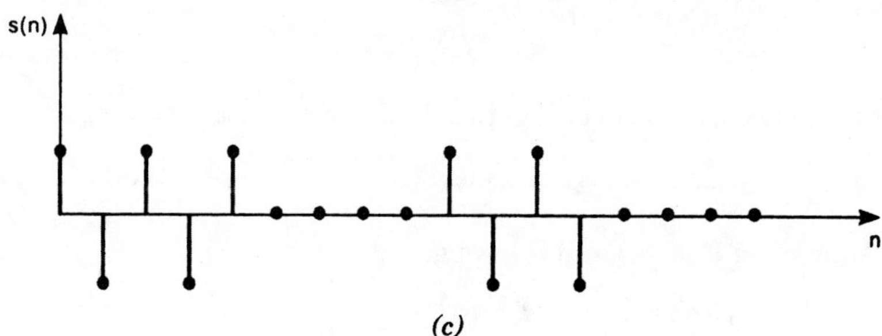

(c)

Fig. 4.10

Solution. (*a*) For the sequence $s(n)$ in Fig. 7.10 (*a*), we find $N = 8$ and that,

$s(1) = -s(7)$

$s(2) = -s(6)$

$s(3) = -s(5)$,

except $s(0)$ and $s(4)$

Therefore, it obeys symmetry

$$s(n) = -s(N-m)$$

Hence the Fourier coefficients $S(K)$ will be imaginary.

(*b*) For the Fig. 7.10 (*b*), we find that

$$s(1) = s(7)$$

$$s(2) = s(6)$$

and $\quad s(3) = s(5)$,

thus it obeys the symmetry $s(n) = s(N-n)$. Hence the fourier coefficients $S(K)$ are real.

(*c*) Sequence $s(n)$ of Fig. 7.10 (*c*) does not obey any symmetry, thus the Fourier coefficients $S(K)$ will be complex numbers.

Example 7.11 Show for an arbitrary K, that

(*a*) $S(-K) = S^*(K)$

(*b*) $S\left(\frac{N}{2} - K\right) = S^*\left(\frac{N}{2} + K\right)$

where $S(K)$ is the DFT of $s(n)$

$S^*(K)$ is the complex conjugate of $S(K)$. Here $s(n)$ is real and N is even.

Solution. (*a*) DFT of $s(n)$ is given by

$$S(K) = \sum_{n=0}^{N-1} s(n)\ W_N^{-nK} \qquad ...(1)$$

where $W_N = e^{j2\pi/N}$

Substituting $K = -K$ in Eq. (1), we get

$$S(-K) = \sum_{n=0}^{N-1} s(n)\ W_N^{nK} \qquad ...(2)$$

Taking the complex conjugate of both sides of Eq. (2), we get

$$S^*(-K) = \sum_{n=0}^{N-1} s(n)\ W_N^{-nK} \qquad ...(3)$$

Comparing Eqs. (1) and (3), we get

$$S(K) = S^*(-K) \text{ Proved}$$

(*b*) Since $s(n)$ is real, then

$$S(K) = S^*(-K)$$

It holds symmetry in its Fourier coefficients (DFTs)

$$S\left(\frac{N}{2} - K\right) = \sum_{n=0}^{N-1} s\left(\frac{N}{2} - n\right) W_N^{(N/2-n)K} \quad ...(4)$$

$$S\left(\frac{N}{2} + K\right) = \sum_{n=0}^{N-1} s\left(\frac{N}{2} - n\right) W_N^{(N/2-n)(-K)}$$

or $$S\left(\frac{N}{2} + K\right) = \sum_{n=0}^{N-1} s(N/2 - n)\ W_N^{(N/2-n)K} \quad ...(5)$$

Taking the complex conjugate of both sides of Eq. (5), we get

$$S^*\left(\frac{N}{2} + K\right) = \sum_{n=0}^{N-1} s\left(\frac{N}{2} - n\right) W_N^{-(N/2-n)K} \quad ...(6)$$

because sequence $s(n)$ is a real sequence.

$$s^*(n) = s(n)$$

Comparing Eqs. (4) and (6), we get

$$S\left(\frac{N}{2} - K\right) = S^*\left(\frac{N}{2} + K\right) \text{ Proved}$$

Example 7.12 If sequence $s_1(n)$ is a periodic sequence with period N and it is also periodic with $2N$. Assume that $S_1(K)$ is the DFT of $s_1(n)$ with period N and $S_2(K)$ is the DFT of $s_1(n)$ with period $2N$. Determine $S_2(K)$ in terms of $S_1(K)$.

Solution. DFT of $s_1(n)$ with period N is given by

$$S_1(K) = \sum_{n=0}^{N-1} s_1(n)\ e^{-(j2\pi/N)nK} \quad ...(1)$$

Substituting $W_N = e^{-j2\pi/N}$ in above Eq. (1), we get

$$S_1(K) = \sum_{n=0}^{N-1} s(n)\ W_N^{nK}. \quad ..(2)$$

Similarly, DFT of $s_1(n)$ with period $2N$ is given by

$$S_2(K) = \sum_{n=0}^{2N-1} s_1(n)\, e^{-(j2\pi/2N)nK} \quad ...(3)$$

Substituting $W_{2N} = e^{-j2\pi/2N}$ in Eq. (3), we get

$$S_2(K) = \sum_{n=0}^{2N-1} s_1(n)\ W_{2N}^{nK} \quad ...(4)$$

Now we want to express $S_2(K)$ in terms of $S_1(K)$, where $s(n)$ is a periodic sequence and it satisfies the condition

$$s(n) = s(n + N) = s(n + 2N) \quad ...(5)$$

$$W_{2N}^{nK} = W_N^{nK/2} = e^{-j(2\pi/2N)nK} = e^{-j(2\pi/N)nK/2} \quad ...(6)$$

Substituting Eq. (6) in Eq. (4), we get

$$S_2(K) = \sum_{n=0}^{2N-1} s_1(n)\, W_{2N}^{nK}$$

$$= \sum_{n=0}^{2N-1} s_1(n)\, W_N^{nK/2} \qquad ...(7)$$

$$= \sum_{n=0}^{N-1} s(n)\, W_N^{nK/2} + \sum_{n=N}^{2N-1} s(n+N)\, W_N^{(n+N)K/2}$$

$$= S_1\left(\frac{K}{2}\right) + \left[\sum_{n=N}^{2N-1} s(n+N)\right] W_N^{(n+N)K/2}$$

$$\text{where } S_1\left(\frac{K}{2}\right) = \sum_{n=0}^{N-1} s(n)\, W_N^{nK/2}$$

$$= S_1\left(\frac{K}{2}\right) + \left[\sum_{n=0}^{N-1} s(n)\, W_N^{(n)K/2}\right] W_N^{NK/2}$$

$$= S_1\left(\frac{K}{2}\right) + S_1\left(\frac{K}{2}\right) W_N^{NK/2}$$

or
$$S_2(K) = \left(1 + W_N^{NK/2}\right) S_1\left(\frac{K}{2}\right) \qquad ...(8)$$

This is the desired expression.

Note. If K is even then $W_N^{NK/2} = 1$

and if K is odd then $W_N^{NK/2} = (-1)^K$, therefore, for K is even,

$$S_2(K) = 2S_1\left(\frac{K}{2}\right)$$

for K is odd,

$$S_2(K) = [1 + (-1)^K]\, S_1\left(\frac{K}{2}\right) = [1-1]\, S_1\left(\frac{K}{2}\right)$$

$$= [0]\, S_1\left(\frac{K}{2}\right) = 0$$

Example 7.13 Consider two periodic sequences, $s(n)$ is periodic with period N and $y(n)$ is periodic with period M. The sequence $z(n)$ is defined as $z(n) = s(n) + y(n)$,

(*a*) Show that $z(n)$ is periodic with period MN.

(*b*) DFTs of $s(n)$ and $y(n)$ will be periodic with period N and M respectively.

Find $Z(L)$ in the terms $S(L)$ and $Y(L)$. Where $Z(L)$, $S(L)$ and $Y(L)$ are the DFTs of $z(n)$, $s(n)$ and $y(n)$ respectively.

Solution. Given :

$$z((n))_K = s((n))_N + y((n))_M \qquad ...(1)$$

Here double brackets are used as it illustrates a sequence with its time-period.

For linearity both sequences $s(n)$ and $y(n)$ should have same period.

$$z((n))_K = s((n))_K + y((n))_K \qquad ...(2)$$

In Eqn. (2) all the sequences have same period, *i.e.*, K.

(*a*) Now we want to determine the relation between K, M and N.

$$\text{DFT of } z(n) = Z(L) = \sum_{n=0}^{K-1} z(n)\, e^{-j(2\pi/K)nL} \qquad ...(3)$$

where L is the discrete frequency index.

We know that $Z(L)$ is periodic with period K. Hence

$$Z(0) = Z(L)$$

$$Z(1) = Z(L+1)$$

$$Z(2) = Z(L+2) \text{ etc.}$$

Substituting Twiddle factor $W_N = e^{-j2\pi/N}$ in Eq. (3), we get

$$Z((L))_K = \sum_{n=0}^{K-1} z(n)\, W_K^{nL} = \sum_{n=0}^{K=1} [s(n) + y(n)]\, W_K^{nL}$$

$$= \sum_{n=0}^{K-1} s((n))\, W_N^{nL} + \sum_{n=0}^{K-1} y((n))\, W_N^{nL}$$

or $$Z((L))_K = \sum_{n=0}^{K-1} s((n))\, e^{-(j2\pi/N)\,nL} + \sum_{n=0}^{K-1} y((n))\, e^{-(j2\pi/M)\,nL}$$

or $$= \sum_{n=0}^{K-1} s((n))\, e^{-(j2\pi/MN)\,MnL} + \sum_{n=0}^{K-1} y((n))\, e^{-(j2\pi/MN)\,NnL}$$

(We know that $M \times 2\pi = N \times 2\pi = K \times 2\pi = 2\pi$. Hence $K = MN$)

$$Z((L))_K = \sum_{n=0}^{K-1} s((n))\, e^{-(j2\pi/K)\,nL} + \sum_{n=0}^{K-1} y((n))\, e^{-(j2\pi/K)\,nL}$$

or $$Z((L))_K = \sum_{n=0}^{K-1} s(n)\, W_K^{nL} + \sum_{n=0}^{K-1} y(n)\, W_K^{nL} \qquad ...(4)$$

The relation of Eq. (4) show that if $K = MN$, then only the sum of eqn. $z((n))_K = s((n))_N + y((n))_M$ is possible and relevent, and that LCM of M and N is MN, where $M \neq N$.

(*b*) Now we are required to determine the relation between $Z((L))_K$, $S((L))_K$ and $Y((L))_K$, where $K = MN$.

$$Z((L))_K = \sum_{n=0}^{K-1} z((n))\, W_K^{nL}$$

$$= \sum_{n=0}^{MN-1} s((n))\, W_{MN}^{nL} + \sum_{n=0}^{MN-1} y((n))\, W_{MN}^{nL}$$

or $$Z((L))_K = S_2((L))_K + Y_2((L))_K \qquad ...(5)$$

Now $S_2((L))_K$ can be expressed in term of $S((L))_N$.

Where,

$$S((L)) = \sum_{n=0}^{N-1} s((n))\, W_N^{nL} \qquad \text{...(6)}$$

Since $s((n))$ is also periodic with period MN

$$s((n)) = s((n+N)) = s((n+2N)) = \ldots = S((n+(M-1)\,N)) \qquad \text{...(7)}$$

Double brackets is used to show the periodicity of a particular sequence.

From above two equations (6) and (7), we can expand $S((L))$ as

$$\sum_{n=0}^{N-1} s((n))\, W_{NM}^{nL} + \sum_{n=0}^{N-1} s((n))\, W_{NM}^{(n+N)L} + \sum_{n=0}^{N-1} s((n))\, W_{NM}^{(n+2N)L} + \ldots + \sum_{n=0}^{N-1} s((n))\, W_{NM}^{[n+(M-1)N]L} \qquad \text{...(8)}$$

Rewriting Eqn. (8) in convenient from we get

$$\sum_{n=0}^{N-1} s((n))\, W_N^{nL/M} + W_N^{NL/M} \sum_{n=0}^{N-1} s((n))\, W_N^{nL/M} + \ldots + W_N^{(M-1)\,NL/M} \sum_{n=0}^{N-1} s((n))\, W_N^{nL/M}$$

Now we can write an expression tor $S_2((L))$ as

$$S_2((L)) = 1 + W_N^{NL/M} + W_N^{N(L+1)/M} + \ldots + W_N^{N\,(M-1)\,L/M} \sum_{n=0}^{N-1} s((n))\, W_N^{nL/M}$$

$$= F.S\left(\left(\frac{L}{M}\right)\right) \qquad \text{...(9)}$$

Where,

$$F = 1 + W_N^{NL/M} + W_N^{N\,(L+1)/M} + W_N^{N\,(L+2)/M} + \ldots + W_N^{N\,(M-1)\,L/M}$$

$$= \sum_{P=0}^{M-1} W_N^{PN} = \sum_{P=0}^{M-1} 1 = M \qquad \text{...(10)}$$

where P is any integer

$\left[\text{Since } W_N^{PN} = e^{-j2\pi/N\,.\,PN} = 1\right]$

$$F = M \qquad \text{...(11)}$$

Substituting $F = M$ in Eq. (11), we get

$$S_2((L)) = FS\left(\left(\frac{L}{M}\right)\right) = MS\left(\left(\frac{L}{M}\right)\right) \qquad \text{...(12)}$$

Let $Y((L)) = \sum_{n=0}^{M-1} y((n))\, W_M^{nL}$

Since $y((n))$ is also periodic in NM

$$y((n)) = y((n+M)) = y((n+2M)) + \ldots + y[n+(N-1)M] \quad \ldots(13)$$

$$Y_2((L)) = \sum_{n=0}^{M-1} y((n))\, W_{NM}^{nL} + \sum_{n=0}^{M-1} y((n))\, W_{NM}^{(n+M)L}$$

$$+ \sum_{n=0}^{M-1} y((n))\, W_{NM}^{(n+M+1)L} + \ldots + \sum_{n=0}^{M-1} y((n))\, W_{NM}^{[n+(N-1)M]L}$$

$$= \sum_{n=0}^{M-1} y((n))\, W_{NM}^{nL} + W_{NM}^{ML} \sum_{n=0}^{M-1} y((n))\, W_{NM}^{nL} + \ldots$$

$$+ W_{NM}^{(N-1)ML} \sum_{n=0}^{M-1} y((n))\, W_{NM}^{nL}$$

$$= \left[1 + W_{NM}^{ML} + W_{NM}^{(M+1)L} + \ldots + W_{NM}^{(N-1)ML}\right] \sum_{n=0}^{M-1} y((n))\, W_{NM}^{nL}$$

$$\left(\text{Putting } W_{NM}^{NL} = W_M^{ML/N} \text{ and } W_{NM}^{nL} = W_M^{nL/N}\right)$$

$$= \left[1 + W_M^{ML/N} + W_M^{(M+1/N)L} + \ldots + W_M^{(N-1/N)ML}\right] \sum_{n=0}^{M-1} y((n))\, W_M^{nL/N}$$

$$= N \sum_{n=0}^{M-1} y((n))\, W_N^{nL/N} = NY\left(\frac{L}{N}\right) \quad \ldots(14)$$

where $\left[1 + W_M^{ML/N} + W_M^{(M+1/N)L} + \ldots + W_M^{(N-1/N)ML} = N\right]$ already determined.

$$V((L))_K = S_2((L))_K + Y_2((L))_K \quad \ldots(15)$$

$$\boxed{S_2((L)) = MS\left(\left(\frac{L}{M}\right)\right) \quad \text{and} \quad Y_2((L)) = NY\left(\left(\frac{L}{M}\right)\right)}$$

Substituting the values of $S_2((L))$ and $Y_2((L))$ in Eq. (15), we get

$$V((L))_K = S_2((L))_K + Y_2((L))_K = MS\left(\left(\frac{L}{M}\right)\right) + NY\left(\left(\frac{L}{N}\right)\right)$$

Hence proved.

REVIEW QUESTIONS

1. Differentiate between Discrete-Time Fourier Transform (DTFT) and Discrete Fourier Transform (DFT) of a sequence.
2. How DFT can be used as a linear transformation tool in signal analysis and Digital Signal Processing ?
3. Discuss various properties of the DFT in brief.
4. Differentiate between circular convolution and linear convolution of two discrete-time sequences.
5. Discuss circular convolution of two discrete-time sequences in detail.
6. What is Parseval's theorem for discrete-time sequences ?
7. What is the relationship between DFT and z-transform ?
8. How linear convolution can be performed uing the DFT ?
9. In what case resultant of linear convolution and circular convolution will be the same ? Discuss..
10. Discuss various problems of pitfalls in using DFT.
11. Write short notes on the following :
    (*i*) Truncation error. (*ii*) Aliasing error.
    (*iii*) Spectral leakage. (*iv*) Improper use of DFT.
    (*v*) Picket-fence effect.
12. Discuss remedies of various problems of pitfalls in using DFT.

NUMERICAL PROBLEMS

1. Evaluate the N-point DFT of the sequences given as follows :

   (*i*) $s(n) = \begin{cases} \frac{1}{3}, & 0 \le n \le 2 \\ 0, & \text{elsewhere} \end{cases}$ (*ii*) $s(n) = (A)^n u(n)$

2. Determine the DFT of the following sequences

   (*i*) $s(n) = \begin{cases} 2, & 2 \le n \le 6 \\ 0, & \text{for } n = 0, 1, 7, 8, 9 \end{cases}$ Given $N = 10$

   (*ii*) $s(n) = \begin{cases} 1, & n = 3, 7 \\ 0, & \text{for } n = 1, 2, 4, 5, 6, 8, 9 \end{cases}$ Given $N = 10$

   (*iii*) $s(n) = \begin{cases} 1, & 0 \le n \le 5 \\ 2, & 6 \le n \le 9 \end{cases}$ Given $N = 10$

3. Determine the circular convolution of the following sequences using the time-domain formula.

   (*i*) $s_1(n) = \{1, 2, 3, 1\}$ and $s_2(n) = \{4, 3, 2, 2\}$
   (*ii*) $s_1(n) = \{1, 1, 1, 1\}$ and $s_2(n) = \{2, 2, 2, 2\}$
   (*iii*) $s_1(n) = \{1, 2, 3, 1\}$ and $s_2(n) = \{1, 1, 1, 1\}$

4. **Determine linear convolution using DFT method**

$$s_1(n) = \begin{cases} 1, & n = 0 \\ \frac{1}{2}, & n = 1 \\ 0, & \text{otherwise} \end{cases} \quad \text{and} \quad s_2(n) = \begin{cases} \frac{1}{2}, & n = 0 \\ 1, & n = 1 \\ 0, & \text{otherwise} \end{cases}$$

5. **Find the circular convolution of two sequences given as follows :**

$$S_1(n) = \begin{cases} 1, & n = 0 \\ \frac{1}{2}, & n = 1 \\ 0, & \text{otherwise} \end{cases} \quad \text{and} \quad S_2(n) = \begin{cases} \frac{1}{2}, & n = 0 \\ 1, & n = 1 \\ 0, & \text{otherwise} \end{cases}$$

6. **Given the two 5-point sequences as**

$$s_1(n) = \{0, 1, 2, 3, 4\}$$
$$\uparrow$$

$$s_2(n) = \{0, 1, 0, 0, 0\}$$
$$\uparrow$$

Determine $s(n) = s_1(n) \, \textcircled{N} \, s_2(n)$ **by DFT method.**

8

Fast Fourier Transform (FFT) Algorithms

8.1 INTRODUCTION

The Fast Fourier Transform (FFT) does not represent a transform different from the DFT but they are special algorithms for speedier implementation of DFT. FFT requires a comparatively smaller number of arithmetic operations such as multiplications and additions than DFT. FFT also requires lesser computational time than DFT. The fundamental principle on which all these algorithms are based upon is that of decomposing the computation of the DFT of a sequence of length N into successively smaller DFTs. The way in which this principle is implemented leads to a variety of different algorithms, all with comparable improvements in computational speed. Thus, we can say that DFT plays an important role in several applications of digital signal processing such as linear filtering, correlation analysis and spectrum analysis.

Direct Computation of the DFT. Discrete Fourier Transform (DFT) for a complex-valued sequence, $s(n)$ may be expressed as

$$S(K) = \sum_{n=0}^{N-1} s(n)\, e^{-j2\pi nK/N} \qquad ...(8.1)$$

Here both $S(K)$ and $s(n)$ are complex-valued discrete- frequency and discrete-time sequences of N points, respectively.

$$S(K) = S_R(K) + j\, S_I(K)$$

$$s(n) = s_R(n) + j\, s_I(n)$$

$$e^{-j2\pi nK/N} = \cos\frac{2\pi nK}{N} - j\sin\frac{2\pi nK}{2}$$

Substituting the values of $S(K)$, $s(n)$ and $e^{-j2\pi nK/N}$ in Eq. (8.1), we get

$$S(K) = \sum_{n=0}^{N-1} s(n)\, e^{-j2\pi nK/N}, \quad K = 0, 1, \ldots, N-1 \qquad \ldots(8.2)$$

$$S_R(K) + j\, S_I(K) = \sum_{n=0}^{N-1} [s_R(n) + j\, s_I(n)] \left[\cos \frac{2\pi nK}{N} - j \sin \frac{2\pi nK}{N} \right]$$

$$= \sum_{n=0}^{N-1} \left[s_R(n) \cos \frac{2\pi nK}{N} + s_I(n) \sin \frac{2\pi nK}{N} \right]$$

$$-j \sum_{n=0}^{N-1} \left[s_R(n) \sin \frac{2\pi nK}{N} - s_I(n) \cos \frac{2\pi nK}{N} \right] \qquad \ldots(8.3)$$

Separating real and imaginary parts of above Eq. (8.3), we get

$$S_R(K) = \sum_{n=0}^{N-1} \left[s_R(n) \cos \frac{2\pi nK}{N} + s_I(n) \sin \frac{2\pi nK}{N} \right] \qquad \ldots(8.4)$$

and $$S_I(K) = -\sum_{n=0}^{N-1} \left[s_R(n) \sin \frac{2\pi nK}{N} - s_I(n) \cos \frac{2\pi nK}{N} \right] \qquad \ldots(8.5)$$

where $s_R(n)$ and $s_I(n)$ are the real and imaginary parts of a discrete-time sequence $s(n)$, respectively.

$S_R(K)$ and $S_I(I)$ are the real and imaginary parts of the DFT of $s(n)$, respectively.

From Eqs. 8.4 and 8.5 it is clear that for each value of K, the direct computation of DFT of $s(n)$, *i.e.*, $S(K)$ requires $4N$ real multiplications and $4(N-1)$ real additions. Since $S(K)$ is computed for N different values of K, the direct computation of the DFT of a sequence $s(n)$ requires $4N \times N = 4N^2$ real multiplications and $4(N-1) \times N = 4N(N-1)$ real additions. In other words, we can say that direct DFT computation requires N^2 complex multiplications and $N(N-1)$ complex additions.

The direct computation of Eqs. 8.4 and 8.5 requires :

(*i*) N^2 Complex multiplications or $4N^2$ Real multiplications

(*ii*) $N(N-1)$ Complex additions or $4N(N-1)$ Real additions

(*iii*) $2N^2$ Evaluations of trigeometric functions such as $\sin \frac{2\pi nK}{4}$ and $\cos \frac{2\pi nK}{N}$.

8.2 GEORTZEL ALGORITHM

In Geortzel algorithm, the periodicity of the phase factors W_N^K is exploited. This algorithm allows us to express the computation of the DFT as linear filtering.

DFT of sequence $s(n)$ is given by

$$S(K) = \sum_{m=0}^{N-1} s(m)\, W_N^{Km} \qquad \ldots(8.6)$$

where $W_N = e^{-j2\pi/N}$ is called phase factor.

Since $W_N^{-KN} = (e^{j(2\pi/N)})^{-KN} = e^{-j2\pi K} = 1$, for $K = 0, 1, 2 \ldots$

Now multiplying above Eqn. (8.6) by W_N^{-KN}, we get

$$S(K) = W_N^{-KN} \sum_{m=0}^{N-1} s(n)\, W_N^{Km} = \sum_{m=0}^{N-1} s(m)\, W_N^{K(m-N)}$$

or
$$S(K) = \sum_{m=0}^{N-1} s(m)\, W_N^{-K(N-m)} \qquad \ldots(8.7)$$

From the inspection of Eq. 8.7 we note that it is in the form of convolution. Now, if we define the discrete-time sequence $y_K(n)$ as

$$y_K(n) = \sum_{m=0}^{N-1} s(m)\, W_N^{-K(n-m)} \qquad \ldots(8.8)$$

then, $y_K(n)$ is the convolution of a sequence $s(n)$ of length N with a filter that has an unit-sample response

$$h_K(n) = W_N^{-Kn} \,.\, u(n) \qquad \ldots(8.9)$$

Comparing Eqs. 8.7 and 8.8, we get

$$S(K) = y_k(n)\Big|_{n=N} \qquad \ldots(8.10)$$

System function $H(z)$ of a filter whose unit-sample response, $h_K(n) = W_N^{-Kn}\, u(n)$.

Taking the z-transform of Eq. (8.9), we get

$$H_K(z) = \mathcal{Z}\left[\, h_K(n) \,\right] = \mathcal{Z}\left[\, W_N^{-Kn} \,.\, u(n) \,\right]$$

$$= \frac{1}{1 - W_N^{-K} z^{-1}} = \frac{z}{z - W_N^{-K}} \qquad \ldots(8.11)$$

or
$$H_K(z) = \frac{z}{z - W_N^{-K}} = \frac{N(z)}{D(z)}$$

where $N(z)$ and $D(z)$ are numerator and denominator polynomial of rational z-transform.

The zeros of a z-transform $H_K(z)$ are the values of z for which $H_K(z) = 0$ and poles of a z-transform $H_K(z)$ are the values of z for which $H_K(z) = \infty$.

In other words, the roots of numerator polynomial $N(z)$ is called zeros and roots of denominator polynomial $D(z)$ is called poles of rational z-transform.

In above z-transform, there is one zero at $z = 0$ and one pole at $z = W_N^{-K}$.

These poles and zeros are illustrated graphically by pole-zero plot as shown in Fig. 8.1.

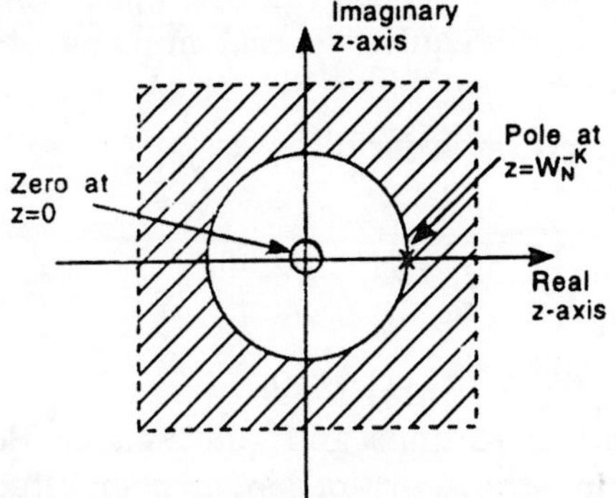

Fig. 8.1 *Pole-zero plot for above z-transform $H_K(z)$.*

This filter has a pole at $z = W_N^{-K}$

$$e^{j\omega_K} = e^{+j2\pi K/N}$$

or
$$\omega_K = \frac{2\pi K}{N}$$

It means it has a pole on the unit circle at the frequency $\omega_K = \dfrac{2\pi K}{N}$.

Thus, the entire DFT can be computed by passing the block of input data into a parallel bank of N single pole filters.

$$H_K(z) = \frac{1}{1 - W_N^{-K} z^{-1}}$$

or
$$\frac{Y_K(z)}{S_K(z)} = \frac{1}{1 - W_N^{-K} z^{-1}}$$

or
$$Y_K(z) = [1 - W_N^{-K} z^{-1}] = S_K(z)$$

or
$$Y_K(z) - W_N^{-K} z^{-1} Y_K(z) = S_K(z)$$

or
$$Y_K(z) = W_N^{-K} z^{-1} Y_K(z) + S_K(z) \qquad ...(8.12)$$

Taking the inverse z-transform of above Eq. (8.12), we get

$$y_K(n) = W_N^{-K} y_K(n-1) + s_K(n) \qquad ...(8.13)$$

When initial conditions are zero, $y(-1) = 0$

The desired output is $S(K) = y_K(n)\Big|_{n=N} = y_K(N)$ ...(8.14)

for $K = 0, 1, \ldots, N-1$

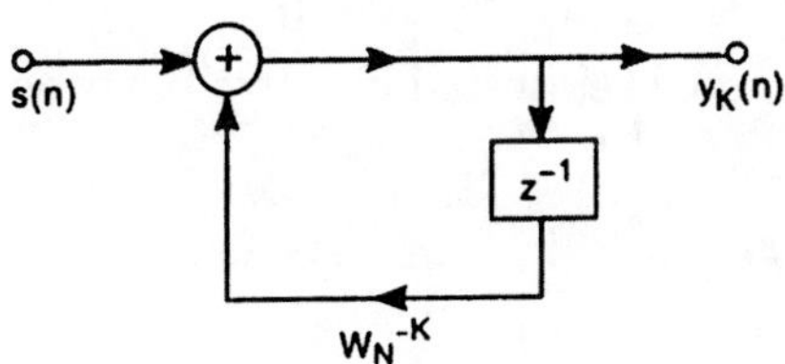

Fig. 8.2 *Flow graph of Ist order complex recursive computation of $S(K)$.*

Since both the input $s(n)$ and W_N^{-K} are complex, the computation of each new value of $y_K(n)$ requires 4 real multiplications and 4 real additions.

System function of first order filter* is

$$H_K(z) = \frac{1}{1 - W_N^{-K} z^{-1}}$$

and its difference equation is

$$y_K(n) = W_N^{-K} y_K(n-1) + s_K(n)$$

The complex multiplications and additions in Eq. (8.13) can be avoided by combining the pairs of single pole filters as it possess complex-conjugate poles. Complex conjugate poles are those poles whose imaginary parts have opposite sign. For example : One pole is given by $p_1 = a + jb$ then its complex conjugate pole will be given by $p_2 = p_1^* = a - jb$.

We can get second order filter from first order filter by multiplying both the numerator and denominator of a system function of first order filter by a factor $(1 - W_N^K z^{-1})$.

$$H_K(z) = \frac{1 - W_N^K z^{-1}}{(1 - W^{-K} z^{-1})(1 - W^K z^{-1})}$$

$$= \frac{1 - W_N^K z^{-1}}{1 - 2\cos\left(\frac{2\pi K}{N}\right) z^{-1} + z^{-2}} \qquad \text{...(8.15)}$$

This is the system function of a second order filter. Above system function can be written as follows :

$$H_K(z) = \frac{Y_K(z)}{S(z)} = \frac{1 - W_N^K z^{-1}}{1 - 2\cos\left(\frac{2\pi K}{N}\right) z^{-1} + z^{-2}} \qquad \text{...(8.16)}$$

Eqn. (8.16) can also be written as

$$\frac{Y_K(z)}{S(z)} = \frac{1 - W_N^K z^{-1}}{1 - 2\cos\left(\frac{2\pi K}{N}\right) z^{-1} + z^{-2}}$$

or
$$\frac{Y_K(z)}{V_K(z)} \cdot \frac{V_K(z)}{S(z)} = \frac{1 - W_N^K z^{-1}}{1 - 2\cos\left(\frac{2\pi K}{N}\right) z^{-1} + z^{-2}}$$

The above Eqn. can be separated into two parts as

$$\frac{Y_K(z)}{V_K(z)} = 1 - W_N^K z^{-1} \qquad \text{...(8.17)}$$

and $$\frac{V_K(z)}{S(z)} = \frac{1}{1 - 2\cos\left(\frac{2\pi K}{N}\right)z^{-1} + z^{-2}}$$

...(8.18)

Taking the inverse z-transform of Eqns. (8.17) and (8.18), we get

$$y_K(n) = v_K(n) - W_N^K\, v_K(n-1) \qquad \text{...(8.19)}$$

where all initial conditions are zero.

$$v_K(n) = 2\cos\frac{2\pi K}{N}\, v_K(n-1) - v_K(n-2) + s(n) \qquad \text{...(8.20)}$$

where initial conditions are zero.

The Direct Form-II Realization of the system function given by Eq.

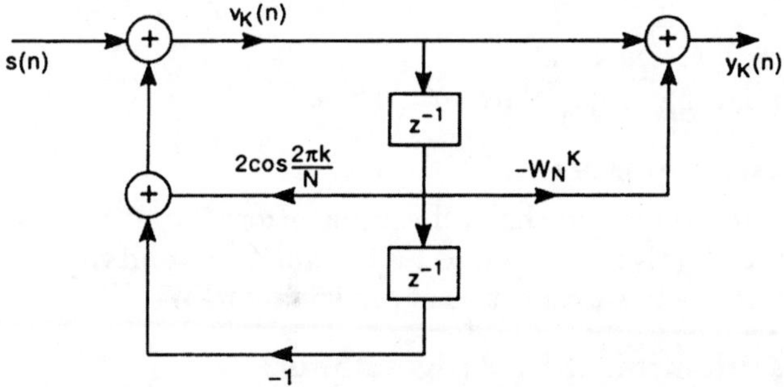

Fig. 8.3 *Flow graph of IInd order recursive computation of S(K). This is an illustration of Goertzel algorithm.*

(8.20) is shown in Fig. 8.3.

Only two multiplications are required to implement the poles of the system given in Eq. (8.20). Since the coefficients are real and (–1) is not counted as a multiplication. Four additions are required for implementing the poles. Complex multiplication by ($-W_N^K$) is required to implement the zero. This operation is not performed at every iteration of the difference equation.

Geortzel algorithm is specially attractive for DFT computation of smaller number of points. Number of complex multiplications required for N-point DFT computation using Goertzel algorithm is N^2 and number of additions required is $2N^2$

From Eqn. (8.16), we get

$$\left[1 - 2\cos\left(\frac{2\pi K}{N}\right)z^{-1} + z^{-2}\right] Y_K(z) = [1 - W_N^K z^{-1}]\, S(z)$$

...(8.21)

Taking inverse z-transform of above Eq. (8.21), we get

$$y_K(n) - 2\cos\left(\frac{2\pi K}{N}\right) y_K(n-1) + y_K(n-2) = s(n) - W_N^K s(n-1) \quad ...(8.22)$$

with initial conditions are zero.

$$y_K(n) = 2\cos\left(\frac{2\pi K}{N}\right) y_K(n-1) - y_K(n-2) + s(n) - W_N^K s(n-1) \quad ...(8.23)$$

The signal flow graph of second order filter is illustrated in Fig. 8.3.

8.3 FAST FOURIER TRANSFORM ALGORITHMS

Direct computation of the DFT is less efficient because it does not exploit the properties of symmetry and periodicity of the phase factor $W_N = e^{-j2\pi/N}$.

These properties are :

Symmetry property : $W_N^{K+N/2} = -W_N^K$

Periodicity property : $W_N^{K+N} = W_N^K$

As we already know that all computationally efficient algorithms for DFT are collectively known as FFT Algorithms and these algorithms exploit the above two properties of phase factor, W_N.

8.3.1 Classification of FFT Algorithms

According to the storage of the components of the intermediate vector, FFT algorithms are classified into two groups.

1. In-Place FFT algorithms
2. Natural Input-Output FFT algorithms.

In-Place FFT Algorithms. In this FFT algorithm, component of an intermediate vector can be stored at the same place as the corresponding component of the previous vector.

In-place FFT algorithms reduce the memory space requirement.

Natural Input-Output FFT Algorithms. In this FFT algorithm, both input and output are in natural order. It means both discrete- time sequence $s(n)$ and is DFT $S(K)$ are in natural order. This type of algorithm consumes more memory space for preservation of natural order of $s(n)$ and $S(K)$.

The disadvantage of an In-place FFT algorithm is that the output appears in an unnatural order necessitating proper shuffling of $s(n)$ or $S(K)$.

In-place FFT algorithms are superior to the Natural Input-output FFT algorithms although it needs shuffling of $s(n)$ or $S(K)$. This

shuffling operation is known as *Scrambling.*

The scrambled value of an integer I is defined as a new number generated by reversing the order of all bits in the equivalent binary number for that integer.

Another classification of FFT algorithms based on Decimation of $s(n)$ or $S(K)$. Decimation means decomposition into decimal parts.

On the basis of decimation process, FFT algorithms are of two types:

1. Decimation-in-Time FFT algorithms.
2. Decimation-in-Frequency FFT algorithms.

Decimation-in-Time (DIT) FFT Algorithms. In DIT FFT algorithms, the sequence $s(n)$ will be broken up into odd numbered and even numbered subsequences.

This algorithm was first proposed by Cooley and Tukey in 1965.

Decimation-in-Frequency (DIF) FFT Algorithms. In DIF FFT algorithms, the sequence $s(n)$ will be broken up into two equal halves. This algorithm was first proposed by Gentlemen and Sande in 1966.

Computation reduction factor of FFT algorithms

$$= \frac{\text{Number of computations required for direct DFT}}{\text{Number of computations required for FFT algorithm}}$$

$$= \frac{N^2}{\frac{N}{2}\log_2(N)}$$

8.3.2 Number of Stages in DFT Computation using FFT Algorithms

Number of stages in DFT computation using FFT algorithms depends upon the total number of points (N) in a given sequence.

For these algorithms, number of points in a discrete-time sequence,

$$N = 2^r \text{ where } r > 0.$$

r is the number of stages required for DFT computation via FFT algorithms.

Let us have a 8-point discrete-time sequence, $N = 8 = 2^3$. It requries three stages for DFT computations.

In Decimation-in-time (DIT) FFT algorithm, input discrete-time sequence $s(n)$ is in Bit-reversed order but output, $S(K)$ is in Natural order for in-place computation. In Decimation-in-frequency (DIF) FFT algorithm, input discrete- time sequence $s(n)$ is in Natural order but its DFT is in Bit-reversed order for in-place computation. For in-place computation smaller memory space is required.

Generally, we use Radix-2 FFT algorithms. In Radix-2 FFT

algorithms, original discrete-time sequence, $s(n)$ is divided in two parts and DFT computation is done on each part separately and resultant of each parts added to get the overall discrete-frequecy sequence.

In DIT FFT algorithm, original sequence $s(n)$ is divided in even-numbered points and odd-numbered points. But in DIF FFT algorithm, original discrete-time sequence $s(n)$ is divided in two parts as first half and second half. Fig. 8.4 illustrates the number of stages required in N-point DFT computation via. DIT FFT algorithm (Here $N = 8$).

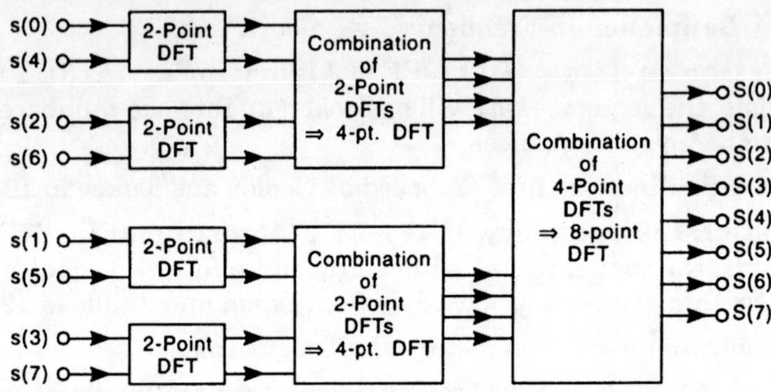

Fig. 8.4 *Three stages in N-point DFT computation via decimation-in-time FFT algorithm ($N = 8$).*

8.3.3 Decimation-in-Time (DIT) FFT Algorithm

This FFT algorithm was first proposed by Cooley and Tukey in 1965. Algorithms in which the decomposition is based on breaking the original sequence $s(n)$ into successively smaller subsequences are called Decimation-in-time FFT algorithms.

The prinicple of decimation-in-time is conveniently explained by considering $N = 2^r$, where $r > 0$. In this algorithm, DFT can be computed by separating original sequence $s(n)$ into two $N/2$-point sequences. These two $N/2$- point sequences consist of Even-numbered and Odd-numbered points in $s(n)$.

DFT of sequence $s(n)$ is given by

$$S(K) = \sum_{n=0}^{N-1} s(n)\, W_N^{nK}, \quad K = 0, 1, \ldots, N-1 \qquad \ldots(8.24)$$

where $W_N = e^{-j2\pi/N}$ is called the Phase Factor.

Now separating $s(n)$ into even and odd-numbered points, we get

$$S(K) = \sum_{n=0}^{N-1} s(n)\, W_N^{nK} = \sum_{n\ \text{Even}} s(n)\, W_N^{nK} + \sum_{n\ \text{Odd}} s(n)\, W_N^{nK} \qquad \ldots(8.25)$$

Substituting $n = 2m$ for even and $n = 2m + 1$ for n odd in Eq. (8.25). we get

$$S(K) = \sum_{m=0}^{\frac{N}{2}-1} s(2m)\, W_N^{2mK} + \sum_{m=0}^{\frac{N}{2}-1} s(2m+1)\, W_N^{(2m+1)K}$$

$$= \sum_{m=0}^{\frac{N}{2}-1} s(2m)\, W_N^{2mK} + \sum_{m=0}^{\frac{N}{2}-1} s(2m+1)\, W_N^{2mK}\, W_N^K$$

$$= \sum_{m=0}^{\frac{N}{2}-1} s(2m)\, W_N^{2mK} + W_N^K \sum_{m=0}^{\frac{N}{2}-1} s(2m+1)\, W_N^{2mK}$$

Since $W_N^2 = (e^{-j2\pi/N})^2 = e^{(-j2\pi)/(N/2)} = W_{N/2}$

Putting the value of W_N^2 in above Eqn., we get

or $$S(K) = \sum_{m=0}^{\frac{N}{2}-1} s(2m)\, W_{N/2}^{mK} + W_N^K \sum_{m=0}^{\frac{N}{2}-1} s(2m+1)\, W_{N/2}^{mK}$$

$$= G(K) + W_N^K\, H(K)$$

ir $$S(K) = G(K) + W_N^K\, H(K) \qquad \text{...(8.26)}$$

where $G(K) = \sum_{m=0}^{\frac{N}{2}-1} s(2m)\, W_{N/2}^{mK}$ = DFT of even-numbered points of sequence $s(n)$

$H(K) = \sum_{m=0}^{\frac{N}{2}-1} s(2m+1)\, W_{N/2}^{mK}$ = DFT of odd-numbered points of sequence $s(n)$

Both $s(n)$ and $S(K)$ are periodic with period N, consequently $G(K)$ and $H(K)$ will be also periodic with period $\frac{N}{2}$.

After computing $N/2$-point DFTs, *i.e.*, $G(K)$ and $H(K)$, they are then combined to produce the N-point DFT, $S(K)$. To illustrate the procedure. Let us considered Eq. (8.26) again

$$S(K) = G(K) + W_N^K\, H(K)\,, \quad K = 0\,, 1\,, \ldots\,, N-1$$

Now we are considering a case where sequence has 8 points or samples.

$K = 0\,,\quad S(0) = G(0) + W_N^0\, H(0)$

$K = 1\,,\quad S(1) = G(1) + W_N^1\, H(1)$

$K = 2\,,\quad S(2) = G(2) + W_N^2\, H(2)$

$K = 3\,,\quad S(3) = G(3) + W_N^3\, H(3)$

$K = 4\,,\quad S(4) = G(4) + W_N^4\, H(4) = G(0) + W_N^4\, H(0)$

Since $G(4) = G(0+4) = G(0)\,;\quad H(4) = H(0+4) = H(0)$

$K = 5\,,\quad S(5) = G(5) + W_N^5\, H(5) = G(1) + W_N^5\, H(1)$

Since $G(5) = G(1 + 4) = G(1)$; $H(5) = H(1 + 4) = H(1)$

$K = 6$, $S(6) = G(6) + W_N^6 H(6) = G(2) + W_N^6 H(2)$

Since $G(6) = G(2 + 4) = G(2)$; $H(6) = H(2 + 4) = H(2)$

$K = 7$, $S(7) = G(7) + W_N^7 H(7) = G(3) + W_N^7 H(3)$

Since $G(7) = G(3 + 4) = G(3)$; $H(7) = H(3 + 4) = H(3)$

Note. $G(K)$ and $H(K)$ are periodic with period $N/2$. Above sequence has 8 points then its DFT also has 8 points. Therefore $G(K)$ and $H(K)$ are also periodic with period $\frac{N}{2}$, *i.e.*, $\frac{8}{2} = 4$

or $$G\left(K + \frac{N}{2}\right) = G(K) \; ; \quad H\left(K + \frac{N}{2}\right) = H(K)$$

Fig. 8.6 illustrates the flow graph of DIT decomposition of N-point DFT computation into two $N/2$-point DFT computation (For $N = 8$).

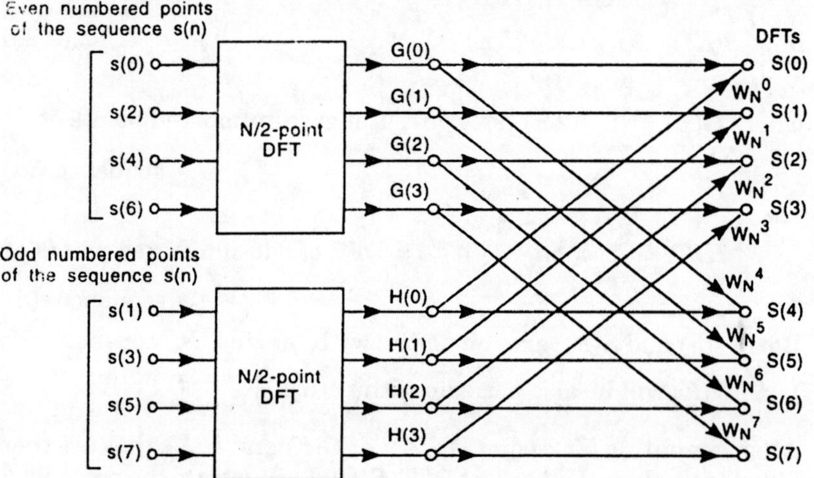

Fig. 8.5 ***Flow graph of the decimation-in-time (DIT) decomposition of an N-point DFT computation into two N/2-point DFT computation (N = 8).***

$G(K)$ and $H(K)$ in Eq. (8.26) can be computed as follows :

$$G(K) = \sum_{m=0}^{N/2-1} g(m)\, W_{N/2}^{mK} = \sum_{m\ Even} g(m)\, W_{N/2}^{mK} + \sum_{m\ Odd} g(m)\, W_{N/2}^{mK}$$

Substituting $m = 2l$ for m even and $m = 2l + 1$ for m odd in above Eqn., we get

$$G(K) = \sum_{l=0}^{N/4-1} g(2l)\, W_{N/2}^{2lK} + \sum_{l=0}^{N/4-1} g(2l+1)\, W_{N/2}^{(2l+1)K}$$

$$= \sum_{l=0}^{N/4-1} g(2l)\, W_{N/4}^{lK} + W_{N/2}^{K} \sum_{l=0}^{N/4-1} g(2l+1)\, W_{N/4}^{lK} \qquad ...(8.27)$$

Similarly,

$$H(K) = \sum_{l=0}^{N/4-1} h(2l)\, W_{N/4}^{lK} + W_{N/2}^{K} \sum_{l=0}^{N/4-1} h(2l+1)\, W_{N/4}^{lK} \qquad \ldots(8.28)$$

4-point DFT is computed according to Eqs. (8.27) and (8.28). This computation is shown in Fig. 8.6.

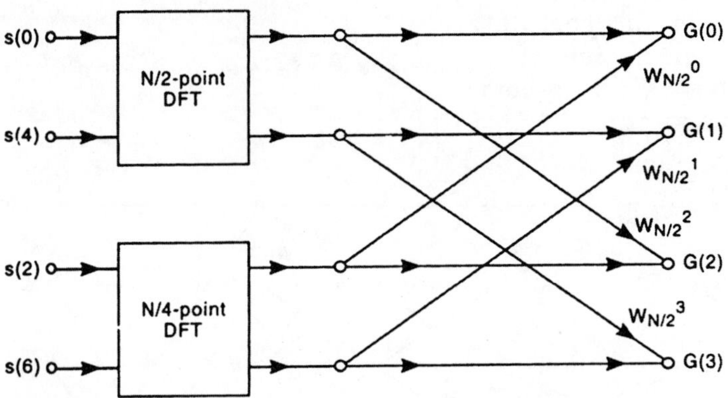

Fig. 8.6 *Flow graph of the decimation-in-time decomposition of an N/2-point DFT computation into two N/4-point DFT computations (where N = 8).*

After inserting Fig. 8.6 into Fig. 8.5, we get Fig. 8.7.

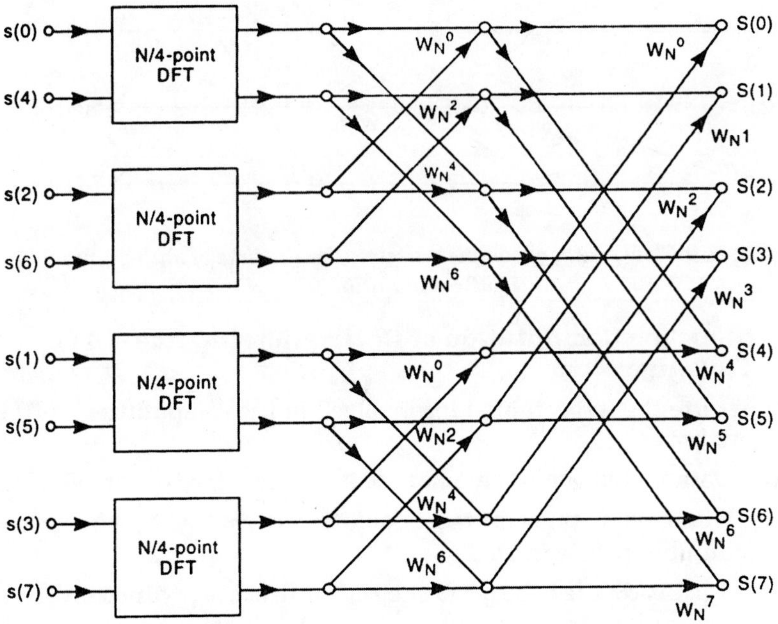

Fig. 8.7 *Result of substituting Fig.* 8.5 *into Fig.* 8.6.

Two-point DFT is computed as given in Fig. 8.8. It is also known as butterfly computation because of its appearance.

Substituting Fig. 8.8 into Fig. 8.7, we get Fig. 8.9 which is a complete decimation-in-time (DIT) decomposition of 8-point DFT computation.

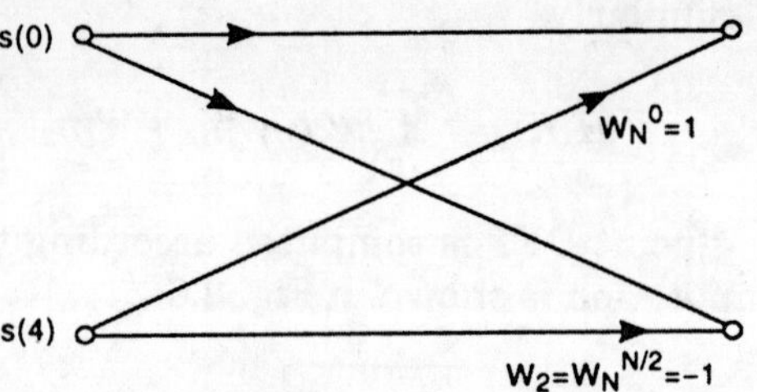

Fig. 8.8 *Flow graph of 2-point DFT.*

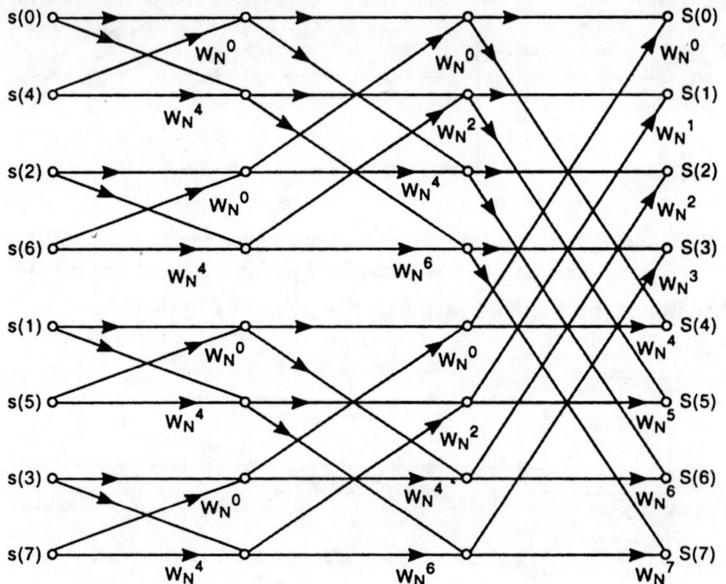

Fig. 8.9 *Flow graph of complete decimation-in-time of an 8-point DFT computation.*

8.3.4 Steps for Computation of Decimation-in-Time (DIT) FFT Algorithm

Followings are the important steps involved in the computation of DIT FFT algorithms :

1. Data shuffling must be done first. It is a performed by Bit Reversal.
2. Number of stages in the calculation is $r = \log_2 N$, where N = number points in DFT.
3. 2-point DFT is to be performed at the first stage which involves no multiplication, but it requires some addition and subtraction.

4. General formulation for Decimation-in-time (DIT) FFT algorithm provides us a ready-made formula for calculation of phase rotation factors required for each butterfly.

Example 8.1 Find the DFT of the following discrete-time sequence

$$s(n) = \{1, -1, -1, -1, 1, 1, 1, -1\}$$

using Radix-2 decimation-in-time (DIT) FFT algorithm.

Solution. The Twiddle factor or phase rotation factor $W_N = e^{-j2\pi/N}$ involved in the FFT calculation are found out as follows for $N = 8$.

W_N = Phase roration factor $= e^{-j2\pi/N}$

$$W_8^0 = e^{-j(2\pi/8).0} = e^0 = 1$$

$$W_8^1 = e^{-j(2\pi/8).1} = e^{-j\pi/4} = \cos\frac{\pi}{4} - j\sin\frac{\pi}{4} = \frac{1-j}{\sqrt{2}}$$

$$W_8^3 = e^{-j(2\pi/8).2} = e^{-j\pi/2} = \cos\frac{\pi}{2} - j\sin\frac{\pi}{2} = -j$$

In Radix-2 decimation-in-time (DIT) FFT algorithm, original sequence $s(n)$ is decomposed in two parts even-numbered point and odd numbered point sequence. This is done very easily by bit reversal of original discrete-time sequence. Flow graph of Radix-2 decimation-in-time FFT algorithm is given in Fig. 8.10. Determination of DFT using Radix-2 DIF FFT algorithm requires three stages because the number of points in a given sequence is 8, *i.e.*, $2^3 = 2^r = N = 8$, where r is number of stages required = 3.

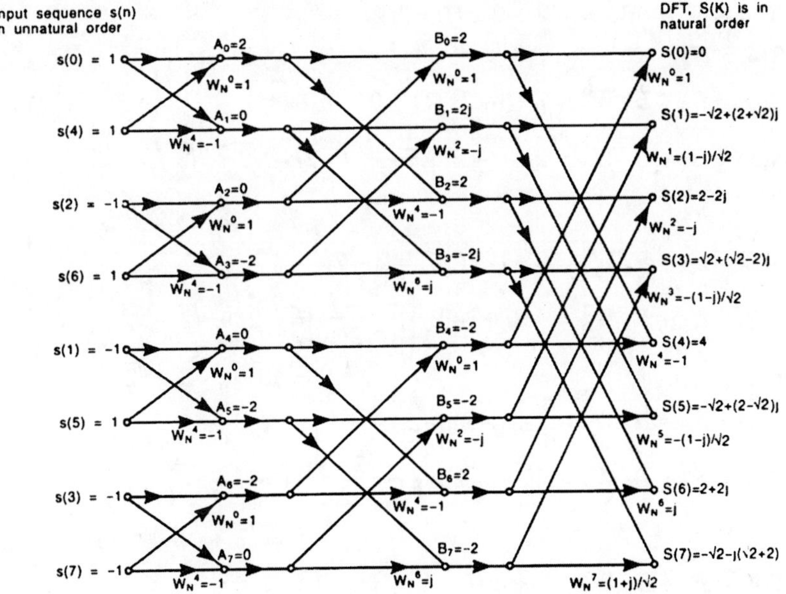

Fig. 8.10 *Flow graph of Radix-2 decimation-in-time FFT algorithm for N = 8.*

Stage I :

$A_0 = s(0) + s(4)\, W_N^0 = 1 + 1 \times 1 = 2$

$A_1 = s(0) + s(4)\, W_N^4 = 1 + 1 \times (-1) = 0$

$A_2 = s(2) + s(6)\, W_N^0 = -1 + 1 \times (1) = 0$

$A_3 = s(2) + s(6)\, W_N^4 = -1 + 1 \times (-1) = -2$

$A_4 = s(1) + s(5)\, W_N^0 = -1 + 1 \times (1) = 0$

$A_5 = s(1) + s(5)\, W_N^4 = -1 + 1 \times (-1) = -2$

$A_6 = s(3) + s(7)\, W_N^0 = -1 + (-1) \times 1 = -2$

$A_7 = s(3) + s(7)\, W_N^4 = -1 + (-1) \times (-1) = 0$

Stage II :

$B_0 = A_0 + A_2\, W_N^0 = 2 + 0 \times 1 = 2$

$B_1 = A_1 + A_3\, W_N^2 = 0 + (-2)\,(-j) = 2j$

$B_2 = A_0 + A_2\, W_N^4 = 2 + 0 \times (-1) = -2$

$B_3 = A_1 + A_3\, W_N^6 = 0 + (-2)\, j = -2j$

$B_4 = A_4 + A_6\, W_N^0 = 0 + (-2)\,(1) = 2$

$B_5 = A_5 + A_7\, W_N^2 = -2 + 0 \times (-j) = -2j$

$B_6 = A_4 + A_6\, W_N^4 = 0 + (-2)\,(-1) = 2$

$B_7 = A_5 + A_7\, W_N^6 = -2 + 0 \times (j) = -2j$

Stage III :

$S(0) = B_0 + B_4\, W_N^0 = 2 + (-2)\,(1) = 0$

$$S(1) = B_1 + B_5\, W_N^1 = 2j + (-2)\left(\frac{1-j}{\sqrt{2}}\right) = 2j - \sqrt{2} + \sqrt{2}\, j$$

$$= -\sqrt{2} + (2 + \sqrt{2})\, j$$

$S(2) = B_2 + B_6\, W_N^2 = 2 + (2)\,(-j) = 2 - 2j$

$$S(3) = B_3 + B_7\, W_N^3 = -2j + (-2)\left(\frac{-1-j}{\sqrt{2}}\right) = -2j + \sqrt{2} + \sqrt{2}\, j$$

$$= \sqrt{2} + (-2 + \sqrt{2})\, j$$

$S(4) = B_0 + B_4\, W_N^4 = 2 + (-2)\,(-1) = 4$

$$S(5) = B_1 + B_5\, W_N^5 = 2j + (-2) \times \left(\frac{-(1-j)}{\sqrt{2}}\right) = 2j + \sqrt{2} - \sqrt{2}\, j$$

$$= \sqrt{2} + (2 - \sqrt{2})\, j$$

$S(6) = B_2 + B_6\, W_N^6 = 2 + 2\,(j) = 2 + 2j$

$$S(7) = B_3 + B_7\, W_N^7 = -\,2j + (-\,2)\,\frac{(1+j)}{\sqrt{2}} = -\,2j - \sqrt{2} - \sqrt{2}\,j$$
$$= -\,\sqrt{2} - (2+\sqrt{2})\,j$$

Resultant discrete-frequency sequency will be

$$S(K) = (\,S(0)\,,\,S(1)\,,\,S(2)\,,\,S(3)\,,\,S(4)\,,\,S(5)\,,\,S(6)\,,\,S(7)\,\}$$
$$= \{\,0\,,\,-\sqrt{2} + (2+\sqrt{2})\,j\,,\,2-2j\,,\,\sqrt{2} + (-\,2+\sqrt{2})\,j\,,\,4\,,$$
$$\{\,\sqrt{2} + (2-\sqrt{2})\,j\,,\,2+2j\,,\,-\sqrt{2} - (2+\sqrt{2})\,j\,\}$$

Table 8.1 Phase rotation factors for quick computation.

Number of points in DFT, N	*Stage 1*	*Stage 2*	*Stage 3*	*Stage 4*	*Stage 5*
4 No. of stages = 2	Twiddle factor not required	W_4^0, W_4^1	—	—	—
8 No. of stages = 3	Twiddle factor not required	W_8^0, W_8^2	W_8^0, W_8^1 W_8^2, W_8^3	—	—
16 No. of stages = 4	Twiddle factor not required	W_{16}^0, W_{16}^4	W_{16}^0, W_{16}^2 W_{16}^4, W_{16}^6	W_{16}^0, W_{16}^1 W_{16}^2, W_{16}^3 W_{16}^4, W_{16}^5 W_{16}^6, W_{16}^7	—
32 No. of Stages = 5	Twiddle factor not required	W_{32}^0, W_{32}^8	W_{32}^0, W_{32}^4 W_{32}^8, W_{32}^{12}	W_{32}^0, W_{32}^2 .. W_{32}^{14}	W_{32}^0, W_{32}^1, .. W_{32}^{15}

8.3.5 Decimation-in-Frequency FFT Algorithm

In decimation-in-frequency FFT algorithm, the output DFT sequence $S(K)$ is broken into smaller and smaller subsequences. For the derivation of this algorithm, the number of points or samples in a given sequence should be $N = 2^r$ where $r > 0$. For this purpose we can first-divide the input sequence into the first-half and the second-half of the points.

DFT of sequence $s(n)$ is given by

$$S(K) = \sum_{n=0}^{N-1} s(n)\, W_N^{nK}\,, \quad K = 0\,,\,1\,,\,\ldots\,N-1 \qquad \ldots(8.29)$$

where $W_N = e^{-j2\pi/N}$ is called the Phase Factor.

$$= \sum_{\text{Ist Half}} s(n)\, W_N^{nK} + \sum_{\text{IInd Half}} s(n)\, W_N^{nK}$$

$$= \sum_{n=0}^{N/2-1} s(n)\, W_N^{nK} + \sum_{n=N/2}^{N-1} s(n)\, W_N^{nK}$$

Putting $n = n + \frac{N}{2}$ in the second part of above equation, we get

$$S(K) = \sum_{n=0}^{N/2-1} s(n)\, W_N^{nK} + \sum_{n=0}^{N/2-1} s\left(n + \frac{N}{2}\right) W_N^{(n+N/2)K}$$

or $$S(K) = \sum_{n=0}^{N/2-1} s(n)\, W_N^{nK} + W_N^{(N/2).K} \sum_{n=0}^{N/2-1} s\left(n + \frac{N}{2}\right) W_N^{nK}$$

Since $W_N^{(N/2).K} = (-1)^K$. Substituting the value of $W_N^{NK/2}$ in above Eqn., we get

$$S(K) = \sum_{n=0}^{N/2-1} s(n)\, W_N^{nK} + (-1)^K \sum_{n=0}^{N/2-1} s\left(n + \frac{N}{2}\right) W_N^{nK} \qquad \text{...(8.30)}$$

or $$S(K) = \sum_{n=0}^{N/2-1} \left[s(n) + (-1)^K s\left(n + \frac{N}{2}\right)\right] W_N^{nK} \qquad \text{...(8.31)}$$

Putting, $S(2m)$ for K Even — It is for even numbered points of DFT

$S(2m + 1)$ for K Odd — It is for odd numbered points of DFT

in Eq. (8.31), we get

$$S(2m) = \sum_{n=0}^{N/2-1} \left[s(n) + s\left(n + \frac{N}{2}\right)\right] W_N^{2mn} = \sum_{n=0}^{N/2-1} g(n)\, W_{N/2}^{mn} \qquad \text{...(8.32)}$$

$$S(2m+1) = \sum_{n=0}^{N/2-1} \left[s(n) - s\left(n + \frac{N}{2}\right)\right] W_N^{(2m+1)n}$$

$$= \sum_{n=0}^{N/2-1} \left[s(n) - s\left(n + \frac{N}{2}\right)\right] W_N^{2mn}\, W_N^{n}$$

or $$S(2m+1) = \sum_{n=0}^{N/2-1} [h(n)]\, W_N^{n}\, W_{N/2}^{mn} \qquad \text{...(8.33)}$$

where $g(n) = s(n) + s\left(n + \frac{N}{2}\right)$

$h(n) = s(n) - s\left(n + \frac{N}{2}\right)$

$W_N^2 = W_{N/2}$

$m = 0, 1, \ldots, \left(\frac{N}{2} - 1\right)$

Eqns. (8.32) and (8.33) can be recognized as $N/2$-point DFTs. The DFT can be computed by first forming the sequences $g(n)$ and $h(n)$, then computing the $N/2$-point DFTs of these two sequences that are called the even-numbered output points and odd-numbered points, respectively.

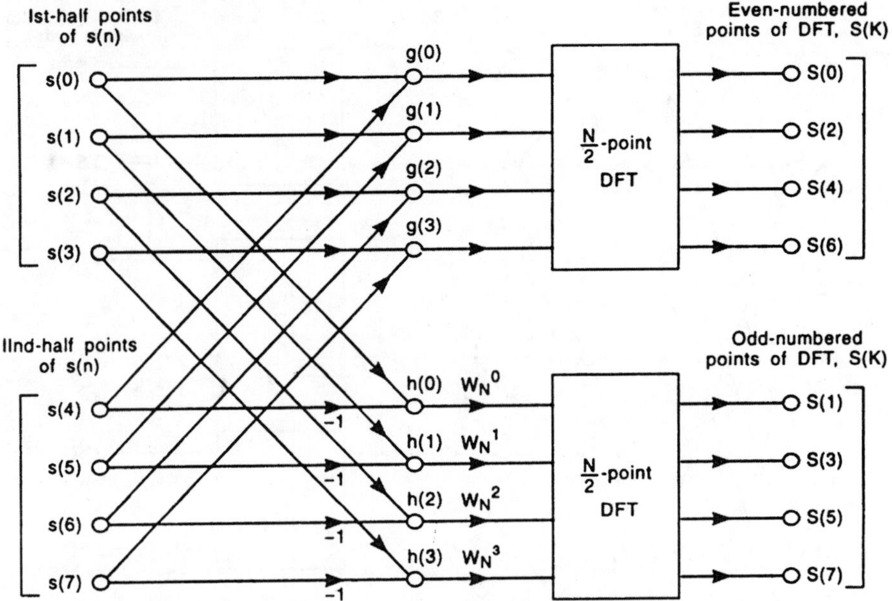

Fig. 8.11 *Flow graph of the decimation-in-frequency decomposition of an N-point DFT computation into two N/2-point DFT computation. Here, N = 8.*

Eqns. (8.32) and (8.33) are illustrated in Fig. 8.11 for 8-point DFT. 8-point DFT is broken into two 4-point DFTs. The first 4-point DFT is for even-numbered points and second 4-point DFTs for odd-numbered points of $S(K)$. Now 4-point DFTs is further broken into two 2-point DFTs in the same manner as given above. It is illustrated in Fig. 8.12.

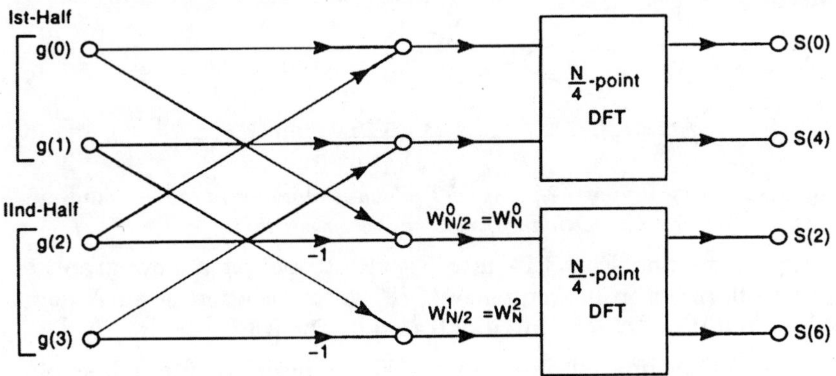

Fig. 8.12 *Flow graph of the decimation-in-frequency decomposition N/2-point DFT compuation into two N/4-point DFT compuations. Here, N = 8.*

After inserting Fig. 8.12 into Fig. 8.11, we get Fig. 8.13 and this is illustrated as follows.

Now 2-point DFT is computed as follows in Fig. 8.14.

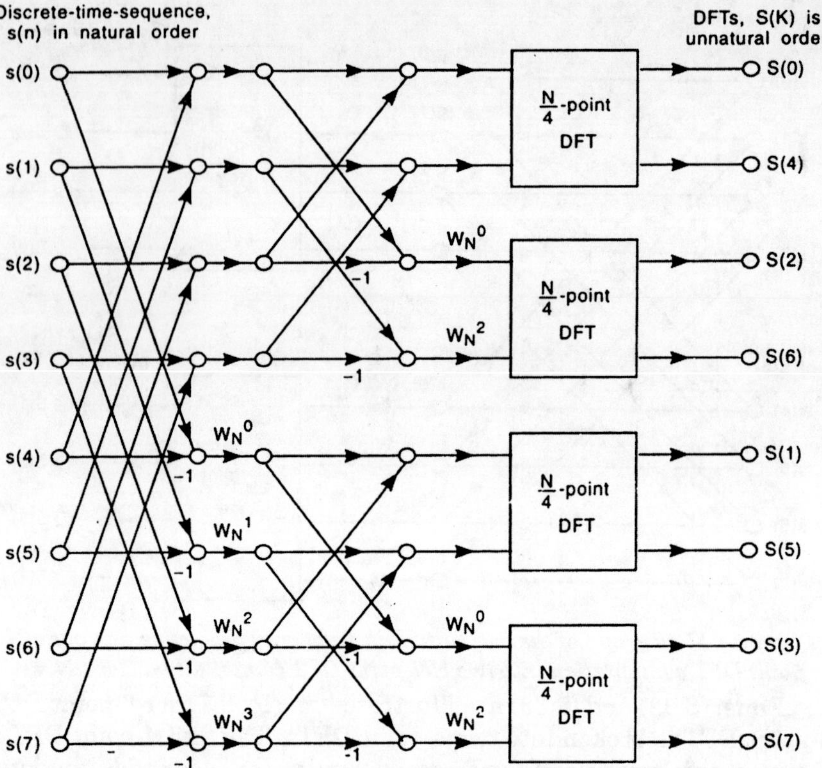

Fg. 8.13 *Flow graph after inserting Fig.* 8.12 *into Fig.* 8.11.

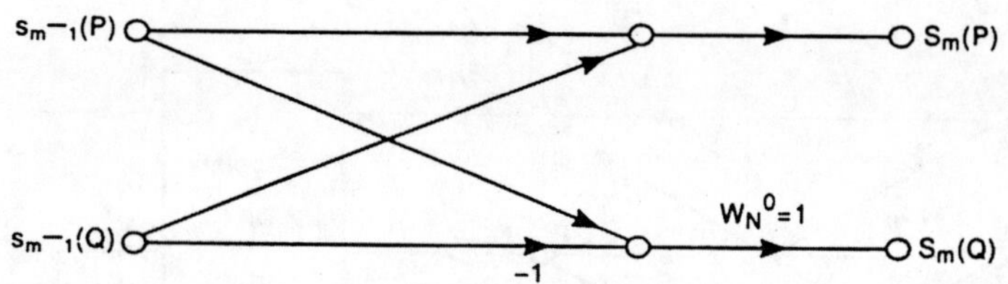

Fig. 8.14 *Flow graph of a 2-point DFT which is required in the last stage of decimation-in-frequency decomposition.*

After inserting Fig. 8.14 into Fig. 8.13, we get a flow graph of complete decimation-in-frequency (DIF) decomposition of an N-point DFT computation ($N = 8$) and illustrated in Fig. 8.15.

Note. In decimation-in-frequency decomposition, input discrete-time sequence $s(n)$ is in natural order and output DFT sequence $S(K)$ is in Bit-reversed order for an in-place computation. But in decimation-in-time decomposition, input discrete-time sequence, $s(n)$ is in Bit-reversed order and output DFT sequence, $S(K)$ is in natural order for an in-place computation.

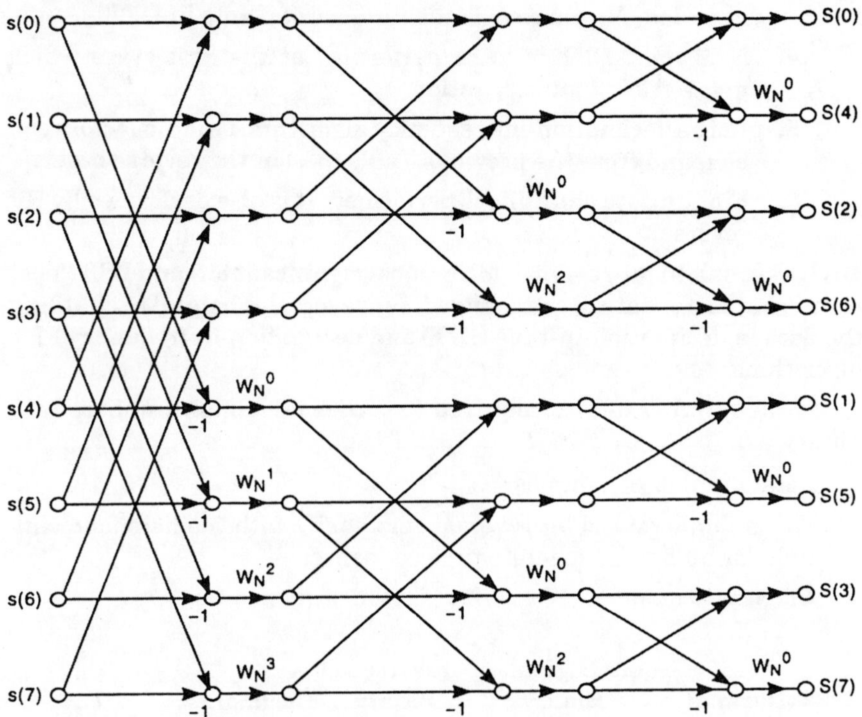

Fig. 8.15 *Flow graph of complete decimation-in-frequency (DIF) decomposition of an N-point DFT computation (N = 8).*

8.3.6 Steps for Computation of Decimation-in-Frequency (DIF) FFT Algorithm

Given below are the important steps for the computation of DIF FFT algorithms.

1. Data shuffling is not requried but whole sequence is divided in two parts : first half and second half. From these we calculate $g(n)$ and $h(n)$ as follows :

$$g(n) = s(n) + s\left(n + \frac{N}{2}\right)$$

and $h(n) = s(n) - s\left(n + \frac{N}{2}\right)$

where $n = 0, 1, \ldots, \frac{N}{2} - 1$

2. From these two sequences, we compute $N/2$-point DFTs. Before computation of $N/2$-point DFT for $h(n)$, it is multiplied by W_N^n.

3. Number of stages required in the calculation of N-point DFT is $r = \log_2 N$, where N is the number of points in DFT.
4. $N/2$-point DFT is to be performed at the first stage which involves $N/2$ multiplication.
5. In the decimation-in-frequency algorithm, twiddle factors can be found from the previous Table (8.1) in the reverse order.
6. Finally data shuffling is performed. It is also performed by Bit reversal.

Bit Reversal. In bit-reversal, MSB (most significant bit) and LSB (least significant bit) of data are exchanged. Bit reversal is helpful in shuffling the data in decimation-in-time (DIT) and decimation-in-frequency FFT algorithms.

If the number of bits required to convert decimal numbers into binary = 3.

Total number of samples = $2^3 = 8$.

We can understand bit reversal very easily with the help following example for an 8 point discrete time sequence.

Original sequence   Bit reversed sequence

Decimal		Original Binary		Bit reversed Binary		Decimal
$s(0)$	$\Rightarrow$	$s(000)$	$\rightarrow$	$s(000)$	$\Rightarrow$	$s(0)$
$s(1)$	$\Rightarrow$	$s(001)$	$\rightarrow$	$s(100)$	$\Rightarrow$	$s(4)$
$s(2)$	$\Rightarrow$	$s(010)$	$\rightarrow$	$s(010)$	$\Rightarrow$	$s(2)$
$s(3)$	$\Rightarrow$	$s(011)$	$\rightarrow$	$s(110)$	$\Rightarrow$	$s(6)$
$s(4)$	$\Rightarrow$	$s(100)$	$\rightarrow$	$s(001)$	$\Rightarrow$	$s(1)$
$s(5)$	$\Rightarrow$	$s(101)$	$\rightarrow$	$s(101)$	$\Rightarrow$	$s(5)$
$s(6)$	$\Rightarrow$	$s(110)$	$\rightarrow$	$s(011)$	$\Rightarrow$	$s(3)$
$s(7)$	$\Rightarrow$	$s(111)$	$\rightarrow$	$s(111)$	$\Rightarrow$	$s(7)$

Example 8.2 Find the DFT of the following discrete-time sequence

$$s(n) = \{1, -1, -1, -1, 1, 1, 1, -1\}$$

using Radix-2 decimation-in-frequency (DIF) FFT algorithm.

Solution. The Twiddle factor or phase rotation factor $W_N = e^{-j2\pi/N}$ involved in the FFT calculation are found out as follows for $N = 8$.

W_N = Phase roation factor = $e^{-j2\pi/N}$

$W_8^0 = e^{-(j2\pi/8)0} = e^0 = 1$

$$W_8^1 = e^{-j(2\pi/8)1} = e^{-j\pi/4} = \frac{1-j}{\sqrt{2}}$$

$$W_8^2 = e^{-j(2\pi/8)1} = e^{-j\pi/2} = -j$$

$$W_8^3 = e^{-j(2\pi/8)3} = e^{-j3\pi/4} = \frac{-(1+j)}{\sqrt{2}}$$

In Radix-2 decimation-in-frequency (DIF) FFT algorithm, original sequence $s(n)$ is decomposed into two subsequences as first half and second half of a sequence. There is no need of reordering (shuffling) the original sequence as in Radix-2 decimation-in-time (DIT) FFT algorithm. But resultant discrete frequency sequence $S(K)$ is shuffled (reordered) into natural order because these are obtained in unnatural order. Flow graph of Radix-2 decimation-in-frequency (DIF) FFT algorithm for $N = 8$ is shown in Fig. 8.11. Determination of DFT using Radix-2 DIF FFT algorithm requires three stages because the number of points in a given sequence is 8, *i.e.*, $2^3 = 2^r = N = 8$, where r is number of stages required = 3.

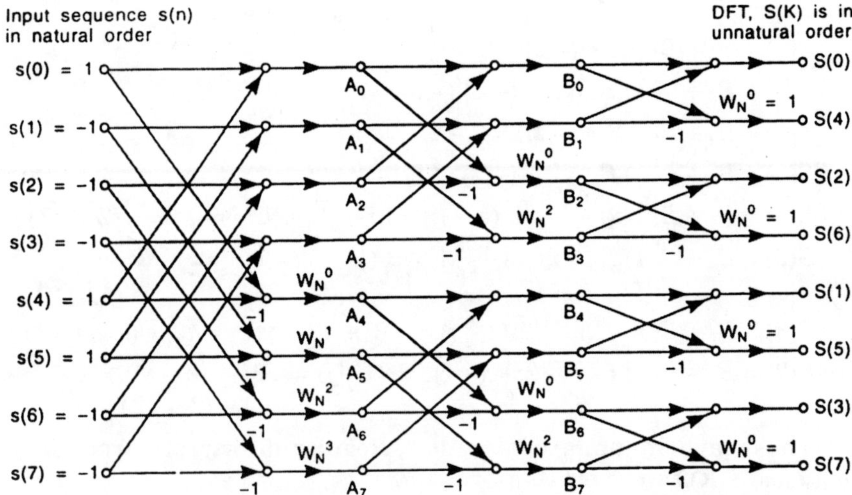

Fig. 8.16 *Flow graph of Radix-2 decimation-in-frequency (DIF) FFT algorithm N = 8.*

Stage I :

$A_0 = s(0) + s(4) = 1 + 1 = 2$

$A_1 = s(1) + s(5) = -1 + 1 = 0$

$A_2 = s(2) + s(6) = -1 + 1 = 0$

$A_3 = s(3) + s(7) = -1 - 1 = -2$

$A_4 = [s(0) + (-1)\, s(4)]\, W_8^0 = [1 + (-1)\,(1)] \times 1 = 0$

$$A_5 = [s(1) + (-1)\,s(5)]\,W_8^1 = [-1 + (-1)\,(1)]\,\frac{(1-j)}{\sqrt{2}} = -\sqrt{2}\,(1-j)$$

$$A_6 = [s(2) + (-1)\,s(6)]\,W_8^2 = [-1 + (-1) \times 1]\,(-j) = 2j$$

$$A_7 = [s(3) + (-1)\,s(7)]\,W_8^3 = [-1 + (-1)\,(-1)]\left\{\frac{-(1+j)}{\sqrt{2}}\right\} = 0$$

Stage II :

$B_0 = A_0 + A_2 = 2 + 0 = 2$

$B_1 = A_1 + A_3 = 0 + (-2) = -2$

$B_2 = [A_0 + (-1)\,A_2]\,W_8^0 = [2 - 0] \times 1 = 2$

$B_3 = [A_1 + (-1)\,A_3]\,W_8^2 = [0 + (-1)\,(-2)] \times (-j) = -2j$

$B_4 = A_4 + A_6 = 0 + 2j = 2j$

$B_5 = A_5 + A_7 = [-\sqrt{2}\,(1-j)] + 0 = -\sqrt{2}\,(1-j)$

$B_6 = [A_4 + (-1)\,A_6]\,W_8^0 = [0 + (-1)\,2j] \times 1 = -2j$

$B_7 = [A_5 + (-1)\,A_7]\,W_8^2 = [-\sqrt{2}\,(1-j) + (-1) \times 0] \times (-j) = \sqrt{2}\,(1+j)$

Stage III :

$S(0) = B_0 + B_1 = 2 + (-2) = 0$

$S(4) = B_0 + (-1)\,B_1 = 2 + (-1)\,(-2) = 4$

$S(2) = B_2 + B_3 = 2 + (-2j) = 2 - 2j$

$S(6) = B_2 + (-1)\,B_3 = 2 + (-1)\,(-2j) = 2 + 2j$

$S(1) = B_4 + B_5 = 2j + [-\sqrt{2}\,(1-j)] = 2j - \sqrt{2} + \sqrt{2}\,j = -\sqrt{2} + (2+\sqrt{2})\,j$

$$S(5) = B_4 + (-1)\,B_5 = 2j + (-1)\,[-\sqrt{2}\,(1-j)] = 2j + \sqrt{2} - \sqrt{2}\,j$$
$$= \sqrt{2} + (2-\sqrt{2})\,j$$

$S(3) = B_6 + B_7 = -2j + \sqrt{2}\,(1+j) = -2j + \sqrt{2} + \sqrt{2}\,j = \sqrt{2} + (-2+\sqrt{2})\,j$

$$S(7) = B_6 + (-1)\,B_7 = -2j + (-1)\,\sqrt{2}\,(1+j) = -2j - \sqrt{2} - \sqrt{2}\,j$$
$$= -\sqrt{2} - (2+\sqrt{2})\,j$$

These are in unnatural order. Resultant discrete frequency sequence $S(K)$ can be written as

$$S(K) = \{S(0), S(1), S(2), S(3), S(4), S(5), S(6), S(7)\}$$

or $S(K) = \{0, -\sqrt{2} + (2+\sqrt{2})\,j, 2 - 2j, \sqrt{2} + (-2+\sqrt{2})\,j, 4,$
$\sqrt{2} + (2-\sqrt{2})\,j, 2 + 2j, -\sqrt{2} - (2+\sqrt{2})\,j\}$

8.3.7 Number of Complex Multiplications Required in Decimation-in-Frequency (DIF) FFT Algorithm

Number of complex multiplications required in decimation-in-frequency (DIF) FFT algorithm are the same as that required in decimation-in-time (DIT) FFT algorithm. Number of complex multiplication required in these FFT algorithms are $N/2 \log_2 N$, where

$N = 2^r$, $r > 0$ and N is the total number of points (or samples) in a discrete-time sequence $s(n)$. Thus the total computations (number of multiplication and addition operations) are the same in both FFT algorithms.

Now we will compare the computational complexity for the direct computation of the DFT and FFT algorithm. This comparison is given in Table 8.2.

Table 8.2.

No. of points (or samples) in a sequence $s(n)$, N	*Complex multiplications in direct computation of DFT =* $N \times N = A$	*Complex multiplications in FFT algorithms* $N/2 \log_2 N = B$	*Speed improvement Factor =* A/B
$2^2 = 4$	16	4	$\frac{16}{4} = 4.0$
$2^3 = 8$	64	12	$\frac{64}{12} = 5.3$
$2^4 = 16$	256	32	$\frac{256}{32} = 8.0$
$2^5 = 32$	1024	80	$\frac{1024}{80} = 12.8$
$2^6 = 64$	4096	192	$\frac{4096}{192} = 21.3$
$2^7 = 128$	16384	448	$\frac{16384}{448} = 36.6$
$2^8 = 256$	65536	1024	$\frac{65536}{1024} = 64.0$
$2^9 = 512$	262144	2304	$\frac{262144}{2304} = 113.8$

REVIEW QUESTIONS

1. What is Fast Fourier Transformation (FFT) ?
2. Explain Goertzel algorithm.
3. Define following terms :
   (*a*) Decimation-in-time (*b*) Decimation-in-frequency
   (*c*) Phase rotation factor or twiddle factor.
4. Distinguish between decimation-in-time and decimation-in-frequency algorithms.
5. Give classification of FFT algorithms.
6. Give various steps required for computation of decimation-in-time (DIT) FFT algorithms and decimation-in-frequency (DIF) FFT algorithms.

7. Discuss the number of complex multiplication required in FFT algorithms.

8. Give flow graph of complete decimation-in-time (DIT) FFT algorithm ($N = 8$).

9. Give flow graph of complete decimation-in-frequency (DIF) FFT algorithm ($N = 8$), where N is the total number of samples in DFT.

10. Find the DFT of the following sequences using decimation-in-time (DIT) and decimation-in-frequency (DIF) FFT algorithms.

(*a*) $s(n) = \{1, 1, 1, 1, 1, 1, 1, 1\}$

(*b*) $s(n) = \{1, 0, 0, 0, 1, 1, 1, 0\}$

(*c*) $s(n) = \{1, 0, 0, 1, -1, 1\}$

(*d*) $s(n) = \{1, 1, 1, 1, 0, 0, 0, 0\}$

9

The Laplace Transform

9.1 INTRODUCTION TO LAPLACE TRANSFORM

In the preceding chapters, we have studied Fourier analysis tools such as Fourier series and Fourier transforms. These tools are used for the analysis of signals and LTI systems. A broad class of signals can be represented as a linear combination of periodic complex exponentials and these complex exponentials are eigen functions of LTI systems. The Continuous-Time Fourier Transform (CTFT) is used to represent signals as a linear combination of complex exponentials of the form e^{st} with $s = j\omega$. Since we know that the eigen function property and many of its consequences can be applied for arbitrary values of s and not only to those values which are purely imaginary. This observation leads to a generalization of the Continuous-time Fourier Transform (CTFT) known as the Laplace transform. In this chapter we will develop the Laplace transform. As we will also see, that Laplace transform has many of properties that make Fourier analysis useful. Laplace transform can be used to analyse a broad class of signals and systems and in some cases it can be used where Fourier transform cannot be used. Laplace transform can be applied to the analysis of many stable systems. It plays an important role in the investigation of the stability or instability of systems. Also, Laplace transform can be used to analyse feedback systems.

9.2 DEFINITION OF LAPLACE TRANSFORM

We know that the response of a continuous-time LTI system with impulse response $h(t)$ to a complex exponential input of the form e^{st} is given by

$$y(t) = H(s)e^{st} \qquad ...(9.1)$$

where

$$H(s) = \int_{-\infty}^{\infty} h(t)e^{-st}\,dt \qquad ...(9.2)$$

If $s = j\omega$ then the integral of Eq. 9.2 is called the CTFT of $h(t)$. For general values of the complex variable s, it is referred to as the Laplace transform of the impulse response $h(t)$. Laplace transform of a general signal $s(t)$ is defined as

$$S(s) = \int_{-\infty}^{\infty} s(t)\, e^{-st}\, dt \qquad \text{...(9.3)}$$

The function of the independent complex variable s corresponding to the complex variable in the exponent of e^{-st}. The complex variable s can be written as $s = \sigma + j\omega$, where σ and ω are the real and imaginary parts of complex variable s respectively. For convenience, we denote the Laplace transform pair as

$$s(t) \xleftrightarrow{\text{LT}} S(s) \qquad \text{...(9.4)}$$

where $s(t)$ = continuous-time signal

$S(s)$ = Laplace transform of $s(t)$

and s = complex variable = $\sigma + j\omega$

When $s = j\omega$ then Eq. 9.3 becomes

$$S(j\omega) = \int_{-\infty}^{\infty} s(t)\, e^{-j\omega t}\, dt \qquad \text{...(9.5)}$$

Eq. 9.5 is called the Continuous-time Fourier Transform (CTFT) of signal $s(t)$.

$$\text{CTFT}\,\{s(t)\} = S(s)\,\big|_{S=j\omega} = \text{LT}\,\{s(t)\}\,\big|_{S=j\omega} \qquad \text{...(9.6)}$$

The Laplace transform also has a direct relationship with the continuous-time Fourier transform (CTFT) when the complex variable s is not purely imaginary.

Substituting $s = \sigma + j\omega$ in Eq. 9.3, we get

$$S(\sigma + j\omega) = \int_{-\infty}^{\infty} s(t)\, e^{-(\sigma + j\omega)t}\, dt$$

or $$S(\sigma + j\omega) = \int_{-\infty}^{\infty} s(t)\, e^{-\sigma t}\, e^{-j\omega t}\, dt$$

or $$S(\sigma + j\omega) = \int_{-\infty}^{\infty} [s(t)\, e^{-\sigma t}]\, e^{-j\omega t}\, dt \qquad \text{...(9.7)}$$

Now we can say that the Laplace transform of signal $s(t)$ can be interpreted as the Continuous-time Fourier transform of $s(t)$ after multiplication by a real exponential signal $e^{-\sigma t}$. This real exponential signal $e^{-\sigma t}$ may be decaying or growing with time which depends on the value of σ. If σ is positive valued then the exponential signal $e^{-\sigma t}$ will be decaying type and if σ is negative valued then the exponential signal $e^{-\sigma t}$ will be growing type.

We can understand the relationship between Laplace transform and Continuous-time Fourier transform (CTFT) of a continuous-time signal by considering following numerical examples.

Example 9.1 Determine the Laplace transform and CTFT of a continuous-time exponential signal $s(t) = e^{-At}u(t)$. Also illustrate the relationship of Laplace transform with CTFT.

Solution. Continuous-time Fourier transform (CTFT) of a signal $s(t)$ can be determined as

$$S(j\omega) = \text{CTFT}\,\{s(t)\} = \int_{-\infty}^{\infty} s(t)\, e^{-j\omega t}\, dt$$

$$= \int_{-\infty}^{\infty} e^{-At}\, u(t)\, e^{-j\omega t}\, dt \qquad \text{...(1)}$$

Unit step function $u(t)$ is defined as

$$u(t) = \begin{cases} 1\,, & t > 0 \\ 0\,, & t < 0 \end{cases} \qquad \text{...(2)}$$

Substituting the value of $u(t)$ in Eq. 1, we get

$$S(j\omega) = \int_{0}^{\infty} e^{-At}\,.1\, e^{-j\omega t}\, dt$$

$$= \int_{0}^{\infty} e^{-At}\, e^{-j\omega t}\, dt = \int_{0}^{\infty} e^{-(A+j\omega)t}\, dt$$

$$= \left[\frac{1}{-(A+j\omega)}\, e^{-(A+j\omega)t}\right]_{0}^{\infty} = \frac{1}{j\omega + A}\,, \quad A > 0$$

or $$S(j\omega) = \frac{1}{j\omega + A}\,, \quad A > 0 \qquad \text{...(3)}$$

Laplace transform of a signal $s(t)$ can be determined as

$$S(s) = \int_{-\infty}^{\infty} s(t)\, e^{-st}\, dt = \int_{-\infty}^{\infty} e^{-At}\, u(t)\, e^{-st}\, dt$$

$$= \int_{0}^{\infty} e^{-At}\, 1\, e^{-st}\, dt = \int_{0}^{\infty} e^{-(A+s)t}\, dt$$

$$= \left[\frac{1}{-(A+s)}\, e^{-(A+s)t}\right]_{0}^{\infty} = \frac{1}{A+s}$$

or $$S(s) = \frac{1}{s+A}, \; s > 0 \qquad \text{...(4)}$$

Substituting $s = \sigma + j\omega$ in Eq. 4, we get

$$S(\sigma + j\omega) = \frac{1}{(\sigma + j\omega) + A}$$

$$= \frac{1}{(\sigma + A) + j\omega}, \; \sigma + A > 0$$

$$= \frac{1}{(\sigma + A) + j\omega}, \; \sigma > -A$$

Since $$s = \sigma + j\omega$$

$$S(s) = \frac{1}{s+A}, \; \sigma > -A \qquad \text{...(5)}$$

That is $$e^{At}u(t) \xleftrightarrow{\text{LT}} \frac{1}{s+A}, \; \sigma > -A \qquad \text{...(6)}$$

If $A = 0$ then we can determine the Laplace transform of unit step function $u(t)$ as

$$e^{0t}u(t) \xleftrightarrow{\text{LT}} \frac{1}{s+0}, \; \sigma > 0$$

$$u(t) \xleftrightarrow{\text{LT}} \frac{1}{s}, \; \sigma > 0 \qquad \text{...(7)}$$

where σ and ω are the real and imaginary parts of complex variable s respectively.

Example 9.2 Find the Laplace transform of a continuous-time signal

$$s(t) = -e^{-At}u(-t)$$

Solution. Laplace transform of a signal $s(t)$ can be determined as

$$S(s) = \int_{-\infty}^{\infty} s(t)e^{-st}\, dt$$

$$= \int_{-\infty}^{\infty} [-e^{-At}u(-t)]\, e^{-st}\, dt$$

or $$S(s) = -\int_{-\infty}^{\infty} e^{-At}\, u(-t)e^{-st}\, dt \qquad \text{...(1)}$$

Since $$u(-t) = \begin{cases} 1, & t \le 0 \\ 0, & t > 0 \end{cases} \qquad \text{...(2)}$$

Substituting Eq. 2 in Eq. 1, we get

$$S(s) = -\int_{-\infty}^{0} e^{-At}\, 1\, e^{-st}\, dt = -\int_{-\infty}^{0} e^{-(s+A)t}\, dt$$

$$= -\left[\frac{1}{-(s+A)} e^{-(s+A)t} \right]_{-\infty}^{0} = \frac{1}{s+A}$$

or $$S(s) = \frac{1}{s+A}, \quad s + A < 0$$

or $$S(s) = \frac{1}{s+A}, \quad s < -A \qquad \text{...(3)}$$

Laplace transform of a continuous-time signal $s(t) = -e^{-At}u(-t)$ can be written as

$$-e^{-At}u(-t) \xleftrightarrow{\text{LT}} \frac{1}{s+A}, \quad \sigma < -A \qquad \text{...(4)}$$

σ is the real part of the complex variable s.

Comparing the Laplace transform which are determined in Examples 9.1 and 9.2, we see that the algebraic expressions for the Laplace transforms are identical for both the signals. However, for same expressions of Laplace transforms, the set of values of s for which the expressions are valid, are different.

In general, the range of values of s for which the integral of equation $S(s) = \int_{-\infty}^{\infty} s(t)\, e^{-st}\, dt$ converges is referred to as the region of convergence (ROC) of the Laplace transform.

9.3 THE REGION OF CONVERGENCE FOR THE LAPLACE TRANSFORMS

The range of values of s for which the integral of equation

$$S(s) = \text{LT}\,\{s(t)\} = \int_{-\infty}^{\infty} s(t)\, e^{-st}\, dt$$

converges is called the region of convergence (ROC) of the Laplace transform (LT) of the continuous-time signal $s(t)$. That is, the region of convergence (ROC) consists of those values of complex variable $s = \sigma + j\omega$ for which the Continuous-time Fourier transform (CTFT) of $s(t)\, e^{-\sigma t}$ converges. The region of convergence (ROC) can be displayed on s-plane, which is a complex variable. The coordinate axes are Real $\{s\}$ along the horizontal axis and Imaginary $\{s\}$ along the vertical axis. The horizontal and vertical axis are also called the σ-axis and $j\omega$-axis, respectively. The region of convergence (ROC) for Laplace transforms of two continuous-time signals is given as

$$e^{At}u(t) \xleftrightarrow{\text{LT}} \frac{1}{s+A}, \quad \text{ROC: } \sigma > -A$$

and $$-e^{-At}u(-t) \xleftrightarrow{\text{LT}} \frac{1}{s+A}, \quad \text{ROC: } \sigma < -A$$

The ROC for above Laplace transforms are shown by shaded portion in Fig. 9.1.

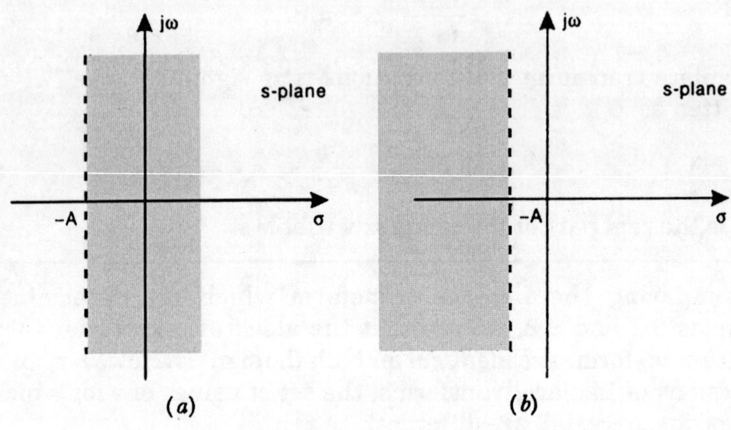

Fig. 9.1

Example 9.3 Determine the Laplace transform of a continuous-time signal

$$s(t) = 2e^{-3t}u(t) - e^{-2t}u(t)$$

Also find the ROC of Laplace transform of above signal.

Solution. Laplace transform of a signal $s(t)$ is determined as

$$S(s) = \text{LT}\{s(t)\} = \int_{-\infty}^{\infty} s(t)\, e^{-st}\, dt$$

$$= \int_{-\infty}^{\infty} [2e^{-3t}u(t) - e^{-2t}u(t)\,]\, e^{-st}\, dt$$

$$= 2\int_{-\infty}^{\infty} e^{-3t}u(t)\, e^{-st}\, dt - \int_{-\infty}^{\infty} e^{-2t}u(t)\, e^{-st}\, dt$$

From the definition unit step function $u(t) = \begin{cases} 1, & t > 0 \\ 0, & t < 0 \end{cases}$

$$S(s) = 2\int_{0}^{\infty} e^{-3t}\, 1\, e^{-st}\, dt - \int_{0}^{\infty} e^{-2t}\, 1\, e^{-st}\, dt$$

$$= 2\int_{0}^{\infty} e^{-(s+3)t}\, dt - \int_{0}^{\infty} e^{-(s+2)t}\, dt$$

$$= 2\left[\frac{1}{-(s+3)}e^{-(s+3)t}\right]_0^\infty - \left[\frac{1}{-(s+2)}e^{-(s+2)t}\right]_0^\infty$$

or $$S(s) = \frac{1}{s+3} - \frac{1}{s+2} \qquad ...(1)$$

This is the Laplace transform of the signal $s(t) = 2e^{-3t}u(t) - e^{-2t}u(t)$

The overall ROC of the Laplace transform of the signal given in example 9.3, is shown in Fig. 9.2.

Determination of ROC :

If $$e^{-3t}u(t) \xleftrightarrow{LT} \frac{1}{s+3}, \; R_1 : \sigma > -3$$

and $$e^{-2t}u(t) \xleftrightarrow{LT} \frac{1}{s+2}, \; R_2 : \sigma > -2$$

then $$2e^{-3t}u(t) - e^{-2t}u(t) \xleftrightarrow{LT} \frac{2}{s+3} - \frac{1}{s+1}$$

$$= \frac{(s-1)}{(s+3)(s+1)}$$

$$R = R_1 \cap R_2 = \sigma > -2$$

where R_1, R_2 and R are the ROCs.

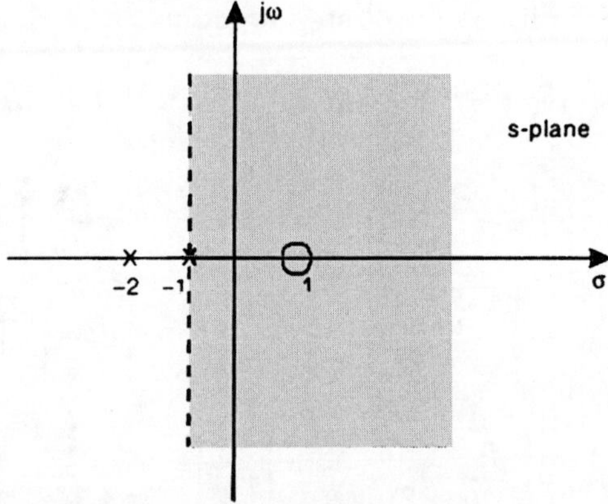

Fig. 9.2 *s-plane representation of the Laplace transform for example 9.3. The shaded portion indicates the ROC of the Laplace transform of signal s(t). Here circle (O) shows the location of zero and cross (X) shows the location of pole in the Laplace transform S(s).*

Example 9.4 Determine the Laplace transform of the continuous-time signal

$$s(t) = e^{-2t}u(t) + e^{-t}\cos(3t)u(t)$$

Also determine the ROC of the Laplace transform of the above signal.

Solution. By using Euler's relation, signal $s(t)$ can be written as

$$s(t) = e^{-2t}u(t) + e^{-t}\cos(3t)u(t)$$

$$= e^{-2t}u(t) + e^{-t}\left[\frac{e^{j3t} + e^{-j3t}}{2}\right]u(t)$$

$$= e^{-2t}u(t) + \frac{1}{2}e^{(3j-1)t}u(t) + \frac{1}{2}e^{(-3j-1)t}u(t)$$

$$s(t) = e^{-2t}u(t) + \frac{1}{2}e^{-(1-3j)t}u(t) + \frac{1}{2}e^{-(1+3j)t}u(t) \quad ...(1)$$

Now Laplace transform of signal $s(t)$ can be determined as

$$S(s) = \text{LT}\,\{s(t)\} = \int_{-inf}^{\infty} s(t)\,e^{-st}\,dt$$

$$= \int_{-\infty}^{\infty}\left[e^{-2t}u(t) + \frac{1}{2}e^{-(1-3j)t} + \frac{1}{2}e^{-(1+3j)t}u(t)\right]e^{-st}\,dt$$

or

$$S(s) = \int_{-\infty}^{\infty} e^{-2t}u(t)\,e^{-st}\,dt + \frac{1}{2}\int_{-\infty}^{\infty} e^{-(1-3j)t}u(t)\,e^{-st}\,dt$$

$$+ \frac{1}{2}\int_{-\infty}^{\infty} e^{-(1+3j)t}u(t)\,e^{-st}\,dt$$

Using the definition of unit step function

$$S(s) = \int_{0}^{\infty} e^{-2t}\,1\,e^{-st}\,dt + \frac{1}{2}\int_{0}^{\infty} e^{-(1-3j)t}\,1\,e^{-st}\,dt$$

$$+ \frac{1}{2}\int_{0}^{\infty} e^{-(1+3j)t}\,1\,e^{-st}\,dt$$

$$= \int_{0}^{\infty} e^{-(s+2)t}\,dt + \frac{1}{2}\int_{0}^{\infty} e^{-(s+1-3j)t}\,dt + \frac{1}{2}\int_{0}^{\infty} e^{-(s+1+3j)t}\,dt$$

$$= \left[-\frac{1}{(s+2)}e^{-(s+2)t}\right]_0^{\infty} + \frac{1}{2}\left[-\frac{1}{(s+1-3j)}e^{-(s+1-3j)t}\right]_0^{\infty}$$

$$+ \frac{1}{2}\left[-\frac{1}{(s+1+3j)}e^{-(s+1+3j)t}\right]_0^{\infty}$$

$$= \frac{1}{s+2} + \frac{1}{2}\left(\frac{1}{(s+1-3j)}\right) + \frac{1}{2}\left(\frac{1}{(s+1+3j)}\right)$$

or

$$S(s) = \frac{1}{s+2} + \frac{1}{2}\left(\frac{1}{s+(1-3j)}\right) + \frac{1}{2}\left(\frac{1}{s+(1+3j)}\right) \quad ...(2)$$

Determination of ROCs :

If $$e^{-2t}u(t) \xleftrightarrow{\text{LT}} \frac{1}{s+2},\ R_1 : \sigma > -2$$

$$e^{-(1-3j)t}su(t) \xleftrightarrow{\text{LT}} \frac{1}{s+(1-3j)},\ R_2 : \sigma > -1$$

$$e^{-(1+3j)t}u(t) \xleftrightarrow{\text{LT}} \frac{1}{s+(1+3j)},\ R_3 : \sigma > -1$$

then $$e^{-2t}u(t) + \frac{1}{2}e^{-(1-3j)t}u(t) + \frac{1}{2}e^{-(1+3j)t}u(t)$$

$$\xleftrightarrow{\text{LT}} \frac{1}{(s+2)} + \frac{1}{2}\left(\frac{1}{s+(1-3j)}\right) + \frac{1}{2}\left(\frac{1}{s+(1+3j)}\right)$$

$$= \frac{2s^2+5s+12}{(s^2+2s+10)(s+2)},$$

$$R : R_1 \cap R_2 \cap R_3 = \sigma > -1$$

where R_1, R_2, R_3 and R are the ROCs of the Laplace transform. The overall ROC of the Laplace transform of the signal given in example 9.4 is shown in Fig. 9.3.

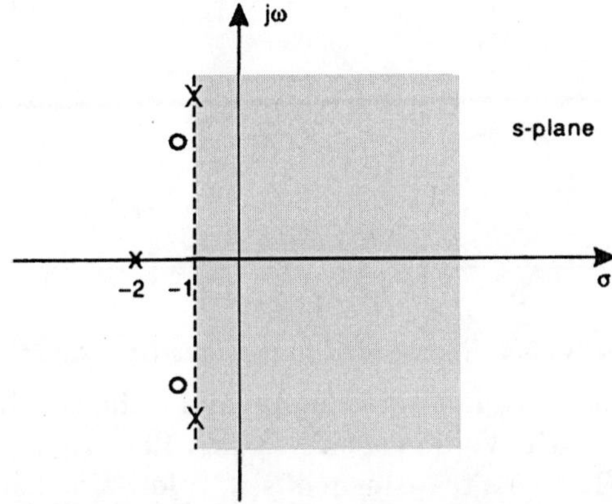

Fig. 9.3 *s-plane representation of the Laplace transform for example 9.4. The shaded portion indicates the ROC of the Laplace transform of the signal s(t).*

9.3.1 Properties of ROCs for Laplace Transforms

The Laplace transform of a continuous-time signal $s(t)$ is specified by not only the algebraic expression for $S(s)$, but also by its ROC. In some cases, Laplace transform expressions are same but ROCs are different. In this subsection, we will study various properties of ROCs of the Laplace transforms. The Laplace transforms can olny be distinguished by their region of convergence (ROC).We can reconstruct the ROC from the knowledge of only the algebraic expression for Laplace transform of

$s(t)$, *i.e.*, $S(s)$ and certain general characteristics of $s(t)$ in the time-domain. Now let us discuss some properties of ROC :

Property 1. The ROC of $S(s)$ consists of strips parallel to the imaginary axis, *i.e.*, $j\omega$-axis in the s-plane.

The ROC of the Laplace transform of signal $s(t)$ consists of those values of s for which $s(t)\, e^{-\sigma t}$ is absolutely integrable.

$$\int_{-\infty}^{\infty} |s(t)|\, e^{-\sigma t}\, dt < \infty \qquad \text{...(9.8)}$$

Property 2. The ROC for rational Laplace transforms does not contain any poles.

Property 3. If $s(t)$ is a finite-duration signal and is absolutely integrable, then ROC of Laplace transform of $s(t)$ is the entire s-plane.

Property 4. If $s(t)$ is a right-sided signal and if the line Real $\{s\} = \sigma_0$ is in the ROC, then all the values of s for which Real $\{s\} > \sigma_0$ will also be in the ROC.

A right-sided signal is a signal for which $s(t) = 0$ prior to some finite time T_1. For a right-sided signal, the ROC is commonly referred to as a right-half plane.

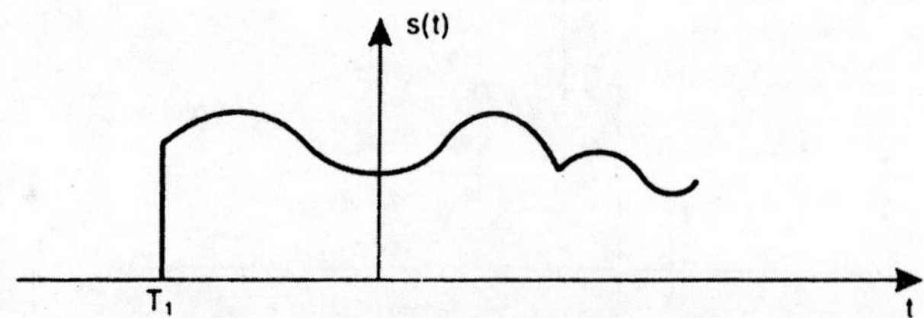

Fig. 9.4 *Right-sided continuous-time signal.*

Property 5. If $s(t)$ is a left-sided signal and if the line Real $\{s\} = \sigma_0$ is in the ROC, then all the values of s for which Real $\{s\} < \sigma_0$ will also be in the ROC of the Laplace transform of $s(t)$. A left-sided signal is a signal for which $s(t) = 0$ after some finite time T_2. The ROC for left-sided signal is commonly referred to as a left-half plane.

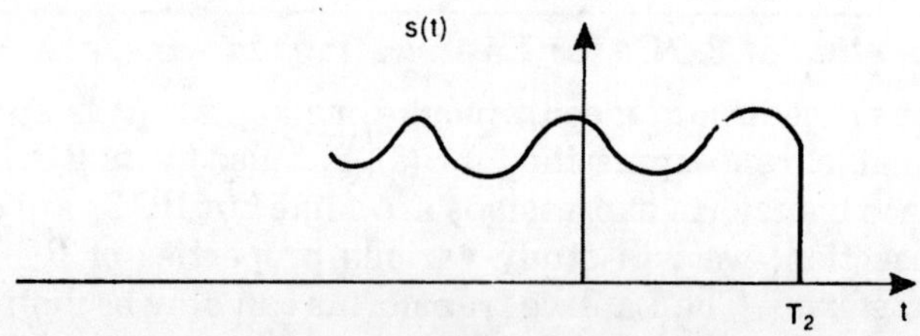

Fig. 9.5 *Left-sided continuous-time signal.*

Property 6. If $s(t)$ is a two-sided signal, and if the line Real $\{s\} = \sigma_0$ is in the ROC, then the ROC of the Laplace transform of $s(t)$ will consist of a strip in the s-plane which will include the line Real $\{s\} = \sigma_0$.

A two-sided signal is a signal which is of infinite extent for both $t > 0$ and $t < 0$. Two-sided signal is shown in Fig. 9.6 (*a*).

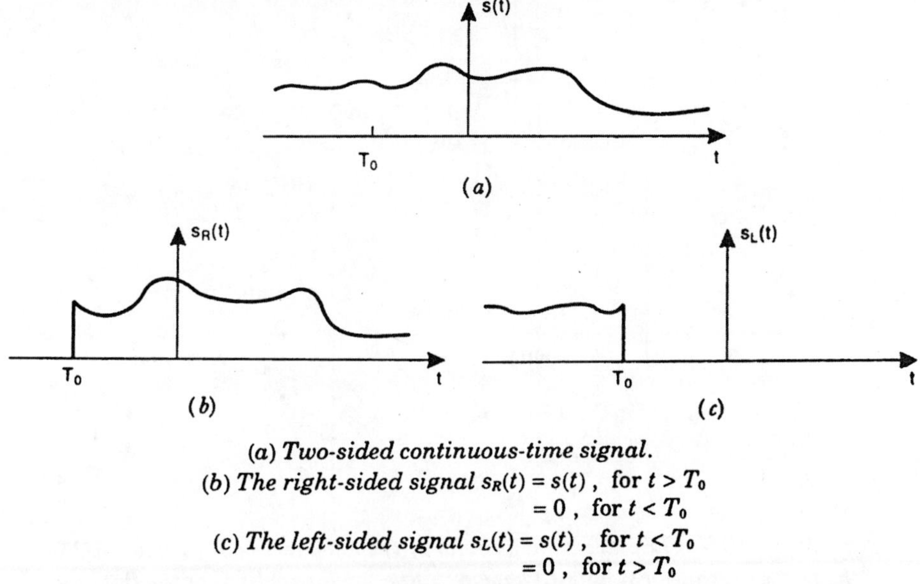

(*a*) *Two-sided continuous-time signal.*

(*b*) *The right-sided signal* $s_R(t) = s(t)$, for $t > T_0$
$= 0$, for $t < T_0$

(*c*) *The left-sided signal* $s_L(t) = s(t)$, for $t < T_0$
$= 0$, for $t > T_0$

Fig. 9.6 *Division of two-sided signal* $s(t)$ *into the sum of right-sided signal* $s_R(t)$ *and left-sided signal* $s_L(t)$.

The ROC of Laplace transform of two-sided signal $s(t)$ can be examined by choosing an arbitrary time T_0 and dividing the signal $s(t)$ into the sum of a right-sided signal $s_R(t)$ and a left-sided signal $s_L(t)$. These two signals are shown in Fig. 9.6 (*b*) and Fig. 9.6 (*c*).

The Laplace transform of two-sided signal $s(t)$ converges for those values of s for which the Laplace transforms of both $s_R(t)$ and $s_L(t)$ converge. We know from property 4, the ROC of Laplace transform of $s_R(t)$ consists of a half-plane Real $\{s\} > \sigma_R$ for some value of σ_R and from property 5, the ROC of Laplace transform of $s_L(t)$ consists of a half-plane Real $\{s\} < \sigma_L$ for some value of σ_L. The ROC of Laplace transform of $s(t)$ is then the overlap of these two half-planes. These are shown in Fig. 9.7.

The ROCs of Laplace transforms of right-sided signal $s_R(t)$, left-sided signal $s_L(t)$ and two-sided signal $s(t)$ is shown in Fig. 9.7. This assumes, that $\sigma_R < \sigma_L$, so that there is some overlap. If there is no overlap, then even though the Laplace transforms of $s_R(t)$ and $s_L(t)$ individually exist but Laplace transform of two-sided signal $s(t)$ cannot exist.

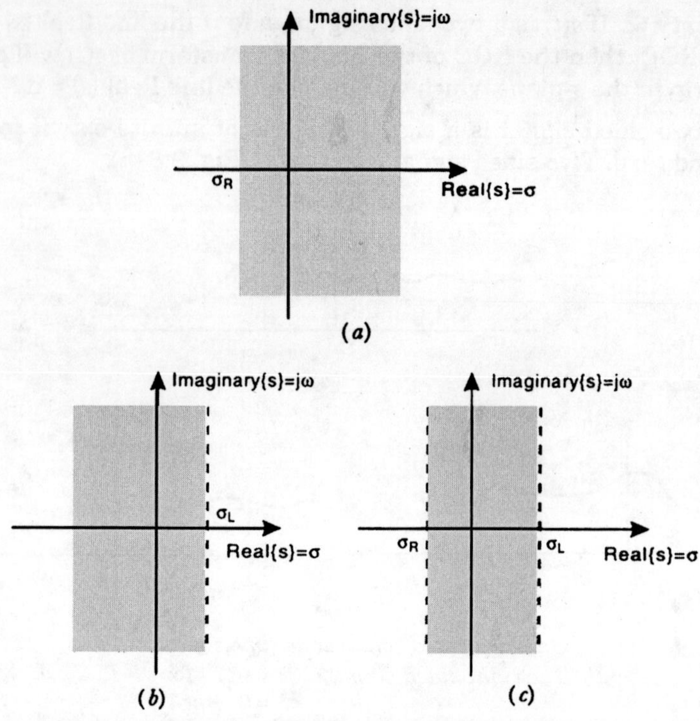

Fig. 9.7 (*a*) *ROC of Laplace transform for $s_R(t)$ shown in Fig.* 9.6 (*a*). (*b*) *ROC of the Laplace transform for $s_L(t)$ shown in Fig.* 9.6 (*c*). (*c*) *ROC of the Laplace transform for two-sided signal s(t) shown in Fig.* 9.6 (*a*). *$s(t) = s_R(t) + s_L(t)$, assuming that the ROCs in* (*a*) *and* (*b*) *overlap.*

Property 7. If the Laplace transform $S(s)$ of signal $s(t)$ is a rational function then its ROC is bounded by poles or extends to infinity. In addition, no poles of $S(s)$ are contained in the ROC of Laplace transform of signal $s(t)$. The validity of this property results from the fact that a signal with a rational Laplace transform consists of a linear combination of exponential functions and the ROC for the Laplace transform of individual terms in this combination must have this property.

Property 8. If the Laplace transform $S(s)$ of the signal $s(t)$ is a rational function, then if signal $s(t)$ is right-sided, the ROC is the region in the s-plane to the right of the right most pole.

Example 9.5 Determine the ROC of the Laplace transform

$$S(s) = \frac{1}{(s+2)\,(s+3)}$$

Also draw the associated pole-zero pattern for Laplace transform $S(s)$.

Solution. Determination of poles and zeros of $S(s)$. There are no zeros in $S(s)$.

There are two poles determined as

$$s + 2 = 0 \quad \text{or} \quad s = -2$$

$$s + 3 = 0 \quad \text{or} \quad s = -3$$

Pole-zero pattern for $S(s)$ is shown in Fig. 9.8 (*a*).

There are three possible ROCs that can be associated with this algebraic expression, corresponding to three different signals. The signal associated with the pole-zero pattern given in Fig. 9.8 (*b*) is right-sided signal. Because the ROC includes the imaginary axis ($j\omega$), the Continuous-time Fourier transform (CTFT) of this signal converges. The pole-zero pattern of Fig. 9.8 (*c*) corresponds to the left-sided signal. The pole-zero pattern of Fig. 9.8 (*d*) corresponds to the two- sided signal neitheı the left-sided nor the two-sided signals have CTFT because their ROCs do not include the imaginary axis ($j\omega$).

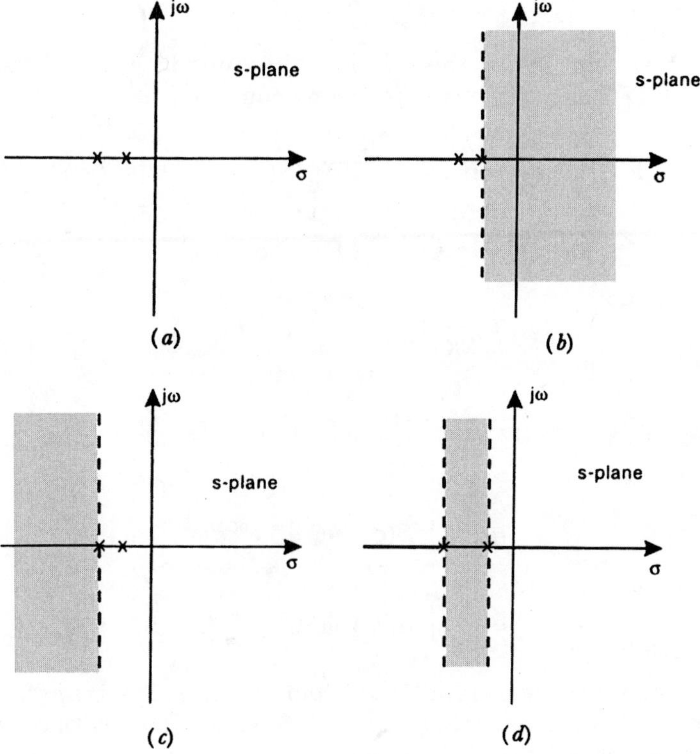

Fig. 9.8 (*a*) *Pole-zero pattern for Laplace transform* $S(s) = \dfrac{1}{(s+2)(s+3)}$ (*b*) *ROC corresponding to a right-sided signal* (*c*) *ROC corresponding to a left-sided signal* (*d*) *ROC corresponding to a two-sided signal.*

9.4 INVERSE LAPLACE TRANSFORM

We have already studied that Laplace transform of a continuous-time signal $s(t)$ is interpreted as Continuous-time Fourier transform (CTFT) of an exponentially weighted version of the signal $s(t)$.

Laplace transform of a signal $s(t)$ is defined as

$$S(s) = \text{LT}\{s(t)\} = \int_{-\infty}^{\infty} s(t)\, e^{-st}\, dt \qquad \text{...(9.9)}$$

Substituting $s = \sigma + j\omega$ in the ROC in Eq. 9.9, we get

$$S(\sigma + j\omega) = \int_{-\infty}^{\infty} s(t)\, e^{-(\sigma + j\omega)t}\, dt$$

$$= \int_{-\infty}^{\infty} [s(t)\, e^{-\sigma t}]\, e^{-j\omega t}\, dt$$

$$= \text{CTFT}\{s(t)\, e^{-\sigma t}\} \qquad \text{...(9.10)}$$

Now we can invert the relationship given in Eq. 9.10 using the inverse CTFT as given by the following equation.

$$s(t) = \text{Inverse CTFT}\{S(j\omega)\} = \frac{1}{2\pi} \int_{-\infty}^{\infty} S(j\omega)\, e^{j\omega t}\, d\omega$$

or $$s(t)\, e^{-\sigma t} = \text{Inverse CTFT}\{S(\sigma + j\omega)\}$$

$$= \frac{1}{2\pi} \int_{-\infty}^{\infty} S(\sigma + j\omega)\, e^{j\omega t}\, d\omega \qquad \text{...(9.11)}$$

or $$s(t) = \frac{1}{2\pi} e^{\sigma t} \int_{-\infty}^{\infty} S(\sigma + j\omega)\, e^{j\omega t}\, d\omega$$

$$= \frac{1}{2\pi} \int_{-\infty}^{\infty} S(\sigma + j\omega)\, e^{j\omega t}\, e^{\sigma t}\, d\omega$$

or $$s(t) = \frac{1}{2\pi} \int_{-\infty}^{\infty} S(\sigma + j\omega)\, e^{(\sigma + j\omega)t}\, d\omega \qquad \text{...(9.12)}$$

We can recover signal $s(t)$ from its Laplace transform $S(s)$ determined for a set of values of $s = \sigma + j\omega$ in the ROC, with σ fixed and ω varying from $-\infty$ to $+\infty$.

Now to recover $s(t)$ from $S(s)$ we change the variable of integration in Eq. 9.12 from ω to $s = \sigma + j\omega$, with σ fixed and ω varying from $-\infty$ to $+\infty$ and $ds = jd\omega$. This results in the inverse Laplace transform which is given by

$$s(t) = \frac{1}{2\pi} \int_{s=\sigma-j\infty}^{\sigma+j\infty} S(s)\, e^{st} \left(\frac{ds}{j}\right)$$

or
$$s(t) = \frac{1}{2\pi j} \int_{s=\sigma-j\infty}^{\sigma+j\omega} S(s)\, e^{st}\, ds \quad ...(9.13)$$

Eq. 9.13 is called the inverse Laplace transformation equation. It states that $s(t)$ can be represented as a weighted integral of complex exponential functions. The contour of integration in inverse Laplace transformation equation is a straight line in the s-plane corresponding to all points s which satisfy Real $\{s\} = \sigma$. This line is parallel to the imaginary axis, *i.e.*, $j\omega$-axis. Also we can choose any value of σ such that $S(\sigma + j\omega)$ converges. The formal determination of the integral given in Eq. 9.13 for a general $S(s)$ requires the use of contour integration in the complex plane (s-plane). The inverse Laplace transform for a rational Laplace transform can be determined by using partial fraction expansion method that was used to determine the inverse CTFT in chapter 5. The partial fraction expansion method is based on expanding the rational algebraic expression into a linear combination of lower order terms.

Example 9.6 Find the inverse Laplace transform of

$$S(s) = \frac{1}{(s+2)(s+3)}, \quad \text{ROC : Real } \{s\} > -2$$

using partial fraction expansion method.

Solution. To obtain inverse Laplace transform, we first perform a partial-fraction expansion as

$$S(s) = \frac{1}{(s+2)(s+3)} \quad ...(1)$$

$$S(s) \frac{1}{(s+2)(s+3)} = \frac{\alpha_1}{(s+2)} + \frac{\alpha_2}{(s+3)} \quad ...(2)$$

$s + 2 = 0$ or $s = -2$', $\alpha_1 = \dfrac{1}{s+3} = \dfrac{1}{-2+3} = 1$

$s + 3 = 0$ or $s = -3$', $\alpha_2 = \dfrac{1}{s+2} = \dfrac{1}{-3+2} = -1$

Substituting the values of α_1 and α_2 in Eq. 2, we get

$$S(s) = \frac{1}{s+2} + \frac{(-1)}{s+3} \quad ...(3)$$

Taking the inverse Laplace transform of Eq. 3, we get

$$\text{Inverse LT } \{S(s)\} = \text{Inverse LT} \left\{\frac{1}{s+2} - \frac{1}{s+3}\right\}$$

or $\quad s(t) = \text{Inverse LT}\left\{\dfrac{1}{s+2}\right\} - \text{Inverse LT}\left\{\dfrac{1}{s+3}\right\}$

or $\quad s(t) = e^{-2t}u(t) - e^{-3t}u(t) = (e^{-2t} - e^{-3t})u(t) \quad \text{...(4)}$

where $\quad e^{-2t}u(t) \xleftrightarrow{\text{LT}} \dfrac{1}{s+2}$, ROC : Real $\{s\} > -2$

and $\quad e^{-3t}u(t) \xleftrightarrow{\text{LT}} \dfrac{1}{s+3}$, ROC : Real $\{s\} > -3$

then $\quad e^{-2t}u(t) - e^{-3t}u(t) \xleftrightarrow{\text{LT}} \dfrac{1}{s+2} - \dfrac{1}{s+3} = \dfrac{1}{(s+1)\,(s+2)}$

ROC : Real $\{s\} > -2$

Construction of the ROCs for the individual terms in the partial fraction expansion of $S(s)$ is shown in Fig. 9.9.

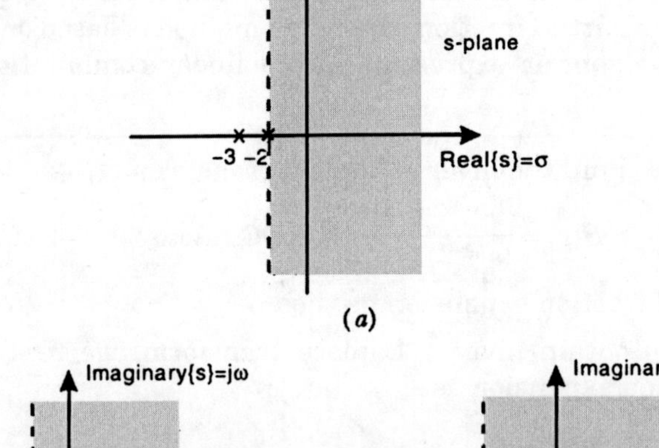

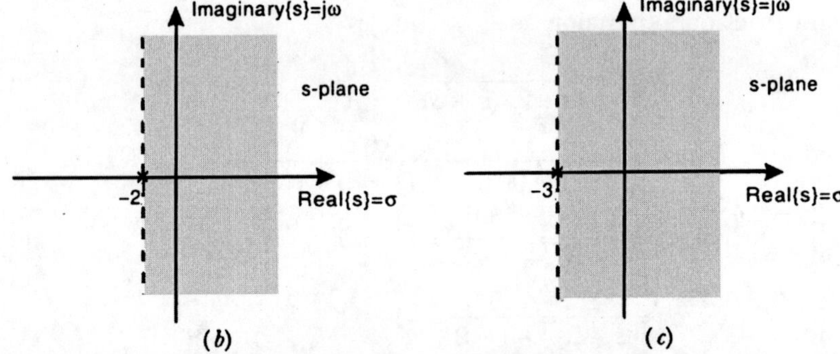

Fig. 9.9 *Construction of the ROCs for the individual terms in the partial fraction expansion of $S(s)$. (a) Pole-zero plot and ROC for $S(s)$ (b) Pole at $s = -2$ and its ROC (c) Pole at $s = -3$ and its ROC.*

9.5 PROPERTIES OF THE LAPLACE TRANSFORM

In chapter 5, we have studied the Continuous-time Fourier transform (CTFT) and its properties. Here we will discuss various properties of the Laplace transform. The derivations of many of the properties of Laplace transform are analogous to those of the corresponding properties for

CTFT. So there is no need to give derivations for various properties of Laplace transform.

We will discuss following properties of Laplace transform

1. Linearity of Laplace transform.
2. Time-shifting of a continuous-time signal.
3. Shifting in the *s*-domain.
4. Time scaling of continuous-time signal.
5. Complex conjugation.
6. Convolution property for Laplace transform.
7. Differentiation of a continuous-time signal in time-domain.
8. Differentiation in the *s*-domain.
9. Integration of continuous-time signal in time-domain.
10. The initial value theorem.
11. The final value theorem..

9.5.1 Linearity of the Laplace Transform

Laplace transform also satisfies the principle of superposition *i.e.*, it is a linear transformation tool. Let us consider two continuous-time signals $s_1(t)$ and $s_2(t)$ with Laplace transforms $S_1(s)$ and $S_2(s)$, respectively.

If $$s_1(t) \xleftrightarrow{\text{LT}} S_1(s)\ ,\ \text{ROC}: R_1$$

and $$s_2(t) \xleftrightarrow{\text{LT}} S_2(s)\ ,\ \text{ROC}: R_2$$

then $$s(t) = A_1 s_1(t) + A_2 s_2(t) \xleftrightarrow{\text{LT}} S(s) = A_1 S_1(s) + A_2 S_2(s)$$

$$\text{ROC}: R = R_1 \cap R_2 \qquad \text{...(9.14)}$$

where R_1, R_2 and R are ROCs for $S_1(s)$, $S_2(s)$ and $S(s)$, respectively. A_1 and A_2 are the constants.

The region of convergence (ROC) of $S(s)$ is atleast the intersection of R_1 and R_2. If intersection of R_1 and R_2, *i.e.*, $R_1 \cap R_2$ is empty then $S(s)$ has no region of convergence (ROC). It means signal $s(t)$ has no Laplace transform. The ROC associated with a linear combination of terms can be constructed by using the properties of ROC of $S(s)$. From the intersection of the ROCs for individual terms, we can obtain a line or strip that is in the ROC of the linear combination.

9.5.2 Time-Shifting of a Continuous-time Signal

The purpose of time shifting is to study a particular signal in its past time or in future time by delaying or advancing the signal in time, respectively. Consider a continuous-time signal $s(t)$ and its Laplace transform $S(s)$.

If $$s(t) \xleftrightarrow{\text{LT}} S(s) \quad \text{with ROC} = R$$

then Laplace transform pair of delayed version of signal $s(t)$ by time t_0 is given as

$$s(t - t_0) \xleftrightarrow{\text{LT}} e^{-st_0} S(s) \quad \text{with ROC} = R. \qquad ...(9.15)$$

9.5.3 Shifting of Laplace Transform S(s) in the s-Domain

Consider a continuous-time signal $s(t)$ and its Laplace transform $S(s)$. The delayed version of $S(s)$ by s_0 is given by $S(s - s_0)$.

If $$s(t) \xleftrightarrow{\text{LT}} S(s - j\omega_0) \quad \text{with ROC} = R$$

then $$e^{s_0t} s(t) \xleftrightarrow{\text{LT}} S(s - s_0) \quad \text{with ROC} = R + \sigma_0 \qquad ...(9.16)$$

where $s_0 = \sigma_0 + j\omega_0$ is a complex constant and σ_0 is the real part of s_0.

The ROC associated with $S(s - s_0)$ is equal to $R + \sigma_0$, where R is the ROC of $S(s)$. A special case of Eq. 9.16 is given by $s_0 = j\omega_0$. It is used to modulate a periodic complex exponential $e^{j\omega_0t}$.

$$e^{j\omega_0t} s(t) \xleftrightarrow{\text{LT}} S(s - j\omega_0) \quad \text{with ROC} = R \qquad ...(9.17)$$

The right hand side of Eq. 9.17 can be interpreted as a shift in the s-plane parallel to the $j\omega$-axis.

9.5.4 Time Scaling of Continuous-Time Signal

Consider a continuous-time signal $s(t)$ and its Laplace transform $S(s)$. Time scaled version of $s(t)$ is given by $s(At)$.

If $$s(t) \xleftrightarrow{\text{LT}} S(s) \quad \text{with ROC} = R$$

then $$s(At) \xleftrightarrow{\text{LT}} \frac{1}{|A|} S\left(\frac{s}{A}\right) \quad \text{with ROC ; } R_1 = \frac{R}{A} \qquad ...(9.18)$$

That is, for any value s in R, the value s/A will be in R_1 for a positive value of $A > 1$.

Note. For $A > 1$, there is a compression in the size of the ROC of $S(s)$ by a factor $1/A$, and for $0 < A < 1$, there is an expansion in the size of the ROC of $S(s)$ by a factor $1/A$. If A is negative, the ROC undergoes a reversal plus a scaling.

For $0 < A > -1$ it involves a reversal about the imaginary axis ($j\omega$). There is also a change in the size of the ROC by a factor of $1/|A|$. Thus, time-reversal of signal $s(t)$ results in a time-reversal of the ROC.

$$s(-t) \xleftrightarrow{\text{LT}} S(-s) \quad \text{with ROC} = -R \qquad ...(9.19)$$

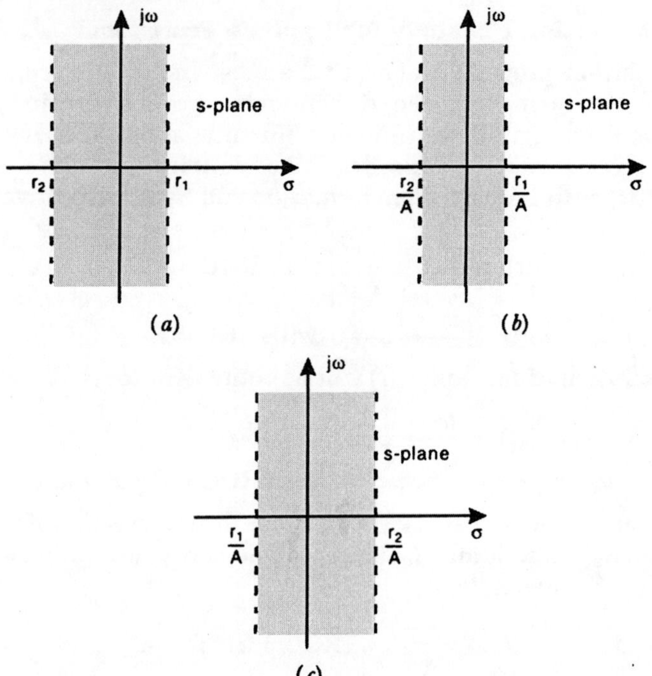

Fig. 9.10 *Illustration of effect on the ROC of time scaling* (*a*) *ROC of* $S(s)$ (*b*) *ROC of* $\frac{1}{|A|} S\left(\frac{s}{A}\right)$ *for* $A > 1$ (*c*) *ROC of* $\frac{1}{|A|} S\left(\frac{s}{A}\right)$ *for* $0 < A < -1$.

9.5.5 Complex Conjugation

Consider a complex continuous-time signal $s(t)$, its Laplace transform $S(s)$ is also a complex function. If $s(t) = s_R(t) + js_I(t)$ then its complex conjugate is given by $s^*(t) = s_R(t) - js_I(t)$ and if $S(s) = S_R(s) + jS_I(s)$ then its complex conjugate is given by $S^*(s) = S_R(s) - jS_I(s)$.

If $$s(t) \xleftrightarrow{\text{LT}} S(s) \quad \text{with ROC} = R_1$$

then $$s^*(t) \xleftrightarrow{\text{LT}} S^*(s^*) \quad \text{with ROC} = R_1 \qquad ...(9.20)$$

where s is a complex variable and is given as $s = \sigma + j\omega$. Complex conjugate of $s = s^* = \sigma - j\omega$.

If signal $s(t)$ is real then

$$S(s) = S^*(s^*) \qquad ...(9.21)$$

Consequently, if signal $s(t)$ is real and if $S(s)$ has a pole or zero at $s = s_0$ then $S(s)$ also has a pole or zero at the complex conjugate point $s = s_0^*$.

9.5.6 Convolution Property for Laplace Transform

The convolution property is one of the most important properties of Laplace transform because it converts the convolution of two continuous-time signals in time-domain into multiplication of their Laplace transforms. Consider two continuous-time signals $s_1(t)$ and $s_2(t)$ with Laplace transform $S_1(s)$ and $S_2(s)$, respectively.

If $\quad s_1(t) \xleftrightarrow{\text{LT}} S_1(s) \quad \text{with ROC} = R_1$

and $\quad s_2(t) \xleftrightarrow{\text{LT}} S_2(s) \quad \text{with ROC} = R_2$

then convolution of $s_1(t)$ and $s_2(t)$ will be equivalent to

$$s(t) = s_1(t) * s_2(t) \xleftrightarrow{\text{LT}} S(s) = S_1(s)S_2(s)$$

$$\text{with ROC} = R = R_1 \cap R_2 \qquad \text{...(9.22)}$$

It means that ROC of $S_1(s)S_2(s)$ includes the intersection of the ROCs of $S_1(s)$ and $S_2(s)$ and it may be larger if pole-zero cancellation occurs in the product. For example, if

$$S_1(s) = \frac{s+2}{s+3}, \quad \text{ROC} = R_1 : \sigma > -3$$

and
$$S_2(s) = \frac{s+3}{s+2}, \quad \text{ROC} = R_2 : \sigma > -2$$

then
$$S_1(s)S_2(s) = \frac{s+2}{s+3} \cdot \frac{s+3}{s+2} = 1$$

and its ROC is the entire s-plane.

9.5.7 Differentiation of a Continuous-Time Signal in Time-Domain

Consider a continuous-time signal $s(t)$ and its Laplace transform $S(s)$.

If $\quad s(t) \xleftrightarrow{\text{LT}} S(s) \quad \text{with ROC} = R$

then $\quad \dfrac{ds(t)}{dt} \xleftrightarrow{\text{LT}} sS(s) \quad \text{with ROC containing } R \qquad \text{...(9.23)}$

This property followed by differentiating both sides of inverse Laplace transformation equation

$$s(t) = \frac{1}{2\pi j} \int_{S=\sigma-j\infty}^{\sigma+j\infty} S(s)\, e^{st}\, ds$$

$$\frac{ds(t)}{dt} = \frac{1}{2\pi j} \int_{s=\sigma-j\infty}^{\sigma+j\infty} S(s)\, (se^{st})\, ds \qquad \text{...(9.24)}$$

Consequently, $\frac{ds(t)}{dt}$ is the inverse Laplace transform of $sS(s)$. The ROC of $sS(s)$ includes the ROC of $S(s)$ and ROC may be larger if $S(s)$ has first order pole at $s = 0$ that is cancelled by the multiplication by s.

9.5.8 Differentiation in s-Domain

Consider a continuous-time signal $s(t)$ and its Laplace transform $S(s)$.

Differentiating both sides of the Laplace transform equation w.r.t. to s

$$S(s) = \int_{-\infty}^{\infty} s(t)\, e^{-st}\, dt \qquad \text{...(9.25)}$$

$$\frac{dS(s)}{ds} = \int_{-\infty}^{\infty} s(t)\,(-t\, e^{-st})\, dt$$

$$= \int_{-\infty}^{\infty} (-t)\, s(t)\, e^{-st}\, dt \qquad \text{...(9.26)}$$

Consequently, if

$$s(t) \xleftrightarrow{\text{LT}} S(s) \text{ with ROC} = R$$

then

$$-t\, s(t) \xleftrightarrow{\text{LT}} \frac{dS(s)}{ds}, \text{ with ROC} = R \qquad \text{...(9.27)}$$

Example 9.7 Find the Laplace transform of a continuous-time signal $te^{-At}u(t)$.

Solution. We know that Laplace transform of $e^{-At}u(t)$ is equal to $\frac{1}{s+A}$ with ROC : $\sigma > -A$, *i.e.*,

$$s(t) = e^{-At}u(t) \xleftrightarrow{\text{LT}} S(s) = \frac{1}{s+A}, \text{ ROC} : \sigma > -A \qquad \text{...(1)}$$

From the differentiation property of Laplace transform in s-domain we know that

$$-ts(t) \xleftrightarrow{\text{LT}} \frac{dS(s)}{ds}, \text{ with same ROC as } S(s) \qquad \text{...(2)}$$

Substituting the value $s(t)$ and $S(s)$ in Eq. 2, we get

$$-te^{At}u(t) \xleftrightarrow{\text{LT}} \frac{d}{ds}\left\{\frac{1}{s+A}\right\}, \text{ ROC} : \sigma > -A$$

or

$$-te^{At}u(t) \xleftrightarrow{\text{LT}} \frac{(s+A)\,(0) - (1+0)}{(s+A)^2}, \text{ ROC} : \sigma > -A$$

or

$$-te^{At}u(t) \xleftrightarrow{\text{LT}} \frac{-1}{(s+A)^2}, \text{ ROC} : \sigma > -A$$

or $$te^{-At}u(t) \xleftrightarrow{LT} \frac{1}{(s+A)^2}, \quad \text{ROC}: \sigma > -A$$

Similarly, we can determine the Laplace transform of signal $\frac{t^2}{2}e^{-At}u(t)$, using the differentiation property of Laplace transform in s-domain as

$$\frac{t^{3-1}}{(3-1)!}e^{-At}u(t) \xleftrightarrow{LT} \frac{1}{(s+A)^3}, \quad \text{ROC}: \sigma > -A$$

In a similar manner, we can determine the Laplace transforms of $\frac{t^3}{3!}e^{-At}u(t)$, $\frac{t^4}{4!}e^{-At}u(t)$, ... $\frac{t^n}{n!}e^{-At}u(t)$.

$$\frac{t^{4-1}}{(4-1)!}e^{-At}u(t) \xleftrightarrow{LT} \frac{1}{(s+A)^4}, \quad \text{ROC}: \sigma > -A$$

$$\vdots$$

$$\frac{t^{n-1}}{(n-1)!}e^{-At}u(t) \xleftrightarrow{LT} \frac{1}{(s+A)^n}, \quad \text{ROC}: \sigma > -A \qquad ...(9.28)$$

9.5.9 Integration of a Continuous-Time Signal in the Time-Domain

This property is the inverse of the differentiation property of the Laplace transform.

Consider a continuous-time signal $s(t)$ and its Laplace transform as $S(s)$. Integration of signal $s(t)$ is given as

If $$s(t) \xleftrightarrow{LT} S(s) \quad \text{with ROC} = R$$

then $$\int_{-\infty}^{t} s(\tau)d\tau \xleftrightarrow{LT} \frac{1}{s}S(s) \quad \text{with ROC}: R \cap \{\text{Real}\{S\} > 0 \qquad ...(9.29)$$

This property can be derived by using the convolution property of Laplace transform.

Convolution of $s(t)$ and $u(t)$ is given by

$$s(t) * u(t) = \int_{\tau=-\infty}^{\infty} s(\tau)\, u(t-\tau)\, d\tau \qquad ...(9.30)$$

But $$u(t-\tau) = \begin{cases} 1, & t-\tau > 0 \\ 0, & t-\tau < 0 \end{cases} = \begin{cases} 1, & \tau < t \\ 0, & \tau > t \end{cases} \qquad ...(9.31)$$

Substituting Eq. 9.31 in Eq. 9.30, we get

$$s(t) * u(t) = \int_{\tau=-\infty}^{t} s(\tau)\, d\tau \qquad ...(9.32)$$

We know that the Laplace transform of $u(t)$ is $1/s$, *i.e.*,

$$u(t) \xleftrightarrow{\text{LT}} \frac{1}{s}, \text{ Real } \{s\} > 0$$

or
$$U(s) = \frac{1}{s} \qquad ...(9.33)$$

Thus, from convolution property of Laplace transform

$$s(t) * u(t) \xleftrightarrow{\text{LT}} S(s)U(s) \qquad ...(9.34)$$

Substituting the value of $U(s)$ in Eq. 9.34, we get

$$s(t) * u(t) \xleftrightarrow{\text{LT}} S(s)\, U(s) = S(s)\frac{1}{s} = \frac{1}{s} S(s) \qquad ...(9.35)$$

Its ROC will contain the intersection of the ROC of $S(s)$ and the ROC of $U(s)$, where $U(s)$ is the Laplace transform $u(t)$.

9.5.10 Initial Value Theorem for Laplace Transform

Consider a continuous-time signal $s(t)$, its Laplace transform is $S(s)$.

$$s(t) \xleftrightarrow{\text{LT}} S(s) \quad \text{with ROC} : R$$

If the Laplace transforms of signal $s(t)$ and $ds(t)/dt$ both exist and also $\lim_{s \to \infty} sS(s)$ exists then from initial value theorem for Laplace transform

$$\lim_{s \to \infty} sS(s) = \lim_{t \to 0^+} s(t) = s(0^+) \qquad ...(9.36)$$

Initial value theorem of Laplace transform is used to determine the initial value of the signal.

Example 9.8 Find the initial value of the signal corresponding to the Laplace transform $S(s) = \dfrac{s+1}{s\,(s+2)}$ by using initial value theorem of Laplace transform. Also verify that the answer obtained is correct.

Solution. We can determine the initial value of the signal $s(t)$ with Laplace transform $S(s)$ using initial value theorem as

$$s(0^+) = \text{Initial value of } s(t) = \lim_{s \to \infty} [sS(s)]$$

$$= \lim_{s \to \infty} \left[\frac{s+1}{s+2}\right] = \lim_{s \to \infty} \left[\frac{1 + 1/s}{1 + 2/s}\right]$$

$$= \frac{1 + 1/\infty}{1 + 2/\infty} = \frac{1+0}{1+0} = 1$$

or
$$s(0^+) = 1$$

This is the initial value of the signal $s(t)$.

Laplace transform of signal $s(t)$ is given as

$$S(s) = \frac{s+1}{s\,(s+2)} = \frac{\alpha_1}{s} + \frac{\alpha_2}{s+2}$$

By partial fraction expansion

$$S(s) = \frac{1/2}{s} + \frac{1/2}{s+2} \qquad ...(1)$$

Taking the inverse Laplace transform of Eq. 1, we get

$$s(t) = \text{Inverse Laplace transform } \{S(s)\}$$

$$= \text{Inverse LT}\left\{\frac{1/2}{s} + \frac{1/2}{s+2}\right\}$$

$$= \frac{1}{2}\text{ Inverse LT}\left\{\frac{1}{s}\right\} + \frac{1}{2}\text{ Inverse LT}\left(\frac{1}{s+2}\right\}$$

or
$$s(t) = \frac{1}{2}u(t) + \frac{1}{2}e^{-2t}\,u(t) \qquad ...(2)$$

$$S(0^+) = s(t)\,\Big|_{t=0} = \left[\frac{1}{2}u(t) + \frac{1}{2}e^{-2t}\,u(t)\right]_{t=0}$$

$$= \frac{1}{2}u(0) + \frac{1}{2}e^{-2\times 0}\,u(0) = \frac{1}{2}\,.\,1 + \frac{1}{2}\,.\,1\,.\,1 = 1$$

or
$$s(0^+) = 1$$

which verifies the result obtained with the initial value theorem.

9.5.11 Final Value Theorem for Laplace Transform

Let us consider a continuous-time signal $s(t)$, its Laplace transform $S(s)$ is denoted as

$$s(t) \xleftrightarrow{\text{LT}} S(s), \text{ with ROC} : R$$

If the Laplace transforms of signal $s(t)$ and $ds(t)/dt$ both exist then from final value theorem

$$s(\infty) = \lim_{t \to \infty} s(t) = \lim_{s \to 0} [s\,S(s)] \qquad ...(9.37)$$

Final value theorem of Laplace transform is used to find out the final value of a continuous-time signal $s(t)$.

Example 9.9 Find the final value of a continuous-time signal $s(t) = [2 + e^{-3t}]\,u(t)$.

Solution. Determination of Laplace transform of $s(t)$,

$$S(s) = \text{LT}\,\{s(t)\} = \text{LT}\,\{(2 + e^{-3t})\,u(t)\}$$

$$= \int_{-\infty}^{\infty} (2 + e^{-3t})\, u(t)\, e^{-st}\, dt \qquad \text{...(1)}$$

Using the definition of unit step function

$$u(t) = \begin{cases} 1\,, & t > 0 \\ 0\,, & t < 0 \end{cases} \qquad \text{...(2)}$$

Substituting the value of $u(t)$ in Eq. 1, we get

$$S(s) = \int_{0}^{\infty} (2 + e^{-3t})\, 1\, e^{-st}\, dt$$

$$= \int_{0}^{\infty} [2\, e^{-st} + e^{-(s+3)t}]\, dt$$

$$= 2 \int_{0}^{\infty} e^{-st}\, dt + \int_{0}^{\infty} e^{-(s+3)t}\, dt$$

$$= 2\left[\frac{1}{-s} e^{-st}\right]_0^{\infty} + \left[\frac{1}{-(s+3)} e^{-(s+3)t}\right]_0^{\infty}$$

$$= \frac{2}{s} + \frac{1}{s+3} = \frac{2\,(s+3) + s}{s\,(s+3)}$$

or $$S(s) = \frac{3s+6}{s\,(s+3)} = \frac{3\,(s+2)}{s\,(s+3)} \qquad \text{...(3)}$$

Determination of final value of the signal $s(t)$ using final value theorem of the Laplace transform

$$s(\infty) = \lim_{t \to \infty} s(t) = \lim_{s \to \infty} [s\, S(s)] \qquad \text{...(4)}$$

Substituting the value of $S(s)$ in Eq. 4, we get

$$s(\infty) = \lim_{s \to \infty} \left[s \,.\, \frac{3\,(s+2)}{s\,(s+3)} \right]$$

$$= \lim_{s \to \infty} \left[\frac{3\,(s+2)}{s+3)} \right] = \frac{3\,(0+2)}{(0+3)} = 2$$

Final value of the signal $s(t)$ is 2.

9.6 TABULATION OF PROPERTIES OF THE LAPLACE TRANSFORM

In this section, we summarize important properties of Laplace transform. These important properties of Laplace transform are given in Table 9.1. Many of these properties are used in applying the Laplace transform for the analysis and characterisation of continuous-time LTI systems. The various properties of Laplace transforms and their ROCs can provide us with considerable information about the signal and its Laplace transform.

Table 9.1 Summary Properties of the Laplace Transforms

Name of properties of Laplace transform	*Continuous-time signals*	*Laplace transform*	*Region of convergence ROC*
	$s_1(t)$ $s_2(t)$ $s_{(t)}$	$S_1(s)$ $S_2(s)$ $S_{(s)}$	R_1 R_2 $Rsub$
Linearity	$A_1\, s_1(t) + A_2\, s_2(t)$	$A_1\, S_1(s) + A_2\, S_2(s)$	ROC will be atleast intersection of R_1 and R_2
Time-shifting	$s(t - t_0)$	$e^{-st_0}\, S(s)$	R
Shifting of the Laplace transform in the s-domain	$e^{s_0t}\, s(t)$	$S(s - s_0)$	ROC will be shifted version of R, *i.e.*, S is in the ROC of if $s - s_0$ is in R
Time-scaling	$s(At)$	$\frac{1}{\lvert A \rvert} S\left(\frac{s}{A}\right)$	Scaled version of ROC, *i.e.*, s is in the ROC if (s/A) is in R
Complex conjugation	$s^*(t)$	$S^*(s^*)$	R
Convolution integral	$s_1(t) * s_2(t)$	$S_1(s)\, S_2(s)$	ROC will be atleast intersection of R_1 and R_2
Differentiation of a continuous-time signal $s(t)$ in time-domain	$\frac{d}{dt} s(t)$	$s\, S(s)$	Atleast R

Contd...

Table 9.1 Contd...

Name of properties of Laplace transform	*Continuous-time signals*	*Laplace transform*	*Region of convergence ROC*
Differentiation of $S(s)$ in s-domain	$-t\,s(t)$	$\frac{d}{ds}S(s)$	R
Integration of a continuous-time signal $s(t)$ in the time-domain	$\int_{-\infty}^{t} s(\tau)\,d\tau$	$\frac{1}{s}S(s)$	ROC wil be atleast intersection of R and real part of $s > 0$
Initial-value Theorem for Laplace Transform : If signal $s(t) = 0$ for $t < 0$ and $s(t)$ contains no impulses or higher order singularities at $t = 0$ then from initial value theorem of Laplace transform, the initial value of the signal, $s(0^+) = \lim_{s \to \infty} \{s\,S(s)\}$.			
Final-value Theorem for Laplace Transform : If signal $s(t) = 0$ for $t < 0$ and $s(t)$ contains no impulses or higher order singularities at $t = 0$ then from final-value theorem of Laplace transform, the final value of signal, $\lim_{t \to \infty} s(t) = \lim_{s \to 0} \{s\,S(s)\}$			

9.7 TABULATION OF SOME LAPLACE TRANSFORM PAIRS

We will summarize important Laplace transform pairs given in Table 9.2. Many of these important transform pairs are used to simplify the analysis and characterization of continuous-time LTI systems.

Table 9.2 Summary of Some Important Laplace Transform Pairs

Continuous-time signal $s(t)$	*Laplace transform* $S(s)$	*Region of convergence* $ROC = R$
$\delta(t)$	1	All the values of s, where s is the complex variable $s = \sigma + j\omega$, σ and ω are the real and imaginary parts of s
$u(t)$	$\frac{1}{s}$	$\sigma > 0$
$-u(-t)$	$\frac{1}{s}$	$\sigma < 0$
$\frac{t^{n-1}}{(n-1)!}u(t)$	$\frac{1}{s^n}$	$\sigma > 0$
$-\frac{t^{n-1}}{(n-1)!}u(-t)$	$\frac{1}{s^n}$	$\sigma < 0$
$e^{-At}\, u(t)$	$\frac{1}{s+A}$	$\sigma > -A$
$-e^{-At}\, u(-t)$	$\frac{1}{s+A}$	$\sigma < -A$
$\frac{t^{n-1}}{(n-1)!}$	$\frac{1}{(s+A)^n}$	$\sigma > -A$
$\frac{-t^{n-1}}{(n-1)!}e^{-At}\, u(-t)$	$\frac{1}{(s+A)^n}$	$\sigma < -A$
$\delta(t - T_0)$	e^{-sT_0}	All values of s
$(\cos \omega_0 t)\, u(t)$	$\frac{s}{s^2 + \omega_0^2}$	$\sigma > 0$
$(\sin \omega_0 t)\, u(t)$	$\frac{\omega_0}{s^2 + \omega_0^2}$	$\sigma > 0$
$(e^{-At} \cos \omega_0 t)\, u(t)$	$\frac{s+A}{(s+A)^2 + \omega_0^2}$	$\sigma > -A$
$(e^{-At} \sin \omega_0 t)\, u(t)$	$\frac{\omega_0}{(s+A)^2 + \omega_0^2}$	$\sigma > -A$

Contd...

Continuous-time signal s(t)	*Lapace transform S(s)*	*Region of convergence ROC = R*
$u_n(t) = \frac{d^n\,\delta(n)}{dt}$	s^n	All values of s
$u_{-n}(t) = u(t) * \ldots * u(t)$ n times	$\frac{1}{s^n}$	$\sigma > 0$

9.8 APPLICATIONS OF LAPLACE TRANSFORM IN ANALYSIS AND CHARACTERIZATION OF CONTINUOUS-TIME LTI SYSTEMS

One of the important applications of Laplace transform is in the anlaysis and characterization of continuous-time LTI systems. Laplace transform is also used to simplify the convolution integral of two continuous-time signals. Its role for continuous-time LTI systems stems directly from the convolution property. Let us consider a continuous-time LTI system with impulse response $h(t)$.

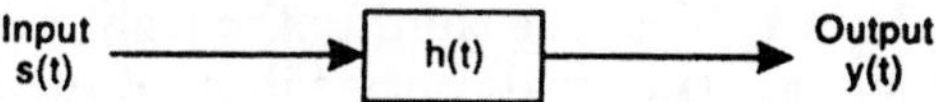

Fig. 9.11 *Continuous-time LTI systems.*

Specifically, the Laplace transforms of input and output of a continuous-time LTI system are related through multiplication by the Laplace transform of the impulse response of continuous-time LTI system. Output of continuous-time LTI system is determined by convolution of input and impulse response of the system as

$$y(t) = \int_{-\infty}^{\infty} s(\tau)\, h(t-\tau)\, d\tau \qquad \text{...(9.38)}$$

On taking the Laplace transform of both sides of Eq. 9.38, we get

$$Y(s) = H(s)\, S(s) \qquad \text{...(9.39)}$$

where $S(s)$, $Y(s)$ and $H(s)$ are the Laplace transforms of the input $s(t)$, output $y(t)$, and impulse response $h(t)$, respectively. In fact, for $s = j\omega$, each of the Laplace transform of Eq. 9.39 reduces to the respective continuous-time Fourier transform (CTFT). The response of a continuous-time LTI system to complex exponential e^{st} is an eigen function of the system with eigen value equal to the Laplace transform of the impulse response $h(t)$. In other words, if the input to a continuous-time LTI system is $s(t) = e^{st}$, then the output will be $H(s)\, e^{st}$

The frequency response of a continuous-time LTI system is equal to the Laplace transform $H(s)$ at $s = j\omega$. The Laplace transform of impulse response $h(t)$, *i.e.*, $H(s)$ is commonly called the *system function* or *transfer function*. Many properties of continuous-time LTI systems are closely related to the characteristics of the sytsem function $H(s)$ in s-plane.

We will discuss various applications of Laplace transform in analysis and characterization of the following properties of continuous-time LTI systems.

1. Causality of a continuous-time LTI system.
2. Stability of a continuous-time LTI system.
3. Characterization of continuous-time LTI system by linear constant-coefficient differential equations.
4. Determination of transfer function of a continuous-time LTI system.

9.8.1 Causality of a Continuous-Time LTI System

If the impulse response $h(t)$ of a continuous-time LTI system is zero for $t < 0$, *i.e.*, impulse response is right-sided then the LTI system is called causal LTI system. Consequently, we see that the ROC associated with the transfer function $H(s)$ for a causal LTI system is a right half plane. On the other hand, an ROC to the right to the rightmost pole does not guarantee that a system is causal; rather it guarantees only that the impulse response $h(t)$ of a continuous-time LTI system is right sided. We can find whether the LTI system is causal simply by checking if its ROC is a right half plane. In other words, for a continuous-time LTI system with rational transfer function $H(s)$, the causality of the LTI system is equivalent to the ROC being the right-half plane to the right of the rightmost pole.

Example 9.10 Check the causality of a continuous-time LTI system with impulse response

$$h(t) = e^{-t}u(t)$$

Solution. Given : Impulse response of the continuous-time LTI system

$$h(t) = e^{-t}u(t) \quad ...(1)$$

From the definition of unit-step function

$$u(t) = \begin{cases} 1\ , & t > 0 \\ 0\ , & t < 0 \end{cases} \quad ...(2)$$

Substituting the values of $u(t)$ in Eq. 1, we get

$$h(t) = \begin{cases} e^{-t}\ , & t > 0 \\ 0\ , & t < 0 \end{cases} \quad ...(3)$$

Since $h(t) = 0$ for $t < 0$, therefore the system is causal.

Determination of Transfer Function of Continuous-Time Systems

Transfer function $H(s)$ = Laplace transform of $h(t)$

$$H(s) = \text{Laplace transform } \{h(t)\}$$

$$= \text{Laplace transform } \{e^{-t}u(t)\}$$

$$= \int_{-\infty}^{\infty} e^{-t}u(t)\, e^{-st}\, dt$$

$$= \int_{0}^{\infty} e^{-t}\, e^{-st}\, dt = \int_{0}^{\infty} e^{-(1+s)t}\, dt$$

$$= \left[\frac{1}{-(1+s)} e^{-(1+s)t}\right]_{0}^{\infty}$$

or $$H(s) = \frac{1}{s+1}, \quad \text{ROC} : \sigma > -1 \qquad \text{...(4)}$$

where $s = \sigma + j\omega$, σ and ω are real and imaginary parts of complex variable s.

In this example, the transfer function $H(s)$ is rational and the ROC in Eq. 4 is to the right of the rightmost pole, $s = -1$. It means that causality for LTI systems with rational transfer functions is equivalent to the ROC being to the right of the rightmost pole.

Example 9.11 Check the causality of a continuous-time LTI system with impulse response

$$h(t) = e^{-|t|}$$

Solution. Given : Impulse response of the continuous-time LTI system

$$h(t) = e^{-|t|} \qquad \text{...(1)}$$

Since $h(t) \neq 0$ for $t < 0$, this system is not causal.

Determination of Transfer Function $H(s)$:

$$H(s) = \text{Laplace transform } \{h(t)\} = \text{LT } \{e^{-|t|}\}$$

$$= \int_{-\infty}^{\infty} e^{-|t|} e^{-st}\, dt$$

$$= \int_{-\infty}^{0} e^{t} e^{-st}\, dt + \int_{0}^{\infty} e^{-t} e^{-st}\, dt$$

$$= \int_{-\infty}^{0} e^{(1-s)t}\, dt + \int_{0}^{\infty} e^{-(1+s)t}\, dt$$

$$= \left[\frac{1}{1-s} e^{(1-s)t}\right]_{-\infty}^{0} + \left[\frac{1}{-(1-s)} e^{-(1-s)t}\right]_{0}^{\infty}$$

$$= \frac{1}{1-s} + \frac{1}{1+s} = \frac{1+s+1-s}{1-s^2}$$

or $$H(s) = \frac{-2}{s^2 - 1}, \quad \text{ROC} : -1 < \sigma < +1 \qquad ...(2)$$

Thus, transfer function $H(s)$ is rational and has an ROC that is not to the right of the rightmost pole, therefore, the system is not causal.

9.8.2 Stability of a Continuous-Time LTI System

The region of a convergence (ROC) of the transfer function $H(s)$ of a continuous-time LTI system can also be related to the stability of the system. The stability of an LTI system is equivalent to its impulse response $h(t)$ being absolutely integrable. When impulse response $h(t)$ is absolutely integrable, its Continuous-time Fourier transform (CTFT) converges. Since CTFT of a signal $s(t)$ equals Laplace transform determined along the imaginary axis ($j\omega$-axis), we conclude that a continuous-time LTI system is stable if and only if the ROC of its transfer function $H(s)$ includes the imaginary axis ($j\omega$-axis), *i.e.*, Real $\{s\} = \sigma = 0$.

For systems with rational transfer functions, stability is easily interpreted in terms of poles of the system. For pole-zero plot, the stability corresponds to the choice of an ROC that is in between the two poles, so that the imaginary axis ($j\omega$- axis) is contained in the ROC.

Stability for causal continuous-time LTI system can be characterized very simply in terms of the locations of the poles. Specifically, we consider a causal LTI system with rational transfer function $H(s)$. Since the system is causal, its ROC is to the right of the rightmost pole. Consequently, this system will be stable if and only if the rightmost pole of $H(s)$ lies to the left of the imaginary axis ($j\omega$-axis).

A causal continuous-time LTI system with rational transfer function $H(s)$ is stable if and only if all the poles of $H(s)$ lie in the left-half of the s-plane (*i.e.*, all the poles will have negative real parts).

Example 9.12 Check the stability of a continuous-time LTI system with transfer function

$$H(s) = \frac{(s-2)}{(s+2)\,(s-3)} \qquad ...(1)$$

Solution. Since the ROC of $H(s)$ has not been specified, hence, there will be several different ROCs and consequently, there will be several different system impulse responses that can be associated with the algebraic expression for $H(s) = \dfrac{(s-2)}{(s+2)\,(s-3)}$.

If we have information about the causality or stability of the system then the appropriate ROC can be identified. If we assume that the system is causal then the ROC will be right-sided.

Determination of impulse response $h(t)$ for causal continuous-time LTI systems.

$$h(t) = \text{Inverse Laplace transform } \{H(s)\}$$

$$= \text{Inverse LT}\left\{\frac{(s-2)}{(s+2)(s-3)}\right\}$$

From partial fraction expansion

$$h(t) = \text{Inverse LT}\left\{\frac{\alpha_1}{(s+2)} + \frac{\alpha_2}{(s-3)}\right\}$$

Determination of values of α_1 and α_2

or $\begin{matrix} s+2=0 \\ s=-2 \end{matrix}$, $\quad \alpha_1 = \dfrac{s-2}{s-3} = \dfrac{-2-2}{-2-3} = \dfrac{-4}{-5} = \dfrac{4}{5}$

or $\begin{matrix} s-3=0 \\ s=3 \end{matrix}$, $\quad \alpha_2 = \dfrac{s-2}{s+2} = \dfrac{3-2}{3+2} = \dfrac{1}{5}$

or $$H(s) = \text{Inverse LT}\left\{\frac{4/5}{(s+2)} + \frac{1/5}{(s-3)}\right\}$$

$$= \frac{4}{5}\text{ Inverse LT}\left\{\frac{1}{s+2}\right\} + \frac{1}{5}\text{ Inverse LT}\left\{\frac{1}{s-3}\right\}$$

or $$h(t) = \frac{4}{5}e^{-2t}\,u(t) + \frac{1}{5}e^{3t}\,u(t) \qquad \text{...(2)}$$

Note that this particular choice of ROC does not include the imaginary axis ($j\omega$-axis), and therefore, the corresponding system is unstable (becuase this impulse response $h(t)$ is not absolutely integrable).

The ROC for causal system given by Eq. 1 is shown in Fig. 9.12 (*a*). If the above system is stable then ROC for system described by Eq. 1 is shown in Fig. 9.12 (*b*).

The corresponding impulse response is given by

$$h(t) = \frac{4}{5}e^{-2t}\,u(t) - \frac{1}{5}e^{3t}\,u(-t) \qquad \text{...(3)}$$

Impulse response given in Eq. 3 is absolutely integrable.

If the above system is assumed to be anticausal and unstable then ROC for the system described by Eq. 1 is shown in Fig. 9.12 (*c*).

The corresponding impulse response is given by

$$h(t) = -\frac{4}{5}e^{-2t}\,u(-t) - \frac{1}{5}e^{3t}\,u(-t)$$

$$= -\left(\frac{4}{5}e^{-2t} + \frac{1}{5}e^{3t}\right)u(-t) \qquad \text{...(4)}$$

Given transfer function, $H(s) = \dfrac{(s-2)}{(s+2)(s-3)}$. There is one zero at $s = 2$ and two poles at $s = -2$ and $s = 3$.

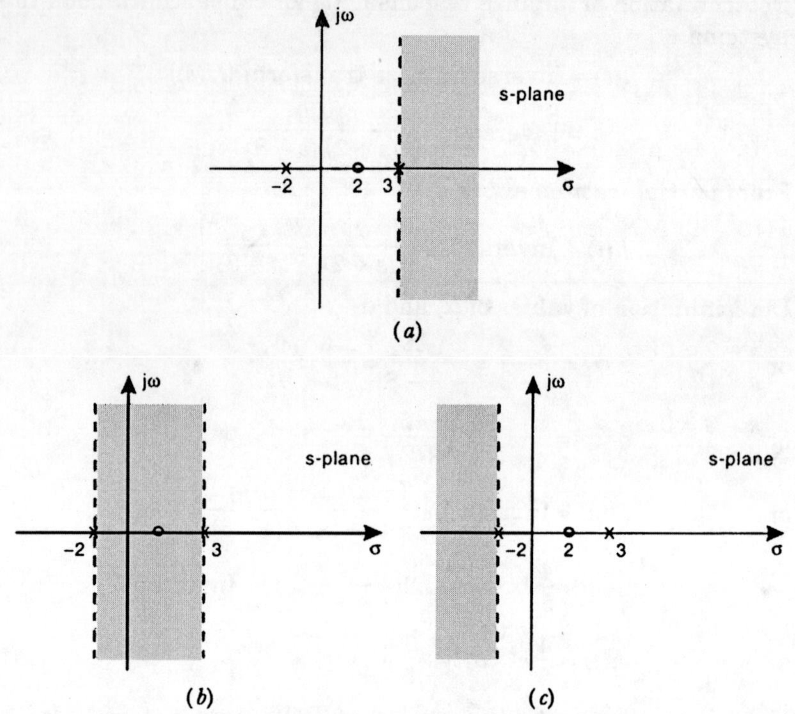

Fig. 9.12 *Different possible ROCs for the system function* $H(s) = \dfrac{(s-2)}{(s+2)(s-3)}$ *with poles at* $s = -2$ *and* $s = 3$ *and zeros at* $s = 1$. *(a) causal and unstable system (b) non-causal and stable system (c) anticausal and unstable system.*

9.8.3 Characterisation of Continuous-Time LTI Systems by Linear Constant-Coefficient Differential Equations

We have already discussed in chapter 5, the use of CTFT to obtain the frequency response of a continuous-time LTI system characterized by a linear constant-coefficient differential equation without first obtaining the impulse response *i.e.,* time-domain solution. In this subsection we will exploit Laplace transform to obtain the transfer function of a continuous-time LTI system characterized by a linear constant-coefficient differenetial equation.

Let us determine the transfer function $H(s)$ for a general linear constant-coefficient differential equation given by

$$\sum_{k=0}^{N} A_k \frac{d^k y(t)}{dt^k} = \sum_{k=0}^{M} B_k \frac{d^k s(t)}{dt^k} \qquad \text{...(9.40)}$$

where $s(t)$ and $y(t)$ are the input and output of the continuous-time LTI system. A_k and B_k are constant-coefficients.

Taking the Laplace transform of both sides of Eq. 9.40, we get

$$\left\{ \sum_{k=0}^{N} A_k s^k \right\} Y(s) = \left\{ \sum_{k=0}^{M} B_k s^k \right\} S(s)$$

or
$$\frac{Y(s)}{S(s)} = \frac{\left\{ \sum_{k=0}^{M} B_k s^k \right\}}{\left\{ \sum_{k=0}^{N} A_k s^k \right\}} \qquad ...(9.41)$$

Transfer function $H(s)$ of a continuous-time LTI system (initially relaxed) is the ratio of Laplace transform of output $y(t)$ and Laplace transform of input $s(t)$. It is defined as

$$H(s) = \frac{Y(s)}{S(s)} = \frac{\left\{ \sum_{k=0}^{M} B_k s^k \right\}}{\left\{ \sum_{k=0}^{N} A_k s^k \right\}} \qquad ...(9.42)$$

Thus, the transfer function $H(s)$ of a continuous-time LTI system specified by a differential equation is always rational.

Zeros of the transfer function $H(s)$ are given as

$$\sum_{k=0}^{M} B_k s^k = 0 \qquad ...(9.43)$$

and poles of the transfer function $H(s)$ are given as

$$\sum_{k=0}^{N} A_k s^k = 0 \qquad ...(9.44)$$

Equation 9.42 does not include a specification of the ROC of $H(s)$, since the linear constant coefficient differential equation by itself does not constrain the ROC. However, we can infer the ROC by using additional information such as the knowledge about the stability or causality of the LTI system. If we impose the condition of initial rest on the system, so that it is causal, ROC will be to the right of the rightmost pole. If all the initial conditions for a continuous-time LTI system are zero then this system is referred to as a initially relaxed system.

Example 9.13 Find the transfer function of a continuous-time LTI system that is described by the following linear constant- coefficient differential equation

$$\frac{dy(t)}{dt} + 4y(t) = 2s(t) \qquad ...(1)$$

where $s(t)$ and $y(t)$ are the input and output of the system respectively.

Solution. We can determine the transfer function of a system by applying the Laplace transform on both the sides of Eq. 1, as

$$\text{LT}\left\{ \frac{dy(t)}{dt} + 4y(t) \right\} = \text{LT}\{2s(t)\}$$

or $$\text{LT}\left\{\frac{dy(t)}{dt}\right\} + 4\ \text{LT}\ \{y(t)\} = 2\ \text{LT}\ \{s(t)\}$$

or $$sY(s) + 4Y(s) = 2S(s)$$

or $$\frac{Y(s)}{S(s)} = \frac{2}{s+4} \qquad \text{...(2)}$$

From the definition of transfer function of the system

$$H(s) = \frac{Y(s)}{S(s)} \qquad \text{...(3)}$$

Therefore, transfer function $H(s)$ is given as

$$H(s) = \frac{2}{s+4} \qquad \text{...(4)}$$

Equation 4 provides the algebraic expression for the transfer function of the system, but not the region of convergence. Infact, the differential equation itself is not a complete specification of the continuous-time LTI system. Therefore, in general, there are different impulse responses for the same differential equation. If for a differential equation, the system is causal, then the ROC can be inferred to be to the right of the rightmost pole. The impulse response $h(t)$ can be determined by applying inverse Laplace transform as

$$h(t) = \text{Inverse LT}\{H(s)\} = \text{Inverse LT}\left\{\frac{2}{s+4}\right\}$$

$$= 2e^{-4t}u(t) \qquad \text{...(5)}$$

It is the impulse response for causal LTI system.

ROC for causal system will be ROC : Real $\{s\} = \sigma > -4$

If for a differential equation, the system is non-causal, then the ROC can be inferred to be the left of the leftmost pole. ROC for a non-causal system will be ROC : Real $\{s\} = \sigma < -4$.

In this case impulse response will be given by

$$h(t) = -\,2e^{-4t}\,u(-t) \qquad \text{...(6)}$$

Example 9.14 Determine the transfer function $H(s)$ for an *RLC* circuit shown in Fig. 9.13 whose capacitor voltage and inductor current are initially zero.

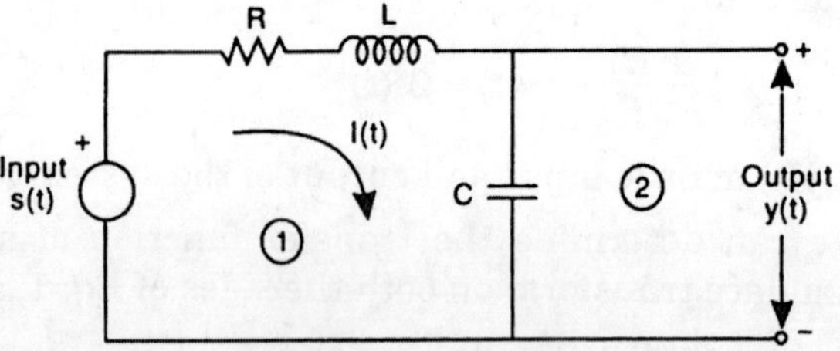

Fig. 9.13 *RLC circuit.*

Solution. Applying *KVL* in mesh (1), we get

$$s(t) = RI(t) + L\frac{dI(t)}{dt} + \frac{1}{C}\int_{0}^{\tau} I(t)\,dt \qquad ...(1)$$

Applying *KVL* in mesh (2), we get

$$y(t) = \frac{1}{C}\int_{0}^{\tau} I(t)\,dt \qquad ...(2)$$

Eq. 2 becomes

$$Cy(t) = \int_{0}^{\tau} I(t)\,dt \qquad ...(3)$$

Differentiating Eq. 3 w.r.t. '*t*', we get

$$C\frac{dy(t)}{dt} = \frac{d}{dt}\left[\int_{0}^{\tau} I(t)\,dt\right] = I(t) \qquad ...(4)$$

Now, differentiating Eq. 4 w.r.t. '*t*', we get

$$C\frac{d^2y(t)}{dt^2} = \frac{dI(t)}{dt} \qquad ...(5)$$

Substituting the values of $\int_{0}^{\tau} I(t)\,dt$, $I(t)$, and $\frac{dI(t)}{dt}$ in Eq. 1, we get

$$s(t) = RI(t) + L\frac{dI(t)}{dt} + \frac{1}{C}\int_{0}^{\tau} I(t)\,dt$$

$$= RC\frac{dy(t)}{dt} + LC\frac{d^2y(t)}{dt} + \frac{C}{C}y(t)$$

or $$LC\frac{d^2y(t)}{dt^2} + RC\frac{dy(t)}{dt} + y(t) = s(t) \qquad ...(6)$$

It is the linear constant-coefficient differential equation for the *RLC* circuit which is a continuous-time LTI system.

Transfer function $H(s)$ for the *RLC* circuit described by Eq. 6 is obtained as follows.

Applying Laplace transform to both sides of Eq. 6 with all initial conditions zero, we get

$$\text{Laplace transform}\left\{LC\frac{d^2y(t)}{dt^2} + RC\frac{dy(t)}{dt} + y(t)\right\}$$

$$= \text{Laplace transform }\{s(t)\}$$

or $$LCs^2\,Y(s) + RCs\,Y(s) + Y(s) = S(s)$$

or $$Y(s)\,[LCs^2 + RCs + 1] = S(s)$$

or $$\frac{Y(s)}{S(s)} = \frac{1}{LCs^2 + RCs + 1}$$

or $$H(s) = \frac{1}{s^2 + \left(\frac{R}{L}\right)s + \left(\frac{1}{LC}\right)} \quad ...(7)$$

This is the transfer function of the *RLC* circuit (continuous-time LTI system).

If the values of passive components such as *R, L* and *C* are positive then the poles of this transfer function *H*(*s*) will have negative real parts and therefore, the system will be stable.

9.8.4 System Behaviour Related to the Transfer Function of the System

We have already studied that the system properties such as causality and stability can be directly related to the transfer function of the system and its characteristics. Infact, each of the properties of Laplace transforms can be used to describe the behaviour of a system which is related to the transfer function of the system.

Example 9.15 Find the transfer function of an LTI system with input $s(t) = e^{-4t}u(t)$ and output $y(t) = [e^{-2t} - e^{-3t}]\, u(t)$. Also discuss the system behaviour which is related to transfer function of the system.

Solution. Transfer function *H*(*s*) of a continuous-time LTI system can be determined as

$$H(s) = \frac{Y(s)}{S(s)} \quad ...(1)$$

where *Y*(*s*) and *S*(*s*) are the Laplace transforms of output *y*(*t*) and input *s*(*t*), respectively.

$$S(s) = \text{Laplace transform } \{y(t)\} = \int_{-\infty}^{\infty} s(t)\, e^{-st}\, dt$$

$$= \text{LT } \{e^{-4t}u(t)\} = \int_{-\infty}^{\infty} (e^{-4t}\, u(t))\, e^{-st}\, dt$$

$$= \int_{0}^{\infty} (e^{-4t} .1)\, e^{-st}\, dt = \int_{0}^{\infty} e^{-(s+4)t}\, dt$$

$$= \left[\frac{1}{-(s+4)} e^{-(s+4)t}\right]_0^{\infty} = \frac{1}{s+4}, \quad \text{ROC}: \sigma > -4$$

or $$S(s) = \frac{1}{s+4}, \quad \text{ROC}: \sigma > -4 \quad ...(2)$$

$$Y(s) = \text{LT}\{y(t)\} = \int_{-\infty}^{\infty} y(t)\, e^{-st}\, dt$$

$$= \int_{-\infty}^{\infty} [e^{-2t} - e^{-3t}]\, u(t)\, e^{-st}\, dt$$

$$= \int_{0}^{\infty} e^{-2t}\, e^{-st}\, dt - \int_{0}^{\infty} e^{-3t}\, e^{-st}\, dt$$

$$= \int_{0}^{\infty} e^{-(s+2)t}\, dt - \int_{0}^{\infty} e^{-(s+3)t}\, dt$$

$$= \left[\frac{1}{-(s+2)}\, e^{-(s+2)}\right]_0^{\infty} - \left[\frac{1}{-(s+3)}\, e^{-(s+3)t}\right]_0^{\infty}$$

$$= \frac{1}{s+2} - \frac{1}{s+3} = \frac{1}{(s+2)\,(s+3)},$$

ROC : $\sigma > -2 \cap \sigma > -3$

or $$Y(s) = \frac{1}{(s+2)\,(s+3)}, \quad \text{ROC : } \sigma > -2 \qquad ...(3)$$

Substituting Eqs. 2 and 3 in Eq. 1, we get

$$H(s) = \frac{Y(s)}{S(s)} = \frac{\dfrac{1}{(s+2)\,(s+3)}}{\dfrac{1}{(s+4)}} = \frac{(s+4)}{(s+2)\,(s+3)}$$

or $$H(s) = \frac{(s+4)}{(s+2)\,(s+3)} \qquad ...(4)$$

From the knowledge of input and output, we have determined the transfer function of this system. We can also deduce a number of other important properties of the system.

We can also determine the ROC for this system. We know from convolution property of Laplace transform that the ROC of $Y(s)$ must include atleast the intersections of ROCs of $S(s)$ and $H(s)$. There are three possible choices for the ROCs of $H(s)$.

These are : to the left of the pole at $s = -3$

: between the poles at $s = -3$ and $s = -2$, and

: to the right of the pole at $s = -2$

The only choice consistent with the ROCs of $S(s)$ and $Y(s)$ is Real $\{s\} = \sigma > -2$. Since this is to the right of rightmost pole of $H(s)$. Now we can conclude that $H(s)$ is causal, and since both poles of $H(s)$ have negative real parts, hence this system is stable.

Now we can find input-output relationship for the system characterized by

$$H(s) = \frac{Y(s)}{S(s)} = \frac{(s+4)}{(s+2)(s+3)} = \frac{s+4}{s^2+5s+6}$$

or $$Y(s)[s^2+5s+6] = S(s)[s+4]$$

or $$s^2Y(s) + 5sY(s) + 6y(s) = sS(s) + 4S(s) \qquad ...(5)$$

Taking inverse Laplace transform of both sides of Eq. 5, we get

$$\frac{d^2y(t)}{dt^2} + 5\frac{dy(t)}{dt} + 6y(t) = \frac{ds(t)}{dt} + 4s(t) \qquad ...(6)$$

This is the linear constant-coefficient differential equation of the system with the condition of initial rest.

Example 9.16 Find the transfer function of the continuous-time LTI system for which the following information is given

(*i*) The system is causal.

(*ii*) The system function is rational and has only two poles at $s = -2$ and $s = 4$.

(*iii*) If $s(t) = 1$, then $y(t) = 0$.

(*iv*) The impulse response $h(t) = 4$ at $t = 0$.

Solution. From (*i*) and (*ii*) we infer that the system is unstable. Since it is causal and has pole at $s = 4$ with positive real part. Therefore, the transfer function will be given as

$$H(s) = \frac{p(s)}{(s+2)(s-4)} = \frac{p(s)}{s^2-2s-8} \qquad ...(1)$$

where $p(s)$ is a polynomial in s. Because the output *i.e.*, response $y(t)$ to the input $s(t) = 1 = e^{0.t}$ is equal to $H(s)\, e^{s.t}\big|_{s=0} = H(0)\, e^{0.t}$, we conclude from (*iii*), that $p(0) = 0$, *i.e.*, $p(s)$ will be at rest at $s = 0$. Thus

$$p(s) = s\, q(s) \qquad ...(2)$$

where $q(s)$ is another polynomial of s.

Finally, from (*iv*) and initial-value thorem of Laplace transform

$$h(0^+) = \lim_{s\to\infty} [sH(s)] = \lim_{s\to\infty}\left[s\left(\frac{s\,q(s)}{s^2-2s-8}\right)\right]$$

or $$h(0^+) = \lim_{s\to\infty} \frac{q(s)}{1-\frac{2}{s}-\frac{8}{s^2}}$$

or $$4 = \frac{q(s)}{1-\frac{2}{\infty}-\frac{8}{\infty^2}}$$

or $$4 = q(s) \qquad ...(3)$$

Substituting $q(s) = 4$ in Eq. 2, we get

$$p(s) = s.4 = 4s \quad ...(4)$$

Substituting Eq. 4 in Eq. 1, we get

$$H(s) = \frac{4s}{s^2 - 2s - 8} \quad ...(5)$$

9.9 BLOCK DIAGRAM REPRESENTATION FOR CONTINUOUS-TIME LTI SYSTEMS

Laplace transform is used to replace time-domain operations such as differentiation, convolution, time-shifting, and so on, with algebraic operations such as additions, multiplications etc. In this section, we will emphasize the block diagram representations for continuous-time LTI systems and use of system function algebraic. Here we will discuss the analysis of interconnections of LTI systems and synthesis of LTI systems as interconnections of elementary system building blocks. Transfer function of a system is also called system function.

9.9.1 Transfer Function of Interconnections of Continuous-Time LTI Systems

In this subsection we will study various interconnections of LTI systems. These interconnections are

1. Parallel interconnection of two LTI systems.
2. Series interconnection of two LTI systems.
3. Feedback interconnection of two LTI systems.

1. Parallel interconnection of two LTI systems. A parallel interconnection of two continuous-time LTI systems is shown in Fig. 9.14.

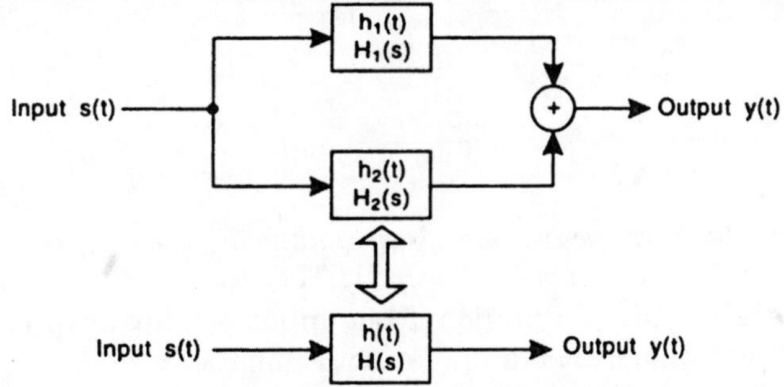

Fig. 9.14 *Parallel interconnection of two continuous-time LTI systems and its equivalent (overall) LTI system.*

The impulse response of the overall system of parallel interconnection of two continuous-time LTI systems is given as

$$h(t) = h_1(t) + h_2(t) \quad ...(9.45)$$

and from property of linearity of Laplace transform

$$H(s) = H_1(s) + H_2(s) \qquad ...(9.46)$$

2. Series interconnection of two LTI systems. Series or cascade interconnection of two continuous-time LTI systems is as shown in Fig. 9.15.

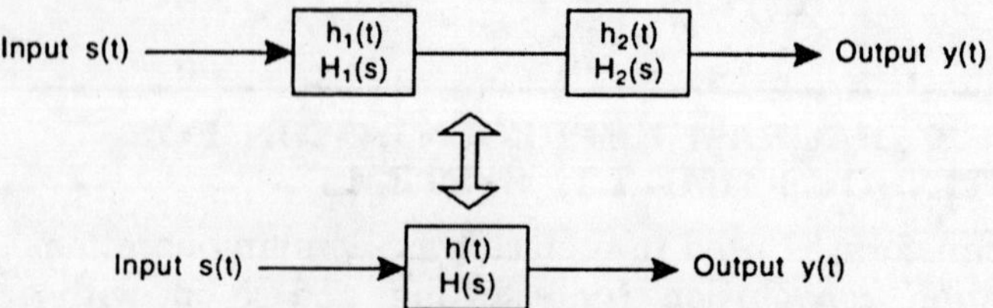

Fig. 9.15 *Series interconnection of two continuous-time LTI systems and its equivalent (overall) LTI system.*

The impulse response of the overall system of series interconnection of two continuous-time LTI systems can be determined by convolving $h_1(t)$ and $h_2(t)$.

$$h(t) = h_1(t) * h_2(t) \qquad ...(9.47)$$

and from convolution property of Laplace transform

$$H(s) = H_1(s)H_2(s) \qquad ...(9.48)$$

3. Feedback interconnection of two LTI systems. Feedback interconnection of two continuous-time LTI systems is shown in Fig. 9.16.

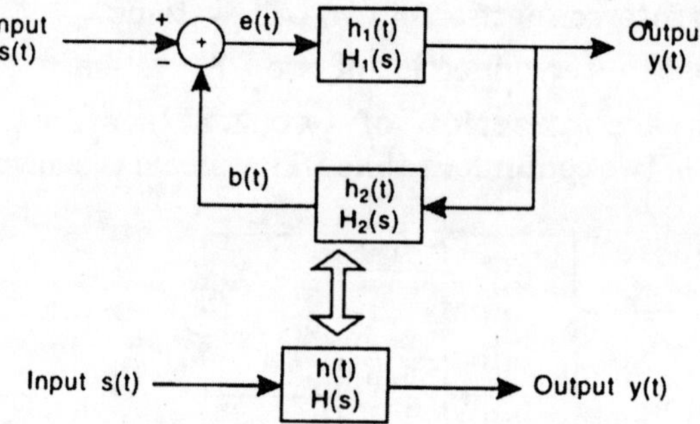

Fig. 9.16 *Feedback interconnection of two continuous-time LTI systems and its equivalent (overall) LTI system.*

The overall system function from input $s(t)$ to output $y(t)$ is straight-forward algebraic manipulation. From Fig. 9.16

Output $\quad y(t) = e(t) * h_1(t) \qquad ...(9.49)$

where $\quad e(t) = \text{Error signal} = s(t) - b(t) \qquad ...(9.50)$

and $\quad b(t) = \text{Feedback signal} = y(t) * h_2(t) \qquad ...(9.51)$

Taking Laplace transform of both sides of Eqs. 9.49, 9.50 and 9.51, we get

$$Y(s) = E(s)\, H_1(s) \qquad ...(9.52)$$

$$E(s) = S(s) - B(s) \quad \text{...(9.53)}$$

$$B(s) = Y(s)\, H_2(s) \quad \text{...(9.54)}$$

Substituting Eq. 9.54 in Eq. 9.53, we get

$$E(s) = S(s) - B(s)$$

or
$$E(s) = S(s) - Y(s)\, H_2(s) \quad \text{...(9.55)}$$

Substituting Eq. 9.55 in Eq. 9.52, we get

$$Y(s) = E(s)\, H_1(s) = [S(s) - Y(s)\, H_2(s)]\, H_1(s)$$

$$= S(s)\, H_1(s) - Y(s)\, H_2(s)\, H_1(s)$$

or
$$Y(s) + Y(s)\, H_1(s)\, H_2(s) = S(s)\, H_1(s)$$

or
$$Y(s)\,[1 + H_1(s) H_2(s)\,] = H_1(s)\, S(s)$$

or
$$\frac{Y(s)}{S(s)} = \frac{H_1(s)}{1 + H_1(s)\, H_2(s)}$$

or
$$H(s) = \frac{Y(s)}{S(s)} = \frac{H_1(s)}{1 + H_1(s)\, H_2(s)} \quad \text{...(9.56)}$$

This is the overall transfer function of the feedback interconnection of two continuous-time LTI systems. The utility of the Laplace transform in representing various combinations of linear systems through algebraic operations extends to more complex combinations (interconnections) than the simple parallel, series and feedback combinations.

9.9.2 Block Diagram Representation of Causal Continuous-Time LTI Systems described by Differential Equations and Rational Transfer Function of the System

Block diagram representation of a continuous-time LTI system is described by a first-order differential equation using the basic operations such as addition, multiplication by a coefficient and integration. These three basic operations can also be used to build block diagrams for higher order continuous-time LTI systems.

The methods for block diagram representation for causal continuous-time LTI systems described by differential equations and rational transfer function $H(s)$ can be understood by following example. These methods can be applied equally to higher order systems. These methods are

1. Direct-form.
2. Parallel-form.
3. Cascade-form.

Example 9.17 Sketch the block diagram representation for a continuous-time LTI system characterized by the transfer function

$$H(s) = \frac{1}{(s+2)(s+3)} = \frac{1}{s^2+5s+6} \quad ...(1)$$

Solution. Determination of differential equation of the LTI system

$$H(s) = \frac{Y(s)}{S(s)} = \frac{\text{Laplace transform of output } y(t)}{\text{Laplace transform of input } s(t)} \quad ...(2)$$

From Eqs. 1 and 2, we get

$$\frac{Y(s)}{S(s)} = \frac{1}{s^2+5s+6}$$

or

$$Y(s)\,[s^2+5s+6] = S(s) \quad ...(3)$$

Taking the inverse Laplace transform of both sides of Eq. 3, we get

$$\frac{d^2y(t)}{dt^2} + 5\frac{d^2y(t)}{dt} + 6\,y(t) = s(t) \quad ...(4)$$

It is a differential equation for the LTI system given by transfer function $H(s) = \frac{1}{s^2+5s+6}$.

Direct Form of Block Diagram Representation of Transfer Function $H(s)$

It is described by the differential equation given by Eq. 4 and is shown in Fig. 9.17 (*a*).

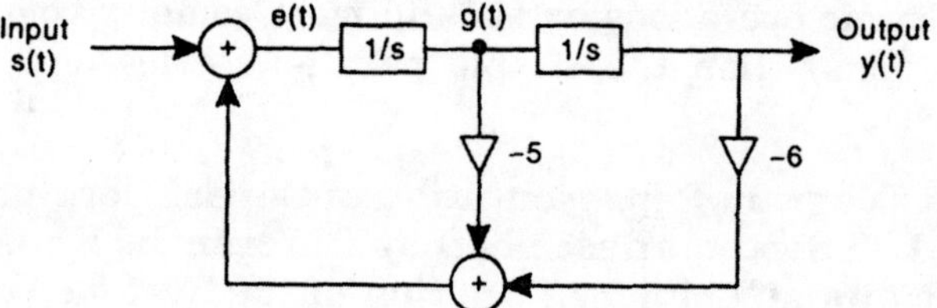

Fig. 9.17 (*a*) *Direct form of block diagram representation for the system function given by Eq.* 1.

Cascade Form of Block Diagram Representation of Transfer Function $H(s)$

$$H(s) = \frac{1}{(s+2)(s+3)} = H_1(s)\,H_2(s) \quad ...(5)$$

Cascade form of block diagram representation of transfer function $H(s)$ is shown in Fig. 9.17 (*b*).

Parallel Form of Block Diagram Representation of Transfer Function $H(s)$

$$H(s) = \frac{1}{s^2+5s+6} = \frac{1}{(s+2)(s+3)}$$

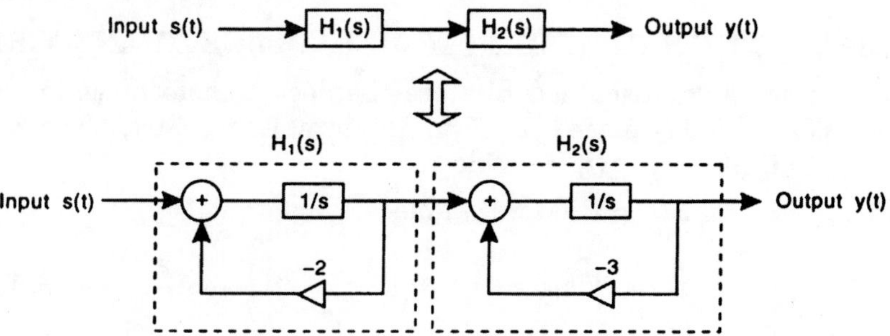

Fig. 9.17 (*b*) *Cascade form of block diagram representation for the system function*

$$H(s) = \frac{1}{s^2 + 5s + 6} = \frac{1}{(s+2)(s+3)}$$

$$= \frac{\alpha_1}{(s+2)} + \frac{\alpha_2}{(s+3)}$$

(By using partial fraction expansion)

or $$H(s) = \frac{1}{(s+2)} - \frac{1}{(s+3)} = H_1(s) + H_2(s) \qquad ...(6)$$

Parallel form of block diagram representation of transfer function given in Eq. 6 is shown in Fig. 9.17 (*c*).

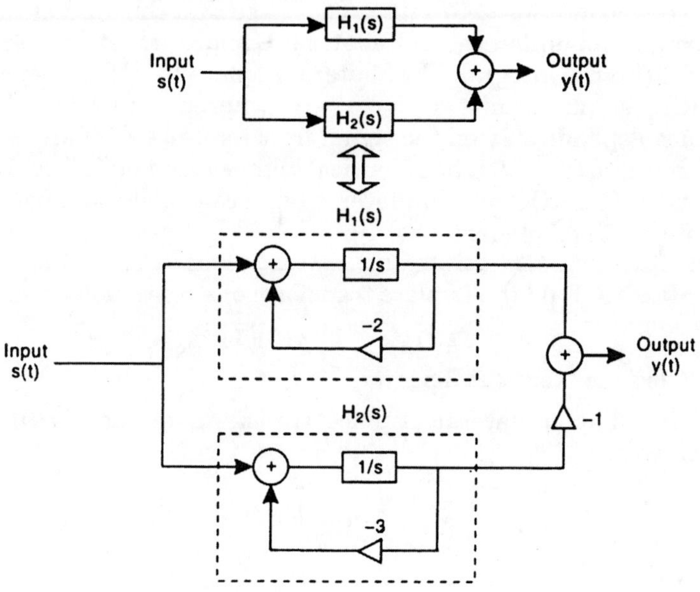

Fig. 9.17 (*c*) *Parallel form of block diagram representation of transfer function given in Eq. 6.*

9.10 DEFINITION OF UNILATERAL LAPLACE TRANSFORM

Upto now we have discussed Bilateral Laplace transform. It is also called as two-sided Laplace transform. Bilateral Laplace transform of a continuous-time signal $s(t)$ is defined as

$$S(s) = \text{Bilateral Laplace transform of } s(t)$$

$$= \int_{-\infty}^{\infty} s(t)\, e^{-st}\, dt \qquad \text{...(9.57)}$$

Unilateral Laplace transform is a special case of Bilateral Laplace transform. In unilateral Laplace transform the lower limit of integration is zero. It signifies that we include in the interval of integration any impulses or higher order singualrity functions concentrated at $t = 0$.

The Unilateral Laplace transform of a continuous-time signal $s(t)$ is defined as

$$S_u(s) = \int_{0}^{\infty} s(t)\, e^{-st}\, dt \qquad \text{...(9.58)}$$

We can use a short hand notation for a continuous-time signal $s(t)$ and its unilateral Laplace transform $S_u(s)$ for convenience given as

$$s(t) \xleftrightarrow{\text{U LT}} S_u(s) \qquad \text{...(9.59)}$$

By comparing Eqs. 9.57 and 9.58, we see that the difference in the definition of the unilateral and bilateral Laplace transform lies in the lower limit on the integral. The bilateral Laplace transform depends on the entire signal from $t = -\infty$ to $+\infty$, whereas unilateral Laplace transform depends only on the signal from $t = 0$ to $+\infty$. If any signal is identically zero for $t < 0$ it has identical bilateral and unilateral Laplace transforms. The ROC for a unilateral Laplace transform must always be a right-half of s-plane.

Example 9.18 Find the Laplace transform of a continuous-time signal

$$s(t) = e^{-A(t+1)}\, u(t+1)$$

where A is a constant coefficient.

Solution. The unilateral Laplace transform of the signal $s(t)$ is determined as

$$S_u(s) = \int_{0}^{\infty} s(t)\, e^{-st}\, dt = \int_{0}^{\infty} e^{-A(t+1)}\, u(t+1)\, e^{-st}\, dt$$

$$= \int_{0}^{\infty} e^{-A(t+1)}\, e^{-st}\, dt = e^{-A} \int_{0}^{\infty} e^{-(s+A)t}\, dt$$

$$= e^{-A}\left[\frac{1}{-(s+A)} e^{-(s+A)t}\right]_0^\infty$$

or $$S_u(s) = e^{-A}\left(\frac{1}{s+A}\right), \text{ ROC : Real } \{s\} = \sigma > -A \quad ...(1)$$

The bilateral Laplace transform of signal $s(t)$ is determined as

$$S(s) = \int_{-\infty}^{\infty} s(t)\, e^{-st}\, dt = \int_{-\infty}^{\infty} e^{-A(t+1)}\, u(t+1)\, e^{-st}\, dt$$

$$= e^{-s}\left(\frac{1}{s+A}\right)$$

(using shifting property of Laplace transform)

$$S(s) = \frac{e^{s}}{s+A}, \text{ ROC : Real } \{s\} = \sigma > -A \quad ...(2)$$

The unilateral and bilateral Laplace transforms given by Eqs. 1 and 2 are different.

Example 9.19 Find the inverse unilateral Laplace transform of

$$S_u(s) = \frac{1}{(s+2)(s+3)}$$

Solution. The bilateral Laplace transform of a continuous-time signal will have several ROCs. But for the unilateral Laplace transform, the ROC must be the right-half of the s-plane to the right of the rightmost pole of $S_u(s)$, *i.e.*, in this case the ROC consists of all points of s, Real $\{s\} = \sigma > -2$.

$$s(t) = \text{Inverse unilateral Laplace transform } \{S_u(s)\}$$

$$= \text{Inverse ULT}\left\{\frac{1}{(s+2)(s+3)}\right\}$$

$$= \text{Inverse ULT}\left\{\frac{\alpha_1}{s+2} + \frac{\alpha_2}{s+3}\right\}$$

$$= \text{Inverse ULT}\left\{\frac{1}{s+2} - \frac{1}{s+3}\right\}$$

(using partial fraction expansion)

$$= \text{Inverse ULT}\left\{\frac{1}{s+2}\right\} - \text{Inverse ULT}\left\{\frac{1}{s+3}\right\}$$

$$= e^{-2t}u(t) - e^{-3t}u(t)$$

or $$s(t) = (e^{-2t} - e^{-3t})\, u(t) \quad \text{for } t > 0 \quad ...(1)$$

Thus, we see that unilateral Laplace transform provides us with information about signals only for $t > 0$.

9.11 PROPERTIES OF THE UNILATERAL LAPLACE TRANSFORM

Unilateral Laplace transform has a number of important properties which are same as bilateral Laplace transform. However, there are some properties which are different. We will not discuss properties of unilateral Laplace transform in detail. We have summarized important properties of unilateral Laplace transform in Table 9.3. There is no need to include the column for ROC for unilateral Laplace transform in Table 9.3 because ROC for a unilateral. Laplace transform is always a right-half of s-plane. The ROC for a rational unilateral laplace transform is always to the right of the rightmost pole. The system analysis tool and block diagram representations for the transfer function developed and used for bilateral transfer function can be extended for unilateral Laplace transform without any change.

It is important to note that the convolution property for unilateral Laplace transforms is applied for only causal continuous-time signals *i.e.*, $s_1(t)$ and $s_2(t)$ are both zero for $t < 0$. The Bilateral Laplace transform of $s_1(t) * s_2(t)$ is always equal to the product of bilateral Laplace transforms of $s_1(t)$ and $s_2(t)$. But, unilateral Laplace transform of $s_1(t) * s_2(t)$ is not equal to the product of unilateral Laplace transforms of $s_1(t)$ and $s_2(t)$. This is valid for causal signals, *i.e.*, both $s_1(t)$ and $s_2(t)$ are zero for $t < 0$.

An important difference between the properties of the unilateral and bilateral Laplace transforms is the differentiation property. Now we are considering a continuous-time signal $s(t)$ with unilateral Laplace transform $S(s)$.

Determination of Unilateral Laplace Transform of $\frac{ds(t)}{dt}$

Unilateral Laplace transform of $\frac{ds(t)}{dt}$ is given as

$$\int_0^{\infty} \frac{ds(t)}{dt} e^{-st}\, dt \qquad \text{...(9.60)}$$

By integrating Eq. 9.60 by parts, we get

$$\int_0^{\infty} \frac{ds(t)}{dt} e^{-st}\, dt = \Big[s(t)\, e^{-st} \Big]_0^{\infty} + \int_0^{\infty} s(t)\, e^{-st}\, dt$$

$$= -s(0)\, e^{-s.0} + s \int_0^{\infty} s(t)\, e^{-st}\, dt$$

or $$\int_0^{\infty} \frac{ds(t)}{dt} e^{-st}\, dt = s\, S(s) - s(0) \qquad \text{...(9.61)}$$

Similarly, we can find the unilateral Laplace transform of $\frac{d^2s(t)}{dt^2}$.

$$\int_0^{\infty} \frac{d^2s(t)}{dt^2} e^{-st}\, dt = s^2\, S(s) - s\, s(0) - s'(0) \qquad \text{...(9.62)}$$

where $s'(0) = \frac{ds(0)}{dt}$ = Derivative of $s(t)$ evaluated at $t = 0$.

In similar manner, we can find unilateral Laplace transform of higher derivatives.

9.11.1 Tabulation of Properties of Unilateral Laplace Transform

We have summarized important properties of unilateral Laplace transform in Table 9.3.

Table 9.3 Summary of Important Properties of Unilateral Laplace Transform

Properties of unilateral Laplace transform	*Continuous-time signal*	*Unilateral Laplace transform*
	$s_1(t)$, $s_2(t)$, $s(t)$	$S_1(s)$, $S_2(s)$, $S(s)$
Linearity of Laplace transform	$A_1\, s_1(t) + A_2\, s_2(t)$	$A_1\, S_1(s) + A_2\, S_2(s)$
Shifting of Laplace transform in s-domain	$e^{js_0}\, s(t)$	$S(s - s_0)$
Time scaling of a continuous-time signal	$s(At)$, $A > 0$	$\frac{1}{A} S\left(\frac{s}{A}\right)$
Complex conjugation	$s^{*(t)}$	$S^*(s)$
Convolution integral for two continuous- time signals (Now we have assumed that $\left.\begin{matrix} s_1(t) = 0 \\ s_2(t) = 0 \end{matrix}\right]$ for $t < 0$	$s_1(t) * s_2(t)$	$S_1(s)\, S_2(s)$

Contd...

Properties of unilateral Laplace transform	*Continuous-time signal*	*Unilateral Laplace transform*
Differentiation of a continuous-time signal in time-domain	$\frac{d}{dt} s(t)$	$sS(s) - s(0^-)$
Differentiation of Laplace transform in s-domain	$-t\, s(t)$	$\frac{d}{ds} S(s)$
Integration of a continuous-time signal in time-domain	$\int_0^t s(\tau)\, d\tau$	$\frac{1}{s} S(s)$

Initial-value Theorem for Unilateral Laplace Transform.
If continuous-time signal $s(t)$ contains no impulses or higher order singularities at $t = 0$ then from initial-value theorem of Laplace transform, initial value of the signal can be determined as $s(t)\big|_{t=0^+} = s(0^+) = \lim_{s \to \infty} [sS(s)]$

Final-value Theorem for Unilateral Laplace Transform.
If continuous-time signal $s(t)$ contains no impulses or higher order singularities at $t = 0$ then from final-value theorem of Laplace transform, the final value of the signal can be determined as

$$\lim_{t \to \infty} s(t) = \lim_{s \to \infty} [sS(s)]$$

9.12 APPLICATIONS OF UNILATERAL LAPLACE TRANSFORM IN SOLVING DIFFERENTIAL EQUATIONS

The unilateral Laplace transform is primarily used to obtain the solution of linear constant-coefficient differential equations with non-zero initial conditions.

Example 9.20 Find the transfer function of a causal continuous-time LTI system described by differential equation

$$\frac{d^2y(t)}{dt^2} + 5\frac{d\,y(t)}{dt} + 6\,y(t) = s(t) \qquad \text{...(1)}$$

together with the condition of initial rest.

Also find the response of the system to input $s(t) = 2u(t)$.

Solution. *Determination of transfer function of the system*. Taking the unilateral Laplace transform of both sides of Eq. 1 keeping all initial conditions equal to zero

$$s^2\,Y(s) + 5s\,Y(s) + 6\,Y(s) = S(s) \qquad \text{...(2)}$$

or $$Y(s)\,[s^2 + 5s + 6] = S(s)$$

or $$H(s) = \frac{Y(s)}{S(s)} = \frac{1}{s^2 + 5s + 6} \qquad ...(3)$$

This is the transfer function of the system described by Eq. 1.

Now we have

input, $s(t) = 2\,u(t)$ ...(4)

Taking the Laplace transform of both sides of Eq. 4, we get

$$S(s) = 2\,\frac{1}{s} \qquad ...(5)$$

Laplace transform of output $y(t)$

$$Y(s) = H(s)\,S(s) \qquad ...(6)$$

Substituting the value of $H(s)$ and $S(s)$ in Eq. 6, we get

$$Y(s) = \left(\frac{1}{s^2 + 5s + 6}\right)\left(\frac{2}{s}\right) = \frac{2}{s(s^2 + 5s + 6)}$$

or $$Y(s) = \frac{2}{s(s+2)\,(s+3)} = \frac{\alpha_1}{s} + \frac{\alpha_2}{(s+2)} + \frac{\alpha_3}{(s+3)} \qquad ...(7)$$

Using partial fraction expansion, we get

$s = 0,$ $$\alpha_1 = \frac{2}{(s+2)\,(s+3)} = \frac{2}{(0+2)\,(0+3)} = \frac{1}{3}$$

or $\begin{matrix} s+2=0 \\ s=-2 \end{matrix}$, $$\alpha_2 = \frac{2}{s(s+3)} = \frac{2}{(-2)\,(-2+3)} = -1$$

or $\begin{matrix} s+3=0 \\ s=-3 \end{matrix}$, $$\alpha_3 = \frac{2}{s(s+2)} = \frac{2}{(-3)\,(-3+2)} = \frac{2}{3}$$

Substituting the values of α_1 , α_2 and α_3 in Eq. 7, we get

$$Y(s) = \frac{1/3}{s} + \frac{(-1)}{(s+2)} + \frac{2/3}{(s+3)} \qquad ...(8)$$

Taking inverse unilateral Laplace transform of both sides of Eq. 8, we get

$$\text{Inverse ULT }\{Y(s)\} = \text{Inverse ULT}\left\{\frac{(1/3)}{s} - \frac{1}{(s+2)} + \frac{2/3}{(s+3)}\right\}$$

or $$y(t) = \frac{1}{3}\,u(t) - e^{-2t}\,u(t) + \frac{2}{3}\,e^{-3t}\,u(t) \qquad ...(9)$$

This is the response of the system described by Eq. 1 to input $s(t) = 2\,u(t)$.

Example 9.21 Find the response of a continuous-time LTI system described by the differential equation

$$\frac{d^2y(t)}{dt^2} + 5\,\frac{dy(t)}{dt} + 6\,y(t) = s(t) \qquad ...(1)$$

with initial conditions $y(0) = 2$ and $\frac{dy(0)}{dt} = y'(0) = 1$

Also given input $s(t) = 2\,u(t)$.

Solution. Taking unilateral Laplace transform of both sides of Eq. 1, we get

$$\text{ULP}\left\{\frac{d^2y(t)}{dt^2} + 5\frac{dy(t)}{dt} + 6\,y(t)\right\} = \text{ULP}\,\{s(t)\}$$

But we have given input $s(t) = 2\,u(t)$

or $$\text{ULP}\left\{\frac{d^2y(t)}{dt^2} + 5\frac{dy(t)}{dt} + 6\,y(t)\right\} = \text{ULP}\,\{2\,u(t)\}$$

or $$\text{ULP}\left\{\frac{d^2y(t)}{dt^2}\right\} + 5\,\text{ULP}\left\{\frac{dy(t)}{dt}\right\} + 6\,\text{ULP}\,\{y(t)\} = 2\,\text{ULP}\,\{u(t)\}$$

or $$[s^2\,Y(s) - s\,y(0) - y'(0)] + 5\,[s\,Y(s) - y(0)] + 6\,Y(s) = 2\frac{1}{s} \qquad ...(2)$$

Substituting the value of initial conditions such as $y(0) = 2$ and $y'(0) = 1$ in Eq. 2, we get

$$[s^2\,Y(s) - 2s - 1] + 5\,[s\,Y(s) - 2] + 6\,Y(s) = \frac{2}{s}$$

or $$s^2\,Y(s) - 2s - 1 + 5s\,Y(s) - 10 + 6\,Y(s) = \frac{2}{s}$$

or $$Y(s)\,[s^2 + 5s + 6] = \frac{2}{s} + 2s + 11$$

or $$Y(s) = \frac{2}{s(s^2 + 5s + 6)} + \frac{2s}{(s^2 + 5s + 6)} + \frac{11}{(s^2 + 5s + 6)} \qquad ...(3)$$

where $Y(s)$ is the unilateral Laplace transform of $y(t)$.

Eq. 3 can be written as

$$Y(s) = \frac{2s^2 + 11s + 2}{s(s^2 + 5s + 6)} = \frac{2s^2 + 11s + 2}{s(s+2)\,(s+3)}$$

$$= \frac{\alpha_1}{s} + \frac{\alpha_2}{(s+2)} + \frac{\alpha_3}{(s+3)} \qquad ...(4)$$

(using partial fraction expansion)

$s = 0$, $$\alpha_1 = \frac{2s^2 + 11s + 2}{(s+2)\,(s+3)} = \frac{2(0)^2 + 11 \times 0 + 2}{(0+2)\,(0+3)} = \frac{1}{3}$$

$$\text{or } \begin{matrix} s+2=0 \\ s=-2 \end{matrix}, \quad \alpha_2 = \frac{2s^2+11s+2}{s(s+3)} = \frac{2(-2)^2+11(-2)+2}{(-2)\,(-2+3)}$$

$$= \frac{8-22+2}{-2} = 6$$

$$\text{or } \begin{matrix} s+3=0 \\ s=-3 \end{matrix}, \quad \alpha_3 = \frac{2s^2+11s+2}{s(s+2)} = \frac{2(-3)^2+11(-3)+2}{(-3)\,(-3+2)}$$

$$= \frac{18-33+2}{3} = \frac{-13}{3}$$

Substituting the values of α_1 , α_2 and α_3 in Eq. 4, we get

$$Y(s) = \frac{(1/3)}{s} + \frac{6}{(s+2)} + \frac{(-13/3)}{(s+3)} \qquad ...(5)$$

Taking the inverse unilateral Laplace transform of both sides of Eq. 5, we get

$$y(t) = \frac{1}{3}\,u(t) + 6\,e^{-2t}u(t) - \frac{13}{3}\,e^{-3t}u(t)$$

$$= \left(\frac{1}{3} + 6\,e^{-2t} - \frac{13}{3}\,e^{-3t}\right)u(t) \text{ for } t > 0 \qquad ...(6)$$

This is the desired output of the system described by Eq. 1 with non-zero initial conditions for input $s(t) = 2\,u(t)$.

Example 9.22 Find the' Laplace transform of the following continuous-time funcdtions

(*a*) $s(t) = t^3 + 3t - 6t + 4$ (*b*) $s(t) = \cos^3 3t$

(c) $s(t) = \sin At \cos Bt$ (*d*) $s(t) = t \sin At$

(*e*) $s(t) = \dfrac{1-e^t}{t}$ (*f*) $s(t) = \delta(t^2 - 3t + 2)$

Solution. Laplace transform of a continuous-time signal $s(t)$ is defined as

$$S(s) = \text{LT}\,\{s(t)\} = \int_0^\infty s(t)\,e^{-st}\,dt$$

$$= \int_0^\infty (t^3 + 3t^2 - 6t + 4)\,e^{-st}\,dt$$

$$= \int_0^\infty t^3\,e^{-st}\,dt + 3\int_0^\infty t^2\,e^{-st}\,dt - 6\int_0^\infty t\,e^{-st}\,dt$$

$$+ 4\int_0^\infty e^{-st}\,dt$$

$$= \frac{6}{s^4} + \frac{6}{s^3} - \frac{6}{s^2} + \frac{4}{s}$$

(*b*) Given : $s(t) = \cos^3 3t$ ...(1)

We know that

$$\cos 3\theta = 4 \cos^3 \theta - 3 \cos \theta$$

Eq. 1 can be written as

$$s(t) = \frac{\cos 9t + 3 \cos 3t}{4}$$

$$S(s) = \text{LT } \{s(t)\} = \int_0^\infty s(t)\, e^{-st}\, dt$$

$$= \int_0^\infty \left(\frac{\cos 9t + 3 \cos 3t}{4} \right) e^{-st}\, dt$$

$$= \frac{1}{4} \int_0^\infty \cos 9t\, e^{-st}\, dt + \frac{3}{4} \int_0^\infty \cos 3t\, e^{-st}\, dt$$

$$= \frac{1}{4}\left[\frac{s}{s^2 + (9)^2} \right] + \frac{3}{4}\left[\frac{s}{s^2 + (3)^2} \right]$$

$$= \frac{1}{4}\left[\frac{s}{s^2 + 81} \right] + \frac{3}{4}\left[\frac{s}{s^2 + 9} \right]$$

(*c*) Given : $s(t) = \sin At \cos Bt$ ...(2)

Eq. 2 can be written as

$$s(t) = \frac{1}{2} [\sin (A + B)\, t + \sin (A - B)\, t]$$

$$S(s) = \text{LT } \{s(t)\} = \int_0^\infty s(t)\, e^{-st}\, dt$$

$$= \int_0^\infty \left\{ \frac{1}{2} [\sin (A + B)\, t + \sin (A - B)\, t] \right\} e^{-st}\, dt$$

$$= \frac{1}{2} \int_0^\infty \sin (A + B)\, t\, e^{-st}\, dt + \frac{1}{2} \int_0^\infty \sin (A - B)\, e^{-st}\, dt$$

$$= \frac{1}{2}\left[\frac{(A + B)}{s^2 + (A + B)^2} \right] + \frac{1}{2}\left[\frac{(A - B)}{s^2 + (A - B)^2} \right]$$

(*d*) Given : $s(t) = t \sin At$

$$S(s) = \text{LT } \{t \sin At\} = -\frac{d}{ds} [\text{LT } \{\sin At\}]$$

$$= -\frac{d}{ds}\left[\frac{A}{s^2 + A^2} \right] = \frac{2As}{(s^2 + A^2)^2}$$

(e) Given : $s(t) = \left[\dfrac{1-e^t}{t}\right]$

$$S(s) = \text{LT}\{s(t)\} = \text{LT}\left\{\frac{1-e^t}{t}\right\}$$

$$= \int_s^\infty [\text{LT}\{1-e^t\}]\, ds = \int_s^\infty \left[\frac{1}{s} - \frac{1}{s-1}\right] ds$$

$$= [\log s - \log(s-1)]_s^\infty = \left[\log\left(\frac{s}{s-1}\right)\right]_s^\infty$$

$$= \left[-\log\left(\frac{s-1}{s}\right)\right]_0^\infty = \left[-\log\left(1-\frac{1}{s}\right)\right]_s^\infty$$

$$= \log\left(1-\frac{1}{s}\right) = \log\left(\frac{s-1}{s}\right)$$

(f) Given : $s(t) = \delta(t^2 - 3t + 2) = \delta[(t-1)(t-2)]$

$$= \delta(t-1)\, u(t-1) + \delta(t-2)\, u(t-2)$$

$$= \delta(t-1) + \delta(t-2)$$

Laplace tranform of $s(t)$

$$= S(s) = \text{LT}\{\delta(t-1) + \delta(t-2)\} = e^{-s} + e^{-2s}$$

Example 9.23 Find the Laplace transform of the rectangular pulse shown in Fig. 9.18.

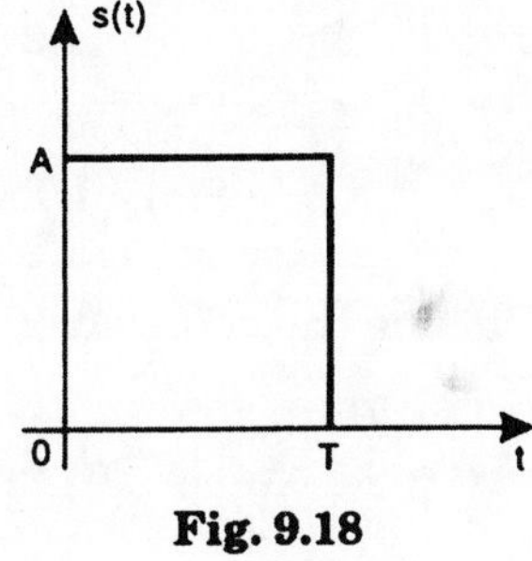

Fig. 9.18

Solution. Laplace transform at the rectangular pulse can be determined by

$$S(s) = \text{LT}\{s(t)\} = \int_0^\infty s(t)\, e^{-st}\, dt$$

$$= \int_0^T A\, e^{-st}\, dt = \left[\frac{Ae^{-st}}{-s}\right]_0^T$$

$$= \frac{A}{s}[1 - e^{-sT}]$$

Example 9.24 Find the Laplace transform of a single sawtooth pulse shown in Fig. 9.19.

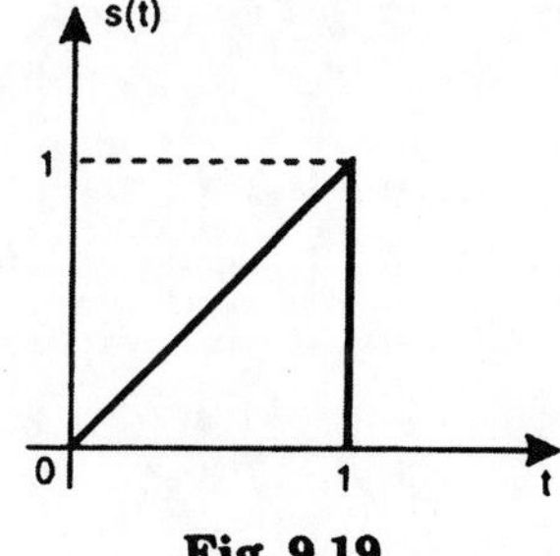

Fig. 9.19

Solution. The function $s(t)$ can be analytically given as

$$s(t) = \begin{cases} t, & \text{for } 0 < t < 1 \\ 0, & \text{for } t > 1 \end{cases} \quad \text{...(1)}$$

$$S(s) = \text{LT}\,\{s(t)\} = \int_0^\infty s(t)\, e^{-st}\, dt \qquad ...(2)$$

Putting the value of $s(t)$ from Eq. 1 in Eq. 2, we get

$$S(s) = \int_0^1 t\, e^{-st}\, dt = \left[\frac{te^{-st}}{-s} - \frac{1\, e^{-st}}{s^2}\right]_0^1$$

$$= \frac{1}{s^2} - e^{-st}\left[\frac{1}{s} + \frac{1}{s^2}\right]$$

Example 9.24 Find the Laplace transform of the triangular pulse shown in Fig. 9.20.

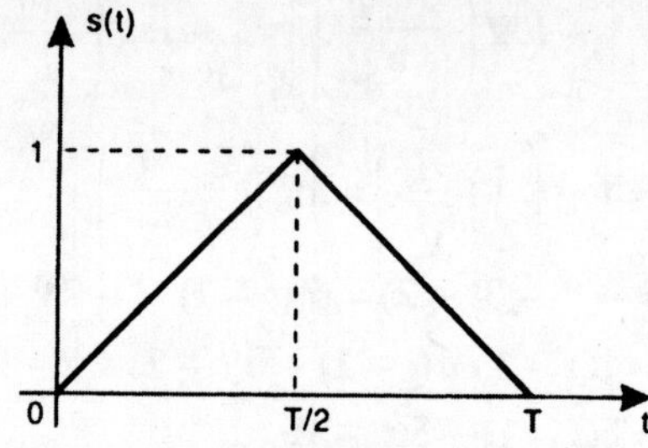

Fig. 9.20

Solution. The function $s(t)$ can be analytically given as

$$s(t) = \begin{cases} \dfrac{2}{T}t\,, & \text{for } 0 \le t \le \dfrac{T}{2} \\ 2 - \dfrac{2}{T}t\,, & \text{for } \dfrac{T}{2} \le t \le T \end{cases} \qquad ...(1)$$

$$S(s) = \text{LT}\,\{s(t)\} = \int_0^\infty s(t)\, e^{-st}\, dt \qquad ...(2)$$

Putting the value of $s(t)$ from Eq. 1 into Eq. 2, we get

$$S(s) = \int_0^{T/2} \left(\frac{2}{T}t\right) e^{-st}\, dt + \int_{T/2}^{T} \left(2 - \frac{2}{T}t\right) e^{-st}\, dt$$

$$= \frac{2}{T}\int_0^{T/2} t\, e^{-st}\, dt + 2\int_{T/2}^{T} e^{-st}\, dt - \frac{2}{T}\int_{T/2}^{T} t\, e^{-st}\, dt$$

$$= \frac{2}{T}\left[\left\{\frac{te^{-st}}{-s}\right\} - \left\{\frac{e^{-st}}{s^2}\right\}\right]_0^{T/2} + 2\left[\frac{e^{-st}}{-s}\right]_{T/2}^{T} - \frac{2}{T}\left[\left\{\frac{te^{-st}}{-s}\right\} - \left\{\frac{e^{-st}}{s^2}\right\}\right]_{T/2}^{T}$$

$$= \frac{2}{T}\frac{1}{s^2} - \frac{4}{T}\frac{e^{-sT/2}}{s^2} + \frac{2}{T}\frac{e^{-sT}}{s^2}$$

Example 9.25 Find the Laplace transform of the square wave shown in Fig. 9.21.

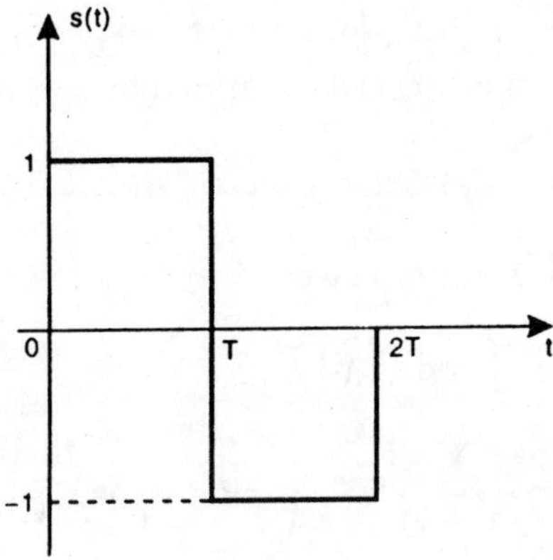

Fig. 9.21

Solution. Function $s(t)$ can be analytically given as

$$s(t) = \begin{cases} 1, & \text{for } 0 < t < T \\ -1, & \text{for } T < t < 2T \end{cases} \qquad \text{...(1)}$$

Laplace transform of $s(t)$ is determined as

$$S(s) = \text{LT}\ \{s(t)\} = \int_0^\infty s(t)\, e^{-st}\, dt \qquad \text{...(2)}$$

Substituting Eq. 1 in Eq. 2, we get

$$S(s) = \int_0^T (1)\, e^{-st}\, dt + \int_T^{2T} (-1)\, e^{-st}\, dt$$

$$= \left[\frac{e^{-st}}{-s}\right]_0^T - \left[\frac{e^{-st}}{-s}\right]_T^{2T}$$

$$= -\frac{1}{s}\,[e^{-st} - 1 - e^{-2sT} + e^{-sT}]$$

$$= \frac{1}{s}\,(1 - e^{-sT})^2$$

Example 9.26 Find the Laplace transform for the waveform shown in Fig. 9.22.

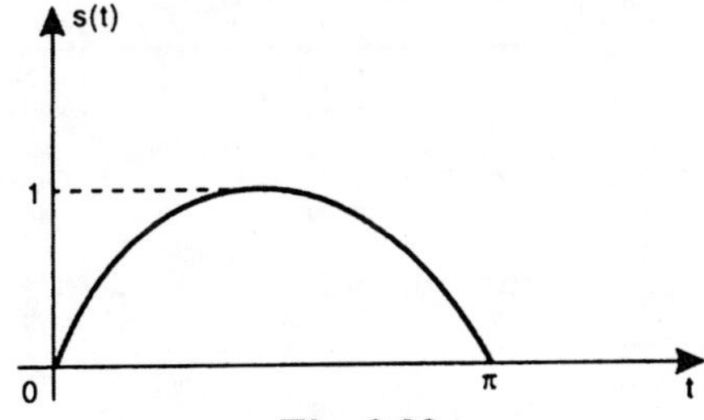

Fig. 9.22

Solution. The function for the given waveform is

$$s(t) = \begin{cases} \sin t, & \text{for } 0 < t < \pi \\ 0, & \text{for } t > \pi \end{cases} \qquad ...(1)$$

Laplace transform of $s(t)$ is determined by

$$S(s) = \text{LT}\,\{s(t)\} = \int_0^{\infty} s(t)\, e^{-st}\, dt \qquad ...(2)$$

Substituting Eq. 1 in Eq. 2, we get

$$S(s) = \int_0^{\pi} \sin t\, e^{-st}\, dt$$

$$= \frac{1}{(s^2+1)} \left[e^{-st}\,(-s \sin t - \cos t)\right]_0^{\pi}$$

$$= \frac{e^{-s\pi}+1}{(s^2+1)}$$

Example 9.27 Find the convolution integral for the following continuous-time functions $s_1(t) = e^{-2t}$ and $s_2(t) = 2t$.

Solution. Convolution integral of $s_1(t)$ and $s_2(t)$ is defined as

$$s(t) = s_1(t) * s_2(t)$$

$$= \int_0^t s_1(\tau)\, s_2(t-\tau)\, d\tau$$

$$= \int_0^t e^{-2\tau}\, 2(t-\tau)\, d\tau$$

$$= 2t \int_0^t e^{-2\tau}\, d\tau - 2 \int_0^t \tau\, e^{-2\tau}\, d\tau$$

$$= \left[t - \frac{1}{2} + \frac{e^{-2t}}{2}\right] u(t)$$

Example 9.28 Find the unit-step response of the *RC* network shown in Fig. 9.23 by using convolution integral.

Solution. The transfer function of the network is given by

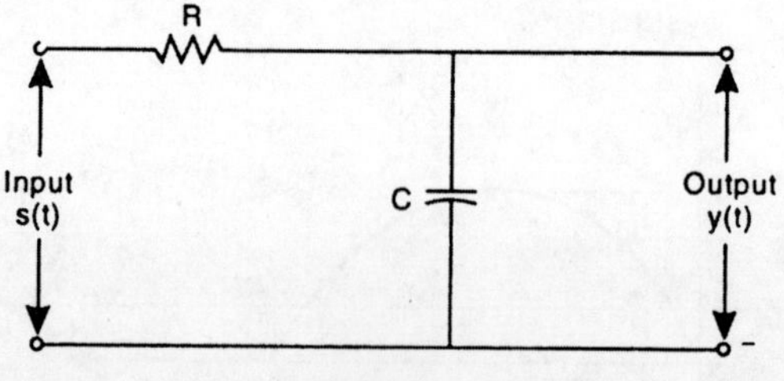

Fig. 9.23

$$H(s) = \frac{Y(s)}{S(s)} = \frac{1/Cs}{R + 1/Cs} + \frac{1}{RCs + 1}$$

or
$$H(s) = \frac{1}{RC}\left[\frac{1}{s + \frac{1}{RC}}\right] \qquad ...(1)$$

Therefore, the impulse response of the network can be determined by taking inverse Laplace transform of Eq. 1,

$$h(t) = \text{Inverse Laplace transform } \{H(s)\}$$

$$= \text{Inverse Laplace transform } \left\{\frac{1}{RC}\left[\frac{1}{s + 1/RC}\right]\right\}$$

or
$$h(t) = \frac{1}{RC} e^{-(1/RC)t} \text{ for } t \geq 0$$

$$= \frac{1}{RC} e^{-(1/RC)t}\, u(t) \qquad ...(2)$$

Unit-step response is the output of the network when input $s(t) = u(t)$. If input $s(t) = u(t)$ then $y(t)$ is called unit-step response denoted by $g(t)$.

Convolution of input $s(t)$ and impulse response $h(t)$ is defined as

$$y(t) = \int_{-\infty}^{\infty} s(\tau)\, h(t - \tau)\, d\tau$$

For $s(t) = u(t)$, $y(t) = g(t)$, we get

$$g(t) = \int_{-\infty}^{\infty} u(\tau)\, h(t - \tau)\, d\tau \qquad ...(3)$$

Substituting Eq. 1 in Eq. 3, we get

$$g(t) = \int_{-\infty}^{\infty} u(\tau) \frac{1}{RC} e^{-(t-\tau)/RC}\, u(t - \tau)\, d\tau \qquad ...(4)$$

$$\left[\text{Since } u(\tau) = \begin{cases} 1, & \tau > 0 \\ 0, & \tau < 0 \end{cases} \text{ and } u(t - \tau) = \begin{cases} 1, & t - \tau \geq 0 \rightarrow \tau \leq t \\ 0, & t - \tau < 0 \rightarrow \tau > t \end{cases}\right]$$

Eq. 4 can be written as

$$g(t) = \int_{0}^{t} \frac{1}{RC} e^{-(t-\tau)/RC}\, d\tau = \frac{e^{-t/RC}}{RC} \int_{0}^{t} e^{\tau/RC}\, d\tau$$

$$= \frac{e^{-t/RC}}{RC}\left[\frac{e^{t/RC}}{\frac{1}{RC}}\right]_0^t = e^{-t/RC}\left[e^{t/RC} - e^{0/RC}\right]$$

$$= 1 - e^{-t/RC} \text{ for } t \geq 0$$

or $$g(t) = \left[1 - e^{-t/RC}\right] u(t)$$

This is the step response of the network given in Fig. 9.23.

Example 9.29 Find the output response $y(t)$ of the RC low-pass network shown in Fig. 9.24 due to an input $s(t) = t\, e^{-t/RC}$ by using convolution integral.

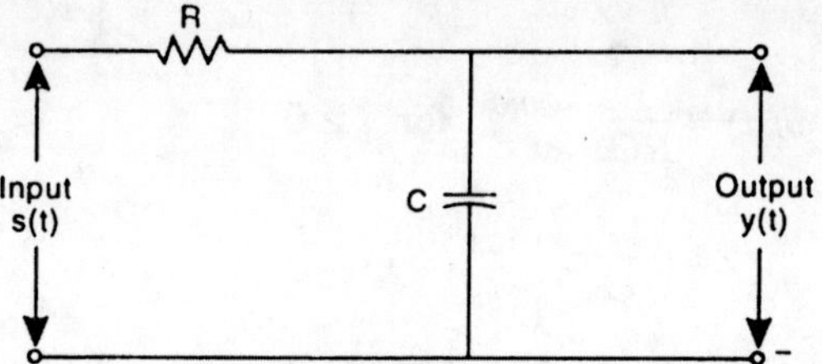

Fig. 9.24

Solution. The transfer function of the network shown in Fig. 9.24 is given by

$$H(s) = \frac{Y(s)}{S(s)} = \frac{1/Cs}{R + 1/Cs} = \frac{1}{(RCs + 1)}$$

$$= \frac{1}{RC}\frac{1}{(s + 1\text{RC})} \qquad \text{...(1)}$$

The input function is $s(t) = t\, e^{-t/RC}$ Taking the Laplace transform of $s(t) = t\, e^{-t/RC}$, we get

$$S(s) = \text{LT}\,\{s(t)\} = \text{LT}\left\{t\, e^{-t/RC}\right\}$$

$$= \frac{1}{(s + 1/RC)^2} \qquad \text{...(2)}$$

We know from convolution theorem of Laplace transform that

$$y(t) = s(t) * h(t) \xleftrightarrow{\text{Laplace transform}} Y(s) = S(s)\,H(s)$$

or $$Y(s) = S(s)\,H(s) \qquad \text{...(3)}$$

Substituting Eqs. 1 and 2 in Eq. 3, we get

$$Y(s) = S(s)\,H(s) = \frac{1}{(s + 1/RC)^2}\frac{1}{RC}\frac{1}{(s + 1/RC)}$$

or $$Y(s) = \frac{1}{RC}\frac{1}{(s + 1/RC)^3} \qquad \text{...(4)}$$

Taking the inverse Laplace transform of Eq. 4, we get

$$y(t) = \text{Inverse Laplace transform } \{Y(s)\}$$

$$= \text{Inverse LT}\left\{\frac{1}{RC}\frac{1}{(s+1/RC)^3}\right\}$$

or
$$y(t) = \frac{1}{RC}\text{ Inverse LT}\left\{\frac{1}{(s+1/RC)^3}\right\}$$

$$y(t) = \frac{1}{RC}\frac{t^2 e^{-t/RC}}{2}u(t)$$

Example 9.30 Draw the pole-zero plot for the Laplace transform of current in a network given by

$$I(s) = \frac{3s}{(s+2)(s+4)}$$

Also obtain the current $i(t)$ from $I(s)$.

Solution. The zero occurs at $s = 0$ and poles at $s = -2$ and $s = -4$. These are shown in Fig. 9.25.

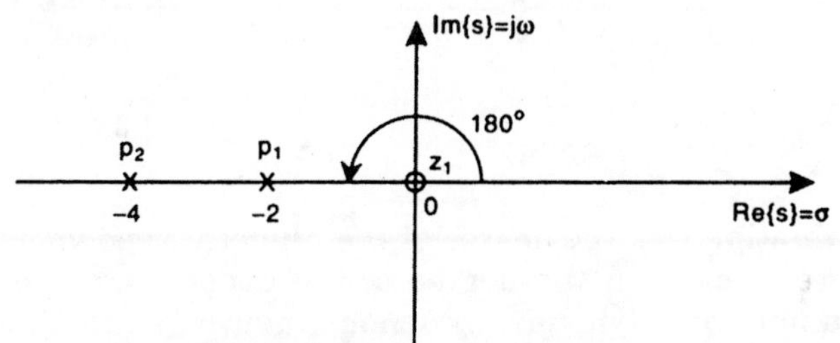

Fig. 9.25 *Pole-zero plot for I(s).*

Inverse Laplace transform of $I(s)$ can be determined by using partial fraction expansion method.

$$I(t) = \text{Inverse Laplace transform } \{I(s)\}$$

$$= \text{Inverse LT}\left\{\frac{3s}{(s+2)(s+4)}\right\}$$

$$= \text{Inverse LT}\left\{\frac{A_1}{(s+2)} + \frac{A_2}{(s+4)}\right\}$$

$$= \text{Inverse LT}\left\{\frac{-3}{s+2} + \frac{6}{s+4}\right\}$$

$$= \text{Inverse LT}\left\{\frac{-3}{s+2}\right\} + \text{Inverse LT}\left\{\frac{6}{s+4}\right\}$$

$$= -3 \times \text{Inverse LT}\left\{\frac{1}{s+2}\right\} + 6 \times \text{Inverse LT}\left\{\frac{1}{s+4}\right\}$$

or
$$I(t) = -3\,e^{-2t} + 6\,e^{-4t}$$

Example 9.31 Draw the pole-zero plot of

$$Y(s) = 4\,\frac{(s+1)\,(s+3)}{(s+2)\,(s+4)}$$

Also obtain $y(t)$.

Solution. The zeros occurs at

$$s + 1 = 0 \Rightarrow s = -1$$

$$s + 3 = 0 \Rightarrow s = -3$$

The poles occurs at

$$s + 2 = 0 \Rightarrow s = -2$$

$$s + 4 = 0 \Rightarrow s = -4$$

The pole-zero plot of $Y(s)$ is shown in Fig. 9.26.

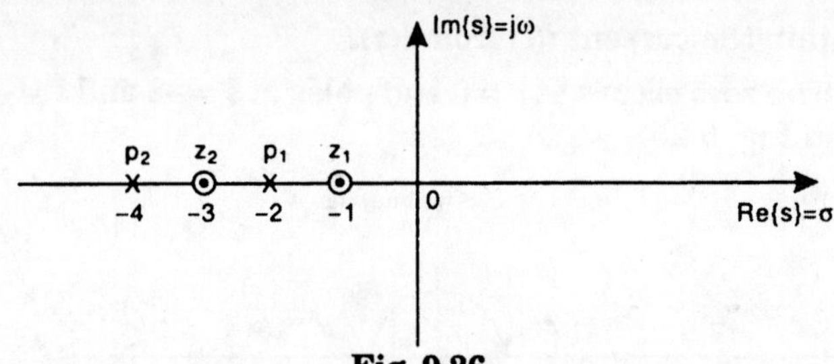

Fig. 9.26

For evaluating $y(t)$, the degree of the numerator polynomial must be one degree less than the degree of the denominator polynomial. Dividing numerator polynomial by denominator polynomial, we get

$$Y(s) = 4\,\frac{(s+1)\,(s+3)}{(s+2)\,(s+4)} = 4\left(\frac{s^2+4s+3}{s^2+6s+8}\right)$$

$$= 4\left[1 - \frac{(2s+5)}{s^2+6s+8}\right] = 4\left[1 - \frac{(2s+5)}{(s+2)\,(s+4)}\right]$$

or $$Y(s) = 4 - 8\left[\frac{s+2.5}{(s+2)\,(s+4)}\right] \quad \text{...(1)}$$

$$Y(s) = 4 - 8\left[\frac{s+2.5}{(s+2)\,(s+4)}\right] = 4 - 8\left[\frac{A_1}{(s+2)} + \frac{A_2}{(s+4)}\right]$$

(using partial fraction expansion)

or $$Y(s) = 4 - 8\left[\frac{1/4}{(s+2)} + \frac{3/4}{(s+4)}\right] \quad \text{...(2)}$$

Taking the inverse Laplace transform of Eq. 2, we get

$$y(t) = \text{Inverse Laplace transform } \{Y(s)\}$$

$$= \text{Inverse LT}\left\{4 - 8\left[\frac{1}{4(s+2)} + \frac{3}{4(s+4)}\right]\right\}$$

$$= \text{Inverse LT } \{4\} - \text{Inverse LT}\left\{\frac{2}{s+2}\right\}$$

$$- \text{Inverse LT}\left\{\frac{6}{s+4}\right\}$$

$$y(t) = 4\,\delta(t) - 2e^{-2t} - 6e^{-4t}$$

Example 9.32 Draw the pole-zero plot for the transfer function of a system given by

$$H(s) = \frac{10s}{s^2 + 2s + 2}$$

Also determine impulse response $h(t)$.

Solution. Given transfer function

$$H(s) = \frac{10s}{s^2 + 2s + 2} = \frac{10s}{(s + 1 - j1)\,(s + 1 + j1)}$$

The zero occurs at $s = 0$

The pole occurs at $s + 1 - j1 = 0 \;\Rightarrow\; s = -1 + j1$

and $\quad s + 1 + j1 = 0 \;\Rightarrow\; s = -1 - j1$

The pole-zero plot for $H(s)$ is shown in Fig. 9.27.

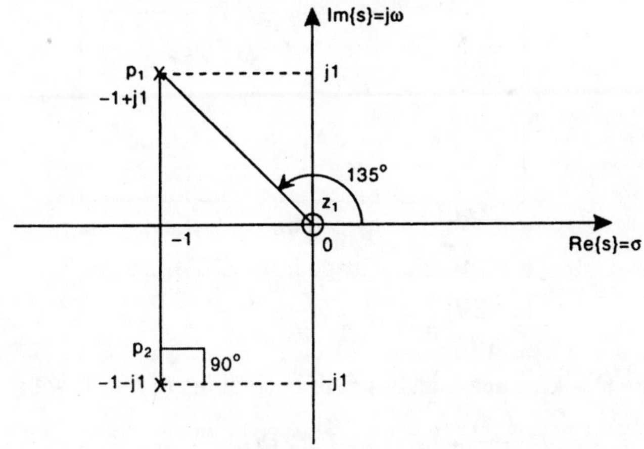

Fig. 9.27

Impulse response $h(t)$ can be determined by taking inverse Laplace transform of transfer function $H(s)$.

$$h(t) = \text{Inverse Laplace transform } \{H(s)\}$$

$$= \text{Inverse LT}\left\{\frac{10s}{(s + 1 - j1)\,(s + 1 + j1)}\right\}$$

$$= \text{Inverse LT}\left\{\frac{A}{(s + 1 - j1)} + \frac{A_2}{(s + 1 + j1)}\right\}$$

(using partial fraction expansion)

$$= \text{Inverse LT}\left\{\frac{7.07\, e^{j\pi/4}}{(s+1-j1)} = \frac{7.07\, e^{-j\pi/4}}{(s+1+j1)}\right\}$$

$$= 7.07\, e^{j\pi/4}\ \text{Inverse LT}\left\{\frac{1}{s+1-j1}\right\}$$

$$+ 7.07\, e^{-j\pi/4}\ \text{Inverse LT}\left\{\frac{1}{s+1+j1}\right\}$$

$$= 7.07\, e^{j\pi/4}\, e^{-(1-j1)t} + 7.07\, e^{-j\pi/4}\, e^{-(1+j1)t}$$

$$= 7.07\, e^{-t}\left\{ e^{j[1+(\pi/4)]t} + e^{-j[1+(\pi/4)]t}\right]$$

$$= 2 \times 7.07\, e^{-t}\left[\frac{e^{j[1+(\pi/4)]t} + e^{-j[1+(\pi/4)]t}}{2}\right]$$

$$= 14.14\, e^{-t} \cos\left(\frac{\pi}{4} + 1\right)t$$

or $$h(t) = 14.14\, e^{-t} \cos\left(\frac{\pi}{4} + 1\right)t$$

Example 9.33 Determine the current $i(t)$ in the series RL circuit shown in Fig. 9.28.

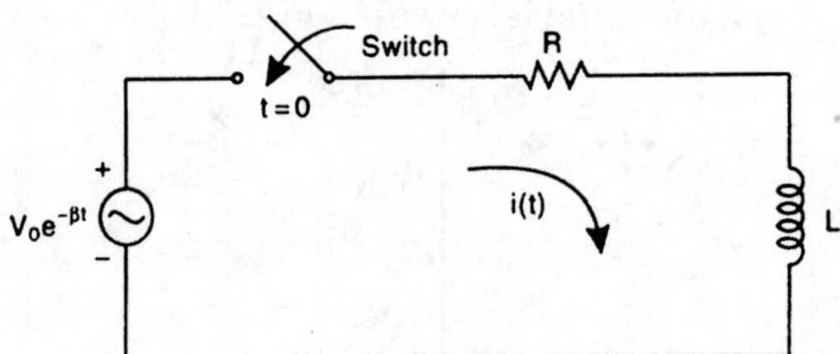

Fig. 9.28

Solution. Using Kirchhoff's voltage law (KVL), we get

$$L\frac{di(t)}{dt} + Ri(t) = V_0 e^{-\beta t} \qquad ...(1)$$

Taking the Laplace transform of both sides of Eq. 1, we get

$$L.T.\left\{ L\frac{di(t)}{dt} + Ri(t)\right\} = L.T.\ \{V_0\, e^{-\beta t}\}$$

or $$L\, sI(s) + RI(s) = V_0\left[\frac{1}{s+\beta}\right]$$

or $$I(s)\,(Ls + R) = V_0\left[\frac{1}{s+\beta}\right]$$

or $$I(s) = V_0\, \frac{1}{(s+\beta)\,(Ls+R)}$$

or $$I(s) = \frac{V_0}{L} \frac{1}{(s+\beta)\left(s+\frac{R}{L}\right)} \qquad ...(2)$$

Case-I : When $\beta \neq \frac{R}{L}$

$$I(s) = \frac{V_0}{L} \frac{1}{(s+\beta)\left(s+\frac{R}{L}\right)} = \frac{V_0}{R}\left[\frac{A_1}{(s+\beta)} + \frac{A_2}{\left(s+\frac{R}{L}\right)}\right]$$

(using partial fraction expansion)

or $$I(s) = \frac{V_0}{L}\left[\frac{\frac{1}{\left(-\beta+\frac{R}{L}\right)}}{(s+\beta)} - \frac{\frac{1}{\left(-\beta+\frac{R}{L}\right)}}{\left(s+\frac{R}{L}\right)}\right]$$

or $$I(s) = \frac{V_0}{L} \frac{1}{\left(\frac{R}{L}-\beta\right)}\left[\frac{1}{s+\beta} - \frac{1}{\left(s+\frac{R}{L}\right)}\right] \qquad ...(3)$$

Taking inverse Laplace transform of Eq. 3, we get

$$i(t) = \text{Inverse LT } \{I(s)\}$$

$$= \text{Inverse LT}\left\{\frac{V_0}{L} \frac{1}{\left(\frac{R}{L}-\beta\right)}\left[\frac{1}{s+\beta} - \frac{1}{\left(s+\frac{R}{L}\right)}\right]\right\}$$

$$= \frac{V_0}{L} \frac{1}{\left(\frac{R}{L}-\beta\right)}[e^{-\beta t} - e^{-(R/L)t}]$$

or $$i(t) = \frac{V_0}{(R-\beta L)}[e^{-\beta t} - e^{-(R/L)t}]$$

Case-II : When $\beta = \frac{R}{L}$

$$I(s) = \frac{V_0}{L} \frac{1}{(s+\beta)^2} \qquad ...(4)$$

Taking the inverse Laplace transform of Eq. 4, we get

$$i(t) = \text{Inverse LT } \{I(s)\} = \text{Inverse LT}\left\{\frac{1}{(s+\beta)^2}\right\}$$

or $$i(t) = \frac{V_0}{L} t\, e^{-\beta t}$$

Example 9.34 Given the circuit shown in Fig. 9.29. Determine

(*a*) Currents $i_1(t)$ and $i_2(t)$ and the output voltage across 5Ω resistor when the switch is closed.

(*b*) Initial and final values of current.

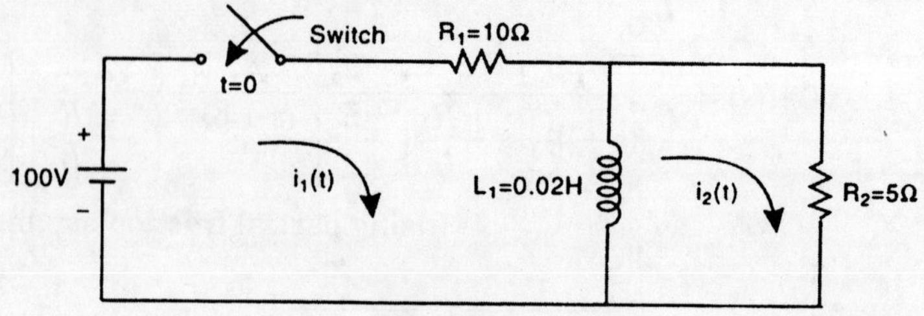

Fig. 9.29

Solution. (*a*) By applying KVL to the circuit, we have the following differential equations

$$R_1 i_1(t) + L_1 \frac{di_1(t)}{dt} - L_1 \frac{di_2(t)}{dt} = 100$$

or $$10\, i_1(t) + 0.02 \frac{di_1(t)}{dt} - 0.02 \frac{di_2(t)}{dt} = 100 \qquad \text{...(1)}$$

$$L_1 \frac{di_2(t)}{dt} - L_1 \frac{di_1(t)}{dt} + R_2\, i_2(t) = 0$$

$$0.02 \frac{di_2(t)}{dt} - 0.02 \frac{di_1(t)}{dt} + 5\, i_2(t) = 0 \qquad \text{...(2)}$$

Taking Laplace transform of Eqs. 1 and 2, we get

$$\text{Laplace transform} \left\{ 10 i_1(t) + 0.02 \frac{di_1(t)}{dt} - 0.02 \frac{di_2(t)}{dt} \right\}$$
$$= \text{Laplace transform } \{100\}$$

or $$10 I_1(s) + 0.02\, s I_1(s) - 0.02\, s I_2(s) = 100 \times \frac{1}{s}$$

or $$(10 + 0.02s)\, I_1(s) - 0.02\, I_2(s) = \frac{100}{s} \qquad \text{...(3)}$$

$$\text{Laplace transform} \left\{ 0.02 \frac{di_2(t)}{dt} - 0.02 \frac{di_1(t)}{dt} + 5\, i_2(t) \right\}$$
$$= \text{Laplace transform } \{0\}$$

or $$0.02\, s I_2(s) - 0.02\, s I_1(s) + 5 I_2(s) = 0$$

or $$(0.02s + 5)\, I_2(s) = 0.02\, s I_1(s)$$

or $$I_2(s) = I_1(s) \frac{(0.02s)}{(0.02s + 5)}$$

or $$I_2(s) = I_1(s) \left[\frac{s}{s + 250} \right] \qquad \text{...(4)}$$

Substituting Eq. 4 in Eq. 3, we get

$$(10 + 0.02s)\, I_1(s) - 0.02\, I_1(s)\left[\frac{s}{s+250}\right] = \frac{100}{s}$$

or $$I_1(s)\left[10 + 0.02s - \frac{0.02s}{s+250}\right] = \frac{100}{s}$$

or $$I_1(s)\left[\frac{10s + 2500 + 0.02s^2 + 5s}{s+250}\right] = \frac{100}{s}$$

$$I_1(s)\left[\frac{0.02s^2 + 15s + 2500}{(s+250)}\right] = \frac{100}{s}$$

or $$I_1(s) = \frac{100\,(s+250)}{s\,(0.02s^2 + 15s + 2500)} = \frac{20}{3}\left\{\frac{s+250}{s\,(s+500/3)}\right\}$$

or $$I_1(s) = \frac{10}{s} - \frac{10/3}{s + \frac{500}{3}} \qquad ...(5)$$

Substituting Eq. 5 in Eq. 4, we get

$$I_2(s) = \frac{20}{3}\left(\frac{1}{s+500/3}\right) \qquad ...(6)$$

Taking inverse Laplace transform of Eqs. 5 and 6, we get

$$i_1(t) = 10 - \frac{10}{3}\, e^{-(500/3)t} \qquad ...(7)$$

and $$i_2(t) = \frac{20}{3}\, e^{-(500/3)t} \qquad ...(8)$$

The voltage across 5 Ω resistor is

$$5\, i_2(t) = 5 \times \frac{20}{3}\, e^{-(500/3)t} = \frac{100}{3}\, e^{-(500/3)t} \text{ Volts}$$

(*b*) Determination of the initial and final values of $i_1(t)$ and $i_2(t)$.

The initial value of $i_1(t)$ is

$$i_1(0^+) = \lim_{s \to \infty} [sI_1(s)]$$

$$= \lim_{s \to \infty}\left[s\,\frac{20}{3}\left\{\frac{s+250}{s\,(s+500/3)}\right\}\right] = \frac{20}{3} \text{ Ampere}$$

The final value of $i_1(t)$ is

$$i_1(\infty) = \lim_{s \to 0} [sI_1(s)]$$

$$= \lim_{s \to 0}\left[s\,\frac{20}{3}\left\{\frac{s+250}{s\,(s+500/3)}\right\}\right] = 10 \text{ Ampere}$$

The initial value of $i_2(t)$ is

$$i_2(0^+) = \lim_{s \to \infty} [sI_1(s)]$$

$$= \lim_{s \to \infty} \left[s \frac{20}{3} \left(\frac{1}{s + 500/3} \right) \right] = \frac{20}{3} \text{ Ampere}$$

The final value of $i_2(t)$ is

$$i_2(\infty) = \lim_{s \to 0} [sI_2(s)]$$

$$= \lim_{s \to 0} \left[s \frac{20}{3} \left(\frac{1}{s + 500/3} \right) \right] = 0$$

Example 9.35 Determine the current $i(t)$ in the RL circuit shown in Fig. 9.30. The input voltage applied is $v(t) = 10 \sin 25t$.

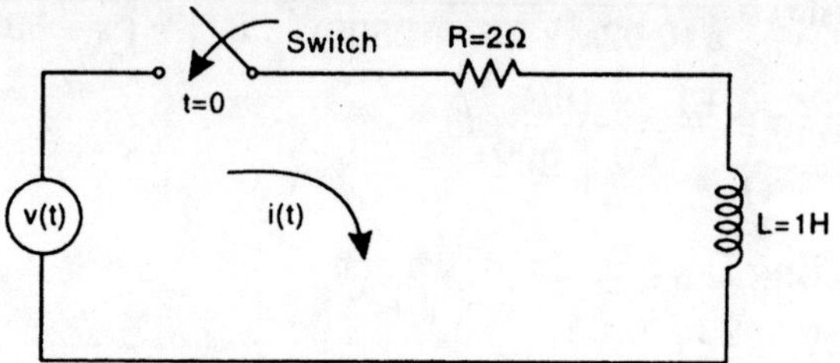

Fig. 9.30

Solution. Applying KVL, the loop equation can be written as

$$Ri(t) + L\frac{di(t)}{dt} = v(t)$$

or
$$2i(t) + 1 \times \frac{di(t)}{dt} = 10 \sin 25t \qquad ...(1)$$

Taking Laplace transform of Eq. 1, we get

$$\text{LT}\left[2i(t) + \frac{di(t)}{dt} \right] = \text{LT } \{10 \sin 25t\}$$

or
$$2I(s) + [sI(s) - i(0)] = 10 \frac{25}{s^2 + (25)^2} \qquad ...(2)$$

where $i(0)$ is the initial current passing through the circuit, as the inductor does not allow sudden changes in current, $i(0) = 0$.

Putting $i(0) = 0$ in Eq. 2, we get

$$I(s) = \frac{250}{(s^2 + 625)\,(s + 2)} \qquad ...(3)$$

Eq. 3 can we written as

$$I(s) = \frac{250}{(s + j25)\,(s - j25)\,(s + 2)}$$

$$= \frac{A_1}{(s + j25)} + \frac{A_2}{s - j25)} + \frac{A_3}{(s + 2)}$$

(using partial fraction expansion)

or $$I(s) = \frac{\left(\frac{-5}{25+j2}\right)}{(s+j25)} + \frac{\left(\frac{-5}{25-j2}\right)}{(s-j25)} + \frac{\left(\frac{250}{629}\right)}{(s+2)} \qquad \text{...(4)}$$

Taking inverse Laplace transform of Eq. 4, we get

$$i(t) = \text{Inverse Laplace transform } \{I(s)\}$$

$$= -\frac{5}{25+j2}e^{-j25t} - \frac{5}{25-j2}e^{j25t} + \frac{250}{629}e^{-2t}$$

On simplification, we get

$$i(t) = -\frac{250}{629}\cos 25t + \frac{20}{629}\sin 25t + \frac{250}{629}e^{-2t}$$

Example 9.36 Determine the resultant current $i(t)$ for series RL circuit shown in Fig. 9.31 (*b*) for applied input voltage shown in Fig. 9.31 (*a*).

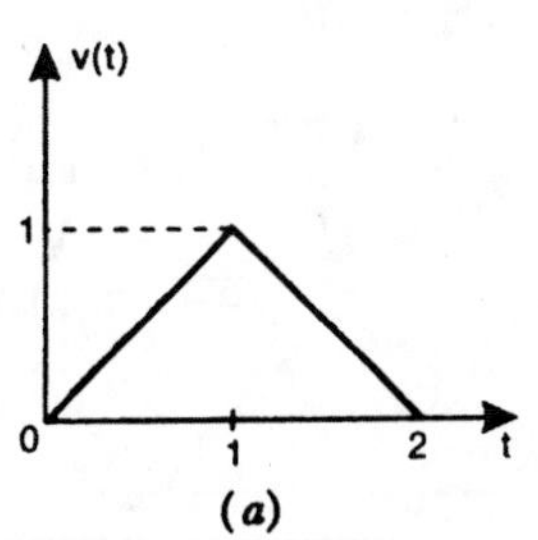

(*a*)

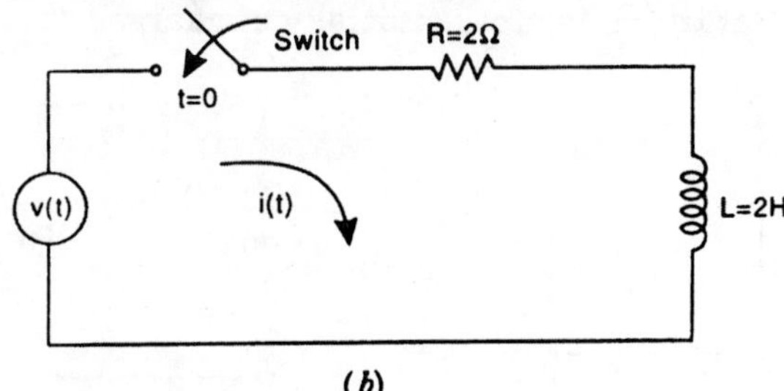

(*b*)

Fig. 9.31

Solution. By using KVL, we get differential equation as

$$Ri(t) + L\frac{di(t)}{dt} = v(t)$$

or $$2i(t) + 2\frac{di(t)}{dt} = v(t) \qquad \text{...(1)}$$

Taking the Laplace transform of Eq. 1, we get

$$2I(s) + [2sI(s) - 2i(0^+)] = V(s) \qquad [\text{Since } i(0^+) = 0]$$

or $$I(s) = \frac{V(s)}{2(s+1)} \qquad \text{...(2)}$$

Now we determine Laplace transform of $v(t)$ shown in Fig. 9.31 (*a*).

$$V(s) = \frac{1 - 2e^{-s} + e^{-2s}}{s^2} \qquad \text{...(3)}$$

Substituting Eq. 3 in Eq. 2, we get

$$I(s) = \frac{V(s)}{2(s+1)} = \frac{1 - 2e^{-s} + e^{-2s}}{2s^2(s+1)}$$

$$= \frac{A_1}{s^2} + \frac{A_2}{s} + \frac{A_3}{(s+1)}$$

(By partial fraction expansion)

or $$I(s) = \frac{1}{s^2} + \frac{(-1)}{s} + \frac{1}{(s+1)} \quad ...(4)$$

Taking inverse Laplace transform of Eq. 4, we get

$$i(t) = \text{Inverse Laplace transform } \{I(s)\}$$

$$= \text{Inverse Laplace transform } \left\{ \frac{1}{s^2} - \frac{1}{s} + \frac{1}{s+1} \right\}$$

$$= t - 1 + e^{-t}$$

or $$I(t) = t - 1 + e^{-t}$$

Example 9.37 A series *RC* circuit is shown in Fig. 9.32 (*b*). A rectangular voltage pulse of height 2 cm and duration of 2 seconds is applied to a series *RC* circuit, as shown in Fig. 9.32 (*a*). Determine the resultant circuit *i*(*t*) in the capacitor as a function of time. Also assume that the capacitor is initially uncharged.

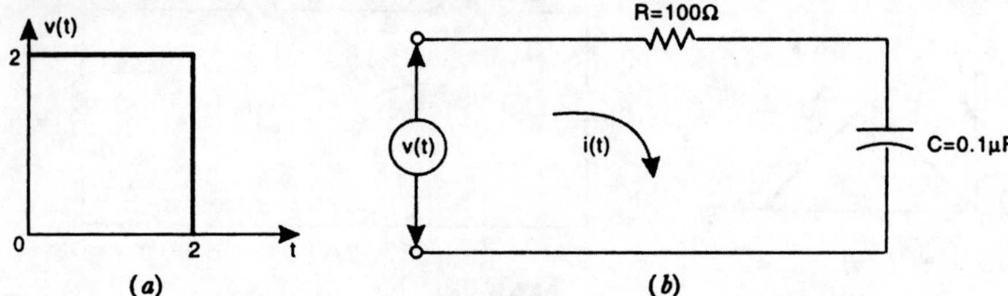

Fig. 9.32

Solution. Applying KVL to the circuit, we get

$$Ri(t) + \frac{1}{C} \int_{-\infty}^{t} i(t)\, dt = v(t) \quad ...(1)$$

Taking Laplace transform of both sides of Eq. 1, we get

$$RI(s) + \frac{1}{C}\left[\frac{I(s)}{s} + \frac{q(0)}{s} \right] = V(s) \quad ...(2)$$

Since the initial charge on the capacitor is zero, $q(0) = 0$.

Therefore, $I(s)\,[\,R + 1/Cs\,] = V(s)$

or $$I(s) = \frac{V(s)}{(R + 1/Cs)} \quad ...(3)$$

The input voltage can be written as the combination of two steps, *i.e.*,

$$v(t) = 2[u(t) - u(t-2)] \quad ...(4)$$

Taking Laplace transform of both sides of Eq. 4, we get

$$V(s) = \text{Laplace transform } \{2[u(t) - u(t-2)]\}$$

$$= 2\left[\frac{1}{s} - \frac{e^{-2s}}{s} \right]$$

or $$V(s) = \frac{2}{s}(1 - e^{-2s}) \qquad ...(5)$$

Substituting Eq. 5 in Eq. 4, we get

$$I(s) = \frac{V(s)}{(R + 1/Cs)} = \frac{\frac{2}{s}(1 - e^{-2s})}{(R + 1/Cs)}$$

$$= \frac{2}{R}\frac{(1 - e^{-2s})}{(s + 1/RC)}$$

or $$I(s) = \frac{2}{R}\left[\frac{1}{(s + 1/RC)} - \frac{e^{-2s}}{(s + 1/RC)}\right] \qquad ...(6)$$

Taking inverse Laplace transform of both sides of Eq. 6, we get

$$i(t) = \frac{2}{R}\{e^{-t/RC}\,u(t) - e^{-(t-2)/RC}\,u(t-2)\} \qquad ...(7)$$

Substituting $R = 100\Omega$ and $C = 0.1\ \mu F$ in Eq. 7, we get

$$i(t) = \frac{2}{100}\left\{e^{-t/100 \times 10^{-7}}\,u(t) - e^{-(t-2)/100 \times 10^{-7}}\,u(t-2)\right\}$$

or $$i(t) = \frac{1}{50}\left\{e^{-10^5 t}\,u(t) - e^{-10^5(t-2)}\,u(t-2)\right\}$$

Example 9.38 Determine the resulting current $i(t)$ in the circuit shown in Fig. 9.33. The switch S is closed at $t = 0$ and there is no initial charge on either of the capacitors. Also determine the initial and final values of the current $i(t)$.

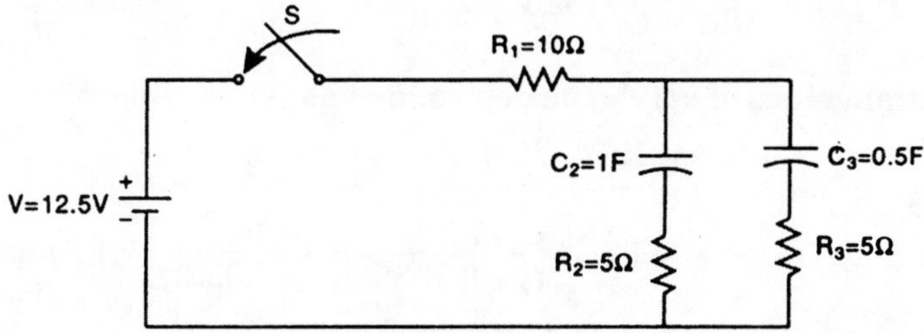

Fig. 9.33

Solution. The equivalent impedance of the network in the s-domain is given by

$$Z(s) = R_1 + \frac{Z_2(s)\,Z_3(s)}{Z_2(s) + Z_3(s)} \qquad ...(1)$$

where $R_1 = 10\ \Omega$

$$Z_2(s) = R_2 + X_{C2} = 5 + \frac{1}{s}$$

$$Z_3(s) = R_3 + X_{C3} = 5 + \frac{2}{s}$$

Substituting above values in Eq. 1, we get

$$Z(s) = 10 + \frac{(5 + 1/s)(5 + 2/s)}{(5 + 1/s + 5 + 2/s)}$$

or $$Z(s) = \frac{125s^2 + 45s + 2}{s(10s + 3)} \quad ...(2)$$

Therefore, $$I(s) = \frac{V(s)}{Z(s)} = \frac{12.5/s}{\dfrac{125s^2 + 45s + 2}{s(10s + 3)}}$$

$$= \frac{(10s + 3)}{125s^2 + 45s + 2}$$

or $$I(s) = \frac{s + 0.3}{(s + 0.308)(s + 0.052)}$$

$$= \frac{A_1}{(s + 0.308)} + \frac{A_2}{(s + 0.052)}$$

(By using partial fraction expansion)

or $$I(s) = \frac{1/32}{(s + 0.308)} + \frac{31/32}{(s + 0.052)} \quad ...(3)$$

Taking inverse Laplace transform of Eq. 3, we get

$$i(t) = \text{Inverse Laplace transform } \{I(s)\}$$

$$= \text{Inverse LT}\left\{\frac{1/32}{(s + 0.308)} + \frac{31/32}{(s + 0.052)}\right\}$$

or $$i(t) = \frac{1}{32}e^{-0.308t} + \frac{31}{32}e^{-0.052t}$$

Initial values of $i(t)$ can be determined as

$$i(0^+) = \lim_{s \to \infty} [sI(s)]$$

$$= \lim_{s \to \infty}\left[\frac{1/32\, s}{(s + 0.308)} + \frac{31/32\, s}{(s + 0.052)}\right] = 1 \text{ Ampere}$$

Final value of $i(t)$ can be determined as

$$i(\infty) = \lim_{s \to 0} [sI(s)]$$

$$= \lim_{s \to 0}\left[\frac{1/32\, s}{(s + 0.308)} + \frac{31/32\, s}{(s + 0.052)}\right] = 0$$

Example 9.39 Determine the current $i(t)$ resulting from the voltage source $v(t)$ in the series *RLC* circuit shown in Fig. 9.34. Here we have assumed that there is an initial current i_L flowing in the inductor and initial voltage v_c across the capacitor when the switch is closed at $t = 0$.

Solution. By applying KVL, we get differential equation for the series *RLC* circuit as

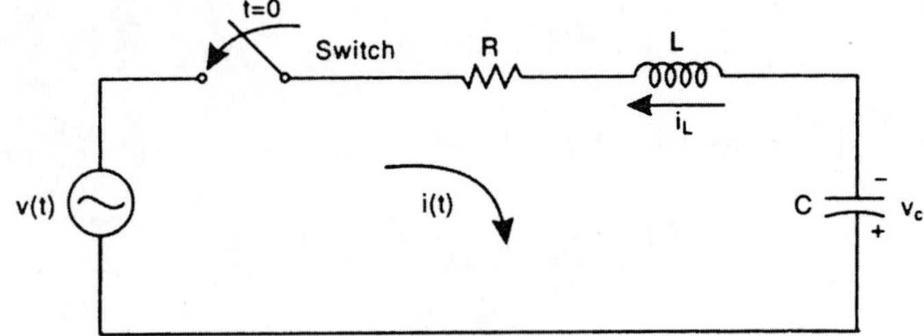

Fig. 9.34

$$v(t) = Ri(t) + L\frac{di(t)}{dt} + \frac{1}{C}\int_{-\infty}^{t} i(t)\, dt \qquad \text{...(1)}$$

Taking Laplace transform of both sides of Eq. 1, we get

$$V(s) = RI(s) + L\ [sI(s) + i(0^+)]\ + \frac{1}{C}\left[\frac{I(s)}{s} - \frac{v_c(0^+)}{s}\right] \qquad \text{...(2)}$$

where $v_c(0^+)$ = voltage across the capacitor after the switch closes at $t = 0^+$

The polarity of the initial conditions depends on the assumed direction for current $i(t)$ w.r.t. the flow of the current in inductance L and the voltage across the capacitor C.

Rearranging terms in Eq. 2, we get

$$I(s)\left[R + Ls + \frac{1}{Cs}\right] = V(s) + \frac{v_c(0^+)}{Cs} - L\, i(0^+)$$

$$\text{Therefore, } I(s) = \frac{V(s) + \dfrac{v_c(0^+)}{Cs} + L\, i(0^+)}{R + Ls + \dfrac{1}{Cs}} \qquad \text{...(3)}$$

$$i(t) = \text{Inverse Laplace transform } \{I(s)\}$$

Example 9.40 Determine the resulting current if the switch s is closed at $t = 0$ in a series RLC circuit shown in Fig. 9.35.

Solution. By applying KVL, we get in differential equation of the RLC circuit as

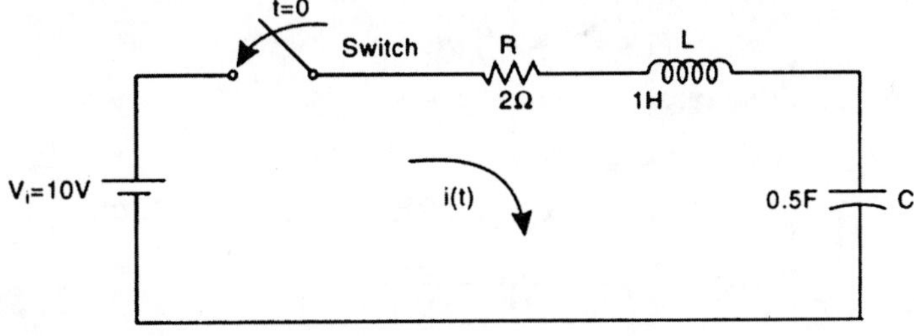

Fig. 9.35

$$Ri(t) + L\frac{di(t)}{dt} + \frac{1}{C}\int_0^t i(t)\,dt = v_i(t) \qquad ...(1)$$

Putting the values of R, L, C and $v_i(t)$ in Eq. 1, we get

$$2i(t) + 1 \times \frac{di(t)}{dt} + \frac{1}{0.5}\int_0^t i(t)\,dt = 10$$

or
$$2i(t) + \frac{di(t)}{dt} + 2\int_0^t i(t)\,dt = 10 \qquad ...(2)$$

Taking Laplace transform of Eq. 2, we get

$$2I(s) + [sI(s) - i(0^+)] + 2\left[\frac{I(s)}{s} - \frac{v_c(0^+)}{s}\right] = \frac{10}{s} \qquad ...(3)$$

There is no initial charge on the capacitor, therefore, $v_c(0^+) = 0$

As $i(0^+) = 0$, Eq. 3, can be written as

$$RI(s) + LsI(s) + \frac{1}{Cs}I(s) = \frac{10}{s}$$

or
$$I(s) = \frac{10}{s^2 + 2s + 2} = \frac{10}{(s+1)^2 + 1} \qquad ...(4)$$

Taking the inverse Laplace transform of Eq. 4, we get

$$i(t) = \text{Inverse Laplace transform } \{I(s)\}$$

$$= \text{Inverse LT}\left\{\frac{10}{(s+1)^2 + 1}\right\}$$

or
$$i(t) = 10\,e^{-t}\sin t \qquad \text{(using table)}$$

Example 9.41 Determine the responses of the system when the excitation (input) is $s(t) = (1 + e^{-3t} - e^{-t})\,u(t)$. Transfer function of the system is

$$H(s) = \frac{1}{(s+1)\,(s^2 + s + 1)}$$

Solution. We have given input function

$$s(t) = (1 + e^{-3t} - e^{-t})\,u(t) \qquad ...(1)$$

Taking Laplace transform of Eq. 1, we get

$$S(s) = \text{Laplace transform } \{s(t)\}$$

$$= \text{Laplace transform }\{(1 + e^{-3t} - e^{-t})\,u(t)\}$$

or
$$S(s) = \frac{1}{s} + \frac{1}{(s+3)} - \frac{1}{(s+1)} \qquad ...(2)$$

We also have given transfer function of the system as

$$H(s) = \frac{1}{(s+1)(s^2+s+1)}$$

$$= \frac{A}{(s+1)} + \frac{Bs+C}{(s^2+s+1)}$$

(By using partial fraction expansion)

$$= \frac{(1)}{(s+1)} + \frac{(-1)\,s+(0)}{(s^2+s+1)}$$

or $$H(s) = \frac{1}{(s+1)} - \frac{s}{(s^2+s+1)} \quad ...(3)$$

We know that $Y(s) = H(s)\,S(s)$ ...(4)

Substituting Eqs. 2 and 3 in Eq. 4, we get

$$Y(s) = \left[\frac{1}{s+1} - \frac{s}{s^2+s+1}\right]\left[\frac{1}{s} + \frac{1}{s+3} - \frac{1}{s+1}\right]$$

$$= \frac{1}{s(s+1)} + \frac{1}{(s+1)(s+3)} - \frac{1}{(s+1)^2} - \frac{1}{(s^2+s+1)} - \frac{s}{(s+3)(s^2+s+1)} + \frac{s}{(s+1)(s^2+s+1)}$$

(By using partial fraction expansion)

or $$Y(s) = \left[\frac{1}{s} - \frac{1}{s+1}\right] + \left[\frac{1}{2(s+1)} - \frac{1}{2(s+3)}\right] - \left[\frac{1}{(s+1)^2}\right] - \left[\frac{1}{s^2+s+1}\right] + \left[\frac{3/7}{s+3} - \frac{(3/7)\,s+1/7}{s^2+s+1}\right] + \left[\frac{-1}{(s+1)} + \frac{s+1}{s^2+s+1}\right] \quad ...(5)$$

Taking the inverse Laplace transform of both sides of Eq. 5, we get

$$y(t) = 1 - \left[\frac{3}{2} + t\right]e^{-t} - \frac{2\sqrt{3}}{7}\sin\frac{\sqrt{3}}{2}t\,e^{-t/2} + \frac{4}{7}\cos\frac{\sqrt{3}}{2}t\,e^{-t/2}$$

REVIEW QUESTIONS

1. Distinguish between Laplace transform and continuous-time Fourier transform (CTFT).
2. Define Laplace transform. Give the relationship between Laplace transform and CTFT.
3. What do you mean by region of convergence for Laplace transform ?
4. Discuss various properties of ROCs for Laplace transform.
5. What is inverse Laplace transform ?

6. Discuss important properties of Laplace transform in brief.

7. Give the statement for the following theorems for Laplace transform
   (*a*) Initial value theorem
   (*b*) Final value theorem

8. Tabulate important properties of Laplace transform.

9. Tabulate important Laplace transform pairs.

10. Discuss important applications of Laplace transform analysis and characterization of continuous-time LTI systems.

11. State and explain the following :
    (*a*) Causality of a continuous-time LTI system.
    (*b*) Stability of a continuous-time LTI system.

12. Describe constant-coefficient differential equations.

13. Describe system behaviour related to the transfer function of system.

14. Discuss block diagram representation for continuous-time LTI systems.

15. Find the transfer function of the following types of interconnections of continuous-time LTI systems.
    (*a*) Parallel interconnection
    (*b*) Series interconnection
    (*c*) Feedback interconnections

16. Discuss and draw the following types of block diagram representations :
    (*a*) Direct-form (*b*) Parallel form
    (*c*) Cascade form

17. Define unilateral or one-sided Laplace transform.

18. Discuss some special properties of unilateral Laplace transform.

19. Tabulate important properties of unilateral Laplace transform.

NUMERICAL PROBLEMS

1. Determine the Laplace transforms of the following functions :
   (*a*) e^{10t} (*b*) e^{-5t}
   (*c*) $\delta(t)$ (*d*) $2 - 2e^{t} + 0.5 \sin 4t$
   (*e*) $e^{-t} \sin 5t$ (*f*) $e^{2t} + 2e^{-2t} - t^2$

   Ans. (*a*) $\frac{1}{s+10}$ (*b*) $\frac{1}{s-5}$ (*c*) 1 (*d*) $\frac{-2\,(s+16)}{s(s-1)\,(s^2+16)}$ (*e*) $\frac{5}{(s+1)^2+(5)^2}$
   (*f*) $\frac{3s^4 - 2s^3 - 2s^2 + 8}{s^3\,(s^2-4)}$

2. Determine the Laplace transform of the continuous-time function

$$s(t) = \sin \alpha t \sin \beta t$$

3. Deterine the bilateral Laplace transform of

$$s(t) = \begin{cases} e^{-3t}, & \text{for} \geq 0 \\ 0, & \text{for } t < 0 \end{cases}$$

Also find its region of convergence.

4. Find the Laplace transform of $s(t) = \sin t \cosh (t)$.

5. Find the Laplace transform of $s(t) = \cos t \sinh (t)$.

6. Find the Laplace transform of the continuous-time signal

$$s(t) = e^{-At} \cos (\omega_c t + \theta)$$

Ans. $S(s) = \dfrac{\cos \theta \,(s + A)}{(s + A)^2 + \omega_c^2} - \dfrac{\omega_c \sin \theta}{(s + A)^2 + \omega_c^2}$

7. Determine the inverse Laplace of the following :

(a) $\dfrac{s - 1}{s(s + 1)}$ (b) $\dfrac{s^3 + 1}{s(s + 1)(s + 2)}$

(c) $\dfrac{s - 1}{(s + 1)(s^2 + 2s + 5)}$

Ans. (a) $-1 + e^{-t}$ (b) $\delta(t) + 0.5 - 3.5\, e^{-2t}$
(c) $-0.5\, e^{-t} + 0.5\, e^{-t} \cos 2t + 0.5\, e^{-t} \sin 2t$

8. Find the initial value of the continuous-time signal if its Laplace transform is given as $S(s) = \dfrac{2s + 1}{s^2 - 1}$ **Ans.** $s(0^+) = 2$

9. Find the final value of the following continuous-time signals for which Laplace transforms are

(a) $S(s) = \dfrac{1}{(s - 1)}$ (b) $S(s) = \dfrac{(s - 2)}{s(s + 1)}$

(c) $S(s) = \dfrac{(s - 1)}{s(s^2 - 1)}$ (d) $S(s) = \dfrac{1}{(s + j)(s - j)}$

Ans. (a) $s(\infty) = \infty$ (b) $s(\infty) = -2$ (c) $s(\infty) = 1$ (d) $s(\infty) \Rightarrow$ undefined

10. Find the initial and final values of function if its Laplace is given by

$$S(s) = \frac{17s^3 + 7s^2 + s + 6}{s^5 + 3s^4 + 5s^3 + 4s^2 + 2s}$$

Ans. $s(0^+) = 0$; $s(\infty) = 3$

11. Find the inverse Laplace transform of $S(s) = \dfrac{2s^2 + 3s + 3}{(s + 1)(s + 3)^2}$

12. Determine the inverse Laplace transform of

$$S(s) = \frac{s}{(s^2+1)^2}$$

Ans. $s(t) = \frac{1}{2} t \sin t$

13. Determine the inverse Laplace transform of following Laplace transform

$$S(s) = \frac{s^2+5}{s^3+2s^2+4s}$$

Ans. $s(t) = \frac{5}{4} e^{-t}\left[\frac{1}{4}\cos\sqrt{3}\,t + \frac{9}{4\sqrt{3}}\sin\sqrt{3}\,t\right]u(t)$

14. We have given transform function of a network

$$H(s) = \frac{3s}{(s+2)\,(s^2+s+2)}$$

(*a*) Draw pole-zero plot for $H(s)$ (*b*) Find its impulse response, $h(t)$

Ans. (*b*) $h(t) = j3e^{-t}\, t\left[\frac{e^{jt}-e^{-jt}}{j2}\right] - \frac{3}{2}e^{-j2t}$

15. Laplace transform for the current is given by

$$I(s) = \frac{15s}{(s+1)\,(s+3)}$$

(*a*) Draw the pole-zero plot for $I(s)$

(*b*) Determine current $i(t)$

16. Determine the poles and zeros of the impedance $Z(s)$ of the network shown in Fig. 9.36.

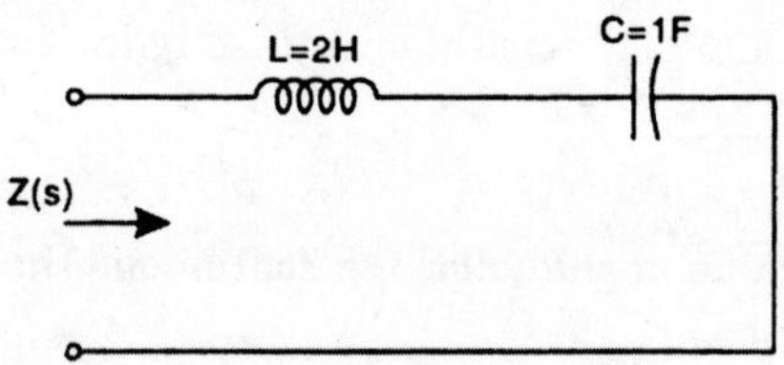

Fig. 9.36

17. Given the Laplace transform for current

$$I(s) = \frac{4s}{s^2+2s+2}$$

(*a*) Draw pole-zero plot for $I(s)$

(*b*) Determine the amplitude and phase response of $I(s)$

18. Find the Laplace transform of $v(t) = e^{-5t}\,u(t) - e^{-5(t-1)}\,u(t-1)$. If this voltage is applied to a network whose impedance is

$$z(s) = \frac{s^2+4s+3}{(s(s^2+6s+8)}$$

Determine (*a*) Laplace transform of current $i(t)$

(*b*) Current $i(t)$

19. A sinusoidal voltage $v(t) = 25 \sin t$ is applied at the instant $t = 0$ to an RL circuit with $R = 2\Omega$ and $L = 1\text{H}$. Determine current $i(t)$ by using Laplace transform method.

20. A circuit diagram is shown in Fig. 9.37. The initial current through L is 2 A and initial voltage across C is 1 volt and input excitation is $s(t) = \cos 2t$. Find the resultant current $i(t)$ and also find $v(t)$ across capacitor C.

Ans. $i(t) = -\frac{18}{5} e^{-t} + 5e^{-2t} + \frac{3}{5} \cos 2t + \frac{1}{5} \sin 2t$

$v(t) = \frac{18}{5} e^{-t} - \frac{5}{2} e^{-2t} + \frac{3}{10} \sin 2t - \frac{1}{10} \cos 2t$

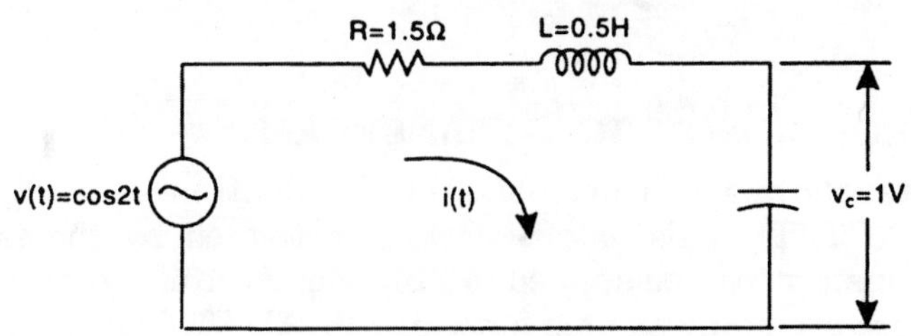

Fig. 9.37

10

z-Transform

10.1 INTROUDCTION TO z-TRANSFORM

Laplace transform is an extension of the Continuous-time Fourier transform (CTFT). This extension was motivated by the fact that Laplace transform can be applied to a broader class of signals than the CTFT. There are many signals for which the CTFT does not converge but it does for Laplace transform.

z-transform for discrete-time signals or sequences is the discrete-time counterpart of the Laplace transform for continuous-time signals. Both Laplace transform and *z*- transform have a similar relationship to their corresponding Fourier transforms (CTFT and DTFT). The movitation for introducing generalization (*z*-transform) of Discrete-time Fourier transform (DTFT) are

1. The DTFT does not converge for all discrete-time signals or sequences but its generalization (*z*-transform) does. Therefore, *z*-transform can be applied to a broader class of signals.
2. In analytical problems, the *z*-transform notation is often more conveniently used than the DTFT.

It is worth noting that transform techniques are used for analysis of various types of signals and linear time-invariant (LTI) systems.

10.2 DEFINITION OF z-TRANSFORM

We know that the response $y(n)$ of a discrete-time LTI system with impulse response $h(n)$ to a complex exponential input $s(t)$ of the form z^n is given as

$$y(n) = H(z)z^n \qquad ...(10.1)$$

where

$$H(z) = \sum_{n=-\infty}^{\infty} h(n)\, z^{-n} \qquad ...(10.2)$$

For $z = e^{j\omega}$ with ω real or with $|z| = 1$, the summation of Eq. 10.2 is equal to the DTFT of $h(n)$. If $|z| \neq 1$, the summation is referred to as *z*-transform of impulse response $h(n)$.

z-transforms are of two types

1. Bilateral or two-sided *z*-transform.
2. Unilateral or one-sided *z*-transform.

The two-sided *z*-transform of a discrete-time signal $s(n)$ is deinfed as a power series given as

$S(z)$ = *z*-transform of discrete-time signal or sequence $s(n)$

$$\mathcal{Z}\,[s(n)] = \sum_{n=-\infty}^{\infty} s(n)\, z^{-n} \qquad \text{...(10.3)}$$

where z is a complex variable and it is expressed as

$$z = re^{j\omega}$$

r and ω are the magnitude and angle of the z, respectively.

For convenience, a discrete-time signal $s(n)$ and its *z*- transform $S(z)$ is denoted by *z*-transform-pair given as

$$s(n) \overset{\mathcal{Z}}{\longleftrightarrow} S(z) \qquad \text{...(10.4)}$$

We have already studied that there are a number of important relationships between the Laplace transform and CTFT. Similarly, there are a number of important relationships between the *z*-transform and DTFT for discrete-time signals. Now, we will develop a relationship between *z*-transform and discrete-time Fourier transform (DTFT) of a discrete-time signal.

Substituting $z = re^{j\omega}$ in Eq. 10.3, we get

$$S(re^{j\omega}) = \sum_{n=-\infty}^{\infty} s(n)\, [re^{j\omega}]^{-n}$$

or

$$S(re^{j\omega}) = \sum_{n=-\infty}^{\infty} [s(n)\, r^{-n}]\, e^{-j\omega n}$$

$$= \text{DTFT}\,\{[s(n)\, r^{-n}]\} \qquad \text{...(10.5)}$$

From Eq. 10.5, we see that $S(re^{j\omega})$ is the Discrete-time Fourier transform of discrete-time signal $s(n)$ multiplied by a real exponential r^{-n}.

$$S(re^{j\omega}) = \text{DTFT}\,\{[s(n)\, r^{-n}]\} \qquad \text{...(10.6)}$$

If $r > 1$, then exponential weighing r^{-n} will be decaying with increasing n.

If $r < 1$, then exponential weighing r^{-n} will be growing with increasing n.

If $r = 1$ or $|z| = 1$, then Eq. 10.5 reduces to the DTFT of $s(n)$.

$$S(z)\Big|_{z=e^{j\omega}} = S(e^{j\omega}) = \text{DTFT}\{s(n)\}$$

$$= \sum_{n=-\infty}^{\infty} s(n)\, z^{-n} \qquad \text{...(10.7)}$$

As already stated the relationship between z-transform and DTFT for discrete-time signals parallels closely with the relationship between the Laplace transform and CTFT for continuous-time signals, but there exist some important differences also. These are

Laplace transform of continuous-time signals reduces to the CTFT when the real part of transform variable s is zero. In other words, we can say that Laplace transform reduces to the CTFT on the imaginary axis ($j\omega$-axis). But the z- transform of discrete-time signals reduces to the DTFT when the magnitude of the transform variable z, *i.e.*, $|z|$ is unity. Therefore, the z-transform of a discrete-time signal $s(n)$ reduces to the DTFT on the contour in the complex z-plane which corresponds to a circle of radius equal to unity. The circle of radius unity in the z-plane is called *unit circle*. The unit circle plays a very important role in z-transform. It is similar to the role of imaginary axis ($j\omega$) in the s-plane for Laplace transform of continuous-time signals.

Unit circle in the z-plane is shown in Fig. 10.1.

In general, the z-transform of a discrete-time signal $s(n)$ has association with its range of values of z for which z-transform $S(z)$ converges. This range of values is referred to as Region of Convergence (ROC) of $S(z)$. If the ROC includes the unit circle, $|z| = 1$, in the z-plane, then the DTFT of $s(n)$ also converges.

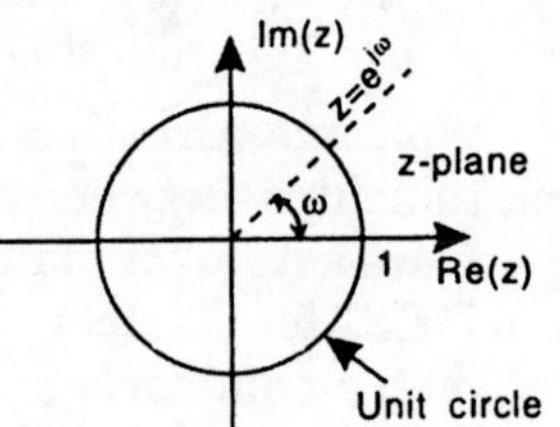

Fig. 10.1 *Illustration of unit circle in the complex z-plane.*

One-sided or unilateral z-transform of a discrete-time signal $s(n)$ is defined as

$$S(z) = \mathcal{Z}u\,[s(n)] = \sum_{n=0}^{\infty} s(n)\, z^{-n} \qquad \text{...(10.9)}$$

The difference between two-sided and one-sided z-transform is that the lower limit of summation is zero for one-sided z-transform and it is $-\infty$ for two-sided z-transfrom. Generally, one-sided z-transform is used for solving difference equations of discrete-time LTI systems with non-zero initial conditions. For causal sequences or discrete-time signals, both the two-sided and one- sided z-transforms are equivalent. The two-sided z-transform can be defined only for a particular region of convergence (ROC). The ROC is not important for one-sided z-transform.

10.3 REGION OF CONVERGENCE FOR THE z-TRANSFORM OF A DISCRETE-TIME SIGNAL

We have already studied that the specific properties of the region of convergence of the Laplace transform for different classes of continuous-time signals. In the same manner, in this section we will study a number of important properties of the region of convergence (ROC) for the z-transform. Most of the properties of ROC for z-transform of discrete-time signal are parallel to the properties of ROC for Laplace transform of continuous-time signals.

10.3.1 Definition of Region of Convergence (ROC)

For all the values of z, the power series of equation $S(z) = \sum_{n=-\infty}^{\infty} s(n)\, z^{-n}$ does not converge to a finite value. The portion of the z-plane for which the power series $S(z) = \sum_{n=-\infty}^{\infty} s(n)\, z^{-n}$ converges is called the region of convergence (ROC). The ROC depends upon the magnitude of z. The ROC cannot contain any poles, since the power series becomes infinite at the poles. The ROC can be a circle, interior of a circle, the exterior of a circle, an annulus, or the entire z-plane.

10.3.2 Possible Configurations of the ROC for the z- Transform

There are following possible configurations of the ROC for the z-transform.

1. A circle [Fig. 10.2 (*a*)]
2. Interior of a circle [Fig. 10.2 (*b*)]
3. Exterior of a circle [Fig. 10.2 (*c*)]
4. An annulus [Fig. 10.2 (*d*)]
5. The entire z-plane [Fig. 10.2 (*e*)].

10.3.3 Properties of ROC for the z-Transform

In this subsection, we will explore a number of important properties of the region of convergence for the z-transform of a discrete-time signal.

Property 1. The ROC of z-transform $S(z)$ consists of a ring (annulus region) in the z-plane centered around the origin $(\sigma\,, j\omega) = (0\,,0)$.

This property of ROC is shown in Fig. 10.3 and follows from the fact that the ROC consists of those values of $z = re^{j\omega}$ for which discrete-time signal $s(n)\, r^{-n}$ has a CTFT that converges. In other words, we can say that the ROC of the z-transform of $s(n)$ contains the values of z for which signal $s(n)\, r^{-n}$ is absolutely summable.

$$\sum_{n=-\infty}^{\infty} |s(n)|\, r^{-n} < \infty = \text{Finite value} \qquad ...(10.10)$$

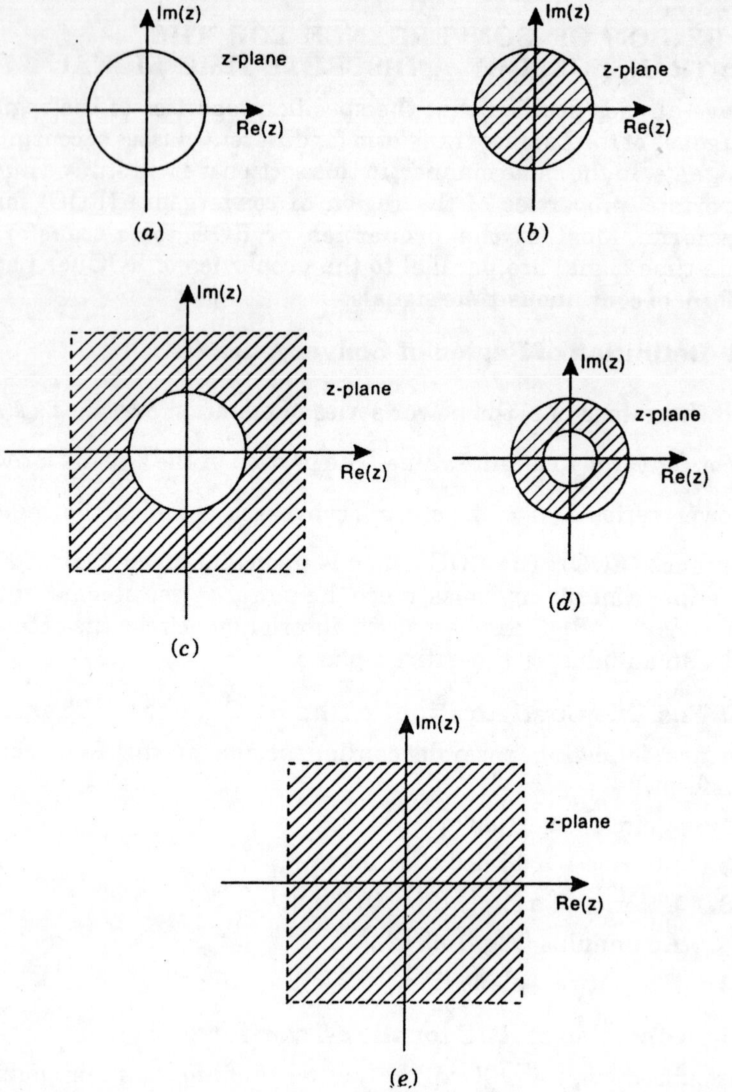

Fig. 10.2 *Possible configurations of the ROC for the z- transform (a) a circle (b) interior of a circle (c) exterior of a circle (d) an annulus (e) the entire z-plane.*

Note that in some cases the radius of inner circle can extend inward to the origin and in this case ROC becomes a disc. In other cases, the radius of outer circle can extend outward to infinity. The convergence of $S(z)$ depends upon $r = |z|$ and not on ω. If a specific value of z is in the ROC then all the values of z on

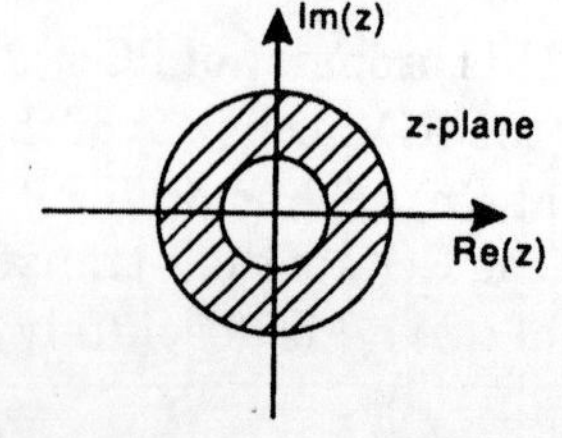

Fig. 10.3 *ROC as an annulus region, i.e., a ring in the z- plane.*

the same circle will be in the ROC. This by itself guarantees that the ROC will consist of concentric rings (annulus region). ROC will consist of only a single ring.

Property 2. The ROC of z-transform of a discrete-time signal does not contain any poles. This property results from the fact that at poles z-transform $S(z)$ is infinite and therefore, $S(z)$ does not converge.

Property 3. If discrete-time signal $s(n)$ is of finite duration then the ROC of $S(z)$ will be entire z-plane except possibly $z = 0$ and/or $z = \infty$. A finite-duration discrete-time signal or sequence has only a finite number of non- zero values. It is defined as $s(n) \neq 0$, $N_1 \leq n \leq N_2$.

z-transform for finite-duration discrete-time signal is defined as

$$S(z) = \sum_{n=N_1}^{N_2} s(n)\, z^{-n} \qquad \text{...(10.11)}$$

For $z \neq 0$ or ∞, each term in the summation of Eq. 10.11 will be finite and consequently, the z- transform $S(z)$ will converge.

1. If N_1 is negative and N_2 positive then the discrete-time signal $s(n) \neq 0$ for $n < 0$ and $n > 0$.

   Also $S(z)$ will include the terms with positive powers of z and negative powers of z.

   As $|z|$ tends to zero, the terms of negative powers of z become unbounded.

   As $|z|$ tends to infinity, the terms of positive powers of z become unbounded.

   Consequently, for N_1 negative and N_2 positive, the ROC of $S(z)$ does not include $z = 0$ or ∞.

2. If $N_1 = 0$ or positive, then there are only negative powers of z in Eq. 10.11 and consequently, the ROC of $S(z)$ will include $z = \infty$.

3. If $N_2 = 0$ or negative, then there are only positive powers of z in Eq. 10.11 and consequently, the ROC of $S(z)$ will include $z = 0$.

Property 4. If discrete-time signal $s(n)$ is a right-sided signal $s(n)$, and if ROC of its z-transform $S(z)$ contains a circle $|z| = r_0$, then all finite values of z for which $|z| > r_0$ will also be in the ROC.

A right-sided sequence is zero prior to some value of n (say N_1). If the circle $|z| = r_0$ is in the ROC then discrete-time signal $[s(n)\, r_0^{-n}]$ is absolutely summable. We have a circle of radius $|z| = r_1$ with $r_1 > r_0$ so that exponential sequence r_1^{-n} decays more quickly than exponential sequence r_0^{-n} for increasing n. A right-sided sequence is shown in Fig. 10.4.

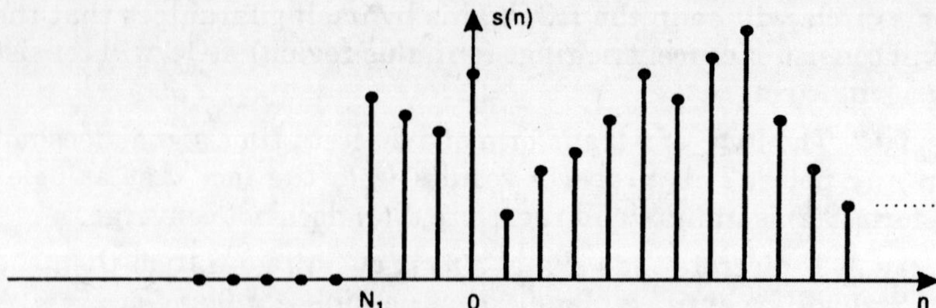

Fig. 10.4 *A right-sided discrete-time signal.*

Since discrete-time signal $s(n)$ is right-sided, *i.e.*,

$$s(n) = 0\ ,\ n < N_1 \qquad \neq 0\ ,\ n \geq N_1 \qquad ...(10.12)$$

In particular,

$$s(n)\, z^{-n} = 0\ ,\ \text{for } n < N_1 \qquad \neq 0\ ,\ \text{for } n \geq N_1 \qquad ...(10.13)$$

Consequently, sequence $s(n)\, r_1^{-n}$ will be absolutely summable, *i.e.*,

$$\sum_{n=N_1}^{\infty} s(n)\, r_1^{-n} < \infty \qquad ...(10.14)$$

z-transform for right-sided signal or sequence $s(n)$ is given as

$$S(z) = \sum_{n=N_1}^{\infty} s(n)\, z^{-n} \qquad ...(10.15)$$

where N_1 is finite. It may be either positive or negative. If N_1 is negative, then the summation of Eq. 10.15 will include positive powers of z, which become unbounded as $|z|$ approaches infinity. In general, the ROC of right-sided sequence does not include infinity. The ROC of a special case of right-sided sequence, *i.e.*, causal right- sided sequence will include infinity. For causal right-sided sequence $N_1 = 0$.

Property 5. If a discrete-time signal $s(n)$ is a left-sided signal, and, if ROC of its z-transform $S(z)$ contains a circle $|z| = r_0$, then all the values of z for which $0 < |z| < r_0$ will also be in the ROC.

A left-sided sequence is shown in Fig. 10.5. It is zero after some value of n (say N_2).

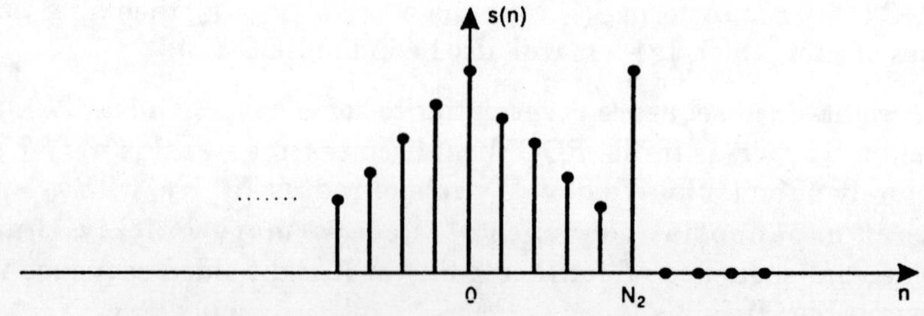

Fig. 10.5 *A left-sided discrete-time signal or sequence.*

A left-sided sequence is defined as

$$\begin{aligned} s(n) &\neq 0\,, \quad n \le N_2 \\ &= 0\,, \quad n > N_2 \end{aligned} \qquad \text{...(10.16)}$$

z-transform for left-sided discrete-time signal $s(n)$ is given as

$$S(z) = \sum_{n=-\infty}^{N_2} s(n)\, z^{-n} \qquad \text{...(10.17)}$$

where N_2 may be positive or negative. If N_2 is positive then Eq. 10.17 includes negative powers of z, which become unbounded as $|z|$ approaches zero. In general, the ROC of the left-sided signal $s(n)$ will not include $z = 0$. In particular, if $N_2 \le 0$, then the ROC of $S(z)$ will include $z = 0$.

Property 6. If a discrete-time signal $s(n)$ is a two-sided signal, and, if ROC of its z-transform $S(z)$ contains a circle $|z| = r_0$, then the ROC will consist of an annulus region, *i.e.*, ring in the z-plane that includes the circle $|z| = r_0$.

A two-sided discrete-time signal $s(n)$ is shown in Fig. 10.6.

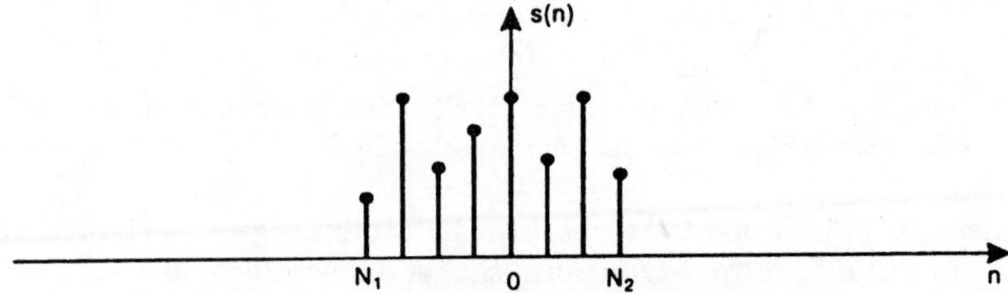

Fig. 10.6 *A two-sided discrete-time signal.*

A two-sided discrete-time signal is defined as

$$\begin{aligned} s(n) &\neq 0\,, \quad N_1 \le n \le N_2 \\ &= 0\,, \quad \text{elsewhere} \end{aligned} \qquad \text{...(10.18)}$$

The ROC for a two-sided signal $s(n)$ can be examined by expressing $s(n)$ as the sum of right-sided and left-sided signal. The ROC for a right-sided signal (or right-sided component of two-sided signal) is a region bounded on the inside by a circle and extending outward to infinity. The ROC for a left-sided signal (or left-sided component of two-sided signal) is a region bonded on the outside by a circle and extending inward to the origin. The ROC for a combined signal (two-sided) includes the intersection of ROCs of right-sided and left-sided components.

ROCs for right-sided, left-sided and two-sided signals are shown in Fig. 10.7.

Property 7. If the z-transform $S(z)$ of a discrete-time signal $s(n)$ is rational, then its ROC is bounded by poles or extends to infinity.

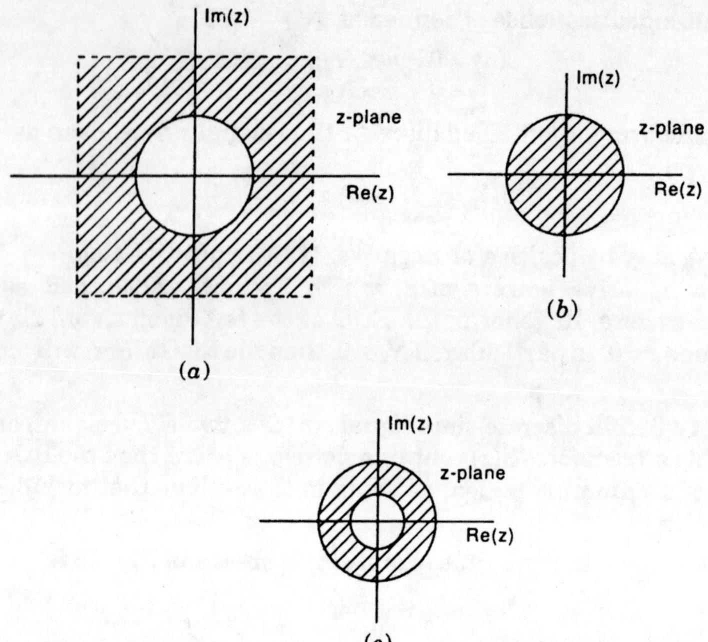

Fig. 10.7 (*a*) *ROC for right-sided signal* (*b*) *ROC for left-sided signal* (*c*) *ROC for two-sided signal that is the sum of the right-sided and left-sided signals.*

Property 8. This property is obtained by combining property 7 with properties 4 and 5. If the z-transform $S(z)$ of a discrete-time signal $s(n)$ is rational, and if $s(n)$ is right-sided, then the ROC is the region in the z-plane outside the outermost pole.

In other words, ROC of a rational z-transform for right- side signal $s(n)$ will be outside the circle of radius equal to largest magnitude of the poles of $S(z)$.

The ROC for z-transform of causal right-sided signal will include $z = 0$.

Property 9. The ROC for rational z-transform $S(z)$ for left-sided signal $s(n)$ will be inside the circle of radius equal to the smallest magnitude of the poles of $S(z)$ other than any at $z = 0$. In particular, if signal $s(n)$ is non- causal, then the ROC also includes $z = 0$.

10.3.4 Tabulation of Properties of ROC for z-Transform

The summary of properties of ROC for z-transform of a discrete-time signal $s(n)$ is given in Table 10.1.

Table 10.1 Summary of Properties of ROC for z-Transform

Discrete-time signal or sequence $s(n)$	ROC for its z-plane $S(z)$
Unit-sample sequence $s(n) = \delta(n) = \begin{cases} 1, & n = 0 \\ 0, & n \neq 0 \end{cases}$	Entire z-plane
Finite duration causal sequence	Entire z-plane except $z = 0$
Finite-duration sequence with $s(n) \neq 0$, for some $n < 0$ $s(n) = 0$, for all $n > 0$	Entire z-plane except for $z = \infty$
Finite-duration sequence with $s(n) \neq 0$, for some $n < 0$ $s(n) = 0$, for some $n > 0$	Entire z-plane except $z = 0$ and $z = \infty$
Right-sided sequence $s(n) = 0$ for some $n < 0$	Outward from outermost pole
Right-sided sequence $s(n) \neq 0$ for some $n < 0$	Outward from outermost pole, $z = \infty$ is excluded
Left-sided sequence $s(n) = 0$ for all $n > 0$	Inward from innermost pole
Left-sided sequence $s(n) \neq 0$ for some $n > 0$	Inward from innermost pole, $z = 0$ is excluded
Two-side sequence	Ring or annulus region

Example 10.1 Determine the z-transforms of the following finite-duration sequences

(a) $s(n) = \{\underset{\uparrow}{2}, 4, 5, 6, 1\}$ (b) $s(n) = \{1, 2, \underset{\uparrow}{3}, 4, 5\}$

(c) $s(n) = \{\underset{\uparrow}{0}, 0, 1, 2, 3\}$ (d) $s(n) = \delta(n)$

(e) $s(n) = \delta(n - k), k > 0$ (f) $s(n) = \delta(n + k), k > 0$

Solution. (a) z-transform is defined as

$$S(z) = \mathcal{Z}[s(n)] = \sum_{n=-\infty}^{\infty} s(n) z^{-n}$$

Given $s(n) = \{\underset{\uparrow}{2}, 4, 5, 6, 1\}$

or $$S(z) = \sum_{n=-\infty}^{\infty} s(n)\, z^{-n} = \sum_{n=0}^{4} s(n)\, z^{-n}$$

$$= s(0)\, z^{-0} + s(1)\, z^{-1} + s(2)\, z^{-2} + s(3)\, z^{-3} + s(4)\, z^{-4}$$

$$= 2.1 + 4\, z^{-1} + 5\, z^{-2} + 6\, z^{-3} + 1\, z^{-4}$$

$$= 2 + 4\, z^{-1} + 5\, z^{-2} + 6\, z^{-3} + z^{-5}$$

ROC : Entire z-plane except $z = 0$

(*b*) **Given** $s(n) = \left\{ 1 , 2 , \underset{\uparrow}{3} , 4 , 5 \right\}$

z-transform is defined as

$$S(z) = \mathcal{Z}\, [\, s(n)\,] = \sum_{n=-\infty}^{\infty} s(n)\, z^{-n} = \sum_{n=-2}^{2} s(n)\, z^{-n}$$

$$= s(-2)\, z^{2} + s(-1)\, z^{1} + s(0)\, z^{0} + s(1)\, z^{-1} + s(2)\, z^{-2}$$

$$= 1.z^{2} + 2\, z^{1} + 3.1 + 4\, z^{-1} + 5\, z^{-2}$$

$$= z^{2} + 2\, z + 3 + 4\, z^{-1} + 5\, z^{-2}$$

ROC : Entire z-plane excpet $z = \infty$ and $z = 0$

(*c*) **Given** $s(n) = \left\{ \underset{\uparrow}{0} , 0 , 1 , 2 , 3 \right\}$

z-transform is defined as

$$S(z) = \mathcal{Z}\, [\, s(n)\,] = \sum_{n=-\infty}^{\infty} s(n)\, z^{-n} = \sum_{n=0}^{4} s(n)\, z^{-n}$$

$$= s(0)\, z^{-0} + s(1)\, z^{-1} + s(2)\, z^{-2} + s(3)\, z^{-3} + s(4)\, z^{-4}$$

$$= 0.\, z^{-0} + 0.\, z^{-1} + 1.\, z^{-2} + 2.\, z^{-3} + 3\, z^{-4}$$

$$= z^{-2} + 2\, z^{-3} + 3\, z^{-4}$$

ROC : Entire z-plane except $z = 0$.

(*d*) **Given** $s(n) = \delta(n) = \begin{cases} 1 , n = 0 \\ 0 , n \neq 0 \end{cases}$

z-transform is defined as

$$S(z) = \mathcal{Z}\, [\, s(n)\,] = \sum_{n=-\infty}^{\infty} s(n)\, z^{-n}$$

$$= \ldots + s(-2)\, z^{2} + s(-1)\, z^{1} + s(0)\, z^{0} + s(1)\, z^{-1} + s(2)\, z^{-2} + \ldots$$

$$= \ldots + 0\, z^{2} + 0.\, z^{1} + 1.\, z^{0} + 0.\, z^{-1} + 0.\, z^{-2} + \ldots = 1$$

or $S(z) = 1$

ROC : Entire z-plane.

(*e*) **Given** $s(n) = \delta(n - k)$, $k > 0$

z-transform is defined as

$$S(z) = \mathcal{Z}[\,s(n)\,] = \sum_{n=-\infty}^{\infty} s(n)\, z^{-n}$$

$$= \sum_{n=-\infty}^{\infty} \delta(n-k)\, z^{-n}$$

Substituting $m = n - k$

or $\qquad n = m + k$, we get

$$S(z) = \sum_{m=-\infty}^{\infty} \delta(m)\, z^{-(m+k)} = \sum_{m=-\infty}^{\infty} \delta(m)\, z^{-m} z^{-k}$$

$$= z^{-k} \sum_{m=-\infty}^{\infty} \delta(m)\, z^{-m} = z^{-k} \,.\, 1 = z^{-k}$$

where $\sum_{m=-\infty}^{\infty} \delta(m)\, z^{-m} = 1$

$$S(z) = \mathcal{Z}\,[\,\delta\,(n-k)\,] = z^{-k}$$

ROC : Entire z-plane except $z = 0$.

(*f*) Given $s(n) = \delta\,(n+k)\,,\, k > 0$

z-transform is defined as

$$S(z) = \mathcal{Z}\,[\,\delta(n+k)\,] = \sum_{n=-\infty}^{\infty} \delta(n+k)\, z^{-n}$$

Substituting $m = n + k$

or $\qquad n = m - k$, we get

$$= \sum_{m=-\infty}^{\infty} \delta(m)\, z^{-(m-k)} = \sum_{m=-\infty}^{\infty} \delta(m)\, z^{-m} z^{k}$$

$$= z^{k} \sum_{m=-\infty}^{\infty} \delta(m)\, z^{-m} = z^{k} \,.\, 1 = z^{K}$$

Where $\sum_{m=-\infty}^{\infty} \delta(m)\, z^{-m} = 1$

$$S(z) = \mathcal{Z}\,[\,\delta(n+k)\,] = z^{k}$$

ROC : Entire z-plane except $z = 0$.

Example 10.2 Determime the z-transform of the following discrete-time signals

(*a*) $s(n) = u(n)$ (*b*) $s(n) = A^n u(n)$

(*c*) $s(n) = -A^n u(-n-1)$ (*d*) $s(n) = \cos\,(\omega_0 n) u(n)$

(*e*) $s(n) = \sin\,(\omega_0 n) u(n)$ (*f*) $s(n) = A^n u(n) + B^n u(-n-1)$

(*g*) $s(n) = n u(n)$ (*h*) $s(n) = u(n-1)$

Solution. (*a*) Given $s(n) = u(n)$ and $u(n)$ is defined as

$$u(n) = \begin{cases} 1, n \geq 0 \\ 0, n < 0 \end{cases}$$

z-transform is defined as

$$S(z) = \mathcal{Z}[s(n)] = \sum_{n=-\infty}^{\infty} s(n) z^{-n}$$

$$= \sum_{n=-\infty}^{\infty} u(n) z^{-n} = \sum_{n=0}^{\infty} 1. z^{-n}$$

$$= z^{-0} + z^{-1} + z^{-2} + ... + z^{-\infty}$$

$$= \frac{1}{1 - z^{-1}}$$

[Sum of geometric series for infinite number of points is given by

$$S_{n=\infty} = \frac{\text{First point}}{1 - \text{common ratio}} \;]$$

This is sum of geometric series for infinite number of points.

ROC : $|z^{-1}| < 1$

or $|z| > 1$

(*b*) Given $s(n) = A^n u(n)$

z-transform is defined as

$$S(z) = \mathcal{Z}[s(n)] = \sum_{n=-\infty}^{\infty} s(n) z^{-n} = \sum_{n=-\infty}^{\infty} A^n u(n) z^{-n}$$

$$= \sum_{n=0}^{\infty} A^n z^{-n} = \sum_{n=0}^{\infty} (Az^{-1})^n,$$

(This is a geometric series for infinite number of points)

$$= (Az^{-1})^0 + (Az^{-1})^1 + (Az^{-2})^2 + ... + (Az^{-1})^{\infty}$$

$$= \frac{1}{1 - Az^{-1}}$$

ROC : $\left|Az^{-1}\right| < 1$

or $|z| > |A|$

where A is a scalar quantity.

ROC of the z-transform of the sequence

$s(n) = A^n u(n)$ is $|z| > |A|$

This is a causal sequence. The ROC is the exterior of a circle having radius $|A|$. ROC of the z-transform of the sequence $s(n) = A^n u(n)$ can be shown graphically on z-plane and is illustrated in Fig. 10.8.

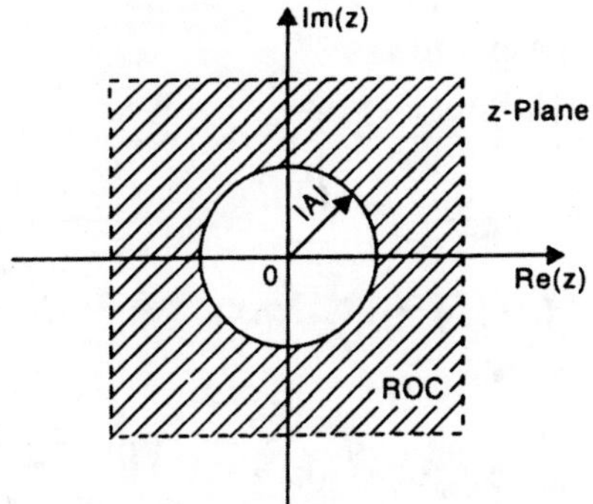

Fig. 10.8 *ROC of the z-transform of the sequence $s(n) = A^n u(n)$.*

(*b*) Given

$$s(n) = -A^n u(-n-1)$$

$$= \begin{cases} 0\,, & n \geq 0 \\ -A^n\,, & n < 0 \end{cases}$$

z-transform is defined as

$$S(z) = \mathcal{Z}[\,s(n)\,] = \sum_{n=-\infty}^{\infty} s(n)\, z^{-n}$$

$$= \sum_{n=-\infty}^{\infty} -A^n u(-n-1)\, z^{-n}$$

$$= \sum_{n=-\infty}^{-1} (-A^n)\, z^{-n} = -\sum_{n=-\infty}^{-1} (A^{-1} z^1)^{-n}$$

Substituting $l = -n$

$$S(z) = -\sum_{l=\infty}^{1} (A^{-1}z)^l = -\sum_{l=1}^{\infty} (A^{-1}z)^l$$

$$= -[(A^{-1}z)^1 + (A^{-1}z)^2 + (A^{-1}z)^3 + \ldots..]$$

$$= \frac{-A^{-1}z}{1 - A^{-1}z}$$ Sum of geometric series for inifinite number of points.

ROC : $|A^{-1} z| < 1$

or $\quad |z| < |A|$

The ROC is interior of a circle having radius $|A|$.

Above sequence,

$$s(n) = -A^n u(-n-1)$$

is a non-causal sequence. ROC of the *z*-transform of above sequence can be shown graphically on the *z*-plane. Its ROC is shown in Fig. 10.9.

(*d*) Given $s(n) = \cos \omega_0 n u(n)$

$$= \left\{ \frac{e^{j\omega_0 n} + e^{-j\omega_0 n}}{2} \right\} u(n)$$

z transform is given as

$$S(z) = \mathbf{Z}[\, s(n) \,] = \sum_{n=-\infty}^{\infty} s(n)\, z^{-n}$$

$$= \sum_{n=-\infty}^{\infty} \cos(\omega_0 n) u(n)\, z^{-n}$$

$$= \sum_{n=-\infty}^{\infty} \left\{ \frac{e^{j\omega_0 n} + e^{-j\omega_0 n}}{2} \right\} u(n)\, z^{-n}$$

Fig. 10.9 *ROC of the z-transform of above sequence* $s(n) = -A^n u(n-1)$.

$$= \frac{1}{2} \sum_{n=0}^{\infty} e^{j\omega_0 n} z^{-n} + \frac{1}{2} \sum_{n=0}^{\infty} e^{-j\omega_0 n} z^{-n}$$

$$= \frac{1}{2} \sum_{n=0}^{\infty} \left(e^{j\omega_0} z^{-1} \right)^n + \frac{1}{2} \sum_{n=0}^{\infty} \left(e^{-j\omega_0} z^{-1} \right)^n$$

$$= \frac{1}{2} \left[\frac{1}{1 - e^{j\omega_0} z^{-1}} \right] + \frac{1}{2} \left[\frac{1}{1 - e^{-j\omega_0} z^{-1}} \right]$$

$$= \frac{1 - z^{-1} \cos \omega_0}{1 - 2z^{-1} \cos \omega_0 + z^{-2}}, \quad \text{ROC} : |Z| > 1$$

ROC : $\left| e^{j\omega_0} z^{-1} \right| < 1 \Rightarrow |z| > \left| e^{j\omega_0} \right| > 1$

and $\left| e^{-j\omega_0} z^{-1} \right| < 1 \Rightarrow |z| > \left| e^{-j\omega_0} \right| > 1$

If we set $\left| e^{\pm j\omega_0} \right| = 1$

ROC of the *z*-transform of the sequence $s(n) = \cos(\omega_0 n)\, u(n)$ is exterior to the unit circle. It can be shown graphically on *z*-plane. Its ROC is shown in Fig. 10.10.

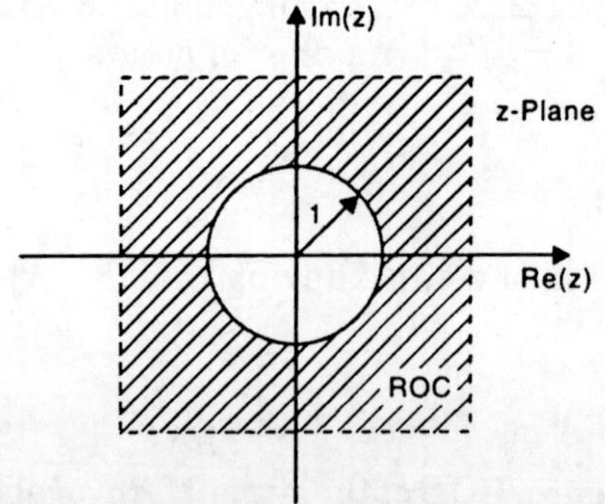

Fig. 10.10 *ROC of the transform of the sequence* $s(n) = \cos(\omega_0 n)\, u(n)$.

(e) Given $s(n) = \sin(\omega_0 n)\, u(n) = \left[\dfrac{e^{j\omega_0 n} - e^{-j\omega_0 n}}{2j}\right] u(n)$

z-transform is given as

$$S(z) = \mathcal{Z}[\, s(n)\,] = \sum_{n=-\infty}^{\infty} s(n)\, z^{-n}$$

$$= \sum_{n=-\infty}^{\infty} \sin(\omega_0 n)\, u(n)\, z^{-n}$$

$$= \sum_{n=-\infty}^{\infty} \left[\frac{e^{j\omega_0 n} - e^{-j\omega_0 n}}{2j}\right] u(n)\, z^{-n}$$

$$= \frac{1}{2j} \sum_{n=0}^{\infty} \left(e^{j\omega_0} z^{-1}\right)^n - \frac{1}{2j} \sum_{n=0}^{\infty} \left(e^{-j\omega_0} z^{-1}\right)^n$$

$$= \frac{1}{2j}\left[\frac{1}{1 - e^{j\omega_0} z^{-1}}\right] - \frac{1}{2j}\left[\frac{1}{1 - e^{-j\omega_0} z^{-1}}\right]$$

$$= \frac{z^{-1} \sin \omega_0}{1 - 2z^{-1} \cos \omega_0 + z^{-2}}, \text{ ROC : } |z| > 1$$

Its ROC is also shown in Fig. 3.4.

(f) Given $s(n) = A^n u(n) + B^n u(-n-1)$

z-transform is defined as

$$S(z) = \mathcal{Z}[\, s(n)\,] = \sum_{n=-\infty}^{\infty} s(n)\, z^{-n}$$

or $$S(z) = \sum_{n=-\infty}^{\infty} \left[A^n u(n) + B^n u(-n-1) \right] z^{-1}$$

or $$S(z) = \sum_{n=-\infty}^{\infty} A^n z^{-n} u(n) + \sum_{n=-\infty}^{\infty} B^n z^{-n} u(-n-1)$$

$$u(-n-1) = \begin{cases} 1, & n \le -1 \\ 0, & n \ge 0 \end{cases}$$

Substituting these values in above Eq., we get

$$S(z) = \sum_{n=0}^{\infty} A^n z^{-1} + \sum_{n=-\infty}^{-1} B^n z^{-n}$$

or $$S(z) = \underbrace{\sum_{n=0}^{\infty} (Az^{-1})^n}_{\text{1st part}} + \underbrace{\sum_{n=-\infty}^{-1} (Bz^{-1})^n}_{\text{2nd part}}$$

In 2nd part, putting $m = -n$

$$S(z) = \sum_{n=0}^{\infty} (Az^{-1})^n + \sum_{m=\infty}^{1} (Bz^{-1})^{-m}$$

or $$S(z) = \sum_{n=0}^{\infty} (Az^{-1})^n + \sum_{m=1}^{\infty} (Bz^{-1})^{-m}$$

or $$S(z) = \sum_{n=0}^{\infty} (Az^{-1})^n + \sum_{m=1}^{\infty} (B^{-1}z)^m$$

Ist series IInd series

The first power series converges if $|Az^{-1}| < 1 \Rightarrow |z| > |A|$

The second power series converges if $|B^{-1}z| < 1 \Rightarrow |z| < |B|$

For convergence of $S(z)$, we consider two different cases

Case I. If $|B| < |A|$. In this case the two ROCs above do not overlap, as shown in Fig. 10.11 (a). Consequently, we cannot find values of z for which both power series converge simultaneously.

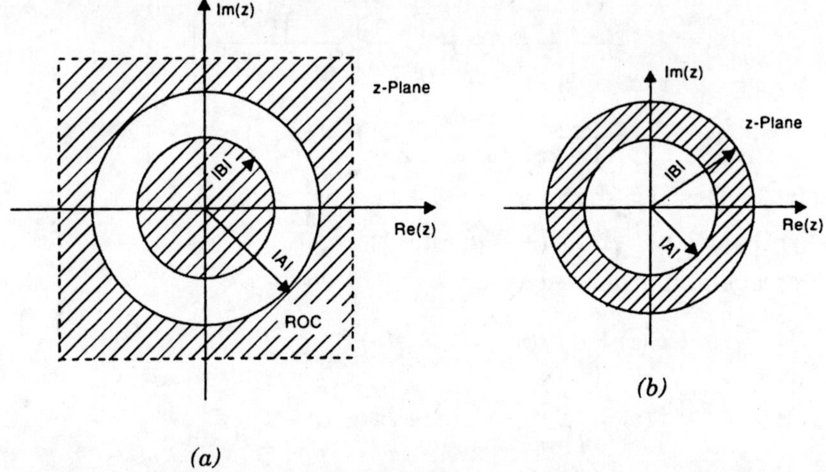

Fig. 10.11 (a) *If* $|B| < |A|$ *then* $S(z)$ *does not exist* (b) *ROC for a z-transform of the sequence* $s(n) = A^n u(n) + B^n u(-n-1)$ *if* $|B| > |A|$.

In this case, we cannot determine $S(z)$.

Case II. If $|B| > |A|$. In this case there is an annulus region in the z-plane where both power series converge simultaneously as shown in Fig. 10.11 (b).

In this case we can obtain $S(z)$.

$$S(z) = \frac{1}{1 - Az^{-1}} + \frac{B^{-1}z}{1 - B^{-1}z}$$

$$= \frac{1 - B^{-1}z + B^{-1}z - AB^{-1}}{1 - Az^{-1} - B^{-1}z + AB^{-1}}$$

$$= \frac{B - A}{B - ABz^{-1} - z + A} = \frac{B - A}{B + A - z - ABz^{-1}}$$

The ROC of the $S(z)$ is $|A| < |z| < |B|$

(g) Given $s(n) = r(n)$. It is a ramp sequence and given as $s(n) = nu(n)$

z-transform is defined as

$$S(z) = \mathcal{Z}[\,s(n)\,] = \sum_{n=-\infty}^{\infty} s(n)\, z^{-n} = \sum_{n=-\infty}^{\infty} r(n)\, z^{-n}$$

$$= \sum_{n=-\infty}^{\infty} nu(n)\, z^{-n}$$

From the definition of $u(n) = \begin{cases} 1, & n \geq 0 \\ 0, & n < 0 \end{cases}$, substituting this above Eqn., we get,

$$S(z) = \sum_{n=0}^{\infty} nz^{-n}$$

Now expanding the above power series, we get

$$S(z) = 0\, z^{-0} + 1\, z^{-1} + 2\, z^{-2} + 3\, z^{-3} + 4\, z^{-4} + \ldots$$

$$z^{-1}S(z) = 0z^{-1} + 1\, z^{-2} + 2\, z^{-3} + 3\, z^{-4} + \ldots$$

Subtracting

$$S(z) - z^{-1}\, S(z) = z^{-1} + z^{-2} + z^{-3} + z^{-4} + \ldots$$

$$S(z)\,[1 - z^{-1}] = \frac{z^{-1}}{1 - z^{-1}}$$ [Sum of geometric series of infinite number of points]

$$S(z)\,[1 - z^{-1}] = \frac{z^{-1}}{1 - z^{-1}}$$

or $$S(z) = \frac{z^{-1}}{(1 - z^{-1})^2} = \frac{z}{(z - 1)^2}$$

ROC : $|z^{-1}| < 1$

or $|z| > 1$

(h) Given $s(n) = u(n - 1)$

z-transform is defined as

$$S(z) = \mathcal{Z}[\,s(n)\,] = \sum_{n=-\infty}^{\infty} s(n)\, z^{-n} = \sum_{n=-\infty}^{\infty} u(n-1)\, z^{-n}$$

Substituting $m = n - 1$

or $n = m + 1$

$$= \sum_{m=-\infty}^{\infty} u(m)\, z^{-(m+1)} = \sum_{m=-\infty}^{\infty} u(m)\, z^{-m}\, z^{-1}$$

$$= z^{-1} \sum_{m=-\infty}^{\infty} u\,(m)\, z^{-m}, \text{ for } u(m) = \begin{cases} 1, & m \geq 0 \\ 0, & m < 0 \end{cases}$$

$$= z^{-1}\left[\sum_{m=0}^{\infty} z^{-m}\right] = z^{-1}\left[z^{-0} + z^{-1} + z^{-2} + z^{-3} + \ldots\right]$$

$$= z^{-1} \frac{1}{1 - z^{-1}} = \frac{z^{-1}}{1 - z^{-1}} = \frac{1}{z - 1}$$

ROC : $|z^{-1}| < 1$ or $|z| > 1$

10.4 PROPERTIES OF z-TRANSFORM

We already know that z-transform is a very powerful tool for the analysis of discrete-time and LTI systems. Derivation of its properties are analogous to the derivations of property of other transforms. It possesses the following important properties.

1. Linearity
2. Time shifting of a discrete-time sequence
3. Scaling in the z-domain
4. Time reversal of a discrete-time sequence
5. Time expansion
6. Differentiation in the z-domain
7. Convolution of two discrete-time sequences
8. Correlation of two discrete-time sequences
9. Multiplication of two discrete-time sequences
10. Parseval's theorem for z-transform
11. Complex conjugation
12. The initial value theorem.

10.4.1 Linearity

A discrete-time system is said to be linear if it is homogenous and satisfies the principle of superposition. z-transform is also a linear transformation tool similar to Laplace transform.

If the z-transform for two discrete-time sequences is given as

$$s_1(n) \overset{Z}{\longleftrightarrow} S_1(z) \text{ with ROC} = R_1$$

and
$$s_2(n) \overset{Z}{\longleftrightarrow} S_2(z)\text{, with ROC} = R_2$$

then according to property of linearity of z-transform.

$$s(n) = A_1\, s_1(n) + A_2\, s_2(n) \overset{Z}{\longleftrightarrow} S(z) = A_1\, S_1(z) + A_2\, S_2(z)$$

$$\text{with ROC} = R\text{ , } R = R_1 \cap R_2 \qquad \text{...(10.19)}$$

The ROC of z-transform of a linear combination of two discrete-time sequences is atleast the intersection of R_1 and R_2. For discrete-time sequence with rational z- transform, if the poles of $A_1S_1(z) + A_2S_2(z)$ consist of all of the poles of $S_1(z)$ and $S_2(z)$, then the ROC will be exactly equal to the overlap of individual ROCs. If the linear combination has

introduced some zeros then it cancels some poles. In this case the ROC may be larger.

Example 10.3 **Find the z-transform and ROC of the signal sequence**

$$s(n) = [4(2^n) - 5(3^n)]\, u(n)$$

Solution. **Given :** $s(n) = 4[2^n u(n)] - 5[3^n u(n)]$ ...(1)

Now the combined signal is given by,

$$s(n) = As_1(n) + Bs_2(n) \quad ...(2)$$

After comparing Eqs. (1) and (2), we get two sequences as

$$s_1(n) = (2^n)\, u(n) \text{ and}$$

$$s_2(n) = (3^n)\, u(n)$$

Taking the z-transform of both sides of Eq. (2), we get

$$S(z) = AS_1(z) + BS_2(z)$$

$$= 4S_1(z) - 5S_2(z) \quad ...(3)$$

$$\mathcal{Z}\,[s_1(n)] = S_1(z) = \sum_{n=-\infty}^{\infty} s_1(n)\, z^{-n} = \sum_{n=-\infty}^{\infty} (2)^n\, u(n)\, z^{-n}$$

$$\left[\text{From } u(n) = \begin{cases} 1\,, n \geq 0 \\ 0\,, n < 0 \end{cases} \right]$$

$$= \sum_{n=0}^{\infty} (2)^n\, z^{-n} = \sum_{n=0}^{\infty} (2z^{-1})^n = \frac{1}{1 - 2z^{-1}}$$

ROC : $|z| > 2$

$$\mathcal{Z}\,[s_2(n)] = S_2(z) = \sum_{n=-\infty}^{\infty} s_2(n)\, z^{-n} = \sum_{n=-\infty}^{\infty} 3^n\, u(n)\, z^{-n}$$

$$= \sum_{n=0}^{\infty} 3^n z^{-n} = \sum_{n=0}^{\infty} (3z^{-1})^n = \frac{1}{2 - 3z}$$

ROC : $|z| > 3$

After substituting the value of $S_1(z)$ and $S_2(z)$ in Eq. 3, we get

$$S(z) = \frac{4}{1 - 2z^{-1}} - \frac{5}{1 - 3z^{-1}} \qquad \text{ROC} : |z| > 3$$

10.4.2 Time Shifting of a Discrete-Time Sequence

The purpose of time shifting is that we can see a particular signal in past time or future time by delaying the signal in time or advancing the signal in time, respectively. Consider a discrete-time sequence $s(n)$ and its z-transfom $S(z)$.

If $s(n) \stackrel{\mathcal{Z}}{\longleftrightarrow} S(z)$ with ROC $= R$

For delayed version of sequence $s(n)$ by k units of time

$$s(n-k) \overset{Z}{\longleftrightarrow} z^{-k} S(k) \qquad ...(10.20)$$

with ROC = R, except for the possible addition or deletion of the origin or infinity.

Because of multiplication of z^{-k}, for $k > 0$, poles will be introduced at $z = 0$, which may cancel corresponding zeros of $S(z)$ at $z = 0$. Consequently, $z = 0$ is a pole of $z^{-k} S(z)$ while it is not a pole of $S(z)$. In this case the ROC for $z^{-k} S(z)$ is equal to ROC for $S(z)$ but with origin deleted.

For $k < 0$, zeros will be introduced at $z = 0$, which cancels poles of $S(z)$ at $z = 0$. Consequently, $z = 0$ is zero of $z^{-k}S(z)$ while it is not a zero of $S(z)$. Therefore, in this case $z = \infty$ is a pole of $z^{-k}S(z)$ and thus the ROC for $z^{-k}S(z)$ equals the ROC of $S(z)$ but with the $z = \infty$ deleted.

Example 10.4 Find z-transform of signal

$$s_2(n) = \{1, 2, \underset{\uparrow}{3}, 4, 0, 1\}$$

from the z-transform of signal

$$s_1(n) = \{\underset{\uparrow}{1}, 2, 3, 4, 0, 1\}$$

by using time-shifting property of z-transform.

Solution. Above two signal sequences $s_2(n)$ and $s_1(n)$ can be related in time-domain by

$$s_2(n) = s_1(n + 2) \qquad ...(1)$$

Taking z-transform of Eq. (1), we get

$$S_2(z) = z^2 S_1(z) \qquad ...(2)$$

But $$S_1(z) = \sum_{n=-\infty}^{\infty} s_1(n) z^{-n}$$

or $$S_1(z) = \sum_{n=0}^{5} s_1(n) z^{-1} = s_1(0) z^0 + s_1(1) z^{-1} + s_1(2) z^{-2} + s_1(3) z^{-3} + s_1(4) z^{-4} + s_1(5) z^{-5}$$

or $$S_1(z) = 1 + 2 z^{-1} + 3 z^{-2} + 4 z^{-3} + 0 z^{-4} + 1 z^{-5}$$

or $$S_1(z) = 1 + 2 z^{-1} + 3 z^{-2} + 4 z^{-3} + z^{-5} \qquad ...(3)$$

Substituting Eq. (3) in Eq. (2), we get

$$S_2(z) = z^2 S_1(z) = z^2 [1 + 2 z^{-1} + 3 z^{-2} + 4 z^{-4} + z^{-5}]$$

$$= z^2 + 2z + 3 + 4 z^{-1} + z^{-3}$$

ROC : All values of z except $z = \infty$.

10.4.3 Scaling in the z-Domain

Scaling in the z-domain means we want to multiply or divide the z-domain parameter by any constant A. By scaling in z-domain, we can increase or decrease the value of z. Scaling in the z-domain (frequency-domain) by a scalar quantity A is equivalent to scaling in time-domain by a scalar quantity 1/A.

$$\text{If} \qquad s(n) \xleftrightarrow{\mathcal{Z}} S(z) \text{ with ROC} = R$$

$$\text{then} \qquad A^n\, s(n) \xleftrightarrow{\mathcal{Z}} S\left(\frac{z}{A}\right) \text{ with ROC} = |A|\,R \qquad \text{...(10.21)}$$

where ROC = $|A|R$ is the scaled versin of R. A is a real or complex constant. That is, if z is a point in the ROC of $S(z)$, then the point $|A|z$ is in the ROC of $S\left(\frac{z}{A}\right)$

10.4.4 Time Reversal of a Discrete-TIme Sequence

Time reversal of a discrete-time sequence means taking the mirror image of the signal or folding the signal.

If we have a discrete-time signal $s(n)$, then its mirror image will be $s(-n)$.

$$\text{If} \qquad s(n) \xleftrightarrow{\mathcal{Z}} S(z) \text{ with ROC} = R$$

$$\text{then} \qquad s(-n) \xleftrightarrow{\mathcal{Z}} S\left(\frac{1}{z}\right) \text{ with ROC} = \frac{1}{R} \qquad \text{...(10.22)}$$

That is, if z_0 is in the ROC for $s(n)$, then $\frac{1}{z_0}$ is in the ROC for $s(-n)$.

Example 10.5 Find the z-transform of sequence $s(n) = u(-n)$.

Solution. We know that, z-transform of unit-step sequence $u(n)$ is given by

$$S(z) = \mathcal{Z}\,[u(n)] = \frac{1}{1 - z^{-1}} \qquad \text{ROC} : |z| > 1$$

By using property of time reversal of z-transform

$$\mathcal{Z}\,[s(-n)] = S(z^{-1})$$

$$\text{or} \qquad \mathcal{Z}\,[u(-n)] = S(z^{-1}) = \frac{1}{1 - z^{-1}}\Big|_{z = z^{-1}}$$

$$- \frac{1}{1 - z} \qquad \text{ROC} : |z| < 1$$

10.4.5 TIme Expansion

Time expansion of a signal is also termed as time scaling of a signal by a factor greater than unity. The concept of time scaling does not directly extend to discrete-time signals because the discrete-time indcx n is defined only for integer values. We know that time scaling operation can be directly applied to continuous-time signals because the time variable t is defined for all values. The discrete-time concept of time expansion can be understood by inserting a number of zero between successive values of a discrete-time signal $s(n)$. Time expansion of a discrete-time signal $s(n)$ plays an important role in discrete-time signal and system analysis.

Time expanded, *i.e.*, scaled discrete-time signal is defined as

$$s_{(k)}(n) = \begin{cases} s(n/k)\,, & \text{if } n \text{ is a multiple of } k \\ 0\,, & \text{if } n \text{ is not a multiple of } k \end{cases} \qquad ...(10.23)$$

Here in the time expanded signal, we have inserted $(k - 1)$ zeros between successive values of the original signal. In this case,

$$\text{If} \qquad s(n) \overset{Z}{\longleftrightarrow} S(z)\,, \text{ with ROC} = R$$

$$\text{then} \qquad s_{(k)}(n) \overset{Z}{\longleftrightarrow} S(z^k)\,, \text{ with ROC} = R^{1/k} \quad ...(10.24)$$

If z_0 is in the ROC of $S(z)$ then the point $z_o^{1/k}$ is in the ROC of $S(z^k)$.

If z-transform $S(z)$ has a pole or zero at $z = A$, then $S(z^k)$ has a pole or zero at $z = A^{1/k}$.

10.4.6 Differentiation in the z-domain

Here we discuss differentiation of z-transform of a discrete-time sequence $s(n)$ with respect to variable 'z', *i.e.*, $\dfrac{dS(z)}{dz}$.

$$\text{If} \qquad s(n) \overset{Z}{\longleftrightarrow} S(z)$$

$$\text{then} \qquad ns(n) \overset{Z}{\longleftrightarrow} -z\,\frac{dS(z)}{dz}$$

In this case both transforms will have the same region of convergence (ROC).

Example 10.6 Find the sequence $s(n)$ whose z-transform is given by

$$S(z) = \log\,(1 + Az^{-1})\,, \quad |z| > |A|$$

Solution. Taking first derivative of $S(z)$, we get

$$\frac{dS(z)}{dz} = \frac{d}{dz}\,[\log\,(1 + Az^{-1})\,]$$

or $$\frac{dS(z)}{dz} = \frac{1}{(1+Az^{-1})}(-Az^{-2})$$

or $$\frac{dS(z)}{dz} = \frac{-Az^{-2}}{(1+Az^{-1})}$$

or $$-z\frac{dS(z)}{dz} = \frac{+Az^{-1}}{(1+Az^{-1})} \quad ...(1)$$

From the property of differentiation in z-domain of z-transform.

$$-z\frac{dS(z)}{dz} = \mathcal{Z}\,[ns(n)] \quad ...(2)$$

Comparing Eqs. (1) and (2), we get

$$\mathcal{Z}\,[ns(n)] = \frac{Az^{-1}}{(1+Az^{-1})} \quad ...(3)$$

From Eq. (3), we get

$$ns(n) = \mathcal{Z}^{-1}\left[\frac{Az^{-1}}{1+Az^{-1}}\right] = A\,\mathcal{Z}^{-1}\left[\frac{z^{-1}}{1+Az^{-1}}\right]$$

$$= A\,(-A)^{n-1}\,u(n-1)$$

or $$s(n) = (-1)^{n-1}\,\frac{A^n}{n}\,u(n-1)$$

10.4.7 Convolution of the Two Discrete-time Sequences

The convolution property is one of the most important property of the z-transform because it is used to convert the convolution of two discrete-time signals in time-domain into multiplication of their z-transform.

Let us consider two discrete-time sequences $s_1(n)$ and $s_2(n)$.

If $$s_1(n) \overset{\mathcal{Z}}{\longleftrightarrow} S_1(z)$$

and $$s_2(n) \overset{\mathcal{Z}}{\longleftrightarrow} S_2(z)$$

Then $$s(n) = s_1(n) * s_2(n) = \sum_{k=-\infty}^{\infty} s_1(k)\,s_2(n-k) \overset{\mathcal{Z}}{\longleftrightarrow} S(z)$$

$$= S_1(z)\,.\,S_2(z)$$

* (asterix) shows the convolution operation between sequences $s_1(n)$ and $s_2(n)$.

The ROC of the z-transform of $s(n) = s_1(n) * s_2(n)$ is at least the intersection of $S_1(z)$ and $S_2(z)$.

We now enumerate the steps involved in computation of the convolution of two signals using z- transform.

Step I. **Compute the z-transform of individual signals which are to be convolved.**

$$S_1(z) = \mathcal{Z}\,[s_1(n)]$$

$$S_2(z) = \mathcal{Z}\,[s_2(n)]$$

Step II. **Multiply the two z-transforms**

$$S(z) = S_1(z)\,.\,S_2(z)$$

Step III. **Find the inverse z-transform of $S(z)$**

$$s(n) = \mathcal{Z}^{-1}\,[S(z)]$$

Example 10.7 **Find the convolution of sequences**

$$s_1(n) = \{\underset{\uparrow}{1}, -2, 1\}$$

an $$s_2(n) = \{\underset{\uparrow}{1}, 1, 1\}$$

Solution. ***Step I.*** **Determine z-transforms of individual signal sequences.**

$$S_1(z) = \mathcal{Z}\,[s_1(n)] = \sum_{n=0}^{2} s_1(n)\,z^{-n}$$

$$= s_1(0)\,z^0 + s_1(1)\,z^{-1} + s_1(2)\,z^{-2}$$

$$= 1 - 2\,z^{-1} + 1\,z^{-2} = 1 - 2\,z^{-1} + z^{-2}$$

$$S_2(z) = \mathcal{Z}\,[s_2(n)] = \sum_{n=0}^{\infty} s_2(n)\,z^{-n}$$

$$= s_2(0)\,z^0 + s_2(1)\,z^{-1} + s_2(2)\,z^{-2}$$

$$= 1 \times 1 + 1 \times z^{-1} + 1 \times z^{-2}$$

$$= 1 + z^{-1} + z^{-2}$$

Step II. **Multiplcation of $S_1(z)$ and $S_2(z)$**

$$S(z) = S_1(z)\,.\,S_2(z)$$

$$= (1 - 2\,z^{-1} + z^{-2})\,(1 + z^{-1} + z^{-2})$$

$$= 1 - 3\,z^{-1} - 2\,z^{-3} + z^{-4}$$

Step III. **Taking inverse z-transform of $S(z)$**

$$s(n) = \mathcal{Z}^{-1}\,[S(z)] = \mathcal{Z}^{-1}\,[1 - 3\,z^{-1} - 2\,z^{-3} + z^{-4}]$$

$$= \{\underset{\uparrow}{1}, 3, 0, 0, -2, 1\}$$

10.4.8 Correlation of two Discrete-time Sequences

Correlation of two signals means a degree of similarity between two similar or dissimilar signals. Correlation of two discrete-time sequences

is equivalent to multiplication of z-transform of $s_1(n)$, $S_1(z)$ and z-transform of second sequence $s_2(n)$, $S_2(z)$ at $z = \frac{1}{z}$.

If $$s_1(n) \overset{\mathcal{Z}}{\longleftrightarrow} S_1(z)$$

$$s_2(n) \overset{\mathcal{Z}}{\longleftrightarrow} S_2(z)$$

then $$R_{s_1 s_2}(l) = \sum_{n=-\infty}^{\infty} s_1(n)\, s_2(n-l) \overset{\mathcal{Z}}{\longleftrightarrow} R_{S_1 S_2}(z)$$

$$= S_1(z)\, S_2\left(\frac{1}{z}\right)$$

We now enumerate the steps involved in computation of the correlation of two sequences using z-transform.

Step I. Compute the z-transform of individual sequences which are to be correlated

$$S_1(z) = \mathcal{Z}\,[s_1(n)] \text{ and } S_2(z) = \mathcal{Z}\,[s_2(n)]$$

Step II. Determine $S_2\left(\frac{1}{z}\right)$

$$S_2\left(\frac{1}{z}\right) = S_2(z)\Big|_{z=1/z}$$

Step III. Multiply two z-transforms

$$S(z) = S_1(z)\,.\,S_2\left(\frac{1}{z}\right)$$

Step IV. Find the inverse z-transform of $S(z)$

$$s(n) = \mathcal{Z}^{-1}\,[S(z)]$$

where $s(n)$ is the resultant of correlation between sequences $s_1(n)$ and $s_2(n)$.

Example 10.8 Find the auto-correlation sequence of the signal

$$s_1(n) = A^n u(n), \quad -1 < A < 1$$

Solution. Auto-correlation sequence is given by

$$r_{s_1 s_1}(l) = \sum_{n=-\infty}^{\infty} s_1(n)\, s_1(n-l)$$

This sequence can be determined by z-transform and it can be represented as

$$R_{s_1 s_1}(z) = \mathcal{Z}\,[r_{s_1 s_2}(l)] = \mathcal{Z}\left[\sum_{n=-\infty}^{\infty} s_1(n)\, s_1(n-l)\right]$$

$$= S_1(z)\, S_1\left(\frac{1}{z}\right) \qquad \text{...(1)}$$

$$S_1(z) = \mathcal{Z}\,[s_1(n)] = \sum_{n=-\infty}^{\infty} s_1(n)\, z^{-n}$$

$$= \sum_{n=-\infty}^{\infty} A^n u(n)\, z^{-n} = \sum_{n=0}^{\infty} A^n z^{-n}$$

$$= \sum_{n=0}^{\infty} [Az^{-1}]^n$$

or $$S_1(z) = \frac{1}{1 - Az^{-1}} \qquad \text{ROC} : |z| > |A| \qquad \text{...(2)}$$

It is a causal signal.

Also we can determine

$$S_1\left(\frac{1}{z}\right) = S_1(z)\Big|_{z=1/z} = \frac{1}{1 - Az} \quad \text{ROC} : |z| = \frac{1}{|A|} \quad \text{...(3)}$$

It is a non-causal signal.

Substituting the Eqns. (3) and (2) in Eqn. (1), we get

$$R_{S_1 S_1}(z) = S_1(z)\, S_1\left(\frac{1}{z}\right) = \left(\frac{1}{1 - Az^{-1}}\right)\left(\frac{1}{1 - Az}\right)$$

$$= \frac{1}{1 - A\,(z + z^{-1}) + A^2}, \quad \text{ROC} : |A| < |z| < \frac{1}{|A|}$$

Since the ROC of $R_{S_1 S_1}(z)$ is a ring, $r_{s_1 s_1}(l)$ is a two-sided signal, even if $s_1(n)$ is causal.

Auto-correlation sequence,

$$r_{s_1 s_1}(l) = \mathcal{Z}^{-1}\,[R_{S_1 S_1}(z)] = \mathcal{Z}^{-1}\left[\frac{1}{1 - A\,(z + z^{-1}) + A^2}\right]$$

$$= \mathcal{Z}^{-1}\left[\left(\frac{1}{1 - Az^{-1}}\right)\left(\frac{1}{1 - Az}\right)\right]$$

$$= \mathcal{Z}^{-1}\left[\frac{1}{(1 - A^2)\,(1 - Az^{-}1)} + \frac{1}{(1 - A^2)\,(1 - Az)}\right]$$

$$= \frac{1}{1 - A^2}\,\mathcal{Z}^{-1}\left[\frac{1}{1 - Az^{-1}} + \frac{1}{1 - A^2}\right]$$

$$= \frac{1}{1 - A^2}\, A^{|l|}$$

10.4.9 Multiplication of Two Discrete-time Sequences

The z-transform of a multiplication of two discrete-time sequences is the complex convolution of the z-transforms of the individual sequences.

If $$s_1(n) \overset{\mathcal{Z}}{\longleftrightarrow} S_1(z)$$

$$s_2(n) \overset{Z}{\longleftrightarrow} S_2(z)$$

then $s(n) = s_1(n) \, . \, s_2(n) \overset{Z}{\longleftrightarrow} S(z) = \dfrac{1}{2\pi j} \oint_c S_1(v)\, S_2\left(\dfrac{z}{v}\right) v^{-1}\, dv$

where v = other variable in frequency-domain

z = standard variable in frequency domain for z-transform

$j = \sqrt{-1}$

π = standard parameter.

This is called complex convolution of $S_1(z)$ and $S_2(z)$.

Where c is the closed contour that encloses the origin and lies within the ROC common to both $S_1(v)$ and $S_2\left(\dfrac{1}{v}\right)$.

This property of z-transform is not generally used, because the complexity of computations increase.

10.4.10 Parseval's Theorem

If we have two discrete-time sequences $s_1(n)$ and $s_2(n)$ and both are complex-valued sequences. Then, according to parseval's theorem

$$\sum_{n=-\infty}^{\infty} s_1(n)\, s_2^*(n) = \frac{1}{2\pi j} \oint_c S_1(v)\, S_2^*\left(\frac{1}{v^*}\right) v^{-1}\, dv \qquad ...(3.3)$$

ROC of $S_1(z)$: $R_1 l < |z| < R_1 u$

ROC of $S_2(z)$: $R_2 l < |z| < R_2 u$

For the above theorem the following condition should be satisifed.

$$R_1 l \, R_2 l < 1 < R_1 u \, R_2 u$$

where $s_2^*(n)$ is the complex conjugate of $s_2(n)$ and $s_2(n)$ is a complex-valued sequence if

$$s_2(n) = s_{2R}(n) + js_{2I}(n)$$

then $\qquad s_2^*(n) = s_{2R}(n) - js_{2I}(n)$

where $s_{2R}(n)$ and $s_{2I}(n)$ are the real part and imaginary part of signal $s_2(n)$, respectively.

$S_2^*\left(\dfrac{1}{v^*}\right)$ is the complex conjugate of z-transform of $s_2(n)$ at $z = \dfrac{1}{v^*}$, where v is another frequency varilable and v^* is the comlex conjugate of v.

$j = \sqrt{-1}$

l for lower limit, u for upper limit

$\oint_c$ represents a contour integration.

Parseval equations relate energy in a signal to energy in its frequency spectrum.

10.4.11 Conjugation of a Complex Sequence

Let us take a sequence $s(n)$ which is complex valued.

If $$s(n) \xleftrightarrow{Z} S(z)$$

then $$s^*(n) \xleftrightarrow{Z} S^*(z^*)$$

ROC for both $S(z)$ and $S^*(z^*)$ will be the same.

Where $s^*(n)$ and $S^*(z)$ are the complex conjugates of $s(n)$ and $S(z)$ respectively and z^* is complex conjugate of z.

10.4.12 Initial Value Theorem

If a discrete-time sequence $s(n) = 0$ for $n < 0$, *i.e.*, $s(n)$ is a causal sequence;

then $s(n)$ at $n = 0$

$$= s(0) = \lim_{z \to 0} \left\{ S(z) \right\} \qquad ...(3.4)$$

This is called *initial value theorem.*

10.5 TABULATION FOR SOME IMPORTANT PROPERTIES OF z-TRANSFORM

Some important properties of z-transform are summarized in Table 10.2.

Table 10.2 Properties of z-transform

Property	*Signal sequences* $s(n)$, $s_1(n)$, $s_2(n)$	*z-transform of sequences* $S(z)$, $S_1(z)$, $S_2(z)$	*ROC* R, R_1, R_2
Linearity	$As_1(n) + Bs_2(n)$	$AS_1(z) + BS_2(z)$	At least the intersection of R_1 and R_2
Time shifting	$s[n - n_0]$	$z^{-n_0} S(z)$	R, except for the possible addition or deletion of origin ($z = 0$)
Scaling of z-domain	$A^n s(n)$	$S(z/A)$	Scaled version of R, *i.e.*, $\|A\|R$
Time reversal	$s(-n)$	$S(z^{-1})$	Inverted version of R (*i.e.*, R^{-1} = the set of points z^{-1}, where z is in R)

Contd...

Property	*Signal sequences* $s(n)$, $s_1(n)$, $s_2(n)$	*z-transform of sequences* $S(z)$, $S_1(z)$, $S_2(z)$	*ROC* R, R_1, R_2
Time expansion	$s(mn)$	$S(z^m)$	$R^{1/m}$ (*i..e.*, the set of points $z^{1/m}$, where z in R)
Conjugation	$s^*(n)$	$S^*(z^*)$	R
Complex convolution	$s_1(n) * s_2(n)$	$S_1(m)\, S_2(m)$	At least the intersection of R_1 and R_2
Correlation	$r_{s_1s_2}(l) = s_1(l) * s_2(-l)$	$R_{S_1S_2}(z) = S_1(z)\, S_2(1/z)$	At least, the intersection of R_1 and R_2^{-1}
First difference	$s(n) - s(n-1)$	$(1 - z^{-1})\, S(z)$	At least the intersection of R and $\|z\| > 0$
Accumulation	$\sum_{m=-\infty}^{\infty} s(m)$	$\frac{1}{1-z^{-1}} S(z)$	At least the intersection of R and $\|z\| > 1$
Differentiation in the z-domain	$ns(n)$	$-z\, dS(z)/dz$	R

10.6 TABULATION FOR SOME COMMON z-TRANSFORM PAIRS

Some commonly used discrete-time signals and their z-transforms are given in Table 10.3.

Table 10.3 Some common z-transform pairs

Signal $s(n)$	*z-transform of signal,* $S(z)$	*Region of convergence ROC*
$\delta(n)$	1	All values of z
$u(n)$	$\frac{1}{1-z^{-1}}$	$\|z\| > 1$
$r(n) = nu(n)$	$\frac{z^{-1}}{(1-z^{-1})^2}$	$\|z\| > 1$
$A^n u(n)$	$\frac{1}{1-Az^{-1}}$	$\|z\| > \|A\|$
$nA^n u(n)$	$\frac{Az^{-1}}{(1-Az^{-1})^2}$	$\|z\| > \|A\|$
$-A^n u(-n-1)$	$\frac{1}{1-Az^{-1}}$	$\|z\| < \|A\|$
$-nA^n u(-n-1)$	$\frac{Az^{-1}}{(1-Az^{-1})^2}$	$\|z\| < \|A\|$

Contd...

Signal s(n)	*z-transform of signal, S(z)*	*Region of convergence ROC*
$(\cos \omega_0 n)\, u(n)$	$\dfrac{1 - z^{-1} \cos \omega_0}{1 - 2z^{-1} \cos \omega_0 + z^{-2}}$	$\lvert z \rvert > 1$
$(\sin \omega_0 n)\, u(n)$	$\dfrac{z^{-1} \sin \omega_0}{1 - 2z^{-1} \cos \omega_0 + z^{-2}}$	$\lvert z \rvert > 1$
$(A^n \cos \omega_0 n)\, u(n)$	$\dfrac{1 - Az^{-1} \cos \omega_0}{1 - 2Az^{-1} \cos \omega_0 + A^2 z^{-2}}$	$\lvert z \rvert > \lvert A \rvert$
$(A^n \sin \omega_0 n)\, u(n)$	$\dfrac{Az^{-1} \sin \omega_0}{1 - 2Az^{-1} \cos \omega_0 + A^2 z^{-2}}$	$\lvert z \rvert > \lvert A \rvert$
$u(n-1)$	$z^{-1}\left(\dfrac{1}{1 - z^{-1}}\right)$	$\lvert z \rvert > 1$
$\delta(n-k)$	$z^{-K} . 1 = z^{-k}$	All values of z except $z = 0$
$\delta(n+k)$	$z^{K} . 1 = z^{k}$	All values of z except $z = \infty$

10.7 THE INVERSE z-TRANSFORM

The mechanism for transforming $S(z)$ back to a discrete-time sequence $s(n)$ is called the *inverse* z-transform.

Inverse z-transform is formally given as

$$s(n) = \frac{1}{2\pi j} \oint_c S(z)\, z^{n-1}\, dz \qquad \text{...(10.27)}$$

Equation 10.27 represents a contour integral over a closed path c that encloses the origin and lies within the ROC of $S(z)$.

Contour integration is a complex integration over a closed path in the region of convergence. It encompasses the origin $z = 0$ in the z-plane, exactly once in the anticlockwise direction, a circle of radius $c > R_1$. Where R_1 is the radius of convergence of z-transform.

From Cauchy integral theorem, we know that

$$\frac{1}{2\pi j} \oint_c z^{n-1-m}\, dz = \begin{cases} 1\,, & m = n \\ 0\,, & m \neq n \end{cases} \qquad \text{...(10.28)}$$

where the integral is over any simple contour c encircling the origin.

Fig. 10.12 shows the closed contour c.

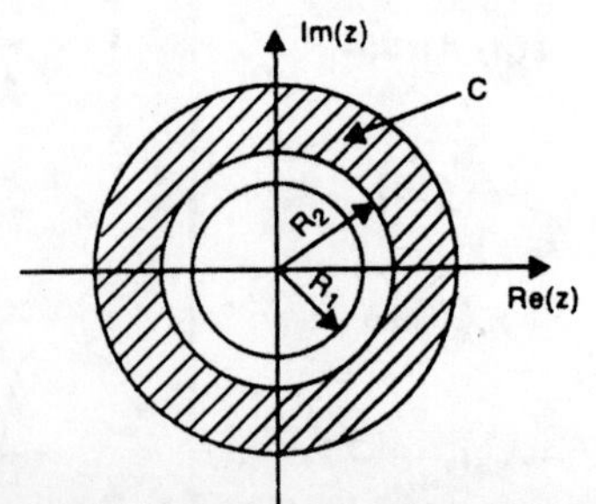

Fig. 10.12 *Closed contour c taken for the integral, in the ROC of S(z) and encircling the orizin z = 0 exactly in the anticlockwise direction.*

From the definition of z-transform

$$S(z) = \sum_{m=-\infty}^{\infty} s(m)\, z^{-m} \qquad ...(10.29)$$

Multiplying Eq. 10.29 by z^{n-1} and integrating over a closed contour with the ROC and enclosing the origin, we get

$$\oint_c S(z)\, z^{n-1}\, dz = \oint_c \sum_{m=-\infty}^{\infty} s(m)\, z^{n-1-m}\, dz \qquad ...(10.30)$$

where contour c is taken exactly once in the anticlockwise direction.

Now order of integration and summation can be interchanged as the series is convergent on this contour. It yields

$$\oint_c S(z)\, z^{n-1}\, dz = \sum_{m=-\infty}^{\infty} s(m) \oint_c z^{n-1-m}\, dz \qquad ...(10.31)$$

$$= 2\pi j\, s(n) \text{ (on applying Couchy integral theorem)}$$

Three methods that are often used for the evaluation of the inverse z-transform in practice. These are

1. The inverse z-transform by contour integration method.
2. The inverse z-transform by power series expansion method.
3. The inverse z-transform by partial fraction expansion method.

10.7.1 The Inverse z-Transform by Contour Integration Method

In this method, we use the Cauchy Residue Theorem for determining the inverse z-transform directly from the contour integral.

Cauchy Residue Theorem. Let us consider $S(z)$ which is a function of complex variable z and c is a closed path in the z-plane. Here we assume that $dS(z)/dz$ exists on and inside the contour c and $S(z)$ has no poles at $z = z_0$.

In general, Cauchy Residue Theorem is stated as

$$\frac{1}{2\pi j} \oint_c \frac{S(z)}{(z - z_0)^k}\, dz = \begin{cases} \dfrac{1}{k-1} \left. \dfrac{d^{k-1} S(z)}{dz^{k-1}} \right|_{z=z_0} & \text{if } z_0 \text{ is inside } c \\ 0, & \text{if } z_0 \text{ is outside } c \end{cases} \qquad ....(10.32)$$

$$\text{For } k = 1 \quad \frac{1}{2\pi j} \oint_c \frac{S(z)}{(z - z_0)}\, dz = \begin{cases} S(z)\,|_{z=0} & \text{if } z_0 \text{ is inside } c \\ 0 & \text{if } z_0 \text{ is outside } c \end{cases} \qquad ...(10.33)$$

The values on the RHS of Eqs. 10.32 and 10.33 are called *residues of pole at* $z = z_0$.

Inverse z-transform of $S(z)$ is given by

$$s(n) = \frac{1}{2\pi j} \oint_c S(z)\, z^{n-1}\, dz$$

$$= \sum_{\text{All poles inside } c} [\text{Residue of } S(z)\, z^{n-1} \text{ at } z = z_0]$$

$$= \sum_i (z - z_i)\; S(z)\, z^{n-1} \Big|_{z = z_i} \qquad \text{...(10.34)}$$

where $\{ z_i \}$ are simple poles.

Note. If $S(z)\, z^{n-1}$ has no poles inside c for one or more values of n. Then, $s(n) = 0$ for these values.

Example 10.9 Evaluate the inverse z-transform of

$$S(z) = \frac{z}{z - A}, \quad |z| > |A|$$

Using contour integration method.

Solution. Given $S(z) = \dfrac{z}{z - A}$, $|z| > |A|$

The contour integration formula is given as

$$s(n) = \frac{1}{2\pi j} \oint_c S(z)\, z^{n-1}\, dz$$

$$= \sum_i (z - z_i)\, S(z)\, .\, z^{n-1} \Big|_{z = z_i}$$

Putting the value of $S(z)$

$$s(n) = \sum_i (z - z_i)\, .\, \frac{z}{z - A}\, .\, z^{n-1} \Big|_{z = z_i}$$

or

$$= \sum_i (z - z_i) \frac{z^n}{z - A} \Big|_{z = z_i} \qquad \text{...(1)}$$

For $n \geq 0$: there is only one pole at $z = A$

$$s(n) = (z - A) \frac{z^n}{(z - A)} \Big|_{z = A} = z^n \Big|_{z = A} = (A)^n \qquad \text{...(2)}$$

For $n < 0$, $n = -1$, Eqn. (1) becomes

$$s(-1) = \sum_i (z - z_i) \frac{z^{-1}}{z - A} \Big|_{z = z_i}$$

$$= \sum_i (z - z_i) \frac{1}{(z)\,(z - A)} \Big|_{z = z_i}$$

$$= (z - 0) \frac{1}{(z)\,(z - A)} \Big|_{z = 0} + (z - A) \frac{1}{(z)\,(z - A)} \Big|_{z = A}$$

$$= \frac{1}{z-A}\bigg|_{z=0} + \frac{1}{z}\bigg|_{z=A} = -\frac{1}{A}+\frac{1}{A} = 0$$

For $n = -2$, Eq. (1) becomes

$$s(-2) = \sum_i (z-z_i)\frac{z^{-2}}{z-A}\bigg|_{z=z_i}$$

$$= \sum_i (z-z_i)\frac{1}{z^2(z-A)}\bigg|_{z=z_i}$$

$$= \frac{d}{dz}\left\{\frac{1}{z-A}\right\}\bigg|_{z=0} + \frac{1}{z^2}\bigg|_{z=A}$$

$$= -\frac{1}{(z-A)^2}\bigg|_{z=0} + \frac{1}{z^2}\bigg|_{z=A} = -\frac{1}{A^2}+\frac{1}{A^2} = 0$$

Similarly for $n = -3$, $s(-3) = 0$

$n = -4$, $s(-4) = 0$ and so on.

Now we can show that $s(n) = 0$, $n < 0$.

Thus $s(n) = A^n u(n)$.

Example 10.10 Determine the inverse z-transform of

$$S(z) = \frac{5z}{(z-1)(z-2)}$$

Using the contour integration method.

Solution. The contour integral formula is given by

$$s(n) = \frac{1}{2\pi j}\oint_c S(z)\, z^{n-1}\, dz$$

$$= \sum_{\text{All poles inside c}} \text{Residue of } S(z)\, z^{n-1} \text{ at poles of } S(z)$$

$$= \sum_i (z-z_i)\, S(z)\, z^{n-1}\bigg|_{z=z_i}$$

$$= \sum_i (z-z_i)\frac{5z}{(z-1)(z-2)} z^{n-1}\bigg|_{z=z_i}$$

$$= \sum_i (z-z_i)\frac{5z^n}{(z-1)(z-2)}\bigg|_{z=z_i}$$

$$= \frac{5z^n}{(z-2)}\bigg|_{z=1} + \frac{5z^n}{(z-1)}\bigg|_{z=2}$$

$$= -5(1)^n + 5(2)^n = 5(-1+2^n), \quad n \geq 0$$

or $\qquad s(n) = 5(-1+2^n)$, for $n \geq 0$

For $n < 0$, $n = -1$, The Eq. (1) becomes

$$s(n) = \sum_i (z - z_i) \frac{5z^n}{(z-1)(z-2)}\Bigg|_{z=z_i} \quad ...(1)$$

$$s(-1) = \sum_i (z - z_i) \frac{5z^{-1}}{(z-1)(z-2)}\Bigg|_{z=z_i}$$

$$= 5 \sum_i (z - z_i) \frac{1}{z(z-1)(z-2)}\Bigg|_{z=z_i}$$

$$= 5\left[\frac{1}{(z-1)(z-2)}\Bigg|_{z=0} + \frac{1}{z(z-2)}\Bigg|_{z=1} + \frac{1}{z(z-1)}\Bigg|_{z=2} \right]$$

$$= 5\left[\frac{1}{(0-1)(0-2)} + \frac{1}{1(1-2)} + \frac{1}{2(2-1)} \right]$$

$$= 5\left[\frac{1}{2} - 1 + \frac{1}{2} \right] = 5\,[1 - 1] = 0$$

For $n = -2$, $s(-2) = \sum_i (z - z_i) \dfrac{5z^{-2}}{(z-1)(z-2)}\Bigg|_{z=z_i}$

$$= 5 \sum_i (z - z_i) \frac{1}{z^2 (z-1)(z-2)}\Bigg|_{z=z_i}$$

$$= 5\left[\frac{d}{dz}\left\{ \frac{1}{(z-1)(z-2)} \right\}\Bigg|_{z=0} + \frac{1}{z^2 (z-2)}\Bigg|_{z=1} + \frac{1}{z^2 (z-1)}\Bigg|_{z=2} \right]$$

$$= \frac{3}{4} - 1 + \frac{1}{4} = 1 - 1 = 0$$

For $n = -3$, $s(-3) = 0$

$n = -4$, $s(-4) = 0$ and so on.

Hence, $s(n) = 0$, $n < 0$

But $s(n) = 5\,(2^n - 1)$, for $n \geq 0$

$\therefore$ $s(n) = 5\,(2^n - 1)\, u(n)$

10.7.2 The Inverse z-Transform by Power Series Expansion Method

For rational z-transforms a power series expansion can be obtained by long division. For this method, we have a z-transform $S(z)$ with its corresponding ROC.

Now we can expand $S(z)$ into a power series of the form

$$S(z) = \sum_{n=-\infty}^{\infty} C_n z^{-n} \qquad ...(10.35)$$

Which converges in the given ROC.

Inverse z-transforms of $S(z)$ is given by

$$s(n) = \mathcal{Z}^{-1} [\ S(z)\ ] = C_n \text{ for all } n \qquad ...(10.36)$$

Example 10.11 Determine the inverse z-transform of

$$S(z) = \frac{2}{2 - 3z^{-1} + z^{-2}} \text{ by long division method.}$$

When (a) ROC : $|z| > 1$

(b) ROC : $|z| < \frac{1}{2}$

Solution. (a) In this part ROC is $|z| > 1$. Since the ROC is the exterior of a circle whose radius is 1. Now $s(n)$ will be causal sequence and it requires a power series expansion in negative power of z.

Long division is performed as :

$$2 - 3z^{-1} + z^{-2} \;)\; 2 \;\left(\; 1 + \frac{3}{2}z^{-1} + \frac{7}{4}z^{-2} + \ldots\right.$$

$$\begin{array}{l}
2 - 3z^{-1} + z^{-2} \\
- \quad + \quad - \\
\hline
3z^{-1} - z^{-2} \\
3z^{-1} - \frac{9}{2}z^{-2} + \frac{3}{2}z^{-3} \\
- \quad + \quad - \\
\hline
\frac{7}{2}z^{-2} - \frac{3}{2}z^{-3} \\
\frac{7}{2}z^{-2} - \frac{21}{4}z^{-3} + \frac{7}{4}z^{-4} \\
- \quad + \quad - \\
\hline
\frac{15}{4}z^{-3} - \frac{7}{4}z^{-4}
\end{array}$$

$$S(z) = \frac{2}{2 - 3z^{-1} + z^{-2}} = 1 + \frac{3}{2} z^{-1} + \frac{7}{4} z^{-2} + \ldots$$

But $\quad S(z) = \sum_{n=0}^{\infty} s(n)\, z^{-n} = s(0)\, z^{-0} + s(1)\, z^{-1} + s(2)\, z^{-2} + \ldots$

$$s(n) = \{ s(0)\,,\, s(1),\, s(2)\,,\, \ldots \} = \left\{ \underset{\uparrow}{1}\,,\, \frac{3}{2}\,,\, \frac{7}{4}\,,\, \ldots \right\}$$

(*b*) In this case the ROC is interior of a circle of radius 1/2. Consequently, the signal $s(n)$ is non- causal and it requires a power series expansion in positive powers of z.

Now long division is performed as

$$
\begin{array}{l}
z^{-2} - 3z^{-1} + 2 \;\big)\; 2 \qquad\qquad (\; 2z^2 + 6z^3 + 14z^4 + \ldots \\
\qquad\qquad\qquad 2 - 6z + 4z^2 \\
\qquad\qquad\qquad - \;\; + \;\; - \\
\hline
\qquad\qquad\qquad\qquad 6z - 4z^2 \\
\qquad\qquad\qquad\qquad 6z - 18z^2 + 12z^3 \\
\qquad\qquad\qquad\qquad - \;\; + \;\; - \\
\hline
\qquad\qquad\qquad\qquad\qquad 14z^2 - 12z^3 \\
\qquad\qquad\qquad\qquad\qquad 14z^3 - 42z^3 + 28z^4 \\
\qquad\qquad\qquad\qquad\qquad - \;\; + \;\; - \\
\hline
\qquad\qquad\qquad\qquad\qquad\qquad 30z^3 - 28z^4
\end{array}
$$

$$S(z) = 2\, z^2 + 6\, z^3 + 14\, z^4 + \ldots \qquad \ldots(1)$$

z-transform for $n \le 0$

$$S(z) = s(0)\, z^{-0} + s(-1)\, z^1 + s(-2)\, z^2 + s(-3)\, z^3 + s(-4)\, z^4 + \ldots$$

Comparing Eq. (1) with expansion of z-transform for $n \le 0$, we get the sequence

$$s(n) = \{ \ldots..\,,\, 14\,,\, 6\,,\, 2\,,\, 0\,,\, \underset{\uparrow}{0} \}$$

This is a non-causal sequence.

Example 10.12 Find $s(0)$, $s(1)$, $s(2)$, $s(3)$, $s(4)$ for

$$S(z) = \frac{z^2 - 1}{z^3 + 2z + 4} \text{ by long division method.}$$

Solution. Long division is performed as follows :

$$
\begin{array}{l}
z^3 + 2z + 4 \;\big)\; z^2 - 1 \qquad (\; z^{-1} - 3z^{-3} - 4z^{-4} \\
\qquad\qquad\qquad z^2 + 2 + 4z^{-1} \\
\qquad\qquad\qquad - \;\; - \;\; - \\
\hline
\qquad\qquad\qquad\qquad -3 - 4z^{-1} \\
\qquad\qquad\qquad\qquad -3 \qquad -6z^{-2} - 12z^{-3} \\
\qquad\qquad\qquad\qquad + \qquad\;\; + \qquad + \\
\hline
\end{array}
$$

$$\begin{array}{r} -4z^{-1}+6z^{-2}+12z^{-3} \\ -4z^{-1} \qquad\quad -8z^{-3}-16z^{-4} \\ + \qquad\qquad\quad + \qquad + \\ \hline 6z^{-2}+20z^{-3}+16z^{-4} \end{array}$$

Thus $\quad S(z) = z^{-1} - 3\,z^{-3} - 4\,z^{-4} + \ldots \qquad$ ...(1)

By the definition of z-transform

$$S(z) = \sum_{n=-\infty}^{\infty} s(n)\, z^{-n}$$

$$= \ldots s(-4)\, z^{4} + s(-3)\, z^{3} + s(-2)\, z^{2} + s(-1)\, z^{1} + s(0)\, z^{0} + s(1)\, z^{-1} + s(2)\, z^{-2} + s(3)\, z^{-3} + s(4)\, z^{-4} + \ldots. \qquad \text{...(2)}$$

Comparing Eqns. (1) and (2), we get

$s(0) = 0$

$s(1) = 1$

$s(2) = 0$

$s(3) = -3$

$s(4) = -4$

10.7.3 The Inverse z-Transform by Partial Fraction Expansion Method

In this method, we attempt to express the function $S(z)$ as a linear combination of $S_1(z)$, $S_2(z)$, $S_3(z)$ where $S_1(z)$, $S_2(z)$ ... are additive parts of function $S(z)$ recovered by partial fraction. It is shown in Eq. 10.37.

$$S(z) = A_1\, S_1(z) + A_2\, S_2(z) + \ldots + A_k\, S_k(z) \qquad \text{...(10.37)}$$

where $S_1(z)$, $S_2(z)$, ... $S_k(z)$ are the expressions with inverse z-transforms $s_1(n)$, $s_2(n)$, ... $s_k(n)$. Then inverse z- transform of $S(z)$ will be

$$s(n) = \mathcal{Z}^{-1}\,[\,S(z)\,]$$

$$= \mathcal{Z}^{-1}\left[\,A_1S_1(z) + A_2S_2(z) + \ldots + A_kS_k(z)\,\right]$$

$$= \mathcal{Z}^{-1}\left[\,A_1S_1(z)\,\right] + \mathcal{Z}^{-1}\left[\,A_2S_2(z)\,\right] + \ldots + \mathcal{Z}^{-1}\left[\,A_kS_k(z)\,\right]$$

$$= A_1s_1(n) + A_2s_2(n) + \ldots + A_ks_k(n) \qquad \text{...(10.38)}$$

This method is particularly useful if $S(z)$ is a rational function given as follows

$$S(z) = \frac{N(z)}{D(z)} = \frac{B_0 + B_1\, z^{-1} + \ldots + B_M\, z^{-M}}{A_0 + A_1\, z^{-1} + \ldots + A_N\, z^{-N}} \qquad \text{...(10.39)}$$

Here A_0 should be equal to 1.

The partial fraction method is applicable only for proper rational transfer functions. A rational transfer function is called proper if

$$A_N \neq 0 \text{ and } M < N.$$

Example 10.13 Determine the inverse z-transform by using partial fraction expansion method. Given

$$S(z) = \frac{z}{\left(z - \frac{1}{2}\right)\left(z - \frac{1}{4}\right)}$$

Solution. Partial fraction expansion of $S(z)$ is given as

$$S(z) = \frac{z}{\left(z - \frac{1}{2}\right)\left(z - \frac{1}{4}\right)} = \frac{\alpha_1}{\left(z - \frac{1}{2}\right)} + \frac{\alpha_2}{\left(z - \frac{1}{4}\right)} \quad \text{...(1)}$$

$$z = \frac{1}{2}, \quad \alpha_1 = \frac{z}{z - \frac{1}{4}} = \frac{\frac{1}{2}}{\frac{1}{2} - \frac{1}{4}} = 2$$

$$z = \frac{1}{4}, \quad \alpha_2 = \frac{z}{z - \frac{1}{2}} = \frac{\frac{1}{4}}{\frac{1}{4} - \frac{1}{2}} = -1$$

Substituting the value of α_1 and α_2 in Eq. (1), we get

$$S(z) = \frac{2}{\left(z - \frac{1}{2}\right)} - \frac{1}{\left(z - \frac{1}{4}\right)} \quad \text{...(2)}$$

Taking the inverse z-transform of both sides of Eq. (2), we get

$$s(n) = \mathcal{Z}^{-1}\left[\frac{2}{z - \frac{1}{2}}\right] - \mathcal{Z}^{-1}\left[\frac{1}{z - \frac{1}{4}}\right]$$

$$= \mathcal{Z}^{-1}\left[\frac{2z^{-1}}{1 - \frac{1}{2}z^{-1}}\right] - \mathcal{Z}^{-1}\left[\frac{z^{-1}}{1 - \frac{1}{4}z^{-1}}\right]$$

$$= 2\left(\frac{1}{2}\right)^{n-1} u(n-1) - \left(\frac{1}{4}\right)^{n-1} u(n-1)$$

$$= 4\left[\left(\frac{1}{2}\right)^{n} - \left(\frac{1}{4}\right)^{n}\right] u(n-1)$$

Example 10.14 Obtain inverse z-transform using partial fraction expansion method. Given

$$S(z) = \frac{4z^2 - 2z}{z^3 - 5z^2 + 8z - 4}$$

Solution. **Here $s(z)$ is a proper rational function. So we can apply partial fraction expansion method for inverse z- transform.**

$$S(z) = \frac{4z^2 - 2z}{z^3 - 5z^2 + 8z - 4} = \frac{4z^2 - 2z}{(z-1)(z-2)^2}$$

$$= \frac{\alpha_1}{(z-1)} + \frac{\alpha_2}{(z-2)} + \frac{\alpha_3}{(z-2)^2} \qquad ...(1)$$

$z - 1 = 0$, or $z = 1$ $\quad \alpha_1 = \frac{4z^2 - 2z}{(z-2)^2} = \frac{4(1)^2 - 2 \times 1}{(1-2)^2} = 2$

$z - 2 = 0$, or $z = 2$ $\quad \alpha_2 = \frac{d}{dz}\left[\frac{4z^2 - 2z}{z-1}\right] = \frac{(z-1)(8z-2) - (4z^2 - 2z)(1)}{(z-1)^2}$

$$= \frac{(2-1)(8 \times 2 - 2) - [4(2)^2 - 2 \times 2]}{(2-1)^2} = \frac{14 - 12}{1} = 2$$

$z - 2 = 0$, or $z = 2$, $\quad \alpha_3 = \frac{4z^2 - 2z}{z-1} = \frac{4(2)^2 - 2 \times 2}{2-1} = \frac{16-4}{1} = 12$

Putting the values of α_1, α_2 and α_3 in Eq. (1), we get

$$S(z) = \frac{2}{(z-1)} + \frac{2}{(z-2)} + \frac{12}{(z-2)^2} \qquad ...(2)$$

$$\Rightarrow \qquad S(z) = \frac{2z^{-1}}{1 - z^{-1}} + \frac{2z^{-1}}{1 - 2z^{-1}} + \frac{12z^{-2}}{(1 - 2z^{-1})^2} \qquad ...(3)$$

Taking the inverse z-transform of both sides of Eqn. (3), we get

$$s(n) = \mathcal{Z}^{-1}\left[\frac{2z^{-1}}{1 - z^{-1}}\right] + \mathcal{Z}^{-1}\left[\frac{2z^{-1}}{1 - 2z^{-1}}\right] + \mathcal{Z}^{-1}\left[\frac{12z^{-2}}{(1 - 2z^{-1})^2}\right]$$

$$= 2(1)^{n-1} + 2(2)^{n-1} + 6(n-1)(2)^{n-1}, \quad n \geq 0$$

Table 10.4 Inverse z-transform

z-transform $S(z)$	*Inverse z-transform* $s(n)$	*ROC*
1	$\delta(n)$	All z
$\frac{z}{z-1}$	$u(n)$, $n \geq 0$	$\|z\| > 1$
$\frac{z}{z-1}$	$-u(n)$, $n < 0$	$\|z\| > 1$
$\frac{z}{z-A}$	A^n, $n \geq 0$	$\|z\| > \|A\|$

Contd...

z-transform $S(z)$	Inverse z-transform $s(n)$	ROC
$\frac{z}{z-A}$	$-A^n$, $n<0$	$\|z\| > \|A\|$
$\frac{z}{z-A}$	$A^n u(n)$	$\|z\| > \|A\|$
$\frac{Az}{(z-A)^2}$	$nA^n u(n)$	$\|z\| > \|A\|$
$\frac{z}{z-A}$	$-A^n u(-n-1)$	$\|z\| < \|A\|$
$\frac{Az}{(z-A)^2}$	$-nA^n u(-n-1)$	$\|z\| < \|A\|$
$\frac{z^2 - z\cos\omega_0}{z^2 - 2z\cos\omega_0 + 1}$	$(\cos\omega_0 n)\, u(n)$	$\|z\| > 1$
$\frac{z\sin\omega_0}{z^2 - 2z\cos\omega_0 + 1}$	$(\sin\omega_0 n)\, u(n)$	$\|z\| > 1$
$\frac{z^2 - Az\cos\omega_0}{z^2 - 2Az\cos\omega_0 + A^2}$	$(A^2\cos\omega_0 n)\, u(n)$	$\|z\| > \|A\|$
$\frac{Az\sin\omega_0}{z^2 - 2Az\cos\omega_0 + A^2}$	$(A^2\sin\omega_0 n)\, u(n)$	$\|z\| > \|A\|$

10.8 APPLICATIONS OF z-TRANSFORM IN ANALYSIS OF DISCRETE-TIME LTI SYSTEMS

The z-transform plays an important role in the analysis and representation of discrete-time LTI systems. From the convolution property of z-transform.

$$y(n) = h(n) * s(n) \xleftrightarrow{Z} Y(z) = H(z)\, S(z)$$

or $$Y(z) = H(z)\, S(z) \qquad \text{...(10.40)}$$

where $S(z)$, $Y(z)$ and $H(z)$ are the z-transforms of system input, output and impulse response, respectively. $H(z)$ is called the transfer function of discrete-time LTI system. It is also called system function. For z evaluated at unit circle ($z = e^{j\omega}$), the transfer function $H(z)$ reduces to frequency response of the discrete-time LTI system. We assume that unit circle is in the ROC for $H(z)$. Also we know that if the input to a discrete-time LTI system is a complex exponential signal $s(n) = z^n$ then the output of the system will be $H(z)z^n$. In other words, we can say that z^n is an eigen function of the system with eigen value given by $H(z)$.

Many important properties of a system are directly related to characteristics of the poles, zeros and ROC of the system function $H(z)$.

In this section we will discuss applications of z-transform in

1. Causality of discrete-time LTI system.
2. Stability of discrete-time LTI system.
3. Characterization of discrete-time LTI systems by linear constant- coefficient difference equations.
4. Determination of poles and zeros of a rational z-transform.
5. System behaviour related to the system function of a discrete-time LTI system.

10.8.1 Causality of Discrete-Time LTI System

The impulse response of a causal discrete-time LTI system $h(n) = 0$ for $n < 0$ and therefore, $h(n)$ is right- sided. Transfer function $H(z)$ is the z-transform of $h(n)$. The ROC of $H(z)$ of a causal discrete-time LTI system is the exterior of a circle in the z-plane. If impulse response $h(n)$ for LTI system is equal to $\delta(n)$ then transfer function for this system will be unity, *i.e.*, $H(z) = 1$. The ROC for $H(z) = 1$ can extend all the way into z-plane and possibly include the origin. In general, ROC of z-transform of right-sided impulse response may or may not include infinity.

z-transform of impulse response of causal LTI system is given as

$$H(z) = \sum_{n=0}^{\infty} h(n)\, z^{-n} \qquad \text{...(10.41)}$$

This transfer function does not include any positive powers of z. Consequently, the ROC includes infinity.

A discrete-time LTI system is causal if and only if the ROC of its transfer function is the exterior of a circle, including infinity.

A discrete-time LTI system which has a rational transfer function $H(z)$ will be causal if and only if

1. the ROC is the exterior of a circle outside the outermost pole and
2. with $H(z)$ expressed as a ratio of polynomials in z, the order of the numerator should be smaller than order of denominator.

Example 10.15 Check for causality for the discrete-time LTI system with transfer function

$$H(z) = \frac{z^3 - 3z^2 + 2z}{z^2 + \frac{1}{2}z - \frac{1}{4}}$$

Solution. Causality of the discrete-time LTI system described by transfer function

$$H(z) = \frac{z^3 - 3z^2 + 2z}{z^2 + \frac{1}{2}z - \frac{1}{4}} \qquad \text{...(1)}$$

can be checked directly by inspection without knowing the ROC. By inspection of the transfer function given in Eq. 1, we can conclude that the system is not causal because the order of numerator is higher than the order of the denominator of $H(z)$.

Example 10.16 Check for causality for the discrete-time LTI system with transfer function

$$H(z) = \frac{1}{\left(1 - \frac{1}{2}z^{-1}\right)} + \frac{1}{(1 - 2z^{-1})} \quad \text{ROC} : |z| > 2 \qquad \text{...(1)}$$

Also determine impulse response $h(n)$ for this system.

Solution. Since the ROC for this system function $H(z)$ is the exterior of a circle outside the outermost pole, $|z| > 2$, so the impulse response is right-sided. Now we need to check other conditions required for causality. So we express $H(z)$ as a ratio of polynomials in z.

$$H(z) = \frac{1}{\left(1 - \frac{1}{2}z^{-1}\right)} + \frac{1}{(1 - 2z^{-1})}$$

$$= \frac{2z}{(2z-1)} + \frac{z}{(z-2)} = \frac{z\,[2(z-2) + (2z-1)]}{(2z-1)\,(z-2)}$$

$$= \frac{z\,[2z - 4 + 2z - 1]}{(2z-1)\,(z-2)} = \frac{4z^2 - 5z}{2z^2 - 5z + 2} \qquad \text{...(2)}$$

Both the numerator and denominator polynomials of transfer function $H(z)$ given in Eq. 2 have same order, *i.e.*, 2 and consequently, we conclude that the system given in Eq. 1 is causal.

Determination of impulse response of the system

$$H(z) = \text{Inverse } z\text{-transform } \{H(z)\}$$

$$= \mathcal{Z}^{-1}\,[H(z)] = \mathcal{Z}^{-1}\left[\frac{1}{1 - \frac{1}{2}z^{-1}} + \frac{1}{1 - 2z^{-1}}\right]$$

$$= \mathcal{Z}^{-1}\left[\frac{1}{1 - \frac{1}{2}z^{-1}}\right] + \mathcal{Z}^{-1}\left[\frac{1}{1 - 2z^{-1}}\right]$$

$$= \left(\frac{1}{2}\right)^n u(n) + (2)^n u(n)$$

or

$$h(n) = \left[\left(\frac{1}{2}\right)^n + (2)^n\right] u(n) \qquad \text{...(3)}$$

or $$h(n) = \left[\left(\frac{1}{2}\right)^n + (2)^n \right], \quad n > 0$$
$$= 0, \quad n < 0$$

Since $h(n) = 0$ for $n < 0$, we can confirm that the system is a causal discrete-time LTI system.

10.8.2 Stability for Discrete-Time LTI Systems

The stability of a discrete-time LTI system is equivalent to its impulse response $h(n)$ being absolutely summable, *i.e.*,

$$S \equiv \sum_{k=-\infty}^{\infty} |h(k)| < \infty \qquad \text{...(10.42)}$$

In this case the DTFT of $h(n)$ converges and the ROC of transfer function $H(z)$ must include the unit circle $|z| = 1$.

A discrete-time LTI system is stable if and only if the ROC of its transfer function $H(z)$ includes the unit circle $|z| = 1$.

It is perfectly possible for a system to be stable but not causal. For causal systems, stability can easily be checked by examining the locations of poles in transfer function $H(z)$. For a causal discrete-time LTI system with rational transfer function $H(z)$, the ROC is outside the outermost pole. A causal discrete-time LTI system with rational transfer function $H(z)$ is stable if and only if all of the poles of $H(z)$ lie inside the unit circle $|z| = 1$.

Example 10.17 Check for stability of the causal discrete-time LTI system with transfer function

$$H(z) = \frac{1}{1 - Az^{-1}}$$

Solution. Transfer function $H(z) = \dfrac{1}{1 - Az^{-1}}$ has a pole at $z = A$. This system will be stable if its pole is inside the unit circle $|z| = 1$, i.e., $|A| < 1$.

Determination of Impulse Response $h(n)$

$$h(n) = \text{Inverse } z\text{-transform } \{H(z)\}$$
$$= \mathcal{Z}^{-1}\,[H(z)] = \mathcal{Z}^{-1}\left[\frac{1}{1 - Az^{-1}}\right] = A^n\, u(n)$$

or $$h(n) = A^n\, u(n)$$

If $h(n)$ is absolutely summable for $|A| < 1$, then the system is stable.

10.8.3 Characterization of Discrete-Time LTI Systems by Linear Constant-Coefficient Difference Equations

The z-transform is a powerful tool or procedure for obtaining the transfer function, frequency response or time-domain response (or impulse response) of a system described by linear constant- coefficient difference equations.

Let us consider a more general case of an Nth order difference equation given as

$$\sum_{k=0}^{N} A_k\, y(n-k) = \sum_{k=0}^{M} B_k\, s(n-k) \qquad ...(10.43)$$

To determine the transfer function $H(z)$ of the discrete-time LTI system described by Eq. 10.43, we proceed as follows. Take the z-transform of both sides of Eq. 10.43 and using the property of linearity and property of time shifting of z-transform, we get

$$Z\left[\sum_{k=0}^{N} A_k\, y(n-k)\right] = Z\left[\sum_{k=0}^{M} B_k\, s(n-k)\right]$$

or $$\sum_{k=0}^{N} A_k\, z^{-k}\, Y(z) = \sum_{k=0}^{M} B_k\, z^{-k}\, S(z)$$

or $$\frac{Y(z)}{S(z)} = \frac{\left[\sum_{k=0}^{M} B_k\, z^{-k}\right]}{\left[\sum_{k=0}^{N} A_k\, z^{-k}\right]}$$

$$\text{Transfer function } H(z) = \frac{Y(z)}{S(z)} = \frac{\left[\sum_{k=0}^{M} B_k\, z^{-k}\right]}{\left[\sum_{k=0}^{N} A_k\, z^{-k}\right]} \qquad ...(10.44)$$

The z-transform of an impulse response of discrete-time system is called the transfer function of the system. It is expressed as

$$H(z) = Z\,[h(n)] = \sum_{n=-\infty}^{\infty} h(n)\, z^{-n} \qquad ...(10.45)$$

In other words, system function or transfer function can be defined as the ratio of z-transform of output and z- transform of input keeping all initial conditions zeros, *i.e.*, initially the system is relaxed.

$$\text{Transfer function } H(z) = \frac{Y(z)}{S(z)} = \frac{z\text{–transform of output}}{z\text{–transform of input}}$$

The transfer function $H(z)$ represents the z-domain characterization of the system, whereas impulse response $h(n)$ is the corresponding time-domain characterization of the system.

The transfer function of a system satisfying a linear constant-coefficient difference equation is always rational. The difference equation by itself does not provide information about the ROC associated with algebraic expression $H(z)$. For specifying the ROC of $H(z)$, we require an additional constraint such as causality or stability of the system. If we know in addition that the system is causal, the ROC of the system will be outside the outermost pole. If the system is stable then the ROC of $H(z)$ will include the unit circle $|z| = 1$.

Example 10.18 Determine the system function, $H(z)$ and unit-sample response $h(n)$ of the system whose difference equation is

$$y(n) = \frac{1}{2} y(n-1) + 2s(n)$$

where $y(n)$ and $s(n)$ are the output and input of the system, respectively.

Solution. Taking the z-transform of above difference equation

$$Y(z) = \frac{1}{2} z^{-1} Y(z) + 2S(z)$$

or $$Y(z)\left[1 - \frac{1}{2} z^{-1}\right] = 2S(z)$$

or $$H(z) = \frac{Y(z)}{S(z)} = \frac{2}{1 - \frac{1}{2} z^{-1}}$$

This system function has a pole at $z = \frac{1}{2}$ and zero at $z = 0$.

$$H(z) = \frac{2}{1 - \frac{1}{2} z^{-1}}$$

$$h(n) = \text{Inverse } z\text{-transform of } H(z) = Z^{-1}\left[\frac{2}{1 - \frac{1}{2} z^{-1}}\right]$$

$$= 2\left(\frac{1}{2}\right)^n u(n)$$

This is the unit-sample response of the system.

Example 10.19 Determine the transfer function and impulse response of a discrete-time LTI system described by linear constant-coefficient difference equation given as

$$y(n) = \frac{1}{2} y(n-1) + s(n) + \frac{1}{3} s(n-1) \qquad ...(1)$$

Solution. To determine the transfer function or system function of the above difference equation, we proceed as follows. Take z-transform of both sides of Eq. 1.

$$\mathcal{Z}\,[y(n)] = \mathcal{Z}\left[\frac{1}{2}y(n-1) + s(n) + \frac{1}{3}s(n-1)\right]$$

or
$$Y(z) = \frac{1}{2}z^{-1}\,Y(z) + S(z) + \frac{1}{3}z^{-1}\,S(z)$$

or
$$Y(z)\left[1 - \frac{1}{2}z^{-1}\right] = \left(1 + \frac{1}{3}z^{-1}\right)S(z)$$

or
$$\frac{Y(z)}{S(z)} = \frac{\left(1 + \frac{1}{3}z^{-1}\right)}{\left(1 - \frac{1}{2}z^{-1}\right)} \qquad \text{...(2)}$$

Transfer function of the discrete-time LTI system is defined as

$$H(z) = \frac{Y(z)}{S(z)} = \frac{\left(1 + \frac{1}{3}z^{-1}\right)}{\left(1 - \frac{1}{2}z^{-1}\right)} \qquad \text{...(3)}$$

This transfer function $H(z)$ provides the algebraic expression for $Y(z)/S(z)$, but not the ROC. In fact, there are two different impulse responses that are consistent with the difference equation given in Eq. 1, one is right-sided sequence and the other is the left-sided sequence. Therefore, there will be two different choices for the ROC related to algebraic expression of $H(z)$ given by Eq. 3. One ROC will be $|z| > 1/2$ which gives that $h(n)$ is right-sided and the other ROC will be $|z| < 1/2$ which gives that $h(n)$ is left- sided.

Determination of Impulse Response $h(n)$

(*i*) Now let us consider a choice of ROC equal to $|z| > 1$

$$H(z) = \frac{\left(1 + \frac{1}{3}z^{-1}\right)}{\left(1 - \frac{1}{2}z^{-1}\right)} = \left(1 + \frac{1}{3}z^{-1}\right)\left(\frac{1}{1 - \frac{1}{2}z^{-1}}\right) \qquad \text{...(4)}$$

Taking inverse z-transform of both sides of Eq. 4 using together with linearity and time-shifting properties, we get

$$\mathcal{Z}\,[H(z)] = \mathcal{Z}^{-1}\left[\left(1 + \frac{1}{3}z^{-1}\right)\left(\frac{1}{1 - \frac{1}{2}z^{-1}}\right)\right]$$

or
$$h(n) = \mathcal{Z}^{-1}\left[\left(\frac{1}{1 - \frac{1}{2}z^{-1}}\right) + \left(\frac{\frac{1}{3}z^{-1}}{1 - \frac{1}{2}z^{-1}}\right)\right]$$

$$= \mathcal{Z}^{-1}\left[\frac{1}{1-\frac{1}{2}z^{-1}}\right] + \frac{1}{3}\mathcal{Z}^{-1}\left[\frac{z^{-1}}{1-\frac{1}{2}z^{-1}}\right]$$

or $$h(n) = \left(\frac{1}{2}\right)^n u(n) + \frac{1}{3}\left(\frac{1}{2}\right)^{n-1} u(n-1) \qquad ...(5)$$

(*ii*) **Now let us consider another choice of ROC equal to $|z| < 1$.**

Taking inverse z-transform of both the sides of Eq. 4 and using it together with linearity and time-shifting properties of z-transform, we get

$$h(n) = -\left(\frac{1}{2}\right)^n u(-n-1) - \frac{1}{3}\left(\frac{1}{2}\right)^{n-1} u(-n) \qquad ...(6)$$

In this case, the system is non-causal and unstable.

$$h(n) = 0 \text{ for } n > 0$$

10.8.4 Determination of Poles and Zeros of Rational z-Transform

An important family of z-transforms are those for which $S(z)$ is a rational function. Rational transfer function is a ratio of two polynomials of z, *i.e.*, $S(z) = \frac{P(z)}{Q(z)}$. The zeros of a z-transform $S(z)$ are the values of z for which $S(z) = 0$. The poles of a z-transform $S(z)$ are the values of z for which $S(z) = \infty$.

We can represent $S(z)$ graphically by pole-zero plot in the complete z-plane. Pole is located by × and zero by 0. From the definition of ROC, the ROC of a z-transform should not contain any pole.

Example 10.20 Determine the pole-zero plot for the signal $s(n) = (2)^n u(n)$.

Solution $$S(z) = \mathcal{Z}[\,s(n)\,] = \sum_{n=-\infty}^{\infty} s(n)\,z^{-n}$$

$$= \sum_{n=-\infty}^{\infty} (2)^n u(n)\,z^{-n} = \frac{1}{1-2z^{-1}}, \quad \text{ROC}: |z| > |2|$$

or $$S(z) = \frac{z}{z-2}$$

Thus $S(z)$ has one zero at $z = 0$ and one pole at $z = 2$. The pole-zero plot is shown in Fig. 10.13.

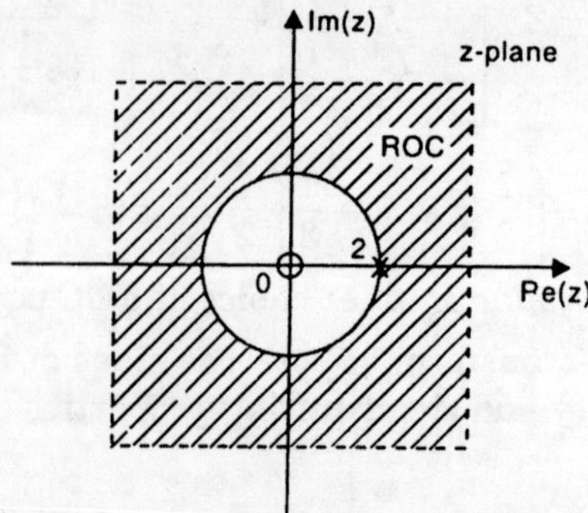

Fig. 10.13 *Pole-zero plot for the sequence $s(n) = (s)^n u(n)$.*

10.8.5 System Behaviour Related to the Transfer Function of Discrete-Time LTI Systems

Many important properties of discrete-time LTI systems can be directly related to their transfer functions and their characteristics. Here we will further use the z-transform for analyzing discrete-time LTI system.

Example 10.21 Determine the transfer function $H(z)$ of a discrete-time LTI system for which the following information is given. Also determine the value of A and deduce other properties of the system.

(*i*) If the input to the system is $s_1(n) = \left(\frac{1}{6}\right)^n u(n)$ then the output is

$$y_1(n) = \left[A\left(\frac{1}{2}\right)^n + 10\left(\frac{1}{3}\right)^n\right]u(n)$$

where A is real number.

(*ii*) If the input to the system is $s_2(n) = (-1)^n$ then the output is

$$y_2(n) = \frac{7}{4}(-1)^n$$

Solution. (*i*) Taking the z-transform of discrete-time signals given in the first part of information as

$$S_1(z) = \mathcal{Z}[s_1(n)] = \sum_{n=-\infty}^{\infty} s_1(n)\, z^{-n} = \sum_{n=-\infty}^{\infty} \left(\frac{1}{6}\right)^n u(n)\, z^{-n}$$

Now using definition of $u(n)$

$$\text{or} \qquad S_1(z) = \sum_{n=0}^{\infty} \left(\frac{1}{6}\right)^n z^{-n} = \sum_{n=0}^{\infty} \left(\frac{1}{6} z^{-1}\right)^n = \frac{1}{\left(1 - \frac{1}{6} z^{-1}\right)}$$

$$\text{with ROC} : |z| > \frac{1}{6}$$

or $$S_1(z) = \frac{1}{1-\frac{1}{6}z^{-1}}, \quad \text{ROC} : |z| > \frac{1}{6} \quad ...(1)$$

$$Y_1(z) = Z[y_1(n)] = \sum_{n=-\infty}^{\infty} y_1(n) z^{-n}$$

$$= \sum_{n=-\infty}^{\infty} \left[A\left(\frac{1}{2}\right)^n + 10\left(\frac{1}{3}\right)^n \right] u(n) z^{-n}$$

Using the definition of $u(n)$

or $$Y_1(z) = \sum_{n=0}^{\infty} \left[A\left(\frac{1}{2}\right)^n + 10\left(\frac{1}{3}\right)^n \right] 1\, z^{-n}$$

$$= A \sum_{n=0}^{\infty} \left(\frac{1}{2}z^{-1}\right)^n + 10 \sum_{n=0}^{\infty} \left(\frac{1}{3}z^{-1}\right)^n$$

or $$Y_1(z) = A\frac{1}{\left(1-\frac{1}{2}z^{-1}\right)} + 10\frac{1}{\left(1-\frac{1}{3}z^{-1}\right)}$$

$$\text{ROC} : \left\{ |z| > \frac{1}{2} \right\} \cap \left\{ |z| > \frac{1}{3} \right\} = |z| > \frac{1}{2} \quad ...(2)$$

Transfer function of the system

$$H(z) = \frac{Y_1(z)}{S_1(z)} = \frac{\frac{A}{(1-1/2z^{-1})} + \frac{10}{(1-1/3z^{-1})}}{\frac{1}{(1-1/6z^{-1})}}$$

$$= \frac{A(1-1/3\,z^{-1})(1-1/6\,z^{-1}) + 10(1-1/2\,z^{-1})(1-1/6\,z^{-1})}{(1-1/2\,z^{-1})(1-1/3\,z^{-1})}$$

or $$H(z) = \frac{[(A+10)-(5+A/3)\,z^{-1}]\,[1-1/6\,z^{-1}]}{(1-1/2\,z^{-1})(1-1/3\,z^{-1})} \quad ...(3)$$

(*ii*) In the second part of question, we have given input,

$$s_2(n) = (-1)^n \quad ...(4)$$

output $$y_2(n) = \frac{7}{4}(-1)^n \quad ...(5)$$

If we apply input $s(n) = z^n$ to the discrete-time LTI system then output to this system will be given by

$$y(n) = H(z)\, z^n \quad ...(6)$$

where $$s(n) = z^n \quad ...(7)$$

For $z = -1$ Eqs. 7 and 6 will be written as

$$s_1(n) = (-1)^n \quad ...(8)$$

$$y_1(n) = H(-1)\,(-1)^n \quad ...(9)$$

On comparing Eqs. 5 and 9, we get

$$H(-1) = \frac{7}{4} \quad ...(10)$$

Substituting $z = -1$ in Eq. 3, we get

$$H(z) = \frac{[(A+10) - (5 + A/3)\,z^{-1}]\,[1 - 1/6\,z^{-1}]}{(1 - 1/2\,z^{-1})\,(1 - 1/3\,z^{-1})}$$

$$H(-1) = \frac{[(A+10) - (5 + A/3)\,(-1)^{-1}]\,[1 - 1/6\,(-1)^{-1}]}{(1 - 1/2\,(-1)^{-1})\,(1 - 1/3\,(-1)^{-1})}$$

$$= \frac{[(A+10) + (5 + A/3)]\,[1 + 1/6]}{[1 + 1/2]\,[1 + 1/3]}$$

or $$H(-1) = \frac{(4A/3 + 15)\,(7/6)}{(3/2)\,(4/3)} = \left(\frac{4A}{3} + 15\right)\frac{7}{12} \quad ...(11)$$

From Eqs. 10 and 11, we get

$$\frac{7}{4} = \left(\frac{4A}{3} + 15\right)\frac{7}{12}$$

or $$\frac{4A}{3} + 15 = \frac{7}{4} \times \frac{12}{7} = 3$$

or $$\frac{4A}{3} = 3 - 15 = -12$$

$$A = \frac{-12 \times 3}{4} = -9 \quad ...(12)$$

Substituting the value of A in Eq. 3, we get

$$H(z) = \frac{\left[(A+10) - \left(5 + \frac{A}{3}\right)z^{-1}\right]\left(1 - \frac{1}{6}z^{-1}\right)}{\left(1 - \frac{1}{2}z^{-1}\right)\left(1 - \frac{1}{3}z^{-1}\right)}$$

$$= \frac{\left[(-9+10) - \left(5 + \frac{(-9)}{3}\right)z^{-1}\right]\left(1 - \frac{1}{6}z^{-1}\right)}{\left(1 - \frac{1}{2}z^{-1}\right)\left(1 - \frac{1}{3}z^{-1}\right)}$$

or $$H(z) = \frac{(1 - 2z^{-1})\left(1 - \frac{1}{6}z^{-1}\right)}{\left(1 - \frac{1}{2}z^{-1}\right)\left(1 - \frac{1}{3}z^{-1}\right)}$$

$$= \frac{z^2 - \frac{13}{6}z + \frac{1}{3}}{\left(z - \frac{1}{2}\right)\left(z - \frac{1}{3}\right)}$$

or $$H(z) = \frac{z^2 - \frac{13}{6}z + \frac{1}{3}}{z^2 - \frac{5}{6}z + \frac{1}{6}} \quad ...(13)$$

This is the transfer function of the discrete-time LTI system.

From the convolution property of z-transform, we know that the ROC of $Y_1(z)$ must include atleast the intersections of ROCs of $S_1(z)$ and $H_1(z)$. There will be three possible ROCs for $H(z)$. These ROCs are

$$|z| < \frac{1}{3},\ \frac{1}{2} < |z| < \frac{1}{3} \text{ and } |z| > \frac{1}{2}$$

Here we find that the only choice that is consistent with ROCs of $S_1(z)$ and $Y_1(z)$ is $|z| > \frac{1}{2}$.

Since the ROC for the system includes the unit circle, therefore the system is stable. The transfer function $H(z)$ viewed as a ratio of polynomials in z, the order of the numerator does not exceed that of the denominator and thus we can say that the LTI system is causal.

$$H(z) = \frac{Y(z)}{S(z)} = \frac{z^2 - \frac{13}{6}z + \frac{1}{3}}{z^2 - \frac{5}{6}z + \frac{1}{6}} = \frac{1 - \frac{13}{6}z^{-1} + \frac{1}{3}z^{-2}}{1 - \frac{5}{6}z^{-1} + \frac{1}{6}z^{-2}}$$

or $$Y(z)\left[1 - \frac{5}{6}z^{-1} + \frac{1}{6}z^{-2}\right] = S(z)\left[1 - \frac{13}{6}z^{-1} + \frac{1}{3}z^{-2}\right]$$

or $$Y(z) - \frac{5}{6}z^{-1}Y(z) + \frac{1}{6}z^{-2}Y(z) = S(z) - \frac{13}{6}z^{-1}S(z) + \frac{1}{3}z^{-2}Y(z)$$

Taking inverse z-transform of both sides of above equation and assuming that the system is initially relaxed, we get

$$y(n) - \frac{5}{6}y(n-1) + \frac{1}{6}y(n-2) = s(n) - \frac{13}{6}s(n-1) + \frac{1}{3}s(n-2)$$

This is the difference equation of the discrete-time LTI system characterized by a transfer function given in Eq. 13.

10.9 BLOCK DIAGRAM REPRESENTATION FOR DISCRETE-TIME LTI SYSTEMS

We already know that z-transform is used to replace time-domain operations such as convolution and time-shifting with algebraic operations for discrete-time signals and LTI systems. We have already studied that we can replace the difference equations of a discrete-time LTI system with algebraic descriptions. z-transform is also used to convert system description to algebraic equations that are helpful in analysis of inter-connections of LTI systems. These algebraic equations are also helpful in representing and synthesizing LTI systems as inter-connections of basic system building blocks.

10.9.1 Transfer Function of Inter-connection of Discrete-Time LTI Systems

The transfer function algebra for analyzing discrete-time block diagrams such as series or cascade, parallel and feedback inter-connections is exactly the same as that for corresponding continuous-time LTI systems. In this subsection, we will discuss about the following interconnections of discrete-time LTI systems.

1. Series or cascade inter-connection of two discrete-time LTI systems.
2. Parallel inter-connection of two discrete-time LTI systems.
3. Feedback inter-connection of two discrete-time LTI systems.

Now let us discuss each and every interconnection in detail.

(*i*) Series or Cascade Inter-connection of Two Discrete-Time LTI Systems

Series interconnection of two discrete-time LTI systems is shown in Fig. 10.14 (*a*).

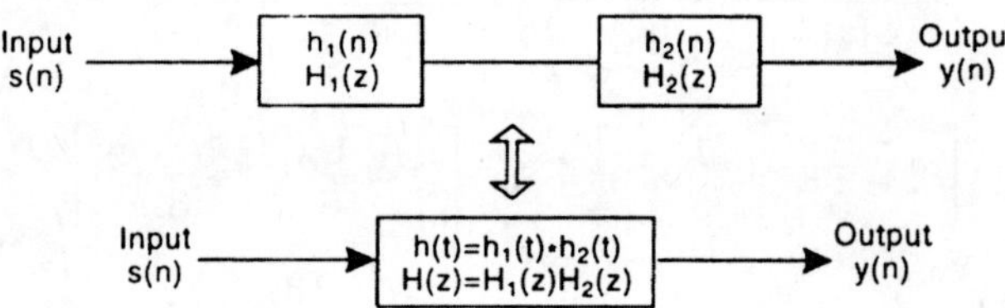

Fig. 10.14 (*a*) *Series interconnection of two discrete-time LTI systems.*

The overall impulse response of series interconnection of two discrete-time LTI systems can be determined by using convolution sum between $h_1(n)$ and $h_2(n)$, where $h_1(n)$ and $h_2(n)$ are the impulse responses of two discrete-time LTI sytems.

The overall impulse response of series inter-connection of two discrete-time LTI systems is given as

$$h(n) = h_1(n) * h_2(n) \qquad ...(10.46)$$

The transfer function of series interconnection of two discrete- time LTI systems can be determined by taking z-transform of both sides of Eq. 10.46 and using convolution property of z-transform,

$$\mathcal{Z}\ [h(n)] = \mathcal{Z}\ [h_1(n) * h_2(n)]$$

or $$H(z) = H_1(z)\, H_2(z) \quad \text{...(10.47)}$$

Thus we see that transfer function $H(z)$ for the cascade of two discrete-time LTI systems is the product of transfer functions of individual systems.

(*ii*) Parallel Inter-connection of Two Discrete-Time LTI Systems

Parallel inter-connection of two discrete-time LTI systems is shown in Fig. 10.14 (*b*).

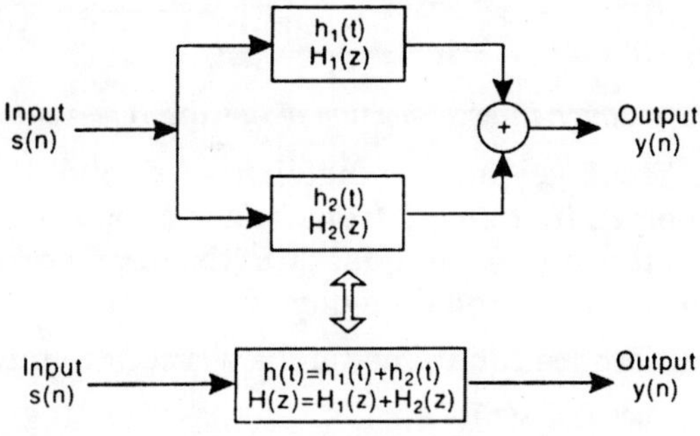

Fig. 10.14 (*b*) *Parallel interconnection of two discrete- time LTI systems.*

The overall impulse response of parallel inter-connection of two discrete-time LTI systems can be determined by addition of individual impulse responses of the two systems.

The overall impulse response of parallel inter-connection of two of the above discrete-time LTI systems is given as

$$h(n) = h_1(n) + h_2(n) \quad \text{...(10.48)}$$

The transfer function of parallel inter-connection of two discrete-time LTI systems can be determined by taking z-transform of both sides of Eq. 10.48 and using property of linearity of z-transform,

$$\mathcal{Z}\,[h(n)] = \mathcal{Z}\,[h_1(n) + h_2(n)]$$

or $$\mathcal{Z}\,[h(n)] = \mathcal{Z}\,[h_1(n)] + \mathcal{Z}\,[h_2(n)]$$

or $$H(z) = H_1(z) + H_2(z) \quad \text{...(10.49)}$$

The transfer function $H(z)$ for parallel inter-connection of two discrete-time LTI systems is the sum of transfer functions of individual systems in parallel. Here $H(z)$ is the transfer function of parallel interconnection of two discrete-time LTI systems.

(*iii*) Feedback Inter-connection of Two Discrete-Time LTI Systems

The feedback interconnection of two discrete-time LTI systems is shown in Fig. 10.14 (*c*).

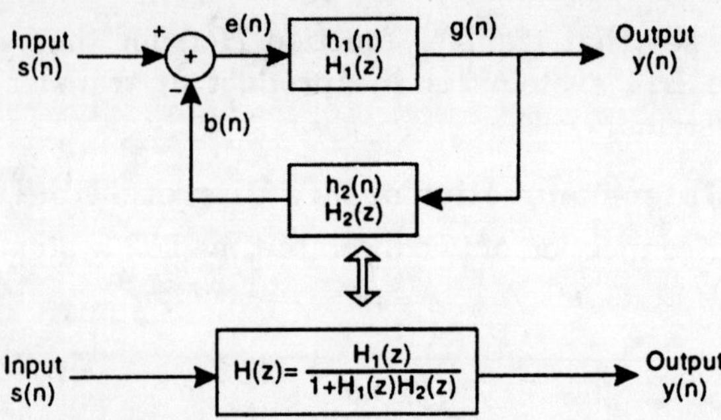

Fig 10.14 (c) *Feedback interconnection of two discrete- time LTI systems.*

In case of feedback interconnection of two discrete-time LTI systems, the determination of transfer function $H(z)$ involves the determination of the difference equation of the impulse response for the overall system working in the time-domain.

Following difference equations can be written from Fig. 10.14 (c).

$$e(n) = s(n) - b(n) \quad \text{...(10.50)}$$

$$y(n) = e(n) * h_1(n) \quad \text{...(10.51)}$$

$$b(n) = y(n) * h_2(n) \quad \text{...(10.52)}$$

Substituting Eq. 10.52 in Eq. 10.50, we get

$$e(n) = s(n) - y(n) * h_2(n) \quad \text{...(10.53)}$$

Substituting Eq. 10.53 in Eq. 10.51, we get

$$y(n) = e(n) * h_1(n)$$

$$= [s(n) - y(n) * h_2(n)] * h_1(n)$$

or $$y(n) = s(n) * h_1(n) - y(n) * h_2(n) * h_1(n) \quad \text{...(10.54)}$$

Taking z-transform of both sides of Eq. 10.54, we get

$$\mathcal{Z}\,[y(n)] = \mathcal{Z}\,[s(n) * h_1(n) - y(n) * h_2(n) * h_1(n)]$$

or $$\mathcal{Z}\,[y(n)] = \mathcal{Z}\,[s(n) * h_1(n)] - \mathcal{Z}\,[y(n) * h_2(n) * h_1(n)]$$

(using convolution property of z-transform)

$$Y(z) = S(z)\,H_1(z) - Y(z)\,H_1(z)\,H_2(z)$$

or $$Y(z) + Y(z)\,H_1(z)\,H_2(z) = S(z)\,H_1(z)$$

or $$\frac{Y(z)}{S(z)} = \frac{H_1(z)}{1 + H_1(z)\,H_2(z)}$$

or $$H(z) = \frac{Y(z)}{S(z)} = \frac{H_1(z)}{1 + H_1(z)\,H_2(z)} \quad \text{...(10.55)}$$

This is the overall transfer function of the feedback inter-connection of two discrete-time LTI systems.

10.9.2 Block Diagram Representation of Causal Discrete-Time LTI System described by Difference Equations and Rational Transfer Functions

Causal discrete-time LTI systems described by linear constant-coefficient difference equations can be represented by block diagrams involving three basic operations. The operations are addition, multiplication by a constant-coefficient, and a unit delay. We also know that causal continuous-time LTI systems described by linear constant-coefficient difference equations can also be represented by block diagrams involving the above three basic algebraic operations. Now we will discuss some more examples to illustrate the basic ideas in constructing block diagram representation.

Example 10.22 Draw the block diagram representation in direct form, cascade form and parallel form for a discrete-time LTI system described by the transfer function

$$H(z) = \frac{1}{\left(1 + \frac{1}{3}z^{-1}\right)\left(1 - \frac{1}{6}z^{-1}\right)} \qquad ...(1)$$

Solution. We know that transfer function $H(z)$ is defined as

$$H(z) = \frac{Y(z)}{S(z)} \qquad ...(2)$$

From Eqs. 1 and 2, we get

$$\frac{Y(z)}{S(z)} = \frac{1}{\left(1 + \frac{1}{3}z^{-1}\right)\left(1 - \frac{1}{6}z^{-1}\right)}$$

or
$$\frac{Y(z)}{S(z)} = \frac{1}{1 + \frac{1}{6}z^{-1} - \frac{1}{18}z^{-2}}$$

or
$$Y(z)\left[1 + \frac{1}{6}z^{-1} - \frac{1}{18}z^{-2}\right] = S(z)$$

or
$$Y(z) + \frac{1}{6}z^{-1}Y(z) - \frac{1}{18}z^{-2}Y(z) = S(z) \qquad ...(3)$$

Taking the inverse z-transform of both sides of Eq. 3 with zero initial conditions, we get

$$y(n) + \frac{1}{6}y(n-1) - \frac{1}{18}y(n-2) = s(n) \qquad ...(4)$$

Direct form block diagram representation of a discrete-time LTI system described by difference equation given by Eq. 4

$$y(n) + \frac{1}{6}y(n-1) - \frac{1}{18}y(n-2) = s(n)$$

or $$y(n) = -\frac{1}{6}y(n-1) + \frac{1}{18}y(n-2) + s(n) \quad ...(5)$$

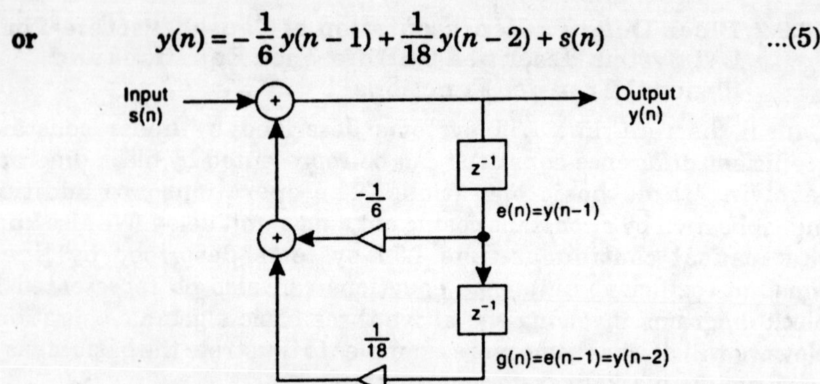

Fig. 10.15 (*a*) *Direct form block diagram representation of a discrete-time LTI system described by transfer function given by Eq.* 1.

Cascade form block diagram representation of a discrete-time LTI system described by transfer function given by Eq. 1

$$H(z) = \frac{1}{\left(1+\frac{1}{3}z^{-1}\right)\left(1-\frac{1}{6}z^{-1}\right)} = \left[1+\frac{1}{3}z^{-1}\right]^{-1}\left[1-\frac{1}{6}z^{-1}\right]^{-1}$$

$$= \left(1-\frac{1}{3}z^{-1}\right)\left(1+\frac{1}{6}z^{-1}\right) = H_1(z)\,H_2(z) \quad ...(6)$$

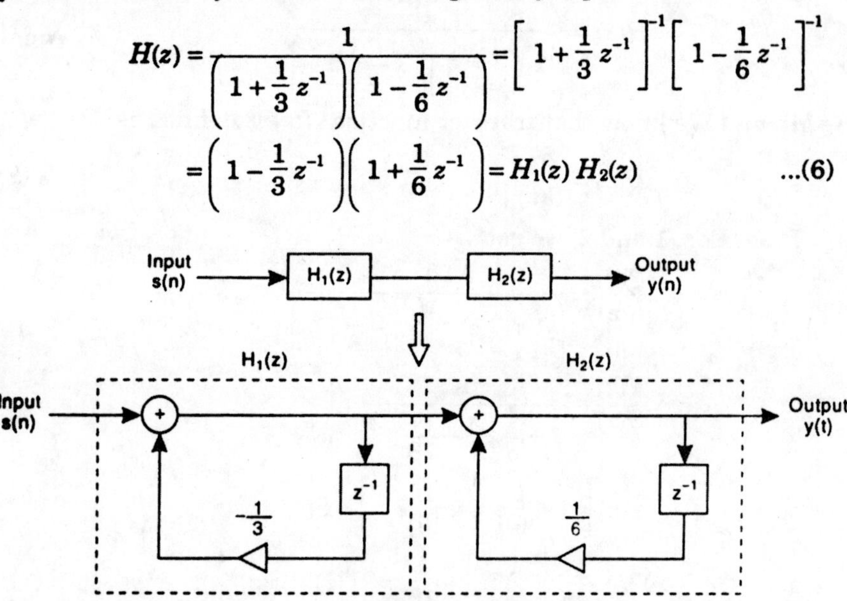

Fig. 10.15 (*b*) *Cascade form block diagram representation of a discrete-time LTI system described by transfer function given by Eq. 6.*

Parallel form block diagram representation of a discrete-time LTI system described by Eq. 1

$$H(z) = \frac{1}{\left(1+\frac{1}{3}z^{-1}\right)\left(1-\frac{1}{6}z^{-1}\right)} = \frac{\alpha_1}{\left(1+\frac{1}{3}z^{-1}\right)} + \frac{\alpha_2}{\left(1-\frac{1}{6}z^{-1}\right)}$$

(using partial fraction expansion)

$$= \frac{2/3}{\left(1 + \frac{1}{3} z^{-1}\right)} + \frac{1/3}{\left(1 - \frac{1}{6} z^{-1}\right)} = H_1(z) + H_2(z) \qquad ...(7)$$

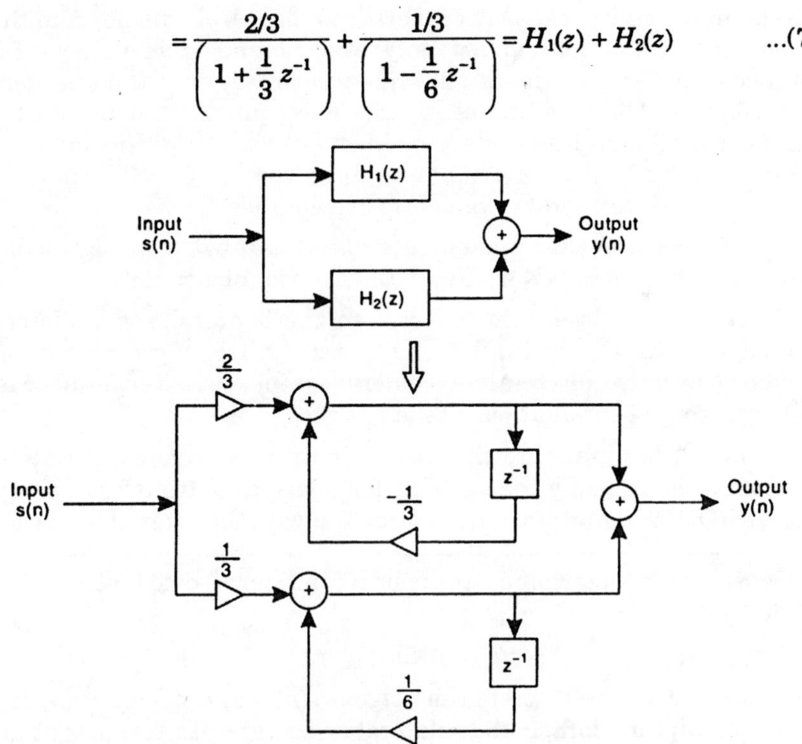

Fig. 10.15 (c) *Parallel form block diagram representation of a discrete-time LTI system described by transfer function given in Eq. 7.*

10.10 DEFINITION OF UNILATERAL z-TRANSFORM

Upto now in this chapter we have discussed bilateral (two-sided) z-transform. As in the case of Laplace transform, we have an alternative form of z-transform, *i.e.*, unilateral (one-sided) z-transform. Unilateral z-transform is particularly useful in analysis of causal discrete-time LTI systems that are described by linear constant-coefficient difference equations with non-zero initial conditions. *All discrete-time LTI systems that are not initially at rest, poses non-zero initial conditions.* In this section, we will discuss various properties of unilateral z-transform. We will also discuss various applications of unilateral z-transform.

The unilateral z-transform of a discrete-time signal $s(n)$ is defined as

$$S(z) = \mathcal{Z}_u\,[s(n)] = \sum_{n=0}^{\infty} s(n)\, z^{-n} \qquad ...(10.56)$$

We can use a convenient short hand notation for a discrete-time signal $s(n)$ and its z-transform $S(z)$ as

$$s(n) \overset{\mathcal{Z}_u}{\longleftrightarrow} S(z) = \mathcal{Z}_{\cdot}\ [\qquad ...(10.57)$$

The unilateral z-transform differs from bilateral transform in that the summation is carried out only over non-negative values of n, whether or not $s(n) = 0$ for $n < 0$. Thus we can say that the unilateral z-transform of a discrete-time sequence $s(n)$ is equivalent to the bilateral z-transform of a discrete-time sequence $s(n)u(n)$. In particular, for a discrete-time sequence $s(n)$ that is zero for $n < 0$, the unilateral z-transform and bilateral z-transform are identical.

We also know that sequence $s(n)u(n)$ is always a right-sided sequence, the ROC of $S(z)$ is always the exterior of a circle.

As there is a close relation between the bilateral and unilateral z-transforms, the determination of unilateral z- transform proceeds in the same way as the bilateral z- transform except that in the former case the lower limit of summation is taken as zero.

In a similar manner, the determination of inverse unilateral z-transform is basically the same as for bilateral z- transform, except that the ROC for unilateral z-transform is always the exterior of a circle.

Example 10.23 Determine z-transform (unilateral) of a discrete-time signal

$$s(n) = A^n u(n)$$

Solution. Since the discrete-time signal $s(n) = 0$, $n < 0$, *i.e.*, $s(n)$ is a causal signal, therefore, both unilateral and bilateral z-transforms are identical.

$$S(z) = Z\,[s(n)] = \sum_{n=-\infty}^{\infty} s(n) z^{-n}$$

$$= \sum_{n=-\infty}^{\infty} A^n\, u(n) z^{-n}$$

Now using the defintion of unit step sequence $u(n)$, we get

$$S(z) = \sum_{n=0}^{\infty} A^n 1 z^{-n} = \sum_{n=0}^{\infty} (Az^{-1})^n$$

or
$$S(z) = \frac{1}{1 - Az^{-1}}, \quad \text{ROC}: \; |z| > |A|$$

Here $S(z)$ is unilateral as well as bilateral z-transform of the signal

$$s(n) = A^n\, u(n)$$

where A is a scalar constant.

Example 10.22 Determine the unilateral and bilateral z- transform a discrete-time signal

$$s(n) = A^{n+1}\, u(n+1)$$

Solution. In this case the unilateral and bilateral z-transform will not be equal since $s(-1) = 1 \neq 0$. The bilateral (or two-sided) z-transform can be determined by using time-shifting property of z-transform as

$$S(z) = \mathcal{Z}\,[s(n)] = \sum_{n=-\infty}^{\infty} s(n)\, z^{-n} = \sum_{n=-\infty}^{\infty} A^{n+1}\, u(n+1)\, z^{-n} \qquad ...(1)$$

Substituting $n + 1 = m$ or $n = m - 1$ in Eq. 1, we get

$$S(z) = \sum_{m=-\infty}^{\infty} A^m\, u(m)\, z^{-(m-1)} = \sum_{m=-\infty}^{\infty} A^m\, u(m)\, z^{-m}\, z^1$$

or $$S(z) = z \sum_{m=-\infty}^{\infty} A^m\, u(m)\, z^{-m} \qquad ...(2)$$

Using the definition of unit step sequence in Eq. 2 becomes

$$S(z) = z \sum_{m=0}^{\infty} A^m\, 1\, z^{-m} = z \sum_{m=0}^{\infty} (Az^{-1})^m = z\left(\frac{1}{1 - Az^{-1}}\right)$$

or $$S(z) = \frac{z}{1 - Az^{-1}}, \quad \text{ROC}: |z| > |A|$$

This is the bilateral z-transform of a discrete-time signal $s(n) = A^{n+1}u(n+1)$.

Determination of Unilateral z-Transform

$$S(z) = Z_u\,[s(n)] = \sum_{n=0}^{\infty} s(n)\, z^{-n}$$

$$= \sum_{n=0}^{\infty} A^{n+1}\, z^{-n} = \sum_{n=0}^{\infty} A\, (Az^{-1})^n$$

or $$S(z) = A\left(\frac{1}{1 - Az^{-1}}\right) = \frac{A}{1 - Az^{-1}}, \quad \text{ROC}: |z| > |A|$$

This is the unilateral z-transform of a discrete-time signal $s(n) = A^{n+1}u(n+1)$.

Example 10.24 Determine inverse z-transform for the unilateral z-transform given as

$$S_u(z) = \frac{\frac{1}{6}(18 - 5z^{-1})}{(1 - 1/4\, z^{-1})\,(1 - 1/3\, z^{-1})} \qquad ...(1)$$

Solution. We know that the inverse z-transform for a bilateral z-transform $S(z)$ of the form of Eq. 1 will have several different ROCs. In the case of unilateral z- transform, the ROC for $S_u(z)$ will be exterior of the circle of radius equal to the largest magnitude of the poles of $S_u(z)$. In this example, largest magnitude of the poles of $S_u(z)$ is $|z| > \frac{1}{3}$.

$$S_u(z) = \frac{\frac{1}{6}(18 - 5z^{-1})}{(1 - 1/4\, z^{-1})\,(1 - 1/3\, z^{-1})}$$

$$= \frac{\alpha_1}{(1 - 1/4\, z^{-1})} + \frac{\alpha_2}{(1 - 1/3\, z^{-1})} \qquad ...(2)$$

(By using partial fraction expansion)

or $1 - \frac{1}{4}z^{-1} = 0$, $z^{-1} = 4$

$$\alpha_1 = \frac{\frac{1}{6}(18 - 5z^{-1})}{(1 - 1/3\, z^{-1})} = \frac{\frac{1}{6}(18 - 5 \times 4)}{(1 - 1/3 \times 4)} = \frac{(-2/3)}{(-1/3)} = 1$$

or $\alpha_1 = 1$

or $1 - \frac{1}{3}z^{-1} = 0$, $z^{-1} = 3$

$$\alpha_2 = \frac{\frac{1}{6}(18 - 5z^{-1})}{(1 - 1/4\, z^{-1})} = \frac{\frac{1}{6}(18 - 5 \times 3)}{(1 - 1/4 \times 3)} = \frac{(1/2)}{(1/4)} = 2$$

Substituting the value α_1 and α_2 in Eq. 2, we get

$$S_u(z) = \frac{1}{(1 - 1/4\, z^{-1})} + \frac{2}{(1 - 1/3\, z^{-1})} \qquad ...(3)$$

Taking the inverse z-transform of both sides of Eq. 3, we get

$$\mathcal{Z}\ [S_u(z)] = \mathcal{Z}^{-1}\left[\frac{\text{time}}{1 - 1/4\, z^{-1}} + \frac{2}{1 - 1/3\, z^{-1}}\right]$$

or
$$s(n) = \mathcal{Z}^{-1}\left[\frac{1}{1 - 1/4\, z^{-1}}\right] + 2\,\mathcal{Z}^{-1}\left[\frac{1}{1 - 1/3\, z^{-1}}\right]$$

or
$$s(n) = \left(\frac{1}{4}\right)^n u(n) + 2\left(\frac{1}{3}\right)^n u(n) \quad \text{for } n \geq 0 \qquad ...(4)$$

Eq. 4, leads us to the conclusion that inverse unilateral z-transform provides us with information about signal $s(n)$ only for $n \geq 0$.

10.11 PROPERTIES OF UNILATERAL *z*-TRANSFORM

The unilateral z-transform has many properties, which are identical to its bilateral (two-sided) counterpart and there are some properties which differ significantly. In the table of properties of unilateral z-transform we have not included a column for ROCs since the ROC of any unilateral z-transform is always the exterior of a circle. We know that the ROC for a rational unilateral Laplace transform is always outside the outermost pole.

In particular, several properties of unilateral z-transform such as linearity, scaling of $S(z)$ in z-domain, time expansion or time scaling, complex conjugation, and differentiation of $S(z)$ in z-domain are identical to bilateral z-transform of a discrete-time signal $s(n)$.

Specifically, the initial value theorem is fundamentally a unilateral z-transform property, since it needs $s(n) = 0$ for $n \leqslant 0$. One specific bilateral z- transform property is time-reversal property and their is no meaningful counterpart for it in the unilateral z- transform. Other properties differ in important ways between the bilateral and unilateral z-transform cases. These properties are

1. Convolution property of unilateral z-transform.
2. Accumulation or summation property of unileteral z- transform.
3. Time-shifting property or time-delay property of unilateral z-transform.

Now we will discuss these properties in detail.

10.11.1 Convolution Property of Unilateral z-Transform

The convolution sum of two causal discrete-time signals $s_1(n)$ and $s_2(n)$ is given as

$$s_1(n) * s_2(n) \xleftrightarrow{Z_u} S_1(z)\, S_2(z) \quad ...(10.58)$$

where $$s_1(n) = s_2(n) = 0 \text{ for all } n < 0 \quad ...(10.59)$$

In this case (Eq. 10.59) the unilateral z-transform and bilateral z-transform are identical for each of these signals given in Eq. 10.59. The system analysis and system function algebra which was previously developed and used, can be applied to unilateral z-transform without any change. For a causal discrete-time LTI system, both the input and output are causal. Impulse response for causal discrete-time LTI system is also causal *i.e.*, $h(n) = 0\ ,\ n < 0$.

10.11.2 Accumulation or Summation Property for Unilateral z-Transform

If $s(n)$ is zero for $n < 0$, then accumulation of a discrete-time signal $s(n)$ is given as

$$\sum_{k=0}^{n} s(k) = s(n) * u(n) \xleftrightarrow{Z_u} S(z)\, U(z)\, S(z) \frac{1}{(1 - z^{-1})} \quad ...(10.60)$$

Example 10.25 Solve the difference equation of a causal discrete- time LTI system which is given as

$$y(n) + 3\, y(n) = s(n)$$

The system is initially relaxed.

Solution. Transfer function $H(z)$ can be determined by taking unilateral z-transform of both sides of Eq. 1.

$$y(n) + 3\, y(n-1) = s(n) \quad ...(1)$$

$$Z_u\, [y(n) + 3\, y(n-1)] = Z_u\, [s(n)]$$

or $$Y(z) + 3z^{-1}\, Y(z) = S(z) \quad ...(2)$$

The system is initially relaxed, *i.e.*, all the initial conditions are zero for this system.

From Eq. 2,

$$H(z) = \frac{Y(z)}{S(z)} = \frac{1}{(1+3z^{-1})} \quad ...(3)$$

This is the transfer function of the causal LTI system described by Eq. 1.

Now we assume that input,

$$s(n) = u(n) \quad ...(4)$$

Taking z-transform of Eq. 4, we get

$$S(z) = U(z) = \frac{1}{1-z^{-1}} \quad ...(5)$$

We know that

$$Y(z) = H(z)\, S(z) = \frac{1}{(1+3z^{-1})\,(1-z^{-1})}$$

or $$Y(z) = \frac{1}{(1+3z^{-1})\,(1-z^{-1})}$$

(using partial fraction expansion)

$$= \frac{\alpha_1}{1+3z^{-1}} + \frac{\alpha_2}{(1-z^{-1})}$$

or $$Y(z) = \frac{3/4}{(1+3z^{-1})} + \frac{1/4}{(1-z^{-1})} \quad ...(6)$$

Taking inverse unilateral z-transform of both sides of Eq. 6, we get

$$Z_u^{-1}\,[Y(z)] = Z_u^{-1}\left[\frac{3/4}{1+3z^{-1}} + \frac{1/4}{1-z^{-1}}\right]$$

$$= \frac{3}{4} Z_u^{-1}\left[\frac{1}{1+3z^{-1}}\right] + \frac{1}{4} Z_u^{-1}\left[\frac{1}{1-z^{-1}}\right]$$

or $$y(n) = \frac{3}{4}(-3)^n\, u(n) + \frac{1}{4}(1)^n\, u(n)$$

or $$y(n) = \left(\frac{1}{4} + \frac{3}{4}(-3)^n\right) u(n) \quad ...(7)$$

This the solution for difference equation given in Eq. 1.

The convolution property for unilateral z-transform is applied only for causal discrete-time signals.

The bilateral z-transforms of $s_1(n) * s_2(n)$ equals the product of bilateral z-transforms of $s_1(n)$ and $s_2(n)$, whereas the unilateral z- transform of $s_1(n) * s_2(n)$ is not equal to multiplication of unilateral z-transforms of $s_1(n)$ and $s_2(n)$ if $s_1(n)$ and $s_2(n)$ are non-zero for $n < 0$. The unilateral z-transform is applied in analysis of causal discrete-time LTI systems which are characterized by linear constant-coefficient difference equations with non-zero initial conditions.

10.11.3 Time-Shifting Property of Unilateral z- Transform

Time shifting property of unilateral z-transform is also very useful in analysis and determination of solutions for discrete-time LTI systems characterized by linear constant- coefficient difference equations with non-zero initial conditions. Whereas the time shifting property of bilateral z- transform is used for discrete-time LTI systems characterized by linear constant-coefficient difference equation with initial conditions equal to zero.

The time-delayed version of discrete-time signal is given as

$$y(n) = s(n-1) \qquad \text{...(10.61)}$$

Taking the unilateral z-transform of both sides of Eq. 10.61, we get

$$\mathcal{Z}_u\,[y(n)] = \mathcal{Z}\,[s(n-1)] = \sum_{n=0}^{\infty} s(n-1)\,z^{-n}$$

or
$$Y(z) = s(-1) + \sum_{n=1}^{\infty} s(n-1)\,z^{-n}$$

or
$$Y(z) = s(-1) + \sum_{m=0}^{\infty} s(m)\,z^{-(m+1)} \qquad \text{(Substituting } m = n-1)$$

$$= s(-1) + z^{-1} \sum_{m=0}^{\infty} s(m)\,z^{-m}$$

or
$$Y(z) = s(-1) + z^{-1}\,S(z) \qquad \text{...(10.62)}$$

By repeated application of Eq. 10.62, the unilateral z-transform

$$w(n) = y(n-1) = s(n-2) \qquad \text{...(10.63)}$$

Taking unilateral z-transform of both sides of Eq. 10.63, we get

$$\mathcal{Z}_u\,[w(n)] = \mathcal{Z}_u\,[y(n-1)] = \mathcal{Z}_u\,[s(n-2)]$$

$$= \sum_{n=0}^{\infty} s(n-2)\,z^{-n}$$

or
$$W(z) = s(-2) + s(-1)\,z^{-1} + \sum_{n=2}^{\infty} s(n-2)\,z^{-n}$$

(Substituting $m = n - 2$)

$$W(z) = s(-2) + s(-1)\,z^{-1} + \sum_{m=0}^{\infty} s(m)\,z^{-(m+2)}$$

$$= s(-2) + s(-1)\, z^{-1} + z^{-2} \sum_{m=0}^{\infty} s(m)\, z^{-m}$$

or $$W(z) = s(-2) + s(-1)\, z^{-1} + z^{-2}\, S(z) \qquad \text{...(10.64)}$$

By using this iterative method, we can determine the unilateral z-transform of $s(n-k)$ for any positive value of k.

This property is also called the time-delay property since $y(n)$ is a delayed version of $s(n)$. We also have a time advance property of unilateral z-transform which relates the z-transform of advanced version of $s(n)$ to $S(z)$.

Time advance version of $s(n)$ is given as

$$y(n) = s(n+1)$$

$$s(n+1) \xleftrightarrow{Z_u} z\, S(z) - z\, s(0) \qquad \text{...(10.65)}$$

10.11.4 Tabulation of Properties of Unilateral z- Transforms

We have summarized important properties of unilateral z- transforms in Table 10.5.

Table 10.5 Summary of Important Properties of Unilateral z-Transforms

Properties of unilateral z-transform	*Discrete-time signal*	*Unilateral z-transform*
	$s_1(n), s_2(n), s(n)$	$S_1(z), S_2(z), S(z)$
Linearity of two discrete-time signals	$A_1\, s_1(n) + A_2\, s_2(n)$	$A_1\, S_1(z) + A_2\, S_2(z)$
Time delay of a discrete-time signal	$s(n-1)$	$z^{-1}\, S(z) + s(-1)$
Time advance of a discrete-time signal	$s(n+1)$	$z\, S(z) - z\, s(0)$
Scaling in the z-domain	$A^n\, s(n)$	$S\left(\frac{z}{A}\right)$
Time expansion of a discrete-time signal	$s_k(n) = \begin{cases} s(m), & n = mk \\ 0, & n \neq mk \end{cases}$ for any value of m	$S(z^k)$
Complex conjugation of discrete-time signal	$s^*(n)$	$S^*(z^*)$

Contd...

Properties of unilateral z-transform	*Discrete-time signal*	*Unilateral z-transform*
Convolution of two causal discrete-time signals	$s_1(n) * s_2(n)$	$S_1(z)\, S_2(z)$
First-difference of a discrete-time signal	$y(n) = s(n) - s(n-1)$	$Y(z) = S(z) - [z^{-1} S(z) + s(-1)]$ $= (1 - z^{-1}) S(z) - s(-1)$
Accumulation of a discrete-time signal	$\sum_{k=0}^{n} s(k)$	$\left(\dfrac{1}{1 - z^{-1}}\right) S(z)$
Differentiation of $S(z)$ in z-domain	$n\, s(n)$	$-z \dfrac{d\, S(z)}{dz}$

Initial-value Theorem for Unilateral z-Transform. By using this theorem, we can determine initial value of discrete-time signal using z-transform as

$$s(0) = \lim_{z \to \infty} [S(z)]$$

10.12 APPLICATION OF UNILATERAL z-TRANSFORM IN SOLVING DIFFERENCE EQUATIONS OF DISCRETE-TIME LTI SYSTEMS

The unilateral z-transform is used for solving difference equations of discrete-time LTI systems with non-zero initial conditions. By using unilateral z-transforms and the time-delay properties of unilateral z-transform, we can solve linear constant-coefficient difference equations with non-zero initial conditions. The difference equations which relate the two time-domain signals can be converted into an equivalent algebraic equation with the help of one-sided z-transform (or unilateral z-transform). This algebraic equation can be easily solved. The signal in the time-domain is obtained by taking inverse unilateral z-transform of the resulting unilateral z-transform. These difference equations can be solved by using unilateral z-transform and inverse unilateral z-transform.

Example 10.26 Determine the step response of a system given by

$$y(n) = Ay(n-1) + s(n)\,, \quad -1 < A < 1$$

When the initial condition is $y(-1) = 1$.

Solution. By taking one-sided z-transform of both sides of the above difference equation $y(n) = Ay(n-1) + s(n)$. We get

$$Y(z) = A\,[\,z^{-1}\, Y(z) + y(-1)\,] + S(z)$$

or $$Y(z) = A\,[\,z^{-1}\, Y(z) + 1\,] + S(z)$$

or $$Y(z)\,[\,1 - Az^{-1}\,] = A + S(z) \qquad ...(1)$$

But for step response, $s(n) = u(n)$

Then, $$S(z) = \mathcal{Z}\,[\,u(n)\,] = \frac{1}{1 - z^{-1}} \qquad ...(2)$$

Substituting Eq. (2) in Eq. (1), we get

$$Y(z)\,[\,1 - Az^{-1}\,] = A + \frac{1}{1 - z^{-1}}$$

or $$Y(z) = \frac{A}{1 - Az^{-1}} + \frac{1}{(1 - z^{-1})\,(1 - Az^{-1})} \qquad ...(3)$$

or $$Y(z) = \frac{A}{1 - Az^{-1}} + \frac{B_1}{1 - z^{-1}} + \frac{B_2}{1 - Az^{-1}} \qquad ...(4)$$

[By partial fraction expansion]

$1 - z^{-1} = 0$ or $z = 1$ $$B_1 = \frac{1}{1 - Az^{-1}} = \frac{1}{1 - A}$$

$1 - Az^{-1} = 0$ or $z^{-1} = 1/A$ $$B_2 = \frac{1}{1 - z^{-1}} = \frac{1}{1 - \frac{1}{A}} = \frac{A}{A - 1} = -\frac{A}{1 - A}$$

Substituting the values of B_1 and B_2 in Eq. 4, we get

$$Y(z) = \frac{A}{1 - Az^{-1}} + \frac{\left(\frac{1}{1 - A}\right)}{1 - z^{-1}} + \frac{\left(\frac{-A}{1 - A}\right)}{1 - Az^{-1}} \qquad ...(5)$$

Taking the inverse z-transform of Eq. 5, we get

$$y(n) = A^{n+1}u(n) + \frac{1}{(1 - A)}u(n) + \frac{-1}{(1 - A)}A^{n+1}u(n)$$

$$= A^{n+1}\,u(n) + \left(\frac{1 - A^{n+1}}{1 - A}\right)u(n)$$

$$= \frac{1}{1 - A}\,[\,1 - A^{n+2}\,]\,u(n)$$

Example 10.27 Solve the following difference equation by using z-transform method

$$s(n + 2) + 3s(n + 1) + 2s(n) = 0$$

Initial conditions are $s(0) = 0$ and $s(1) = 1$.

Solution. Given the difference equation

$$s(n + 2) + 3s(n + 1) + 2s(n) = 0 \qquad ...(1)$$

Taking the z-transform of both sides of the above equation, we get

$$[z^2S(z) - z^2s(0) - zs(1)\,] + 3\,[zS(z) - zs(0)] + 2\,[s(z)] = 0$$

or $$[z^2S(z) - z^20 - z1] + 3\,[zS(z) - z0] + 2\,[S(z)] = 0$$

or $$[z^2S(z) - z] + 3zS(z) + 2S(z) = 0$$

or $$S(z)\,[z^2 + 3z + 2] = z$$

$$S(z) = \frac{z}{z^2 + 3z + 2} \qquad ...(2)$$

We now take the inverse z-transform of above z-transform by partial fraction expansion method

$$S(z) = \frac{z}{z^2 + 3z + 2} = \left\{ \frac{A}{(z+1)} + \frac{B}{(z+2)} \right\} z$$

$$= \frac{z}{z+1} - \frac{z}{z+2} = \frac{1}{1+z^{-1}} - \frac{1}{1+2z^{-1}} \qquad ...(3)$$

$$S(z) = \frac{1}{1+z^{-1}} - \frac{1}{1+2z^{-1}} \qquad ...(4)$$

Taking the inverse z-transform of Eq. 4, we get

$$s(n) = \mathcal{Z}^{-1}\left[\frac{1}{1+z^{-1}}\right] - \mathcal{Z}^{-1}\left[\frac{1}{1+2z^{-1}}\right]$$

$$= (-1)^n u(n) - (-2)^n u(n)$$

Example 10.28 Find the response of the system

$$s(n+2) - 3\,s(n+1) + 2\,s(n) = \delta(n) \qquad ...(1)$$

When the all the initial conditions are zero.

Solution. Taking one-sided z-transform of both sides of above equation, we get

$$z^2S(z) - 3\,z^1S(z) + 2\,S(z) = 1$$

or $$S(z)\,[z^2 - 3z + 2] = 1$$

or $$S(z) = \frac{1}{z^2 - 3z + 2} = \frac{1}{(z-2)\,(z-1)} = \frac{A}{z-2} + \frac{B}{z-1}$$

or $$S(z) = \frac{1}{z-2} - \frac{1}{z-1} \qquad ...(2)$$

(By partial fraction expansion)

Taking inverse z-transform of both sides of Eq. 2, we get

$$s(n) = \mathcal{Z}^{-1}\left[\frac{1}{z-2}\right] - \mathcal{Z}^{-1}\left[\frac{1}{z-1}\right]$$

$$= \mathcal{Z}^{-1}\left[\frac{z^{-1}}{1-2z^{-1}}\right] - \mathcal{Z}^{-1}\left[\frac{z^{-1}}{1-z^{-1}}\right]$$

$$= (2)^{n-1} - (1)^{n-1} = -1 - (2)^{n-1}$$

Example 10.29 Find the z-transform of the following discrete-time signals. Also determine the ROC for each of the following cases;

(a) $s(n) = 2^n\, u(n) + 3\left(\frac{1}{2}\right)^n u(n)$

(b) $s(n) = 3\left(-\frac{1}{2}\right)^n u(n) - 2\,(3)^n\, u(-n-1)$

Solution. (a) Two-sided (bilateral) z-transform of $s(n)$ is defined as

$$S(z) = \mathcal{Z}\,[s(n)] = \sum_{n=-\infty}^{\infty} s(n)\, z^{-n}$$

$$= \sum_{n=-\infty}^{\infty} \left[2^n u(n) + 3\left(\frac{1}{2}\right)^n u(n) \right] z^{-n}$$

$$= \sum_{n=-\infty}^{\infty} 2^n u(n)\, z^{-n} + 3 \sum_{n=-\infty}^{\infty} \left(\frac{1}{2}\right)^n u(n)\, z^{-n}$$

$$\left[\text{Since } u(n) = \begin{cases} 1, & n \geq 0 \\ 0, & n < 0 \end{cases}\right]$$

$$= \sum_{n=0}^{\infty} 2^n\,(1)\, z^{-n} + 3 \sum_{n=0}^{\infty} \left(\frac{1}{2}\right)^n (1)\, z^{-n}$$

$$= \sum_{n=0}^{\infty} (2z^{-1})^n + 3 \sum_{n=0}^{\infty} \left(\frac{1}{2} z^{-1}\right)^n$$

$$= \underbrace{\frac{1}{(1-2z^{-1})}}_{\text{I-part}} + \underbrace{3\frac{1}{\left(1-\frac{1}{2}z^{-1}\right)}}_{\text{II-part}}$$

$$= \frac{\left(1-\frac{1}{2}z^{-1}\right) + 3\,(1-2z^{-1})}{(1-2z^{-1})\left(1-\frac{1}{2}z^{-1}\right)}$$

$$S(z) = \frac{4-13z^{-1}}{(1-2z^{-1})\left(1-\frac{1}{2}z^{-1}\right)} \qquad \text{...(1)}$$

ROC of I-part: $|1-2z^{-1}| > 0$

or $|2z^{-1}| < 1$

or $|z^{-1}| < \frac{1}{2}$

or $\left|\frac{1}{z}\right| < \frac{1}{2}$

or $$|z| > 2 \quad \text{...(2)}$$

ROC of II-part : $\left| 1 - \frac{1}{2} z^{-1} \right| > 0$

or $$\left| \frac{1}{2} z^{-1} \right| < 1$$

or $$|z^{-1}| < 2$$

or $$\left| \frac{1}{z} \right| < 2$$

or $$|z| > \frac{1}{2} \quad \text{...(3)}$$

ROC of $S(z)$:

$$\text{ROC} : \{|z| > 2\} \cap \left\{ |z| > \frac{1}{2} \right\} \Rightarrow \text{ROC} : \{|z| > 2\}$$

(*b*) Given $s(n) = 3\left(-\frac{1}{2}\right)^n u(n) - 2(3)^n u(-n-1)$

Two-sided z-transform of $s(n)$ is defined as

$$S(z) = \mathcal{Z}[s(n)] = \sum_{n=-\infty}^{\infty} s(n) z^{-n}$$

$$= \sum_{n=-\infty}^{\infty} \left[3\left(-\frac{1}{2}\right)^n u(n) - 2(3)^n u(-n-1) \right] z^{-n}$$

$$= 3 \sum_{n=-\infty}^{\infty} \left(-\frac{1}{2}\right)^n u(n) z^{-n} - 2 \sum_{n=-\infty}^{\infty} (3)^n u(-n-1) z^{-n}$$

$$\left[\text{Since } u(n) = \begin{cases} 1, & n \geq 0 \\ 0, & n < 0 \end{cases} \text{ and } u(-n-1) = \begin{cases} 1, & (-n-1) \geq 0 \rightarrow n \leq -1 \\ 0, & (-n-1) < 0 \rightarrow n > -1 \end{cases} \right]$$

$$= 3 \sum_{n=0}^{\infty} \left(-\frac{1}{2}\right)^n (1) z^{-n} - 2 \sum_{n=-\infty}^{-1} (3)^n (1) z^{-n}$$

$$= 3 \sum_{n=0}^{\infty} \left(-\frac{1}{2}\right)^n z^{-n} - 2 \sum_{n=-\infty}^{-1} (3)^n z^{-n}$$

$$= 3 \sum_{n=0}^{\infty} \left(-\frac{1}{2} z^{-1}\right)^n - 2 \sum_{n=-\infty}^{-1} (3z^{-1})^n$$

$$= 3 \sum_{n=0}^{\infty} \left(-\frac{1}{2} z^{-1}\right)^n - 2 \sum_{n=1}^{\infty} (3z^{-1})^{-n}$$

or $$S(z) = 3 \frac{1}{1 - \left(-\frac{1}{2} z^{-1}\right)^n} - 2 \frac{(3z^{-1})^{-1}}{1 - (3z^{-1})^{-1}}$$

or $$S(z) = \frac{3}{1+\frac{1}{2}z^{-1}} - \frac{2/3z^{-1}}{1-\frac{1}{3z^{-1}}}$$

$$= \frac{3}{1+\frac{1}{2}z^{-1}} - \frac{2}{3z^{-1}-1}$$

or $$S(z) = \frac{3}{1+\frac{1}{2}z^{-1}} - \frac{2}{1-3z^{-1}} \quad \text{...(4)}$$

I-part II-part

or $$S(z) = \frac{2(1-3z^{-1}) + 2\left(1+\frac{1}{2}z^{-1}\right)}{\left(1+\frac{1}{2}z^{-1}\right)(1-3z^{-1})}$$

or $$S(z) = \frac{4-5z^{-1}}{\left(1+\frac{1}{2}z^{-1}\right)(1-3z^{-1})}$$

ROC of I-part : $$\left|1+\frac{1}{2}z^{-1}\right| < 0$$

or $$\left|\frac{1}{2}z^{-1}\right| < 1$$

or $$|z^{-1}| < 2$$

or $$|z| > \frac{1}{2} \quad \text{...(5)}$$

ROC of II-part : $$|1-3z^{-1}| > 0$$

or $$|3z^{-1}| > 1$$

or $$|z| < 3 \quad \text{...(6)}$$

ROC of $S(z)$:

$$\text{ROC}: \left\{|z| > \frac{1}{2}\right\} \cap \{|z| < 3\} = \frac{1}{2} < |z| < 3$$

Example 10.30 Given two discrete-time signals

$$s_1(n) = \left(\frac{1}{3}\right)^n u(n) \quad \text{and} \quad s_2(n) = \left(\frac{1}{5}\right)u(n)$$

Convolution of $s_1(n)$ and $s_2(n)$ is given by

$$s(n) = s_1(n) * s_2(n)$$

(*a*) Find $S(z)$ using convolution property of z-transform.

(*b*) Find $s(n)$ by taking inverse z-transforms of $S(z)$ by using *partial fraction expansion method*.

Solution. (*a*) Given : $s(n) = s_1(n) * s_2(n)$

From convolution property of *z*-transform, we know

$$s(n) = s_1(n) * s_2(n) \overset{\mathcal{Z}}{\longleftrightarrow} S(z) = S_1(z)\, S_2(z)$$

or $$S(z) = S_1(z)\, S_2(z) \qquad ...(1)$$

where $$S_1(z) = \mathcal{Z}\,[s_1(n)] = \mathcal{Z}\left[\left(\frac{1}{3}\right)^n u(n)\right]$$

$$= \frac{1}{1 - \frac{1}{3}z^{-1}}, \quad \text{ROC} : |z| > \frac{1}{3} \qquad ...(2)$$

and $$S_2(z) = \mathcal{Z}\,[s_2(n)] = \mathcal{Z}\left[\left(\frac{1}{5}\right)^n u(n)\right]$$

$$= \frac{1}{1 - \frac{1}{5}z^{-1}}, \quad \text{ROC} : |z| > \frac{1}{5} \qquad ...(3)$$

Substituting Eqs. 2 and 3 in Eq. 1, we get

$$S(z) = S_1(z)\, S_2(z) = \frac{1}{\left(1 - \frac{1}{3}z^{-1}\right)} \frac{1}{\left(1 - \frac{1}{5}z^{-1}\right)}$$

or $$S(z) = \frac{1}{\left(1 - \frac{1}{3}z^{-1}\right)} \frac{1}{\left(1 - \frac{1}{5}z^{-1}\right)} \qquad ...(4)$$

$$\text{ROC} : \left\{ |z| > \frac{1}{3} \right\} \cap \left\{ |z| > \frac{1}{5} \right\} = |z| > \frac{1}{3}$$

(*b*) From Eq. 4,

$$S(z) = \frac{z^2}{\left(z - \frac{1}{3}\right)\left(z - \frac{1}{5}\right)}$$

or $$\frac{S(z)}{z} = \frac{z}{\left(z - \frac{1}{3}\right)\left(z - \frac{1}{5}\right)} = \frac{A_1}{\left(z - \frac{1}{3}\right)} + \frac{A_2}{\left(z - \frac{1}{5}\right)}$$

(using partial fraction expansion)

or $$\frac{S(z)}{z} = \frac{5/2}{\left(z - \frac{1}{3}\right)} + \frac{-3/2}{\left(z - \frac{1}{5}\right)}$$

or $$S(z) = \frac{5/2z}{\left(z - \frac{1}{3}\right)} - \frac{3/2z}{\left(z - \frac{1}{5}\right)}$$

or $$S(z) = \frac{5/2}{\left(1 - \frac{1}{3}z^{-1}\right)} - \frac{3/2}{\left(1 - \frac{1}{5}z^{-1}\right)} \qquad ...(3)$$

Since both the poles are surrounded by ROC they give only positive time sequence terms.

Taking inverse z-transform of both sides of Eq. 3, we get

$$s(n) = \mathcal{Z}^{-1}[S(z)] = \mathcal{Z}^{-1}\left[\frac{5/2}{\left(1 - \frac{1}{3}z^{-1}\right)} - \frac{3/2}{\left(1 - \frac{1}{5}z^{-1}\right)}\right]$$

or $$s(n) = \frac{5}{2}\left(\frac{1}{3}\right)^n u(n) - \frac{3}{2}\left(\frac{1}{5}\right)^n u(n)$$

Example 10.31 Find the z-transform of the sequence given by

$$s(n) = \begin{cases} 2^n, & n < 0 \\ (1/2)^n, & n = 0, 2, 4 \\ (1/3)^n, & n = 1, 3, 5 \end{cases} \qquad ...(1)$$

Solution. z-transform of $s(n)$ is given by

$$S(z) = \mathcal{Z}[s(n)] = \sum_{n=-\infty}^{\infty} s(n)\, z^{-n} \qquad ...(2)$$

Substituting Eq. 2 in Eq. 3, we get

$$S(z) = \sum_{n=-\infty}^{-1} 2^n z^{-n} + \sum_{\substack{n=0 \\ (n\text{-even})}}^{\infty} \left(\frac{1}{2}\right)^n z^{-n} + \sum_{\substack{n=0 \\ (n\text{-odd})}}^{\infty} \left(\frac{1}{3}\right)^n z^{-n}$$

$$= \sum_{m=1}^{\infty} 2^{-m} z^{m} + \sum_{p=0}^{\infty} \left(\frac{1}{2}\right)^{2p} z^{-2p} + \sum_{q=0}^{\infty} \left(\frac{1}{3}\right)^{2q-1} z^{-(2q+1)}$$

where $m = -n$; $p = \frac{n}{2}$ and $q = \left(\frac{n-1}{2}\right)$

Therefore, $S(z) = S_1(z) + S_2(z) + S_3(z)$

or $$S(z) = \frac{z/2}{\left(1 - \frac{z}{2}\right)} + \frac{z^2}{\left(z^2 - \frac{1}{4}\right)} + \frac{z/3}{\left(z^2 - \frac{1}{9}\right)} \qquad ...(3)$$

ROC for $S_1(z)$: $|z| < 2$

ROC for $S_2(z)$: $|z| > \frac{1}{2}$

ROC for $S_3(z)$: $|z| > \frac{1}{3}$

Hence, ROC for $S(z)$: $\{|z| < 2\} \cap \left\{|z| > \frac{1}{2}\right\} \cap \left\{|z| > \frac{1}{3}\right\}$

$$= \frac{1}{2} < |z| < 2$$

Example 10.32 Find the convolution $s(n)$ of the signals

$$s_1(n) = \{\underset{\uparrow}{4}, -2, 1\}$$

$$s_2(n) = \begin{cases} 1, & 0 \le n \le 5 \\ 0, & \text{otherwise} \end{cases}$$

Solution. $S_1(z) = \mathcal{Z}[s_1(n)] = \sum_{n=-\infty}^{\infty} s_1(n)\, z^{-n}$

$$= \ldots + s_1(-2)\, z^2 + s_1(-1)\, z^1 + s_1(0)\, z^0 + s_1(1)\, z^{-1} + s_1(2)\, z^{-2} + s_1(3)\, z^{-3} + s_1(4)\, z^{-4} + \ldots$$

or $$S_1(z) = 4 - 2z^{-1} + z^{-2} \qquad \ldots(1)$$

$$S_2(z) = \mathcal{Z}[s_2(n)] = \sum_{n=-\infty}^{\infty} s_2(n)\, z^{-n}$$

or $$S_2(z) = 1 + z^{-1} + z^{-2} + z^{-3} + z^{-4} + z^{-5} \qquad \ldots(2)$$

From convolution property of z-transform, we know that

$$S(z) = S_1(z)\, S_2(z) \qquad \ldots(3)$$

Substituting Eqs. 1 and 2 in Eq. 3, we get

$$S(z) = (4 - 2z^{-1} + z^{-2})\,(1 + z^{-1} + z^{-2} + z^{-3} + z^{-4} + z^{-5})$$

or $$S(z) = 4 + 2z^{-1} + 3z^{-2} + 3z^{-3} + 3z^{-4} + 3z^{-5} + z^{-6} + z^{-7} \qquad \ldots(4)$$

Taking the inverse z-transform of Eq. 4, we get

$$s(n) = \mathcal{Z}^{-1}\,[S(z) = \{\underset{\uparrow}{4}, 2, 3, 3, 3, 3, -1, 1\}$$

Example 10.33 Determine z-transform of the discrete-time signal

$$s(n) = 2^n u(n-2)$$

Solution. We know that the

$$\mathcal{Z}[u(n)] = \frac{1}{1 - z^{-1}}$$

Hence the z-transform of $u(n-2)$ can be determined as

$$\mathcal{Z}[u(n-2)] = z^{-2}\, \mathcal{Z}[u(n)]$$

$$= z^{-2}\left(\frac{1}{1 - z^{-1}}\right) \qquad \ldots(1)$$

Therefore, $\mathcal{Z}[2^n\, u(n-2)] = \mathcal{Z}[u(n-2)]\,\big|_{z^{-1} \to 2z^{-1}}$

$$= \frac{z^{-2}}{1 - z^{-1}}\Bigg|_{z^{-1} \to 2z^{-1}} = \frac{(2z^{-1})^2}{1 - 2z^{-1}} = \frac{4z^{-2}}{1 - 2z^{-1}}$$

Example 10.34 Find the cross-correlation sequence $r_{s_1 s_2}(l)$ of the sequences

$$s_1(n) = \{\underset{\uparrow}{1}, 2, 3, 4\}$$

$$s_2(n) = \{ \underset{\uparrow}{4}, 3, 2, 1\}$$

Solution. Cross-correlation sequence can be obtained by using the correlation property of z-transform

$$R_{s_1s_2}(z) = S_1(z)\, S_2(z^{-1}) \qquad ...(1)$$

Taking z-transforms of sequence $s_1(n)$ and $s_2(n)$, we get

$$S_1(z) = \mathcal{Z}\, [s_1(n)] = 1 + 2z^{-1} + 3z^{-2} + 4z^{-3} \qquad ...(2)$$

$$S_2(z) = \mathcal{Z}\, [s_2(n)] = 4 + 3z^{-1} + 2z^{-2} + z^{-3} \qquad ...(3)$$

Substituting $z = z^{-1}$ in Eq. 3, we get

$$S_2(z^{-1}) = 4 + 3z + 2z^2 + z^3 \qquad ..(4)$$

Substituting Eqs. 2 and 4 in Eq. 1, we get

$$R_{s_1s_2}(z) = S_1(z)\, S_2(z^{-1})$$

$$= (1 + 2z^{-1} + 3z^{-2} + 4z^{-3})\, (4 + 3z + 2z^2 + z^3)$$

or $$R_{s_1s_2}(z) = (z^3 + 4z^2 + 10z + 20 + 25z^{-1} + 24z^{-2} + 16z^{-3}) \qquad ...(5)$$

Taking the inverse z-transform of Eq. 5, we get

$$r_{s_1s_2}(l) = \mathcal{Z}^{-1}\, [R_{s_1s_2}(z)] = \{ 1, 4, 10, \underset{\uparrow}{20}, 25, 24, 16\}$$

Example 10.35 Find the initial and final values of the corresponding sequence, $s(n)$. Given

$$S(z) = 2 + 3z^{-1} + 4z^{-2}$$

Solution. Initial value of the sequence $s(n)$

$$s(0) = \lim_{|z| \to \infty} [2 + 3z^{-1} + 4z^{-2}]$$

$$= 2 + \frac{3}{\infty} + \frac{4}{\infty} = 2$$

Final value of the sequence $s(n)$

$$s(\infty) = \lim_{|z| \to 1} [(1 - z^{-1})\, (2 + 3z^{-1} + 4z^{-2})]$$

$$= \lim_{|z| \to 1} [2 + z^{-1} + z^{-2} - 4z^{-3}]$$

$$= 2 + (1)^{-1} + (1)^{-2} - 4(1)^{-3}$$

$$= 2 + 1 + 1 - 4 = 0$$

Initial and final values of the signals are 2 and 0 respectively.

Example 10.36 Determine the causal discrete-time signal $s(n)$ for $S(z)$ given by

$$S(z) = \frac{1 + 2z^{-1}}{1 - 2z^{-1} + 4z^{-2}}$$

Solution. Given : $S(z) = \dfrac{1 + 2z^{-1}}{1 - 2z^{-1} + 4z^{-2}}$ ...(1)

Eq. 1 can be written as

$$S(z) = \frac{z^2 + 2z}{z^2 - 2z + 4} \quad ...(2)$$

We know that

$$\mathcal{Z}\,[A^n \cos \omega_0 n] = \frac{z(z - A \cos \omega_0)}{z^2 - 2Az \cos \omega_0 + A^2}$$

and $\mathcal{Z}\,[A^n \sin \omega_0 n] = \dfrac{Az \sin \omega_0}{z^2 - 2Az \cos \omega_0 + A^2}$

The denominator of $S(z)$ gives

$$z^2 - 2z + 4 = z^2 - 2Az \cos \omega_0 + A^2$$

$$A \cos \omega_0 = 1 \quad \text{and} \quad A = 2$$

$$2 \cos \omega_0 = 1$$

or $$\cos \omega_0 = \frac{1}{2}$$

or $$\cos \omega_0 = \cos 60°$$

or $$\omega_0 = 60° = \frac{\pi}{3}$$

$$\sin \omega_0 = \sin 60° = \frac{\sqrt{3}}{2}$$

The numerator of $S(z)$ gives

$$z^2 + 2z = z^2 - Az \cos \omega_0 + \alpha\,(Az \sin \omega_0)$$

or $$z^2 + 2z = z^2 - 2z\left(\frac{1}{2}\right) + \alpha\left(2z\,\frac{\sqrt{3}}{2}\right)$$

or $$\alpha = \sqrt{3}$$

Therefore, $$S(z) = \frac{\left[z^2 - 2z \cos \frac{\pi}{3}\right] + \sqrt{3}\left[2z \sin \frac{\pi}{3}\right]}{z^2 - 2z\left(2 \cos \frac{\pi}{3}\right) + 4}$$

Hence $$s(n) = 2^n\left[\cos \frac{n\pi}{3} + \sqrt{3} \sin \frac{n\pi}{3}\right] u(n)$$

Example 10.37 Impulse response of a discrete-time LTI system is given by

$$h(n) = \{\underset{\uparrow}{1}, 2, 3\}$$

Determine the input sequence $s(n)$ for output response

$$y(n) = \{\underset{\uparrow}{1}, 1, 2, -1, 3\}$$

Solution. Taking the z-transform of $h(n)$ and $y(n)$, we get

$$H(z) = \mathcal{Z}\,[h(n)] = \sum_{n=0}^{2} h(n)\, z^{-n}$$

$$= h(0)\, z^{-0} + h(1)\, z^{-1} + h(2)\, z^{-2}$$

$$= 1 + 2z^{-1} + 3z^{-2} \qquad ...(1)$$

$$Y(z) = \mathcal{Z}\,[y(n)] = \sum_{n=0}^{4} y(n)\, z^{-n}$$

$$= y(0)\, z^{-0} + y(1)\, z^{-1} + y(2)\, z^{-2} + y(3)\, z^{-3} + y(4)\, z^{-4}$$

$$= 1 + z^{-1} + 2z^{-2} + z^{-3} + 3z^{-4} \qquad ...(2)$$

We know that

$$Y(z) = S(z)\, H(z)$$

or $$S(z) = \frac{Y(z)}{H(z)} \qquad ...(3)$$

Substituting Eqs. 1 and 2 in Eq. 3, we get

$$S(z) = \frac{1 + z^{-1} + 2z^{-2} - z^{-3} + 3z^{-4}}{1 + 2z^{-1} + 3z^{-2}} \qquad ...(4)$$

$$1 + 2z^{-1} + 3z^{-2} \;\big)\; 1 + z^{-1} + 2z^{-2} - z^{-3} + 3z^{-4} \;\big(\; 1 - z^{-1} + z^{-2}$$

$$\underset{-\quad\;\; -\quad\;\;\; -}{1 + 2z^{-1} + 3z^{-2}}$$

$$-z^{-1} - z^{-2} - z^{-3}$$

$$\underset{+\qquad +\qquad +}{-z^{-1} - 2z^{-2} - 3z^{-3}}$$

$$z^{-2} + 2z^{-3} + 3z^{-4}$$

$$\underset{-\qquad -\qquad -}{z^{-2} + 2z^{-3} + 3z^{-4}}$$

$$0$$

Therefore, $$S(z) = 1 - z^{-1} + z^{-2} \qquad ...(5)$$

Taking the inverse z-transform Eq. 5, we get

$$s(n) = \mathcal{Z}^{-1}\,[S(z)] = \mathcal{Z}^{-1}\,[1 - z^{-1} + z^{-2}] = \{\underset{\uparrow}{1}, -1, 1\}$$

Example 10.38 Find the inverse z-transform of the following $S(z)$ by the partial fraction expansion method

$$S(z) = \frac{z+2}{2z^2 - 7z + 3}$$

if the ROCs are

(a) $|z| > 3$ (b) $|z| < \frac{1}{2}$

(c) $\frac{1}{2} < |z| < 3$

Solution. We desire the partial fraction expansion of $\frac{S(z)}{z}$

$$S'(z) = \frac{S(z)}{z} = \frac{z+2}{z(2z^2 - 7z + 3)}$$

$$= \frac{z+2}{2z\left(z - \frac{1}{2}\right)(z-3)}$$

$$= \frac{A_1}{z} + \frac{A_2}{\left(z - \frac{1}{2}\right)} + \frac{A_3}{(z-3)}$$

(using partial fraction expansion)

$$S(z) = \frac{2}{3} - \frac{z}{z - \frac{1}{2}} + \frac{z/3}{z-3} \quad ...(1)$$

Here, the given function $S(z)$ has two poles $p_1 = 1/2$ and $p_2 = 3$ and it will have the following three inverse z-transforms.

(a) In the region $|z| > 3$, all poles are interior, *i.e.*, the signal $s(n)$ is causal, and therefore,

$$s(n) = \frac{2}{3}\delta(n) - \left(\frac{1}{2}\right)^n u(n) + \frac{1}{3}(3)^n u(n)$$

(b) In the region $|z| < \frac{1}{2}$, both the poles are exterior, *i.e.*, signal $s(n)$ is anticausal and hence

$$s(n) = \frac{2}{3}\delta(n) - \left(\frac{1}{2}\right)^n u(-n-1) - \frac{1}{3}(3)^n u(-n-1)$$

(c) In the region, $\frac{1}{2} < |z| < 3$, the pole $p_1 = \frac{1}{2}$ is interior and $p_2 = 3$ is exterior, and hence

$$s(n) = \frac{2}{3}\delta(n) - \left(\frac{1}{2}\right)^n u(n) - \frac{1}{3}(3)^n u(-n-1)$$

Example 10.39 A low-pass RC network is shown in Fig. 10.16. Find the equivalent discrete-time expressions for the circuit output respones $y(n)$, when the input is $s(t) = e^{-2t}$ and the sampling frequency is $f_s = 50$ Hz.

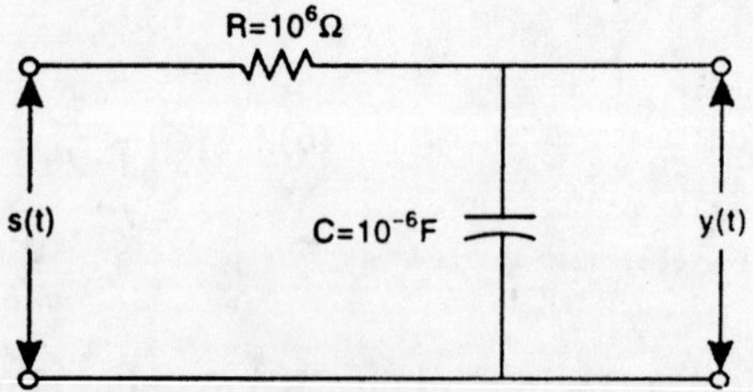

Fig. 10.16

Solution. The transfer function of the given network in the s-domain can be expressed as

$$H(s) = \frac{Y(s)}{S(s)} = \frac{1/RC}{s + \frac{1}{RC}} = \frac{1}{s+1} \quad \text{...(1)}$$

Taking the inverse Laplace transform, we get

$$h(t) = \text{Inverse Laplace transform } \{H(s)\}$$

$$= \text{Inverse Laplace transform } \left\{\frac{1}{s+1}\right\}$$

$$= e^{-t} \quad \text{...(2)}$$

Now we use z-transform approach.

Putting $t = nT_s$ in Eq. 2, we get

$$h(nT_s) = e^{-nT_s} \quad \text{...(3)}$$

Taking the z-transform of Eq. 3, we get

$$H(z) = \mathcal{Z}\,[h(nT_s)] = \mathcal{Z}\,[e^{-nT_s}]$$

$$= \frac{z}{z - e^{-T_s}} \quad \text{...(4)}$$

We have given input

$$s(t) = e^{-2t} \quad \text{...(5)}$$

Putting $t = nT_s$ in Eq. 5, we get

$$s(nT_s) = e^{-2nT_s} \quad \text{...(6)}$$

Taking z-transform of Eq. 6, we get

$$S(z) = \mathcal{Z}\,[s(nT_s)] = \mathcal{Z}\,[e^{-2nT_s}]$$

$$= \frac{z}{z - e^{-2T_s}} \quad \text{...(7)}$$

We know that

$$Y(z) = H(z)\, S(z) \qquad ...(8)$$

Substituting Eqs. 4 and 7 in Eq. 8, we get

$$Y(z) = \frac{z}{(z - e^{-T_s})} \frac{z}{(z - e^{-2T_s})}$$

or
$$\frac{Y(z)}{z} = \frac{z}{(z - e^{-T_s})(z - e^{-2T_s})}$$

$$= \frac{A_1}{(z - e^{-T_s})} + \frac{A_2}{(z - e^{-2T_s})}$$

(using partial fraction expansion)

or
$$\frac{Y(z)}{z} = \frac{\left(\frac{1}{1 - e^{-T_s}}\right)}{(z - e^{-T_s})} + \frac{\left(\frac{1}{1 - e^{T_s}}\right)}{(z - e^{-2T_s})}$$

or
$$Y(z) = \frac{1}{(1 - e^{-T_s})} \frac{z}{(z - e^{-T_s})} + \frac{1}{(1 - e^{T_s})} \frac{z}{(z - e^{-2T_s})} \qquad ...(9)$$

Taking inverse z-transform of Eq. 9, we get

$$y(nT_s) = \frac{1}{(1 - e^{-T_s})} (e^{-nT_s}) + \frac{1}{(1 - e^{T_s})} (e^{-2nT_s}) \qquad ...(10)$$

Substituting $T_s = \frac{1}{50} = 0.02$ in Eq. 10, we get

$$y(n) = 50.5(0.980)^n - 49.5(0.961)^n$$

Thus, the required output response is

$$y(n) = 50.5(0.980)^n - 49.5(0.961)^n$$

REVIEW QUESTIONS

1. Distinguish z-transform from discrete-time Fourier transform (DTFT).
2. Define z-transform. Give the relationship between z-transform and DTFT.
3. Distinguish z-transform from Laplace transform.
4. What do you mean by region of convergence for the z- transform ?
5. Discuss the following types of z-transform :
   (a) Bilateral or two-sided z-transform
   (b) Unilateral or one-sided z-transform
6. What are various configurations of ROCs for the z- transform ?
7. Write the important properties of ROC for the z-transform.
8. Summarize the important properties of ROC for the z-transform.
9. Discuss important properties of z-transform in brief.
10. Tabulate important properties of z-transform.
11. Tabulate some common z-transform pairs.

12. What is inverse z-transform ?
13. Discuss the following methods of inverse z-transformation :
    (*a*) Contour integration method
    (*b*) Power series expansion method
    (*c*) Partial fraction expansion method
14. Tabulate inverse z-transform.
15. Discuss important properties of z-transform in analysis of the discrete-time LTI systems.
16. State and explain the following :
    (*a*) Causality of discrete-time LTI systems
    (*b*) Stability of discrete-time LTI systems
17. Discuss the characterization of discrete-time LTI systems with the help of linear constant-coefficient difference equations.
18. Discuss determination of poles and zeros of a rational z- transform.
19. Describe the system behaviour related to the transfer function of the discrete-time system.
20. Explain the block diagram representation for discrete-time LTI system.
21. Find the transfer function of the following types of interconnections of discrete-time LTI systems.
    (*a*) Series or cascasde interconnections
    (*b*) Parallel interconnections
    (*c*) Feedback interconnections
22. Define the unilateral z-transform.
23. Discuss special properties of unilateral z-transform.
24. Tabulate important properties of unilateral z-transform.
25. Discusss applications of unilateral z-transform in solving difference equations of discrete-time LTI systems.

NUMERICAL PROBLEMS

1. Find the z-transform of the following discrete-time signals
    (*a*) $s(n) = \sin \omega n$ (*b*) $s(n) = \cos \omega n$
2. Find the z-transform of $s(n) = n^2 e^{-\alpha n}$

    Ans. $S(z) = \dfrac{2[z^{-4} - z^{-3}(1 - e^{-\alpha} z^{-1})]}{(1 - e^{-\alpha} z^{-1})}$
3. Find the z-transform of the discrete-time signal $s(n) = -A^n u(-n-1)$. Also sketch its ROC.

    Ans. $S(z) = \dfrac{1}{1 - Az^{-1}}$, ROC $\Rightarrow |z| < |A|$
4. Find the z-transforms and ROCs for the following discrete- time signals
    (*a*) $s(n) = e^{-3n} u(n-1)$ (*b*) $s(n) = n^2 u(n)$

(c) $s(n) = \begin{cases} 0\,, & \text{for } n < 0 \\ 1\,, & \text{for } 0 \le n \le 5 \\ 2\,, & \text{for } 6 \le n \le 10 \\ 3\,, & \text{for } n > 10 \end{cases}$

Ans. (a) $S(z) = \dfrac{1}{1 - e^{-3}z^{-1}}$, ROC : $|z| > |e^{-3}|$ (b) $S(z) = \dfrac{z^{-2}}{(1 - z^{-1})^2}$, ROC : $|z| > 1$ (c) $S(z) = 1 + z^{-1} + z^{-2} + z^{-3} + z^{-4} + z^{-5} + 2(z^{-6} + z^{-7} + z^{-8} + z^{-9} + z^{-10}) + 3(z^{-11} + z^{-12} + ...)$

5. Find the z-transform of the following discrete-time sequences

(a) $s(n) = u(n-4)$ (b) $s(n) = e^{jn\pi/4}\, u(n)$

(c) $s(n) = \delta(n-5)$ (d) $s(n) = \left(\dfrac{1}{3}\right)^n u(-n)$

Ans. (a) $S(z) = \dfrac{z^{-4}}{1 - z^{-1}}$, $|z| > 1$ (b) $S(z) = \dfrac{1}{1 - e^{jn\pi/4} z^{-1}}$ (c) $S(z) = z^{-5}$

(d) $S(z) = \dfrac{1}{1 - 3z}$, $|z| < \dfrac{1}{3}$

6. Find the bilateral or two-sided z-transform of the sequence

$s(n) = \begin{cases} (1/3)^n\,, & n \ge 0 \\ (-2)^n\,, & n < 0 \end{cases}$ **Ans.** $S(z) = \dfrac{z}{z - 1/3} - \dfrac{z}{z + 2}$

7. Use convolution sum to determine discrete-time signal $s(n)$ if $S(z)$ is given by

$$S(z) = \frac{1}{(1 - 1/2\, z^{-1})\,(1 + 1/4\, z^{-1})}$$

Ans. $s(n) = \dfrac{2}{3}\left(\dfrac{1}{2}\right)^n u(n) + \dfrac{1}{3}\left(-\dfrac{1}{4}\right)^n u(n)$

8. Determine the output $y(n)$ of a discrete-time LTI system using convolution property of z-transform when input $s(n) = \{1, 2, 3, 1, -1, 1\}$ and impulse response $h(n) = \{1, 1, 1\}$

Ans. $y(n) = \{\underset{\uparrow}{1}\,, 3\,, 6\,, 6\,, 3\,, 1\,, 0\,, 1\}$

9. Find the convoluton sum of discrete-time signals $s(n)$ and $h(n)$ using z-transform method where

$$s(n) = \begin{cases} 0\,, & n < 0 \\ \alpha^n\,, & n \ge 0 \end{cases} \quad \text{and} \quad h(n) = \begin{cases} 0\,, & n < 0 \\ \beta^n\,, & n \ge 0 \end{cases}$$

Specify the answers if

(a) $\alpha \ne \beta$ (b) $\alpha = \beta$

Ans. (a) $y(n) = \dfrac{1}{\alpha - \beta}\{\alpha^n - \beta^n\}\, u(n)$ (b) $y(n) = n\, \alpha^{n-1}\, u(n)$

10. Determine the cross-correlation sequence $r_{s_1 s_2}(l)$ of the following discrete-time sequences

$$s_1(n) = \{2, -1, 3, 7, 1, 2, -3\}$$
$$\uparrow$$

$$s_2(n) = \{1, -1, 2, -2, 4, 1, -2, 5\}$$
$$\uparrow$$

Ans. $r_{s_1 s_2}(l) = \{10, -9, 19, 36, -14, 33, 0, 7, 13, -18, 16, -7, 5, -3\}$

11. Find the autocorrelation sequence of the signal

$$s(n) = A^n u(n) \text{ for } -1 < A < 1$$

Ans. $r_{ss}(l) = \dfrac{1}{1-A^2} A^{|l|}$, for $-\infty < l < \infty$

12. Determine the final value of the signal corresponding to the z-transform

$$S(z) = \frac{2z^{-1}}{1 - 1.8z^{-1} + 0.8z^{-2}}$$

Ans. $s(\infty) = 10$

13. Find the inverse z-transform of the following

(a) $S(z) = \dfrac{z^2}{(z-1)(z-0.2)}$ (b) $S(z) = \dfrac{z}{(z-1)^3}$

(c) $S(z) = \dfrac{1}{1-4z^{-1}}$

Ans. (a) $s(n) = [1.25 - 0.25\,(0.2)^n]\, u(n)$ (b) $s(n) = \dfrac{n(n-1)\,u(n)}{2}$

(c) $s(n) = (4)^n\, u(n)$

14. Find the inverse z-transform for $S(z) = \dfrac{z^{-1}}{1-3z^{-1}}$ with ROC : $|z| < 3$.

Ans. $s(n) = (-3)^n\, u(n)$

15. Determine the discrete-time sequence $s(n) \geq 0$ given that

$$S(z) = \frac{1}{(1 - 1/2\, z^{-1})(1 - 1/4\, z^{-1})}$$

Ans. $s(n) = \left[2\left(\dfrac{1}{2}\right)^n - \left(\dfrac{1}{4}\right)^n\right] u(n)$

16. By using partial fraction expansion method, determine $s(n)$ for

$$S(z) = \frac{1 - z^{-1}}{(1 + z^{-1})(1 + 1/2\, z^{-1})}$$

Ans. $s(n) = \left[4(-1)^n - 3\left(-\dfrac{1}{2}\right)^n\right] u(n)$

17. Determine $s(n)$ if $S(z) = \dfrac{z^{-3}}{1 - 1/4\, z^{-1}}$. **Ans.** $s(n) = \left(\dfrac{1}{4}\right)^{n-3} u(n-3)$

18. By using long division method, determine the inverse z- transform of

$$S(z) = \frac{z}{3z^2 - 4z + 1}$$

where the ROC is (a) $|z| > 1$ and (b) $|z| < \frac{1}{3}$

Ans. (a) $s(n) = \left\{ \underset{\uparrow}{0}, \frac{1}{3}, \frac{4}{9}, \frac{13}{27}, \frac{40}{81}, \ldots \right\}$

(b) $s(n) = \{\ldots 121, 40, 13, 4, 1, \underset{\uparrow}{0}\}$

19. By using long division method, find the inverse z-transform of

$$S(z) = \frac{1 + 2z^{-1}}{1 - 2z^{-1} + z^{-2}}$$

if (a) sequence $s(n)$ is causal and (b) sequence $s(n)$ is anticausal.

Ans. (a) $s(n) = \{\underset{\uparrow}{1}, 4, 7, 10, 13, \ldots\}$ (b) $s(n) = \{\ldots, 14, 11, 8, 5, 2, \underset{\uparrow}{0}\}$

20. Find the inverse z-transform by using

(a) partial fraction expansion method

(b) long-division method

(c) residue method of the

$$S(z) = \frac{1.8z\,(z + 0.1667)}{z^2 - 0.4z - 0.05}$$

Also verfiy the results in each case for n in the range $0 \le n \le 3$.

Ans. (a) $s(n) = 2(0.5)^n - 0.2(-0.1)^n$; $s(0) = 1.8$, $s(1) = 1.02$, $s(2) = 0.498$, $s(3) = 0.2502$ (b) $s(0) = 1.8$, $s(1) = 1.02$, $s(2) = 0.498$, $s(3) = 0.2502$ (c) $s(n) = 2(0.5)^n - 0.2(-0.1)^n$; $s(0) = 1.8$, $s(1) = 1.02$, $s(2) = 0.498$, $s(3) = 0.2502$

11

Characterization of Signals and Systems in Time and Frequency-Domain

11.1 INTRODUCTION

The frequency-domain characterization of a (continuous-time/discrete-time) LTI system in terms of its frequency response represents an alternative to the time-domain characterization through convolution operation. We often conveniently utilize the frequency-domain for analyzing LTI systems because differential and difference equations and convolution operations in the time- domain become algebraic operations (such as addition, multiplication etc.) in the frequency-domain. Frequency selective filtering (low-pass filtering, high-pass filtering, band-pass filtering and band-reject filtering) are readily and simply visualized in the frequency-domain. In any system design, we consider both time-domain and frequency-domain characteristics. The significant oscillatory behaviour in the impulse response (time-domain response) of a frequency-selective filter (low-pass, high-pass, band-pass and band reject filter) is undesirable. Consequently, we wish to sacrifice the level of frequency selectivity in a frequency selective filter in order to obtain the required tolerances on impulse response behaviour.

In practice, in most applications, we like to specify certain characteristics of a system in both time-domain and frequency-domain.

Both time-domain and frequency-domain characteristics of a system frequently result in conflicting requirements. Therefore, in system design and analysis, it is important to relate time-domain and frequency-domain characteristics and their trade-offs.

11.2 THE MAGNITUDE-PHASE REPRESENTATION OF CTFT AND DTFT

Both continuous-time Fourier transform (CTFT) and discrete-time Fourier transform (DTFT) are in general complex valued. Hence, both

CTFT and DTFT can be represented in terms of their real and imaginary components. Also, these Fourier transforms can be represented in terms of their magnitude and phase.

Continuous-time Fourier transform (CTFT) of a continuous-time signal $s(n)$ is defined as

$$S(j\omega) = \int_{-\infty}^{\infty} s(t)\, e^{-j\omega t}\, dt \qquad ...(11.1)$$

where $S(j\omega)$ is the CTFT of continuous-time signal $s(n)$ and $S(j\omega)$ is a complex-valued function of ω. It is expressed in terms of real and imaginary components as

$$S(j\omega) = S_R(j\omega) + j\, S_I(j\omega) \qquad ...(11.2)$$

The magnitude-phase representation of the CTFT $S(j\omega)$ is expressed as

$$S(j\omega) = |S(j\omega)|\, e^{j\underline{/S(j\omega)}} \qquad ...(11.3)$$

where $|S(j\omega)|$ and $\underline{/S(j\omega)}$ are the magnitude and phase of $S(j\omega)$, respectively.

Magnitude of $S(j\omega)$ is expressed as

$$|S(j\omega)| = \sqrt{[S_R(j\omega)]^2 + [S_I(j\omega)]^2} \qquad ...(11.4)$$

Phase of $S(j\omega)$ is expressed as

$$\underline{/S(j\omega)} = \tan^{-1}\left[\frac{S_I(j\omega)}{S_R(j\omega)}\right] \qquad ...(11.5)$$

Discrete-time Fourier transform (DTFT) of discrete-time signal $s(n)$ is defined as

$$S(e^{j\omega}) = \sum_{-\infty}^{\infty} s(n)\, e^{-j\omega n} \qquad ...(11.6)$$

where $S(e^{j\omega})$ is the DTFT of the discrete-time signal $s(n)$ is a complex-valued function of ω. It is expressed in terms of real and imaginary components as

$$S(e^{j\omega}) = S_R(e^{j\omega}) + j\, S_I(e^{j\omega}) \qquad ...(11.7)$$

The magnitude-phase representation of the DTFT $S(e^{j\omega})$ is expressed as

$$S(e^{j\omega}) = |S(e^{j\omega})|\, e^{j\underline{/S(e^{j\omega})}} \qquad ...(11.8)$$

where $|S(e^{j\omega})|$ and $\underline{/S(e^{j\omega})}$ are the magnitude and phase of $S(e^{j\omega})$ respectively. Magnitude of $S(e^{j\omega})$ is expressed as

$$|S(e^{j\omega})| = \sqrt{[S_R(e^{j\omega})]^2 + [S_I(e^{j\omega})]^2} \qquad ...(11.9)$$

Phase of $S(e^{j\omega})$ is expressed as

$$\underline{/S(e^{j\omega})} = \tan^{-1}\left[\frac{S_I(e^{j\omega})}{S_R(e^{j\omega})}\right] \qquad ...(11.10)$$

In the remaining part of this section, we will concentrate on the continuous-time case in describing and illustrating several important points related to magnitude-phase representations. These important points can be equally applied to discrete-time case as done above.

Continuous-time Fourier transform (CTFT) synthesis equation $s(t) = \frac{1}{2\pi}\int_{-\infty}^{\infty} S(j\omega)\, e^{j\omega t}\, d\omega$ provides a decomposition of the signal $s(t)$ into sum of complex exponentials at different frequencies. Infact $|S(j\omega)|^2$ is interpreted as the energy-density spectrum (eds) of $s(t)$. In other words, we can say that $\frac{1}{2\pi}|S(j\omega)|^2\, d\omega$ can be thought of as the amount of energy in the continuous-time signal $s(t)$ that lies in the infinitesimal frequency band $d\omega$ between ω and $\omega + d\omega$. The term $|S(j\omega)|$ provides us information about the relative magntiude of the complex exponential functions that make up the signal $s(t)$. Therefore, magnitude $|S(j\omega)|$ describes the basic frequency content of a continuous-time signal $s(t)$. If magnitude of $S(j\omega)$ *i.e.*, $|S(j\omega)|$ is zero outside a small band of frequencies centered at zero, then $s(t)$ will show only relatively low-frequency oscillations.

The phase angle $\underline{/S(j\omega)}$ does not affect the amplitudes of the individual frequency components. It provides us information concerning the relative phases of these complex exponentials. Phase relationships have a significant effect on the nature of the signal $s(t)$. Thus, phase angle contain a substantial amount of information about the signal. If the phases are different but the magnitudes are same then we can obtain very different looking signals. For example, if we have a signal that encounters the superposition of three wave trains, each of which can be modelled as a sinusoidal signal. With fixed magnitudes for these three sinusoidal signals, the amplitude of their sum may be quite small or very large, depending on the relative phases between different components of the signal $s(t)$. In general, changes in the phase function of $S(j\omega)$ *i.e.*, $\underline{/S(j\omega)}$ lead to changes in time-domain characteristics of the continuous- time signal $s(t)$. In many applications, phase distortion is very important and in many other applications, it is not so. For example, the human auditory system is relatively insensitive to phase. Specifically, if the CTFT of a spoken sound (*e.g.*, a vowel) is subjected to distortion such that the phase is changed but the magnitude is unchanged. In this case, the effect can be perceptually negligible although the waveform of spoken sound in time-domain may look considerably different. The mild phase distortions (such as those affecting individual sounds) do not lead to a loss of intelligbility but more severe phase distortions of speech affects intelligibility. If $s(t)$ is a tape recording signal (*i.e.*, sentence), then signal $s(-t)$ will represent the sentence played back. In other words, we can say that $s(-t)$ is a folded version or mirror image of a signal $s(t)$. We assume that signal $s(t)$ is a real-valued signal, the corresponding effect in the frequency-domain is to replace the CTFT phase by its negative that is

$$\text{CTFT}\{s(-t)\} = S(-j\omega) = |S(j\omega)|\, e^{-j\angle S(j\omega)} \qquad ...(11.11)$$

Thus, the spectrum of a spoken sentence played in reverse direction or played back $s(-t)$ has the same magntiude function as the spectrum of the original sentence $s(t)$ but it differs only in phase. This phase change has a significant impact on the intelligibility of the signal that was recorded.

11.3 THE MAGNITUDE-PHASE REPRESENATATION OF THE FREQUENCY RESPONSE OF CONTINUOUS-TIME AND DISCRETE-TIME LTI SYSTEMS

From the convolution property for CTFT, the output of a continuous-time LTI system can be computed as

$$y(t) = h(t) * s(t) \xleftrightarrow{\text{CTFT}} Y(j\omega) = H(j\omega)\, S(j\omega)$$

$$\text{or} \qquad Y(j\omega) = H(j\omega)\, S(j\omega) \qquad ...(11.12)$$

where $Y(j\omega)$ and $S(j\omega)$ are the CTFTs of the output $y(n)$ and input $s(n)$, respectively. $H(j\omega)$ is the frequency response of the continuous-time LTI system, *i.e.*, CTFT of the impulse response $h(t)$ of the continuous-time LTI system. The impulse response of a continuous-time LTI system is the response (output) to a system with input equal to unit- impulse function or Dirac-Delta function $\delta(t)$.

Similarly, from the convolution property of DTFT, the output of a discrete-time LTI system can be computed as

$$y(n) = h(n) * s(n) \xleftrightarrow{\text{DTFT}} Y(e^{j\omega}) = H(e^{j\omega})\, S(e^{j\omega})$$

$$\text{or} \qquad Y(e^{j\omega}) = H(e^{j\omega})\, S(e^{j\omega}) \qquad ...(11.13)$$

where $Y(e^{j\omega})$ and $S(e^{j\omega})$ are the DTFTs of the output $y(n)$ and input $s(n)$, respectively. $H(e^{j\omega})$ is the frequency response of discrete-time LTI system, *i.e.*, DTFT of the impulse response $h(n)$ of the discrete-time LTI system. The impulse response of a discrete-time LTI system is also called unit-sample response. The impulse response of a discrete-time LTI system is the response (output) to a system with input equal to unit-sample sequence or discrete- time unit-impulse function $\delta(n)$.

The effect that an LTI system has on the input is to change the complex amplitude of each of the frequency components of the signal. Here we will look at this effect in terms of the magntiude-phase representation. By doing so, we will understand the nature of this effect in more detail.

Magnitude-phase Representation for Continuous-Time LTI System

Magnitude-phase representation for $S(j\omega)$, $Y(j\omega)$ and $H(j\omega)$ can be expressed as

$$S(j\omega) = |S(j\omega)|\, e^{j\angle S(j\omega)} \quad ...(11.14)$$

$$Y(j\omega) = |Y(j\omega)|\, e^{j\angle Y(j\omega)} \quad ...(11.15)$$

$$H(e^{j\omega}) = |H(j\omega)|\, e^{j\angle H(j\omega)} \quad ...(11.16)$$

Substituting Eqs. 11.14, 11.15 and 11.16 in Eq. 11.12, we get

$$Y(j\omega) = H(j\omega)\, S(j\omega)$$

or $\quad |Y(j\omega)|\, e^{j\angle Y(j\omega)} = \left\{ |H(j\omega)|\, e^{j\angle H(j\omega)} \right\}\left\{ |S(j\omega)|\, e^{j\angle S(j\omega)} \right\}$

or $\quad |Y(j\omega)|\, e^{j\angle Y(j\omega)} = |H(j\omega)|\ |S(j\omega)|\, e^{j[\angle H(j\omega) + \angle S(j\omega)]}$

On comparing both sides, we get

$$|Y(j\omega)| = |H(j\omega)|\ |S(j\omega)| \quad ...(11.17)$$

and $\quad Y(j\omega) = \angle H(j\omega) + \angle S(j\omega) \quad ...(11.18)$

On the same lines we can also obtain an exactly analogous relationship for discrete-time LTI systems. Magnitude-phase representation for $S(e^{j\omega})$, $Y(e^{j\omega})$ and $H(e^{j\omega})$ can be expressed as

$$S(e^{j\omega}) = |S(e^{j\omega})|\, e^{j\angle S(e^{j\omega})} \quad ...(11.19)$$

$$Y(e^{j\omega}) = |Y(e^{j\omega})|\, e^{j\angle Y(e^{j\omega})} \quad ...(11.20)$$

$$H(e^{j\omega}) = |H(e^{j\omega})|\, e^{j\angle H(e^{j\omega})} \quad ...(11.21)$$

Substituting Eqs. 11.19, 11.20 and 11.21 in Eq. 11.13, we obtain

$$Y(e^{j\omega}) = H(e^{j\omega})\, S(e^{j\omega})$$

or $\quad |Y(e^{j\omega})|\, e^{j\angle Y(e^{j\omega})} = \left\{ |H(e^{j\omega})|\, e^{j\angle H(e^{j\omega})} \right\}\left\{ |S(e^{j\omega})|\, e^{j\angle S(e^{j\omega})} \right\}$

or $\quad |Y(e^{j\omega})|\, e^{j\angle Y(e^{j\omega})} = |H(e^{j\omega})|\ |S(e^{j\omega})|\, e^{j[\angle H(e^{j\omega}) + \angle S(e^{j\omega})]}$

On comparing both sides, we get

$$|Y(e^{j\omega})| = |H(e^{j\omega})|\ |S(e^{j\omega})| \quad ...(11.22)$$

and $\quad \angle Y(e^{j\omega}) = \angle H(e^{j\omega}) + \angle S(e^{j\omega}) \quad ...(11.23)$

From Eqs. 11.17 and 11.22, we see that the magnitude of the Fourier transform (CTFT or DTFT) of the output is equal to multiplication of magnitude of the Fourier transform (CTFT or DTFT) of the input with the magnitude of the Fourier transform (CTFT or DTFT) of impulse response. For this reason, $|H(j\omega)|$ or $|H(e^{j\omega})|$ is commoly referred to as the gain of the system. Also, from Eqs. 11.18 and 11.23, we see that the phase of the Fourier transform (CTFT or DTFT) of the output of an LTI system is equal to the sum of its impulse response and its input. For this response $\angle H(j\omega)$ or $\angle H(e^{j\omega})$ is commonly called phase shift of the system.

The phase shift of the system can change the relative phase relationships among the components of the input. The change in relative phase relationships results in significant modifications in the time-domain characteristics of the input even when the gain of the system is constant for all frequencies. The changes in the magnitude and phase which result from the application of an input to an LTI system may be either desirable or undesirable. If the signal is modified in a useful way then the changes in the magntiude and phase are desirable. If the signal is modified in an unwanted manner then the changes in the magnitude and phase are undesirable otherwise the changes are undesirable.

11.3.1 Linear and Non-Linear Phases of LTI Systems

When the phase shift at the frequency ω is a linear function of frequency, there is straight-forward interpretation of the effect in the time-domain. Now let us consider the continuous-time LTI system with frequency response.

$$H(j\omega) = \text{CTFT of } h(t) = e^{-j\omega t_0} \qquad ...(11.24)$$

or

$$H(j\omega) = \cos \omega t_0 - j \sin \omega t_0 \qquad ...(11.25)$$

Magnitude of $H(j\omega)$ is also called gain of the system and is determined as

$$\begin{aligned} |H(j\omega)| &= \sqrt{[H_R(j\omega)]^2 + [H_I(j\omega)]^2} \\ &= \sqrt{[\cos \omega t_0]^2 + [-\sin \omega t_0]^2} \\ &= \sqrt{\cos^2 \omega t_0 + \sin^2 \omega t_0} \\ &= \sqrt{1} = 1 \end{aligned}$$

or

$$|H(j\omega)| = 1 \qquad ...(11.26)$$

Phase-shift of the system is determined as

$$\begin{aligned} \angle H(j\omega) &= \tan^{-1}\left[\frac{H_I(j\omega)}{H_R(j\omega)}\right] \\ &= \tan^{-1}\left[\frac{-\sin \omega t_0}{\cos \omega t_0}\right] \\ &= \tan^{-1}[-\tan \omega t_0] = -\omega t_0 \end{aligned}$$

or

$$\angle H(j\omega) = -\omega t_0 \qquad ...(11.27)$$

The system with frequency response given in Eq. 11.24 has linear phase.

The system with the frequency response characteristc given in Eq. 11.24 produces an output that is simply a time-shift of the input. It is expressed as

$$y(t) = s(t - t_0) \qquad ...(11.28)$$

The effect of linear phase in discrete-time case is similar to that in the continuous-time case when the slope of the linear phase is an integer. Now we have a discrete-time LTI system with frequency response given as

$$H(e^{j\omega}) = \text{DTFT of } h(n) = e^{-j\omega n_0} \quad ...(11.29)$$

or $$H(e^{j\omega}) = \cos \omega n_0 - j \sin n\omega_0 \quad ...(11.30)$$

Magnitude of $H(e^{j\omega})$ is also called gain of the system and is determined as

$$|H(e^{j\omega}) = \sqrt{[H_R(e^{j\omega})]^2 + [H_I(e^{j\omega})]^2}$$

$$= \sqrt{[\cos \omega n_0]^2 + [-\sin \omega n_0]^2}$$

$$= \sqrt{\cos^2 \omega n_0 + \sin^2 \omega n_0}$$

$$= \sqrt{1} = 1$$

or $$|H(e^{j\omega})| = 1 \quad ...(11.31)$$

Phase-shift of the system is determined as

$$H(e^{j\omega}) = \tan^{-1}\left[\frac{H_I(e^{j\omega})}{H_R(e^{j\omega})}\right] = \tan^{-1}\left[\frac{-\sin \omega n_0}{\cos \omega n_0}\right]$$

$$= \tan^{-1}[-\tan \omega n_0] = -\omega n_0$$

or $$\underline{/H(e^{j\omega})} = -\omega n_0 \quad ...(11.32)$$

The system with frequency response given in Eq. 11.29 has linear phase. The system with the frequency response characteristic given in Eq. 11.29 produces an output that is simply a time-shift of the input. It is expressed as

$$y(n) = s(n - n_0) \quad ...(11.33)$$

Thus, a linear phase shift with an integer slope corresponds to a shift of input $s(n)$ by an integer number of samples. When the phase slope is not an integer, the effect in the time-domain is somewhat more complex. The effect is a time-shift of the envelope of the discrete-time sequence $s(n)$ values, but the values themselves may change. A linear phase shift leads to very simple and easily understood and visualised changes in a signal. If an input signal is subjected to a non-linear phase shift that is function of ω, then the complex exponential components of the input at different frequencies will be shifted. It will shift in a manner that results in a change in their relative phases When these exponential functions are superimposed, we can obtain a signal that may look considerably different from the input signal. The effect of phase shift on a continuous-time signal is shown in Fig. 11.1. A continuous-time signal $s(t)$ that is applied as input to three different systems, is shown in Fig. 11.1 (*a*).

Fig. 11.1 (*b*) shows the output when the signal $s(t)$ is applied as input to a continuous-time system with frequency response $H_1(j\omega) = e^{-j\omega t_0}$. This

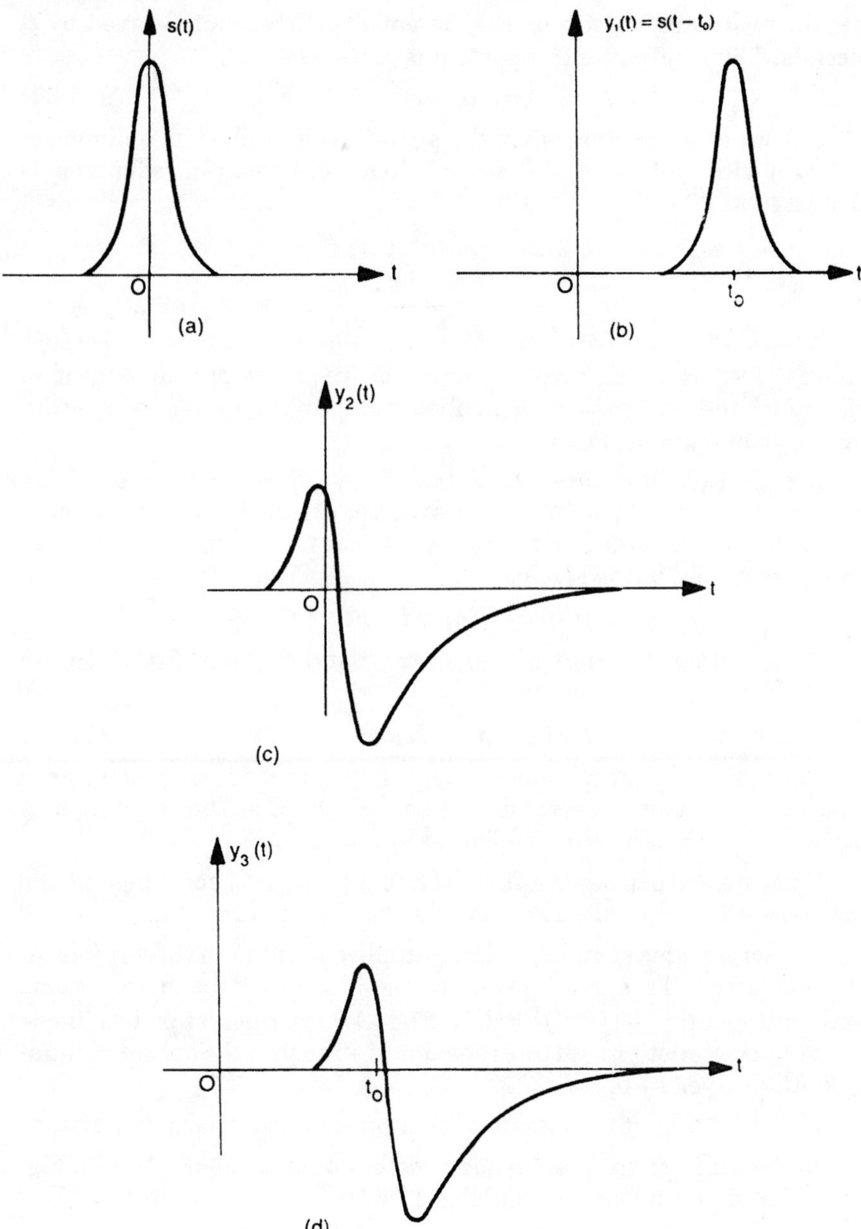

Fig. 11.1 *(a) Continuous-time signal s(t) that is applied as the input to various continuous-time systems. The frequency response of these systems have unity magnitude,* $|H(j\omega)| = 1$. *(b) Response (output) for a system with linear phase. (c) Response (output) for a system with non-linear phase. (d) Response (output) for a system with linear phase of part (b) plus non-linear phase of part (c).*

system results in an output that is equal to the input delayed by t_0 seconds. Thus, output of this system is given as

$$y_1(t) = s_1(t - t_0) \quad \text{...(11.34)}$$

Output of the system when the signal $s(t)$ is applied as an input to a system with unity gain, $|H(j\omega)| = 1$ then non-linear phase function is expressed as

$$H_2(j\omega) = |H_2(j\omega)| \underline{/e^{j\,H_2(j\omega)}}$$

$$= 1\underline{/e^{j\,H_2(j\omega)}}$$

or $$H_2(j\omega) = \underline{/e^{j\,H_2(j\omega)}} \quad \text{...(11.35)}$$

where $\underline{/H_2(j\omega)}$ is a non-linear function of frequency ω. The output of system when the signal $s(t)$ is applied as input to a system with unity gain is shown in Fig. 11.1 (*c*).

Fig. 11.1 (*d*) illustrates the output from another non-linear phase system. For this system, the corresponding frequency response has a phase shift that is obtained by adding a linear phase term to non-linear phase term $\underline{/e^{j\,H_2(j\omega)}}$. It is given as

$$H_3(j\omega) = H_1(j\omega)\, H_2(j\omega) \quad \text{...(11.36)}$$

Substituting the value of $H_1(j\omega) = e^{-j\omega t_0}$ and $H_2(j\omega)$ in Eq. 11.36, we get

$$H_3(j\omega) = e^{-j\omega t_0}\, H_2(j\omega) \quad \text{...(11.37)}$$

Thus, the output of combination of linear phase and non-linear phase continuous-time systems can be thought of as the response to a cascade of the system $H_2(j\omega)$ followed by a time-shift.

Now, we will discuss the effect of linear phase and non-linear phase on discrete-time signals. These effects are shown in Fig. 11.2.

A discrete-time signal $s(n)$ that is applied as input to three different discrete-time LTI systems is shown in Fig. 11.2 (*a*). All the three systems have unity gain *i.e.*, $|H(e^{j\omega})| = 1$. In Fig. 11.2 (*b*), the system has linear phase characteristics with integer slope of -5 so that the output is equal to input delayed by 5.

In Fig. 11.2 (*c*), the system has non-linear phase characteristics.

In Fig. 11.2 (*d*), the system has a combination of linear- phase of Fig. 11.2 (*b*) and non-linear phase of Fig. 11.2 (*c*).

Note that all systems (continuous-time/discrete-time) which were discussed and illustrated in Fig. 11.1 and Fig. 11.2 have unity gain or unity magnitude. Hence magnitude of the Fourier transform of the input $s(t)$ or $s(n)$ to any of these systems is passed through unchanged by the system. That is why, these systems are called *all-pass systems*. The characteristics of an all-pass system are completely determined by its phase-shift characteristics.

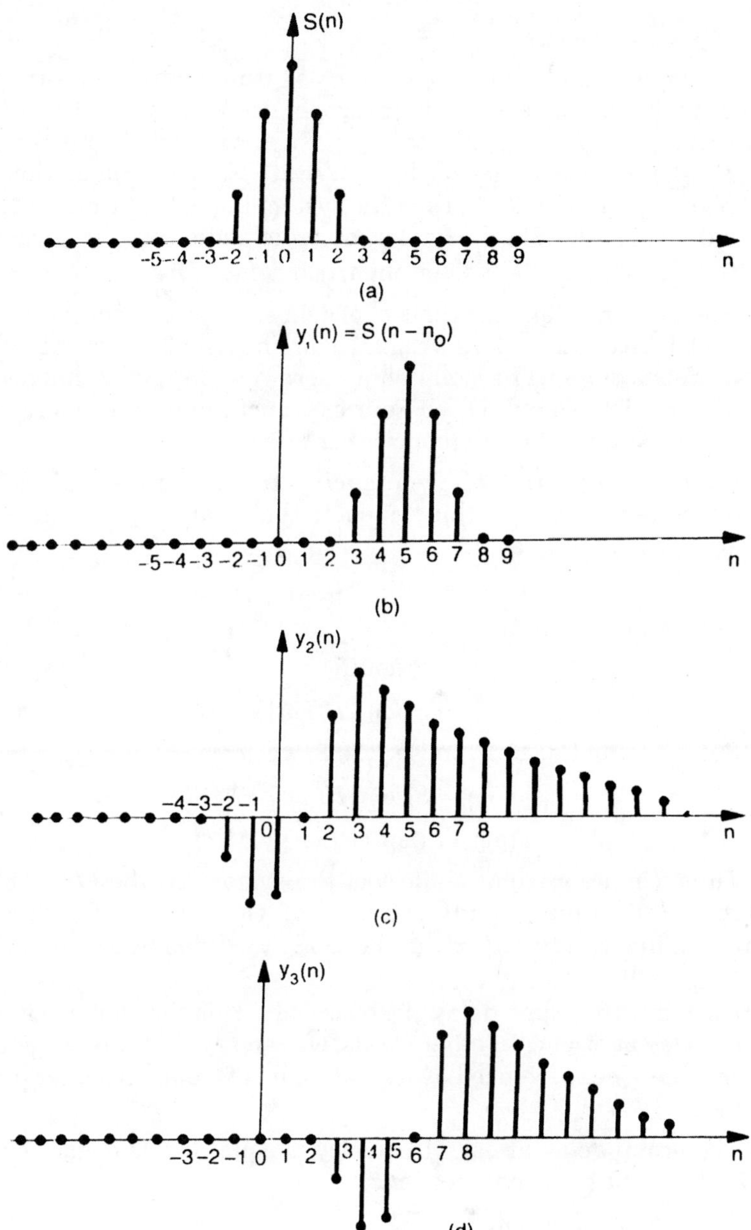

Fig. 11.2 (*a*) *Discrete-time signal s*(*n*) *that is applied as an input to various discrete-time systems. The frequency response of these systems has unity magnitude,* $|H(e^{j\omega})| = 1$. (*b*) *Response (output) for a system with linear phase with slope of* –5. (*c*) *Response (output) for a system with non-linear phase.* (*d*) *Response (output) for a system whose phase characteristic is combination of linear-phase of part* (*b*) *and non-linear phase part* (*c*).

11.3.2 Group Delay

We have already discussed that systems with linear phase characteristics have simple interpretation as time-shifts. From equations $y(t) = s(t - t_0)$ and $y(n) = s(n - n_0)$, we know about the phase slope that tells us the size of the time-shift. For continuous-time case, if phase-shift $H(j\omega) = -\omega t_0$, then the system imparts of time-shift of $-t_0$ or a delay of t_0. Similarly, for discrete-time case, if phase-shift $H(e^{j\omega}) = -\omega n_0$, then the system imparts a delay of n_0.

Now we can extend the concept of delay to a system with non- linear phase characteristics. Here we shall examine the effects of the phase of a continuous-time LTI system on a narrowband input. A narrowband input signal is a signal whose Fourier transform is zero or very small outside a small band of frequencies centered at $\omega = \omega_0$.

Now taking band of frequencies to be very small, we can approximate the phase of this system in these band of frequencies with linear approximation as

$$H(j\omega) \approx -\phi - \omega\alpha \qquad ...(11.37)$$

Also, we know that

$$Y(j\omega) \approx S(j\omega)\, H(j\omega)$$

$$\approx S(j\omega)\, |H(j\omega)|\, e^{j\angle H(j\omega)} \qquad ...(11.38)$$

Substituting Eq. 11.37 in Eq. 11.38, we get

$$Y(j\omega) \approx S(j\omega)\, |H(j\omega)|\, e^{j(-\phi - \omega\alpha)}$$

or

$$Y(j\omega) \approx S(j\omega)\, |H(j\omega)|\, e^{-j\phi}\, e^{-j\omega\alpha} \qquad ...(11.39)$$

Thus, the approximate effect of the system on the CTFT of this narrowband input signal consists of the shape of magnitude corresponding to $|H(j\omega)|$, multiplication by $e^{-j\phi}$ and by $e^{-j\omega\alpha}$, where $e^{-j\phi}$ is an overall constant complex factor and $e^{-j\omega\alpha}$ is a linear phase corresponding to a time-delay of a α seconds. This time delay α is called *group delay* at $\omega = \omega_0$. In other words, we can say that it is the effective common delay experienced by a small band of frequencies centered at $\omega = \omega_0$.

The group delay at each frequency is equal to the negative of the slope of the phase at that frequency.

The group delay is defined as

$$\tau(\omega) = -\frac{d}{d\omega}\{\angle H(j\omega)\} \qquad ...(11.40)$$

The concept of group delay is applied directly to discrete-time systems as well.

11.3.3 Log-Magnitude Plots and Bode Plots

Log-Magnitude Plots

We often conveniently use a logarithmic scale for the graphical display of magnitude of Fourier transform (CTFT or DTFT) and system frequency response in polar form.

The magnitude and phase of the output of a continuous-time LTI system is given as

$$|Y(j\omega)| = |H(j\omega)|\ |S(j\omega)| \qquad ...(11.41)$$

and

$$\angle Y(j\omega) = H(j\omega) + S(j\omega) \qquad ...(11.42)$$

Note that the phase relationships for the CTFT of output of a continuous-time LTI system is additive, while the magnitude for output of a continuous-time LTI system is the multiplication of $|H(j\omega)|$ and $|S(j\omega)|$. Now taking log of both sides of Eq. 11.41, we get

$$\log |Y(j\omega)| = \log \{H(j\omega)\ S(j\omega)\}$$

or

$$\log |Y(j\omega)| = \log H(j\omega) + \log S(j\omega) \qquad ...(11.43)$$

Thus logarithmic scale changes multiplication of $|H(j\omega)|$ and $|S(j\omega)|$ into addition.

We can also obtain an exactly analogous expression in discrete-time.

Consequently, if we have a graph of log magnitude and phase of a Fourier transform (CTFT or DTFT) input ande the frequency response of a continuous-time/discrete-time LTI system, the Fourier transform of the output is obtained by adding the log- magnitude plots and by adding the phase plots. Since we know that the frequency response of the cascade of two LTI systems is the product of the individual frequency responses, *i.e.*, $H(j\omega) = H_1(j\omega)\ H_2(j\omega)$. Therefore, we can obtain a plot of the log-magnitude and phase of the overall frequency response of cascaded systems. It can be done by adding the corresponding log-magnitude and phase plots for each of the component systems.

The plotting of the magnitude of the Fourier transform on a logarithmic scale allows details which can be displayed over a wider range. On a linear scale, the detailed magnitude characteristics in the stop band of a frequency selective filter with high attenuation are not evident. But these detailed characteristics can be easily shown on a logarithmic scale.

A specific logarithmic amplitude scale used is in units of 20 $\log_{10}$. It is called *decibels* and abbreviated as *dB*. Magnitude of frequency response of a continuous-time LTI system in dB is given as

$$|H(j\omega)|_{dB} = 20 \log_{10} |H(j\omega)| \qquad ...(11.44)$$

Bode Plot

For continuous-time LTI systems, it is also common and useful to use logarithmic frequency sacle. Plots of $|H(j\omega)|_{dB}$ vs $\log_{10}(\omega)$ and $\underline{/H(j\omega)}$ vs $\log_{10}(\omega)$ are referred to as Bode plots. A typical Bode plot is shown in Fig. 11.3. If impulse response $h(t)$ is real then magnitude of its frequency response $H(j\omega)$ is an even function of frequency ω and phase of its frequency response is an odd function of frequency ω. The plots for negative frequency ω are superfluous and can be obtained immediately from the plots for positive frequency.

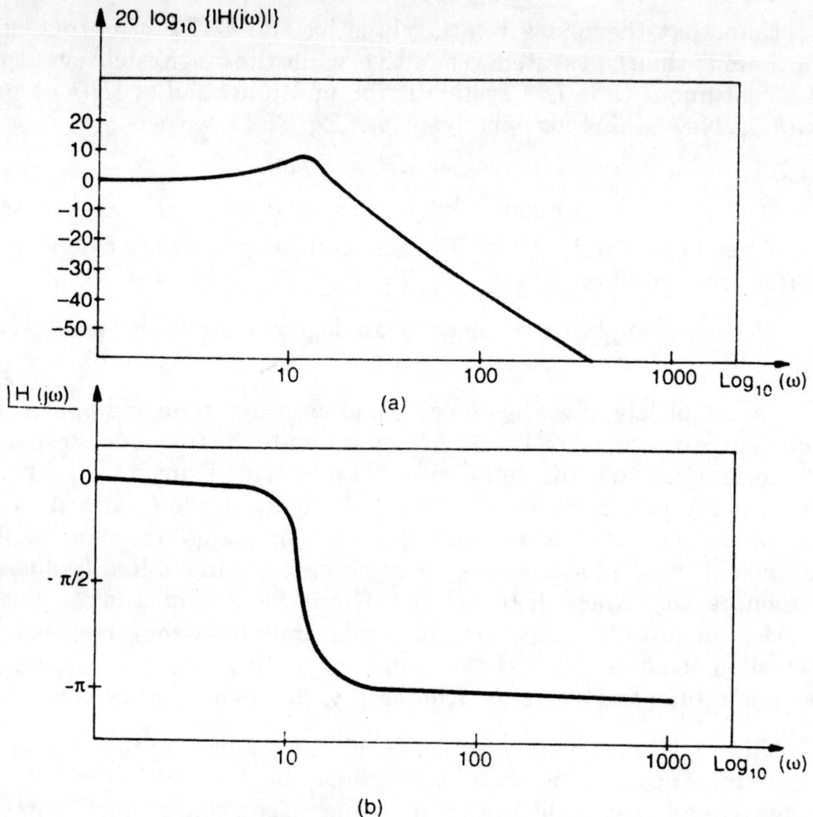

Fig. 11.3 *Bode plots (a) Plot* $|H(j\omega)|_{dB}$ *vs* $\log_{10}(\omega)$. *(b) Plot* $\underline{/H(j\omega)}$ *vs* $\log_{10}(\omega)$.

The use of a logarithmic frequency scale has a number of advantages in continuous-time case. It often permits a much wider range of frequencies to be displayed than does a linear frequency scale. In addition, on a logarithmic frequency scale, the shape of a particular response curve does not change even if the frequency is scaled. Therefore, for continuous-time LTI systems which are described by linear constant-coefficient differential equation, we can draw an approximate sketch of log-magnitude vs log-frequency through the use of asymptotes.

In a similar manner, for discrete-time case, the magnitude of DTFTs and frequency responses are often displayed in dB. This is because they are in continuous-time case. However, logarithmic frequency scale is not used in discrete-time case. The reason being; (*i*) the range of frequencies is always limited, (*ii*) the advantage found for differential equations (having linear asymptotes) does not aply to difference equations. A graphical representation of the magnitude and phase of a discrete-time frequency response $|H(e^{j\omega})|$ is illustrated in Fig. 11.4. In this plot we have sketched phase $\underline{/H(e^{j\omega})}$ in radians and magnitude $|H(e^{j\omega})$ in dB as a function of frequency ω. If impulse response of a discrete-time system is real then we need a plot only for $0 \le \omega \le \pi$. In this case the symmetry property of the DTFT implies that we can then determine frequency response $H(e^{j\omega})$ for $-\pi \le \omega \le 0$ using the following relations.

$$|H(e^{j\omega}) = |H(e^{-j\omega})| \quad ...(11.45)$$

and

$$\underline{/H(e^{j\omega})} = -\underline{/H(e^{j\omega})} \quad ...(11.46)$$

Furthermore, there is a no need to consider values of $|\omega|$ greater than ω because of the periodicity of frequency response $H(e^{j\omega})$.

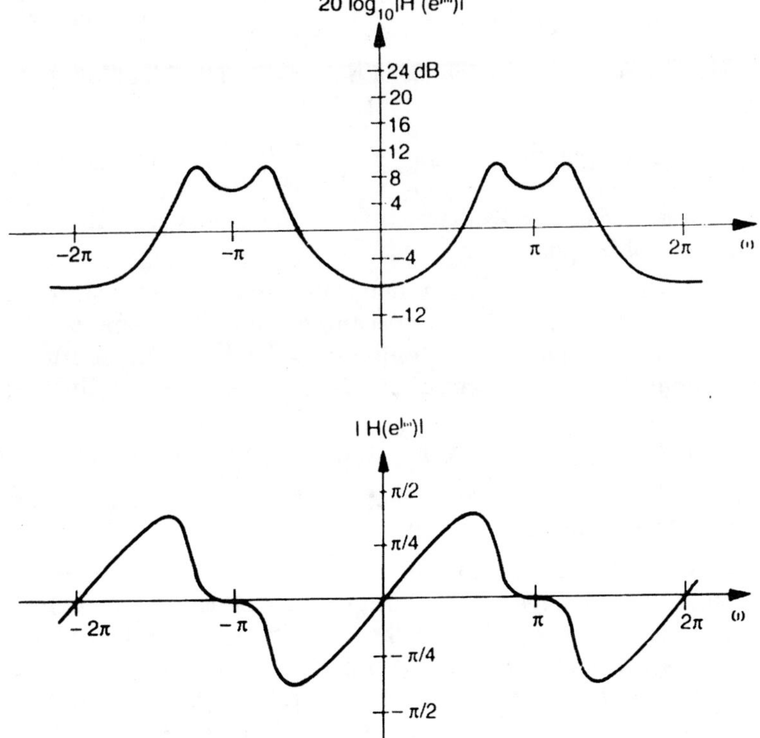

Fig. 11.4 *Graphical representation of (a) magnitude of a discrete-time frequency response $H(e^{j\omega})$ (b) phase of a discrete-time frequency response $H(e^{j\omega})$.*

A logarithmic amplitude scale is often useful and important. But in some situations it is convenient to use a linear amplitude scale. The magntiude of frequency response for ideal frequency selective filters is a non-zero constant over some frequency bands and zero over others, Here a linear amplitude scale is more appropriate. So both linear and logarithmic graphical representations for the magnitude of the Fourier transforms (CTFT or DTFT) can be used.

11.4 PROPERTIES OF IDEAL FREQUENCY SELECTIVE FILTERS IN TIME-DOMAINS

There exists a class of frequency selective filters (or continuous-time/discrete-time LTI systems) with specific frequency responses so chosen so as to pass one or several bonds of frequencies with little or no attenuation and also stop frequencies outside those band of frequencies. There are a number of important issues that arise in frequency-selective filtering applications. These issues are directly related to the characteristics of the frequency-selective filters. On the basis of the range of frequency selection, the frequency selective filters are classified as low-pass, high-pass, band-pass, band- stop and all-pass filters.

A continuous-time ideal low-pass filter has a frequency response

$$H(j\omega) = \text{CTFT}\,\{h(t)\} = \begin{cases} 1, & |\omega| \le \omega_c \\ 0, & |\omega| > \omega_c \end{cases} \qquad ...(11.47)$$

Similarly, the frequency response of a discrete-time ideal low-pass filter is given as

$$H(e^{j\omega}) = \text{DTFT}\,\{h(n)\} = \begin{cases} 1, & |\omega| \le \omega_c \\ 0, & \omega_c < |\omega| \le \pi \end{cases} \qquad ...(11.48)$$

The frequency response of discrete-time ideal low-pass filter is a periodic function of frequency ω.

The frequency responses of continuous-time and discrete-time ideal low pass filters are shown in Fig. 11.4 (*a*) and Fig. 11.4 (*b*), respectively. Ideal low-pass filters have perfect frequency selectivity. These filters can pass through (without attenuation) frequencies of the following characteristics :

(*i*) Those frequencies at or lower than cut-off frequency ω_c.

(*ii*) The filters can completely stop all frequenices higher than cut-off frequency ω_c.

We have already studied that non-linear phase characteristics can lead to significant changes. It can happen so in the time-domain characteristics of a signal even when the magnitude of its CTFT is not changed by the system. A filter with magnitude characteristics as given by Eqs. 11.47 and 11.48 and non-linear phase, produces undesirable effects in some applications. On the other hand, an ideal

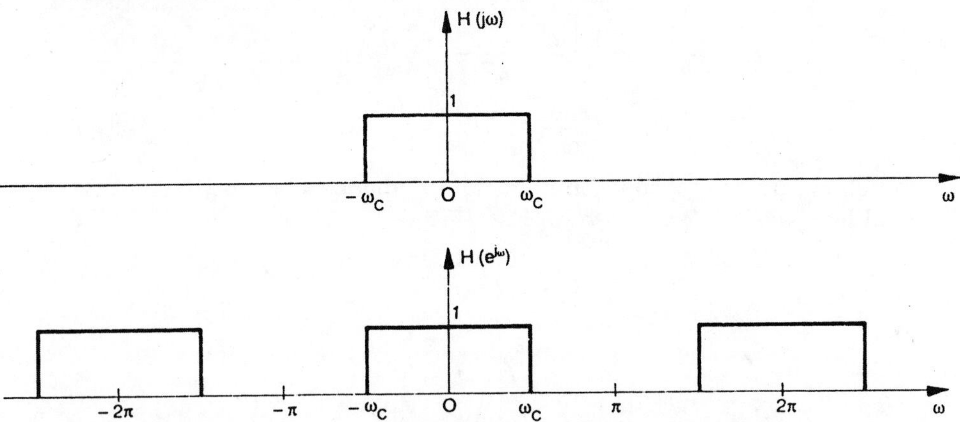

Fig. 11.4 (*a*) *The frequency response of a continuous-time ideal low-pass filter* (*b*) *the frequency response of a discrete-time ideal low pass filter.*

frequency-selective filter having linear phase over the pass band introduces only a simple time-shift relative to the response of the ideal low pass filter with zero phase characteristics. The magnitude and phase characteristics of a linear-phase ideal low pass filter is given in Fig. 11.5.

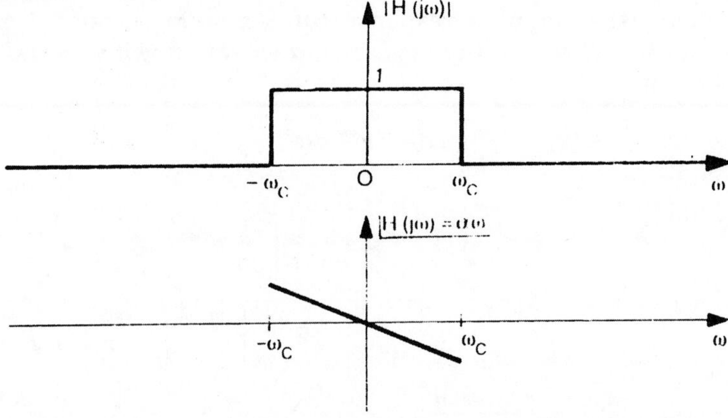

Fig. 11.5 (*a*) *The magnitude characteristics of a linear phase, continuous-time ideal low pass filter* (*b*) *the phase characteristic of a linear phase continuous-time ideal loss pass filter.*

Impulse response of a continuous-time ideal low pass filter is determined by taking inverse CTFT of frequency response given as

$$h(t) = \frac{1}{2\pi}\int_{-\infty}^{\infty} H(j\omega)\, e^{j\omega t}\, d\omega = \frac{1}{2\pi}\int_{-\omega_c}^{\omega_c} 1\, e^{j\omega t}\, d\omega$$

$$= \frac{1}{2\pi}\left[\frac{e^{j\omega t}}{jt}\right]_{-\omega_c}^{\omega_c} = \frac{1}{2\pi jt}\,[e^{j\omega_c t} - e^{-j\omega_c t}]$$

$$= \frac{1}{\pi t}\left[\frac{e^{j\omega_c t} - e^{-j\omega_c t}}{2j}\right] = \frac{1}{\pi t}\sin\omega_c t$$

or $$h(t) = \frac{\sin\omega_c t}{\pi t} \qquad ...(11.49)$$

The impulse response (time-domain response of a continuous-time ideal low pass filter is shown in Fig. 11.6.

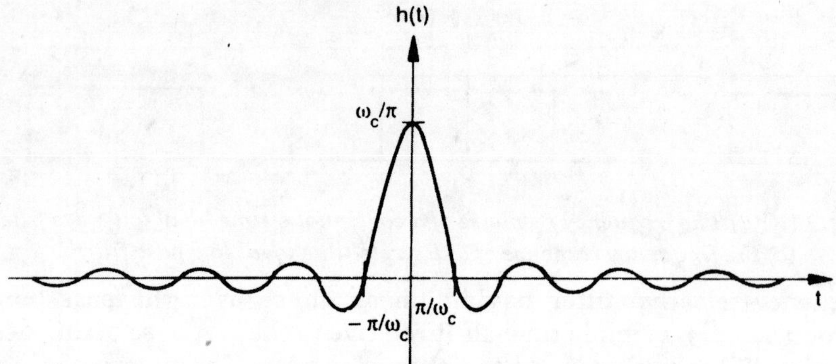

Fig. 11.6 *The impulse response (time-domain response) of the continuous-time ideal low pass filter shown in Fig.* 11.4 (*a*).

Similarly, the impulse response (unit-sample response) of a discrete-time ideal low pass filter is determined by taking inverse DTFT of frequency as

$$h(n) = \frac{1}{2\pi}\int_{2\pi} H(e^{j\omega})\, e^{j\omega n}\, d\omega$$

$$= \frac{1}{2\pi}\int_{-\omega_c}^{\omega_c} 1\, e^{j\omega n}\, d\omega = \frac{1}{2\pi}\left[\frac{1}{jn}e^{j\omega n}\right]_{-\omega_c}^{\omega_c}$$

$$= \frac{1}{2\pi jn}\left[e^{j\omega_c n} - e^{-j\omega_c n}\right] = \frac{1}{2\pi}\left[\frac{e^{j\omega_c n} - e^{-j\omega_c n}}{2j}\right]$$

$$= \frac{\sin\omega_c n}{\pi n}$$

or $$h(n) = \frac{\sin\omega_c n}{\pi n} \qquad ...(11.50)$$

The impulse response (time-domain response) of a discrete-time ideal low pass filter is shown in Fig. 11.7 for $\omega_c = \pi/4$.

For ideal frequency response, *i.e.*, $H(j\omega)$ or $H(e^{j\omega})$ with a linear phase characteristic, the impulse response, *i.e.*, $h(t)$ or $h(n)$ is simply delayed by an amount equal to the negative of the slope of this phase function. The impulse response of a continuous-time low pass filter with magnitude and phase shown as in Fig. 11.5, is depicted in Fig. 11.8. In both continuous-time and discrete-time, the width of the filter passband

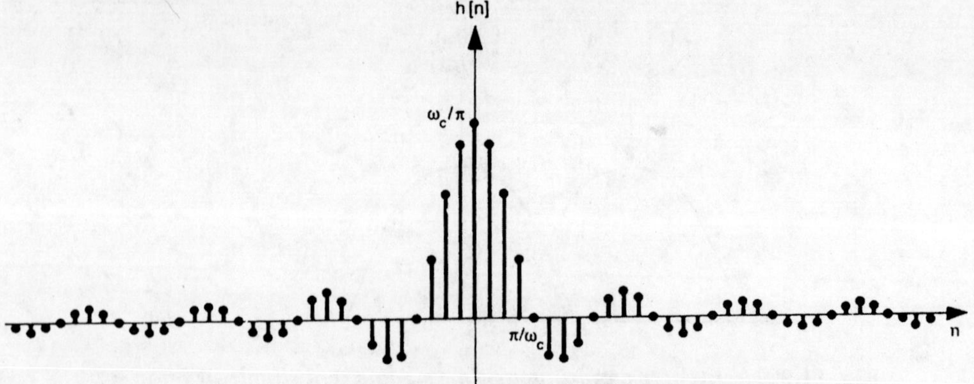

Fig. 11.7 ***The impulse response (time-domain response) of the discrete-time ideal low pass filter shown in Fig.*** **11.4 (*b*)** ***for*** $\omega_c = \pi/4$.

is proportional to the cut-off frequency ω_c, while the width of the main lobe of impulse response is inversely proportional to ω_c. As the bandwidth of the filter increases, the impulse response $h(t)$ becomes narrower, and, vice-versa. This relationship is consistent with the inverse relationship between time and frequency.

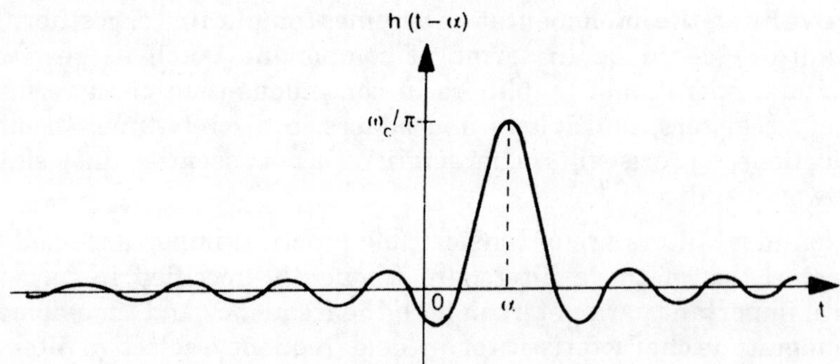

Fig. 11.8 ***Impulse response of a continuous-time ideal low pass filter with magnitude and phase shown in Fig.*** **11.5.**

11.5 TIME-DOMAIN AND FREQUENCY-DOMAIN ASPECTS OF NON-IDEAL FREQUENCY SELECTIVE FILTERS

The characteristics of an ideal frequency selective filter are not always desirable in practice. For example, in many filtering applications, the signals to be separated do not always lie in totally disjoint frequency bands.

We have a situation that is depicted in Fig. 11.9, where the specrtra of two continuous-time signals slightly overlap each other. In such a case, we may wish to trade off the fidelity with which the filter preserves one of these signals [say $s_1(t)$] against the level of which frequency components of the second signal [say $s_2(t)$] are attenutated. For filtering the superposition of two signals with overlapping spectra, we prefer a filter with a gradual transition from passband to stop band.

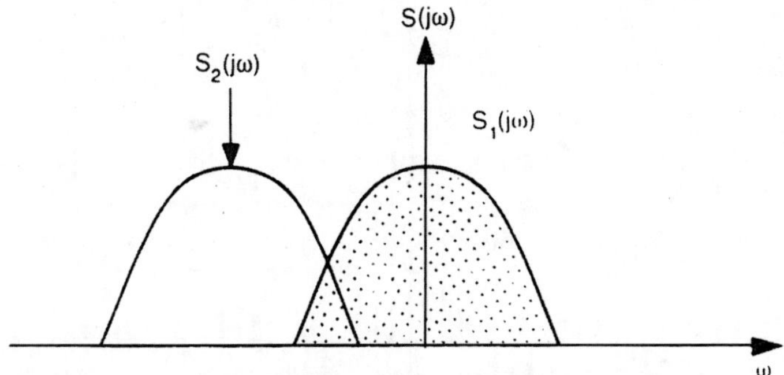

Fig. 11.9 *Spectra of two continuous-time signals that slightly overlap.*

Moreover, in cases where the ideal frequency-selective characteristics are desirable, it is not possible to attain these characteristics. For example an ideal low pass filter is non-causal. If filtering is carried out in real time then causality is a necessary constraint, and this, a causal approximation to the ideal characteristics would be required.

In general, for more precise approximations of an ideal frequency selective filter, the implementation becomes complicated or costlier. The expenditure could be in terms of components such as resistors, capacitors, operational amplifiers in continuous-time or in terms of memory registers, multipliers and adders in discrete-time. In many applications, a precise filter characteristic is not essential and a simple filter would suffice.

Non-ideal filters are of considerable practical importance and the characteristics of these filters are frequently specified in terms of several important parameters in both the frequency and time-domain. The magnitude characteristics of an ideal frequency- selective filter are unachievable or undesirable. It is preferred to allow some flexibility in the behaviour of the filter in the passband and in the stop band, as well as permit a more graudal transition between the two bands. Ideal filters provide abrupt transition between passband and stop band. In the case of low pass filters, the specifications may allow some deviation from unity gain in the passband and from zero gain in the stop band. Also there is a transition band between passband edge and stop band edge. The amount by which the frequency response differs from unity in the passband is called the passband ripple. The amount by which the frequency response differs from zero in the stop band is referred to as the stop band ripple. The frequencies ω_p and ω_s are referred to as the passband edge and stop band edge, respectively. The frequency range from ω_p to ω_s is referred to as transition band and it is given as

$$\omega_t = \omega_s - \omega_p \qquad \text{...(11.51)}$$

Similar definitions are also applicable to discrete-time low pass

filters as well as other continuous-time and discrete-time frequency selective filters such as high-pass, band-pass, band- stop and all-pass filters.

Fig. 11.10 shows tolerances for the magnitude characteristic of a low pass filter. In some cases, it is important to know the specifications of magnitude characteristics in the frequency- domain as well as phase characteristics. We frequently desire that the phase characteristic of a filter over the passband should be linear or nearly linear.

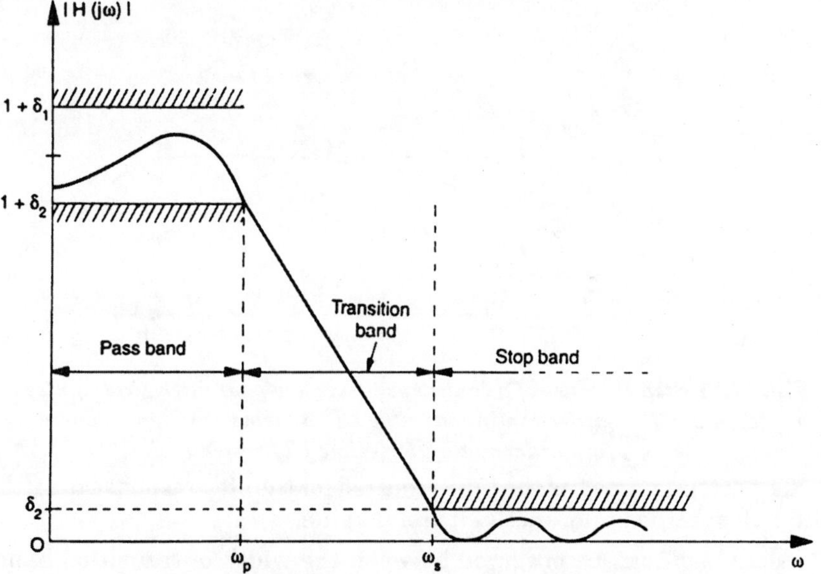

Fig. 11.10 *Magnitude characteristic of a low pass filter*
δ_1 = *allowable passband ripple;* δ_2 = *allowable stop band ripple.*

For controlling the time-domain behaviour, the specifications are frequently imposed on the step response of a filter. The step response of a continuous-time low pass filter is shown in Fig. 11.11. There is one quantity in step response of a continuous-time low pass filter that is called rise time t_r of the step response. It is the time interval over which the step response rises towards its final value. The presence or absence of oscillatory behaviour or ringing in the step response of a filter is often of importance. If ringing is present, then there are three other quantities which are often used to characterize the nature of these oscillations. These three quantities are

(*i*) The overshoot of the final value of the step response and it is denoted by Δ.

(*ii*) Ringing frequency ω_r.

(*iii*) The settling time, t_s.

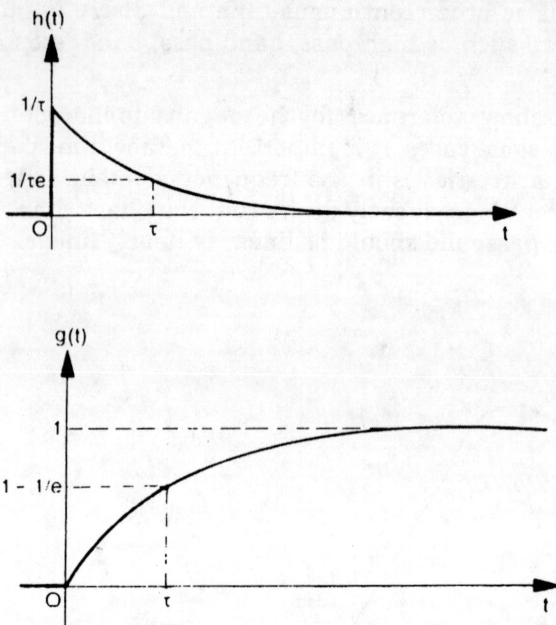

Fig. 11.11 *Step response of a continuous-time low pass filter. (a) Impulse response of a first-order continuous-time LTI systems. (b) Step response of a first order continuous- time LTI systems.*

The settling time t_s is the time required for the step response to settle with in a specified tolerance of its final value.

A trade-off can be observed between the width of transition band ($\omega_t = \omega_s - \omega_p$) and the settling time t_r of the step response for a non-ideal low pass filter. The width of the transition band ω_t represents a frequency response characteristic of the filter, whereas settling time t_r of the step response of a filter represents a time-domain characteristic of the filter. The consideration of the trade-offs between time-domain and frequency-domain characteristics and other issues (such as complexity and cost of filters) form the core of field of filter design.

11.6 LOWER-ORDER CONTINUOUS-TIME LTI SYSTEMS

Continuous-time LTI systems described by linear constant- coefficient differential equations are of great practical importance because many physical systems can be modelled by these differential equations. Also those type of systems can often be conveniently implemented. The higher order continuous-time LTI systems are frequently implemented due to a variety of practical reasons. The higher order systems are represented by combining first-order and second-order systems in cascade (series or parallel) arrangements. Consequently, the properties of first-order and second-order continuous-time LTI systems play an important role in analysis and design of higher order continuous-time

LTI systems. These lower order systems also play an important role in understanding the time-domain and frequency- domain characteristics of higher order continuous-time LTI systems.

11.6.1 First-Order Continuous-Time LTI Systems

The linear constant-coefficient differential equation for the first-order continuous-time LTI system is expressed as

$$\tau \frac{d\,y(t)}{dt} + y(t) = s(t) \qquad ...(11.52)$$

where τ is a constant-coefficient (or parameter) that is called the time-constant of the system. It controls the rate at which the first-order continuous-time LTI system responds.

Taking the Laplace transform of both sides of Eq. 11.52 with zero initial conditions, we obtain

$$\text{LT}\left\{\tau \frac{d\,y(t)}{dt} + y(t)\right\} = \text{LT}\,\{s(t)\}$$

or $$\tau s\, Y(s) + Y(s) = S(s)$$

or $$H(s) = \frac{Y(s)}{S(s)} = \frac{1}{1+\tau s}$$

or Transfer function of the first order system

$$= H(s) = \frac{\text{Laplace transform of output}}{\text{Laplace transform of input}}$$

$$= \frac{\text{LT}\,\{y(t)\}}{\text{LT}\,\{s(t)\}} = \frac{Y(s)}{S(s)} = \frac{1}{s\tau + 1}$$

or $$H(s) = \frac{1}{s\tau + 1} \qquad ...(11.53)$$

Frequency response of a first order continuous-time system can be determined by substituting $s = j\omega$ in Eq. 11.53,

Frequency response $H(j\omega) = H(s)\,\big|_{s=j\omega}$

or $$H(j\omega) = \frac{1}{j\omega\tau + 1} \qquad ...(11.54)$$

Impulse response $h(t)$ of a first order continuous-time system can be determined by taking inverse Laplace transform of Eq. 11.53,

Impulse response, $h(t)$ = Inverse Laplace transform $\{H(s)\}$

$$= \text{Inverse LT}\left\{\frac{1}{1+\tau s}\right\}$$

$$= \text{Inverse LT}\left\{\frac{1/\tau}{s + 1/\tau}\right\}$$

or $$h(t) = \frac{1}{\tau}\, e^{(-1/\tau)t}\, u(t) \qquad ...(11.55)$$

The impulse response $h(t)$ is shown in Fig. 11.11 (*a*).

The step response of the first order continuous-time LTI system is deterined as

$$g(t) = h(t) * u(t)$$

$$= \int_{t_0=-\infty}^{\infty} h(t_0)\, u(t - t_0)\, dt_0$$

$$= \int_{t_0=-\infty}^{\infty} \frac{1}{\tau} e^{-t_0/\tau}\, u(t_0)\, u(t - t_0)\, dt_0$$

From the definition

$$u(t_0) = \begin{cases} 1, & t_0 > 0 \\ 0, & t_0 < 0 \end{cases}$$

and $$u(t - t_0) = \begin{cases} 1, & t - t_0 > 0 \text{ or } t_0 < t \\ 0, & t - t_0 < 0 \text{ or } t_0 > t \end{cases}$$

$$g(t) = \int_{t_0=0}^{t} \frac{1}{\tau} e^{-t_0/\tau}\, d\tau = \left[\frac{1}{\tau} (-\tau)\, e^{-t_0/\tau} \right]_{t_0=0}^{t}$$

$$= (1 - e^{-t/\tau}),\ t > 0$$

$$g(t) = (1 - e^{-t/\tau})\, u(t) \qquad ...(11.56)$$

This is the step response of the first order continuous-time LTI system. The step response of the first order continuous-time LTI system is shown in Fig. 11.11 (*b*). For example, the impulse response at $t = \tau$ has reached $\frac{1}{e}$ times its value at $t = 0$, *i.e.*,

$$h(\tau) = \frac{1}{e}\, h(0) \qquad ...(11.57)$$

The step response is written $\frac{1}{e}$ of its final value.

Therefore, as time constant τ is decreased the impulse response decays more sharply and the rise time of the step response becomes shorter. It means, step response of first order continuous-time LTI system rises more sharply towards its final value. The step response of a first-order system does not exhibit any ringing.

Bode plot of the frequency response of the first order continuous-time LTI system is shown in Fig. 11.12. In this figure we have shown one of the advantages of using a logarithmic frequency scale. We can obtain a useful approximate Bode plot for a continuous-time first order LTI system. Now we shall examine the plot of the log magnitude of the frequency response.

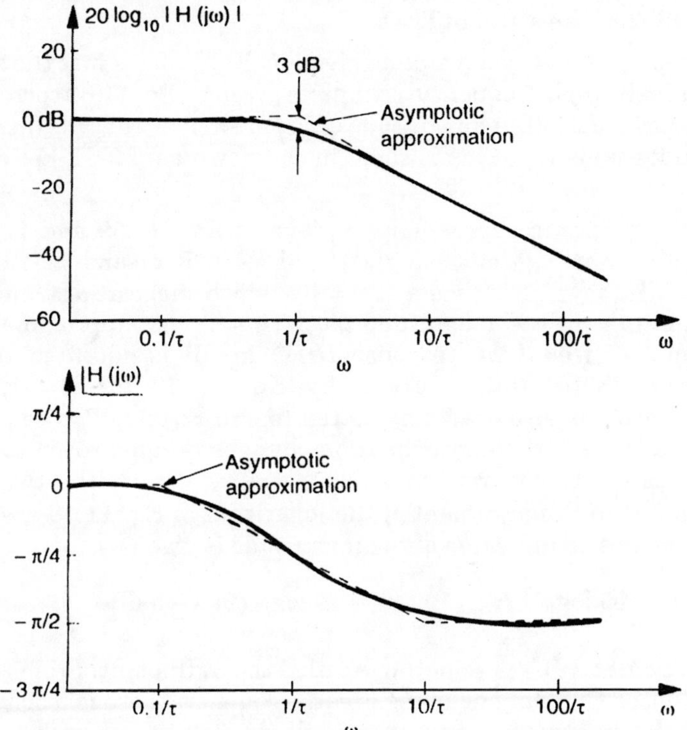

Fig. 11.12 *Bode plot for a continuous-time first order LTI system.*

Frequency response is given as

$$H(j\omega) = \frac{1}{j\omega\tau + 1}$$

$$20 \log_{10} |H(j\omega)| = 20 \log_{10} \frac{1}{\sqrt{(\omega\tau)^2 + 1}}$$

or $$|H(j\omega)|_{dB} = -10 \log_{10} \{(\omega\tau)^2 + 1\} \quad \text{...(11.58)}$$

From Eq. 11.57, for $\omega\tau << 1$, the log magnitude $|H(j\omega)|_{dB}$ is approximately zero, while for $\omega\tau >> 1$, the log magnitude, $|H(j\omega)|_{dB}$ is approxiamtely a linear function of $\log_{10}(\omega)$ that is,

$$20 \log_{10} |H(j\omega)| \approx 0 \quad \text{for} \quad \omega << \frac{1}{\tau} \quad \text{...(11.59)}$$

and $$20 \log_{10} |H(j\omega)| \approx -20 \log_{10} (\omega\tau)$$

$$= -20 \log_{10} (\omega) - 20 \log_{10} (\tau)$$

$$\text{for} \quad \omega >> \frac{1}{\tau} \quad \text{...(11.60)}$$

In other words, we can say that for the first order continuous- time LTI systems, the low- and high-frequency asymptotes of the log magnitude are the straight lines.

The low-frequency asymptote given by Eq. 11.59 is just the zero dB line, while the high-frequency asymptote given by Eq. 11.60 corresponds to a decrease of 20 dB in magnitude of $H(j\omega)$ for every decade (*i.e.,* factor of 10) in frequency ω. This is sometimes known as "20-dB per decade" asymptote.

Two asymptotic approximation given in Eqs. 11.59 and 11.60 are equal to a point $\log_{10}(\omega) = -\log_{10}(\tau)$ or $\omega = 1/\tau$. It means that the two straight- line asymptotes meet at $\omega = 1/\tau$, which suggests a straight-line approximation to the magnitude plot. That is our approximation to magnitude of frequency response $H(j\omega)$ in dB is equal to zero for frequency $\omega \leq 1/\tau$ and is given by Eq. 11.60 for $\omega = 1/\tau$. This approximation is also drawn by dotted line in Fig. 11.12. The point at which the slope of the approximation changes is equal to $\omega = 1/\tau$. This point is called break frequency. Note that at $\omega = 1/\tau$ the two terms $[(\omega\tau)^2$ and 1] in the argument of the logarithm in Eq. 11.58 are equal. Therefore, the actual value of the magnitude is given by

$$20 \log_{10} \left| H\left(j\frac{1}{\tau}\right) \right| = -10 \log_{10}(2) \approx -3 \text{ dB} \qquad ...(11.61)$$

The point $\omega = 1/\tau$ is sometimes called the 3-dB point. From the Fig. 11.12, we see that only near the break frequency, there is a significant error in the straight line approximate Bode plot. So, we wish to obtain more accurate sketch of the Bode plot by modifying the approximation near the break frequency. It is also possible to obtain a useful straight-line approximation to phase of frequency response $H(j\omega)$.

$$H(j\omega) = \text{Phase of } H(j\omega) = \text{Phase of } \frac{1}{j\omega\tau + 1}$$

$$= \tan^{-1}\left[\frac{H_I(j\omega)}{H_R(j\omega)}\right] = \tan^{-1}\left[\frac{-\omega\tau}{1}\right]$$

$$= -\tan^{-1}(\omega\tau) \qquad ...(11.62)$$

Eq. 11.62 can be written as

$$\underline{/H(j\omega)} = \begin{cases} 0, & \omega \leq 0.1\pi \\ -\left(\frac{\pi}{4}\right)[\log_{10}(\omega\tau) + 1], & \frac{0.1}{\tau} \leq \omega \leq \frac{10}{\tau} \\ -\frac{\pi}{2}, & \omega \geq \frac{10}{\pi} \end{cases} \qquad ...(11.63)$$

Note that approximation of phase of $H(j\omega)$ decreases linearily from 0 to $-\pi/2$ as a function of $\log_{10}(\omega)$ in the range 0.1 over $\tau \leq \omega \leq \frac{10}{\tau}$.

In the range from one decade below the break frequency to one decade above the break frequency. The correct asymptotic value of $\angle H(j\omega)$ for $\omega << \frac{1}{\tau}$ is zero, whereas correct asymptotic value of $\angle H(j\omega)$ for $\omega >> \frac{1}{\tau}$ is $-\frac{\pi}{2}$. Furthermore, the approximation agrees with the actual value of $\angle H(j\omega)$ at the break frequency $\omega = \frac{1}{\tau}$.

At break frequency, phase of the frequency response $H(j\omega)$ is given as

$$\angle H\left(j\frac{1}{\tau}\right) = -\frac{\pi}{4} \qquad ...(11.64)$$

From this first-order continuous-time LTI system, we can again see the inverse relationship between time and frequency. If we decrease time constant τ of a first-order system then the time response of the system speeds up and we simultaneously make the break frequency large.

11.6.2 Second-Order Continuous-Time LTI Systems

The second order continuous-time LTI system is described by following linear constant-coefficient differential equations

$$\frac{d^2 y(t)}{dt^2} + 2\zeta\omega_n \frac{d\,y(t)}{dt} + \omega_n^2\, y(t) = \omega_n^2\, s(t) \qquad ...(11.65a)$$

These types of equations arise in many physical systems such as *RLC* circuits and mechanical systems. A mechanical system is composed of a spring, a mass, and a viscous damper or dashpot. A mechanical second order system consisting of a spring and a dashpot attached to a moveable mass and a fixed support is shown in Fig. 11.13.

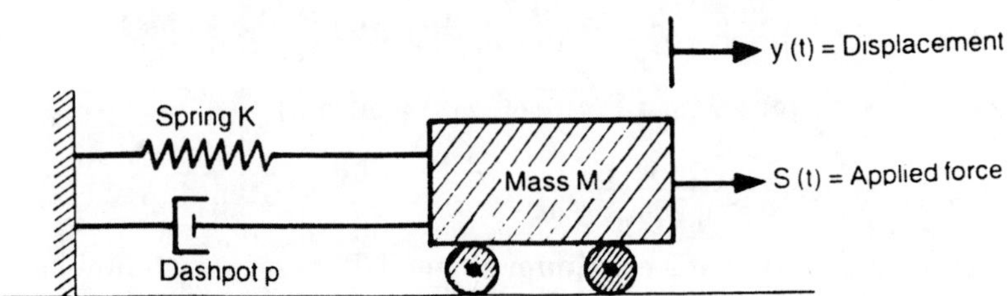

Fig. 11.13 *Mechanical second order conntinuous-time LTI system consisting of a spring and a dashpot. The spring and dashpot are attached to a moveable mass and a fixed support.*

In Fig. 11.13, the input is the applied force $s(t)$ and the output is the displacement of the mass $y(t)$ from some equilibrium position. At equilibrium position, the spring exerts zero restoring force.

The differential equation which describes the input-output relationship of this mechanical system is given as

$$m\frac{d^2y(t)}{dt^2} = s(t) - K\,y(t) - p\,\frac{d\,y(t)}{dt}$$

or
$$m\frac{d^2y(t)}{dt^2} + p\,\frac{d\,y(t)}{dt} + K\,y(t) = s(t)$$

or
$$\frac{d^2y(t)}{dt^2} + \frac{p}{m}\frac{d\,y(t)}{dt} + \frac{K}{m}y(t) = \frac{1}{m}s(t) \qquad \text{...(11.65}b\text{)}$$

Comparing Eqs. 11.65 and 11.66, we get

$$\frac{d^2y(t)}{dt^2} + \left(\frac{p}{m}\right)\frac{d\,y(t)}{dt} + \left(\frac{K}{m}\right)y(t) = \frac{1}{m}s(t) \qquad \text{...(11.66)}$$

$$\frac{d^2y(t)}{dt^2} + 2\zeta\omega_n\frac{d\,y(t)}{dt} + \omega_n^2\,y(t) = \omega_n^2\,s(t) \qquad \text{...(11.67)}$$

$$\omega_n^2 = \frac{K}{m} \quad \text{or} \quad \omega_n = \sqrt{\frac{K}{m}} \qquad \text{...(11.68)}$$

$$2\zeta\omega_n = \frac{p}{m} \quad \text{or} \quad \zeta = \frac{p}{m}\frac{1}{2\omega_n} \qquad \text{...(11.69)}$$

Substituting the value ω_n from Eq. 11.68 in Eq. 11.69

$$\zeta = \frac{p}{m}\frac{1}{2}\sqrt{\frac{m}{K}} \quad \text{or} \quad \zeta = \frac{p}{2\sqrt{mK}} \qquad \text{...(11.70)}$$

Determination of the Transfer Function of a Second Order Continuous-Time LTI System

Here we are assuming that the system is initially relaxed *i.e.,* it has zero initial conditions. Taking the Laplace transform of both sides of Eq. 11.65 with zero initial conditions, we get

$$\text{LT}\left\{\frac{d^2y(t)}{dt^2} + 2\zeta\omega_n\frac{d\,y(t)}{dt} + \omega_n^2\,y(t)\right\} = \text{LT}\,\{\omega_n^2\,s(t)\}$$

or
$$s^2\,Y(s) + 2\zeta\omega_n s\,Y(s) + \omega_n^2\,Y(s) = \omega_n^2\,S(s)$$

or
$$\frac{Y(s)}{S(s)} = \frac{\omega_n^2}{s^2 + 2\zeta\omega_n s + \omega_n^2} \qquad \text{...(11.71)}$$

Transfer function of a continuous-time LTI system is defined as the ratio of Laplace transform of output to Laplace transform of input assuming all initial conditions equal to zero. It is defined as

$$H(s) = \frac{\text{Laplace transform of output}}{\text{Laplace transform of input}} = \frac{Y(s)}{S(s)} \qquad \text{...(11.72)}$$

From Eqs. 11.71 and 11.72, the transfer function of the second order continuous-time LTI system is given by

$$H(s) = \frac{\omega_n^2}{s^2 + 2\zeta\omega_n s + \omega_n^2} \qquad \text{...(11.73)}$$

The frequency response for the second order continuous-time LTI system is determined by substituting $s = j\omega$ in Eq. 11.73. Substituting, we get

$$H(j\omega) = H(s)\Big|_{s=j\omega} = \frac{\omega_n^2}{(j\omega)^2 + 2\zeta\omega_n(j\omega) + \omega_n^2} \quad ...(11.74)$$

The denominator of frequency response $H(j\omega)$ can be factored to yield

$$H(j\omega) = \frac{\omega_n^2}{[j\omega + \zeta\omega_n - \omega_n\sqrt{\zeta^2 - 1}\,]\,[j\omega + \zeta\omega_n + \omega_n\sqrt{\zeta^2 - 1}\,]}$$

or $$H(j\omega) = \frac{\omega_n^2}{(j\omega - c_1)(j\omega - c_2)} \quad ...(11.75)$$

$$\left.\begin{aligned} \text{where } c_1 &= -\zeta\omega_n + \omega_n\sqrt{\zeta^2 - 1} \\ \text{and } c_2 &= -\zeta\omega_n - \omega_n\sqrt{\zeta^2 - 1} \end{aligned}\right\} \quad ...(11.76)$$

Performing partial fraction expansion of frequency response $H(j\omega)$ given in Eq. 11.75 for $\zeta \neq 1$ while c_1 and c_2 being unequal.

$$H(j\omega) = \frac{\omega_n^2}{(j\omega - c_1)(j\omega - c_2)} = \frac{\alpha_1}{(j\omega - c_1)} + \frac{\alpha_2}{(j\omega - c_2)} \quad ...(11.77)$$

or $j\omega - c_1 = 0$, $j\omega = c_1$, $$\alpha_1 = \frac{\omega_n^2}{j\omega - c_2} = \frac{\omega_n^2}{c_1 - c_2} = \frac{\omega_n}{2\sqrt{\zeta^2 - 1}} = M$$

or $j\omega - c_2 = 0$, $j\omega = c_2$, $$\alpha_2 = \frac{\omega_n^2}{j\omega - c_1} = \frac{\omega_n^2}{c_2 - c_1} = \frac{-\omega_n}{2\sqrt{\zeta^2 - 1}} = -M$$

Substituting the value of α_1 and α_2 in Eq. 11.77, we get

$$H(j\omega) = \frac{M}{(j\omega - c_1)} - \frac{M}{(j\omega - c_2)} \quad ...(11.78)$$

where $$M = \frac{\omega_n}{2\sqrt{\zeta^2 - 1}} \quad ...(11.79)$$

Impulse response or time-domain response of a second order continuous-time LTI system can be determined by taking inverse CTFT of both sides of Eq. 11.78,

Impulse response, $h(t)$ = Inverse CTFT $\{H(j\omega)\}$

$$= \text{Inverse CTFT}\left\{\frac{M}{(j\omega - c_1)} - \frac{M}{(j\omega - c_2)}\right\}$$

$$= \text{Inverse CTFT}\left\{\frac{M}{(j\omega - c_1)}\right\} - \text{Inverse CTFT}\left\{\frac{M}{(j\omega - c_2)}\right\}$$

or $$= M\,e^{c_1 t}\,u(t) - M\,e^{c_2 t}\,u(t)$$

or $$h(t) = M[e^{C_t} - e^{C_2 t}\,]\,u(t). \quad ...(11.80)$$

If $\zeta = 1$, then $c_1 = -\zeta\omega_n + \omega_n\sqrt{\zeta^2 - 1} = -1 \times \omega_n + \omega_n\sqrt{(1)^2 - 1} = -\omega_n$

$c_2 = -\zeta\omega_n - \omega_n\sqrt{\zeta^2 - 1} = -1 \times \omega_n - \omega_n\sqrt{(1)^2 - 1} = -\omega_n$

Hence, $c_1 = c_2 = -\omega_n$ ...(11.81)

Substituting $\zeta = 1$ and $c_1 = c_2 = -\omega_n$ in Eq. 11.77, we get

$$H(j\omega) = \frac{\omega_n^2}{(j\omega - c_1)(j\omega - c_2)}$$

$$= \frac{\omega_n^2}{(j\omega + \omega_n)(j\omega + \omega_n)}$$

or $$H(j\omega) = \frac{\omega_n^2}{(j\omega + \omega_n)^2} \quad ...(11.82)$$

Impulse response of the second order continuous-time LTI system for $\zeta = 1$ and $c_1 = c_2 = -\omega_n$ can be determined by taking inverse CTFT of both sides of Eq. 11.82.

Impulse response, $h(t)$ = Inverse CTFT $\{H(j\omega)\}$

$$= \text{Inverse CTFT}\left\{\frac{\omega_n^2}{(j\omega + \omega_n)^2}\right\}$$

or $$h(t) = \omega_n^2\, t\, e^{-\omega_n t}\, u(t) \quad ...(11.83)$$

Note from Eqs. 11.80 and 11.83 that $\frac{h(t)}{\omega_n}$ is a function of $\omega_n t$. Now Eq. 11.74 can be rewritten as

$$H(j\omega) = \frac{\omega_n^2}{(j\omega)^2 + 2\zeta\omega_n(j\omega) + \omega_n^2}$$

or $$H(j\omega) = \frac{1}{\left(\frac{j\omega}{\omega_n}\right)^2 + 2\zeta\left(\frac{j\omega}{\omega_n}\right) + 1} \quad ...(11.84)$$

From Eq. 11.84, we see that the frequency response is a function of ω/ω_n. Thus, variation in ω_n is essentially equivalent to a time and frequency scaling. The parameter ζ is referred to as the damping ratio and parameter ω_n is called *undamped natural frequency*.

From Eq. 11.78, we see that for $0 < \zeta < 1$, c_1 and c_2 are complex and we can rewrite Eq. 11.80 as

$$h(t) = M\,[e^{c_1 t} - e^{c_2 t}]\,u(t)$$

$$= \frac{\omega_n}{2\sqrt{\zeta^2 - 1}}\left\{e^{[-\zeta\omega_n + \omega_n\sqrt{\zeta^2 - 1}]t} - e^{[-\zeta\omega_n - \omega_n\sqrt{\zeta^2 - 1}]t}\right\}u(t)$$

$$= \frac{\omega_n e^{-\zeta\omega_n t}}{2\sqrt{\zeta^2 - 1}}\left\{e^{[\omega_n\sqrt{\zeta^2 - 1}]t} - e^{[-\omega_n\sqrt{\zeta^2 - 1}]t}\right\}u(t)$$

$$= \frac{\omega_n e^{-\zeta\omega_n t}}{2j\sqrt{1 - \zeta^2}}\left\{e^{j\omega_n\sqrt{1 - \zeta^2}\,t} - e^{-j\omega_n\sqrt{1 - \zeta^2}\,t}\right\}u(t)$$

$$= \frac{\omega_n e^{-\zeta\omega_n t}}{\sqrt{1-\zeta^2}} \left\{ \frac{e^{j\omega_n \sqrt{1-\zeta^2}\, t} - e^{-j\omega_n \sqrt{1-\zeta^2}\, t}}{2j} \right\} u(t)$$

or $$h(t) = \frac{\omega_n e^{-\zeta\omega_n t}}{\sqrt{1-\zeta^2}} [\sin (\omega_n \sqrt{1-\zeta^2}\,) t]\, u(t) \qquad ...(11.85)$$

(*i*) Thus, for $0 < \zeta < 1$, the second order continuous- time LTI system has impulse response that shows damped oscillatory behaviour, and in this case the system is referred to as being *under-damped*.

(*ii*) If damping ratio, $\zeta > 1$, both c_1 and c_2 are real and negative. Impulse response $h(t)$ in this case is the difference between two decaying exponentials. In this case, system is called *over damped*.

(*iii*) If the damping ratio, $\zeta = 1$ (where $c_1 = c_2$) then in this case system is called *critically damped system*.

The plot of $\frac{h(t)}{\omega_n}$ for various values of damping ratio ζ is shown in Fig. 11.14.

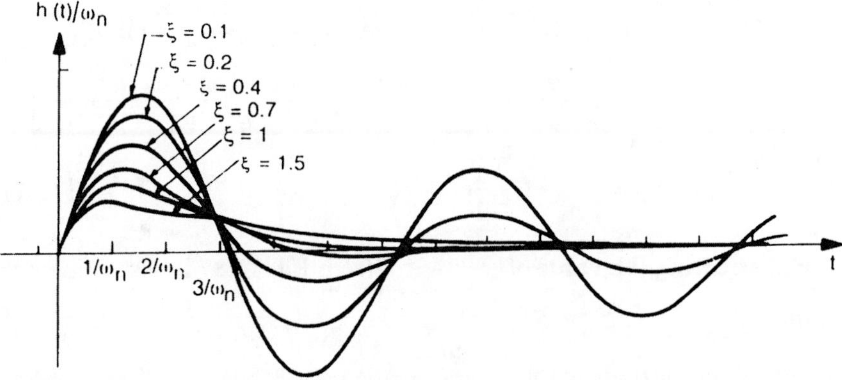

Fig. 11.14 *Impulse response of a second-order continuous-time LTI system with different values of damping ratio ζ.*

Determination of Step Resonse $g(t)$ of a Second Order Continuous-Time LTI System

It is determined by convolution integral between impulse response $h(t)$ and unit step function $u(t)$ and is given as

$$g(t) = h(t) * u(t) = \int_{t_0=-\infty}^{\infty} h(t_0)\, u(t - t_0)\, dt_0$$

$$= \int_{t_0=-\infty}^{\infty} M\, \{e^{c_1 t_0} - e^{c_2 t_0}\}\, u(t_0)\, u(t - t_0)\, dt_0$$

From the definition of $u(t_0)$ and $u(t - t_0)$

$$u(t_0) = \begin{cases} 1, & t_0 > 0 \\ 0, & t_0 < 0 \end{cases}$$

and $$u(t - t_0) = \begin{cases} 1, & t - t_0 > 0 \text{ or } t_0 < t \\ 0, & t - t_0 < 0 \text{ or } t_0 > t \end{cases}$$

or $$g(t) = \int_{t_0=0}^{t} M\{e^{c_1 t_0} - e^{c_2 t_0}\}\, dt_0 = M \int_{t_0=0}^{t} e^{c_1 t_0}\, dt_0 - M \int_{t_0=0}^{t} e^{c_2 t_0}\, dt_0$$

$$= M\left[\frac{1}{c_1} e^{c_1 t_0}\right]_{t_0=0}^{t} - M\left[\frac{1}{c_2} e^{c_2 t_0}\right]_{t_0=0}^{t}$$

$$= M\left[\frac{e^{c_1 t}}{c_1} - \frac{e^{c_1 \times 0}}{c_1}\right] - M\left[\frac{e^{c_2 t}}{c_2} - \frac{e^{c_2 \times 0}}{c_2}\right]$$

or $$g(t) = \frac{M}{c_2} - \frac{M}{c_1} + M\left[\frac{e^{c_1 t}}{c_1} - \frac{e^{c_2 t}}{c_2}\right] \quad \text{...(11.86)}$$

Now, we will determine the value of $\left(\frac{M}{c_2} - \frac{M}{c_1}\right)$ as

$$M\left(\frac{1}{c_2} - \frac{1}{c_1}\right) = \frac{\omega_n}{2\sqrt{\zeta^2 - 1}}\left[\frac{1}{-\zeta\omega_n - \omega_n\sqrt{\zeta^2 - 1}} - \frac{1}{-\zeta\omega_n + \omega_n\sqrt{\zeta^2 - 1}}\right]$$

$$= \frac{\omega_n}{2\sqrt{\zeta^2 - 1}}\left[\frac{-\zeta\omega_n + \omega_n\sqrt{\zeta^2 - 1} + \zeta\omega_n + \omega_n\sqrt{\zeta^2 - 1}}{\zeta^2\omega_n^2 - \zeta^2\omega_n^2 + \omega_n^2}\right]$$

$$= \frac{2\omega_n^2\sqrt{\zeta^2 - 1}}{2\omega_n^2\sqrt{\zeta^2 - 1}} = 1 \quad \text{...(11.87)}$$

Substituting the value of $\left(\frac{M}{c_2} - \frac{M}{C_1}\right)$ from Eq. 11.87 in Eq. 11.86, we obtain

$$g(t) = \left\{1 + M\left[\frac{e^{c_1 t}}{c_1} - \frac{e^{c_2 t}}{c_2}\right]\right\} u(t) \quad \text{...(11.88)}$$

For a critically damped system, $\zeta = 1$, the step response of a second order continuous-time LTI system can be determined by using impulse response $h(t) = \omega_n^2\, t\, e^{-\omega_n t}\, u(t)$.

$$s(t) = h(t) * u(t) = \int_{t_0=-\infty}^{\infty} h(t_0)\, u(t - t_0)\, dt_0$$

$$= \int_{t_0=-\infty}^{\infty} \omega_n^2\, t_0\, e^{-\omega_n t_0}\, u(t_0)\, u(t - t_0)\, dt_0$$

(using the definition of $u(t_0)$ and $u(t - t_0)$)

$$= \int_{t_0=0}^{t} \omega_n^2\, t_0\, e^{-\omega_n t_0}\, dt_0$$

or $$s(t) = [1 - e^{-\omega_n t} - \omega_n\, t\, e^{-\omega_n t}]\, u(t) \qquad ...(11.89)$$

The step response of a second order continuous-time LTI system is shown in Fig. 11.15.

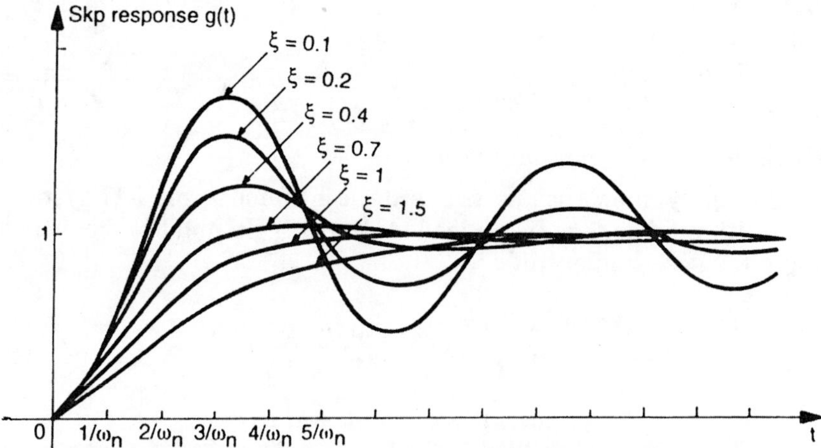

Fig. 11.15 *Step response of a second order continuous-time LTI system with different value of damping ratio* ζ.

From Fig. 11.15, we see that in the under damped case ($\zeta < 1$), the step response $g(t)$ of a second order continuous-time system possesses both overshoot and ringing. The overshoot is the exceed in the step response in its final value. Ringing is the oscillatory behaviour of the step response of a system. For a critically damped case, $\zeta = 1$, the step response has the fastest response that is possible without overhsoot and thus has the shortest settling time. Fastest response provides the shortest rise time. As the damping ratio ζ increases beyond unity, the step response becomes slower. From Eqs. 11.76 and 11.88, we see that as ζ increases, c_1 becomes smaller in magnitude while c_2 increases in magnitude. Therefore, the time constant associated with $e^{c_2 t}$, *i.e.*, $(1/c_2)$ decreases and time constant associated with $e^{c_1 t}$, *i.e.*, $(1/c_1)$ increases. Therefore, the term involving $e^{c_1 t}$ in Eq. 11.88 will take longer time to decay to zero. Thus it is the time constant associated with this term that determines the settling time of the step response of a second order continuous-time LTI system. Consequently, the step response will take longer time to settle for large values of damping ratio ζ.

The undamped natural frequency ω_n essentially controls the time scale of the impulse response $h(t)$ and step response $g(t)$ for a second order continuous-time LTI system. In the underdamped case ($\zeta < 1$), the larger the term ω_n, the more compressed is the impulse response $h(t)$ as a function of t and higher is the frequency of the oscillations or ringing in both impulse response $h(t)$ and step response $g(t)$. The frequency of oscillations in $h(t)$ and $g(t)$ is equal to $\omega_n \sqrt{1 - \zeta^2}$.

This frequency does not increase with increasing undamped natural frequency ω_n. This frequency depends explicitly on the damping ratio ζ and is not equal to ω_n. It is smaller than ω_n except in the undamped case, $\zeta = 0$. It is for this reason that ω_n is called undamped natural frequency.

Bode plot of the frequency response

$$H(j\omega) = \frac{\omega_n^2}{(j\omega)^2 + 2\zeta\omega_n (j\omega) + \omega_n^2}$$

for several values of damping ratio ζ.

We already know that for first order continuous-time LTI system, the logarithmic frequency scale leads to linear high- and low-frequency asymptotes for log magnitude.

$$\text{Frequency response, } H(j\omega) = \frac{\omega_n^2}{(j\omega)^2 + 2\zeta\omega_n (j\omega) + \omega_n^2}$$

or

$$H(j\omega) = \frac{1}{(j\omega/\omega_n)^2 + 2\zeta(j\omega/\omega_n) + 1}$$

$$= \frac{1}{[1 - (\omega^2/\omega_n^2)] + j[2\zeta\,(\omega/\omega_n)]}$$

$$\text{Magnitude of } H(j\omega) = |H(j\omega)| = \frac{1}{\sqrt{[1 - (\omega/\omega_n)^2\,]^2 + 4\zeta^2\,(\omega/\omega_n)^2}}$$

Taking 20 $\log_{10}$ of both sides of above equation, we obtain

$$20\log_{10} |H(j\omega)| = \frac{1}{20\log_{10} \sqrt{[1 - (\omega/\omega_n)^2\,]^2 + 4\zeta^2\,(\omega/\omega_n)^2}}$$

$$= -10\log\,\{\sqrt{[1 - (\omega/\omega_n)^2\,]^2 + 4\zeta^2\,(\omega/\omega_n)^2}\,\} \quad \text{...(11.90)}$$

From the expression given in Eq. 11.90, it can be written that

$$20\log_{10} |H(j\omega)| \approx \begin{cases} 0\,, & \text{for } \omega << \omega_n \\ -40\log_{10}\omega + 40\log_{10}\omega_n\,, & \text{for } \omega >> \omega_n \end{cases} \quad \text{...(11.91)}$$

Therefore, the low-frequency asymptote of the log-magnitude is the zero dB line, whereas high-frequency asymptote has a slope of –40 dB/decade. The meaning of –40 dB/decade is that the magnitude of frequency response, $|H(j\omega)|$ decreases by 40 dB for every increase in ω by a factor of 10. Two straight-line asymptote meet at the point $\omega = \omega_n$. Thus, we can obtain a straight line apprxomation of Eq. 11.91 for $\omega \le \omega_n$. Due to this reason, frequency ω_n is called the break frequency of the second order continuous-time LTI system. This approximation is shown in Fig. 11.16 (*a*) by dottedd line.

We can also obtain a straight-line approximation of the phase of frequency response $H(j\omega)$ of a second-order continuous- time LTI system by using equation of frequency response

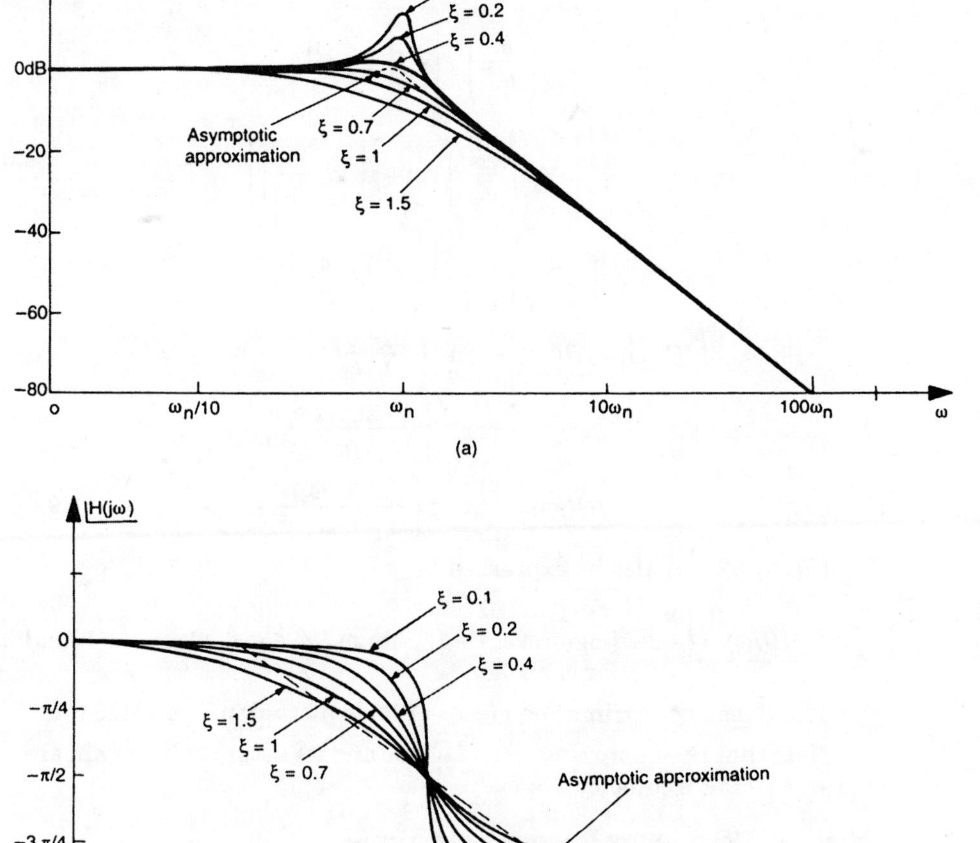

Fig. 11.16 ***Bode plot second order continuous-time LTI systems for various value of damping ratio*** ζ.

$$H(j\omega) = \frac{\omega_n^2}{(j\omega)^2 + 2\zeta\omega_n(j\omega) + \omega_n^2}$$

$$= \frac{1}{-\left(\frac{\omega}{\omega_n}\right)^2 + 2\zeta j\left(\frac{\omega}{\omega_n}\right) + 1}$$

$$= \frac{1}{\left[1 - \left(\frac{\omega}{\omega_n}\right)^2\right] + j\left[2\zeta\frac{\omega}{\omega_n}\right]}$$

$$= \left\{\left[1 - \left(\frac{\omega}{\omega_n}\right)^2\right] + j\left[2\zeta\frac{\omega}{\omega_n}\right]\right\}^{-1}$$

$$= \left[1 - \left(\frac{\omega}{\omega_n}\right)^2\right] - j\left[2\zeta\frac{\omega}{\omega_n}\right]$$

$$\text{Phase of } H(j\omega) = \underline{/H(j\omega)} = \tan^{-1}\left[\frac{H_I(j\omega)}{H_R(j\omega)}\right]$$

$$= \tan^{-1}\left\{\frac{[-2\zeta(\omega/\omega_n)]}{[1-(\omega/\omega_n)^2]}\right\}$$

or
$$\underline{/H(j\omega)} = -\tan^{-1}\left\{\frac{2\zeta(\omega/\omega_n)}{1-(\omega/\omega_n)^2}\right\} \quad ...(11.92)$$

Eq. 11.92 can also be expressed as

$$\underline{/H(j\omega)} = \begin{cases} 0, & \text{for } \omega \le \omega_n/10 \\ -\pi/2\,[\log_{10}(\omega/\omega_n) + 1], & \text{for } \omega_n/10 \le \omega \le 10\omega_n \\ -\pi, & \text{for } \omega \ge 10\omega_n \end{cases} \quad ...(11.93)$$

The phase approximation of Eq. 11.93 is plotted in Fig. 11.16 (*b*).

Note that the approximation of phase and its actual value again are equal at break frequency $\omega = \omega_n$.

Phase of Frequency Response at $\omega = \omega_n$

Substituting $\omega = \omega_n$ in Eq. 11.92, we get

$$\underline{/H(j\omega)} = -\tan^{-1}\left[\frac{2\zeta(\omega/\omega_n)}{1-(\omega/\omega_n)^2}\right]$$

or
$$\underline{/H(j\omega_n)} = -\tan^{-1}\left[\frac{2\zeta(\omega_n/\omega_n)}{1-(\omega_n/\omega_n)^2}\right] = -\tan^{-1}(\infty) = -\frac{\pi}{2}$$

or
$$\underline{/H(j\omega_n)} = -\frac{\pi}{2} \quad ...(11.93)$$

We observe that the asymptotic approximation for magnitude and phase (Eqs. 11.91 and 11.93) for a second order continuous-time LTI system do not depend on the damping ratio ζ, while the actual plots of

$|H(j\omega)|$ and $\underline{/H(j\omega)}$ depend on the damping ratio ζ. Thus, for obtaining an acurate sketch (especially near the break frequency $\omega = \omega_n$) we must modify the approximations to conform more closely to the actual plots. The discrepancy is most pronounced for small values of damping ratio ζ. In this case the actual log magnitude has a peak around $\omega = \omega_n$.

From straight-forward calculations using Eq. 11.90, we will determine maximum value of ω and its magnitude.

If $$\zeta < \frac{\sqrt{2}}{2} = \frac{1}{\sqrt{2}} \approx 0.707$$

then magnitude of $H(j\omega)$ will have maximum value at

$$\omega_{max} = \omega_n \sqrt{1 - 2\zeta^2} \qquad \text{...(11.95)}$$

$$|H(j\omega_{max})| = \frac{1}{2\zeta\sqrt{1-\zeta^2}} \qquad \text{...(11.96)}$$

For $\zeta > 0.707$, the frequency response $H(j\omega)$ decreaes monotonically as frequency increases from zero. The fact that $H(j\omega)$ can have a peak is extremely important in the design of frequency selective filters and amplifiers. In some applications, we prefer to design certain circuits that have a sharp peak in the magnitude of the frequency response at some specified frequency. Therefore, it provides large frequency selective amplification for sinusoids at frequencies within a narrow band. The quality factor Q of such a circuit is defined to be a measure of sharpness of the peak.

The quality factor Q for a second-order circuit described by linear constant-coefficient differential equation (Eq. 11.65) is given as

$$Q = \frac{1}{2\zeta} \qquad \text{...(11.97)}$$

The lesser damping in a system, the sharper is the peak in magnitude of $H(j\omega)$.

Bode Plots for Rational Frequency Responses

The first-order and second-order systems are used as basic building blocks for more complex LTI systems with rational frequency responses. Here we will discuss Bode plots that will essentially provide us with all the information, we need to construict Bode plots for arbitrary rational frequency responses. Here we will obtain the Bode plots for frequency responses of the form

$$H(j\omega) = 1 + j\omega\tau \qquad \text{...(11.98)}$$

and $$H(j\omega) = 1 + 2\zeta\left(\frac{j\omega}{\omega_n}\right) + \left(\frac{j\omega}{\omega_n}\right)^2 \qquad \text{...(11.99)}$$

The Bode plots for Eqs. 11.98 and 11.99 follow directly from Fig. 11.16.

We know that

$$20 \log_{10} |H(j\omega)| = 20 \log_{10} \left| \frac{1}{H(j\omega)} \right| \quad ...(11.100)$$

and $$\angle H(j\omega) = -\frac{1}{\angle H(j\omega)} \quad ...(11.101)$$

Now, we consider a continuous-time system that is a constant gain

$$H(j\omega) = K \quad ...(11.102)$$

K can be written as

$$\left.\begin{aligned} K &= |K|\ e^{j.0} \text{ for } K > 0 \\ \text{and } K &= |K|\ e^{j\pi} \text{ for } K < 0 \end{aligned}\right\} \quad ...(11.103)$$

Eq. 11.102 can be written as

$$20 \log_{10} |H(j\omega)| = 20 \log_{10} |K| \quad ...(11.104)$$

$$\angle H(j\omega) = \begin{cases} 0, & \text{for } K > 0 \\ \pi, & \text{for } K < 0 \end{cases} \quad ...(11.105)$$

Since a rational frequency response can be factored into the product of a constant gain and first-order terms and second-order terms.

Bode plots can be obtaind by adding the plots for each of these terms.

Example 11.1 Find the Bode plot for the frequency response of a second-order continuous-time LTI systems.

Frequency response, $H(j\omega) = \dfrac{2 \times 10^4}{(j\omega)^2 + 100(j\omega) + 10^4}$

Solution. Let us assume that

$$H(j\omega) = 2\, H'(j\omega) \quad ...(1)$$

where $H'(j\omega)$ has the same form as the standard second-order frequency response that is given as

$$H(j\omega) = \frac{\omega_n^2}{(j\omega)^2 + 2\zeta\omega_n(j\omega) + \omega_n^2} \quad ...(2)$$

Determination of Magnitude of Eq. 1

$$|H(j\omega)| = 2|H'(j\omega)| \quad ...(3)$$

Taking the 20 $\log_{10}$ of both sides of Eq. 3, we get

$$20 \log_{10} |H(j\omega)| = 20 \log_{10} \{2\ |H'(j\omega)|\}$$

or $$20 \log_{10} |H(j\omega)| = 20 \log_{10} (2) + 20 \log_{10} H'(j\omega) \quad ...(4)$$

By comparing $H'(j\omega)$ with standard frequency response of second-order continuous-time LTI system given in Eq. 2, we get

$$\omega_n^2 = 10^4 \text{ or } \omega_n = 100$$

and $$2\zeta\omega_n = 100 \text{ or } 2\zeta \times 100 = 100 \text{ or } \zeta = \frac{1}{2}$$

for frequency response $H'(j\omega)$.

Determination of Asymptotes for 20 log$_{10}$ $|H'(j\omega)|$

We will determine asymptotes for 20 log$_{10}$ $|H'(j\omega)|$ by using following expression

$$20\log_{10}|H'(j\omega)| \approx \begin{cases} 0\,, & \text{for } \omega \le \omega_n \\ -40\log_{10}(\omega) + 40\log_{10}(\omega_n)\,, & \text{for } \omega \ge \omega_n \end{cases}$$

$$\approx \begin{cases} 0\,, & \text{for } \omega \le 100 \\ -40\log_{10}(\omega) + 40\log_{10}(100)\,, & \text{for } \omega \ge 100 \end{cases}$$

$$\text{or } 20\log_{10}|H'(j\omega)| \approx \begin{cases} 0\,, & \text{for } \omega \le 100 \\ -40\log_{10}(\omega) + 80\,, & \text{for } \omega \ge 100 \end{cases} \quad ...(5)$$

20 log$_{10}$ $|H(j\omega)|$ will have the same asymptotes except for a constant offset at all frequencies due to the addition of the 20 log$_{10}$(2) ≈ 6 dB. The dotted lines in Fig. 11.17 (*a*) shows these asymptotes. The bold lines in the same figure represents actual Bode plot that is generated by computer for 20 log$_{10}$ $|H(j\omega)|$.

Determination of a Plot of Phase $\angle H(j\omega)$

$$\angle H(j\omega) = \angle H'(j\omega) \quad ...(6)$$

We will determine asymptote for $\angle H'(j\omega)$ by using the following expression

$$\angle H'(j\omega) = \begin{cases} 0\,, & \text{for } \omega \le \omega_n/10 \\ -\pi/2\,[\log_{10}(\omega/\omega_n) + 1]\,, & \text{for } \omega_n/10 \le \omega \le 10\omega_n \\ -\pi\,, & \text{for } \omega \ge 10\omega_n \end{cases}$$

$$= \begin{cases} 0\,, & \text{for } \omega \le 100/10 = 10 \\ -\pi/2\,[\log_{10}(\omega/100) + 1]\,, & \text{for } 100/10 \le \omega \le 10 \times 100 \\ -\pi\,, & \text{for } \omega \ge 10 \times 100 \end{cases}$$

$$\text{or} \quad \angle H'(j\omega) = \begin{cases} 0\,, & \text{for } \omega \le 10 \\ -\pi/2\,[\log_{10}(\omega/100) + 1]\,, & \text{for } 10 \le \omega \le 1000 \\ -\pi\,, & \text{for } \omega \ge 1000 \end{cases} \quad ...(7)$$

The asymptotes of phase of $H(j\omega)$ and actual value for $\angle H(j\omega)$ are plotted in with dotted and bold lines, respectively in Fig. 11.17 (*b*).

Example 11.2 **Find the Bode plot for the frequency response of a continuous-time system given as**

$$H(j\omega) = \frac{100\,(1 + j\omega)}{(10 + j\omega)\,(100 + j\omega)}$$

Solution. **For obtaining the Bode plot, we rewrite the expression for $H(j\omega)$ in the following factored form**

$$H(j\omega) = \frac{100\,(1 + j\omega)}{(10 + j\omega)\,(100 + j\omega)}$$

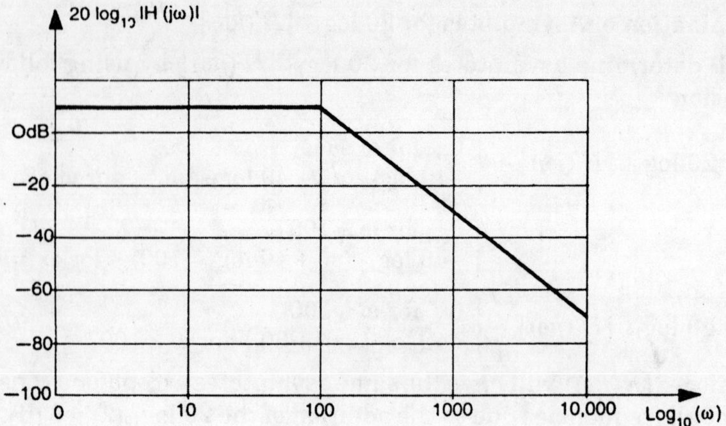

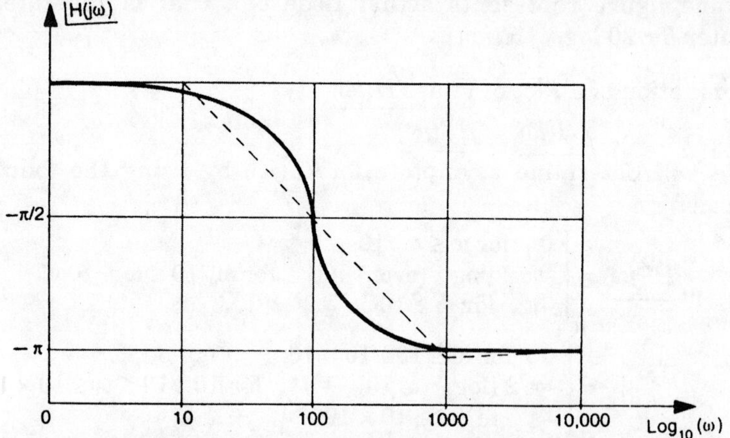

Fig. 11.17 *Bode plot for the frequency response*

$$H(j\omega) = \frac{2 \times 10^4}{(j\omega)^2 + 100(j\omega) + 10^4}$$

(*a*) *Magnitude;* 20 $\log_{10} |H(j\omega)|$ *vs* $\log_{10}(\omega)$ (*b*) *Phase;* $\angle H(j\omega)$ *vs* $\log_{10}(\omega)$

$$= \frac{100\,(1 + j\omega)}{10\left(1 + \dfrac{j\omega}{10}\right)100\left(1 + \dfrac{j\omega}{100}\right)}$$

or
$$H(j\omega) = \underset{\text{I}}{\left(\frac{1}{10}\right)}\underset{\text{II}}{\left(\frac{1}{1 + \dfrac{j\omega}{10}}\right)}\underset{\text{III}}{\left(\frac{1}{1 + \dfrac{j\omega}{100}}\right)}\underset{\text{IV}}{(1 + j\omega)} \qquad ...(1)$$

Here, the first factor is a constant, next two factors have the standard form for the first-order frequency response and fourth factor is the reciprocal of the same first-order standard form.

The Bode plot for 20 $\log_{10} |H(j\omega)|$ is therefore the sum of the Bode plots corresponding to each of the factors. We can obtain the asymptotes for the overall Bode plot by adding the asymptotes corresponding to each factor of the frequency response $H(j\omega)$. The asymptotes corresponding to each factor and the actual values of 20 $\log_{10} |H(j\omega)|$ are shown in Fig. 11.18 (*a*). The constant factor of 1/10 accounts for an offset of –20 dB at each frequuency. The break frequency at $\omega = 1$ corresponds to the $(1 + j\omega)$ factor, which produces the 20 dB/decade rise which starts at

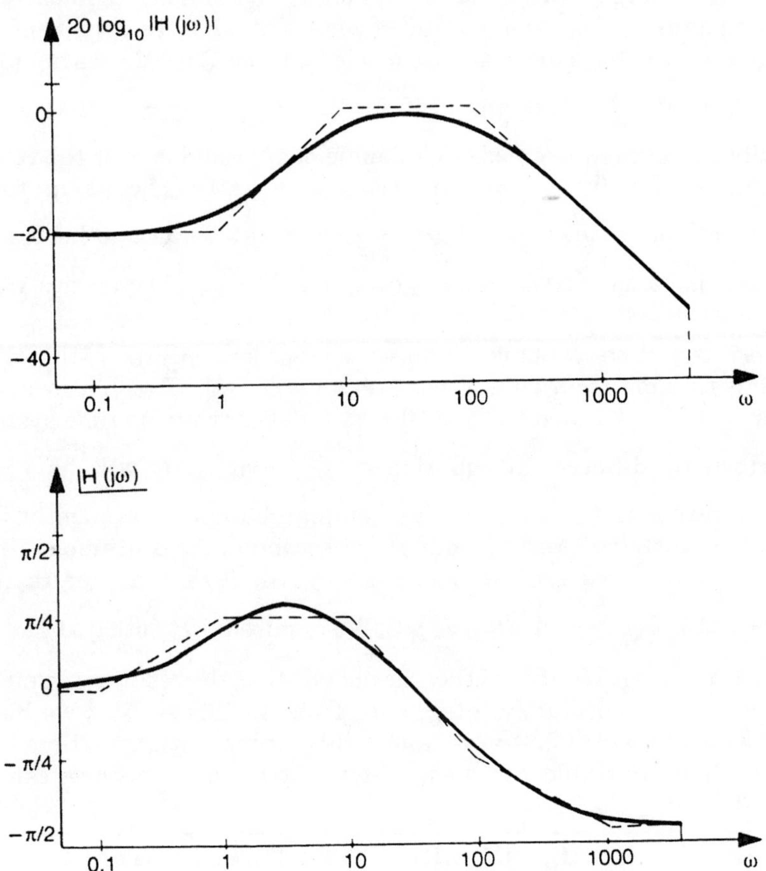

Fig. 11.18 *Bode plot for the frequency response*

$$H(j\omega) = \frac{100\,(1 + j\omega)}{(10 + j\omega)\,(100 + j\omega)}$$

(*a*) *Magnitude;* 20 $\log_{10} |H(j\omega)|$ *vs* $\log_{10}(\omega)$ (*b*) *Phase;* $\angle H(j\omega)$ *vs* $\log_{10}(\omega)$

$\omega = 1$ and is canceled by the 20 dB/decade decay that starts at the break frequency at $\omega = 10$. This 20 dB/decade decay is due to the factor $\frac{1}{[1 + (j\omega/10)]}$.

Finally, the $\frac{1}{[1 + (j\omega/100)]}$ factor of $H(j\omega)$ gives another break frequency at $\omega = 100$ and provides a subsequent decay of 20 dB/decade.

In a similar fashion, we can construct the asymptotic approximation for phase of $H(j\omega)$ from the individual asymptotes of phase for each factor of $H(j\omega)$. The asymptotic approximation of phase of $H(j\omega)$ and the exact value of the phase are plotted in Fig. 11.18 (*b*). The constant factor (1/10) contributes zero to the phase. The factor $(1 + j\omega)$ contributes an asymptotic approximation that is zero for $\omega < 1/10$ and rises linearly as a function of $\log_{10}(\omega)$ from a value of zero at $\omega = 1/10$ to a value of $\pi/2$ radians at $\omega = 10$. This rise is canceled at $\omega = 1$ by the asymptotic approximation for the angle of factor $\frac{1}{[1 + (j\omega/100)]}$. This factor contributes a linear decrease in an angle of $\pi/2$ radians over the range of frequencies from $\omega = 1$ to $\omega = 100$. Finally, the asymptotic approximation for the phase of $\frac{1}{[1 + (j\omega/100)]}$ contributes another linear decrease in phase of $\pi/2$ radians over the range of frequencies from $\omega = 10$ to $\omega = 1000$.

In this section, we have discussed first-order continuous-time LTI systems for time $\tau > 0$. That is, the system is causal. We can also easily check that if $\tau < 0$, then the causal first-order continuous-time system described by differential equation $\tau \frac{d\,y(t)}{dt} + y(t) = s(t)$ will have an impulse response that is not absolutely integrable and consequently, the system is unstable. Also in analysis of second-order continuous-time causal LTI system described by differential equation $\frac{d^2 y(t)}{dt^2} + 2\zeta\omega_n \frac{d\,y(t)}{dt} + \omega_n^2\, y(t) = \omega_n^2\, s(t)$. We required that both ζ and ω_n^2 be positive numbers. If either of these is not positive, the resulting impulse response is not absolutely integrable. Thus, in this section, we have discussed only causal first-order and second-order continuous-time LTI systems that are stable. For these systems, frequency responses can be defined.

11.7 LOWER ORDER DISCRETE-TIME LTI SYSTEMS

In this section, we will examine the important properties of the first-order and second-order discrete-time LTI systems that are described by linear constant-coefficient difference equations. Any discrete-time LTI system that is discribed by a linear constant-coefficient difference equations can be expressed as product or sum of first-order and second-order discrete-time systems. These lower-order discrete-time systems are of considerable value in both implementing and analyzing more complex discrete-time LTI systems.

11.7.1 First-Order Discrete-Time LTI Systems

First-order discrete-time LTI system is described by the linear constant-coefficient difference equation given as

$$y(n) - A\,y(n-1) = s(n) \qquad \text{...(11.106)}$$

with constant $|A| < 1$

Determination of Transfer Function $H(z)$

Taking the z-transform of both sides of Eq. 11.106 and assuming that all the initial conditions of above system are zero.

$$\mathcal{Z}\,[y(n) - Ay(n-1)] = \mathcal{Z}\,[s(n)]$$

or $$\mathcal{Z}\,[y(n)] - A\,\mathcal{Z}\,[y(n-1)] = \mathcal{Z}\,[s(n)]$$

or $$Y(z) - Az^{-1}\,Y(z) = S(z)$$

or $$Y(z)\,[1 - Az^{-1}] = S(z)$$

or $$\frac{Y(z)}{S(z)} = \frac{1}{1 - Az^{-1}} \qquad \text{...(11.107)}$$

Transfer function of a discrete-time LTI system is the ratio of z-transform of output $y(n)$ to the z-transform of input $s(n)$ with zero initial conditions. An initially relaxed system has zero initial conditions.

$$H(z) = \frac{z\text{–transform of output } y(n)}{z\text{–transform of input } y(n)} = \frac{Y(z)}{S(z)} \qquad \text{...(11.108)}$$

From Eqs. 11.107 and 11.108, we get transfer function of the first-order discrete-time LTI system.

$$H(z) = \frac{1}{1 - Az^{-1}} \qquad \text{...(11.109)}$$

Frequency response of a first-order discrete-time LTI system can be determined by substituting $z = e^{j\omega}$ in Eq. 11.109. Substituting, we get

$$H(e^{j\omega}) = H(z)\,\big|_{z=e^{j\omega}} = \left[\frac{1}{1 - Az^{-1}}\right]_{z=e^{j\omega}}$$

or $$H(e^{j\omega}) = \frac{1}{1 - A\,(e^{j\omega})^{-1}} = \frac{1}{1 - Ae^{-j\omega}} \qquad \text{...(11.110)}$$

Impulse response of first-order discrete-time LTI system can be determined by taking inverse z-transform of both sides of Eq. 11.109,

Impulse response, $h(n)$ = Inverse z-transform $\{H(z)\}$

$$= \text{Inverse } z\text{-transform}\left\{\frac{1}{1 - Az^{-1}}\right\}$$

or $$h(n) = A^n\, u(n)\,. \qquad ..(11.111)$$

The step response of a first-order discrete-time LTI system can be determined by convolving impulse response $h(n)$ with unit- step sequence $u(n)$ as

$$\text{Step response, } g(n) = h(n) * u(n)$$

$$= \sum_{k=-\infty}^{\infty} h(k)\, u(n-k) \qquad ...(11.112)$$

Substituting the value of impulse response $h(n)$ from Eq. 11.111 in Eq. 11.112, we obtain

$$g(n) = \sum_{k=-\infty}^{\infty} A^k\, u(k)\, u(n-k)$$

using the definition of $u(k)$ and $u(n-k)$

$$u(k) = \begin{cases} 1\,, & k \geq 0 \\ 0\,, & k < 0 \end{cases}$$

and $$u(n-k) = \begin{cases} 1\,, & n-k \geq 0 \text{ or } k \leq n \\ 0\,, & n-k < 0 \text{ or } k > n \end{cases}$$

or $$g(n) = \sum_{k=0}^{n} A^k \qquad ...(11.113)$$

[It is a geometric progression with first term 1 and common ratio A. There will be $(n + 1)$ number of terms in this GP.]

or $$g(n) = \frac{1\,[1 - A^{n+1}]}{[1 - A]} = \left(\frac{1 - A^{n+1}}{1 - A}\right),\; n \geq 0$$

or $$g(n) = \left(\frac{1 - A^{n+1}}{1 - A}\right) u(n) \qquad ...(11.114)$$

It is the step response of first order discrete-time LTI system.

The magnitude of the parameter A plays a role similar to that of the time constant τ in the first-order continuous- time LTI system. Magnitude of parameter A is denoted by $|A|$. It is used to determine the rate at which the first- order system responds.

From equations

$$h(n) = A^n\, u(n) \text{ and } g(n) = \left[\frac{1 - A^{n+1}}{1 - A}\right] u(n)$$

and Figs. 11.19 and 11.20, we see that impulse response $h(n)$ and step response $g(n)$ converge to their final value at the rate at which $|A|^n$ converges to zero. The impulse response $h(n)$ decays sharply and step response $g(n)$ settles quickly for smaller values of $|A|$. For $|A|$ nearer to unity, the impulse and step responses are slower. The first-order discrete-time LTI system described by Eq. 11.106 displays oscillatory behaviour. But first-order continuous-time LTI system described by linear constant-coefficient differential equation does not display

oscillatory behaviour. This type of oscillatory behaviour occurs where $A < 0$. In this case the step response processes both overshoot of its final value and rining (oscillatory behaviour).

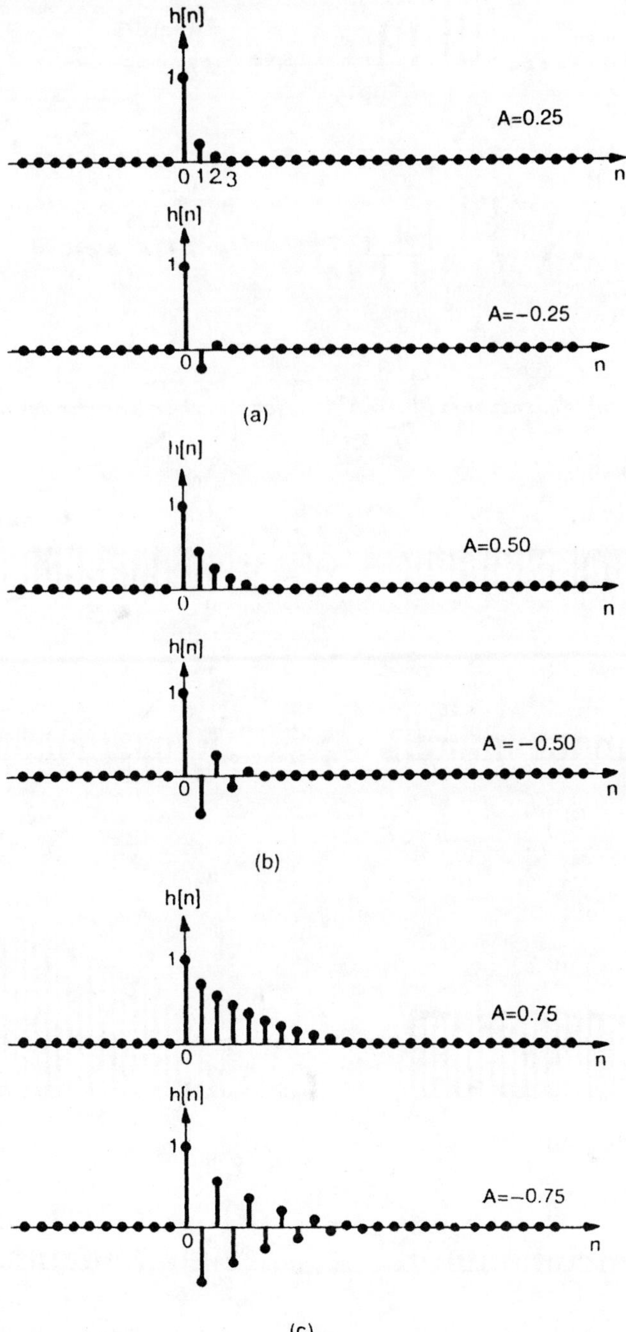

Fig. 11.19 *Impulse response $h(n) = A^n\, u(n)$ for a first-order discrete-time LTI system.*

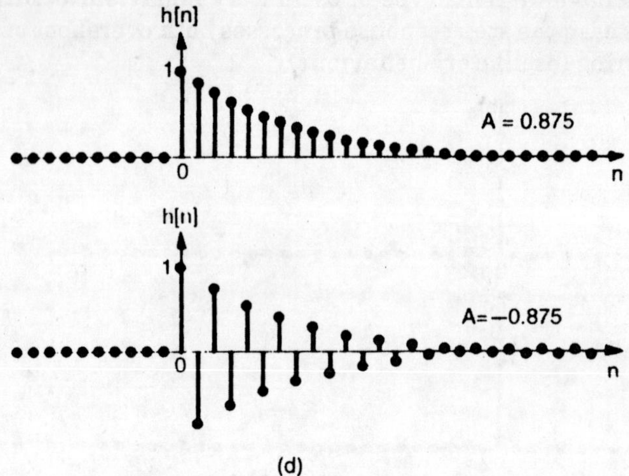

Fig. 11.19 ***Impulse response $h(n) = A^n\, u(n)$ for a first-order discrete-time LTI system.***

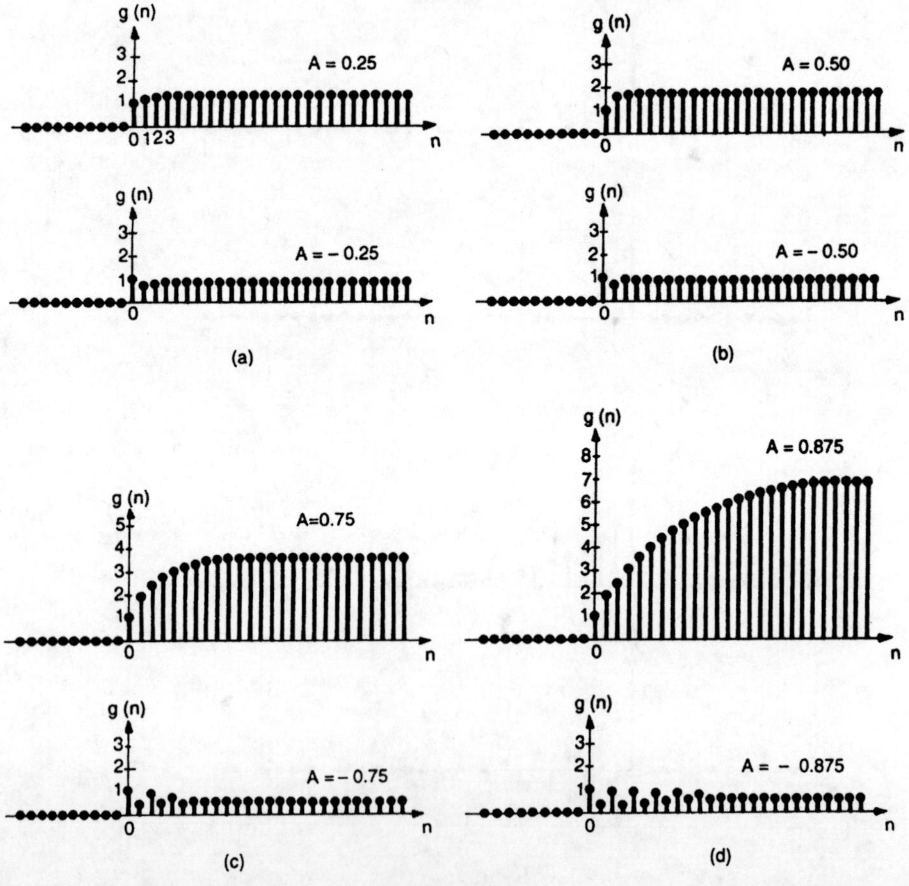

Fig. 11.20 ***Step response $g(n) = \left[\dfrac{1 - A^{n+1}}{1 - A}\right] u(n)$ of a first-order discrete-time LTI system.***

Determination of Magnitude and Phase of the Frequency Response of the First-Order Discrete-Time LTI Systems given by Eq. 11.106

Frequency response, $H(e^{j\omega}) = \dfrac{1}{1 - Ae^{-j\omega}}$

$$= \frac{1}{1 - A\,(\cos\omega - j\sin\omega)}$$

or $$H(e^{j\omega}) = \frac{1}{(1 - A\cos\omega) + j(A\sin\omega)} \quad ...(11.115)$$

Magnitude of $H(e^{j\omega}) = |H(e^{j\omega})| = \dfrac{1}{\sqrt{(1 - A\cos\omega)^2 + (A\sin\omega)^2}}$

or $$|H(e^{j\omega})| = \frac{1}{\sqrt{1 + A^2 - 2A\cos\omega}} \quad ...(11.116)$$

Phase of $H(e^{j\omega}) = \angle H(e^{j\omega}) = \tan^{-1}\left[\dfrac{-A\sin\omega}{1 - A\cos\omega}\right]$

or $$\angle H(e^{j\omega}) = -\tan^{-1}\left[\frac{A\sin\omega}{1 - A\cos\omega}\right] \quad ...(11.117)$$

The log magnitude and phase of the frequency response in Eq. 11.110 are shown in Fig. 11.21 for several values of $A > 0$.

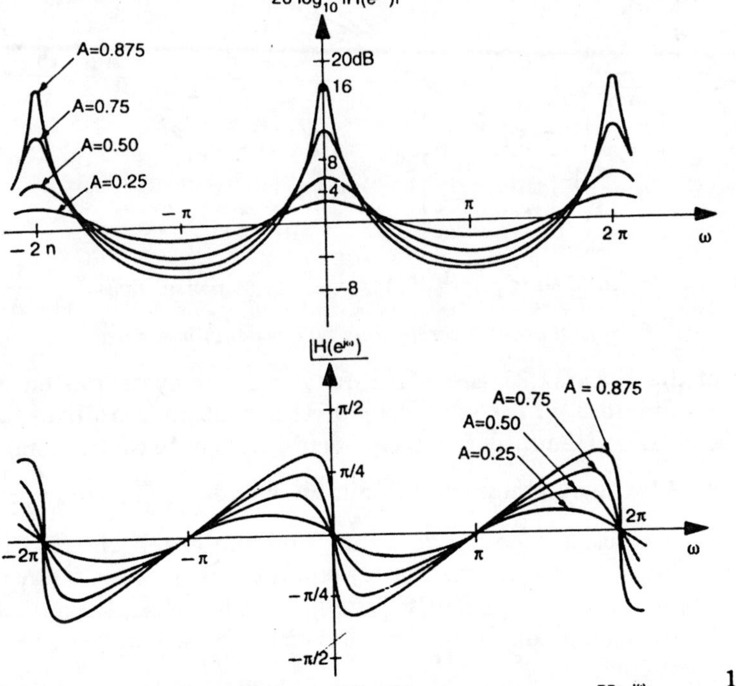

Fig. 11.21 *Magnitude and phase of the frequency response* $H(e^{j\omega}) = \dfrac{1}{(1 - Ae^{-j\omega})}$ *for a first-order discrete-time LTI system for* $A > 0$.

The log magnitude and phase of the frequency response in Eq. 11.110 is shown in Fig. 11.22 for several values of $A < 0$.

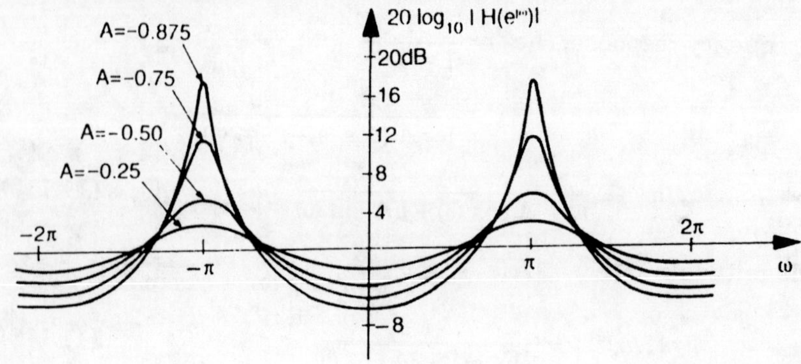

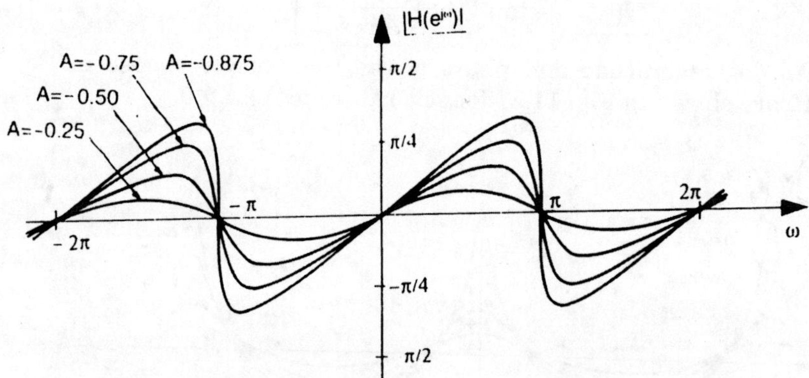

Fig. 11.22 *Magnitude and phase of the frequency response* $H(e^{j\omega}) = \dfrac{1}{(1 - Ae^{-j\omega})}$ *for a first order discrete-time LTI system for* $A < 0$.

From the Figs. 11.21 and 11.22, for $A > 0$, the system attenuates high frequencies, while when $A < 0$, the system amplifies high frequencies and attenuates low frequencies. Also note that for smaller values of $|A|$, the maximum and minimum values, $\dfrac{1}{(1+A)}$ and $\dfrac{1}{(1-A)}$, of $|H(e^{j\omega})|$ are close together in value, and the graph of $|H(e^{j\omega})|$ is relatively flat. For $|A|$ approximately equal to unity, these quantities differ significantly. Consequently $|H(e^{j\omega})|$ is more sharply peaked. It provides filtering and amplification that is more selective over a narrow band of frequencies.

11.7.2 Second-Order Discrete-Time LTI Systems

Second order discrete-time LTI systems is described by a linear constant-coefficient difference equation given as

$$y(n) - A_1\, y(n-1) - A_2\, y(n-2) = s(n) \qquad ...(11.118)$$

where A_1 and A_2 are constant coefficients.

$$A_1 = 2r \cos\theta \;\text{ and }\; A_2 = -r^2$$

with $0 < r < 1$ and $0 \le \theta \le \pi$

Eq. 11.118 can also be written as

$$y(n) - 2\, r \cos\theta\, y(n-1) + r^2\, y(n-2) = s(n) \qquad ...(11.119)$$

Determination of Transfer Function of the Second Order Discrete-time LTI System

It is obtained by taking z-transform of both sides of Eq. 11.119 for an initially relaxed system. All the initial conditions are zero for this system.

$$\mathcal{Z}\,[y(n) - 2\, r \cos\theta\, y(n-1) + r^2\, y(n-2)] = \mathcal{Z}\,[s(n)]$$

or $$\mathcal{Z}\,[y(n)] - 2\, r \cos\theta\, \mathcal{Z}\,[y(n-1)] + r^2\, \mathcal{Z}\,[y(n-2)] = \mathcal{Z}\,[s(n)]$$

or $$Y(z) - 2\, r \cos\theta\, z^{-1}\, Y(z) + r^2\, z^{-2}\, Y(z) = S(z)$$

or $$\frac{Y(z)}{S(z)} = \frac{1}{(1 - 2\, r \cos z^{-1} + r^2\, z^{-2})} \qquad ...(11.120)$$

But transfer function of a discrete-time LTI system is defined as

$$H(z) = \frac{Y(z)}{S(z)} = \frac{z\text{–transform of output } y(n)}{z\text{–transform of input } s(n)} \qquad ...(11.121)$$

From Eqs. 11.120 and 11.121, we obtain transfer function

$$H(z) = \frac{1}{(1 - 2r \cos\theta\, z^{-1} + r^2\, z^{-2})} \qquad ...(11.122)$$

This is the transfer function or system function of the second order discrete-time LTI system described by difference equation in Eq. 11.119.

Frequency response is determined by substituting $z = e^{j\omega}$ in the expression of transfer function of the system of Eq. 11.122.

Frequency $H(e^{j\omega}) = H(z)\,\big|_{z = e^{j\omega}}$

$$= \left[\frac{1}{1 - 2r \cos\theta\, z^{-1} + r^2\, z^{-2}}\right]_{z = e^{j\omega}}$$

$$= \frac{1}{1 - 2r \cos\theta\, z^{-1} + r^2\, z^{-2}}$$

or $$H(e^{j\omega}) = \frac{1}{1 - 2r \cos\theta\, e^{-j\omega} + r^2 e^{-j2\omega}} \qquad ...(11.123)$$

Impulse response is determined by taking inverse z-transform of the transfer function given in Eq. 11.122.

Impulse response, $h(n)$ = Inverse z-transform $\{H(z)\}$

$$= \text{Inverse } z\text{–transform}\left\{\frac{1}{1-2r\cos\theta\, z^{-1}+r^2 z^{-2}}\right\}$$

$$= \text{Inverse } z\text{–transform}\left\{\frac{1}{1-2r\left[\dfrac{e^{j\theta}-e^{-j\theta}}{2}\right]z^{-1}+r^2 z^{-2}}\right\}$$

$$= \text{Inverse } z\text{–transform}\left\{\frac{1}{[1-(re^{j\theta})\,z^{-1}]\,[1-(re^{-j\theta})\,z^{-1}]}\right\}$$

(using partial fraction expansion)

$$= \text{Inverse } z\text{–transform}\left\{\frac{\alpha_1}{[1-(re^{j\theta})\,z^{-1}]}+\frac{\alpha_2}{[1-(re^{-j\theta})\,z^{-1}]}\right\}$$

where $\alpha_1 = \dfrac{e^{j\theta}}{2j\sin\theta}$, $\alpha_2 = \dfrac{-e^{-j\theta}}{2j\sin\theta}$

or $$h(n) = \alpha_1 \text{ Inverse } z\text{–transform}\left\{\frac{1}{[1-(re^{j\theta})\,z^{-1}]}\right\} + \alpha_2 \text{ Inverse } z\text{–transform}\left\{\frac{1}{[1-(re^{-j\theta})\,z^{-1}]}\right\}$$

or $$h(n) = \alpha_1\,(re^{j\theta})^n\,u(n) + \alpha_2\,(re^{-j\theta})^n\,u(n) \qquad ...(11.124)$$

Substituting values of α_1 and α_2 in Eq. 11.124, we obtain

$$h(n) = \alpha_1\,(re^{j\theta})^n\,u(n) + \alpha_2\,(re^{-j\theta})^n\,u(n)$$

$$= \left[\frac{e^{j\theta}}{2j\sin\theta}(re^{j\theta})^n + \frac{-e^{-j\theta}}{2j\sin\theta}(re^{-j\theta})^n\right]u(n)$$

$$= \left[\frac{r^n e^{j(n+1)\theta} - r^n e^{-j(n+1)\theta}}{2j\sin\theta}\right]u(n)$$

$$= r^n\,\frac{1}{\sin\theta}\left[\frac{e^{j(n+1)\theta} - e^{-j(n+1)\theta}}{2j}\right]u(n)$$

or $$h(n) = r^n\,\frac{\sin[(n+1)\,\theta]}{\sin\theta}\,u(n) \qquad ...(11.125)$$

Special case-I : for $\theta = 0°$

Frequency response, $$H(e^{j\omega}) = \frac{1}{1-2r\cos\theta\,e^{-j\omega}+r^2 e^{-j2\omega}}$$

$$= \frac{1}{1-2r\cos(0)\,e^{-j\omega}+r^2 e^{-j2\omega}}$$

$$= \frac{1}{1-2r\times 1\times e^{-j\omega}+r^2 e^{-j2\omega}}$$

$$= \frac{1}{1 - 2r\, e^{-j\omega} + r^2\, e^{-j2\omega}}$$

or $$H(e^{j\omega}) = \frac{1}{(1 - re^{-j\omega})^2} \quad \text{...(11.126)}$$

Impulse response, $h(n) = r^n \dfrac{\sin[(n+1)\,\theta]}{\sin\theta}\, u(n)$

using L' Hospital's rule for determining equation $\dfrac{\sin[(n+1)\,\theta]}{\sin\theta}$ **at** $\theta = 0°$

$$\lim_{\theta \to 0} \frac{\frac{d}{d\theta}\{\sin[(n+1)\,\theta]\}}{\frac{d}{d\theta}\{\sin\theta\}} = \lim_{\theta \to 0}\left[\frac{(n+1)\cos(n+1)\,\theta}{\cos\theta}\right]$$

$$= \frac{(n+1)\cos(n+1) \times 0}{\cos(0)}$$

$$= \frac{(n+1)\cos(0)}{\cos(0)} = n + 1$$

Therefore, impulse response will be given as

$$h(n) = (n+1)\, r^n\, u(n) \quad \text{...(11.127)}$$

Special case II : for $\theta = \pi$

Frequency response, $H(e^{j\omega}) = \dfrac{1}{1 - 2r\cos\theta\, e^{-j\omega} + r^2\, e^{-j2\omega}}$

$$= \frac{1}{1 - 2r\cos(\pi)\, e^{-j\omega} + r^2\, e^{-j2\omega}}$$

$$= \frac{1}{1 + 2r\, e^{-j\omega} + r^2\, e^{-j2\omega}}$$

or $$h(n) = \frac{1}{(1 + re^{-j\omega})^2} \quad \text{...(11.128)}$$

Impulse response, $h(n) = r^n \dfrac{\sin[(n+1)\,\theta]}{\sin\theta}\, u(n)$

$$= r^n \frac{\sin[(n+1)\,\pi]}{\sin(\pi)}\, u(n)$$

$$= r^n\, (n+1)\, (-r)^n\, u(n)$$

or $$h(n) = (n+1)\, (-r)^n\, u(n) \quad \text{...(11.129)}$$

From Eq. 11.125, we see that the rate of decay of impulse response $h(n)$ **is controlled by variable** r**. If** r **is nearly unity then the decay of** $h(n)$ **will be slower. Similarly, the value of** θ **determines the frequency of oscillation. If** $\theta = 0°$**, then there will be no oscillation in the impulse response of the system but if** $\theta = \pi$ **then the oscillation will be very rapid.**

We can also see the effect of different values of r and θ in the step response of the second-order discrete-time LTI system.

Determination of Step Response of the Second-Order Discrete-Time LTI System

Step response $g(t)$ can be determined by convolving impulse response $h(n)$ with unit-step sequence $u(n)$ as

$$g(n) = h(n) * u(n) = \sum_{k=-\infty}^{\infty} h(k)\, u(n-k) \qquad \text{...(11.130)}$$

Substituting the value $h(n)$ from Eq. 11.125 in Eq. 11.130, we get

$$g(n) = \sum_{k=-\infty}^{\infty} \left\{ \frac{r^k \sin[(k+1)\theta]}{\sin\theta}\, u(k)\, u(n-k) \right\}$$

Using the definition of $u(k)$ amd $u(n-k)$

$$u(k) = \begin{cases} 1, & k \geq 0 \\ 0, & k < 0 \end{cases}$$

$$u(n-k) = \begin{cases} 1, & n-k \geq 0 \text{ or } k \leq n \\ 0, & n-k < 0 \text{ or } k > n \end{cases}$$

or
$$g(n) = \sum_{k=0}^{n} r^k \frac{\sin[(k+1)\theta]}{\sin(\theta)} \times 1 \times 1$$

$$= \sum_{k=0}^{n} r^k \frac{\sin[(k+1)\theta]}{\sin(\theta)}$$

$$= r^0 \frac{\sin[(0+1)\theta]}{\sin(\theta)} + r^1 \frac{\sin[(1+1)\theta]}{\sin(\theta)} + r^2 \frac{\sin[(2+1)\theta]}{\sin(\theta)} + \ldots + r^n \frac{\sin(n+1)\theta]}{\sin(\theta)}$$

$$= r^0 \frac{\sin\theta}{\sin\theta} + r\frac{\sin 2\theta}{\sin\theta} + r^2 \frac{\sin 3\theta}{\sin\theta} + r^3 \frac{\sin 4\theta}{\sin\theta} + \ldots + r^n \frac{\sin[(n+1)\theta]}{\sin\theta}$$

$$= \frac{1}{\sin\theta}[\sin\theta + r\sin 2\theta + r^2 \sin 3\theta + \ldots + r^n \sin(n+1)\theta]$$

$$= \frac{1}{\sin\theta}\left[\left(\frac{e^{j\theta} - e^{-j\theta}}{2j}\right) + r\left(\frac{e^{j2\theta} - e^{-j2\theta}}{2j}\right) + r^2\left(\frac{e^{j3\theta} - e^{-j3\theta}}{2j}\right) + \ldots + r^n\left(\frac{e^{j(n+1)\theta} - e^{-j(n+1)\theta}}{2j}\right)\right]$$

$$= \frac{1}{2j\sin\theta}[e^{j\theta} + re^{j2\theta} + r^2 e^{j3\theta} + \ldots + r^n e^{j(n+1)\theta}] - \frac{1}{2j\sin\theta}[e^{-j\theta} + re^{-j2\theta} + r^2 e^{-j3\theta} + \ldots + r^n e^{-j(n+1)\theta}]$$

$$= \frac{1}{2j\sin\theta}\left\{e^{j\theta}\left(\frac{1-(re^{j\theta})^{n+1}}{1-re^{j\theta}}\right)\right\}$$

$$-\frac{1}{2j\sin\theta}\left\{e^{-j\theta}\left(\frac{1-(re^{-j\theta})^{n+1}}{1-re^{-j\theta}}\right)\right\}$$

$$= \frac{e^{j\theta}}{2j\sin\theta}\left(\frac{1-(re^{j\theta})^{n+1}}{1-re^{j\theta}}\right) - \frac{e^{-j\theta}}{2j\sin\theta}\left(\frac{1-(re^{-j\theta})^{n+1}}{1-re^{-j\theta}}\right)$$

$$= \alpha_1\left(\frac{1-(re^{j\theta})^{n+1}}{1-re^{j\theta}}\right) + \alpha_2\left(\frac{1-(re^{-j\theta})^{n+1}}{1-re^{j\theta}}\right) \qquad \text{for } n \geq 0$$

or $$g(n) = \left\{\alpha_1\left(\frac{1-(re^{j\theta})^{n+1}}{1-re^{j\theta}}\right) + \alpha_2\left(\frac{1-(re^{-j\theta})^{n+1}}{1-re^{-j\theta}}\right)\right\}u(n)$$

...(11.130)

where $\alpha_1 = \dfrac{e^{j\theta}}{2j\sin\theta}$ and $\alpha_2 = \dfrac{-e^{-j\theta}}{2j\sin\theta}$

Special case I : Step response $g(n)$ for $\theta = 0°$

$$g(n) = \frac{e^{j\theta}}{2j\sin\theta}\left[\frac{1-(re^{j\theta})^{n+1}}{1-re^{j\theta}}\right]u(n) - \frac{e^{-j\theta}}{2j\sin\theta}\left[\frac{1-(re^{-j\theta})^{n+1}}{1-re^{-j\theta}}\right]u(n)$$

$$= \left\{\frac{e^{j\theta}(1-re^{j\theta})\,[1-(re^{j\theta})^{n+1}] - e^{-j\theta}(1-re^{j\theta}\,[1-(re^{-j\theta})^{n+1}]}{2j\sin\theta\,(1-re^{j\theta})\,(1-re^{-j\theta})}\right\}u(n)$$

$$= \left\{\frac{(e^{j\theta}-r)\,[1-(re^{j\theta})^{n+1}] - (e^{-j\theta}-r)\,[1-(re^{-j\theta})^{n+1}]}{2j\sin\theta\,(1-re^{j\theta})\,(1-re^{-j\theta})}\right\}u(n)$$

$$= \left\{\frac{e^{j\theta}-(r)^{n+1}(e^{j\theta})^{n+2} - r + (r)^{n+2}(e^{j\theta})^{n+1} - e^{-j\theta} + (r)^{n+1}(e^{-j\theta})^{n+2} + r - (r)^{n+2}(e^{-j\theta})^{n+1}}{2j\sin\theta\,(1-re^{j\theta})\,(1-re^{-j\theta})}\right\}u(n)$$

$$= \left\{\frac{\left(\dfrac{e^{j\theta}-e^{-j\theta}}{2j}\right) - \left(\dfrac{e^{j(n+2)\theta}-e^{-j(n+2)\theta}}{2j}\right)(r)^{n+1} + \left(\dfrac{e^{j(n+1)\theta}-e^{-j(n+1)\theta}}{2j}\right)(r)^{n+2}}{\sin\theta\,(1-re^{j\theta})\,(1-re^{-j\theta})}\right\}u(n)$$

$$= \left\{\frac{\sin\theta - \sin[(n+2)\theta]\,(r)^{n+1} + \sin[(n+1)\theta]\,(r)^{n+2}}{\sin\theta\,(1-re^{j\theta})\,(1-re^{-j\theta})}\right\}u(n)$$

or $$g(n) = \left\{\frac{1}{(1-re^{j\theta})(1-re^{-j\theta})} - \frac{(r)^{n+1}}{(1-re^{j\theta})(1-re^{-j\theta})}\frac{\sin[(n+2)\theta]}{\sin\theta}\right.$$

$$\left. + \frac{(r)^{n+2}}{(1-re^{j\theta})(1-re^{-j\theta})}\frac{\sin[(n+1)\theta]}{\sin\theta}\right\}u(n) \quad \text{...(11.131)}$$

Step response $g(n$ for $\theta = 0°$ can be determined by using L' Hospital's rule as

$$g(n)\Big|_{\theta=0} = \lim_{\theta\to 0}\{g(n)\}$$

$$= \lim_{\theta\to 0}\left\{\frac{1}{(1-re^{j\theta})(1-re^{-j\theta})} - \frac{(r)^{n+1}}{(1-re^{j\theta})(1-re^{-j\theta})}\frac{\frac{d}{d\theta}\{\sin(n+2)\theta]\}}{\frac{d}{d\theta}\{\sin\theta\}} + \frac{(r)^{n+2}}{(1-re^{j\theta})(1-re^{-j\theta})} \times \frac{\frac{d}{d\theta}\{\sin[(n+1)\theta]\}}{\frac{d}{d\theta}\{\sin\theta\}}\right\} u(n)$$

$$= \left\{\lim_{\theta\to 0}\left\{\frac{1}{(1-re^{j\theta})(1-re^{-j\theta})}\right\} - \lim_{\theta\to 0}\left\{\frac{(r)^{n+1}}{(1-re^{j\theta})(1-re^{-j\theta})}\frac{(n+2)\cos[(n+2)\theta]}{\cos\theta}\right\} + \lim_{\theta\to 0}\left\{\frac{(r)^{n+2}}{(1-re^{j\theta})(1-re^{-j\theta})} \times \frac{(n+1)\cos[(n+1)\theta]}{\cos\theta}\right\}\right\} u(n)$$

$$= \left\{\frac{1}{(1-r)(1-r)} - \frac{(r)^{n+1}(n+2)}{(1-r)(1-r)} + \frac{(r)^{n+2}(n+1)}{(1-r)(1-r)}\right\} u(n)$$

$$= \left\{\frac{1}{(r-1)^2} - \frac{(r)^{n+1}(n+2)}{(r-1)^2} + \frac{(r)^{n+2}(n+1)}{(r-1)^2}\right\} u(n)$$

$$= \left\{\frac{1}{(r-1)^2} - \frac{(r)^{n+1}}{(r-1)^2} + \frac{r(r-1)(n+1)r^n}{(r-1)^2}\right\} u(n)$$

$$g(n)\Big|_{\theta=0} = \left\{\frac{1}{(r-1)^n} - \frac{r}{(r-1)^n}(r)^n + \frac{r}{(r-1)}(n+1)r^n\right\} u(n) \qquad \text{...(11.132)}$$

Special case II : Step response $g(n)$ for $\theta = \pi$.

In similar manner, we determine step response $g(n)$ for $\theta = \pi$. It is expressed as

$$g(n)\Big|_{\theta=\pi} = \left\{\frac{1}{(r+1)^2} + \frac{r}{(r+1)^2}(-r)^n + \frac{r}{(r+1)}(n+1)(-r)^n\right\} u(n) \qquad \text{...(11.133)}$$

The second-order system described by difference equation $y(n) - 2r\cos\theta\, y(n-1) + r^2 y(n-2) = s(n)$ is the counter-part of underdamped second- order continuous-time LTI system described by differential equation

$$\frac{d^2 y(t)}{dt^2} + 2\zeta\omega_n \frac{d\,y(t)}{dt} + \omega_n^2\, y(t) = \omega_n^2\, s(t)$$

If $\theta = 0°$, then the system is called critically damped system. That is, for any value of θ other than zero, the impulse response $h(n)$ has a damped oscillatory behaviour. The step response $g(t)$ of second order discrete- time LTI system exhibits ringing or oscillatory behaviour and overshoot. The parameter r determines how sharply peaked the frequency response is within this band of frequencies.

Upto here, we have seen that the second order discrete-time LTI system described by its frequency response

$$H(e^{j\omega}) = \frac{1}{1 - 2r\cos\theta\, e^{-j\omega} + r^2\, e^{-j2\omega}}$$

has complex coefficients (unless $\theta = 0$ or π). We can also consider second order systems with real coefficients. Now, frequency response can be expressed as

$$H(e^{j\omega}) = \frac{1}{(1 - c_1\, e^{-j\omega})\,(1 - c_2\, e^{-j\omega})} \qquad \text{...(11.134)}$$

where c_1 and c_2 are both real numbers with $|c_1| < 1$ and $|c_2| < 1$.

This frequency response represents a second order discrete-time LTI system described by difference equation given as

$$y(n) - (c_1 + c_2)\, y(n-1) + c_1c_2\, y(n-2) = s(n) \qquad \text{...(11.135)}$$

Determination of Impulse Response $h(n)$

We can determine impulse response of a second-order discrete-time LTI system by taking inverse DTFT of frequency response given in Eq. 11.134 as

$$h(n) = \text{Inverse DTFT } \{H(e^{j\omega})\}$$

or
$$h(n) = \text{Inverse DTFT} \left\{ \frac{1}{(1 - c_1\, e^{-j\omega})\,(1 - c_2\, e^{-j\omega})} \right\}$$

(using partial fraction expansion)

$$= \text{Inverse DTFT} \left\{ \frac{\beta_1}{(1 - c_1\, e^{-j\omega})} + \frac{\beta_2}{(1 - c_2\, e^{-j\omega})} \right\}$$

where $\beta_1 = \dfrac{c_1}{c_1 - c_2}$ and $\beta_2 = \dfrac{c_2}{c_2 - c_1}$

or
$$h(n) = \text{Inverse DTFT} \left\{ \frac{\beta_1}{(1 - c_1\, e^{-j\omega})} \right\} + \text{Inverse DTFT} \left\{ \frac{\beta_2}{(1 - c_2\, e^{-j\omega})} \right\}$$

$$= \beta_1\, (c_1)^n\, u(n) + \beta_2\, (c_2)^n\, u(n)$$

or $$h(n) = [\beta_1 (c_1)^n + \beta_2 (c_2)^n]\, u(n) \quad ...(11.136)$$

This impulse response $h(n)$ is the sum of two decaying real exponentials.

Determination of Step Response $g(n)$

Step response $g(n)$ can be determined by convolving impulse response $h(n)$ and unit-step function $u(n)$ as

$$g(n) = h(n) * u(n) = \sum_{k=-\infty}^{\infty} h(k)\, u(n-k)$$

$$= \sum_{k=-\infty}^{\infty} [\beta_1 (c_1)^k + \beta_2 (c_2)^k]\, u(k)\, u(n-k)$$

using the definition of $u(k)$ and $u(n-k)$

$$u(k) = \begin{cases} 1, & k \geq 0 \\ 0, & k < 0 \end{cases}$$

and $$u(n-k) = \begin{cases} 1, & n-k \geq 0 \text{ or } k \leq n \\ 0, & n-k < 0 \text{ or } k > n \end{cases}$$

or $$g(n) = \sum_{k=0}^{\infty} [\beta_1 (c_1)^k + \beta_2 (c_2)^k] \times 1 \times 1$$

$$= \beta_1 \sum_{k=0}^{n} (c_1)^k + \beta_2 \sum_{k=0}^{n} (c_2)^k$$

$$= \beta_1 \left[\frac{1-c_1^{n+1}}{1-c_1}\right] + \beta_2 \left[\frac{1-c_2^{n+1}}{1-c_2}\right], \text{ for } n \geq 0$$

or $$g(n) = \left\{\beta_1 \left[\frac{1-c_1^{n+1}}{1-c_1}\right] + \beta_2 \left[\frac{1-c_2^{n+1}}{1-c_2}\right]\right\} u(n) \quad ...(11.137)$$

The discrete-time system with frequency response

$$H(e^{j\omega}) = \frac{1}{(1-c_1 e^{-j\omega})(1-c_2 e^{-j\omega})} = H_1(e^{j\omega})\, H_2(e^{j\omega})$$

corresponds to the cascade of two first-order discrete-time LTI systems with frequency responses $H_1(e^{j\omega})$ and $H_2(e^{j\omega})$. Therefore, we can easily deduce most of its properties from the knowledge of the first- order discrete-time LTI system. The log magnitude and phase plots for frequency response given above can be obtained by adding together the plots for each of the two first-order discrete-time frequency responses terms. Also, we know that for first-order discrete-time systems, the frequency response of the discrete- time system is fast if $|c_1|$ and $|c_2|$ are small. However the system has a long settling time if either of these magnitudes is near unity. If real constants c_1 and c_2 are negative, the response is oscillatory. The case, when c_1 and c_2 are positive, is the counter part of the overdamped case in continuous-time. In this case, both impulse and step responses will settle without oscillation. In this section, we have discussed only causal first-order and second-order discrete-time LTI systems that are stable and for which frequency response can be defined. A causal discrete-time system described by difference equation $y(n) - Ay(n-1) = s(n)$ is unstable for $|A| \geq 1$. Also, a causal discrete-time system described by difference equation $y(n) - 2r \cos\theta\, y(n-1) + r^2 y(n-2) = s(n)$ is unstable if $r \geq 1$. A causal discrete-time system described by difference equation

$$y(n) - (c_1 + c_2)\, y(n-1) + c_1 c_2\, y(n-2) = s(n)$$

(with c_1 and c_2 as real constants) is unstable if either $|c_1|$ or $|c_2|$ exceeds unity.

SOLVED EXAMPLES

Example 11.3 We have been given a continuous-time LTI system with frequency response $H(j\omega) = |H(j\omega)|\, e^{\angle H(j\omega)}$ and real impulse response $h(t)$. We apply input $s(t) = \cos(\omega_0 t + \phi_0)$ to this system. The resulting output of the system is given by

$$y(t) = As(t - t_0)$$

where A is an amplitude scaling factor. It is a non- negative real number and t_0 is a time delay. Express the following :

(*a*) A in terms of $|H(j\omega_0)|$

(*b*) t_0 in terms of $\angle H(j\omega_0)$

Solution. We have been given input

$$s(t) = \cos(\omega_0 t + \phi_0)$$

$$= \frac{e^{j(\omega_0 t + \phi_0)} + e^{-j(\omega_0 t + \phi_0)}}{2} \qquad \text{(using Euler's relation)}$$

$$= \frac{1}{2} e^{j(\omega_0 t + \phi_0)} + \frac{1}{2} e^{-j(\omega_0 t + \phi_0)}$$

$$= s_1(t) + s_2(t) \qquad ...(1)$$

The signal $s(t)$ can be broken up into a sum of the two complex exponental signals

$$s_1(t) = \frac{1}{2} e^{j(\omega_0 t + \phi_0)}$$

and $$s_2(t) = \frac{1}{2} e^{-j(\omega_0 t + \phi_0)}$$

We know that complex exponential signals are Eigen functions of LTI systems. When $s_1(t)$ passes through the LTI system, the output is

$$y_1(t) = s_1(t)\, H(j\omega_0)$$

$$= s_1(t)\, |H(j\omega_0)|\, e^{j\angle H(j\omega_0)}$$

$$= \frac{1}{2} e^{j(\omega_0 t + \phi_0)} \, |H(j\omega_0)| \, e^{j[\omega_0 t + \phi_0 + H(j\omega_0)]}$$

$$= \frac{1}{2} |H(j\omega_0)| \, e^{j[\omega_0 t + \phi_0 + \underline{/H(j\omega_0)}]} \quad ...(2)$$

Similarly, if input $s_2(t)$ passes through the LTI system, the output is

$$y_2(t) = \frac{1}{2} |H(-j\omega_0)| \, e^{-j[\omega_0 t + \phi_0 - \underline{/H(-j\omega_0)}]} \quad ...(3)$$

But since $h(t)$ is given to be real,

$$|H(j\omega_0)| = |H(-j\omega_0)| \quad ...(4)$$

and $$\underline{/H(j\omega_0)} = -\underline{/H(-j\omega_0)} \quad ...(5)$$

Using Eqs. 4 and 5 in Eq. 3, we get

$$y_2(t) = \frac{1}{2} |H(j\omega_0)| \, e^{-j[\omega_0 t + \phi_0 + \underline{/H(j\omega_0)}]} \quad ...(6)$$

Using the principle of linearity to LTI system, when the input to the LTI system is $s(t) = s_1(t) + s_2(t)$, the output will be given by

$$y(t) = y_1(t) + y_2(t)$$

$$= \frac{1}{2} |H(j\omega_0)| \, e^{j[\omega_0 t + \phi_0 + \underline{/H(j\omega_0)}]}$$

$$+ \frac{1}{2} |H(j\omega_0)| \, e^{-j[\omega_0 t + \phi_0 + \underline{/H(j\omega_0)}]}$$

$$= |H(j\omega_0)| \cos [\omega_0 t + \phi_0 + \underline{/H(j\omega_0)}]$$

or $$y(t) = |H(j\omega_0)| \cos \left[\omega_0 \left\{ t - \frac{-\underline{/H(j\omega_0)}}{\omega_0} \right\} + \phi_0 \right] \quad ...(7)$$

But we are given that

$$y(t) = As(t - t_0) \qquad \text{Since } s(t) = \cos(\omega_0 t + \phi_0)$$

Now $y(t)$ can be written as

$$y(t) = A \cos [\omega_0 \{t - t_0\} + \phi_0] \quad ...(8)$$

On comparing Eqs. 7 and 8, we get

(*a*) $A = |H(j\omega_0)|$ (*b*) $t_0 = \dfrac{-\underline{/H(j\omega_0)}}{\omega_0}$

Example 11.4 We have been given a discrete-time LTI system with frequency response $H(e^{j\omega}) = |H(e^{j\omega})| \, e^{j\underline{/H(e^{j\omega})}}$ and real impulse response $h(n)$. We apply input $s(n) = \sin(\omega_0 n + \phi_0)$ to this system. The resulting output to this system is given by

$$y(n) = |H(e^{j\omega_0})| \, s(n - n_0)$$

Determine the relationship between $\underline{/H(e^{j\omega})}$ and ω_0.

Solution. We have been given input

$$s(n) = \sin(\omega_0 n + \phi_0)$$

$$= \frac{e^{j(\omega_0 n + \phi_0)} - e^{-j(\omega_0 n + \phi_0)}}{2j} \quad \text{using Euler's relation}$$

$$= \frac{1}{2j} e^{j(\omega_0 n + \phi_0)} - \frac{1}{2j} e^{-j(\omega_0 n + \phi_0)}$$

$$= s_1(n) + s_2(n) \quad ...(1)$$

The signal $s(n)$ can be broken up into a sum of two complex exponential signals

$$s_1(n) = \frac{1}{2j} e^{j(\omega_0 n + \phi_0)} \quad ...(2)$$

and $$s_2(n) = -\frac{1}{2j} e^{-j(\omega_0 n + \phi_0)} \quad ...(3)$$

Complex exponential signals are Eigen functions of LTI systems. When $s_1(n)$ passes through LTI system, the output is

$$y_1(n) = s_1(n)\, H(e^{j\omega_0})$$

$$= s_1(n)\, |H(e^{j\omega_0})|\, e^{j\angle H(e^{j\omega_0})}$$

Substituting Eq. 2, we get

$$y_1(n) = \frac{1}{2j} e^{j(\omega_0 n + \phi_0)}\, |H(e^{j\omega_0})|\, e^{j\angle H(e^{j\omega_0})}$$

$$= \frac{1}{2j} |H(e^{j\omega_0})|\, e^{j[\omega_0 n + \phi_0 + \angle H(e^{j\omega_0})]} \quad ...(4)$$

Similarly, if input $s_2(n)$ passes through the LTI system, the output is

$$y_2(n) = -\frac{1}{2j} |H(e^{-j\omega_0})|\, e^{-j[\omega_0 n + \phi_0 + \angle H(e^{-j\omega_0})]} \quad ...(5)$$

But since $h(n)$ is given to be real,

$$|H(e^{j\omega_0})| = |H(e^{-j\omega_0})| \quad ...(6)$$

and $$\angle H(e^{j\omega_0}) = -\angle H(e^{-j\omega_0}) \quad ...(7)$$

Using Eqs. 6 and 7, Eq. 5 becomes

$$y_2(n) = -\frac{1}{2j} |H(e^{j\omega_0})|\, e^{-j[\omega_0 n + \phi_0 + \angle H(e^{j\omega_0})]} \quad ...(8)$$

Using the principle of linearity to LTI system, when the input to the LTI system is $s(n) = s_1(n) + s_2(n)$, the output to the system will be given by

$$y(n) = y_1(n) + y_2(n) \quad ...(9)$$

Substituting Eqs. 4 and 8 in Eq. 9, we get

$$y(n) = \frac{1}{2j}\,|H(e^{j\omega_0})|\; e^{j[\omega_0 n + \phi_0 + \angle H(e^{j\omega_0})]}$$

$$- \frac{1}{2j}\,|H(e^{j\omega_0})|\; e^{-j[\omega_0 n + \phi_0 + \angle H(e^{j\omega_0})]}$$

$$= |H(e^{j\omega_0})|\, \sin\,[\omega_0 n + \phi_0 + H(e^{j\omega_0})\,]$$

$$= |H(e^{j\omega_0})|\, \sin\left[\omega_0\left\{n - \frac{-\angle H(e^{j\omega_0})}{\omega_0}\right\} + \phi_0\right] \quad ...(10)$$

But we are given that

$$y(n) = |H(e^{j\omega_0})|\; s(n - n_0) \qquad \text{since } s(n) = \sin\,(\omega_0 n + \phi_0)$$

Now $y(n)$ can be written as

$$y(n) = |H(e^{j\omega_0})|\, \sin\,[\omega_0\,\{n - n_0\} + \phi_0\,] \quad ...(11)$$

On comparing Eqs. 10 and 11, we get

$$n_0 = \frac{-\angle H(e^{j\omega_0})}{\omega_0}$$

It has an integer value.

Therefore, $\angle H(e^{j\omega_0}) = -\,n_0\omega_0$

Now also, note that if we add an integer multiple of 2π to this $\angle H(e^{j\omega_0})$, it does not make any difference. Therefore, we require in general that $\angle H(e^{j\omega_0}) = -\,n_0\,(\omega_0 + 2k\pi)$

Example 11.5 The frequency response for a causal and stable continuous-time LTI system is given by

$$H(j\omega) = \frac{1 - j\omega}{1 + j\omega}$$

(*a*) Find the magnitude of $H(j\omega)$.

(*b*) Determine which of the following statements is true about $\tau(\omega)$, the group delay of the system

1. $\tau(\omega) = 0, \text{ for } \omega > 0$

2. $\tau(\omega) > 0, \text{ for } \omega > 0$

3. $\tau(\omega) < 0, \text{ for } \omega > 0$

Solution. Magnitude of $H(j\omega)$ is determined as

$$|H(j\omega)| = \sqrt{[H_R(j\omega)]^2 + [H_I(j\omega)]^2}$$

where $H_R(j\omega)$ and $H_I(j\omega)$ are the real and imaginary parts of complex frequency response $H(j\omega)$, respectively. Complex frequency response is expressed as

$$H(j\omega) = H_R(j\omega) + jH_I(j\omega)$$

(*a*)

$$|H(j\omega)| = \frac{|1 - j\omega|}{|1 + j\omega|}$$

$$= \frac{\sqrt{(1)^2 + (-\omega)^2}}{\sqrt{(1)^2 + (\omega)^2}} = \frac{\sqrt{1+\omega^2}}{\sqrt{1+\omega^2}} = 1$$

or $\quad |H(j\omega)| = 1$

(*b*) Phase response of $H(j\omega)$ is determined as

$$\angle H(j\omega) = \tan^{-1}\left[\frac{H_I(j\omega)}{H_R(j\omega)}\right]$$

We have been given

$$H(j\omega) = \frac{1-j\omega}{1+j\omega}$$

$$\angle H(j\omega) = \angle \frac{1-j\omega}{1+j\omega}$$

$$= \angle 1-j\omega - \angle 1+j\omega$$

$$= \tan^{-1}\left[\frac{-\omega}{1}\right] - \tan^{-1}\left[\frac{\omega}{1}\right]$$

$$= \tan^{-1}[-\omega] - \tan^{-1}[\omega]$$

$$= -2\tan^{-1}(\omega) \qquad \text{...(1)}$$

Therefore, the group delay is

$$\tau(\omega) = -\frac{d}{d\omega}\angle H(j\omega)$$

$$= -\frac{d}{d\omega}\{-2\tan^{-1}\omega\}$$

$$= 2\frac{d}{d\omega}(\tan^{-1}\omega) = \frac{2}{1+\omega^2}$$

or $$\tau(\omega) = \frac{2}{1+\omega^2} \qquad \text{...(2)}$$

Clearly, $\quad \tau(\omega) > 0$ for $\omega > 0$.

Therefore, statement 2 is true.

Example 11.6 The differential equation (input-output relationship) of a continuous-time causal and stable LTI system is given as

$$\frac{dy(t)}{dt} + 5y(t) = 2\,s(t)$$

(*a*) Determine the step response of the system.

(*b*) Determine the final value of step response.

(*c*) Find the value of t_0 for which

$$g(t_0) = g(\infty)\left[1 - \frac{1}{e^2}\right]$$

where $g(t)$ is the step response of the above mentioned system.

Solution. **To obtain frequency response of the LTI system, we take CTFT of the given differential equation.**

$$\text{CTFT}\left\{\frac{dy(t)}{dt}+5y(t)\right\}=\text{CTFT}\{2s(t)\}$$

or $$j\omega\, Y(j\omega)+5Y(j\omega)=2\,S(j\omega)$$

or $$Y(j\omega)\,[j\omega+5]=2\,S(j\omega)$$

or $$\frac{Y(j\omega)}{S(j\omega)}=\frac{2}{j\omega+5} \quad ...(1)$$

Frequency response $H(j\omega)$ is defined as

$$H(j\omega)=\frac{Y(j\omega)}{S(j\omega)} \quad ...(2)$$

From Eqs. 1 and 2, we get

$$H(j\omega)=\frac{2}{5+j\omega} \quad ...(3)$$

Now taking the inverse CTFT of Eq. 3, we get

$$h(t)=\text{Inverse CTFT}\,\{H(j\omega)\}$$

$$=\text{Inverse CTFT}\left\{\frac{2}{5+j\omega}\right\}$$

or $$h(t)=2e^{-5t}\,u(t) \quad ...(4)$$

(*a*) Step response of an LTI system is the output of the system when we apply input $s(t)=u(t)$. Step response is determined as

$$g(t)=h(t)*u(t)=u(t)*h(t)$$

$$=\int_{\tau=-\infty}^{\infty} h(\tau)\,u(t-\tau)\,d\tau \quad ...(5)$$

We also know that

$$u(t-\tau)=\begin{cases}1, & \text{for } t-\tau>0\rightarrow\tau<t\\ 0, & \text{for } t-\tau<0\rightarrow\tau>t\end{cases} \quad ...(6)$$

Substituting Eqs. 4 and 6 in Eq. 5, we get

$$g(t)=\int_{\tau=-\infty}^{\infty} 2e^{-5\tau}\,u(\tau)\,u(t-\tau)\,d\tau$$

$$=\int_{\tau=0}^{t} 2e^{-5\tau}\,d\tau=2\left[\frac{1}{-5}e^{-5\tau}\right]_{\tau=0}^{t}$$

$$=\frac{2}{5}\,[1-e^{-5t}]\,,\ t>0$$

or $$g(t)=\frac{2}{5}\,[1-e^{-5t}]\,u(t) \quad ...(7)$$

This is the step response of the above mentioned system.

(*b*) The final value of the step response is

$$g(t = \infty) = g(\infty)$$
$$= \frac{2}{5}\,[1 - e^{-5\infty}]\,u(\infty) = \frac{2}{5}$$

(*c*) Step response at $t = t_0$ is

$$g(t_0) = \frac{2}{5}\,[1 - e^{-5t_0}] \quad \text{...(8)}$$

Given step response is

$$g(t_0) = \frac{2}{5}\left[\,1 - \frac{1}{e^2}\,\right] \quad \text{...(9)}$$

Comparing Eqs. 8 and 9, we get

$$\text{Time, } t_0 = \frac{2}{5} \text{ sec.}$$

Example 11.7 We have been given the following second order differential equations for causal and stable continuous-time LTI systems. Determine whether the corresponding impulse response is under-damped, over-damped or critically damped.

(*a*) $\dfrac{d^2y(t)}{dt^2} + 4\dfrac{dy(t)}{dt} + 4y(t) = s(t)$

(*b*) $5\dfrac{d^2y(t)}{dt^2} + 4\dfrac{dy(t)}{dt} + 5y(t) = 7s(t)$

(*c*) $\dfrac{d^2y(t)}{dt^2} + 20\dfrac{dy(t)}{dt} + y(t) = s(t)$

Solution. Frequency response of a continuous-time LTI system is determined by taking CTFT of the differential equation.

(*a*) Frequency response $H(j\omega)$:

$$\text{CTFT}\left\{\frac{d^2y(t)}{dt^2} + 4\frac{dy(t)}{dt} + 4y(t)\right\} = \text{CTFT}\,\{s(t)\}$$

or $$H(j\omega) = \frac{Y(j\omega)}{S(j\omega)} = \frac{1}{(j\omega)^2 + 4j\omega + 4} \quad \text{...(1)}$$

Standard form of frequency response for the second-order continuous-time LTI system is given as

$$H(j\omega) = H(s)\,\Big|_{s=j\omega}$$
$$= \left.\frac{\omega_n^2}{s^2 + 2\,\xi\,\omega_n s + \omega_n^2}\right|_{s=j\omega}$$
$$= \frac{\omega_n^2}{(j\omega)^2 + 2\,\xi\,\omega_n\,(j\omega) + \omega_n^2} \quad \text{...(2)}$$

On comparing Eqs. 1 and 2, we get

$$\omega_n^2 = 4$$

$$\omega_n = 2 \qquad ...(4)$$

and $$2\,\xi\,\omega_n = 4 \qquad ...(5)$$

From Eqs. 4 and 5, we get

$$\xi = 1$$

For this system, damping ratio ξ is unity. Therefore, the system is critically damped.

(*b*) Given differential equation is

$$5\frac{d^2y(t)}{dt^2} + 4\frac{dy(t)}{dt} = 5y(t) = 7s(t)$$

Taking the CTFT of above equation, we get

$$H(j\omega) = \frac{7}{5(j\omega)^2 + 4j\omega + 5} \qquad ...(6)$$

It can be written as

$$H(j\omega) = \frac{7/5}{(j\omega)^2 + \frac{4}{5}j\omega + 1} \qquad ...(7)$$

Comparing Eq. 7 with Eq. 2, we get

$$\xi = \frac{2}{5}$$

Here damping ratio ξ is less than unity. Therefore, the system is under-damped.

(*c*) Given the differential equation is

$$\frac{d^2y(t)}{dt^2} + 20\frac{dy(t)}{dt} + y(t) = s(t)$$

Taking the CTFT of the above equation, we get

$$H(j\omega) = \frac{1}{(j\omega)^2 + 20j\omega + 1} \qquad ...(8)$$

Comparing Eq. 8 with Eq. 2, we get

$$\xi = 10$$

Here the damping ratio ξ is greater than unity. Therefore, the system is overdamped.

Example 11.8 The frequency response $H(j\omega)$ of a causal LTI filter is shown in Fig. 11.23.

Determine the filtered output signal $y(t)$ for following input signals

(*a*) $s(t) = \sin(\omega_0 t)\, u(t)$ (*b*) $S(j\omega) = \frac{1}{2 + j\omega}$

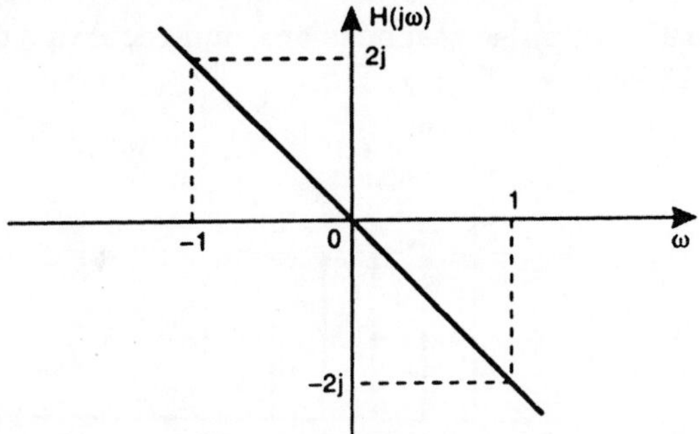

Fig. 11.23

Solution. **From Fig. 11.23, we can find frequency response**

$$H(j\omega) = -2j\omega \qquad ...(1)$$

We also know that

$$Y(j\omega) = H(j\omega)\, S(j\omega) \qquad ...(2)$$

From Eqs. 1 and 2, we get

$$Y(j\omega) = -2j\omega\, S(j\omega) \qquad ...(3)$$

Taking the inverse CTFT of Eq. 3, we get

$$y(t) = -2\frac{ds(t)}{dt} \qquad ...(4)$$

(*a*) Here input $s(t) = \sin(\omega_0 t)\, u(t)$ putting $s(t) = \sin(\omega_0 t)\, u(t)$ in Eq. 4, we get

$$y(t) = -2\frac{d}{dt}[\sin(\omega_0 t)\, u(t)]$$

$$= -2\left[\sin(\omega_0 t)\frac{du(t)}{dt} + \omega_0 \cos(\omega_0 t)\, u(t)\right]$$

$$= -2\,[\sin(\omega_0 t)\,\delta(t) + \omega_0 \cos(\omega_0 t)\, u(t)]$$

$$= -2\,\omega_0 \cos(\omega_0 t)\, u(t)\,. \qquad ..(5)$$

(*b*) Given $S(j\omega) = \dfrac{1}{(2+j\omega)}$ **...(6)**

Taking inverse CTFT of Eq. 6, we get

$$s(t) = e^{-2t}\, u(t) \qquad ...(7)$$

From Eq. 4 and 7, we get

$$y(t) = -2\frac{ds(t)}{dt} = -2\frac{d}{dt}[e^{-2t}\, u(t)]$$

$$= -2\left[e^{-2t}\frac{du(t)}{dt} + (-2)\, e^{-2t}\, u(t)\right]$$

$$= -2\,[e^{-2t}\,\delta(t) - 2e^{-2t}\, u(t)]$$

$$= 4e^{-2t}\, u(t) - e^{-2t}\,\delta(t)$$

Example 11.9 The impulse response of a non-recursive digital filter is shown in Fig. 11.24.

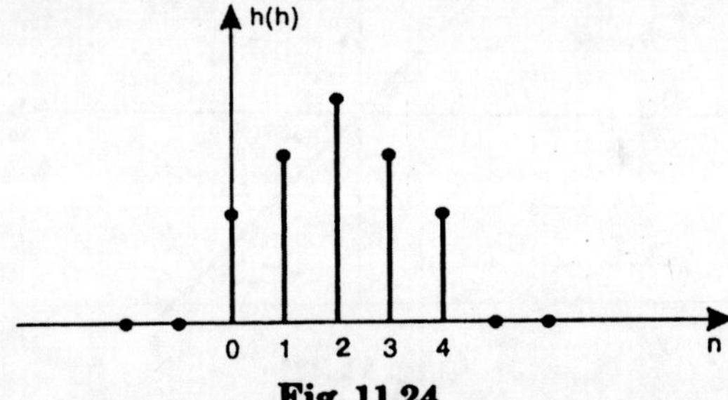

Fig. 11.24

Determine the group delay as a function of frequency for this filter.

Solution. Let us call the impulse response of this non-recursive digital filter $h(n)$. It is easily observed that the signal $h_1(n) = h(n + 2)$ is real and even. Therefore, we know from the properties of DTFT that the DTFT $H_1(e^{j\omega})$ of $h_1(n)$ is real and even. Therefore, $H_1(e^{j\omega})$ has zero phase.

$$h_1(n) = h(n + 2) \qquad \text{...(1)}$$

Taking DTFT of Eq. 1, we get

$$\text{DTFT}\{h_1(n)\} = \text{DTFT}\{h(n + 2)\}$$

or
$$H_1(e^{j\omega}) = e^{j2\omega} H(e^{j\omega})$$

or
$$H(e^{j\omega}) = e^{-j2\omega} H_1(e^{j\omega}) \qquad \text{...(2)}$$

Since $H_1(e^{j\omega})$ has zero phase. Taking phase of Eq. 2, we get

$$\begin{aligned}\angle H(e^{j\omega}) &= \angle e^{-j2\omega} + \angle H_1(e^{j\omega}) \\ &= \angle [\cos 2\omega - j \sin 2\omega] + \angle H_1(e^{j\omega}) \\ &= \tan^{-1}\left[-\frac{\sin 2\omega}{\cos 2\omega}\right] + \angle H_1(e^{j\omega}) \\ &= \tan^{-1}[-\tan 2\omega] + \angle H_1(e^{j\omega}) \\ &= -2\omega + 0\end{aligned}$$

or
$$\angle H(e^{j\omega}) = -2\omega \qquad \text{...(3)}$$

Therefore, the group delay is determined as

$$\begin{aligned}\tau(\omega) &= -\frac{d}{d\omega} \angle H(e^{j\omega}) \\ &= -\frac{d}{d\omega}(-2\omega) = 2\omega\end{aligned}$$

Example 11.10 We are given difference equations of two second- order causal and stable LTI systems. Determine whether or not the step response of the system is oscillatory.

(*a*) $y(n) + y(n-1) + \frac{1}{4} y(n-2) = s(n)$

(*b*) $y(n) - y(n-1) + \frac{1}{4}y(n-2) = s(n)$

Solution. A second-order causal discrete-time LTI system described by

$$y(n) - 2r\cos\theta\, y(n-1) + r^2 y(n-2) = s(n)$$

with $0 < r < 1$ and $0 \le \theta \le \pi$ ...(1)

(*a*) Given difference equation

$$y(n) + y(n-1) + \frac{1}{4}y(n-2) = s(n) \qquad ...(2)$$

Comparing Eqs. 1 and 2, we get

$$-2r\cos\theta = 1 \qquad ...(3)$$

$$r^2 = \frac{1}{4} \text{ or } r = \frac{1}{2} \qquad ...(4)$$

From Eqs. 3 and 4, we get

$$-2r\cos\theta = 1$$

or $$-2 \times \frac{1}{2}\cos\theta = 1$$

or $$\cos\theta = -1 \text{ or } \theta = \pi \qquad ...(5)$$

Therefore, $\theta = \pi$, and system has an oscillatory step response.

(*b*) Given difference equation,

$$y(n) - y(n-1) + \frac{1}{4}y(n-2) = s(n) \qquad ...(6)$$

Comparing Eqs. 1 and 6, we get

$$-2r\cos\theta = 1$$

or $$2r\cos\theta = 1 \qquad ...(7)$$

$$r^2 = \frac{1}{4} \text{ or } r = \frac{1}{2} \qquad ...(8)$$

From Eqs. 7 and 8, we get

$$2r\cos\theta = 1$$

or $$2 \times \frac{1}{2}\cos\theta = 1$$

or $$\cos\theta = 1 = \cos 0^\circ$$

or $$\theta = 0^\circ \qquad ...(9)$$

Therefore, $\theta = 0^\circ$, and hence the system has a non- oscillatory step response.

Example 11.11 A continuous-time LTI system implemented as the *RC* circuit is shown in Fig. 11.25. The voltage source *s*(*t*) is considered as the input to this system. The voltage *y*(*t*) across the capacitor is considered as the system output. Is it possible for the step response of the system to have an oscillatory behaviour ?

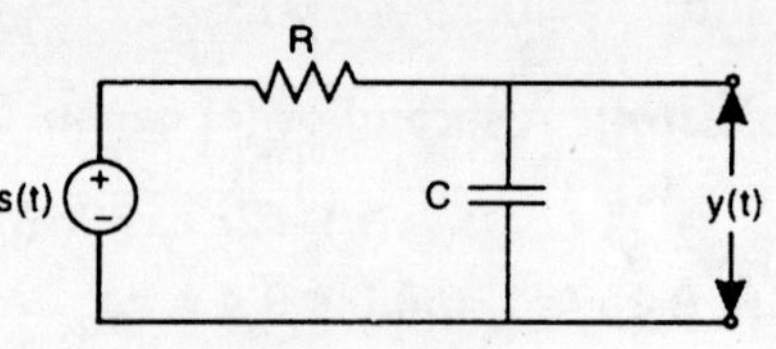

Fig. 11.25

Solution. First of all we have to find the differential equation governing the input and output of this system (or circuit).

Current through resistor *R* = Current through capacitor

$$= C\frac{dy(t)}{dt} \quad ...(1)$$

$$\text{Voltage across the resistor} = RC\frac{dy(t)}{dt} \quad ...(2)$$

Total input voltage

= Voltage across resistor + Voltage across capacitor

$$\text{Therefore, } s(t) = RC\frac{dy(t)}{dt} + y(t) \quad ...(3)$$

Frequency response of this circuit can be determined by taking CTFT of Eq. 3.

$$\text{CTFT}\{s(t)\} = \text{CTFT}\left\{RC\frac{dy(t)}{dt} + y(t)\right\}$$

or $$S(j\omega) = RC\,(j\omega)\,Y(j\omega) + Y(j\omega)$$

or $$\frac{Y(j\omega)}{S(j\omega)} = \frac{1}{RC\,j\omega + 1}$$

or $$H(j\omega) = \frac{Y(j\omega)}{S(j\omega)} = \frac{1}{RC\,j\omega + 1} \quad ...(4)$$

Since this is a first order system, the step response has to be non-oscillatory.

Example 11.12 A continuous-time LTI system implemented as the *RLC* circuit shown in Fig. 11.26. The voltage source *s*(*t*) is considered as the input to this system. The voltage *y*(*t*) across the capacitor is considered as the system output. How should *R*, *L* and *C* be related so that there is no oscillation in the step response ?

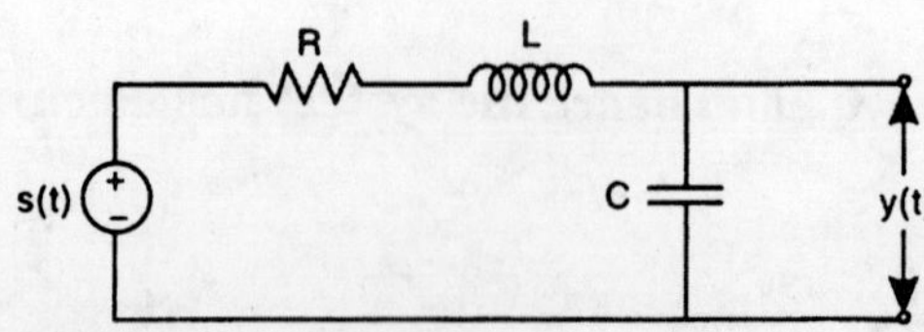

Fig. 11.26

Solution. **First of all we find the differential equation governing the input and output of this circuit.**

Current through resistor and inductor

$$= \text{current through capacitor} = C\frac{dy(t)}{dt}$$

Voltage across resistor $= RC\dfrac{dy(t)}{dt}$

Voltage across inductor $= LC\dfrac{d^2y(t)}{dt^2}$

Total input voltage = voltage across inductor
+ voltage across resistor
+ voltage across capacitor

Therefore,
$$s(t) = LC\frac{d^2y(t)}{dt^2} + RC\frac{dy(t)}{dt} + y(t) \qquad ...(1)$$

Frequency response of this circuit can be determined by taking CTFT of Eq. 1, we get

$$H(j\omega) = \frac{Y(j\omega)}{S(j\omega)}$$

$$= \frac{1}{LC\,(j\omega)^2 + RC\,j\omega + 1} \qquad ...(2)$$

Rewriting Eq. 2,

$$H(j\omega) = \frac{1}{\left(\dfrac{j\omega}{1/\sqrt{LC}}\right)^2 + 2\left(\dfrac{R}{C}\right)\sqrt{C/L}\,\dfrac{j\omega}{1/\sqrt{LC}} + 1}$$

Therefore, the damping constant $\xi = \dfrac{R}{2}\sqrt{\dfrac{C}{L}}$.

In order of the step response to have no oscillations, we must have $\xi \geq 1$. **Tberefore, we require**

$$R \geq 2\sqrt{\frac{L}{C}}$$

REVIEW QUESTIONS

1. How are signals (continuous-time/discrete-time) characterized in time- and frequency domain ?
2. Discuss the magnitude-phase representation of CTFT and DTFT.
3. Also discuss magnitude-phase representation of the frequency response of continuous-time and discrete-time LTI systems.
4. Define linear and non-linear phases of LTI systems.
5. Define and discuss Group delay.

6. Describe Log-magnitude plots and Bode plots.
7. Discuss various properties of ideal frequency selective filters in time-domain.
8. Discuss time-domain and frequency-domain aspects of non-ideal frequency selective filters.
9. Define following quantities :
   (a) The overshoot of the final value of the step response, Δ
   (b) Ringing frequency, ω
   (c) The settling time, t_s
10. Discuss lower-order continuous-time LTI systems.
   (a) First-order continuous-time LTI systems
   (b) Second-order continuous-time LTI systems
11. Discuss lower-order discrete-time LTI systems.
   (a) First-order discrete-time LTI systems
   (b) Second-order discrete-time LTI systems

NUMERICAL PROBLEMS

1. Find the filtered output signal $y(t)$ for each of the input signals given below. Here we have used a causal continuous-time LTI filter with frequncy response $H(j\omega)$ shown in Fig. 11.27.

   (a) $s(t) = e^{jt}$ (b) $S(j\omega) = \dfrac{1}{(j\omega)\,(\sigma + j\omega)}$

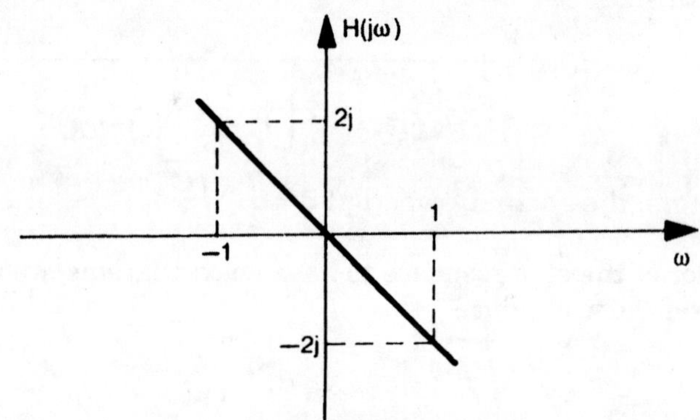

Fig. 11.27

2. Find the filtered output signal $y(t)$ for the input signal $s(t) = \cos(2\pi t + \theta)$. Here we have used a causal continuous-time LTI filter with frequency response shown in Fig. 11.28.
3. Verfiy that each of the following frequency responses have non-linear phase by computing the group delay at selected frequencies.

   (a) $H(j\omega) = \dfrac{1}{j\omega + 1}$ (b) $H(j\omega) = \dfrac{1}{(j\omega + 1)^2}$

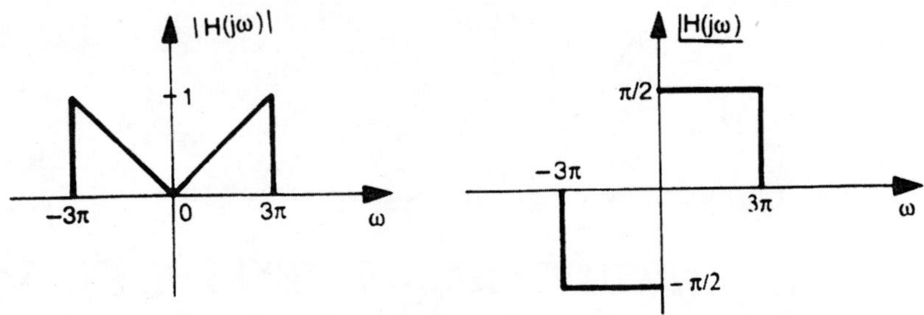

Fig. 11.28

4. Determine the impulse response $h(t)$ of a continuous-time LTI filter with frequency response

$$H(j\omega) = \begin{cases} 1, & |\omega| > \omega_c \\ 0, & \text{otherwise} \end{cases}$$

5. Determine the frequency response of a continuous-time LTI system whose input-output relationship is given by

$$\frac{dy(t)}{dt} + 2y(t) = s(t)$$

Also sketch its Bode plot.

6. Sketch the Bode plots for the following frequency responses

(a) $H(j\omega) = 1 + \left(\frac{j\omega}{10}\right)$ (b) $H(j\omega) = \frac{16}{(j\omega + 2)^4}$

7. Sketch the log magnitude and phase of each of the following frequency responses

(a) $H(e^{j\omega}) = 1 + \frac{1}{2}e^{-j\omega}$ (b) $H(e^{j\omega}) = 1 + 2e^{-j\omega}$

(c) $H(e^{j\omega}) = 1 - 2e^{-j\omega}$

8. We have two discrete-time LTI systems with frequency response

$$H_1(e^{j\omega}) = \frac{1 + \frac{1}{2}e^{-j\omega}}{1 + \frac{1}{4}e^{-j\omega}}$$

$$H_2(e^{j\omega}) = \frac{\frac{1}{2} + e^{-j\omega}}{1 + \frac{1}{4}e^{-j\omega}}$$

Show that both of these frequency responses have the same magnitude function *i.e.*, $|H_1(e^{j\omega})| = |H_2(e^{j\omega})|$, but the group delay of $H_2(e^{j\omega})$ is greater than the group delay of $H_1(e^{j\omega})$ for $\omega > \theta$.

12

Sampling of Continuous-Time Signals

12.1 INTRODUCTION TO SAMPLING PROCESS

In this chapter we will study and develop the concept of sampling of continuous-time signal. Under certain conditions, a continuous-time signal can be completely represented by samples at points equally spaced in time and then recovered from the knowledge of those samples. These certain conditions follow from the basic result that is known as sampling theorem. This theorem is extremely important and useful. For example, in moving pictures, which consist of a sequence of individual frames (each frame represents an instantaneous view) of a continuously changing scene. When these samples (individual frames are equivalent to samples in time) are viewed in sequence at a sufficiently fast rate, we perceive an accurate representation of the originally continuously moving scene.

Much of the importance of the sampling theorem also lies in its role as a bridge between continuous-time signals and discrete-time signals. In other words, sampling theorem is a very important theorem, for converting continuous-time signals into discrete-time signals. In many applications, processing discrete-time signals is more flexible and is often preferable to processing continuous-time signals. The concept of sampling is an extremely attractive and widely employed method for using discrete-time system technology to implement and process continuous-time signals. Sampling process is used to convert a continuous-time signal into a discrete-time signal, process the discrete-time signal using a discrete-time system, and then convert back to continuous-time signal.

12.2 SAMPLING OF CONTINUOUS-TIME SIGNALS

The periodical measurement of a continuous-time signal is called periodic sampling or uniform sampling. By periodic sampling, we can convert a continuous-time signal $s(t)$ into a discrete-time signal $s(nT_s)$ as

$$\text{Continuous-time signal} \xleftrightarrow{\text{Periodic sampling}} \text{Discrete–time signal}$$

$$\text{Discrete-time signal } s(nT_s) = s(t)\Big|_{t=nT_s} \qquad \ldots(12.1)$$

where T_s is the sampling period and reciprocal of sampling period is termed as sampling frequency or sampling rate F_s.

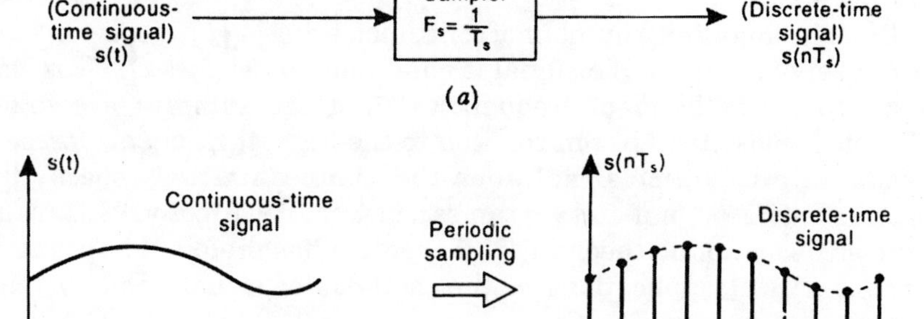

Fig. 12.1 (*a*) *Block diagram of a sampler* (*b*) *periodic sampling of a continuous-time signal.*

Periodic sampling is done according to sampling theorem which is discussed in the next section.

12.3 SAMPLING THEOREM FOR CONTINUOUS-TIME SIGNALS

In general, in the absence of any additional conditions or information, we cannot expect that a continuous-time signal $s(t)$ could be uniquely specified by a sequence of equally spaced samples. The additional condition or information that is imposed on sampling process of a continuous-time signal to get perfect reconstruction of the continuous-time signal from its samples is referred to as sampling theorem. Sampling theorem is stated as :

For perfect reconstruction of a continuous-time signal from its samples, the sampling rate or sampling frequency should be greater than or equal to Nyquist rate of the information-bearing (message) signal.

According to sampling theorem,

$$\text{Sampling rate} \geq \text{Nyquist rate}$$

Nyquist rate is defined as minimum sampling rate required for perfect reconstruction of a continuous-time signal from its samples. Nyquist rate depends upon the highest frequency component of the message signal and it is expressed as

$$\text{Nyquist rate} = 2 \times \text{Highest frequency component of the message signal}$$
$$= 2 \times F_{max}$$

Periodic sampling establishes a relationship between the time variables t and n of a continuous-time signal and discrete-time signal, respectively.

Clearly, an infinite number of continuous-time signals can generate a given set of samples. If a signal is band limited (or if its CTFT is zero outside a finite band of frequencies) and the samples are taken sufficiently close together in relation to the highest frequency present in the continuous-time signal, then the samples uniquely specify the continuous-time signal and we can reconstruct it perfectly. This result is known as sampling theorem. This theorem has profound importance in the practical applications of the methods of signals and systems analysis.

Let us consider a continuous-time signal

$$s(t) = A \cos (2\pi Ft + \theta) \qquad \text{...(12.2)}$$

Sampling the continuous-time signal $s(t)$ given in Eq. 12.2 at a sampling rate F_s samples/second, we get a

Discrete-time signal

$$s(nT_s) = s(t)\Big|_{t = nT_s} \qquad \text{...(12.3)}$$

where $T_s = \dfrac{1}{F_s}$ is the sampling period

$$s(nT_s) = [A \cos (2\pi Ft + \theta)]_{t = nT_s}$$
$$= A \cos (2\pi FnT_s + \theta) = A \cos \left(2\pi Fn \frac{1}{F_s} + \theta \right)$$
$$= A \cos \left(2\pi \frac{F}{F_s} n + \theta \right)$$

or $$s(nT_s) = A \cos (2\pi fn + \theta), \quad -\infty < n < \infty \qquad \text{...(12.4)}$$

where $s(n) = s(nT_s)$ = Discrete-time signal

F = Frequency variable for continuous-time signals

F_s = Sampling frequency in samples/second.

Example 12.1 Given the continuous-time signal

$$s(t) = 5 \cos 200\, \pi t$$

Determine :

(*a*) Minimum sampling rate *i.e.* Nyquist rate required to avoid aliasing.

(*b*) **If sampling frequency $F_s = 400$ Hz. What is the discrete-time signal $s(n)$ or $s(nT_s)$ obtained after sampling ?**

(*c*) **If sampling frequency $F_s = 150$ Hz. What is the discrete-time signal $s(n)$ or $s(nT_s)$ obtained after sampling ?**

(*d*) **What is the frequency $0 < F < F_s/2$ of sinusoidal that yields samples identical to those obtained in part (*c*) ?**

Solution. **(*a*) The highest frequency component of continuous-time signal is $F = 100$ Hz.**

Hence minimum sampling rate required to avoid aliasing is called Nyquist rate and is determined as

$$\text{Nyquist rate} = 2F = 2 \times 100 = 200 \text{ Hz}$$

(*b*) The continuous-time signal $s(t)$ is sampled at $F_s = 400$ Hz. The frequency of the discrete-time signal

$$f = \frac{\text{Frequency of continuous–time signal } F}{\text{Sampling frequency, } F_s}$$

$$= \frac{100}{400} = \frac{1}{4}$$

The discrete-time signal

$$s(n) = 5 \cos 2\pi fn = 5 \cos 2\pi \times \frac{1}{4} n$$

or $$s(n) = 5 \cos \frac{\pi n}{2}$$

(*c*) The continuous-time signal $s(t)$ is sampled at $F_s = 150$ Hz. The frequency of discrete-time signal

$$f = \frac{F}{F_s} = \frac{100}{150} = \frac{2}{3}$$

The discrete-time signal

$$s(n) = A \cos 2\pi fn = 5 \cos 2\pi \left(\frac{2}{3}\right) n = 5 \cos \frac{4\pi}{3} n$$

$$= 5 \cos \left(2\pi - \frac{2\pi}{3}\right) n = 5 \cos \frac{2\pi n}{3}$$

or $$s(n) = 5 \cos \frac{2\pi n}{3}$$

(*d*) For sampling rate of $F_s = 150$ Hz

$$f = \frac{F}{F_s} \text{ or } F = fF_s = \frac{1}{3} \times 150 = 50 \text{ Hz}$$

The sinusoidal signal

$$y(t) = 5 \cos 2\pi Ft = 5 \cos 2\pi \times 50 \times t$$

$$= 5 \cos 100\,\pi t$$

Sampling at $F_s = 150$ Hz, yields identical samples hence $F = 100$ Hz is an alias of $F = 50$ Hz for sampling rate $F_s = 150$ Hz.

Example 12.2 Determine the Nyquist rate for a continuous-time signal

$$s(t) = 5 \cos 50\,\pi t + 20 \sin 300\,\pi t - 10 \cos 100\,\pi t$$

Solution. In a more general form, any continuous-time signal can be written as

$$s(t) = A_1 \cos \omega_1 t + A_2 \cos \omega_2 t + A_3 \cos \omega_3 t$$

On comparing given signal with standard form of a signal, we obtain the frequencies for the given signal as

$$F_1 = \frac{\omega_1}{2\,\pi} = \frac{50\,\pi}{2\,\pi} = 25 \text{ Hz}$$

$$F_2 = \frac{\omega_2}{2\,\pi} = \frac{300\,\pi}{2\,\pi} = 150 \text{ Hz}$$

$$F_3 = \frac{\omega_3}{2\,\pi} = \frac{100\,\pi}{2\,\pi} = 50 \text{ Hz}$$

Highest frequency component of the given message signal

$$F_{max} = 150 \text{ Hz}$$

$$\text{Nyquist rate} = 2\,F_{max} = 2 \times 150 = 300 \text{ Hz}$$

12.3.1 Impulse Train Sampling of a Continuous-Time Signals

Sampling theorem can be developed by using a convenient way in which we can represent the sampling of a continuous-time signal at regular intervals. A useful way to perform sampling is through the use of periodic impulse train multiplied by the continuous-time signal $s(t)$ that we wish to sample. This mechanism is called *impulse train sampling*. Impulse train sampling of a continuous-time signal $s(t)$ is shown in Fig. 12.2. The periodic impulse train $p(t)$ is also called the sampling function. The sampling period of a periodic impulse train is denoted by T_s. The reciprocal of sampling period T_s is called the sampling frequency F_s. The fundamental frequency of the periodic impulse train $p(t)$ is given by $\omega_s = \frac{2\pi}{T_s} = 2\pi F_s$. It is also called angular sampling frequency.

In time-domain $s_p(t) = s(t)\,p(t)$ ...(12.5)

where $p(t)$ is the periodic impulse train. It is given as

$$p(t) = \sum_{n=-\infty}^{\infty} \delta(t - nT_s) \qquad \text{...(12.6)}$$

We know that from the sampling property of the unit-impulse function $\delta(t)$, that multiplying $s(t)$ by a unit impulse samples the value of the signal at the point at which the impulse is located. This property can also be defined as

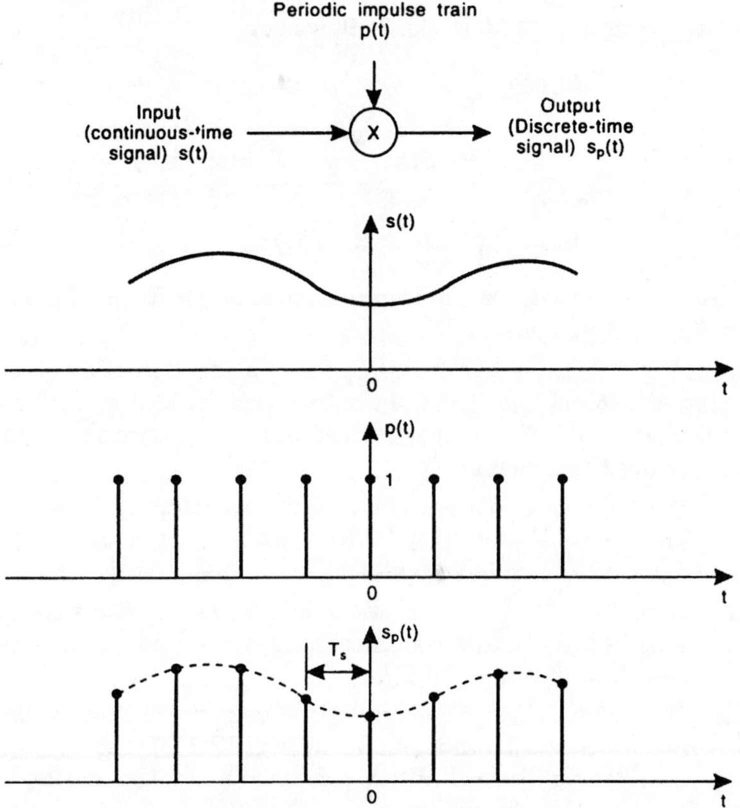

Fig. 12.2 *Illustration of impulse-train sampling of continuous-time signal.*

$$s(t)\ \delta(t - t_0) = s(t_0)\ \delta(t - t_0) \qquad ...(12.7)$$

Applying property of Eq. 12.7 to Eq. 12.5, we see that $s_p(t)$ is an impulse train with the amplitude of the impulses equal to the samples of $s(t)$ at intervals spaced by T_s.

$$s_p(t) = \sum_{n=-\infty}^{\infty} s(nT_s)\ \delta(t - nT_s) \qquad ...(12.8)$$

From the multiplication property of CTFT, we know that

$$s_p(t) = s(t)\, p(t) \overset{\text{CTFT}}{\longleftrightarrow} S_p(j\omega) = \frac{1}{2\pi}\,[S(j\omega) * P(j\omega)]$$

or
$$S_p(j\omega) = \frac{1}{2\pi}\,[S(j\omega) * P(j\omega)] \qquad ...(12.9)$$

Taking the CTFT of both sides of Eq. 12.6, we get

$$\text{CTFT}\ \{p(t)\} = \text{CTFT}\left\{ \sum_{n=-\infty}^{\infty} \delta(t - nT_s) \right\}$$

or
$$P(j\omega) = \frac{2\pi}{T_s} \sum_{k=-\infty}^{\infty} \delta(\omega - k\omega_s) \qquad ...(12.10)$$

Substituting Eq. 12.10 in Eq. 12.9, we get

$$S_p(j\omega) = \frac{1}{2\pi}\,[S(j\omega) * P(j\omega)]$$

$$= \frac{1}{2\pi}\left[S(j\omega) * \frac{2\pi}{T_s} \sum_{k=-\infty}^{\infty} \delta(\omega - k\omega_s) \right]$$

or
$$S_p(j\omega) = \frac{1}{T_s} \sum_{k=-\infty}^{\infty} S(\omega - k\omega_s) \qquad \text{...(12.11)}$$

[Since convolution with impulse simply shifts a signal, *i.e.*, $S(j\omega) * \delta(\omega - \omega_0) = S[j(\omega - \omega_0)]$]

From Eq. 12.11, we know that $S_p(j\omega)$ is a function of frequency ω consisting of a superposition of shifted version (or shifted replicas) of $S(j\omega)$ and scaled by $1/T_s$. The spectrum of $s_p(t)$ *i.e.*, $S_p(j\omega)$ is shown in Fig. 12.3. There are three cases :

1. If $\omega_s > 2\omega_M$, and thus there is no overlapping between the shifted replicas of $S(j\omega)$. In this condition, spectrum of $s_p(t)$ *i.e.*, $S_p(j\omega)$ is shown in Fig. 12.3 (*c*).
2. If $\omega_s = 2\omega_M$, then the spectrum of $s_p(t)$, *i.e.*, $S_p(j\omega)$ is shown by Fig. 12.3 (*d*). In this case, also there is no over-lapping between the shifted replicas of $S(j\omega)$.
3. If $\omega_s < 2\omega_M$, there is an overlap between the shifted replicas of $S(j\omega)$. Its spectrum is shown in Fig. 12.3 (*e*). For the cases 1 and 2, $S(j\omega)$ is faithfully reproduced at integer multiplies of the sampling frequency.

Consequently, if $\omega_s \geq 2\omega_M$, continuous-time signal $s(t)$ can be recovered exactly from $s_p(t)$ by means of a low pass filter with gain T_s. To perform reconstruction of the continuous-time signal $s(t)$, the low pass filter will have a cutoff frequency greater than ω_M and less than $(\omega_s - \omega_M)$. Now we will again state sampling theorem as :

Let us have a continuous-time signal $s(t)$ which is a band-limited signal with $S(j\omega) = 0$ for $|\omega| > \omega_M$. Then $s(t)$ is uniquely determined by its samples $s(nT_s) \neq 0$ for $n = 0, \pm 1, \pm 2, \ldots$ if

$$\omega_s > 2\omega_M$$

where $\omega_s = 2\pi F_s = \dfrac{2\pi}{T_s}$

The frequency $2\omega_M$ is called the Nyquist rate. It is defined as the minimum sampling rate for that we can reconstruct continuous-time signal $s(t)$ from its samples. Ideal filters are generally not used in practice for a variety of reasons. In practical applications, the ideal low pass filter would be replaced by a non-ideal low pass filter that approximates the desired frequency characteristic accurately enough for the problem of interest, *i.e.*,

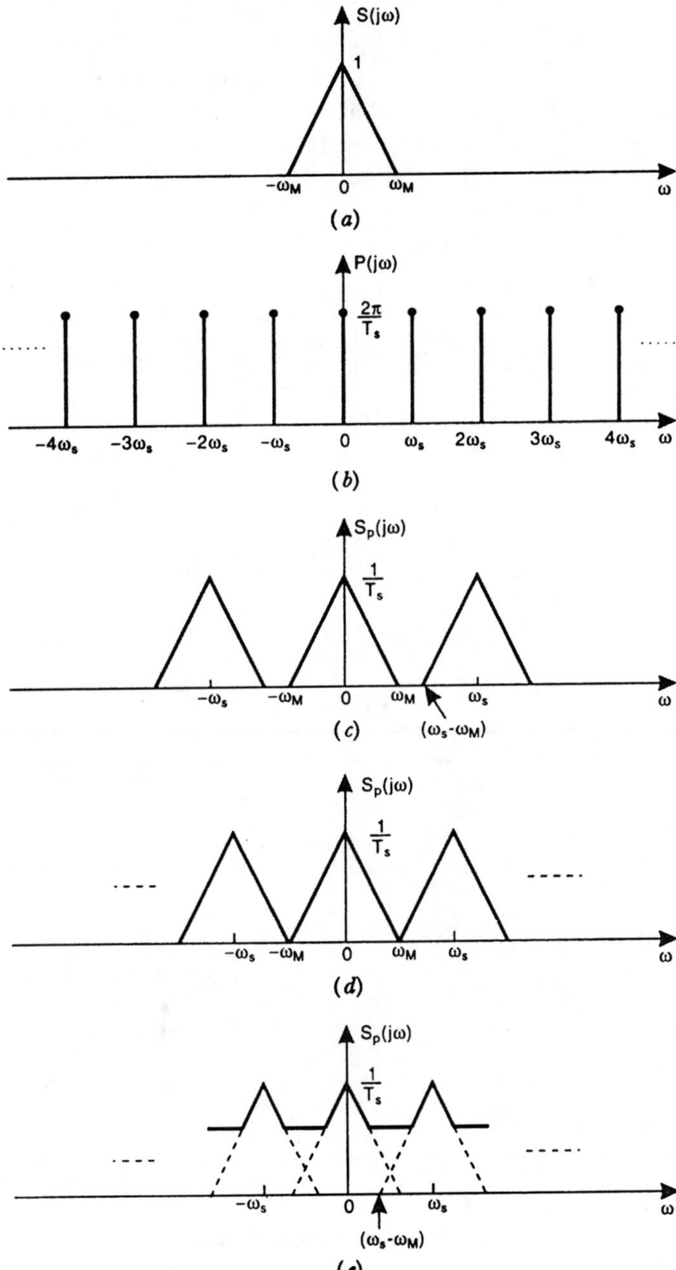

Fig. 12.3 *Illustration of effect in the frequency-domain of sampling in the time domain. (a) Spectrum of original continuous-time signal $s(t)$ (b) spectrum of periodic impulse train or sampling function $p(t)$ (c) spectrum of sampled version of continuous-time signal, $s_p(t)$ with $\omega_s > 2\omega_M$ (d) spectrum of sampled version of continuous-time signal, $s_p(t)$ with $\omega_s = 2\omega_M$ (e) spectrum of sampled version of continuous-time signal, $s_p(t)$ with $\omega_s < 2\omega_M$.*

$$H(j\omega) \approx 1 \text{ for } |\omega| < \omega_M$$

and $$H(j\omega) \approx 0 \text{ for } |\omega| \geq \omega_s - \omega_M$$

For any such approximation in the low pass filter stage, we will see some discrepancy between original signal $s(t)$ and reconstructed signal $s_r(t)$ or equivalently between $S(j\omega)$ and $S_r(j\omega)$. The particular choice of non-ideal filter depends upon the acceptable level of distortion for the application under consideration.

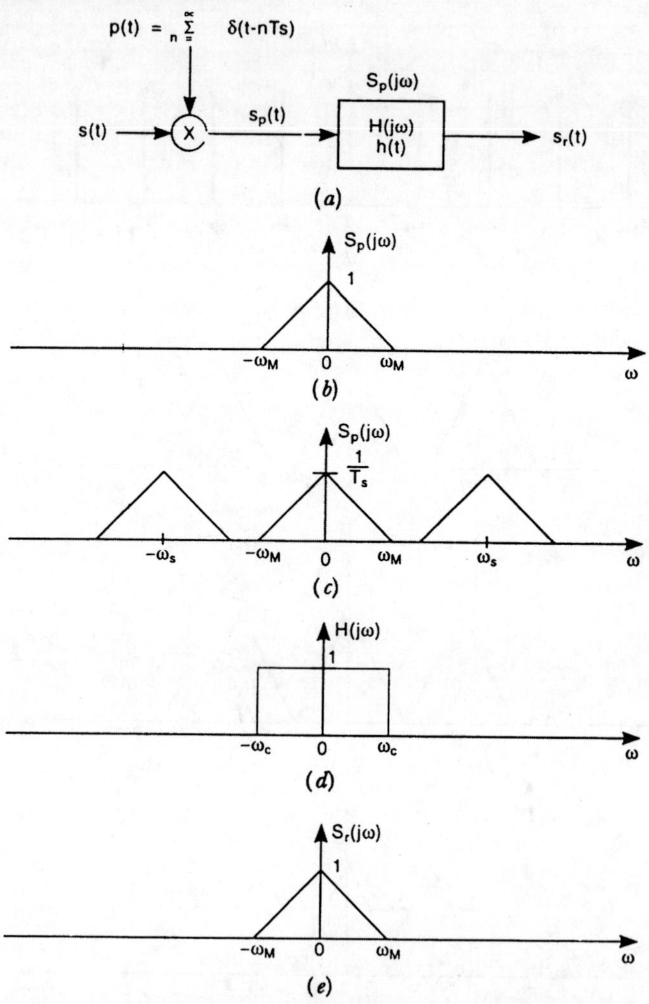

Fig. 12.4 *Illustration of perfect reconstruction of a continuous-time signal from its samples using ideal low pass filter, i.e., reconstruction low pass filter. (a) System for sampling and reconstruction of a continuous-time signal (b) spectrum of original continuous-time signal s(t) (c) spectrum of sampled version of continuous-time signal with* $\omega_s > 2\omega_M$ *(d) frequency response or spectrum of an ideal low pass filter that is used as a reconstruction low pass filter (e) spectrum of reconstructed signal from its samples,* $s_r(t)$.

12.3.2 Zero-Order Hold Sampling

Sampling theorem is most easily explained in terms of impulse-train sampling. Sampling theorem establishes a relationship between band-limited continuous-time signal and its discrete-time version. In practice, it is relatively difficult to generate and transmit narrow, large-amplitude pulses which approximate impulses. It is often more convenient to generate the sampled signal in a form referred to as a zero-order hold. The zero-order hold system samples a continuous-time signal $s(t)$ at a given instant and holds that value until the next instant at which another sample is taken. It is illustrated in Fig. 12.5.

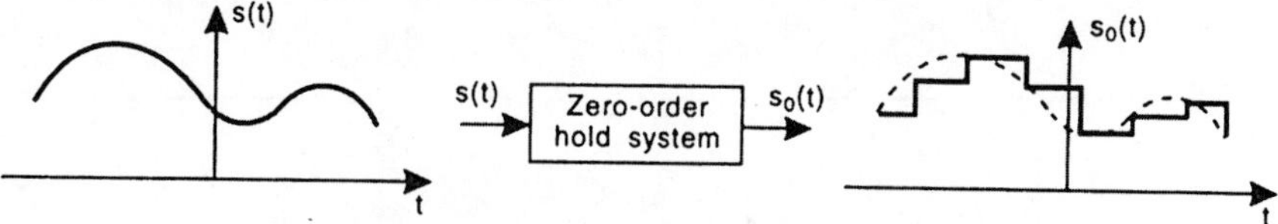

Fig. 12.5 *Illustration of zero-order hold sampling.*

The recovery or reconstruction of a continuous-time signal $s(t)$ from the output of a zero-order hold can again be carried out by low pass filtering. In this case, the required low pass filter no longer has constant gain in the pass band. For developing the required filter characteristic, we first note that the output $s_0(t)$ of the zero-order hold can be generated in principle by impulse train sampling followed by a continuous-time LTI system with rectangular impulse response. It is shown in Fig. 12.6.

For construction of a continuous-time signal $s(t)$ from output of the zero-order hold $s_0(t)$, we consider processing $s_0(t)$ with continuous-time LTI system with impulse response $h_r(t)$. Its frequency response is denoted by $H_r(j\omega)$. The cascade (series) of zero-order hold system with reconstruction low-pass filter of Fig. 12.6 is shown in Fig. 12.7. We wish that output of reconstruction low pass filter should be equal to original continuous-time signal $s(t)$. That is, $s_r(t) = s(t)$. It is only possible when the cascade combination of $h_0(t)$ and $h_r(t)$ are the ideal low pass filters $H(j\omega)$ used in Fig. 12.4.

The impulse response $h_0(t)$ is defined as

$$h_0(t) = \begin{cases} 1, & 0 \le t \le T_s \\ 0, & \text{otherwise} \end{cases} \quad \text{...(12.12)}$$

Frequency response $H_0(j\omega)$ can be determined by taking CTFT of the impulse response $h_0(t)$ as

$$H_0(j\omega) = \text{CTFT}\{h_0(t)\} = \int_{-\infty}^{\infty} h_0(t)\, e^{-j\omega t}\, dt$$

$$= \int_0^{T_s} 1e^{-j\omega t}\, dt = \left[\frac{e^{-j\omega t}}{-j\omega}\right]_0^{T_s}$$

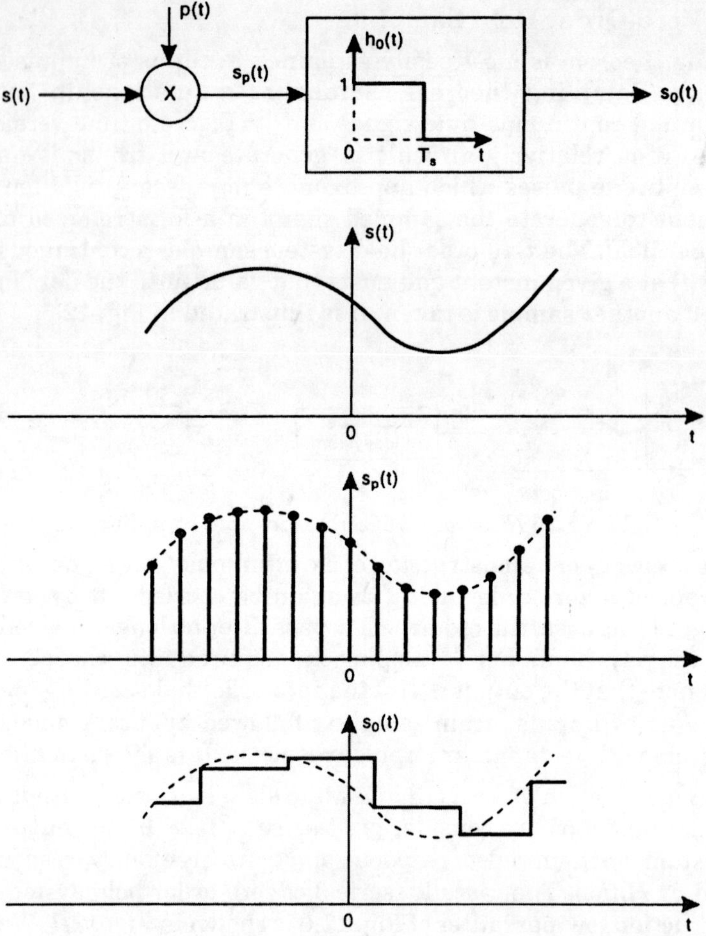

Fig. 12.6 *Illustration of a zero-order hold system as impulse- train sampling followed by an continuous-time LTI system with a rectangular impulse response.*

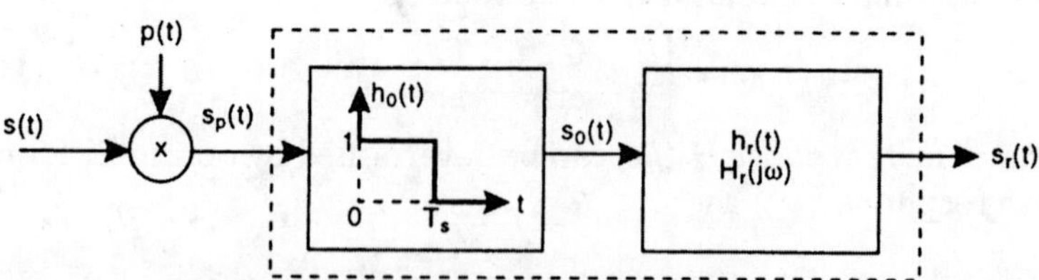

Fig. 12.7 *Cascade combination of zero-order hold system with a reconstruction filter.*

$$= \frac{e^{-j\omega T_s} - e^{-j\omega x(0)}}{-j\omega} = \frac{e^{-j\omega T_s} - 1}{-j\omega}$$

$$= \frac{e^{-j\omega(T_s/2)}}{\omega}\left\{\frac{e^{-j\omega(T_s/2)} - e^{j\omega(T_s/2)}}{-j}\right\}$$

$$= \frac{2e^{-j\omega(T_s/2)}}{\omega}\left\{\frac{e^{j\omega T_s/2} - e^{-j\omega T_s/2}}{2j}\right\}$$

$$= \frac{2e^{-j\omega T_s/2}}{\omega}\sin\left(\frac{\omega T_s}{2}\right)$$

or
$$H_0(j\omega) = e^{-j\omega T_s/2}\left[\frac{2\sin\left(\frac{\omega T_s}{2}\right)}{\omega}\right] \quad \text{...(12.13)}$$

Therefore, frequency response $H_r(j\omega)$ can be determined as

$$H(j\omega) = H_0(j\omega)\, H_r(j\omega)$$

or
$$H_r(j\omega) = \frac{H(j\omega)}{H_0(j\omega)} \quad \text{...(12.14)}$$

Substituting the value $H_0(j\omega)$ from Eq. 12.12 into Eq. 12.13, we get

$$H_r(j\omega) = \frac{H(j\omega)}{H_0(j\omega)} = \frac{H(j\omega)}{e^{-j\omega T_s/2}\left[\dfrac{2\sin \omega T_s/2}{\omega}\right]}$$

or
$$H_r(j\omega) = \frac{e^{j\omega T_s/2}\, H(j\omega)}{\left(\dfrac{2\sin \omega T_s/2}{\omega}\right)} \quad \text{...(12.15)}$$

For example, with a cutoff frequency of $H(j\omega)$ equal to $\omega_s/2$, the ideal magnitude and phase for the reconstruction (low pass) filter following a zero-order hold is shown in Fig. 12.8.

In practice the frequency response given in Eq. 12.15 cannot be realized and thus we design an adequate approximation to it. Infact, in

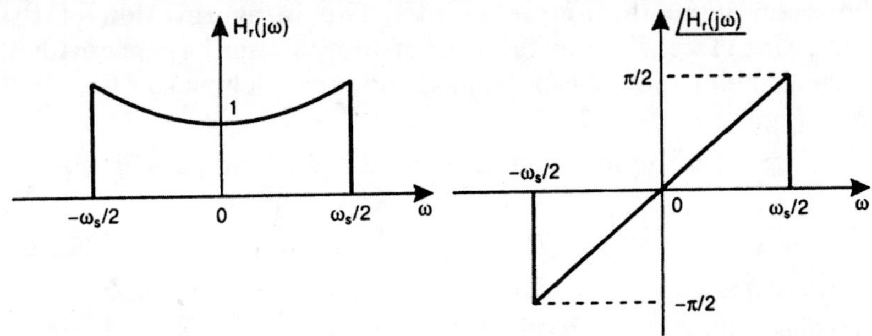

Fig. 12.8 *Magnitude and phase for the reconstruction (low pass) filter for a zero-order hold.*

many situations, the output of a zero-order hold is taken as an adequate approximation to the original continuous-time signal by itself, without using any additional low pass filter. Interpolation is one such method that is used to recover a continuous-time signal from its samples. In some applications, we wish to perform some smoother interpolation between sample values. Interpolation is discussed in the next section.

12.3.3 Reconstruction of a Continuous-Time Signal from its Samples using Interpolation Techniques

Interpolation is the commonly used procedure for reconstructing a continuous-time signal from its sample values. Interpolation gives either approximate or exact reconstruction (or recovery) of the signal. One simple interpolation procedure is zero-order hold. Another useful form of interpolation is the linear interpolation. In linear interpolation the adjacent samples (or sample points) are connected by straight lines as shown in Fig. 12.9. We can also use higher order interpolation formulae for reconstructing the signal from its sample values. In more complicated interpolation formulae, sample points may be connected by higher order polynomials or other mathematical functions.

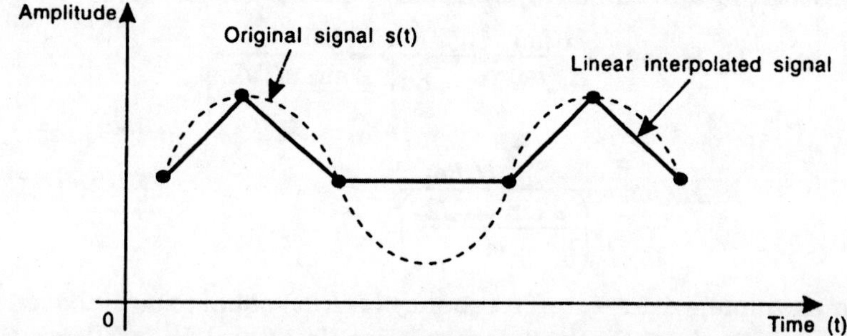

Fig. 12.9 *Linear interpolation between sample points.*

For a band-limited continuous-time signal, if the sampling instants are sufficiently close, then the signal can be reconstructed exactly by using a low pass filter. In this case, an exact interpolation can be carried out between various sample points. The interpretation of the reconstruction of signal $s(t)$ as a process of interpolation becomes evident when we consider the effect in time-domain of a low pass filter. It is shown in Fig. 12.4.

From Fig. 12.4, the output of the reconstruction low pass filter given as

$$s_r(t) = s_p(t) * h(t) \quad \text{...(12.16)}$$

where $h(t)$ is impulse response of a reconstruction low pass filter.

$s_p(t)$ is given as

$$s_p(t) = \sum_{n=-\infty}^{\infty} s(nT_s)\,\delta(t - nT_s) \quad \text{...(12.17)}$$

Substituting Eq. 12.17 in Eq. 12.16, we get

$$s_r(t) = s_p(t) * h(t)$$

$$= \sum_{n=-\infty}^{\infty} s(nT_s)\,\delta(t - nT_s) * h(t)$$

or

$$s_r(t) = \sum_{n=-\infty}^{\infty} s(nT_s)\,h(t - nT_s) \qquad \text{...(12.18)}$$

Eq. 12.18 describes a method to fit a continuous curve between the sample points $s(nT_s)$ and consequently it represents an interpolation formula. Impulse response of an ideal low pass filter is given by

$$h(t) = \omega_c T_s \frac{\sin(\omega_c t)}{\pi \omega_c T_s} \qquad \text{...(12.19)}$$

Substituting the value of $h(t)$ from Eq. 12.19 in Eq. 12.18, we get

$$s_r(t) = \sum_{n=-\infty}^{\infty} s(nT_s) \left\{ \frac{\omega_c T_s}{\pi} \frac{\sin[\omega_c (t - nT_s)]}{\omega_c (t - nT_s)} \right\} \qquad \text{...(12.20)}$$

where ω_c is the cutoff frequency of the low pass filter.

The reconstruction of a signal according to interpolation formula of Eq. 12.20 is shown in Fig. 12.10, where cutoff frequency of low pass filter is equal to $\omega_s/2$.

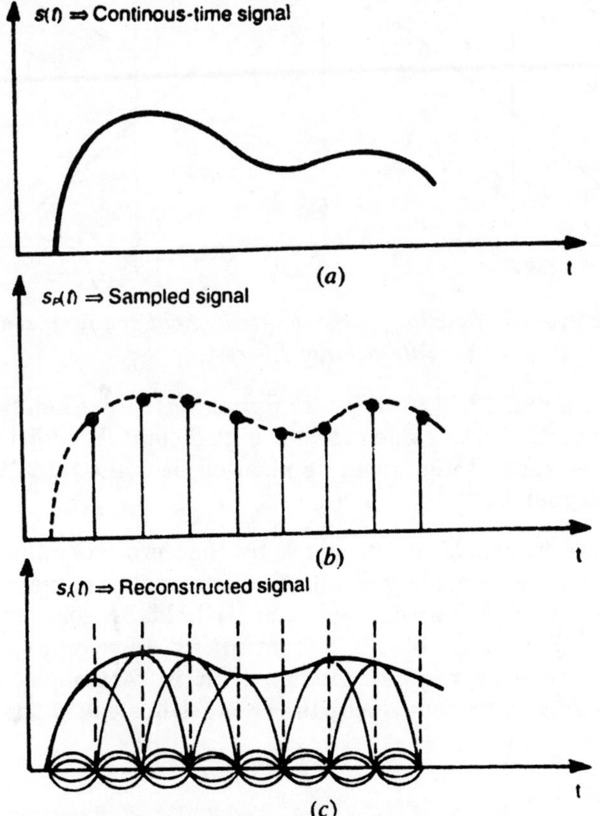

Fig. 12.10 *Ideal band-limited interpolation using the sinc function.*

Interpolation using the impulse response $h(t)$ of an ideal low pass filter given by Eq. 12.20 is called *band-limited interpolation.* It implements exact reconstruction if signal $s(t)$ is band-limited and it satisfies the condition of sampling theorem. According to sampling theorem for perfect reconstruction of a band-limited signal, sampling frequency should be greater than twice the highest frequency component of the signal.

In many cases we prefer simpler interpolating functions such as zero-order hold. The zero-order hold can be viewed as a form of interpolation between sample values in which the interpolating function $h(t)$ is the impulse response $h_0(t)$ shown in Fig. 12.6. In this figure, $s_0(t)$ corresponds to the approximation of continuous-time signal $s(t)$ and $h_0(t)$ represents an approximation of the ideal low pass filter required for exact interpolation. Fig. 12.11 illustrates the magnitude of transfer function or system function of the zero-order hold interpolation filter, superimposed on the desired transfer function of each interpolation filter.

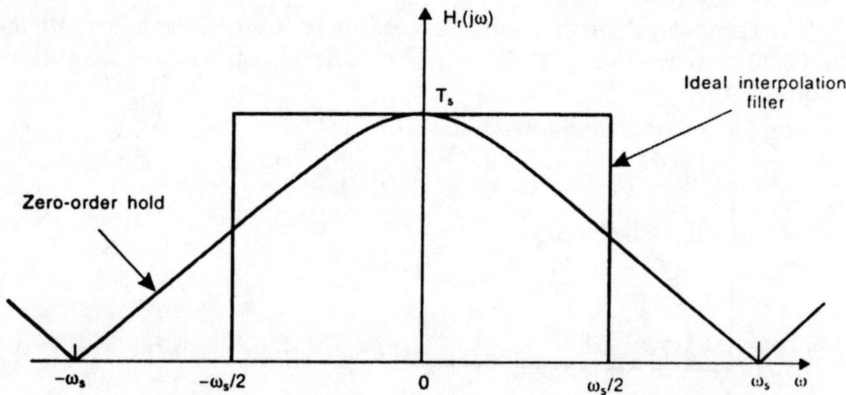

Fig. 12.11 *Transfer function for the zero-order hold and for the ideal interpolation filter.*

We see from Fig. 12.11 and Fig. 12.6 that zero-order hold is a very rough approximation. For some cases, it is sufficient. If additional low pass filtering is applied in a given application, it will tend to improve the overall interpolation.

If the crude interpolation provided by the zero-order hold is not sufficient, we can use a variety of smoother interpolation methods and these are collectively known as higher order holds. A zero-order hold produces an output signal that is discontinuous. In contrast, a linear interpolation produces reconstruction which is continuous and its derivatives are discontinuous due to the changes in slope at the sample points.

Linear interpolation is also referred to as a first-order hold linear interpolation. It is shown in Fig. 12.4 and given by equation $s_r(t) = \sum_{n=-\infty}^{\infty} s(nT_s)\, h(t - nT_s)$ with triangular impulse response shown in Fig. 12.12.

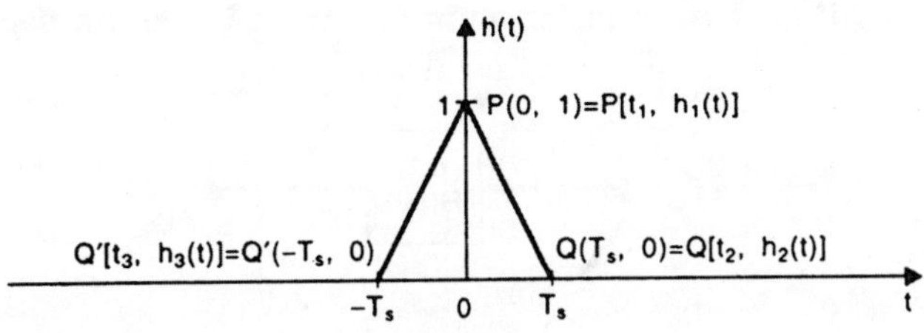

Fig. 12.12

Transfer Function of a Filter with Triangular Impulse Response

Determination of equation of line *PQ*

$$[h(t) - h_1(t)] = \left(\frac{h_2(t) - h_1(t)}{t_2 - t_1}\right)(t - t_1)$$

or $$[h(t) - 1] = \left[\frac{0 - 1}{T_s - 0}\right](t - 0) = \frac{-t}{T_s}$$

or $$h(t) = 1 - \frac{t}{T_s} \qquad \text{...(12.21)}$$

Determination of equation of line *PQ'*

$$[h(t) - h_1(t)] = \left[\frac{h_3(t) - h_1(t)}{t_3 - t_1}\right](t - t_1)$$

or $$[h(t) - 1] = \left[\frac{0 - 1}{-T_s - 0}\right](t - 0)$$

or $$h(t) = 1 + \frac{t}{T_s} \qquad \text{...(12.22)}$$

$$h(t) = \begin{cases} 1 - \dfrac{t}{T_s}, & \text{for } 0 \le t \le T_s \\ 1 + \dfrac{t}{T_s}, & \text{for } -T_s \le t \le 0 \end{cases} \qquad \text{...(12.23)}$$

Frequency response of triangular impulse response given in Eq. 12.23 can be determined by taking CTFT of $h(t)$ as

$$H(j\omega) = \text{CTFT}\,\{h(t)\} = \int_{-\infty}^{\infty} h(t)\, e^{-j\omega t}\, dt$$

$$= \int_{-\infty}^{0} \left(1 + \frac{t}{T_s}\right) e^{-j\omega t}\, dt + \int_{0}^{\infty} \left(1 - \frac{t}{T_s}\right) e^{-j\omega t}\, dt$$

or $$H(j\omega) = \frac{1}{T_s}\left[\frac{\sin(\omega T_s/2)}{\omega/2}\right]^2 \qquad ...(12.24)$$

Eq. 12.24 is the transfer function or frequency response of the low pass filter with triangular impulse response. The transfer function of the first order hold is shown imposed on the transfer function of the ideal interpolation filter. In a similar manner, we can define second and

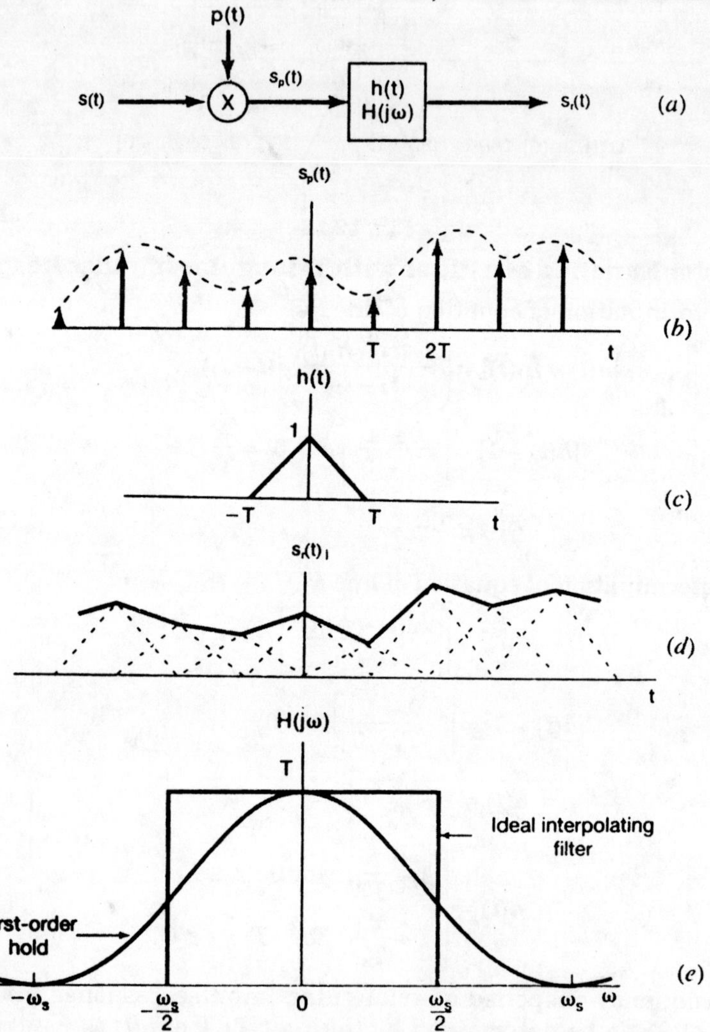

Fig. 12.13 *Illustration of linear interpolation (first-order hold) as impulse train sampling is followed by convolution with a triangular impulse response. (a) System for sampling and reconstruction of a continuous-time signal (b) impulse train of samples of a continuous-time signal (c) impulse response which represents a first-order hold (d) first-order hold is applied to the sampled signal $s_p(t)$ (e) A comparison of transfer function of an ideal interpolation filter and first-order hold.*

higher order holds that produce reconstruction with a higher degree of smoothness. The output of a second-order hold provides an interpolation of sample values that is continuous. In this interpolation, first derivative is also continuous but second derivative is discontinuous.

12.3.4 Aliasing Phenomenon

For perfect reconstruction or recovery of a continuous-time signal from its samples, sampling frequency should be greater or equal to Nyquist rate (we have already discussed Nyquist rate in art. 12.3). Nyquist rate is equal to the minimum rate required for perfect reconstruction of a continuous-time signal from its samples. It is equal to twice the highest frequency component of the message signal. When the sampling frequency is less than the Nyquist rate it is called under sampling. If sampling rate or frequency is less than Nyquist rate aliasing phenomenon occurs.

If sampling rate > Nyquist rate then it is called over sampling and in this case no aliasing phenomenon occurs. Infact this is a suitable and necessary condition for sampling process.

Now, we can say that aliasing phenomenon is caused by under sampling. In other words, aliasing is also called an effect of under sampling. Alasing occurs when sampling rate [$\omega_s < 2\omega_{max}$ or $F_s < 2F_{max}$] is not sufficiently high to prevent the shifting of high frequency information into lower frequencies. Such transference of information from one band of frequencies to another is called aliasing and the resulting frequency response is called an aliased representation of the original continuous-time signal.

Aliasing phenomenon is defined as a phenomenon of high frequency components in a spectrum of a continuous-time signal seemingly taking on the identity of a lower frequency in the spectrum of its sampled version. There are two corrective measures which are used to eliminate aliasing.

1. A pre-alias low pass filter is used before sampling for attenuating those high frequencies that are not essential for the transmission of information.
2. A pre-alias low pass filtered signal is sampled at a rate of frequency slightly higher than the Nyquist rate ($\omega_s > 2\omega_{max}$).

12.4 DISCRETE-TIME PROCESSING OF CONTINUOUS-TIME SIGNALS

In many applications, discrete-time processing of continuous-time signals offers a significant advantage over continuous-time processing of continuous-time signals.

For discrete-timee processing of continuous-time signal, we first convert continuous-time signal into a discrete-time signal. After

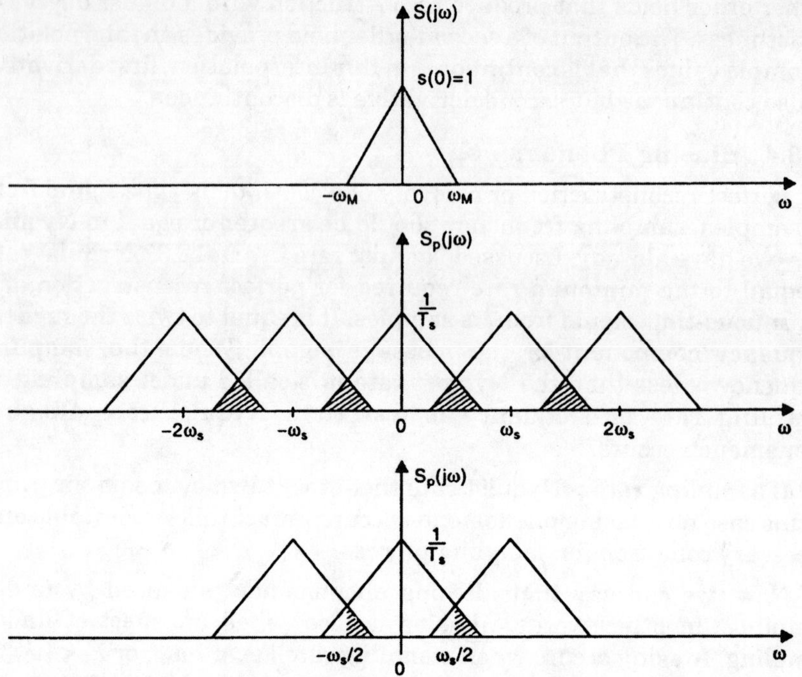

Fig. 12.14 *The effect of under sampling of a continuous-time signal s(t) on its digital frequency response showing aliasing around the folding frequency $\omega_s/2$.*

discrete-time processing, we convert back to a continuous-time signal. The discrete-time signal processing (DTSP) can be implemented with a general or special purpose computer, with microprocessors, or with any of the variety of devices that are specifically oriented toward DTSP.

The discrete-time signal processing (DTSP) can be viewed as the cascade of three operations.

These operations are shown in Fig. 12.15. Here $s_c(t)$ and $y_c(t)$ are continuous-time signals and $s_d(n)$ and $y_d(n)$ are the discrete-time signals corresponding to $s_c(t)$ and $y_c(t)$.

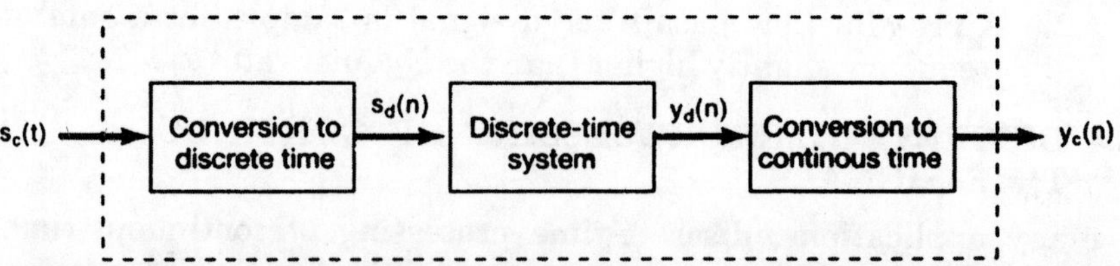

Fig. 12.15 *System for discrete-time processing of continuous- time signals.*

The overall combined system is, of course, a continuous-time system because its input and output are both continuous-time signals.

In discrete-time signal processing (DTSP) or discrete-time processing of continuous-time signals, we convert continuous-time signals to discrete-time signal from its discrete-time representation. If process of periodic sampling satisfies the sampling theorem then the continuous-time signal $s_c(t)$ is exactly represented by a sequence of instantaneous sample values $s_c(nT_s)$; that is, the discrete-time sequence $s_d(n)$. The discrete-time sequence $s_d(n)$ can be related to $s_c(t)$ by

$$s_d(n) = s_c(nT_s) \qquad ...(12.25)$$

The first system in Fig. 12.15 transforms a continuous-time signal into discrete-time signal and is known as continuous-to- discrete-time converter.

The third system in Fig. 12.15 transforms a discrete-time signal into continuous-time signal and known as discrete-to-continuous- time converter. The discrete-to-continuous-time converter performs an interpolation between the sample values provided to it as input. The discrete-to-continuous-time converter produces a continuous-time signal $y_c(t)$ which is related to the discrete-time signal $y_d(n)$ by

$$y_d(n) = y_c(nT_s) \qquad ...(12.26)$$

For digital computers and digital systems, the discrete-time signal is represented in digital form. The device which is commonly used to implement the continuous-time-to-discrete-time conversion is referred to as an analog-to-digital (A/D) converter. The device which is used to implement the discrete-time-to- continuous-time conversion is referred to as an digital-to-analog (D/A) converter.

We can understand the relationship between the continuous-time signal $s_c(t)$ and its discrete-time representation $s_d(n)$ by using the process of periodic sampling followed by a mapping of the impulse train to a sequence. These two steps are shown in Fig. 12.16 (*a*). The first step performs sampling process. The second step performs conversion of impulse train to discrete-time sequence. It can be thought of as as a normalization in time..

The normalization in converting $s_p(t)$ to $s_d(n)$ is evident in Fig. 12.16 (*b*) and 12.16 (*c*). These Fig. 12.16 (*b*) and 12.16 (*c*) are illustrating $s_p(t)$ and $s_d(n)$, respectively, for sampling rates of $T_s = T_1$ and $T_s = 2T_1$.

Fig. 12.17 illustrates the processing stages in frequency domain. Here we shall use CTFT and DTFT for converting continuous-time and discrete-time signals, respectively, in frequency domain. To distinguish the continuous-time and discrete-time frequency variables, we denote continuous-time frequency variable by ω and discrete-time variable by Ω.

The continuous-time Fourier transforms (CTFTs) of $s_c(t)$ and $y_c(t)$ are $S_c(j\omega)$ and $Y_c(j\omega)$, respectively, the discrete-time Fourier transforms (DTFTs) of $s_d(n)$ and $y_d(n)$ are $S_d(e^{j\Omega})$ and $Y_d(e^{j\Omega})$, respectively.

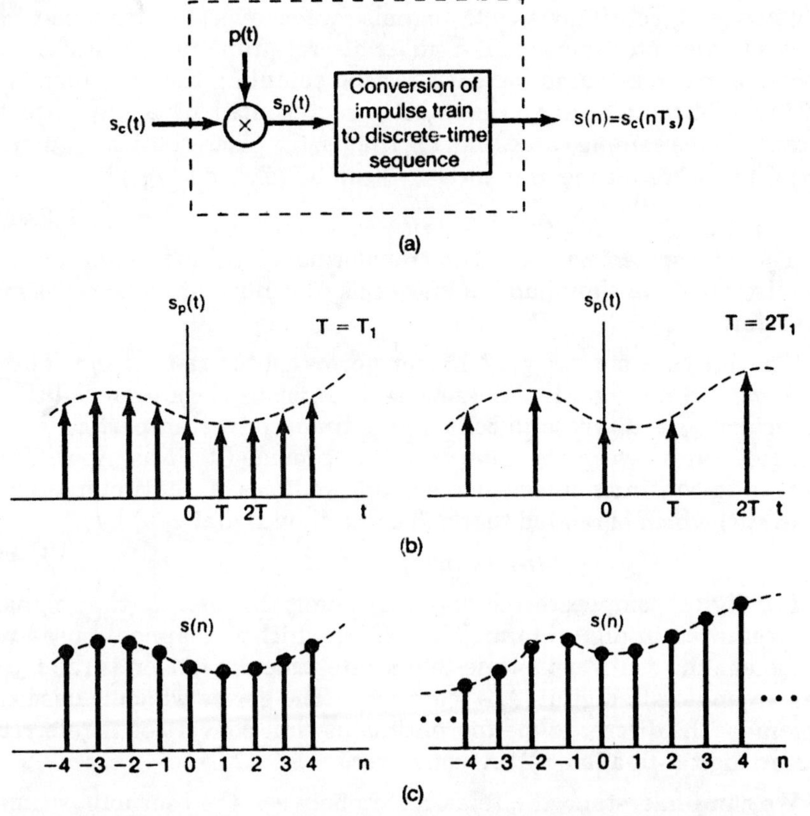

Fig. 12.16 *Illustration of sampling process with a periodic impluse train followed by conversion to a discrete-time sequence. (a) overall system (b) $s_p(t)$ for two sampling periods $T_s = T_1$* and *$T_s = 2T_1$ (c) the output sequence for the two different sampling period.*

We begin by expressing the continuous-time Fourier transform (CTFT) of $s_p(t)$ in terms of sample values of $s_c(t)$ as

$$s_p(t) = \sum_{k=-\infty}^{\infty} s_c(nT_s)\, \delta(t - nT_s) \qquad ...(12.27)$$

Taking CTFT of both sides of Eq. 12.27, we get

$$\text{CTFT}\,[s_p(t)] = \text{CTFT}\left[\sum_{k=-\infty}^{\infty} s_c(nT_s)\, \delta(t - nT_s)\right]$$

$$= \sum_{k=-\infty}^{\infty} s_c(nT_s)\, \text{CTFT}\,[\delta(t - nT_s)]$$

or
$$S_p(j\omega) = \sum_{k=-\infty}^{\infty} s_c(nT_s)\, e^{-j\omega nT_s} \qquad ...(12.28)$$

[Since CTFT of $\delta(t - nT_s)$ is $e^{-j\omega nT_s}$]

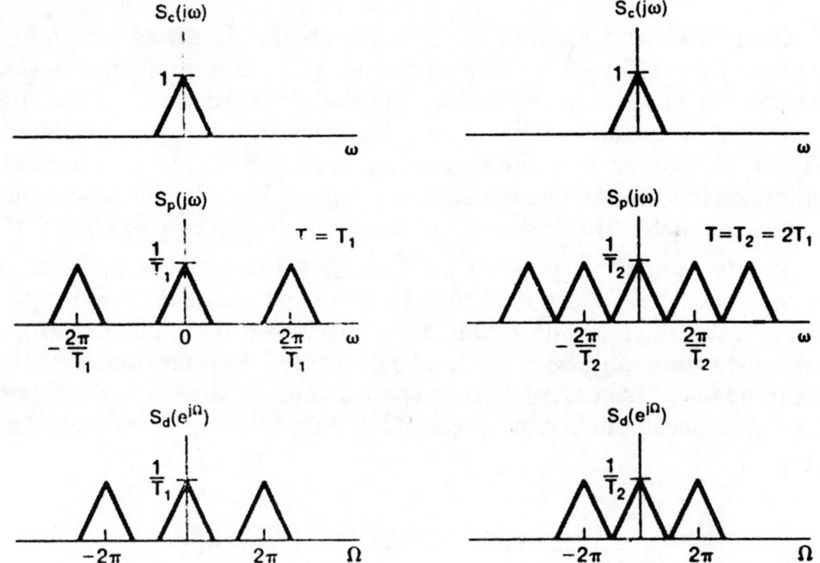

Fig. 12.17 *Illustration of relationship between $S_c(j\omega)$, $S_p(j\omega)$ and $S_d(e^{j\Omega})$ for two different sampling rates.*

The Discrete-time Fourier transform (DTFT) of $s_d(n)$ is given by

$$S_d(e^{j\Omega}) = \sum_{n=-\infty}^{\infty} s_d(n)\, e^{-j\Omega n} \qquad ...(12.29)$$

[Since $s_d(n) = s_c(nT_s)$]

$$S_d(e^{j\Omega}) = \sum_{n=-\infty}^{\infty} s_c(nT_s)\, e^{-j\Omega n} \qquad ...(12.30)$$

The linear frequency scaling can be thought of informally as a consequence of the normalization in time introduced by coverting from the impulse train $s_p(t)$ to the discrete-time sequence $s_d(n)$. The relationship between Ω and ω is given by $\Omega = \omega T_s$. In converting from $s_p(t)$ to $s_d(n)$, the time axis is scaled by $\frac{1}{T_s}$.

By camparing Eqs. 12.28 and 12.30, we see that $S_d(e^{j\Omega})$ and $S_p(j\omega)$ are related through

$$S_d(e^{j\Omega}) = S_p\left(\frac{j\Omega}{T_s}\right) \qquad ...(12.31)$$

We also know that

$$S_p(j\omega) = \frac{1}{T_s} \sum_{k=-\infty}^{\infty} S_c[j(\omega - k\omega_s)] \qquad ...(12.32)$$

Substituting Eq. 12.32 in Eq. 12.31, we get

$$S_d(e^{j\Omega}) = \frac{1}{T_s} \sum_{k=-\infty}^{\infty} S_c\left[j\left(\frac{\Omega - 2\pi k}{T_s}\right)\right] \qquad ...(12.33)$$

The relationship among $S_c(j\omega)$, $S_p(j\omega)$ and $S_d(e^{j\Omega})$ are shown in Fig. 12.17 for two different samping periods. Here is a frequency scaled version of $S_p(j\omega)$, and in particular, is periodic in Ω with period 2π. This periodicity is characteristic of any discrete-time Fourier transform (DTFT). The spectrum of $s_d(n)$ is related to that of $s_c(t)$ through periodic replication followed by linear frequency scaling. The periodic replication is a consequence of the first step in the conversion process in Fig. 12.15.

In the overall system [Fig. 12.15], after processing with a discrete-time system, the resulting discrete-time sequence is converted back to a continuous-time signal. From discrete-time sequence $y_d(n)$, a continuous-time impulse train can be generated. Reconstruction of the continuous-time signal $y_c(t)$ from the continuous-time impulse train $y_p(t)$ is then accomplished by means of low pass filtering. It is illustrated in Fig. 12.18.

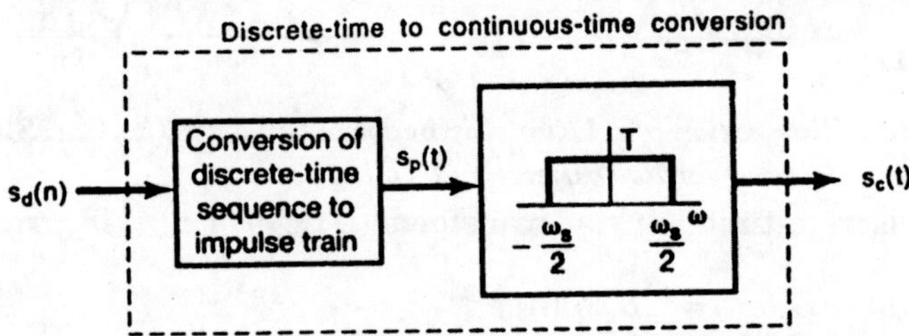

Fig. 12.18 *Conversion of a discrete-time sequence to a continuous-time signal.*

The overall system for filtering a continuous-time signal using discrete-time system or filter is shown in Fig. 12.19. If the discrete-time system is an identity system then, assuming that conditions of sampling theorem are met, the overall system will be an identity system.

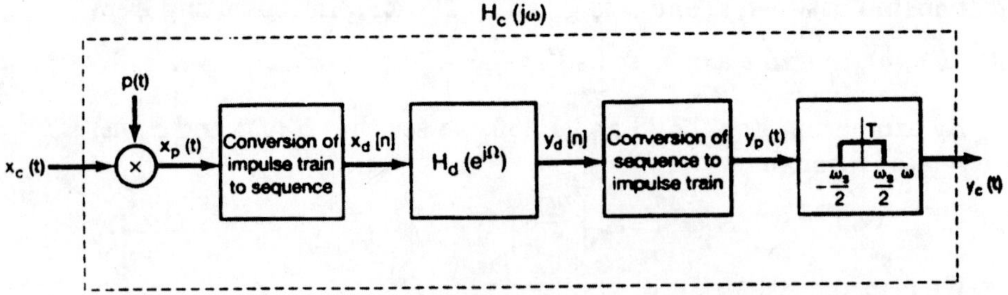

Fig. 12.19 *Block diagram of overall system for filtering a continuous-time signal using a discrete-time system or filter.*

The characteristics of the overall system with a more general frequency response $H_d(e^{j\Omega})$ are perhaps best understood by examining the example depicted in Fig. 12.20.

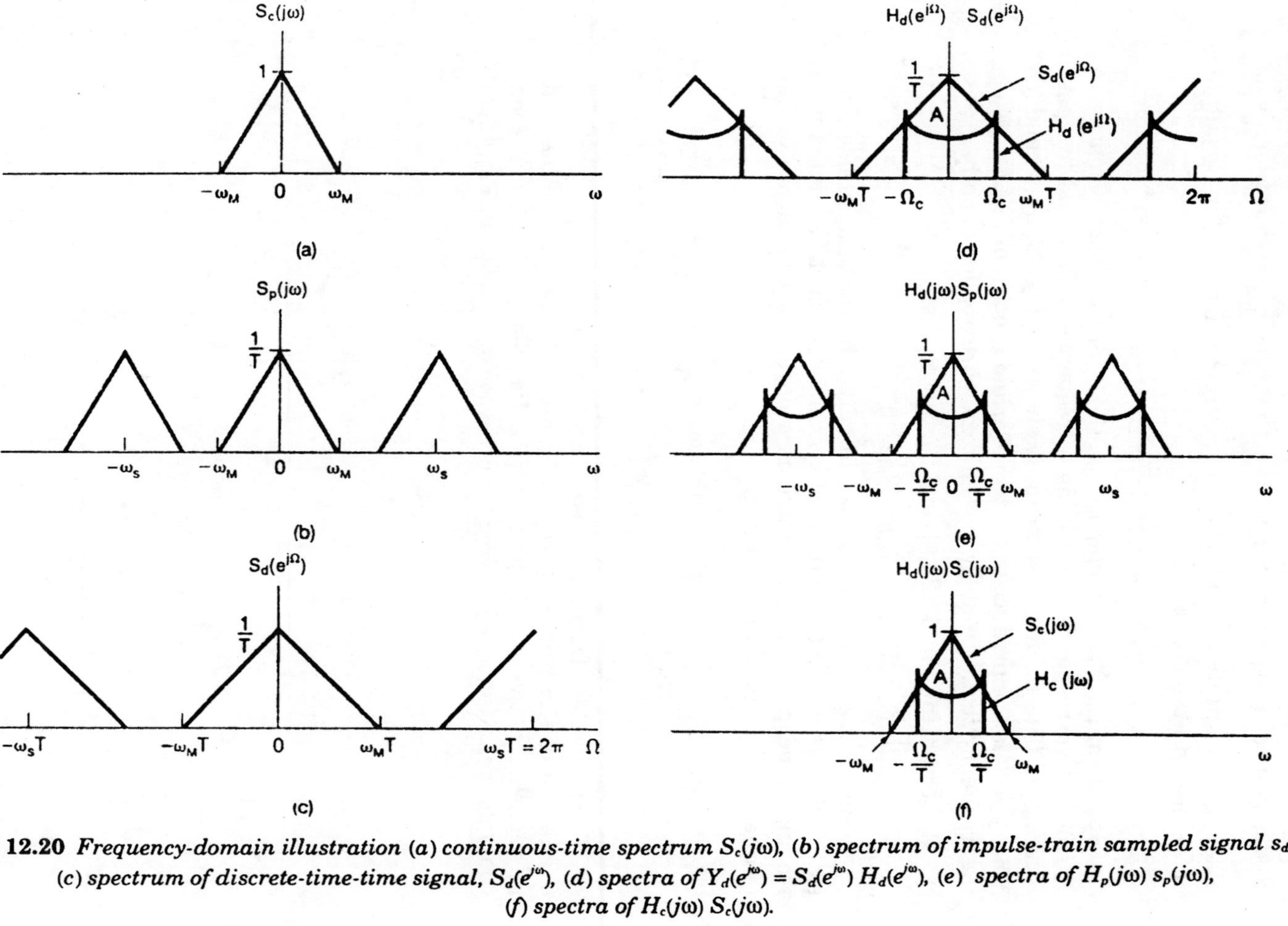

Fig. 12.20 *Frequency-domain illustration (a) continuous-time spectrum $S_c(j\omega)$, (b) spectrum of impulse-train sampled signal $s_d(n)$, (c) spectrum of discrete-time-time signal, $S_d(e^{j\omega})$, (d) spectra of $Y_d(e^{j\omega}) = S_d(e^{j\omega})\, H_d(e^{j\omega})$, (e) spectra of $H_p(j\omega)\, s_p(j\omega)$, (f) spectra of $H_c(j\omega)\, S_c(j\omega)$.*

Fig. 12.20 (*a*) , (*b*) and (*c*) illustrate the spectra of $S_c(j\omega)$, $S_p(j\omega)$ and $S_d(e^{j\Omega})$, respectively, where we assume that $\omega_M < \frac{\omega_s}{2}$, so that there is no aliasing. The spectrum $Y_d(e^{j\Omega})$ is the spectrum corresponding to output of the discrete-time filter.

$$Y_d(e^{j\Omega}) = S_d(e^{j\Omega})\, H(e^{j\Omega})$$

Spectrum $Y_d(e^{j\Omega})$ is depicted in Fig. 12.20 (*d*). The transformation to $Y_c(j\omega)$ then corresponds to applying a frequency scaling and low pass filtering and its resulted spectrum is shown in Fig. 12.20 (*e*) and (*f*). Since $y_d(e^{j\Omega})$ is the product of the two overlaid spectra in Fig. 12.20 (*d*), the frequency scaling and low pass filtering are applied to both.

By comparing Fig. 12.20 (*a*) and (*f*), we see that

$$Y_c(j\omega) = S_c(j\omega)\, H_d(e^{j\omega T_s}) \qquad \text{...(12.34)}$$

If the inputs are sufficiently band-limited then sampling theorem is satisfied. In fact, equivalent to a continuous-time LTI system with frequency response $H_c(j\omega)$ which is related to the discrete-time frequency response $H_d(e^{j\Omega})$ through

$$H_c(j\omega) = \begin{cases} H_d(e^{j\omega T_s}), & |\omega| < \frac{\omega_s}{2} \\ 0, & |\omega| > \frac{\omega_s}{2} \end{cases} \qquad \text{...(12.35)}$$

The equivalent frequency response for the continuous-time filter is one period of the frequency response of the discrete-time filter. The relationship between the discrete-time frequency response and one equivalent continuous-time frequency response is shown in Fig. 12.21.

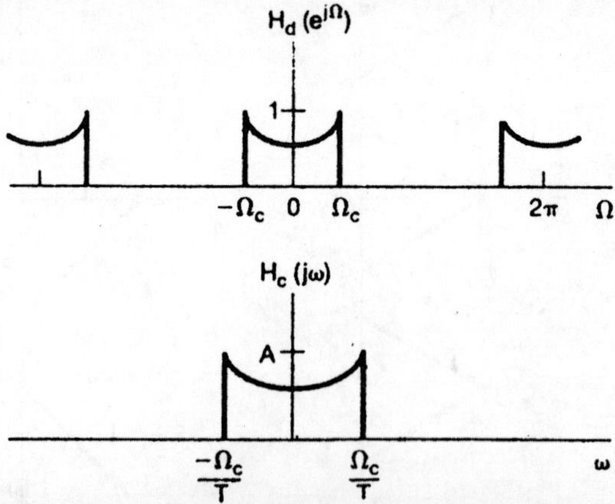

Fig. 12.21 *Discrete-time frequency response and the equivalent continuous-time frequency response for the system in Fig.* 12.19.

The spectra of Fig. 12.20, for band limited input signals with sampling rate are sufficiently high so as to avoid aliasing, the system of Fig. 12.19 is equivalent to a continuous-time LTI system. For such inputs, Fig. 12.19 and Eq. 12.35 provide the conceptual basis for continuous-time processing using discrete- time filters.

Now we shall discuss discrete-time processing of continuous-time signals with some examples. These are

1. Digital differentiator
2. Half-sample delay.

1. Digital Differentiator

Here we shall consider the discrete-time implementation of a continuous- time band-limited differentiating filter.

The frequency response of a continuous-time differentiating filter is given by

$$H_c(j\omega) = j\omega \qquad \text{...(12.36)}$$

The frequency response of a band-limited differentiator with cut- off frequency ω_c is given by

$$H_c(j\omega) = \begin{cases} j\omega, & |\omega| < \omega_c \\ 0, & |\omega| > \omega_c \end{cases} \qquad \text{...(12.37)}$$

This frequency response of Eq. 12.37 is sketched in Fig. 12.22.

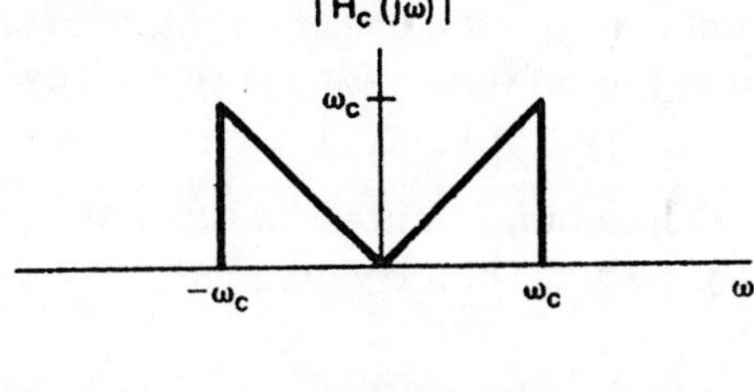

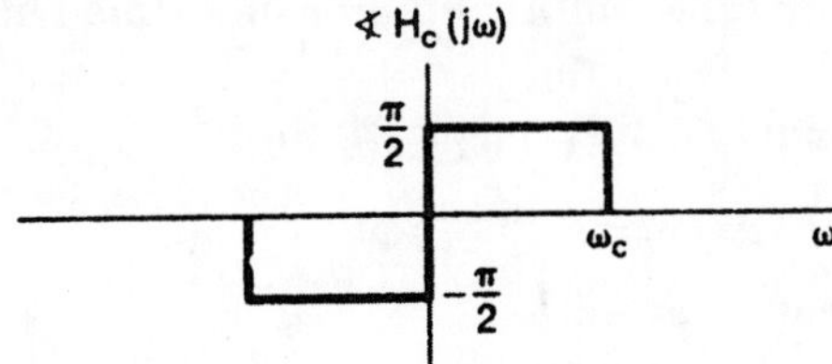

Fig. 12.22 *Frequency response of a continuous-time ideal band- limited differentiator* $H_c(j\omega) = j\omega$, $|\omega| < \omega_c$. *(a) Magnitude response, (b) Phase response.*

Using Eq. 12.35 with a sampling frequency $\omega_s = 2\omega_c$, we see that the corresponding discrete-time transfer function is

$$H_d(e^{j\Omega}) = j\left(\frac{\Omega}{T_s}\right), \quad |\Omega| < \pi \qquad \text{...(12.38)}$$

It is sketched in Fig. 12.23.

Here we have assumed that there is no aliasing, the discrete-time transfer function $H(e^{j\Omega})$ is shown in Fig. 12.23.

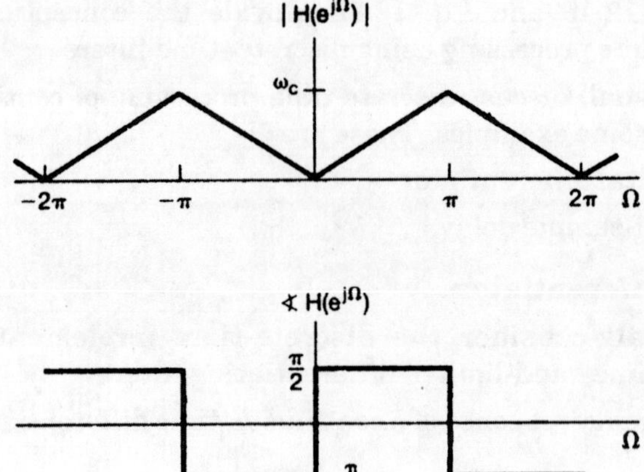

Fig. 12.23 ***Freequency response of discrete-time filter used to implement a continuous-time band limited differentiator. (a) Magnitude response (b) Phase response.***

2. Half-Sample Delay

Now we consider the implementation of a time-shift or delay of a continuous-time signal through the use of a system in the form of Fig. 12.15. The input and output of this system is related by

$$y_c(t) = s_c(t - t_0) \qquad ...(12.39)$$

When the input $s_c(t)$ is band limited and sampling is done at the sampling rate slightly greater than Nyquist rate. The time t_0 is called delay time.

We can determine CTFT of Eq. 12.39 using time-shifting property of CTFT as

$$\text{CTFT }\{y_c(t)\} = \text{CTFT }\{s_c(t - t_0)\}$$

or

$$Y_c(j\omega) = e^{-j\omega t_0} S_c(j\omega) \qquad ...(12.40)$$

Eq. 12.40 can be written as

$$H_c(j\omega) = \frac{Y_c(j\omega)}{S_c(j\omega)} = e^{-j\omega t_0} \qquad ...(12.41)$$

The transfer function for band-limited signal is given by

$$H_c(j\omega) = \begin{cases} e^{-j\omega t_0}, & |\omega| < \omega_c \\ 0, & \text{otherwise} \end{cases} \qquad ...(12.42)$$

where ω_c = the cut-off frequency of the continuous-time filter.

Frequency response $H_c(j\omega)$ corresponds to a time- shift for band-limited signals and rejects all frequencies greater than ω_c. The magnitude and phase response is shown in Fig. 12.24.

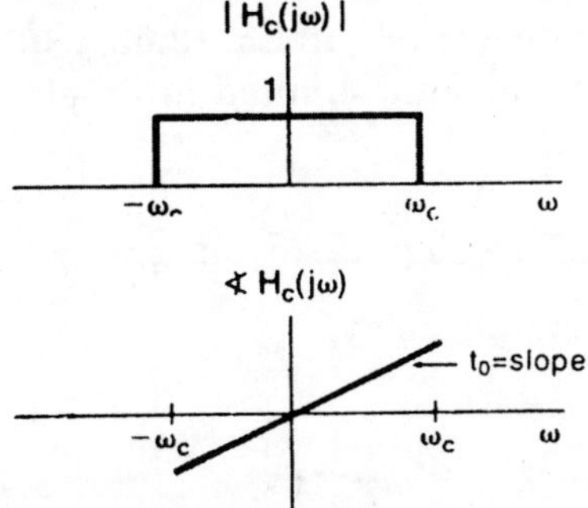

Fig. 12.24 *Frequency response for a continuous-time delay (a) Magnitude response (b) Phase response.*

The corresponding discrete-time frequency response for $\omega_s = 2\omega_c$ is given by

$$H_d(e^{j\Omega}) = e^{-j\Omega t_0/T_s}, \quad |\Omega| < \pi \qquad \text{...(12.43)}$$

This frequency response is shown in Fig. 12.25.

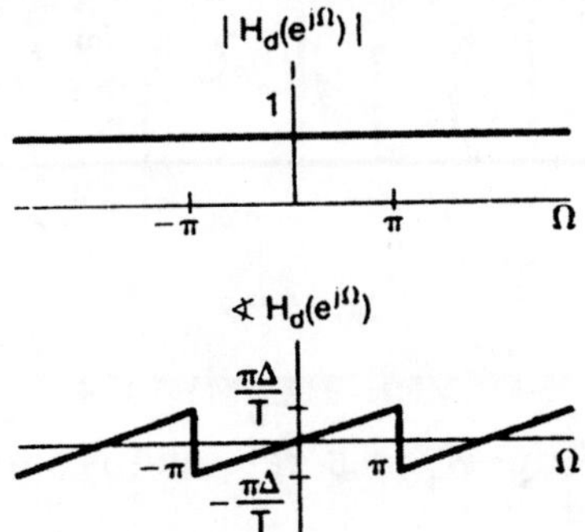

Fig. 12.25 *Frequency response for the corresponding discrete- time delay (a) Magnitude response (b) Phase response.*

For band-limited inputs, the output of the system of Fig. 12.19 with $H_d(e^{j\Omega})$ as in Eq. 12.43 is a delayed replica of the input. For t_0/T_s an integer, the sequence $y_d(n)$ is a delayed replica of $s_d(n)$. It is given by

$$y_d(n) = s_d\left[n - \frac{t_0}{T_s}\right] \qquad \text{...(12.44)}$$

If t_0/T_s is not an integer, then Eq. 12.44 has no meaning, since sequences are defined only at integer values of the index.

However, we can interpret the relationship between $s_d(n)$ and $y_d(n)$ in these cases in terms of band-limited interpolation. The signals

$s_c(t)$ and $s_d(n)$ are related through sampling and band-limited interpolation.

With transfer function $H_d(e^{j\Omega})$ in Eq. 12.43, $y_d(n)$ is equal to samples of a shifted version of the band-limited interpolation of the sequence $s_d(n)$.

This is illustrated in Fig. 12.26 for $\frac{t_0}{T_s} = \frac{1}{2}$, which is sometimes referred to as a half-sample delay.

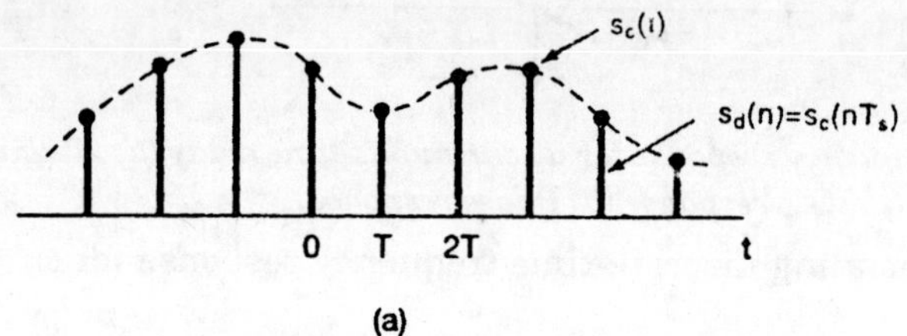

(a)

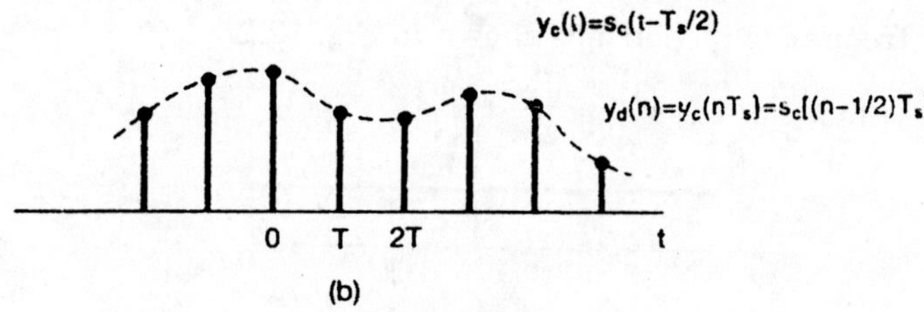

(b)

Fig. 12.26

SOLVED EXAMPLES

Example 12.3 Find the Nyquist rate and Nyquist interval for the continuous-time signal

$$s(t) = \frac{1}{2\pi} \cos(4000\pi t) \cos(1000\pi t)$$

Solution. We are given signal

$$s(t) = \frac{1}{2\pi} \cos(4000\pi t) \cos(1000\pi t)$$

(using trignometric identity for converting above signal into sum of two signals $2 \cos A \cos B = \cos(A + B) + \cos(A - B)$

or $$s(t) = \frac{1}{4\pi} [2 \cos(4000\pi t) \cos(1000\pi t)$$

$$= \frac{1}{4\pi} [\cos(4000\pi t + 1000\pi t) + \cos(4000\pi t - 1000\pi t)]$$

or $$s(t) = \frac{1}{4\pi} [\cos (5000\pi t) + \cos (3000\pi t)] \quad ...(1)$$

Now we presume that there are two frequencies present in the signal. This signal can be represented as

$$s(t) = A_1 \cos \omega_1 t + A_2 \cos \omega_2 t \quad ...(2)$$

where A_1 and A_2 are the amplitudes ω_1 and ω_2 are the angular frequencies.

Comparing Eqs. 1 and 2, we get

$$\omega_1 = 5000\pi \text{ radian/sec}$$

$$\omega_2 = 3000\pi \text{ radian/sec}$$

$$F_1 = \frac{\omega_1}{2\pi} = \frac{5000\pi}{2\pi} = 25000 \text{ Hz}$$

$$F_2 = \frac{\omega_2}{2\pi} = \frac{3000\pi}{2\pi} = 1500 \text{ Hz}$$

Highest frequency component of the given message signal

$$F_{max} = \text{Max } \{F_1 , F_2\}$$

$$= \text{Max } \{2500, 1500\} = 2500 \text{ Hz}$$

$$\text{Nyquist rate} = 2 \times F_{max}$$

$$= 2 \times 2500 = 5000 \text{ Hz}$$

$$\text{Nyquist interval} = \frac{1}{\text{Nyquist rate}}$$

$$= \frac{1}{5000} = 2 \times 10^{-4} = 200 \text{ millisecond}$$

Example 12.4 A real-valued continuous-time signal $s(t)$ is known to be uniquely determined by its samples when the sampling frequency is $\omega_s = 10^4 \pi$ radian/sec. Determine the value of ω when $S(j\omega)$ is guaranteed to be zero.

Solution. From sampling theorem, we know that for perfect reconstruction of the message signal from its samples, sampling frequency or sampling rate should be greater than twice the highest frequency component of the message signal.

If $$S(j\omega) = 0 \text{ for } |\omega| > \frac{\omega_s}{2}$$

then the signal is recovered perfectly from its samples. Therefore,

$$\text{CTFT } \{s(t)\} = S(j\omega) = 0 \text{ for } |\omega| > \frac{\omega_s}{2}$$

or $$|\omega| > \frac{10^4 \pi}{2}$$

or $$|\omega| > 5 \times 10^3 \pi$$

Example 12.5 A continuous-time signal $s(t)$ is obtained at the output of an ideal low pass filter with cutoff frequency $\omega_c = 10^3\,\pi$ radian/sec. If impulse-train sampling is performed on signal $s(t)$, which of the following sampling periods would guarantee that $s(t)$ can be recovered from its sampled version using an appropriate low pass filter ?

(*a*) $T_s = 5 \times 10^{-4}$ second (*b*) $T_s = 2 \times 10^{-3}$ second

(c) $T_s = 1 \times 10^{-4}$ second

Solution. From sampling theorem, we know that the sampling rate (or frequency) must be atleast equal to twice the highest frequency component, *i.e.*, $2\omega_c$.

$$\text{Sampling frequency} \geq 2\omega_c$$

$$\geq 2 \times 10^3\,\pi \text{ radian/sec} \quad ...(1)$$

We are given

(*a*) $$F_s = \frac{1}{T_s} = \frac{1}{5 \times 10^{-4}} = 2 \times 10^3 \text{ Hz}$$

$$\omega_s = 2\pi F_s = 4\pi \times 10^3 \text{ radian/sec}$$

or $$\omega_s = 4 \times 10^3\,\pi \text{ radian/sec} \quad ...(2)$$

(*b*) $$F_s = \frac{1}{T_s} = \frac{1}{2 \times 10^{-3}} = 500 \text{ Hz}$$

$$\omega_s = 2\pi F_s = 2\pi \times 500$$

or $$\omega_s = 10^3\,\pi \text{ radian/sec} \quad ...(3)$$

(c) $$F_s = \frac{1}{T_s} = \frac{1}{1 \times 10^{-4}} = 10^4 \text{ Hz}$$

$$\omega_s = 2\pi F_s$$

or $$\omega_s = 2\pi \times 10^4 = 2 \times 10^4\,\pi \text{ radian/sec} \quad ...(4)$$

From Eqs. 1, 2, 3 and 4, we clearly know that only (*a*) and (*c*) satisfy the sampling theorem condition.

Example 12.6 Determine the Nyquist rate for the continuous-time signal

$$s(t) = \frac{\sin(4 \times 10^3\,\pi t)}{\pi t}$$

Solution. We know that

$$\text{CTFT}\left[\frac{\sin(Wt)}{\pi t}\right] = S(j\omega) = \begin{cases} 1, & |\omega| < W \\ 0, & |\omega| > \omega \end{cases} \quad ...(1)$$

Substituting $W = 4 \times 10^3\,\pi$ radian/sec in Eq. 1, we get

$$\text{CTFT}\left[\frac{\sin(4 \times 10^3\,\pi t)}{\pi t}\right] = S(j\omega) = \begin{cases} 1, & |\omega| < 4 \times 10^3\,\pi \\ 0, & |\omega| > 4 \times 10^3\,\pi \end{cases} \quad ...(2)$$

It means that

$$\text{CTFT}\{s(t)\} = S(j\omega) = 0 \text{ for } |\omega| > 4 \times 10^3 \pi$$

Therefore, the Nyquist rate for this signal is

$$\omega_N = 2\,(4 \times 10^3 \pi) = 8 \times 10^3 \pi \text{ radian/sec}$$

Example 12.7 We are given a continuous-time signal $s(t)$ with Nyquist rate ω_0. Determine the Nyquist rate for the continuous-time signal $s^2(t)$.

Solution. If the continuous-time signal $s(t)$ has a Nyquist rate of ω_0, then its CTFT $S(j\omega) = 0$ for $|\omega| > \frac{\omega_0}{2}$. Let us assume that $y(t) = s^2(t)$. We also know that

$$y(t) = s^2(t) \xleftrightarrow{\text{CTFT}} Y(j\omega) = \frac{1}{2\pi}\,[S(j\omega) * S(j\omega)]$$

Clearly, we can only guarantee that $Y(j\omega) = 0$ for $|\omega| > \omega_0$. Therefore, the Nyquist rate for $y(t)$ is $2\omega_0$.

Example 12.8 We are given a continuous-time signal $s(t)$ with Nyquist rate ω_0. Find the Nyquist rate for the continuous-time signal $y(t) = s(t) \cos \omega_0 t$.

Solution. If the continuos-time signal $s(t)$ has the Nyquist rate of ω_0, then its CTFT $S(j\omega) = 0$ for $|\omega| > \frac{\omega_0}{2}$. We also know that

$$y(t) = s(t) \cos \omega_0 t \xleftrightarrow{\text{CTFT}} = \frac{1}{2} S[j(\omega - \omega_0)] + \frac{1}{2} S[j(\omega + \omega_0)]$$

Clearly, we can guarantee that

$$Y(j\omega) = 0 \text{ for } |\omega| > \omega_0 + \frac{\omega_0}{2}$$

$$> \frac{3\omega_0}{2} \qquad ...(1)$$

Therefore, the Nyquist rate for signal $y(t)$ is $2 \times \frac{3\omega_0}{2}$ *i.e.*, $3\omega_0$.

REVIEW QUESTIONS

1. What is sampling process ?
2. Discuss sampling of continuous-time signals.
3. State and explain sampling theorem for continuous-time signals.
4. Define following terms :
   (*a*) Higher frequency component of the message signal
   (*b*) Nyquist rate
   (*c*) Sampling rate or frequency

5. What do you mean by impluse train sampling of a continuous-time signal?
6. Also discuss zero-order hold sampling.
7. What are interpolation techniques for the reconstruction of a continuous-time signal from its samples ?
8. Describe linear interpolation between sample points.
9. What do you mean by ideal band-limited interpolation using sinc function?
10. Define following terms :
    (*a*) under sampling (*b*) critical sampling
    (*c*) over sampling
11. What is aliasing phenomenon ? How can aliasing phenomena be eliminated ?

NUMERICAL PROBELMS

1. Determine the Nyquist rate corresponding to each of the following signals:
   (*a*) $s(t) = 1 + \cos(2 \times 10^3 \pi t) + \sin(4 \times 10^3 \pi t)$
   (*b*) $s(t) = \dfrac{\sin(4 \times 10^3 \pi t)}{\pi t}$
2. Determine the Nyquist rate for each of the following continuous- time signals if signal $s(t)$ is a signal with Nyquist rate ω_0.
3. A signal $s(t)$ with Fourier transform $S(j\omega)$ undergoes impulse train sampling to generate

$$s_p(t) = \sum_{n=-\infty}^{\infty} s(nT_s)\, \delta(t - nT_s)$$

   where $T_s = 10^{-4}$. For each of the following sets of constraints on $s(t)$ and/or $S(j\omega)$, does the sampling theorem guarantee that $s(t)$ can be recovered exactly from $s_p(t)$?
   (*a*) $S(j\omega) = 0$ for $|\omega| > 5000\pi$
   (*b*) $s(t)$ real and $S(j\omega) = 0$ for $\omega > 5000\pi$
4. The continuous-time signal $y(t)$ is generated by convolving a band-limited signal $s_1(t)$ with another band-limited signal $s_2(t)$,

$$y(t) = s_1(t) * s_2(t)$$

   where $S_1(j\omega) = 0$, for $|\omega| > 10^3 \pi$

$$S_2(j\omega) = 0, \text{ for } |\omega| > 2 \times 10^3 \pi$$

   Impulse train sampling is done on $y(t)$ to obtain

$$y_p(t) = \sum_{n=-\infty}^{\infty} y(nT_s)\, \delta(t - nT_s)$$

   Also specify the range of values for the sampling period T_s which ensures that $y(t)$ is recoverable from $y_p(t)$.

13

Sampling of Discrete-Time Signals

13.1 INTRODUCTION TO SAMPLING PROCESS OF DISCRETE-TIME SIGNALS

In chapter 12, we have already discussed sampling of continuous- time signals. We have also discussed about the analysis that is necessary to understand continuous-time sampling like continuous- time signals. Sampling process can be applied to discrete- time signals too. The periodical measurement of a discrete-time signal is referred to as sampling of discrete-time signals. Impulse-train sampling as well as different order-holds can be used for sampling of discrete-time signals or sequences.

In this chapter we will also study a number of important applications that can be developed for sampling of discrete-time signals.

13.2 IMPULSE-TRAIN SAMPLING OF DISCRETE-TIME SIGNALS

In analogy with sampling of a continuous-time signal as shown in Fig. 12.2, sampling of a discrete-time signal can be represented as shown in Fig. 13.1. Here, the new discrete-time signal $s_p(n)$ resulting from the sampling of a discrete-time signal $s(n)$ is equal to the original discrete-time signal $s(n)$ at integer multiples of the sampling peroid N_0 and is zero at the intermediate samples.

Now, sampled version of $s(n)$ is given as

$$s_p(n) = \begin{cases} s(n)\ , & \text{if } n = \text{an integer multiple of } N_s \\ 0\ , & \text{otherwise} \end{cases} \qquad ...(13.1)$$

where N_s is the sampling period.

As with the sampling of continuous-time signals, the effect in the frequency-domain of sampling of discrete-time signals can be seen by using multiplication property of DTFT and given as

$$s_p(n) = s(n)\, p(n) = \sum_{k=-\infty}^{\infty} s(kN_s)\ \delta(n - kN_s)$$

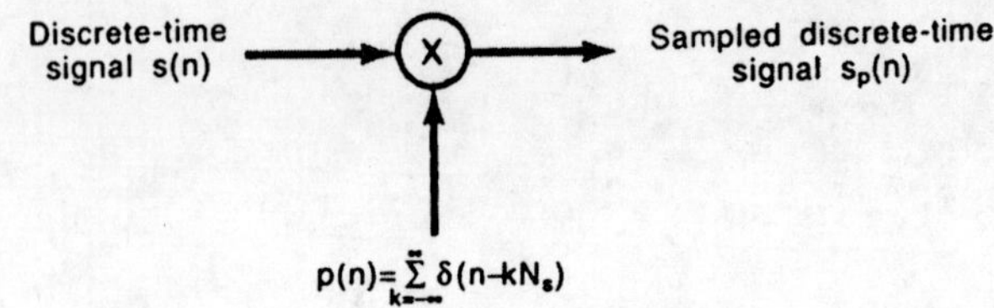

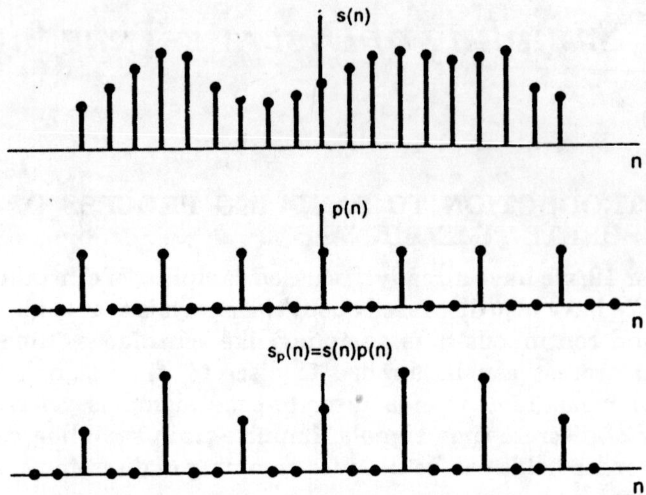

Fig. 13.1 *Illustration of sampling of discrete-time signal.*

$$\xleftrightarrow{\text{DTFT}} S_p(e^{j\omega}) = \text{DTFT}\ \{s_p(n)\}$$

$$= \frac{1}{2\pi} \int_{2\pi} P(e^{j\theta})\, S(e^{j(\omega-\theta)})\, d\theta \qquad \text{...(13.2)}$$

or
$$S_p(e^{j\omega}) = \frac{1}{2\pi} \int_{2\pi} P(e^{j\theta})\, S(e^{j(\omega-\theta)})\, d\theta \qquad \text{...(13.3)}$$

where $P(e^{j\omega})$ is the DTFT of $p(n)$ and $S(e^{j\omega})$ is the DTFT of $s(n)$.

DTFT of the sampling sequence $p(n)$ can be determined as

$$P(e^{j\omega}) = \text{DTFT}\ \{p(n)\}$$

$$= \text{DTFT}\left\{ \sum_{k=-\infty}^{\infty} \delta(n - kN_s) \right\}$$

$$= \sum_{n=-\infty}^{\infty} \left[\sum_{k=-\infty}^{\infty} \delta(n - kN_s) \right] e^{-j\omega n}$$

or
$$P(e^{j\omega}) = \frac{2\pi}{N_s} \sum_{k=-\infty}^{\infty} \delta(\omega - k\omega_s) \qquad \text{...(13.4)}$$

where $\omega_s = \dfrac{2\pi}{N_s}$ is the sampling frequency.

Now combining Eqs. 13.3 and 13.4, we get

$$S_p(e^{j\omega}) = \frac{1}{2\pi}\int_{2\pi} P(e^{j\theta})\, S(e^{j(\omega-\theta)})\, d\theta$$

$$= \frac{1}{2\pi}\int_{2\pi}\left[\frac{2\pi}{N_s}\sum_{k=-\infty}^{\infty}\delta(\theta - k\omega_s)\right] S(e^{j(\omega-\theta)})\, d\theta$$

or $$S_p(e^{j\omega}) = \frac{1}{N_s}\sum_{k=0}^{N_s-1} S(e^{j(\omega - k\omega_s)}) \qquad ...(13.5)$$

Eq. 13.5 represents the counterpart for sampling of discrete-time signal of equation $S_p(j\omega) = \frac{1}{T_s}\sum_{k=-\infty}^{\infty} S[j(\omega - k\omega_s)]$ for sampling of continuous-time signal. The impulse-train sampling of discrete-time signal $s(n)$ in frequency-domain is shown in Fig. 13.2.

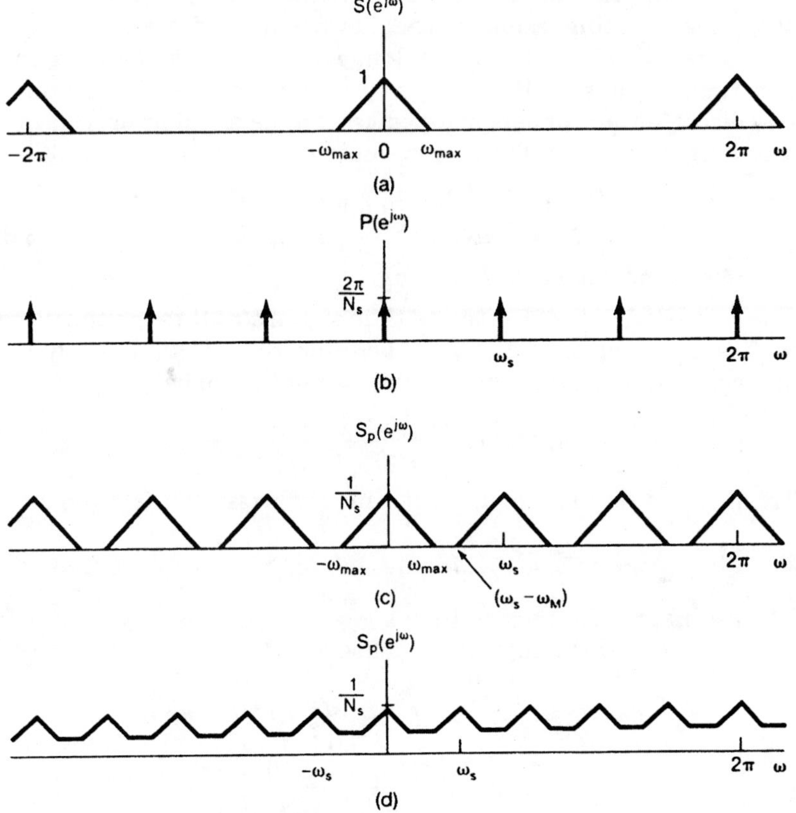

Fig. 13.2 *Illustration of effect of impulse train sampling of a discrete-time signal in frequency-domain. (a) Spectrum of an original discrete-time signal s(n) (b) spectrum of a sampling sequence p(n) (c) spectrum of sampled signal with sampling frequency* $\omega_s > 2\omega_{max}$ *(d) spectrum of sampled signal with sampling frequency* $\omega_s < 2\omega_{max}$.

If sampling rate $\omega_s > 2\omega_{max}$ then there is no aliasing. It means that there are no non-zeros portions of the replicas of $S(e^{j\omega})$ that overlap. This case is shown in Fig. 13.2 (*c*).

If sampling rate $\omega_s < 2\omega_{max}$ then frequency-domain aliasing results. This case is shown in Fig. 13.2 (*d*).

In the absence of aliasing, $S(e^{j\omega})$ is faithfully recovered or reproduced around $\omega = 0$ and integer multiples of 2π. Consequently, signal $s(n)$ can be reconstructed from $s_p(n)$ by means of a low pass filter with gain N_0 and cutoff frequency greater than ω_{max} and less than $(\omega_s - \omega_{max})$. It is shown in Fig. 13.3. In this figure we have specified the cut off frequency of the low pass filters as $\omega_s/2$. Frequency $\omega_s/2$ is also called folding frequency. If the overall system of Fig. 13.3 (*a*) is applied to a discrete-time signal or sequence for which sampling frequency $\omega_s < 2\omega_{max}$ then aliasing results will be shown. In this case, reconstructed discrete-time signal $s_r(n)$ will no longer be equal to the original discrete-time signal $s(n)$. However, as with continuous-time sampling, the two discrete-time signals or sequnces will be equal at multiples of the sampling period (N_s). The corresponding equation to

$$s_r(nT_s) = s(nT_s)\ , \quad n = 0\ , \pm 1\ , \pm 2\ , \ldots$$

$$s_r(kN_s) = s(kN_s)\ , \quad k = 0\ , \pm 1\ , \pm 2\ , \ldots \qquad \ldots(13.6)$$

Eq. 13.6 is independent of aliasing.

The reconstruction of a discrete-time signal $s(n)$ through the use of a low pass filter applied to $s_p(n)$ can be interpreted in time-domain as an interpolation formula similar to the formula given below

$$s_r(t) = \sum_{n=-\infty}^{\infty} s(nT_s) \frac{\omega_c T_s \sin [\omega_c (t - nT_s)}{\pi\, \omega_c (t - nT_s)}$$

The impulse response of discrete-time low pass filter is given as

$$h(n) = \frac{N_s \omega_c}{\pi} \frac{\sin \omega_c n}{\omega_c n} \qquad \ldots(13.7)$$

The reconstructed discrete-time signal or sequence is determined by convolving $s_p(n)$ with impulse response $h(n)$ as

$$s_r(n) = s_p(n) * h(n) = \sum_{k=-\infty}^{\infty} s(kN_s)\, h(n - kN_s)$$

or
$$s_r(n) = \sum_{k=-\infty}^{\infty} s(kN_s) \frac{N_s \omega_c}{\pi} \frac{\sin \omega_c (n - kN_s)}{\omega_c (n - kN_s)} \qquad \ldots(13.8)$$

Eq. 13.8 represents an ideal band-limited interpolation and requires the implementation of an ideal low pass filter. In some typical applications a suitable approximation for the low pass filter is used. For this case the equivalent interpolation formula is given as

$$s_r(n) = \sum_{k=-\infty}^{\infty} s(kN_s)\, h_r(n - kN_s) \qquad \ldots(13.9)$$

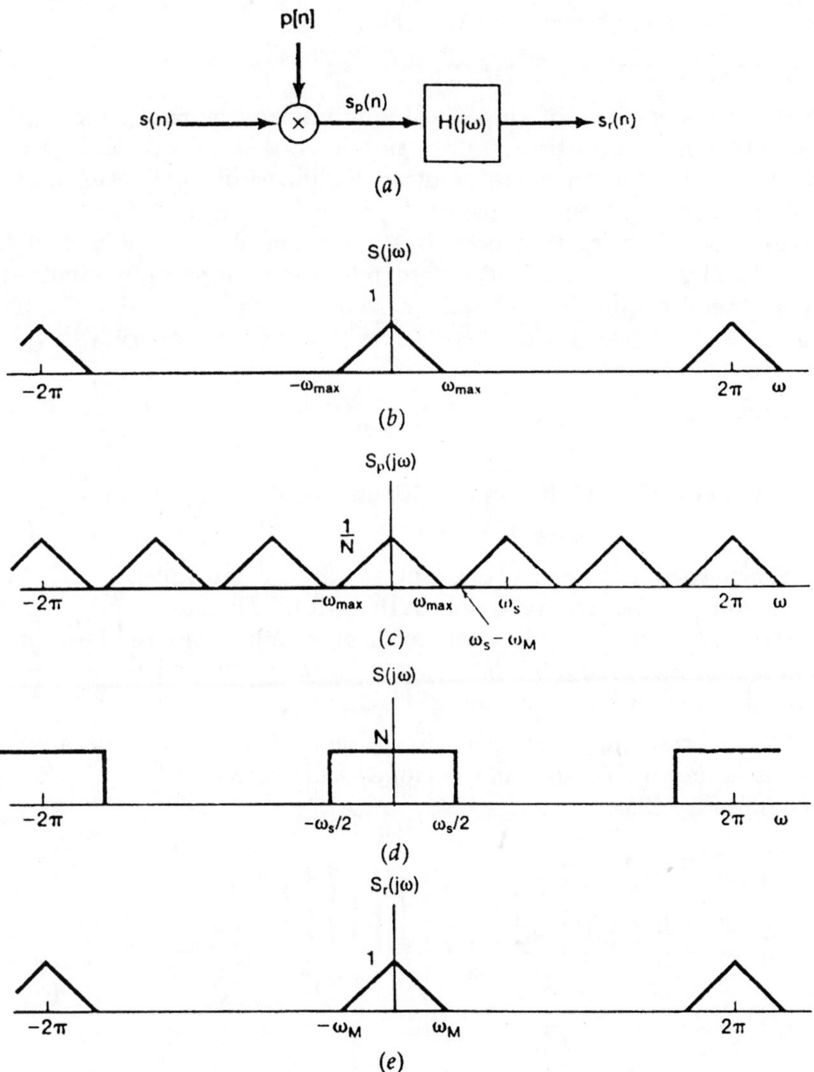

Fig. 13.3 *Perfect reconstruction of a discrete-time signal from its samples using an ideal low pass filter (a) Block diagram for sampling and reconstruction of band-limited discrete- time signal from its samples (b) spectrum of original discrete-time signal s(n) (c) spectrum of sampled version of the original discrete-time signal s(n) (d) frequency response of an ideal low pass filter with cutoff frequency $\omega_s/2$ (e) spectrum of the reconstructed discrete-time signal $s_r(n)$. Here we have taken sampling frequency $\omega_s > 2\omega_{max}$ so that no aliasing occurs and consequently $s_r(n) = s(n)$.*

where $h_r(n)$ is the impulse response of the interpolation filter. Also discrete-time counterpart of zero-order hold and higher-order hold for continuous-time interpolation can exist.

13.3 DISCRETE-TIME DECIMATION

There are a variety of important applications of the principles of discrete-time sampling, such as in digital filter design and implementation or in communiation engineering applications. It is inefficient to represent, transmit or store the sampled discrete- time signal or sequence $s_p(n)$ directly in the form as shown in Fig. 13.1 for many applications, since $s_p(n)$ is zero in between the sampling instants. Thus, the discrete-time signal or sequence is replaced by a new discrete-time signal or sequence $s_d(n)$, which is simply every N_sth value of $s_p(n)$.

It is given as

$$s_d(n) = s_p(nN_s) \qquad ...(13.10)$$

or equivalently Eq. 13.10 can be written as

$$s_d(n) = s(nN_s) \qquad ...(13.11)$$

Since $s_p(n)$ and $s(n)$ are equal at integer multiples of N_s. The operation of extracting every Nth sample is commonly called *Decimation*. In other words, decimation of sampling rate can be defined as : *the process of reducing the sampling rate by an integer factor*. It is also a down-sampling by an integer factor.

The relationship between original discrete-time signal $s(n)$, sampled discrete-time signal $s_p(n)$ and decimated discrete-time signal $s_p'(n)$ is shown in Fig. 13.4.

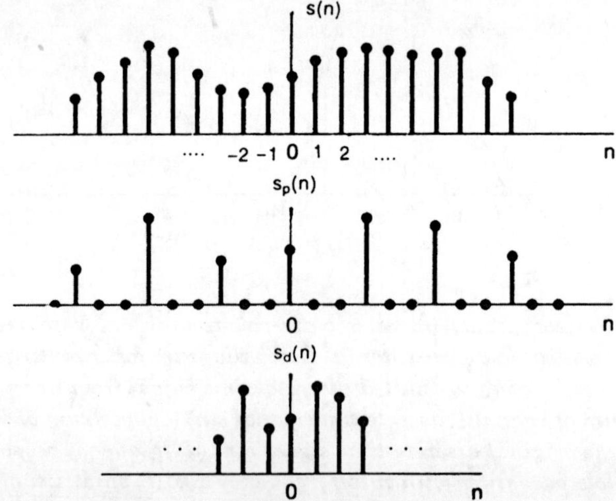

Fig. 13.4 *Illustration of the relationship between $s_p(n)$ corresponding to discrete-time sampling and $s_d(n)$ corresponding to discrete-time decimation.*

Effect of Discrete-time Decimation in Frequency Domain

Here we wish to determine the relationship between $S_d(e^{j\omega})$ and $S(e^{j\omega})$.

where $S_d(e^{j\omega}) = \text{DTFT}\{S_d(n)\}$

and $S(e^{j\omega}) = \text{DTFT}\{s(n)\}$

Taking the DTFT of $s_d(n)$, we get

$$S_d(e^{j\omega}) = \text{DTFT}\{s_d(n)\} = \sum_{k=-\infty}^{\infty} s_d(k)\, e^{-j\omega k} \qquad ...(13.12)$$

But we know that from Eq. 13.10

$$s_d(n) = s_p(nN_s)$$

Substituting the value of $s_d(n) = s_p(nN_s)$ in Eq. 13.12, we get

$$S_d(e^{j\omega}) = \sum_{k=-\infty}^{\infty} s_p(kN_s)\, e^{-j\omega k} \qquad ...(13.13)$$

If we let $n = kN_s$ or $k = \frac{n}{N_s}$, then Eq. 13.13 can be written as

$$S_d(e^{j\omega}) = \sum_{\substack{n = \text{Integer multiple} \\ \text{of } N_s}} s_p(n)\, e^{-j\omega n/k} \qquad ...(13.14)$$

Since $s_p(n) = 0$, where n is not an integer multiple of N_s. Now Eq. 13.14 can be written as

$$S_d(e^{j\omega}) = \sum_{n=-\infty}^{\infty} s_p(n)\, e^{-j\omega n/N_s} \qquad ...(13.15)$$

Futhermore, we can recognize the right-hand side of Eq. 13.15 as the DTFT of $s_p(n)$.

$$\sum_{n=-\infty}^{\infty} s_p(n)\, e^{-j\omega\, n/N_s} = S_p(e^{j\omega/N_s}) \qquad ...(13.16)$$

Thus, from Eq. 13.15 and 13.16, we conclude that

$$S_d(e^{j\omega}) = S_p(e^{j\omega/N_s}) \qquad ...(13.17)$$

The relationship of equation $S_d(e^{j\omega}) = S_p(e^{j\omega/N_s})$ is shown in Fig. 13.5. From this figure, we observe that the spectra for the sampled discrete-time signal (or sequence) and decimated discrete-time signal (or sequence) differs only in a frequency scaling. Frequency scaling is also called normalization of frequency. If the spectrum of the original discrete-time signal $S(e^{j\omega})$ is appropriately band limited then there will be no aliasing. The effect of decimation is to spread spectrum of the original discrete-time signal over larger portion of the frequency band or range of frequencies.

If the original discrete-time signal $s(n)$ is obtained by sampling of continuous-time signal, the process of decimation can be viewed as reducing the sampling on the signal by a integer factor of N_s to avoid

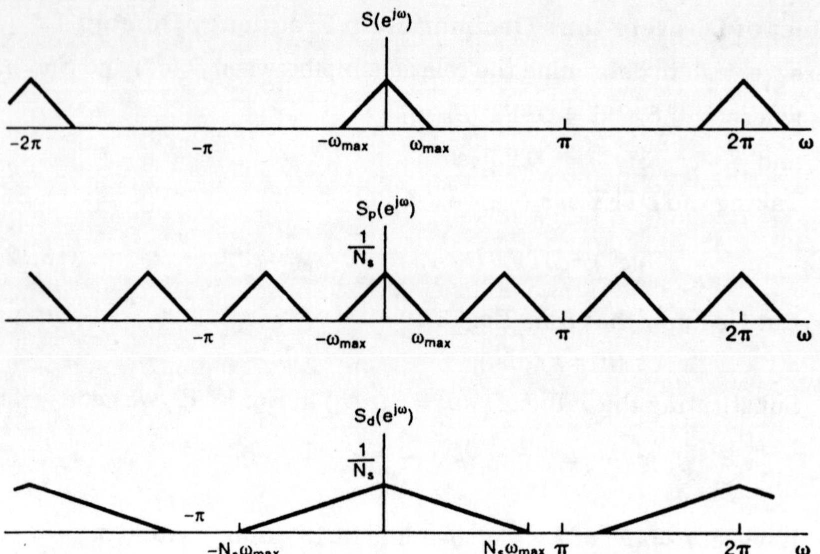

Fig. 13.5 *Illustration of the relationship between sampling and decimation in frequency-domain.*

aliasing, spectrum $S(e^{j\omega})$ cannot occupy the full frequency band. A signal can be decimated without introducing aliasing by sampling a continuous-time signal at a sampling frequency greater than Nyquist rate. In this case, the sampling rate can be reduced without aliasing. If a discrete-time signal $s(n)$ is a sampled version of a continous-time signal then the process of decimation is called *down-sampling*.

In some applications (in which a discrete-time signal is obtained by sampling a continuous-time signal) the original sampling rate can be as low as possible without introducing aliasing But after additional processing and filtering, the range of frequencies (*i.e.*, bandwidth) of the discrete-time signal my be reduced.

13.4 DISCRETE-TIME INTERPOLATION

In some applications it is useful to reduce the sampling rate or to down sample the signal. Similarly, there are some situations in which it is useful to convert a discrete-time signal or sequence to a higher equivalent sampling rate. This process is called *upsampling* or *interpolation*. Upsampling is basically the reverse operation of decimation or down sampling.

The process of increasing the sampling rate by an integer factor is called interpolation. As shown in Fig. 13.4 and Fig. 13.5, in decimation, we first sample and then retain only the discrete- time signal sequence values at sampling instants. For upsampling, we reverse the decimation

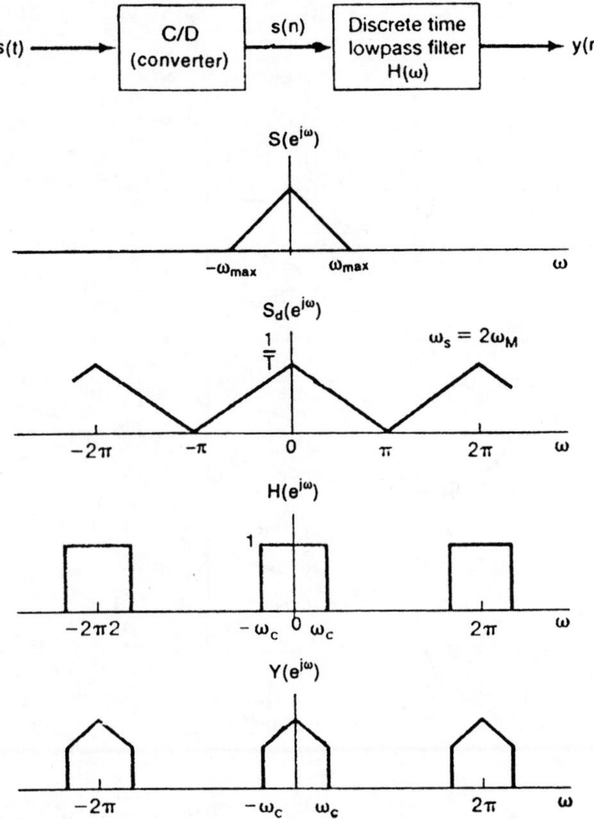

Fig. 13.6 *Illustration of sampling of a continuous-time signal. It is sampled at Nyquist rate. After discrete-time filtering, the resulting discrete-time signal can be further decimated (down-sampled). Here $S(j\omega)$ is the CTFT of the continuous-time signal $s(t)$. $S(e^{j\omega})$ and $Y(e^{j\omega})$ are the DTFTs of $s(n)$ and $y(n)$ respectively. And $H(e^{j\omega})$ is the frequency response of the discrete-time low pass filter shown in figure.*

process. From Fig. 13.4, we consider upsampling the sequence $s_d(n)$ to obtain $s(n)$.

From sequence $s_d(n)$, we can form the sequence $s_p(n)$ by inserting $(N_s - 1)$ sample points with zero amplitude between each of the values in $s_d(n)$. The interpolated discrete-time signal $s(n)$ is then obtained from $s_p(n)$ by low pass filtering. A summary of interpolation of a discrete-time signal is shown in Fig. 13.7.

Example 13.1 We are given a discrete-time signal (or sequence) $s(n)$ whose DTFT $S(e^{j\omega})$ has the property that

$$S(e^{j\omega}) = 0 \text{ or } \frac{2\pi}{9} \leq |\omega| \leq \pi$$

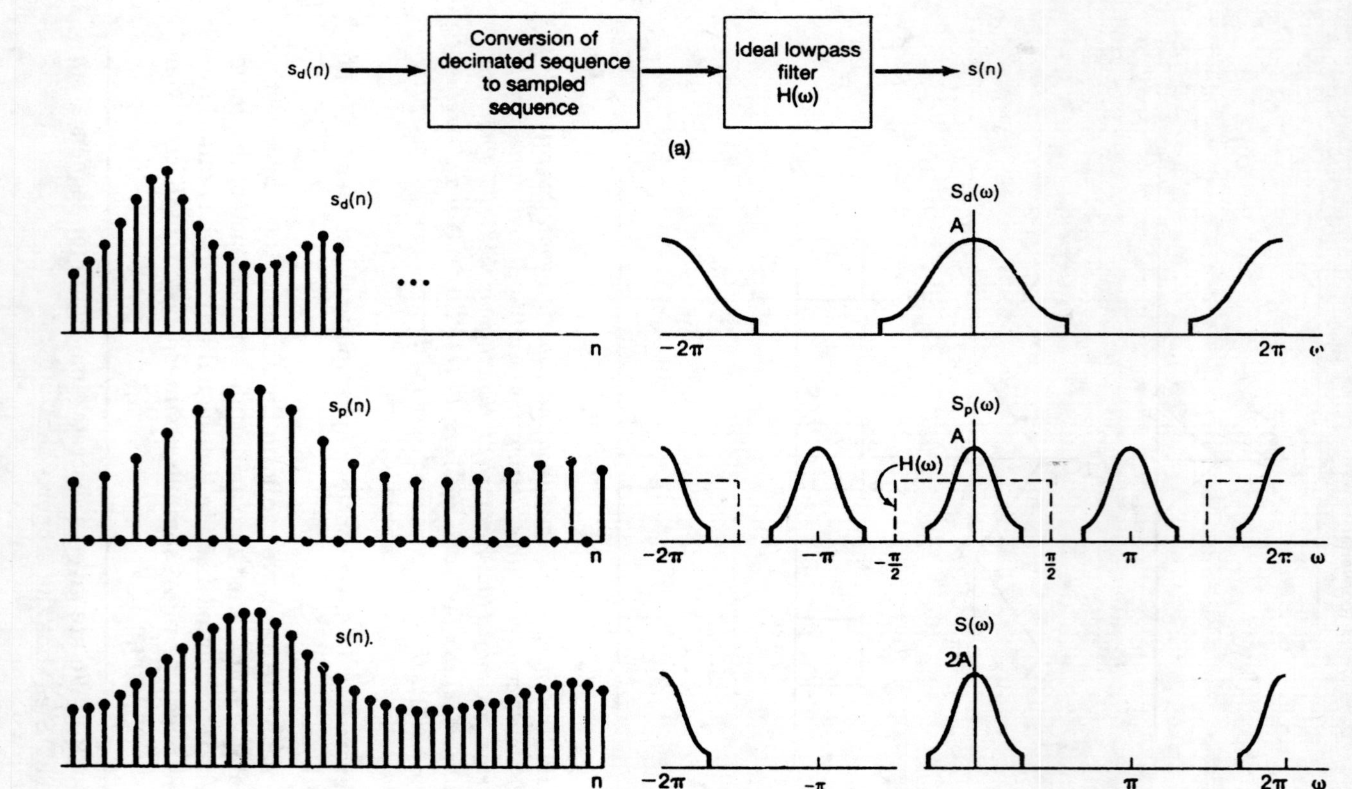

Fig. 13.7 *Illustration of interpolation process (a) overall system for interpolation (b) different associated discrete-time signals and spectra for upsampling by a integer factor of 2.*

Determine the lowest rate at which $s(n)$ may be sampled without the possibility of aliasing.

Solution. Now we determine the largest N.

$$\frac{2\pi}{N} \geq 2\left(\frac{2\pi}{9}\right)$$

or $$N \leq \frac{9}{2}$$

or $$N \leq 4.5$$

We conclude that $N_{max} = 4$ and the corresponding sampling frequency is

$$\frac{2\pi}{4} = \frac{\pi}{2} \text{ radian/sec}$$

Example 13.2 A discrete-time signal $s_d(n)$ has a DTFT $S_d(e^{j\omega})$ with the property that

$$S_d(e^{j\omega}) = 0 \text{ for } \frac{3\pi}{4} \leq |\omega| \leq \pi$$

The signal is convereted into a continuous-time signal

$$s_c(t) = T_s \sum_{n=-\infty}^{\infty} s_d(n) \frac{\sin \frac{\pi}{T_s}(t - nT_s)}{\pi(t - nT_s)}$$

where $T_s = 10^{-3}$ second.

Determine the values of ω for which the CTFT $S_c(j\omega)$ of $s_c(t)$ is guaranteed to be zero.

Solution. We know that discrete and continuous-time frequencies Ω and ω are related by $\Omega = \omega T_s$. Therefore, in this case for $\Omega = \frac{3\pi}{4}$, we find the corresponding value of ω.

$$\omega = \frac{\Omega}{T_s} = \frac{3\pi/4}{10^{-3}} = \frac{3 \times 10^3}{4} \pi = 750\,\pi \text{ radian/sec}$$

REVIEW QUESTIONS

1. What is sampling process for discrete-time signals ?
2. Discuss impulse-train sampling of discrete-time signals.
3. Discuss following :
   (*a*) Discrete-time decimation
   (*b*) Discrete-time interpolation
4. What do you mean by decimation and interpolation ?
5. Discuss effect of discrete-time decimation in frequency-domain.
6. Distinguish discrete-time sampling from continuous-time sampling.

14

Random Signals and Systems (With Random Variables and Processes)

14.1 INTRODUCTION

A signal which is described by some fixed mathematical equations, is known as *deterministic signal.* The behaviour of deterministic signals as well as processing through LTI systems can be determined with the help of mathematical models. These mathematical models represent the complete behaviour of the signal at every instant of time. Hence, we can say that deterministic signals have no degree of uncertainty about the value of any instant of time. It means that deterministic signals are predictable.

There is another class of signals, the behaviour of which cannot be predicted. These signals are called *random signals.* These signals are called random signals because the precise value of these signals cannot be predicted in advance before they actually occur. The example of random signal is noise in communication systems.

Random signals cannot be modelled mathematically. Although the random signals are not predictable in advance precisely, they can be described in terms of their statistical properties. It is possible to analyze the random signals with the help of probability theory.

This chapter begins with the concept of probability theory. In fact, the probability theory is a very essential mathematical tool in the design of digital communication systems.

14.2 BASIC CONCEPTS AND DEFINITIONS

Here we will discuss about probability, mutually exclusive events and statistically independent events.

14.2.1 Probability

Consider an experiment whose outcome varies from trial to trial and is initially unknown. Such an experiment is called a random experiment. Let the event A be one of the possible outcomes. The experiment is repeated n times and n_A is the number of times event A occured, is recorded. The ratio n_A/n is the relative frequency of occurrence of event A for that particular set of experiments, or trials. Now if the experiment has statistical regularity, then ratio n_A/n may approach some definite value as 'n' increases and becomes very large. Assuming that a unique limiting value exists, we define $P(A)$, the probability of event A, as the relative frequency in the limit as $n \to \infty$.

$$P(A) = \lim_{n \to \infty} \left(\frac{n_A}{n} \right) \quad \text{...(14.1)}$$

From the definition it is clear that a probability $P(A)$ is positive number bounded by

$$0 \le P(A) \le 1 \quad \text{...(14.2)}$$

If an event is certain to occur on any trial, its probability is $P(A) = 1$ and if an event is impossible, its probability is $P(A) = 0$.

Out of n, if the outcome of an event is that it has occurred n_A times this means that it has not occured $(n - n_A)$ times. Hence the probability of 'not A' is

$$P(\text{not } A) = \lim_{n \to \infty} \frac{(n - n_A)}{n} = 1 - \lim_{n \to \infty} \left(\frac{n_A}{n} \right)$$

$$= 1 - P(A)$$

Hence, $$P(A) + P(\text{not } A) = 1 \quad \text{...(14.3)}$$

14.2.2 Mutually Exclusive Events

Two possible outcomes of an experiment are defined as being mutually exclusive if the occurrence of one outcome precludes (eliminates) the occurrence of the other. In this case, if the events are A_1 and A_2 with probabilities $P(A_1)$ and $P(A_2)$ then the probability of occurrence of either A_1 or A_2 is written as $P(A_1 \text{ or } A_2)$ and given by

$$P(A_1 \text{ or } A_2) = P(A_1) + P(A_2) \quad \text{...(14.4)}$$

Suppose that in a very large number n of repetitions of the experiment, outcome A_1 had occurred n_1 times and the outcome A_2 had occured n_2 times. Then A_1 and A_2 will have occured $(n_1 + n_2)$ times and

$$P(A_1 \text{ or } A_2) = \lim_{n \to \infty} \frac{n_1 + n_2}{n}$$

$$= \lim_{n \to \infty} \left(\frac{n_1}{n} + \frac{n_2}{n} \right)$$

Therefore, $$P(A_1 \text{ or } A_2) = P(A_1) + P(A_2) \quad \text{...(14.5)}$$

Extension of Eq. 14.5 can be written as

$$P(A_1 \text{ or } A_2 \text{ or } A_3 \text{ or } A_4 \ldots \text{ or } A_n) = \sum_{i=1}^{N} P(A_i) \quad \text{...(14.6)}$$

If all the outcomes have equal probabilities, in which case they are said to be equally likely, then

$$\sum_{i=1}^{N} P(A_i) = N\,P(A_i) = 1$$

Hence $$P(A_i) = \frac{1}{N} \qquad ...(14.7)$$

14.2.3 Statistically Independent Events

Suppose an event B, is independent of A, so the occurrence of A does not influence the occurrence of B. It can be proved that

$$P(AB) = P(A)\,P(B) \qquad ...(14.8)$$

Extension of Eq. 14.8 can be written as

$$P(ABC...) = P(A)\,P(B)\,P(C)... \qquad ...(14.9)$$

Note that statistical independence is quite different from mutual exclusiveness. In fact, if A and B are mutually exclusive, then

$$P(AB) = 0 \qquad ...(14.10)$$

It is quite possible that the occurrence of B depends in some way on the occurrence of A. Given that A has occurred, we should take this information into account in determining the probability that B also has occurred, for that probability may be increased or decreased. To include such additional information we define the conditional probability $P(A/B)$, the probability of B given that A has occurred.

Applying the relative frequency definition

$$P(B/A) = \frac{n_{AB}}{n_A} = \frac{(n_{AB}/n)}{(n_A/n)} = \frac{P(AB)}{P(A)} \qquad ...(14.11)$$

Since n_{BA} is same as n_{AB}, reversing the argument yields

$$P(A/B) = \frac{P(AB)}{P(B)} \qquad ...(14.12)$$

Comparing Eqs. 14.11 and 14.12, we get

$$P(AB) = P(A/B)\,P(B) = P(B/A)\,P(A)$$

or $$P(B/A) = \frac{P(B)\,P(A/B)}{P(A)} \qquad ...(14.13)$$

Eq. 14.13 is known as Baye's Theorem.

14.3 RANDOM VARIABLES

In the study of random signals and communications, we are more concerned with random experiments or processes which have numerical outcomes, *e.g.*, the value of a noise voltage at some instant of time or the number of errors in a digital data message. The analysis and description of such numerical valued random phenomenon is made possible with the help of random variables.

Consider a random process or experiment where the result of any one observation is expressed as a single numerical quantity. A variable is then assigned to the process representing the outcome of a given observation and is called a *Random variable* or *variate*. The random variable X can therefore assume any of the possible values of the process with which it is associated. In other words X is the general symbol for observed values. Now if the process has statistical regularity, we can describe it by probability function $P(X)$ defined in terms of X. Likewise the probability that an observed value is greater than x_k but less than x_l is written as $P(x_k < X \leq x_l)$.

Random variable may be discrete or continuous, depending on the nature of the process. The random variable X is a discrete random variable if X can take on only a finite number of values in any finite observation interval. On the other hand, the random variable X is a continuous random variable if X can take on any value in a whole observation interval.

A random process is often classified according to the type of random variable associated with it. The coin tossing experiment is an example of discrete process. If coins are tossed at a line drawn on the floor, the distance from coin to line is a random number which may be positive or negative, within reasonable limits, and we can assign to the process a continuous random variable, whose range is essentially from $-\infty$ to $+\infty$. Finally, it may be mentioned that discrete random variables are described by frequency functions and continuous random variables are described by density functions. We will define frequency functions and density functions in subsequent subsections.

14.3.1 Distribution Functions and Probability Density Functions

Cumulative Distribution Function. The cumulative distribution function (cdf) $F_X(x)$ of a random variable X takes a value less than or equal to x; that is,

$$F_X(x) = P(X \leq x) \qquad ...(14.14)$$

The cumulative distribution function (cdf) $F_X(x)$ has the following four properties

1. $F_X(x) \geq 0$
2. $F_X(\infty) = 1$
3. $F_X(-\infty) = 0$
4. $F_X(x)$ is a non-decreasing function, that is

$$F_X(x_1) \leq F_X(x_2) \text{ for } x_1 \leq x_2$$

Probability Density Function (pdf). Although distribution function of a random variable completely characterizes the random variable but sometimes it is more convenient to work with another function known as probability density function. It is related to distribution function as

$$\text{Probability density function} = \frac{d}{dx}\{\text{Probability distribution function}\}$$

or

$$f_X(x) = \frac{d}{dx} F_X(x) \qquad \text{...(14.15)}$$

The probability density function $f_X(x)$ is the first derivative of the probability distribution function $F_X(x)$. Note that the first derivative of the probability distribution function may not exist at all points. This is because the probability distribution function may be a discontinuous function for discrete random variables. Here we are assuming that $F_X(x)$ is a continuous function.

We know that the probability density function (pdf) of a random variable completely characterizes the random variable that is

$$F_X(x) = \int_{-\infty}^{x} f_X(x)\, dx \qquad \text{...(14.16)}$$

From the property of probability distribution function

$$F_X(x) = P\{X \le x\} \qquad \text{...(14.17)}$$

From Eqs. 14.16 and 14.17, we get

$$P\{X \le x\} = \int_{-\infty}^{x} f_X(x)\, dx \qquad \text{...(14.18)}$$

Eq. 14.18 relates probability distribution function and probability density function.

Integrating Eq. 14.15 in the limit from x_1 to x_2, we get

$$\int_{x_1}^{x_2} f_X(x)\, dx = F_X(x)\Big|_{x_1}^{x_2}$$

$$= F_X(x_2) - F_X(x_1) \qquad \text{...(14.19)}$$

Also, we know that

$$F_X(x_2) - F_X(x_1) = P\{x_1 \le X \le x_2\} \qquad \text{...(14.20)}$$

From Eqs. 14.19 and 14.20, we get

$$P\{x_1 \le X \le x_2\} = \int_{x_1}^{x_2} f_X(x)\, dx \qquad \text{...(14.21)}$$

We observe from Eqs. 14.16 and 14.21 that the pdf completely characterizes the random variable. If the probability density function of the random variable is known then the probability of the event $\{x_1 \le X \le x_2\}$ is the area under the pdf curve. It is shown in Fig. 14.1.

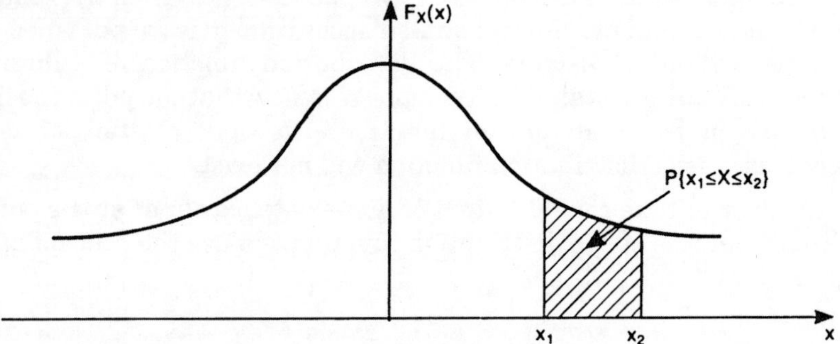

Fig. 14.1 *Probability of an event from its probability density function (pdf), $F_X(x)$.*

Properties of Probability Density Function. The pdf has the following properties

1. The probability density function (pdf) is a non-negative function

$$F_X(x) \geq 0, \text{ for all } x \qquad ...(14.22)$$

Proof : By definition of probability density function (pdf)

$$f_X(x) = \frac{dF_X(x)}{dx}$$

Since $f_X(x)$ is a non-decreasing function, therefore, $f_X(x) \geq 0$ for all x.

2. Area under the pdf curve is unity.

$$\int_{-\infty}^{\infty} f_X(x)\, dx = 1 \qquad ...(14.23)$$

Proof : We know from the definition of pdf,

$$f_X(x) = \frac{dF_X(x)}{dx} \qquad ...(14.24)$$

Integrating both sides and taking the limit from $-\infty$ to $+\infty$, we get

$$\int_{-\infty}^{\infty} f_X(x)\, dx = \Big[F_X(x) \Big]_{-\infty}^{\infty} = F_X(\infty) - F_X(-\infty) = 1 - 0 = 1$$

3. For a continuous random variable the probability of the event $P\{X = x\}$ is zero.

$$P\{X = x\} = 0, \text{ for any real number } x \qquad ...(14.25)$$

Proof: For $x_1 = x_2 = x$, Eq. $P\{x_1 \leq X \leq x_2\} = \int_{x_1}^{x_2} f_X(x)\, dx$ can also be written as

$$P\{X = x\} = \int_{x}^{x} f_X(x)\, dx = 0$$

Upto now, we have discussed the properties of pdf when the random variable is continuous. Now we shall discuss properties of pdf when the random variable is discrete. The distribution function of a discrete random variable is a staircase function. We know that the pdf is the first derivative of the distribution function. Also pdf at the points of discontinuities of distribution function will not exist.

In order to take consideration of non-existence of pdf at the points of discontinuities of the distribution function, we use the concept of an impulse function.

$$\int_{-\infty}^{\infty} f(x)\, \delta(x)\, dx = f(0)$$

where $f(x)$ = any arbitrary function

$\delta(x)$ = impluse function

$$= \begin{cases} 1 \,, & x = 0 \\ 0 \,, & x \neq 0 \end{cases}$$

If we assume that an impulse function $\delta(x)$ exists then derivative of an arbitrary function at a discontinuity x_0 exists and it is equal to $A\delta(x - x_0)$ where A is given by

$$A = f(x_0^{\,+}) - f(x_0^{\,-})$$

where $x_0^{\,+} = x_0 + \in$

and $\quad x_0^{\,-} = x_0 - \in$ for $\in > 0$ and infinitesimal small.

K is known as the measure of the height of discontinuity.

Some Common pdfs. If we know the probability density function of a random variable defined over a given random experiment, then we can find the probability of an event associated with the experiment.

Here we shall introduce some of the commonly occurring pdfs.

These are

1. Uniform pdf
2. Gaussian or normal pdf
3. Exponential pdf.

Now we discuss each of them.

1. Uniform pdf. The pdf of a uniform random variable X, with parameters $a \,.\, b \in R$ has a sample space $S = [a \,, b]$, *i.e.*, $a \leq X \leq b$, is given by

$$f_X(x) = \frac{1}{b - a}, \quad x \in [a \,, b]$$

Uniform pdf is plotted in Fig. 14.2.

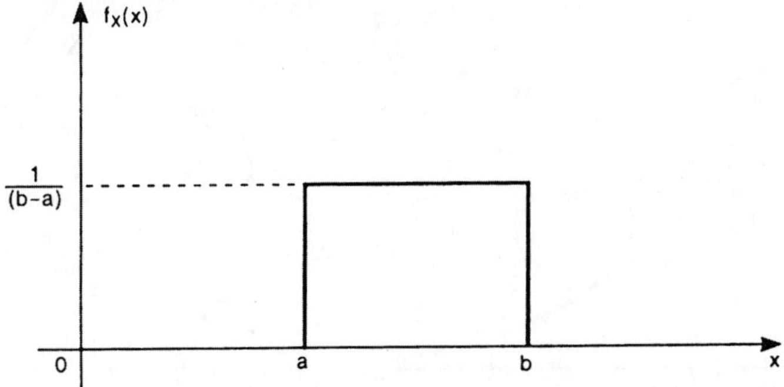

Fig. 14.2 *Uniform pdf.*

2. Gaussian or Normal pdf. The Gaussian random variable X with parameters μ , σ, has sample space $S = (-\infty , \infty)$. Its pdf is given by

$$f_X(x) = \frac{1}{\sqrt{2\pi}\,\sigma} e^{-[(x-\mu)^2]/2\sigma^2}, \quad x \in (-\infty , \infty) \qquad ...(14.26)$$

Gaussian probability density function is the most commonly used model for description of noise in communication channel. Its plot is shown in Fig. 14.3.

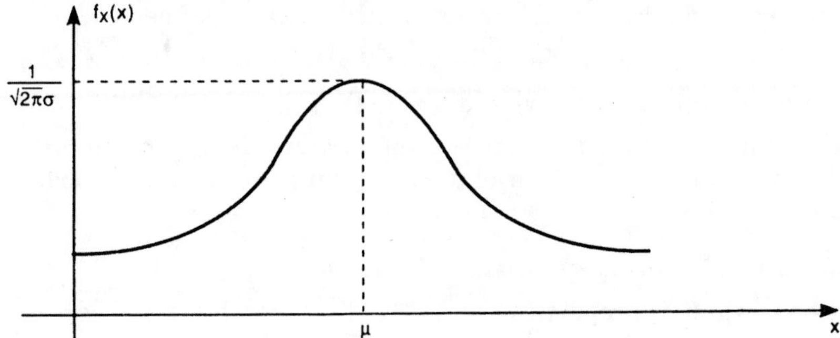

Fig. 14.3 *Gaussian pdf.*

3. Exponential pdf. The exponential random variable X, with parameter γ has a sample space.

$$S = [0 , \infty]$$

Here γ is a positive number. Exponential pdf is given by

$$f(x) = \gamma e^{-\gamma x}, \quad x \in [0 , \infty] \qquad ...(14.27)$$

This is a very widely used model in the analysis and design of computer network systems. Its plot is shown in Fig. 14.4.

There are many other pdfs which are commonly used in different random experiments.

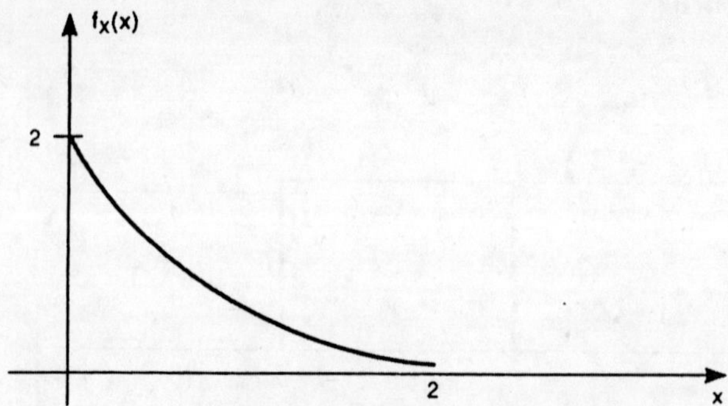

Fig. 14.4 *Exponential pdf for* $\gamma = 2$.

Joint Distribution and Probability Density Functions

We know that a single random variable is completely characterized by either its distribution function or its probability density function. Similarly, the two random variables X and Y are completely characterized by their joint distribution function or joint probability density function.

Joint Distribution Function. The joint distribution function of the two random variables X and Y is denoted by F_{XY} and defined by

$$F_{XY}(x, y) = P[X \leq x, \ Y \leq y] \qquad ...(14.28)$$

where the event $[X \leq x, Y \leq y] = [X \leq x] \cap [Y \leq y]$

The joint distribution function of two random variables is a two-dimensional function. The joint density function depends upon both x and y.

Properties of Joint Distribution Function

It has following properties

1. The joint distribution function of the two random variables completely characterizes the two random variables. It means that both random variables X and Y can be determined from the joint distribution function $F_{XY}(x, y)$.
2. The definition of the joint distribution function can be easily extended to a random vector consisting of N random variables.

$$F_{X_1, X_2, X_3, \ldots X_n}(x_1, x_2, x_3, \ldots x_n)$$

$$= P[X_1 \leq x_1, X_2 \leq x_2, X_3 \leq x_3, \ldots X_n \leq x_n) \qquad ...(14.29)$$

Joint Probability Density Function. The joint probability density function of two random variables X and Y is defined as

$$f_{XY}(x, y) = \frac{\partial F_{XY}(x, y)}{\partial x \, \partial y} \qquad ...(14.30)$$

Properties of Joint Probability Density Function

It has the following properties

1. On integrating both sides of Eq. 14.30, we can get another relation between $f_{XY}(x, y)$ and $F_{XY}(x, y)$.

$$F_{XY}(x, y) = \int_{-\infty}^{x}\int_{-\infty}^{y} f_{XY}(x, y)\, dx\, dy \qquad ...(14.31)$$

2. The joint probability density function of a random vector consisting of N random variables is given by

$$f_{X_1, X_2, ... X_n} = \frac{\partial F_{X_1, X_2, ... X_n}(x_1, x_2, ... x_n)}{\partial x_1\, \partial x_2, ...\, \partial x_n} \qquad ...(14.32)$$

3. The joint probability density function can be used to determine the probabilities in the following way.

$$P[X \le x, Y \le x] = \int_{-\infty}^{x}\int_{-\infty}^{y} f_{XY}(x, y)\, dx\, dx \qquad ...(14.33)$$

$$P[x_1 \le X \le x_2, y_1 \le Y \le y_2] = \int_{x_1}^{x_2}\int_{y_1}^{y_2} f_{XY}(x, y)\, dx\, dy \qquad ...(14.34)$$

   Eqs. 14.33 and 14.34 can also be extended to the case of N random variables.

4. Area of joint probability density function is unity [same as for single random variable].

$$\int_{-\infty}^{\infty}\int_{-\infty}^{\infty} f_{XY}(x, y)\, dx\, dy = 1 \qquad ...(14.35)$$

5. $f_{XY}(x, y)$ completely characterizes the two random variables X and Y.

Marginal Distribution and Probability Density Functions

In the study of two random variables X and Y, $F_X(x)$ and $F_Y(y)$ are known as *marginal distribution functions*. Similarly, $f_X(x)$ and $f_Y(y)$ are known as marginal probability density functions.

Marginal Distribution Function. It can be obtained from the joint distribution function in the following way

As we know that

$$P[X \le x, Y \le x] = F_{XY}(x, y)$$

$$P[X \le x, Y \le \infty] = F_{XY}(x, \infty) \qquad ...(14.36)$$

Now, $\quad [X \le x, Y \le \infty] = [X \le x] \cap [Y \le \infty] = [X \le x] \qquad ...(14.37)$

We can therefore write Eq. 14.36 as

$$P[X \le x] = F_{XY}(x, \infty)$$

Above equation can also be written as

$$F_X(x) = F_{XY}(x\,,\infty) \qquad \text{...(14.38)}$$

Similarly $$F_Y(y) = F_{XY}(\infty\,,y) \qquad \text{...(14.39)}$$

Marginal Probability Density Functions. These density functions can be obtained from following relations

$$F_X(x) = \int_{-\infty}^{x}\int_{-\infty}^{\infty} f_{XY}(x\,,y)\,dx\,dy$$

and $$F_Y(y) = \int_{-\infty}^{\infty}\int_{-\infty}^{y} f_{XY}(x\,,y)\,dx\,dy$$

as $$f_X(x) = \int_{-\infty}^{\infty} f_{XY}(x\,,y)\,dy \qquad \text{...(14.40)}$$

and $$f_Y(y) = \int_{-\infty}^{\infty} f_{XY}(x\,,y)\,dx \qquad \text{...(14.41)}$$

Independent Random Variables. We know the basic theory of probability that two events A and B are independent if the occurrence of one does not affect the probability of occurrence of the other. Mathematically, we say that if $P(A \cap B) = P(A)\,P(B)$, then events A and B are independent.

If events $[X \le x]$ and $[Y \le y]$ are independent for any x and y, then the two random variables X and Y are called independent variables.

For two independent random variables following condition will be satisfied.

$$P[X \le x\,,\,Y \le y] = P[X \le x]\,P[Y \le y] \text{ for all } x \text{ and } y \qquad \text{...(14.42)}$$

or $$F_{XY}(x\,,y) = F_X(x)\,F_Y(y) \qquad \text{...(14.43)}$$

Differentiating Eq. 14.43 wrt. x and y, we get

$$\frac{\partial^2}{\partial x\,\partial y} F_{XY}(x\,,y) = \frac{d}{dx} F_X(x)\,\frac{d}{dy} F_Y(y)$$

or $$f_{XY}(x\,,y) = f_X(x)\,f_Y(y) \qquad \text{...(14.44)}$$

In words we can say that the two random variables X and Y are independent if the product of their marginal distribution (probability density) functions is equal to their joint distribution (probability density) functions.

14.3.2 Several Random Variables

Uptil now, we have discussed situations involving a single random variable. However, we frequently find that the outcome of an experiment requires several random variables for its description.

Here we will discuss the case of two random variables.

The Case of Two Random Variables. Consider two random variables X and Y. We define the joint distribution function $F_{X,Y}(x, y)$ as the probability that the random variables X is less than or equal to a specified value x and that the random variable Y is less than or equal to a specified value y. The variables X and Y may be two separate one-dimensional random variables or the components of a single two-dimensional random variable.

In both cases, the joint sample space is the x-y plane.

The joint distribution function $F_{X,Y}(x, y)$ is the probability that the outcome of an experiment will result in a sample point inside the quadrant $(-\infty < X \leq x, -\infty < Y \leq y)$ of the joint sample space. It is given in Eq. 14.28 as

$$F_{X,Y}(x, y) = P(X \leq x, Y \leq y)$$

The joint distribution function $F_{X,Y}(x, y)$ is continuous everywhere and its partial derivative is given as

$$f_{X,Y}(x, y) = \frac{\partial^2}{\partial x\, \partial y} \{F_{X,Y}(x, y)\} \qquad ...(14.45)$$

Therefore, the partial derivative of $F_{X,Y}(x, y)$ is continuous everywhere. The function $f_{X,Y}(x, y)$ is known as joint probability density function (jpdf) of the random variables X and Y. The joint distribution function $F_{X,Y}(x, y)$ is a monotonic, non-decreasing function of both x and y. The joint probability density function $f_{X,Y}(x, y)$ is always non-negative. The total volume under the graph of a joint probability density function $f_{X,Y}(x, y)$ must be unity. It is given as

$$\int_{-\infty}^{\infty}\int_{-\infty}^{\infty} f_{X,Y}(\alpha, \beta)\, d\alpha\, d\beta = 1 \qquad ...(14.46)$$

The probability density function (pdf) for a single random variable (say X) can be obtained from its joint probability density function (jpdf) with a second random variable (say Y) in the following way

$$F_X(x) = \int_{-\infty}^{\infty}\int_{-\infty}^{\infty} f_{X,Y}(\alpha, \beta)\, d\alpha\, d\beta \qquad ...(14.47)$$

Now, differentiating both sides of Eq. 14.47 with respect to x, we get the following desired relation

$$f_X(x) = \int_{-\infty}^{\infty} f_{X,Y}(x, \beta)\, d\beta \qquad ...(14.48)$$

The probability density function (pdf) can be obtained from the joint probability density function (jpdf) $f_{X,Y}(x, y)$ by integrating it over all possible values of the undesired random variable Y. Similarly, we can find $f_Y(y)$.

$$f_Y(y) = \int_{-\infty}^{\infty} f_{X,Y}(\alpha, y)\, d\alpha \qquad ...(14.49)$$

The functions $f_X(x)$ and $f_Y(y)$ are called marginal probability density functions. Hence, joint probability density function $f_{X,Y}(x, y)$ contains all the possible information about the joint random variable X and Y.

Conditional probability density function of continuous random variable Y for another given continuous random variable $X = x$ is defined by

$$f_{Y/X}(y/x) = \frac{f_{X,Y}(x, y)}{f_X(x)} \qquad ...(14.50)$$

where $f_X(x)$ = marginal density of $X > 0$. The function $f_{Y/X}(y/x)$ is a function of the variable y with variable x arbitrary, but fixed. The conditional probability density function (cpdf) satisfies all the requirements of an ordinary probability density function (pdf), as shown by

$$f_{Y/X}(y/x) \geq 0 \qquad ...(14.51)$$

and $$\int_{-\infty}^{\infty} f_{Y/X}(y/x)\, dy = 1 \qquad ...(14.52)$$

If the random variables X and Y are statistically independent, the knowledge of the outcome of random variable X cannot affect the distribution of random variable Y.

The conditional probability density function (cpdf) $f_{Y/X}(y, x)$ for statistically independent variables reduces to the marginal density $f_Y(y)$. It is given by

$$f_{Y/X}(x, y) = f_Y(y) \qquad ...(14.53)$$

In such a case, joint probability density function (jpdf) of the random variables X and Y is the product of their respective marginal densities. It expressed as

$$f_{X,Y}(x, y) = f_X(x)\, f_Y(y) \qquad ...(14.54)$$

14.3.3 Statistical Averages of Random Variables

In this sub section, we will study various ways of determining the average behaviour of the outcomes arising in random experiments.

Mean Value of Random Variable. The mean value is also called expected value of random variable X. It is defined as

$$m_X = E[X] = \int_{-\infty}^{\infty} x\, f_X(x)\, dx \qquad ...(14.55)$$

where E denotes the expectation operator.

The mean value m_X locates the centre of gravity of the area under the probability density curve of the random variable X.

Similarly, the expected value of function of random variable X is defined as

$$E[g(X)] = \int_{-\infty}^{\infty} g(x)\, f_X(x)\, dx \qquad ...(14.56)$$

Special Case. $g(x) = X^n$, we can obtain the nth moment of the probability distribution of the random variable X.

Expected value of $g(X) = X^n$ is given by

$$E[X^n] = \int_{-\infty}^{\infty} x^n f_X(x)\, dx \qquad ...(14.57)$$

For $n = 1$, Eq. 14.57 gives the mean value of the random variable X.

For $n = 2$, Eq. 14.57 gives the mean square value of random variable X. It is given as

$$E[X^2] = \int_{-\infty}^{\infty} x^2 f_X(x)\, dx \qquad ...(14.58)$$

Central Moments of the Random Variable X. It is simply the moments of the difference between a random variable X and its mean value m_X. The nth central moment of random variable X is

$$E[(X - m_X)^n] = \int_{-\infty}^{\infty} (x - m_X)^n f_X(x)\, dx \qquad ...(14.59)$$

For $n = 1$, the first central moment of random variable is zero.

For $n = 2$, the second central moment of random variable is known as variance of random variable. It is defined as

$$\text{Var}[X] = E[(X - m_X)^2] = \int_{-\infty}^{\infty} (x - m_X)^2 f_X(x)\, dx \qquad ...(14.60)$$

The variance of X is denoted by $\sigma_X{}^2$.

The square root of variance of X is known as standard deviation of random variable. Standard deviation is denoted by σ_X.

In some sense, the variance of a random variable is a measure of the randomness of random variable. By specifying the variance $\sigma_X{}^2$, we constrain the effective width of the probability density function $f_X(x)$ of the random variable X about the mean m_X.

Chebyshev Inequality. Chebyshev inequality states that for any positive number $\in$, we have

$$P(|X - m_X| \geq \in) \leq \frac{\sigma_X{}^2}{\in^2} \qquad ...(14.61)$$

From this statement, we see that the mean and variance of a random variable give a partial description of its probability distribution.

Mean square value of random variable and variance $\sigma_X{}^2$ are related by using Eqs. 14.58 and 14.60.

$$\sigma_X{}^2 = E[(X - m_X)^2] = E[X^2 - 2m_X X + m_X{}^2]$$

$$= E[X^2] - 2m_X E[X] + m_X{}^2 \qquad [\text{Since } m_X = E[X]]$$

or
$$\sigma_X{}^2 = E[X^2] - 2m_X m_X + m_X{}^2$$

$$= E[X^2] - 2m_X{}^2 + m_X{}^2$$

or
$$\sigma_X{}^2 = E[X^2] - m_X{}^2 \qquad \text{...(14.62)}$$

If $m_X = 0$, then the variance $\sigma_X{}^2$ and the mean square value $E[X^2]$ of the random variable X are equal.

Characteristic Function $\Phi_X(v)$ of the Probability Distribution of the Random Variable X

It is defined as the expectation of $e^{(jvX)}$ by

$$\Phi_X(v) = E[e^{(jvX)}] = \int_{-\infty}^{\infty} f_X(x)\, e^{(jvX)}\, dx \qquad \text{...(14.63)}$$

where v is real.

The characteristic function $\Phi_X(v)$ is (except for a sign change in the exponent) the Fourier transform of the probability density function $f_X(x)$.

Inverse relation for Eq. 14.63 can be written as

$$f_X(x) = \frac{1}{2\pi} \int_{-\infty}^{\infty} \Phi_X(v)\, e^{(-jvX)}\, dv \qquad \text{...(14.64)}$$

This relation may be used to determine the probability density function $f_X(x)$ of the random variable X from its characteristic function $\Phi_X(v)$.

Joint Moments of Random Variables

Now let us consider a pair of random variables X and Y. Joint moments are the expected value of X^iY^i, where i and j are assumed positive integer values. Expected value of X^iY^i is given by

$$E[X^iY^j] = \int_{-\infty}^{\infty}\int_{-\infty}^{\infty} X^i\, Y^j f_{X,Y}(x, y)\, dx\, dy \qquad \text{...(14.65)}$$

For $i = j = 1$, Eq. 14.65 defines the correlation, *i.e.*,

$$E[XY] = \int_{-\infty}^{\infty}\int_{-\infty}^{\infty} XY f_{X,Y}(x, y)\, dx\, dy$$

The centered random variables are given by

$$X' = X - E[X]$$
$$Y' = Y - E[Y]$$

The correlation of the centered random variables is called covariance of X and Y. It is defined as

$$\text{Cov}[XY] = E[X' Y'] = E[(X - E[X])\,(Y - E[Y])] \qquad ...(14.66)$$

But we know that

$$m_X = E[X]$$

and $$m_Y = E[Y]$$

Substituting these values in Eq. 14.66, we get

$$\text{Cov}[XY] = E[XY] - m_X\, m_Y \qquad ...(14.67)$$

Let $\sigma_X{}^2$ and $\sigma_Y{}^2$ be the variances of random variables X and Y respectively. The correlation coefficient γ is the ratio of the covariance Cov[xy] to $\sigma_X\,\sigma_Y$. It is given as

$$\gamma = \frac{\text{Cov}[XY]}{\sigma_X \sigma_Y} \qquad ...(14.68)$$

If the random variables X and Y are not correlated then their covariance will be zero, *i.e.*,

$$\text{Cov}[XY] = 0$$

If the random variables X and Y are orthogonal then their correlation will be zero, *i.e.*,

$$E[XY] = 0$$

From Eq. 14.66, we observe that if one of the random variables X and Y or both have zero mean, and they are orthogonal random variables, then they are uncorrelated and vice-versa.

Note If random variables X and Y are statistically independent then they are uncorrelated but its converse is not necessarily true.

Transformation of Random Variables

Now we shall determine the probability density function (pdf) of a random variable Y, which is obtained by a one-to-one transformation of a given random variable X. This is a case when the new random variable Y is monotonic, increasing differentiable function of the random variable X. Transformed random variable Y is given as

$$Y = g(X) \qquad ...(14.69)$$

In this case, distribution function of random variable Y is given by

$$F_Y(y) = P(Y \le y) = P(X \le h(y))$$

where $h(y) = g^{-1}(y)$

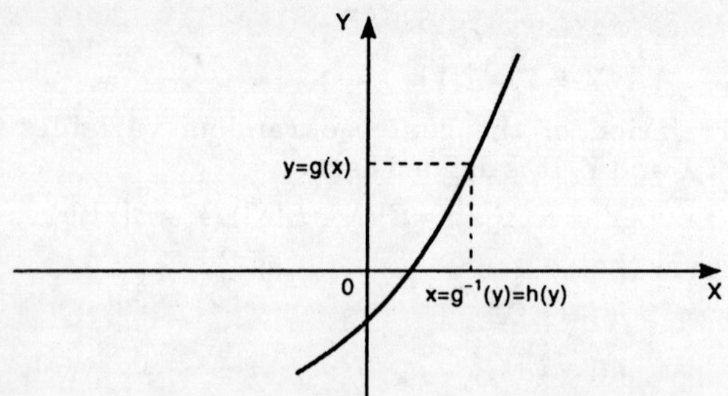

Fig. 14.5 *Illustration of one-to-one transformation of a random variable X.*

This inverse transformation exists for all y, because x and y are related one-to-one. The given random variable X has a probability density function $f_X(x)$.

$$F_Y(y) = \int_{-\infty}^{h(y)} f_X(x)\, dx \qquad \text{...(14.70)}$$

Differentiating both sides of Eq. 14.70 with respect to y, we get

$$\frac{d}{dy} F_Y(y) = \frac{d}{dy} \int_{-\infty}^{h(y)} f_X(x)\, dx$$

or
$$f_Y(y) = f_X(h(y)) \frac{d\,h(y)}{dy} \qquad \text{...(14.71)}$$

Now we have another case when the new random variable Y is monotonic, decreasing differentiable function of the random variable X.

$$F_Y(y) = \int_{h(y)}^{\infty} f_X(x)\, dx \qquad \text{...(14.72)}$$

Differentiating both sides of Eq. 14.72 with respect to y, we get

$$\frac{d}{dy} F_Y(y) = \frac{d}{dy} \int_{h(y)}^{\infty} f_X(x)\, dx$$

or
$$\frac{d}{dy} F_Y(y) = -f_X(h(y)) \frac{d\,h(y)}{dy} \qquad \text{...(14.73)}$$

Since the derivative $\frac{dh(y)}{dy}$ is negative in this case, whereas it is positive in Eq. 14.71.

Results of Eqs. 14.71 and 14.73 are expressed by a single formula which is given as

$$f_Y(y) = f_X(h(y)) \left| \frac{dh(y)}{dy} \right|$$

This is the desired formula for finding the probability density function (pdf) of a one-to-one differentiable function of given random variable.

14.4 RANDOM PROCESS

We know that a random signal is one, whose future values cannot be predicted with certainty at some point of time or over a given interval of time. This nature of random signals can be modelled by Stochastic processes. Random process is also known as *Stochastic process*.

The concept of random process is a logical extension of the random variables. Stochastic is a Greek word, its meaning is "to aim at or to guess". This extension has been necessary for statistical analysis of random signals encountered in communication systems such as voice signal, video signal, digital data and electrical noise. These signals are random as well as functions of time. Many random phenomena that occur in nature also have similar characteristics. Random fluctuation of air temperature and pressure at a particular place is a function of time.

14.4.1 Example of Random Process

Consider a random experiment involving the measurement of air temperature of a certain place at 6 a.m. The temperature is a random variable and is denoted by X. The temperature is a random variable because it takes on different values every day. In order to know the complete statistics of this random variable (say, temperature) we may have to repeat the experiment a large number of times. To determine the of probability distribution function of X we should measure the temperature for many days. We know that temperature at a particular location is a function of time also. The temperature at the same location at different instant of time (say 12 Noon everyday) may have entirely different distribution compared to that at 6 a.m. Thus the random variable X is a function of time and may be expressed as $X(t)$. A random variable that is a function of time is known as a *Random process.* To specify random variable X, we repeat the experiment a large number of times and from the outcomes we find the probability density function (pdf) $f_X(x)$.

In a similar manner, in order to specify the random process $X(t)$, we shall have to carry out the determination of random variables for each value of t.

In this example of random process, to specify the random process $X(t)$, we need to record daily temperatures for each instant of time of the day. This record gives one waveform $X(t\ ,\ s_i)$, where s_i indicates the day for which the record was taken. The same procedure has to be repeated every day for large number of days. The collection of all possible waveforms is known as the ensemble of the random process $X(t)$ and the waveform in this collection is a sample function of the random process.

This is illustrated in Fig. 14.6. Sample function amplitudes at some instant of time $t = t_1$ are the values taken by the random variable $X(t_1) = X_1$ in various trials.

Random variable is the outcome of each trial of the random experiment where the outcome is a number. Random process may be viewed as the outcome of a random experiment where the outcome of each trial is a waveform that is a function of time.

In other words, a random variable X maps the outcomes of a random experiment into numbers along real axis. A random process maps the outcomes into real functions of time. The collection of functions is called ensemble and in this ensemble each member is a sample function.

For statistical analysis of communication systems, it is necessary to characterize the various random signals such as telephone speech signal, television signal etc. These random signals have two important characteristics.

1. These random signals are functions of time defined over some observation interval of time.
2. These signals are random in the sense that before conducting the experiment, it is not possible to exactly describe the waveform at the occurrence of the signal.

On the basis of above discussions, we find that these random signals can only be described by sample points in our sample space each of which is a function of time.

14.4.2 Mathematical Representation of Random Process

Consider a random process $X(t)$ represented by a set of sample functions $\{X_i(t)\}$, $i = 1, 2, 3, ..., n$ and it is shown in Fig. 14.6.

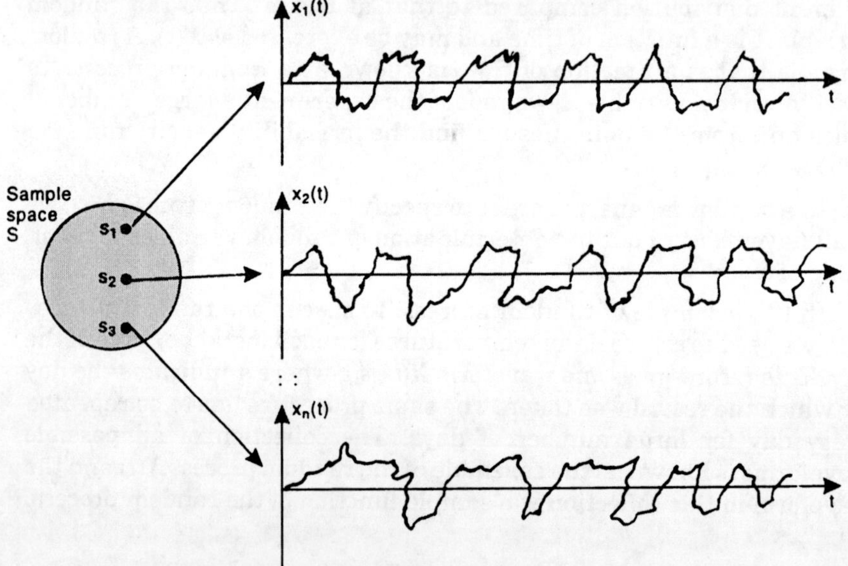

Fig. 14.6 *Illustration of an ensemble of sample functions.*

The sample function $x_1(t)$ corresponds to sample point s_1, $x_2(t)$ corresponds to sample point s_2 of the sample space S and so on for other sample functions. The probability of occurrrence of sample functions are $x_1(t)$, $x_2(t)$,..., $x_n(t)$ and $P(s_1)$, $P(s_2)$,..., $P(s_n)$, respectively.

Let us now observe the set of sample functions $\{x_i(t)\}$, $i = 1, 2, ..., n$ simultaneously at particular instants of time $t = t_1$, $t = t_2$,... etc. Now suppose we observe the set of sample functions $\{x_i(t)\}$, $i = 1, 2, ..., n$, simultaneously at some time instant $t = t_1$. Since each sample point s_i of the sample space S is associated with a number $x_i(t_1)$ and the probability $P(s_i)$, we find that the resulting collection of numbers $\{x_i(t_1)\}$, $i = 1, 2, ...,$ n, forms a random variable. This random variable is denoted by $X(t_1)$. Likewise, for the instant of time $t = t_2$, the set of sample functions $\{x_i(t_2)\}$ define a different random variable $X(t_2)$. In other words, the set of sample functions $\{x_i(t)\}$ defines a different random variable for each choice of observation instant.

By definition of a Stochastic process, $X(t)$ may be viewed as an ensemble of sample functions. Let us consider the values of Stochastic process at any set of time instants, t_1 , t_2, ..., t_n where n is any positive integer. Thus we have n random variables $X(t_1)$, $X(t_2)$, $X(t_3)$, ..., $X(t_n)$.

The joint event may be defined as

$$X(t_1) \leq x_1 , X(t_2) \leq x_2 , \dots , X(t_n) \leq x_n \quad ...(14.74)$$

The joint distribution function is the probability of this joint event and is defined as

$$F_{X(t_1)X(t_2)X(t_n)} (x_1 , x_2 \dots x_n)$$
$$= P[X(t_1) \leq x_1 , X(t_2) \leq x_2 , \dots , X(t_n) \leq x_n] \quad ...(14.75)$$

For the convenience of notation, the joint distribution function is represented as $F_{X(t)}(x)$ where the random vector $X(t)$ is given as

$$X(t) = \begin{bmatrix} X(t_1) \\ X(t_2) \\ \vdots \\ X(t_n) \end{bmatrix} \quad ...(14.76)$$

and dummy vector x is given by

$$x = \begin{bmatrix} x_1 \\ x_2 \\ \vdots \\ x_n \end{bmatrix} \quad ...(14.77)$$

For a particular sample point s_k, the component of random vector $X(t)$ represents the value of the sample function $X_k(t)$ observed at times t_1 , t_2 , t_3 , ... , t_n. It may be noted here that the joint distribution function $F_{X(t)}(x)$ depends on the random process $X(t)$ and the set of time instants $\{t_i\}$, $i = 1, 2, ..., n$.

The joint probability density function of the random vector $X(t)$ is defined as

$$f_{X(t)}(x) = \frac{\partial^n}{\partial x_1\, \partial x_2\, \partial x_3 \ldots \partial x_n} F_{X(t)}(x) \qquad \ldots(14.78)$$

This function is always non-negative and the total volume under its curve is an n-dimensional space and is equal to unity.

14.4.3 Stationary and Non-Stationary Random Processes

A random or stochastic process whose statistical characteristics do not change with time is known as a stationary stochastic process. For a stationary stochastic process a change of time origin cannot be detected and the process will appear to be the same. The example of stationary stochastic process is noise process representing the noise in a channel. It is a stationary process because its statistical properties like mean value and mean square value do not change with time. The random process representing the temperature of a particular location is a non-stationary process. It is because the temperature statistics (mean value and mean square value) depends on the time of the day. Let us consider a random process $X(t)$ for any set of time instants t_1 , t_2 , t_3 , ... , t_n. Then the variables $X(t_1)$, $X(t_2)$, $X(t_3)$, ... , $X(t_n)$ will be obtained frcm the random process $X(t)$. The random process $X(t)$ can be characterized by specifying the joint probability density function $f_{X(t)}(x)$. The random process $X(t)$ is said to be strictly stationary if the joint probability density function is invariant under shift of time origin. A strictly stationary random process will satisfy the following condition

$$f_{X(t)}(x) = f_{X(t+T)}(x) \qquad \ldots(14.79)$$

Eq. 14.79 holds true for every set of time instants $\{t_i\}$, $i = 1, 2, 3, \ldots, n$ and for every time shift T and dummy vector x. The components of random vector $X(t)$ are obtained by observing the random process at time t_1 , t_2 , t_3 , ... t_n. In a similar fashion, the components of random vector $X(t+T)$ are obtained by observing the random process at times $t_1 + T$, $t_2 + T$, $t_3 + T$, ..., $t_n + T$.

The statistics of a stationary random process is time-invariant. In other words, we can say that when the joint probability density functions are different, the stochastic process is non-stationary. The stationary stochastic processes are frequently encountered in practice.

14.4.4 Statistics of Random Process

In this subsection, we shall study about various statistical parameters such as mean, correlation and covariance of stochastic or random process. Mean, correlation and convariance provide a partial description of the distribution of the stochastic or random process. These parameters play key roles to characterize the stochastic process whose probability density function (psd) cannot be determined directly. We have already defined the statistical average or mean of random variable. In the same fashion we can define statistical average or mean for stochastic process. These averages are also known as ensemble averages.

Consider a strictly stationary random process $X(t)$. Here $X(t)$ is a real valued random process.

The mean or average of the stochastic process $X(t)$ is a constant. The mean of stochastic process is defined as

$$m_X(t_j) = E[X(t_j)] \text{ for any time } t_j \qquad ...(14.80)$$

where E denotes the expectation parameter and $X(t_j)$ = Random variable obtained by observing the random process $X(t)$ at time t_j.

$$m_X(t_j) = \int_{-\infty}^{\infty} x\, f_{X(t_j)}(x)\, dx \qquad ...(14.81)$$

The autocorrelation function of the random process $X(t)$ is a function of two time variables t_j and t_i. It is defined as

$$R_X(t_j - t_i) = E[X(t_j)\, X(t_i)] \qquad ...(14.82)$$

for any time t_j and t_i.

where $X(t_j)$ and $X(t_i)$ are random variables obtained by observing the random process $X(t)$ at time t_j and t_i, respectively. The autocorrelation function depends only on the time difference $t_j - t_i$.

Using a single variable τ to represent the time different, the autocorrelation function $R_X(\tau)$ is independent of a shift of time origin.

$$R_X(\tau) = E[X(t)\, X(t - \tau)] \qquad ...(14.83)$$

Both $X(t)$ and $X(t - \tau)$ are random variables. The time variable τ is known as time lag or time delay parameter. For a stationary process, the autocorrelation function $R_X(\tau)$ is independent of shift of time origin.

$$R_X(t_j, t_i) = \int_{-\infty}^{\infty}\int_{-\infty}^{\infty} xy\, f_{X(t_j)\,X(t_i)}(x, y)\, dx\, dy \qquad ...(14.84)$$

The other important characteristics of stationary random process $X(t)$ is the auto-covariance. The auto-covariance function of a stationary random process $X(t)$ is defined as

$$V_X(t_j - t_i) = E[(X(t_j) - m_X)\,(X(t_i) - m_X)]$$

for any time t_j and t_i ...(14.85)

Rewriting Eq. 14.84 for $\tau = t_j - t_i$, we get

$$V_X(\tau) = E[(X(t) - m_X)\,(X(t - \tau) - m_X)] \qquad ...(14.86)$$

This Eq. 14.85 can be written as

$$V_X(\tau) = E[X(t)\, X(t - \tau)] - m_X^{\;2} \qquad ...(14.87)$$

Also, from Eqs. 14.83 and 14.86, we get

$$V_X(\tau) = R_X(\tau) - m_X^{\,2} \qquad ...(14.88)$$

If the random process $X(t)$ has zero mean, then the autocorrelation and autocovariance functions are same. The mean and autocorrelation functions provide only a partial description of the random process. The assumptions of constant mean, time-shift independent autocorrelation function and finite value of zero time lag autocorrelation function are not sufficient to guarantee that the random process is strictly stationary.

14.4.5 Wide-Sense Stationary (WSS) Random Process

A random process may not be strict-sense-stationary (SSS) yet it may have a constant mean value, a constant autocorrelation function that is independent of the shift of time origin and a finite autocorrelation function at zero time lag. Such a random process is called a *Wide-Sense Stationary* (WSS) process. It is also called a weakly stationary process. In other words, we can say that all strictly stationary processes are wide-sense stationary processes but its converse is not necessarily true.

In practice, most random processes are non-stationary processes. This is because every true stationary random process should start at $t = -\infty$ and continue for ever. The eternal sinusoids are the example of stationary process which start at $t = -\infty$ and continue for ever. But true sinusoids (sinusoidal signals) never exist in practice. So we can say that true stationary processes are also non-existent. However, many processes are stationary over a time-interval of interest.

We know that sinusoidal signals simplify the analysis of deterministic signals. Similarly, true random process simplify the modelling of random process.

14.4.6 Ergodic Process

So far we have studied the mean and the autocorrelation function of a random process. These are ensemble average of some kind. For a stationary process, the mean is defined by

$$m_X = E[X(t)] = \int_{-\infty}^{\infty} x\, f_{X(t)}(x)\, dx \qquad ...(14.89)$$

and the autocorrelation function is defined by

$$R_X(\tau) = E[X(t+\tau)\, X(t)] = \int_{-\infty}^{\infty} xy\, f_{X(t+\tau),\, X(t)}(x\,,y)\, dx\, dy \qquad ...(14.90)$$

Computation of ensemble averages such as mean and autocorrelation function is done by taking an average across all the sample functions of the random process. In particular, this computation requires complete knowledge of the first-order and second-order joint probability density functions of the process.

These joint probability density functions are not given in practical situations. But we have given the recording of one of the sample functions of the random process. In such cases, we should also consider time averages of the individual sample functions of the random process.

The time-averaged mean value of the sample function $x(t)$ of a random process $X(t)$ is defined as

$$\langle x(t)\rangle = \lim_{T\to\infty} \frac{1}{T}\int_{-T/2}^{T/2} x(t)\, dt \qquad ...(14.91)$$

where symbol < > stands for the time averaging and T is the observation interval. In a similar way, the time averaged autocorrelation function of sample function $x(t)$ is defined as

$$\langle x(t+\tau)\, x(t)\rangle = \lim_{T\to\infty} \frac{1}{2T}\int_{-T}^{T} x(t+T)\, x(t)\, dt$$

$$= \lim_{T\to\infty} \frac{1}{2T}\int_{-T}^{T} x(t)\, x(t-\tau)\, dt \qquad ...(14.92)$$

The two time averages such as $\langle x(t)\rangle$ and $\langle x(t)\, x(t-\tau)\rangle$ are random variables in the sense that their values depend on that sample function of the random process $X(t)$ which is used in the time-averaging evaluations. On the other hand, the ensemble average m_X is constant and autocorrelation function $R_X(\tau)$ is an ordinary function of time lag τ. For general random process, ensemble averages and time-averages are not equal. A very special class of random processes for which ensemble averages are equal to time-averages of any sample function, is known as ergodic process.

A random process $X(t)$ is said to be ergodic in the most general form if all of its statistical properties can be determined from a sample function representing one possible realization of the process.

If a random process is stationary in strict-sense then it is known as ergodic process. But all stationary processes are not ergodic.

Thus for a random process to be ergodic in mean and autocorrelation function, it must satisfy the following

$$m_X = E[X(t)] = \langle x(t)\rangle \qquad ...(14.93)$$

$$R_X(\tau) = E[X(t)\, X(t-\tau)] = \langle x(t)\, x(t-\tau)\rangle \qquad ...(14.94)$$

The classification of various types of random processes is illustrated in Fig. 14.7.

It is very difficult to check whether a random process is ergodic or not because we must test all possible orders of time and ensemble averages. However, in practice many stationary random processes are

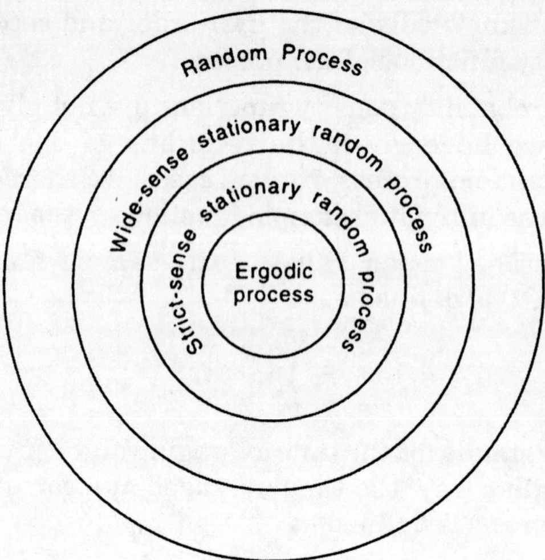

Fig. 14.7 *Classification of random processes.*

usually ergodic w.r.t. at least second-order averages such as mean and autocorrelation function.

14.4.7 Gaussian Process

Now let us consider a random process $X(t)$ for an interval that starts at time $t = 0$ and lasts until $t = T$. Here we weigh the random process $X(t)$ by some function $g(t)$ and then integrate the product $X(t)\, g(t)$ over this observation interval $(0 \le t \le T)$, thereby we obtain a random variable Y which is defined by

$$Y = \int_0^T X(t)\, g(t)\, dt \qquad ...(14.95)$$

where Y is a linear function of $X(t)$.

The random variable Y depends on the course of the argument function $X(t)\, g(t)$ over the observation interval 0 to T. Thus a functional is a quantity that depends on the entire course of one or more functions rather than on a number of discrete variables. Here we assume that the weighing function $g(t)$ is such that the mean square value of the random variable $Y = \int_0^T X(t)\, g(t)\, dt$ is finite. If the random variable Y is a Gaussian-distributed random variable for every value of $g(t)$ then the process $X(t)$ is said to be a Gaussian process.

A random variable Y has a Gaussian distribution if its probability density function has the form

$$f_Y(y) = \frac{1}{\sqrt{2\pi}\,\sigma_y}\, e^{[(-(y-m_Y)^2]/2\sigma_Y^2]} \qquad ...(14.96)$$

where m_Y is the mean and $\sigma_Y{}^2$ is the variance of the random variable Y A Gaussian-distributed random variable is completely characterized by specifying its mean and variance. A plot of Gaussian-distributed probability density function is given in Fig. 14.8 for $m_Y = 0$ and $\sigma_Y{}^2 = 1$.

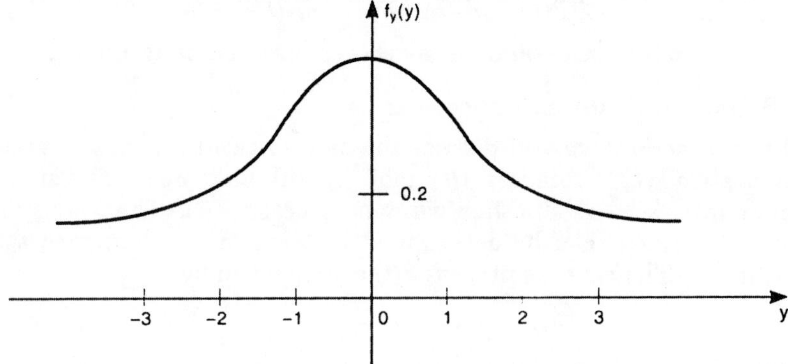

Fig. 14.8 *Normalized Gaussian distributed pdf of a random variable Y.*

A Gaussian process has two main advantages.

1. First, the Gaussian process has many properties that make analytic results possible.
2. Second, the random process produced by physical phenomena are often such that a Gaussian model is appropriate.

The use of a Gaussian model to describe physical phenomena is usually confirmed by experiments. Gaussian process is very important in the study of communication systems. It has the following important properties

1. If a Gaussian process $X(t)$ is applied to a stable linear filter, then the random process $Y(t)$ developed at the output of the filter is also Gaussian.
2. Consider the samples of random process $X(t_1)$, $X(t_2)$, ... , $X(t_n)$, obtained by observing a random process $X(t)$ at time $t = t_1$, $t = t_2$, ..., $t = t_n$. If the random process $X(t)$ is Gaussian, then this set of random variables are jointly Gaussian for any value of n, with their n-fold joint pdf being completely determined by specifying the set of means. Here means used for more than one mean (or average)

$$m_{X(t_i)} = E[X(t_i)] \text{ , } i = 1 \text{ , } 2 \text{ , } \ldots \text{ , } n$$

and the set of autocorrelation function

$$C_X(t_j \text{ , } t_i) = E[\{X(t_j) - m_X(t_j)\} \{X(t_i) - m_X(t_i)\}]$$

$$\text{for } i = j = 1 \text{ , } 2 \text{ , } 3 \text{ , } \ldots \text{ , } n$$

3. If a Gaussian process is wide-sense stationary (WSS), then the process is also stationary in the strict sense.
4. If the set of random variables $X(t_1)$, $X(t_2)$, ... , $X(t_n)$, obtained by sampling a Gaussian process $X(t)$ at times $t = t_1$, $t = t_2$, ..., $t = t_n$, are uncorrelated, *i.e.*,

$$E[\{X(t_j) - m_X(t_j)\}\ \{X(t_i) - m_X(t_i)\}] = 0\ ,\ i \neq j$$

then this set of random variables are statistically independent.

14.4.8 Sum of Random Processes

In this subsection, we will discuss the sum of two random processes. Consider two WSS processes $X(t)$ and $Y(t)$ with zero mean. The sum of random processes is another random process $Z(t)$. The statistical parameters of $Z(t)$ can be determined in terms of random processes $X(t)$ and $Y(t)$. The random process $Z(t)$ can be given by

$$Z(t) = X(t) + Y(t) \qquad ...(14.97)$$

The autocorrelation function of $Z(t)$ can be determined by

$$\begin{aligned} R_Z(\tau) &= E[Z(t)\ Z(t-\tau)] \\ &= E[\{X(t) + Y(t)\}\ \{X(t-\tau) + Y(t-\tau)\}] \\ &= E[X(t)\ X(t-\tau) + X(t)\ Y(t-\tau) \\ &\qquad + X(t-\tau)\ Y(t) + Y(t)\ Y(t-\tau)] \\ &= E[X(t)\ X(t-\tau)] + E[X(t)\ Y(t-\tau)] \\ &\qquad + E[X(t-\tau)\ Y(t)] + E[Y(t)\ Y(t-\tau)] \end{aligned}$$

or

$$R_Z(\tau) = R_X(\tau) + R_{XY}(\tau) + R_{YX}(\tau) + R_Y(\tau) \qquad ...(14.98)$$

The power spectral density of random process $Z(t)$ can be obtained by taking Fourier transform of both sides of Eq. 14.98. Thus,

$$S_Z(\omega) = S_X(\omega) + S_{XY}(\omega) + S_{YX}(\omega) + S_Y(\omega) \qquad ...(14.99)$$

The power spectral density (psd) of the sum of two WSS random processes is equal to the sum of the individual power spectral densities plus the sum of the cross power spectral densities $S_{XY}(\omega)$ and $S_{YX}(\omega)$.

If random processes $X(t)$ and $Y(t)$ are uncorrelated then the cross power spectral densities $S_{XY}(\omega)$ and $S_{YX}(\omega)$ are zero. Now Eq. 14.99 can be written as

$$S_Z(\omega) = S_X(\omega) + S_Y(\omega) \qquad ...(14.100)$$

Therefore ,the power spectral density of the sum of two uncorrelated WSS random processes is equal to the sum of their individual power spectral densities.

14.5 CORRELATION FUNCTION

Correlation function gives a measure of similarity or coherence between a given signal (process) and a replica of the same signal or other signal (process) by a variable amount.

14.5.1 Autocorrelation Function

Autocorrelation function is defined as a measure of similarity between a signal or process and its replica by a variable amount.

The autocorrelation function of a stationary process $X(t)$ is defined as

$$R_{X(t_j - t_i)} = E[X(t_j)\, X(t_i)] \text{ for any } t_j \text{ and } t_i \quad ...(14.101)$$

where $X(t_j)$ and $X(t_i)$ are the random variables obtained by observing the process $X(t)$ at times t_j and t_i, respectively.

The autocorrelation function depends only on the time difference $(t_j - t_i)$. Using $\tau = t_j - t_i$ in Eq. 14.101, we get

$$R_X(\tau) = E[X(t)\, X(t - \tau)] \quad ...(14.102)$$

The expression $X(t)$ and $X(t - \tau)$ are viewed as random variables. The variable τ is known as time-lag or time-delay parameter. The autocorrelation function for stationary random process is independent of a shift of time origin.

14.5.2 Properties of Autocorrelation Function of Random Process

The autocorrelation function $R_X(\tau)$ of a wide-sense stationary (WSS) process has the following important properties.

1. The autocorrelation function of WSS process $X(t)$ is an even function of time-lag. Autocorrelation function $R_X(\tau)$ satistifies the following mathematical relationship.

$$R_X(\tau) = R_X(-\tau) \quad ...(14.103)$$

Proof : From the definition of autocorrelation function of WSS random process

$$R_X(\tau) = E[X(t)\, X(t - \tau)] \quad ...(14.104)$$

Putting $\tau = -\tau$ above Eq. 14.104, we get

$$R_X(-\tau) = E[X(t)\, X(t + \tau)]$$

or

$$R_X(-\tau) = E[X(t + \tau)\, X(t)] \quad ...(14.105)$$

For WSS process, we know that

$$X(t)\, X(t - \tau) = X(t + \tau)\, X(t) \quad ...(14.106)$$

From Eqs. 14.104, 14.105 and 14.106, we can conclude that

$$R_X(\tau) = R_X(-\tau)$$

2. The mean square value of a WSS process is equal to the autocorrelation function of the random process for zero time-lag. It can be written as

$$R_X(0) = R_X(\tau)\Big|_{\tau=0} = E[X(t)\,X(t-\tau)]\Big|_{\tau=0}$$

$$= E[X(t)\,X(t-0)] = E[X^2(t)]$$

or $$R_X(0) = E[X^2(t)] \quad ...(14.107)$$

3. The autocorrelation function of a WSS random process has the maximum magnitude at zero time-lag ($\tau = 0$). This property is stated as

$$|R_X(\tau)| = \text{magnitude of } R_X(\tau)$$

$$= R_X(0) \quad ...(14.108)$$

Proof : We know that the mean square value of the difference between $X(t)$ and $X(t-\tau)$ is always non-negative.

$$E[\{X(t) - X(t-\tau)\}^2] \geq 0 \quad ...(14.109)$$

or $$E[X^2(t) - 2X(t)\,X(t-\tau) + X^2(t-\tau)] \geq 0 \quad ...(14.110)$$

Since expectation operation is a linear operation, we may write Eq. 14.110 as

$$E[X^2(t)] - 2E[X(t)\,X(t-\tau)] + E[X^2(t-\tau)] \geq 0 \quad ...(14.111)$$

For a WSS random process

$$E[X^2(t)] = E[X^2(t-\tau)] = R_X(0) \quad ...(14.112)$$

and $$E[X(t)\,X(t-\tau)] = R_X(\tau) \quad ...(14.113)$$

Substituting Eqs. 14.112 and 14.113 in Eq. 14.111, we get

$$|R_X(\tau)| \leq R_X(0) \quad ...(14.114)$$

The autocorrelation functions $R_X(\tau)$ provide means of decreasing the interdependence of two random variables obtained by observing a random process $X(t)$ τ seconds apart. The autocorrelation function is a measure of the rate of the fluctuation of random process. Various types of random processes are illustrated in Fig. 14.9.

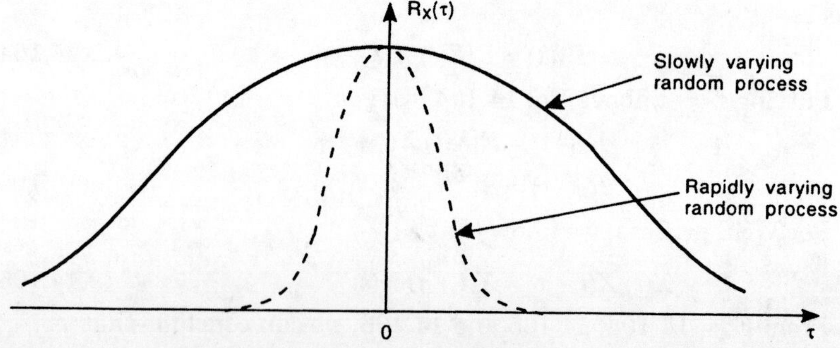

Fig. 14.9 *Illustration of autocorrelation functions of slowly and rapidly varying random processes.*

14.5.2 Cross-Correlation Functions

Autocorrelation function is determined for single random process but cross correlation function is determined for two random processes. Consider two random processes $X(t)$ and $Y(t)$ with autocorrelation functions $R_X(t, u)$ and $R_Y(t, u)$, respectively. There will be two cross-correlation functions between two random processes $X(t)$ and $Y(t)$. These cross-correlations are defined as

$$R_{XY}(t, u) = E[X(t)\, Y(u)] \qquad ...(14.115)$$

and

$$R_{YX}(t, u) = E[Y(t)\, X(u)] \qquad ...(14.116)$$

where t and u are the two values of time at which the processes are observed. The cross-correlation function of two random processes $X(t)$ and $Y(t)$ can be given conveniently in matrix form as

$$R(t, u) = \begin{bmatrix} R_X(t, u) & R_{XY}(t, u) \\ R_{YX}(t, u) & R_Y(t, u) \end{bmatrix} \qquad ...(14.117)$$

$R(t, u)$ is called the correlation matrix of random processes $X(t)$ and $Y(t)$.

Eq. 14.117 can also be written as

$$R(t, u) = \begin{bmatrix} R_X(t-u) & R_{XY}(t-u) \\ R_{YX}(t-u) & R_Y(t-u) \end{bmatrix} \qquad ...(14.118)$$

If both random processes $X(t)$ and $Y(t)$ are WSS processes, then Eq. 14.117 can be written as Eq. 14.118. In addition, both random processes $X(t)$ and $Y(t)$ are said to be jointly wide-sense stationary.

It should be noted that in general, the cross-correlation function is not an even function of τ like an autocorrelation function. It has following symmetry relation

$$R_{XY}(\tau) = R_{YX}(-\tau) \qquad ...(14.119)$$

Moreover, cross-correlation does not have a maximum at the origin like the autocorrelation function. The two random processes $X(t)$ and $Y(t)$ are said to be incoherent or orthogonal if cross-correlation function of $X(t)$ and $Y(t)$ is zero.

$$R_{XY}(\tau) = 0 \qquad ...(14.120)$$

The two random processes are said to be uncorrelated, if their cross-correlation functions $R_{XY}(\tau)$ are equal to the multiplication of mean values.

$$R_{XY}(\tau) = E[X(t)]\, E[Y(t)] \qquad ...(14.121)$$

The incoherent or orthogonal processes are uncorrelated process with $E[X(t)]$ and/or $E[Y(t)] = 0$.

14.6 SPECTRAL DENSITIES

Spectral densities are used to represent random process in frequency-domain. In this section, we shall study frequency-domain description of random processes. Frequency-domain methods are very powerful tools for analysis of various types of signals. We have already discussed some powerful tools such as Fourier series, Fourier transform, Laplace transform for the analysis of deterministic signals. In this section we extend these tools to the stationary random processes.

14.6.1 Power Spectral Density (psd)

Power spectral density $S_X(\omega)$ of a wide-sense stationary (WSS) random process $X(t)$ is defined as

$$S_X(\omega) = \text{Fourier transform } \{R_X(\tau)\}$$

$$= \boxed{\int_{-\infty}^{\infty} R_X(\tau)\, e^{-j\omega\tau}\, d\tau} \quad \text{...(14.122)}$$

where $R_X(\tau)$ = Autocorrelation function of random process $X(t)$.

In words, we can say that the Fourier transform of autocorrelation function $R_X(\tau)$ of a random process $X(t)$ is called power spectral density of random process. Eq. 14.122 was intially developed by Albert Einstein. Later it was developed by N. Weiner and Khenchine. Therefore, this equation is known as Einstein-Weiner-Khenchine (EWK) equation. Power spectral density is used to measure the energy of the signal in the frequency band $[w\ ,\ w + dw]$.

Cross Power Spectral Density (cpsd). The cross power spectral density of two jointly WSS random processes $X(t)$ and $Y(t)$ is defined as

$$S_{XY}(\omega) = \text{Fourier transform } \{R_{XY}(\tau)\}$$

$$= \int_{-\infty}^{\infty} R_{XY}(\tau)\, e^{-j\omega\tau}\, d\tau \quad \text{...(14.123)}$$

where $R_{XY}(\tau)$ = Cross-correlation function of random processes $X(t)$ and $Y(t)$.

Properties of Power Spectral Density

Power spectral density of a random process has the following important properties.

1. Power spectral density of a random process is a real function of frequency ω.

Proof: We know that the autocorrelation function of a random process is an even function

$$R_X(\tau) = R_X(-\tau)$$

Therefore, $S_X(\omega) = \int_{-\infty}^{\infty} R_X(\tau)\, e^{-j\omega\tau}\, d\tau$

$$= \int_{-\infty}^{\infty} R_X(\tau) [\cos \omega\tau - j \sin \omega\tau] \, d\tau$$

$$= \int_{-\infty}^{\infty} R_X(\tau) \cos \omega\tau \, d\tau - j \int_{-\infty}^{\infty} R_X(\tau) \sin \omega\tau \, d\tau \quad ...(14.124)$$

Now we get

$$\int_{-\infty}^{\infty} R_X(\tau) \sin \omega\tau \, d\tau = I \quad ...(14.125)$$

Substituting $t = -\tau$ in Eq. 14.125, we get

$$\int_{-\infty}^{\infty} R_X(-t) \sin (-\omega t) \, (-dt) = I$$

or

$$-\int_{-\infty}^{\infty} R_X(-t) \sin \omega t \, dt = I \quad ...(14.126)$$

Using the even property of autocorrelation function, we get

$$-\int_{-\infty}^{\infty} R_X(\tau) \sin \omega\tau \, d\tau = I \quad ...(14.127)$$

After adding Eqs. 14.125 and 14.127, we get

$$I = 0 \quad ...(14.128)$$

Substituting $\int_{-\infty}^{\infty} R_X(\tau) \sin \omega\tau \, d\tau = 0$ in Eq. 14.124, we get

$$S_X(\omega) = \int_{-\infty}^{\infty} R_X(\tau) \cos \omega\tau \, d\tau \quad ...(14.129)$$

Here $S_X(\omega)$ is a real function of ω.

2. Power spectral density of a random process $X(t)$ is an even function of frequency ω.

$$S_X(\omega) = S_X(-\omega)$$

3. Power spectral density of a random process $X(t)$ is a non- negative function of ω, that is,

$$S_X(\omega) \geq 0 \text{ for all } \omega$$

14.6.2 Energy Spectral Density (esd)

The energy spectral density $\psi_X(\omega)$ is a measure of density of the energy contained in random process $X(t)$ in Joules per Hertz. Note that since the amplitude spectrum of a real-valued random process $X(t)$ is an even function of ω, the energy spectral density of such a signal is symmetrical about the vertical axis passing through the origin.

Total energy of the random process $X(t)$ is defined as

$$E = \frac{1}{2\pi} \int_{-\infty}^{\infty} \psi_X(\omega)\, d\omega$$

14.7 RESPONSE OF LINEAR SYSTEMS TO RANDOM INPUTS

In this subsection, we will study about response of linear systems to random inputs. In other words, we shall study transmission of random process through linear systems or linear filters.

Suppose that a random process $X(t)$ is applied to the input of a linear time-invariant (LTI) system having an impulse response $h(t)$. Let the input random process $X(t)$ produces a random process $Y(t)$ at the system output. The block diagram of this LTI system is shown in Fig. 14.10. Here, we shall determine the mean and the autocorrelation function of the output random process $Y(t)$ in terms of input random process $X(t)$. Here we are assuming that $X(t)$ is a wide-sense stationary (WSS) process.

Fig. 14.10 *Determination of response of a linear system to a random input.*

The mean of the output random process is defined as

$$m_Y(t) = E[Y(t)] = E\left[\int_{-\infty}^{\infty} h(\tau)\, X(t-\tau)\, d\tau\right]$$

$$= \int_{-\infty}^{\infty} h(\tau)\, E[X(t-\tau)]\, d\tau$$

$$= \int_{-\infty}^{\infty} h(\tau)\, m_X(t-\tau)\, d\tau \qquad ...(14.130)$$

When the input random process $X(t)$ is wide-sense stationary, the mean $m_X(t)$ is a constant m_X. Therefore, Eq. 14.130 can be written as

$$m_Y = \int_{-\infty}^{\infty} h(\tau)\, m_X\, d\tau$$

$$= m_X \int_{-\infty}^{\infty} h(\tau)\, d\tau$$

or $$m_Y = m_X\, H(0) \qquad ...(14.131)$$

where $H(0)$ = zero-frequency response of the system.

In words, we can say that the mean of the output random process of a stable linear time-invariant (LTI) system is equal to the mean of the input random process multiplied by the zero-frequency response of the system.

We shall now determine the autocorrelation function of the output random process $Y(t)$. Let t and u be two values of time at which output process is observed. Autocorrelation function of output random process is defined as

$$R_Y(t\,,u) = E[Y(t)\,Y(u)] \qquad ...(14.132)$$

By using convolution integral, Eq. 14.132 can be written as

$$R_Y(t\,,u) = E\left[\int_{-\infty}^{\infty} h(\tau_1)\,X(t-\tau_1)\,d\tau_1 \int_{-\infty}^{\infty} h(\tau_2)\,X(u-\tau_2)\,d\tau_2\right] \qquad ...(14.133)$$

Assuming $E[X^2(t)]$ is finite for all t and the system is stable, we can interchange the order of the expectation operator and integration w.r.t. τ_1 and τ_2 to get

$$R_Y(t\,,u) = \int_{-\infty}^{\infty} h(\tau_1)\,d\tau_1 \int_{-\infty}^{\infty} h(\tau_2)\,d\tau_2\, E[X(t-\tau_1)\,X(u-\tau_2)]$$

$$= \int_{-\infty}^{\infty} h(\tau_1)\,d\tau_1 \int_{-\infty}^{\infty} h(\tau_2)\,d\tau_2\, R_X(t-\tau_1\,,u-\tau_2) \qquad ...(14.134)$$

When the input $X(t)$ is a wide-sense stationary process, the autocorrelation function of $X(t)$ is only a function of the difference between the observation time $t-\tau_1$ and $u-\tau_2$. Substituting $\tau = t-u$ in Eq. 14.134, we get

$$R_Y(\tau) = \int_{-\infty}^{\infty}\int_{-\infty}^{\infty} h(\tau_1)\,h(\tau_2)\,R_X(\tau-\tau_1+\tau_2)\,d\tau_1 d\tau_2 \qquad ...(14.135)$$

$R_Y(\tau)$ is the function of time lag τ.

By combining the result of Eq. 14.135 with Eq. 14.131, we see that if the input to a stable LTI system is a WSS process, then the output of the linear time-invariant (LTI) system is also a WSS process.

Putting $\tau = 0$ in Eq. 14.135, we get

$$R_Y(0) = \int_{-\infty}^{\infty}\int_{-\infty}^{\infty} h(\tau_1)\,h(\tau_2)\,R_X(\tau_2-\tau_1)\,d\tau_1\,d\tau_2 \qquad ...(14.136)$$

Also we know that

$$R_Y(0) = E[Y^2(t)] \qquad ...(14.137)$$

From Eqs. 14.136 and 14.137, we get

$$R_Y(0) = E[Y^2(t)] = \int_{-\infty}^{\infty}\int_{-\infty}^{\infty} h(\tau_1)\, h(\tau_2)\, R_X(\tau_2 - \tau_1)\, d\tau_1\, d\tau_2 \quad ...(14.138)$$

which is a constant.

SOLVED EXAMPLES

Example 14.1 Two coins are tossed together. Find the probability of "match".

Solution. Let H represent the 'head' and T be the tail, then there are four mutually exclusive events. These events are HH, HT, TH, TT. Out of these four events, only two events HH and TT represent a match. Hence

$$P(HH) = \frac{1}{4} \text{ and } P(TT) = \frac{1}{4}$$

The total probability = $P(HH) + P(TT)$

i.e.,
$$P(\text{match}) = \frac{1}{4} + \frac{1}{4} = \frac{2}{4} = \frac{1}{2}$$

or
$$P(\text{match}) = \frac{1}{2}$$

Example 14.2 In a single throw of two dices, determine the probability of a total of 9 or 11.

Solution. Clearly, the events 'a total of 9' and 'a total of 11' are mutually exclusive.

Now, P (a total of 9) $= P(36, 45, 54, 63) = \dfrac{4}{36} = \dfrac{1}{9}$

and $P(\text{a total of } 11) = P(56, 65) = \dfrac{2}{36} = \dfrac{1}{18}$

Thus, P(a total of 9 or 11) = P(a total of 9) + P(a total of 11)

$$= \frac{1}{9} + \frac{1}{18} = \frac{2+1}{18} = \frac{3}{18} = \frac{1}{6}$$

or
$$P(\text{a total of 9 or 11}) = \frac{1}{6}$$

Example 14.3 A dice is tossed twice. Determine the probability of 'a number greater than 4 on each toss'.

Solution. Let us denote by A, the event 'a number greater than 4 on first toss' and by B, the event 'a number greater than 4 on the second toss'. Clearly, events A and B are independent. Thus,

$$P(A \text{ and } B) = P(A)\, P(B) = P(56)\, P(65)$$

$$= \frac{2}{6} \times \frac{2}{6} = \frac{1}{9}$$

or $$P(A \text{ and } B) = \frac{1}{9}$$

Example 14.4 A husband and wife appear in an interview for two vacancies in the same post. The probability of husband's selection is $\frac{1}{7}$ and that of wife's selection is $\frac{1}{5}$. Find the probability

(*a*) Both of them will be selected

(*b*) Only one of them will be selected.

Solution. We let A = Husband is selected

B = Wife is selected

Both events 'A' and 'B' are independent. Also 'A' and 'not B' are independent, 'B' and 'not A' are independent.

(*a*) $$P(A \text{ and } B) = P(A)\,P(B) = \frac{1}{7} \times \frac{1}{5} = \frac{1}{35}$$

(*b*) P(only one of them will be selected)

$$= P(A \text{ and not } B \text{ or } B \text{ and not } A)$$

$P(A$ and not $B) + P(B$ and not $A)$

$$= P(A)\,[1 - P(B)] + P(B)\,[1 - P(A)]$$

$$= \frac{1}{7}\left(1 - \frac{1}{5}\right) + \frac{1}{5}\left(1 - \frac{1}{7}\right)$$

$$= \frac{1}{7} \times \frac{24}{5} + \frac{1}{5} \times \frac{6}{7}$$

$$= \frac{4}{35} + \frac{6}{35} = \frac{10}{35} = \frac{2}{7}$$

Example 14.5 Determine the probability distribution of random variable X, the number of heads, in the two tosses of a coin (or a simultaneous toss of two coins).

Solution. Let X represent the outcome of an event. There can be no head, one head or two heads. Clearly X can have value 0, 1, 2.

Now $P(X = 0) = P$(no head)

$= P$(Tail on the first toss and tail on the second toss)

$= P$(Tail on the first toss) P(tail on the second toss)

$$= \frac{1}{2} \times \frac{1}{2} = \frac{1}{4}$$

$$P(X = 1) = P(1 \text{ head}) = P(HT \text{ or } TH)$$

$$= P(HT) + P(TH) = \frac{1}{4} + \frac{1}{4} = \frac{2}{4} = \frac{1}{2}$$

and $\quad P(X=2) = P(2 \text{ heads}) = P(HH) = \dfrac{1}{4}$

Thus we have

X	0	1	2
$P(X)$	$\frac{1}{4}$	$\frac{1}{2}$	$\frac{1}{4}$

$$\sum_{x=0}^{2} P(x) = P(0) + P(1) + P(2)$$

$$= \frac{1}{4} + \frac{1}{2} + \frac{1}{4} = \frac{1+2+1}{4} = \frac{4}{4} = 1$$

These values are shown in Fig. 14.11.

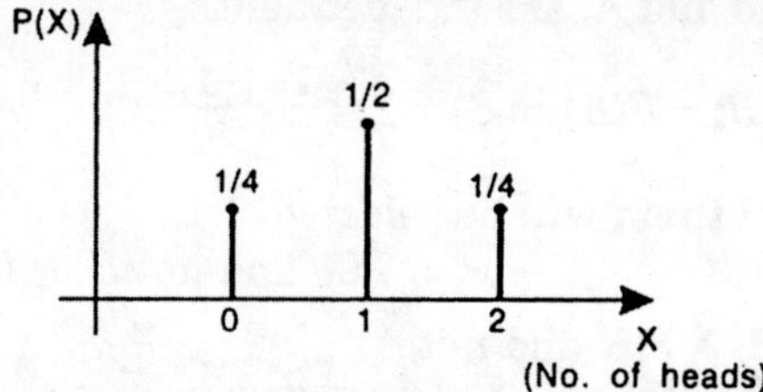

Fig. 14.11

Example 14.6 Five defective bulbs are accidentally mixed with 20 good ones. It is not possible to just look at a bulb and tell whether or not it is defective. Determine the probability distribution of the number of defective bulbs, if four bulbs are drawn at random from this lot.

Solution. We let X, the number of defective bulbs. Clearly X can take a value 0, 1, 2, 3 or 4. Now

$$P(X=0) = P(\text{no defective bulbs})$$

$$= P(\text{all 4 good bulbs})$$

$$= \frac{4C_{20}}{4C_{25}} = \frac{969}{2530}$$

$$P(X=1) = P(\text{defective and 3 good bulbs})$$

$$= \frac{1C_5 \times 3C_{20}}{4C_{25}} = \frac{1140}{2530}$$

$$P(X=2) = P(\text{2 defective and 2 good bulbs})$$

$$= \frac{2C_5 \times 2C_{20}}{4C_{25}} = \frac{380}{2530}$$

$$P(X=3) = P(\text{3 defective and 1 good bulbs})$$

$$= \frac{3C_5 \times 1C_{20}}{4C_{25}} = \frac{40}{2530}$$

and $P(X = 4) = P(\text{All 4 defective bulbs})$

$$= \frac{4C_5}{4C_{25}} = \frac{1}{2530}$$

Thus, we have

X	0	1	2	3	4
$P(X)$	$\frac{969}{2530}$	$\frac{1140}{2530}$	$\frac{380}{2530}$	$\frac{40}{2530}$	$\frac{1}{2530}$

Example 14.7 We have been given a probability density function $f(x) = \alpha\, e^{-\beta|x|}$ where x is a random variable whose allowable values range from $x = -\infty$ to $x = +\infty$. Determine

(*a*) Distribution function

(*b*) The relationship between α and β

(*c*) The probability that the outcome x lies between 1 and 2.

Solution. (*a*) The distribution function is determined as

$$F(x) = P(X \le x) = \int_{-\infty}^{\infty} f(x)\, dx$$

$$= \int_{-\infty}^{\infty} \alpha\, e^{-\beta|x|}\, dx$$

$$= \begin{cases} \dfrac{\alpha}{\beta} e^{\beta x}, & \text{for } x \le 0 \\ \dfrac{\alpha}{\beta} (2 - e^{-\beta x}), & \text{for } x >= 0 \end{cases}$$

(*b*) Also we know that

$$\int_{-\infty}^{\infty} f(x)\, dx = 1$$

or $$\int_{-\infty}^{\infty} \alpha\, e^{-\beta|x|}\, dx$$

or $$\frac{2\alpha}{\beta} = 1$$

or $$\alpha = \frac{\beta}{2}$$

or $$\frac{\alpha}{\beta} = \frac{1}{2}$$

(*c*) The probability that the outcome x lies between 1 and 2 is

$$P(1 \le x \le 2) = \frac{\beta}{2} \int_{1}^{2} e^{-\beta|x|}\, dx = \frac{1}{2}\, [e^{-\beta} - e^{-2\beta}]$$

Example 14.8 The joint probability density function of random variables X and Y is

$$f(x, y) = \frac{1}{4} e^{-|x| - |y|}, \text{ for } -\infty < x < \infty$$
$$-\infty < y < \infty$$

Find the probability that $X \leq 1$ and $Y < = 0$.

Solution. Since $f(x, y)$ can be written as

$$f(x, y) = \frac{1}{2} e^{-|x|} \times \frac{1}{2} e^{-|y|} = f(x) f(y)$$

because we have assumed that random variables X and Y are statistically independent.

Now $$P(X \leq 1, Y \leq 0) = \int_{-\infty}^{1} f(x)\, dx \int_{-\infty}^{0} f(y)\, dy$$

$$= \int_{-\infty}^{1} \frac{1}{2} e^{-|x|}\, dx \int_{-\infty}^{0} \frac{1}{2} e^{-|y|}\, dy$$

$$= \left(\frac{2 + e^{-1}}{2}\right) \frac{1}{2} = \frac{1}{4} (2 + e^{-1})$$

Example 14.9 Find the probability density function $f(z)$ of $Z = X + Y$, in terms of joint probability density function $f(x, y)$.

Solution. Here we assume an arbitrary value of random variable Z and call it z. Then region $Y \leq z - X$ is shown below in Fig. 14.12.

Hence the probability that $Z \leq z$ is the same as the probability that $Y \leq z - X$ independent of the value of X, *i.e.*, for $-\infty \leq X \leq +\infty$.

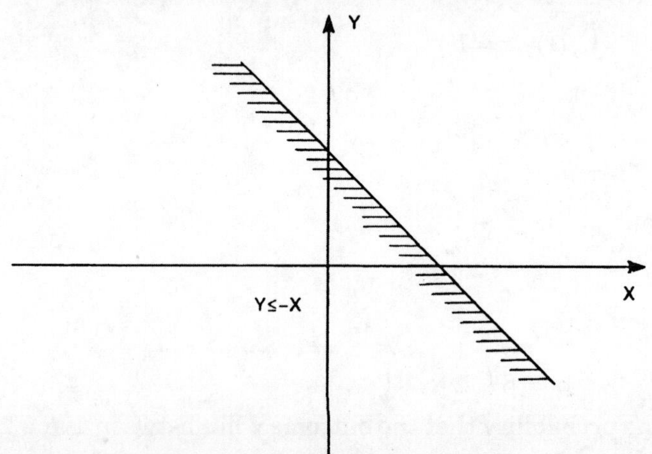

Fig. 14.12

This probability is given by

$$F(z) = P(Z \le z) = P(X \le \infty\, , Y \le z - X)$$

or
$$F(z) = \int_{-\infty}^{\infty} dx \int_{-\infty}^{\infty} f(x\, , y)\, dy$$

Hence $f(z) = \dfrac{d\,F(z)}{dZ}$, we have

$$f(z) = \int_{-\infty}^{\infty} f(x\, , z - x)\, dx$$

If X and Y are independent, *i.e.*,

$$f(x\, , y) = f(x)\, f(y)$$

then
$$f(z) = \int_{-\infty}^{\infty} f(x)\, f(z - x)\, dx$$

Example 14.10 A random variable X follows Gaussian probability density function with parameters μ and σ. Determine its probability distribution function.

Solution. The relationship between probability density function (pdf) and the probability distribution function of a random variable X is given by

$$F_X(z) = \int_{-\infty}^{z} f_X(x)\, dx \qquad \text{...(1)}$$

Since we have given that the random variable X follows Gaussian probability density function, then

$$F_X(z) = \int_{-\infty}^{z} \frac{1}{\sqrt{2\pi}\,\sigma}\, e^{-(x-\mu)/2\sigma^2}\, dx \qquad \text{...(2)}$$

Putting $\dfrac{x - \mu}{\sigma} = y$ in above Eq. 2, we get

$$F_X(z) = \int_{-\infty}^{[(z-\mu)/\sigma]} \frac{1}{\sqrt{2\pi}\,\sigma}\, e^{-y^2/2}\, (\sigma\, dy)$$

$$= \frac{1}{\sqrt{2\pi}} \int_{-\infty}^{(z-\mu)/\sigma} e^{-y^2/2}\, dy$$

$$= \frac{1}{\sqrt{2}} \left[\int_{-\infty}^{0} e^{-y^2/2}\, dy + \int_{0}^{(z-\mu)/2} e^{-y^2/2}\, dy \right]$$

or
$$F_X(z) = \left[- \int_{0}^{\infty} e^{-y^2/2}\, dy + \int_{0}^{[z-\mu)/\sigma]} e^{-y^2/2}\, dy \right] \qquad \text{...(3)}$$

Also error function is defined as

$$erf(x) = \frac{1}{\sqrt{2\pi}} \int_0^x e^{-y^2/2}\, dy \quad ...(4)$$

and $$erf(\infty) = \frac{1}{\sqrt{2\pi}} \int_0^\infty e^{-y^2/2}\, dy = \frac{1}{2} \quad ...(5)$$

Using Eqs. 4 and 5 in Eq. 3, we get

$$F_X(z) = \frac{1}{2} + erf\left(\frac{z-\mu}{\sigma}\right) \quad ...(6)$$

Example 14.11 Prove that for a Gaussian random variable X,

$$P[x_1 \le X \le x_2] = erf\left(\frac{x_2-\mu}{\sigma}\right) - erf\left(\frac{x_1-\mu}{\sigma}\right)$$

where $erf(x)$ = Error function

μ and σ = parameters of Gaussian probability density function

and x_1, x_2 = any real numbers.

Solution. We know that

$$P[x_1 \le X \le x_2] = \int_{x_1}^{x_2} f_X(x)\, dx \quad ...(1)$$

where $$f_X(x) = \frac{1}{\sqrt{2\pi}\,\sigma} e^{(x-\mu)^2/2\sigma^2} \quad ...(2)$$

Substituting Eq. 2 in Eq. 1, we get

$$P[x_1 \le X \le x_2] = \int_{x_1}^{x_2} \frac{1}{\sqrt{2\pi}\,\sigma} e^{(x-\mu)^2/2\sigma^2}\, dx$$

$$= \frac{1}{\sqrt{2\pi}\,\sigma} \int_{x_1}^{x_2} e^{(x-\mu)^2/2\sigma^2}\, dx \quad ...(3)$$

Putting $\left(\frac{x-\mu}{\sigma}\right) = y$ in Eq. 3, we get

$$P[x_1 \le X \le x_2] = \frac{1}{\sqrt{2\pi}\,\sigma} \int_{[(x_1-\mu)/\sigma]}^{[(x_2-\mu)/\sigma]} e^{y^2/2}\, \sigma\, dy$$

$$= \frac{1}{\sqrt{2\pi}} \int_{[(x_1-\mu)/\sigma]}^{[(x_2-\mu)/\sigma]} e^{y^2/2}\, dy$$

$$= \frac{1}{\sqrt{2\pi}} \left[\int_{[(x_1-\mu)/\sigma]}^{0} e^{y^2/2}\, dy + \int_{0}^{[(x_2-\mu)/\sigma]} e^{y^2/2}\, dy \right]$$

$$= \frac{1}{\sqrt{2\pi}} \left[- \int_0^{(x_1-\mu)/\sigma} e^{y^2/2}\, dy + \int_0^{(x_2-\mu)/\sigma} e^{y^2/2}\, dy \right]$$

$$= -\,erf\left(\frac{x_1-\mu}{\sigma}\right) + erf\left(\frac{x_2-\mu}{\sigma}\right) \quad \textbf{Proved.}$$

Example 14.12 The joint probability density function of two random variables X and Y is given by $f_{XY}(x, y) = C(1-x-y)$ for values of x and y for which (x, y) lies within the triangle as shown in Fig. 14.13.

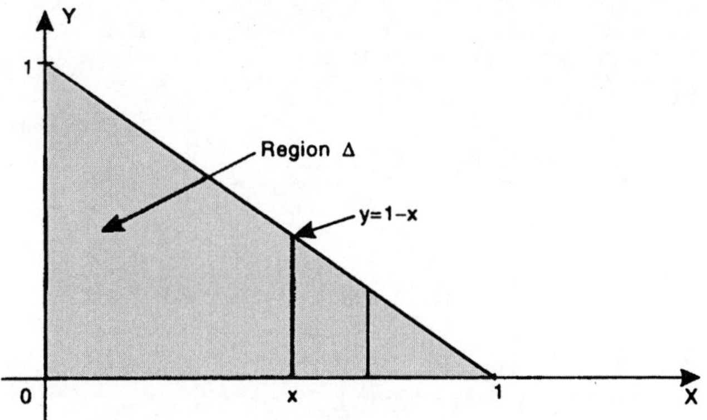

Fig. 14.13

Determine :

(*a*) the value of constant C

(*b*) the marginal probability density function $f_X(x)$ and $f_Y(y)$.

Solution. (*a*) The constant C can be determined by using following condition

$$\iint_{\Delta} f_{XY}(x, y)\, dx\, dy = 1 \qquad ...(1)$$

The triangle is bounded by the region $x = 0$, $x + y = 1$.

Eq. 1 can be written as

$$\int_0^1 \int_0^{1-x} C(1-x-y)\, dx\, dy = 1$$

or
$$\int_0^1 \frac{C(1-x)^2}{2}\, dx = 1$$

or
$$\frac{C}{6} = 1$$

or
$$C = 6$$

(*b*) The marginal probability density function $f_X(x)$ is given

$$f_X(x) = \int_{-\infty}^{+\infty} f_{XY}(x\,,y)\,dy$$

$$= \int_0^{1-x} 6(1-x-y)\,dy$$

$$= 6\left[-\frac{y^2}{2}\right]_0^{1-x} = \begin{cases} -3(1-x)^2\,, & \text{for } 0 \le x \le 1 \\ 0\,, & \text{for } x < 0\,, x \ge 1 \end{cases}$$

$$f_Y(y) = \int_{-\infty}^{\infty} f_{XY}(x\,,y)\,dx$$

$$= \int_0^{1-y} 6(1-x-y)\,dx$$

$$= 6\left[-\frac{x^2}{2}\right]_0^{1-y} = \begin{cases} -3(1-y)^2\,, & \text{for } 0 \le y \le 1 \\ 0\,, & \text{for } y < 0\,, y > 1 \end{cases}$$

Example 14.13 The marginal probability density functions of two random variables X and Y are given by

$$f_X(x) = \begin{cases} 3(1-x)^2\,, & \text{for } 0 \le x \le 1 \\ 0\,, & \text{for } x < 0\,, x > 1 \end{cases}$$

and
$$f_Y(y) = \begin{cases} 3(1-y)^2\,, & \text{for } 0 \le y \le 1 \\ 0\,, & \text{for } y < 0\,, y > 1 \end{cases}$$

Find

(*a*) Mean of random variables X and Y

(*b*) Variance of random variables X and Y.

Solution. (*a*) Mean of $X = \mu_X = E[X]$

$$= \int_{-\infty}^{\infty} x\, f_X(x)\,dx$$

$$= \int_0^1 x\, 3(1-x)^2\,dx = \frac{1}{4}$$

Mean of $Y = \mu_Y = E[Y]$

$$= \int_{-\infty}^{\infty} y\, f_Y(y)\,dy$$

$$= \int_0^1 y\, 3(1-y)^2\,dy = \frac{1}{4}$$

(*b*) Variance of $X = \sigma_X^{\,2} = E[X^2] - \mu_X^{\,2}$ ...(1)

$E[X^2]$ can be determined as

$$E[X^2] = \int_{-\infty}^{\infty} x^2 f_X(x)\, dx = \int_0^1 x^2\, 3(1-x)^2\, dx = \frac{1}{10}$$

Similarly we can determine $E[Y^2]$ as

$$E[Y^2] = \int_{-\infty}^{\infty} y^2 f_Y(y)\, dy = \int_0^1 y^2\, 3(1-y)^2\, dy = \frac{1}{10}$$

Putting the values $E[X^2]$ and μ_X in Eq. 1, we get

$$\sigma_X{}^2 = E[X^2] - \mu_X{}^2 = \frac{1}{10} - \left(\frac{1}{4}\right)^2$$

$$= \frac{8-5}{80} = \frac{3}{80}$$

Similarly, variance of Y is determined as

$$\sigma_Y{}^2 = E[Y^2] - \mu_Y{}^2 \qquad \text{...(2)}$$

Putting the values of $E[Y^2]$ and μ_Y in Eq. 2, we get

$$\sigma_Y{}^2 = \frac{1}{10} - \left(\frac{1}{4}\right)^2 = \frac{3}{80}$$

Example 14.14 We are given two random processes $x(t)$ and $y(t)$ as

$$X(t) = Z_1(t) + 3Z_2(t - \tau)$$

$$Y(t) = Z_2(t + \tau) + 3Z_1(t - \tau)$$

where $Z_1(t)$ and $Z_2(t)$ are independent white noise processes each with variance equal to 0.5. Find, (*a*) autocorrelation functions of $X(t)$ and $Y(t)$, (*b*) cross-correlation functions of $X(t)$ and $Y(t)$.

Solution. (*a*) Autocorrelation function of $X(t)$ is given by

$$R_X(t_1, t_0) = E[X(t_1)\, X(t_0)]$$

$$= E[\{Z_1(t_1) + 3Z_2(t_1 - \tau)\}\, \{Z_1(t_0)\, 3Z_2(t_0 - \tau)\}]$$

$$= E[Z_1(t_1)\, Z_1(t_0)] + 3E[Z_1(t_1) - Z_2(t_0 - \tau)]$$

$$+ 3E[Z_2(t_1 - \tau)\, Z_1(t_0)] + 9[Z_2(t_1 - \tau)\, Z_2(t_0 - \tau)] \qquad \text{...(1)}$$

We know that

$$E[Z_1(t_1), Z_2(t_0 - \tau)] = 0$$

because $Z_1(t)$ and $Z_2(t)$ are independent.

Similarly $\quad E[Z_2(t_1 - \tau)\, Z_1(t_0)] = 0$

For $t_1 \neq t_0$, $\quad E[Z_1(t_1)\, Z_1(t_0)] = 0$

If $t_1 = t_0$ $\quad E[Z_1(t_1)\, Z_1(t_0)] = E[Z_1{}^2(t_1)]$

$$= \sigma^2 + \mu^2 = 0.5\, \delta(t_1 - t_0) \qquad \text{...(2)}$$

Similarly $\quad E[Z_2(t_1-\tau)\,Z_2(t_0-\tau) = 0 \;\text{ for } t_1 \neq t_0$

$$= 0.5 \;\text{ for } t_1 = t_0$$

$$= 0.5\,\delta(t_1 - t_0) \qquad ...(3)$$

Substituting Eqs. 2 and 3 in Eq. 1, we get

$$R_X(t_1\,,\,t_0) = 0.5\,\delta(t_1-t_0) + 4.5\,\delta(t_1-t_0) \qquad ...(4)$$

Similarly $\quad R_Y(t_1\,,\,t_0) = 0.5\,\delta(t_1-t_0) + 4.5\,\delta(t_1-t_0) \qquad ...(5)$

From Eqs. 4 and 5, we observe that the random processes $X(t)$ and $Y(t)$ are wide-sense stationary (WSS).

(*b*) Cross-correlation function of random processes $X(t)$ and $Y(t)$ is given by

$$R_{XY}(t_1\,,\,t_0) = E[X(t_1)\,Y(t_0)]$$

$$= E[\{Z_1(t_1) + 3Z_2(t_1)\}\,\{Z_2(t_0+\tau) + 3Z_1(t_0-\tau)\}]$$

$$= E[Z_1(t_1)\,Z_2(t_0+\tau)] + 3E[Z_1(t_1)\,Z_1(t_0-\tau)]$$

$$+ 3E[Z_2(t_1)\,Z_2(t_0+\tau)] + 9E[Z_2(t_1)\,Z_1(t_0-\tau)] \qquad ...(6)$$

Now $\quad E[Z_1(t_1)\,Z_2(t_0)] = 0$

[because $Z_1(t_1)$ and $Z_2(t_0)$ are independent random variables]

$$E[Z_1(t_1)\,Z_1(t_0-\tau)] = \begin{cases} 0\,, & \text{if } t_1 \neq t_0 - \tau \\ 0.5\,, & \text{if } t_1 = t_0 - \tau \end{cases}$$

$$E[Z_2(t_1-\tau)\,Z_2(t_0)] = \begin{cases} 0\,, & \text{if } t_0 \neq t_1 - \tau \\ 0.5\,, & \text{if } t_0 = t_1 - \tau \end{cases}$$

$$E[Z_2(t_1-\tau)\,Z_1(t_0-\tau)] = 0$$

Substituting above values in Eq. 6, we get

$$R_{XY}(t_1\,,\,t_0) = 1.5\,\delta(t_1 - t_0 + \tau) + 1.5\,\delta(t_0 - t_1 + \tau) \qquad ...(7)$$

Since cross-correlation depends upon the difference of t_1 and t_0 therefore, random processes $X(t)$ and $Y(t)$ are jointly stationary or simply stationary.

Example 14.15 Stationary random process $X(t)$ has the following autocorrelation function

$$R_X(\tau) = \sigma^2\,e^{-\mu|\tau|}$$

where μ and σ^2 are constants. It is passed through a filter whose impulse response is

$$h(\tau) = \alpha\,e^{-\alpha\tau}\,u(\tau)$$

where α is a constant and $u(\tau)$ is a step function.

(*a*) Determine the power spectral density of random signal $X(t)$

(*b*) Determine the power spectral density of the output random signal $Y(t)$.

Solution. (*a*) Power spectral density (psd) of the input is given by

$$S_X(\omega) = \text{CTFT}\ \{R_X(\tau)\}$$

$$= \int_{-\infty}^{\infty} R_X(\tau)\, e^{-j\omega\tau}\, d\tau = \int_{-\infty}^{\infty} \sigma^2\, e^{-\mu|\tau|}\ e^{-j\omega\tau}\ \tau$$

$$= \sigma^2 \int_{-\infty}^{0} e^{(\mu - j\omega)\tau}\, d\tau + \sigma^2 \int_{0}^{\infty} e^{-(\mu + j\omega)\tau}\, d\tau$$

$$= \sigma^2 \left[\frac{1}{\mu - j\omega} + \frac{1}{\mu + j\omega} \right] = \frac{2\mu\sigma^2}{\mu^2 + \omega^2}$$

(*b*) Power spectral density of the output is related to the power spectral density of the input as

$$S_Y(\omega) = |H(\omega)|^2\, S_X(\omega) \qquad \text{...(1)}$$

$$H(\omega) = \text{CTFT}\ \{h(\tau)\}$$

$$= \int_{-\infty}^{\infty} h(\tau)\, e^{-j\omega\tau}\, d\tau = \int_{-\infty}^{\infty} \alpha\, e^{-\alpha\tau}\, u(\tau)\, e^{-j\omega\tau}\, d\tau$$

$$= \alpha \int_{0}^{\infty} e^{-(\alpha + j\omega)\tau}\, d\tau = \frac{\alpha}{\alpha + \omega} \qquad \text{...(2)}$$

Substituting Eq. 2 in Eq. 1, we get

$$S_Y(\omega) = \left| \frac{\alpha}{\alpha + j\omega} \right|^2 S_X(\omega)$$

$$= \frac{\alpha}{\alpha^2 + \omega^2}\, S_X(\omega) = \frac{\alpha}{\alpha^2 + \omega^2}\, \frac{2\mu\sigma^2}{\mu^2 + \omega^2}$$

$$= \frac{2\mu\alpha^2\sigma^2}{(\alpha^2 + \omega^2)\,(\mu^2 + \omega^2)}$$

REVIEW QUESTIONS

1. Distinguish random signals from deterministic signals. Give examples of each signal.
2. Define probability and relative frequency of occurrence of any event.
3. What do you mean by mutually exclusive events and statistically independent events ?
4. Define random variables with examples.
5. Define probability distribution function and probability density function of a random variable.
6. Define cumulative distribution function (cdf) of random variable. Also give properties of cdf.
7. Give important properties of pdf.

8. Discuss following probability density function of a random variable

(*a*) uniform pdf (*b*) Gaussian or normal pdf

(*c*) Exponential pdf.

9. Discuss and define the following

(*a*) Joint distribution function

(*b*) Joint probability density function of two random variables.

10. Give important properties of joint probability density function.

11. Define the following

(*a*) Marginal distribution function

(*b*) Marginal pdf.

12. What is independent random variables ?

13. Define conditional probability density function of a continuous random variable.

14. What do you mean by statistical averages of random variables ?

15. Define

(*a*) Mean value of a random variable

(*b*) Variance of a random variable

(*c*) Chebyshev inequality

(*d*) Characteristic function

(*e*) Joint moments of random variables

(*f*) Central moments of a random variable

(*g*) Covariance of two random variables X and Y

(*h*) Correlation coefficient

16. Discuss transformation of random variables.

17. Define random process with examples.

18. Discuss mathematical representation of random processes.

19. Distinguish non-stationary random processes from stationary random processes.

20. Discuss statistics of random process.

21. Define the following

(*a*) Wide-sense stationary (WSS) process

(*b*) Strict-sense stationary (SSS) process.

22. What is ergodic process ?

23. Give classification of random process.

24. What is Gaussian process ? Give example.

25. Discuss sum of two random processes.

26. What is autocorrelation function of a random process ?

27. Define cross-correlation function of two random processes.

28. Discuss various properties of autocorrelation of a random process.

29. Define :

(*a*) Power spectral density of WSS random process

(*b*) Energy spectral density

(*c*) Cross power spectral density.

30. Discuss properties of power spectral density ?

31. Discuss response of linear systems to random inputs.

NUMERICAL PROBLEMS

1. A box contains 3 red, 4 white and 5 black balls. One ball is drawn at random. Determine the probability that is

(*a*) Red (*b*) not Black

(*c*) Black or white

Ans. (*a*) $P(R) = \frac{1}{4}$ (*b*) $P(\text{Not } B) = \frac{7}{12}$ (*c*) $P(B \text{ or } W) = \frac{3}{4}$

2. Two dices are thrown simultaneously. Find the probability of getting 5.

Ans. $P(5) = \frac{1}{9}$

3. Two cards are drawn at random from an ordinary deck of 52 playing cards. Find the probability that one card is a Diamond and the other card is a Heart.

Ans. $P(A) = \frac{13}{102}$

4. The CDF for a certain random variable is given as

$$fx(x) = \begin{cases} 0\,, & -\infty < x \le 0 \\ qx^2\,, & 0 < x \le 10 \\ 100q\,, & 10 < x < \infty \end{cases}$$

(*a*) Find the value of q

(*b*) Find the value of $P(X \le 5)$

(*c*) Determine the value of $P(5 < X \le 7)$

(*d*) Determine the expression for PDF

Ans. (*a*) $k = \frac{1}{100}$ (*b*) $P(X \le 5) = \frac{1}{4}$ (*c*) $P(5 < X \le 7) = \frac{24}{100}$

$$(d)\ fx(x) = \begin{cases} 0\,, & \text{for } -\infty < x \le 0 \\ 0\,, & \text{for } 10 < x < \infty \\ \frac{x}{50}\,, & \text{for } 0 < x \le 10 \end{cases}$$

5. A continuous random variable has a probability density function (pdf) expressed as

$$fx(x) = 2e^{-2x} \quad \text{for } x \ge 0$$

Find the probability which will take a value between 1 and 3.

Ans. $P(1 < X \le 3) = 0.1328$

6. The random variable Z is the function of another random variable X in such a way that $Z = \cos(X)$ and X is a uniformly distributed random variable distributed in the interval $(-\pi, \pi)$, *i.e.,*

$$f_X(x) = \begin{cases} \frac{1}{2\pi}, & \text{for } -\pi < x < \pi \\ 0, & \text{elsewhere} \end{cases}$$

Determine the value of Z. **Ans.** z has a zero mean

7. A random variable X has the uniform distribution given by

$$f_X(x) = \begin{cases} \frac{1}{2\pi}, & \text{for } 0 \le x \le 2\pi \\ 0, & \text{otherwise} \end{cases}$$

Determine mean, variance and standard deviation of X.

Ans. mean $m_X = \pi$; variance $\sigma_x^2 = \frac{4}{3}\pi^2$; standard deviation $\sigma_X = \frac{\pi}{\sqrt{3}}$

8. Find the cumulative distribution function (CDF) of the Guassian random variable.

Ans. Gaussian CDF, $F_X(x) = \frac{1}{2} erfc\left(\frac{m - x}{\sigma\sqrt{2}}\right)$

SIGNALS AND SYSTEMS

RISHABH ANAND

FEEDBACK FORM

1. Which topics from your examination point of view are inadequately or not covered in this book?

____

____

2. Is there any misprints/ mistakes/ factual inaccuracy in this book? If so, please specify.

____

____

3. What is your assessment of this book as regards the presentation of the subject-matter, expression, precision and price in relation to the other recommended books available on this subject?

____

____

4. Which competing books you regard as better than this? Please mention their author and publishers.

____

____

5. Any further suggestions/comments you would like to make for the improvement of the book?

____

____

PS: If needed, please attach a separate sheet.

Name : ______________________________

College : ______________________________

Address : ______________________________

Email : ______________________________

Mobile : ______________________________

Please mail the feedback form at:

KHANNA BOOK PUBLISHING CO. (P) LTD.

4C/4344, Ansari Road, Daryaganj, New Delhi-110002

You can email this feedback form at contact@khannabooks.com or Whatsapp at +91 – 99109 09320

BEST FEEDBACK WILL BE REWARDED SUITABLY